Acta Numerica 2003

Acta Numerica

Volume 12 2003

CAMBRIDGE
UNIVERSITY PRESS

CAMBRIDGE UNIVERSITY PRESS
Cambridge, New York, Melbourne, Madrid, Cape Town,
Singapore, São Paulo, Delhi, Tokyo, Mexico City

Cambridge University Press
The Edinburgh Building, Cambridge CB2 8RU, UK

Published in the United States of America by Cambridge University Press, New York

www.cambridge.org
Information on this title: www.cambridge.org/9780521174312

First published 2003
First paperback edition 2011

A catalogue record for this publication is available from the British Library

ISBN 978-0-521-82523-8 Hardback
ISBN 978-0-521-17431-2 Paperback

Contents

Acta Numerica (2003), pp. 1–125
DOI: 10.1017/S0962492902000090

Survey of meshless and generalized finite element methods: A unified approach

Ivo Babuška*
Institute for Computational Engineering and Sciences,
ACE 6.412, University of Texas at Austin,
Austin, TX 78712, USA

Uday Banerjee†
Department of Mathematics, 215 Carnegie,
Syracuse University, Syracuse, NY 13244, USA
E-mail: banerjee@syr.edu
http://bhaskara.syr.edu

John E. Osborn†
Department of Mathematics, University of Maryland,
College Park, MD 20742, USA
E-mail: jeo@math.umd.edu
http://www.math.umd.edu/~jeo

In the past few years meshless methods for numerically solving partial differential equations have come into the focus of interest, especially in the engineering community. This class of methods was essentially stimulated by difficulties related to mesh generation. Mesh generation is delicate in many situations, for instance, when the domain has complicated geometry; when the mesh changes with time, as in crack propagation, and remeshing is required at each time step; when a Lagrangian formulation is employed, especially with nonlinear PDEs. In addition, the need for flexibility in the selection of approximating functions (*e.g.*, the flexibility to use non-polynomial approximating functions), has played a significant role in the development of meshless methods. There are many recent papers, and two books, on meshless methods; most of them are of an engineering character, without any mathematical analysis.

In this paper we address meshless methods and the closely related generalized finite element methods for solving linear elliptic equations, using variational principles. We give a unified mathematical theory with proofs, briefly address implementational aspects, present illustrative numerical examples, and provide a list of references to the current literature.

* Supported by NSF grant DMS-98-02367 and ONR grant N00014-99-1-0724.
† Partially supported by the TICAM Visiting Faculty Research Fellowship Program.

The aim of the paper is to provide a survey of a part of this new field, with emphasis on mathematics. We present proofs of essential theorems because we feel these proofs are essential for the understanding of the mathematical aspects of meshless methods, which has approximation theory as a major ingredient. As always, any new field is stimulated by and related to older ideas. This will be visible in our paper.

CONTENTS

1. Introduction

1.1. A brief historical review of the numerical solution of partial differential equations

The numerical solution of partial differential equations has been of central importance for many years. Significant progress has been made in this area, especially in the last 30 years; this progress is directly related to the developments in computer technology. Methods such as, for example, the finite element method, are used in many applications.

Although significant progress has been made, numerical methods for the solution of differential equations are still often based on heuristic ideas, and verified by numerical experiments. Mathematical analysis is often shallow, and fails to address fully important issues that arise in the application of the methods to important problems in engineering and science.

There are three classical families of numerical methods for solving PDEs:

(1) finite difference methods;
(2) finite volume methods;
(3) finite element methods.

These three families have two common, basic features:

(a) they employ a mesh;
(b) they use local approximation by polynomials.

We discuss each of these features in turn.

Mesh generation is often very expensive – especially in human effort. There are several reasons for this effort.

- The domain of the problem can have very complex geometry.
- The domain of the problem may change with time, which requires remeshing at each time step, as for example in the problem of crack propagation or when Lagrangian coordinates are used.
- Adaptive procedures require changes of mesh during computation.

Although great progress has been made in the theory and practice of mesh generation, the construction of the mesh is still a very delicate component of the numerical solution of differential equations. For this reason there is an interest in the development of methods that eliminate or reduce the need for a mesh.

Although polynomials have outstanding approximation properties, there are situations in which they are not effective. We mention problems whose solutions are not smooth, in the sense that they may not have several bounded derivatives. For such problems, there are sometimes other effective approximating functions, which we will refer to as *special*. The classical methods are not flexible in this regard: they do not use these special non-polynomial approximating functions. There is thus an interest in developing and analysing methods that can flexibly use these special approximation functions.

This created the need to develop methods that address both of these issues: the elimination, completely or partially, of the need for meshes; and the effective use of special (nonpolynomial) approximating functions. The inspiration for such methods came mainly from two sources.

The first of these sources is the class of classical particle methods that arise in physical simulation in connection with the Boltzmann equation or with fluid dynamics. Particle methods attempt to describe the motion of the atoms or their averages (or density) in Lagrangian coordinates (see Gingold and Monaghan (1977, 1982), Monaghan (1982, 1988), Nanbu (1980) and Neunzert and Struckmeier (1995), for example).

The other source is the idea of interpolation in the context of general *variational methods* (of Galerkin type). These methods select approximations from a finite-dimensional space, called the *trial space*, and, under certain general conditions, it is known that the error in approximation is no larger than a constant times the error in best approximation by functions in the trial space. Thus the quality of the method is determined by the approximation property of the trial space. It is thus natural to try to find a trial space that has good approximation properties. This property relates directly to interpolation by the approximating functions. For functions in one dimension this is a classical issue in numerical analysis, and, from around 1950, was studied in higher dimensions and for arbitrary distributions of points. It was recognized that the construction of trial spaces could be based on the idea of interpolation.

1.2. Meshless methods

Let us now make the discussion of variational methods more precise. We consider an elliptic PDE, which has the variational or weak form

$$u \in H_1, \qquad B(u,v) = \mathcal{F}(v), \quad \text{for all } v \in H_2, \tag{1.1}$$

where H_1, H_2 are two Hilbert spaces, $B(u,v)$ is a bounded bilinear form on $H_1 \times H_2$, and $\mathcal{F}(v)$ is a continuous linear functional on H_2. Under certain general conditions (the inf-sup or BB condition; see Babuška (1971), Babuška and Aziz (1972)), the solution u is characterized by (1.1). We are interested in approximating u. To that end, we assume we have two finite-dimensional spaces $M_1 \subset H_1, M_2 \subset H_2$ that satisfy the discrete inf-sup condition (see Babuška and Aziz (1972)). The approximate solution u_{M_1} is characterized by

$$u_{M_1} \in M_1, \qquad B(u_{M_1},v) = \mathcal{F}(v), \quad \text{for all } v \in M_2. \tag{1.2}$$

As a consequence of the fact that M_1 and M_2 satisfy the discrete inf-sup condition, we know that the approximation u_{M_1} is quasi-optimal, that is,

$$\|u - u_{M_1}\|_{H_1} \le C \inf_{\chi \in M_1} \|u - \chi\|_{H_1}. \tag{1.3}$$

We note that there are delicate problems related to the discrete inf-sup condition for problems that are not coercive, or where the spaces M_1 and M_2 are different, *e.g.*, in the mixed method or non-self-adjoint problems. Atluri and Shen (2002) use different spaces (without mathematical analysis of the discrete inf-sup condition).

Thus the quality of the approximation, that is, the error $\|u - u_{M_1}\|_{H_1}$, is mainly determined by the approximation properties of the trial space M_1, that is, by

$$E_1 = \inf_{\chi \in M_1} \|u - \chi\|_{H_1}.$$

It is therefore natural to select the trial space M_1 so that E_1 is small. To do this effectively we should use whatever information is available for the solution u. Note that, with a general variational method, as we have formulated it, there is no mention of a mesh. Of course, we may use a mesh to construct a good trial space; that, in fact, is exactly what is done with a finite element method. For example, the trial space might be the space of piecewise linear functions over a mesh.

Meshless methods, however, either avoid the use of a mesh, or use a mesh only minimally, for example, only for the numerical integration. The Petrov–Galerkin method given by (1.2) is a meshless method if the construction of M_1 and M_2 either does not require a mesh or requires a mesh only minimally. Thus, in designing meshless methods within the framework of variational methods, we have two general goals.

(1) The construction of trial spaces M_1 that effectively approximate the solution, and the construction of test spaces M_2 ensuring the inf-sup (stability) condition.

    If the solution has special features, *e.g.*, if it is not smooth, we should have the flexibility to use special approximating functions.

(2) The minimizing of the need for a mesh.

    In meshless methods, there is sometimes a mesh in the background, used for numerical integration, but we may not need a mesh generator.

We note that there are meshless methods that are not of the type given by (1.2), for instance, methods based on collocation, but the construction of approximating space follows the guidelines of the construction of M_1, as mentioned before.

The approximating (trial) spaces can be the spans of specific approximating functions (shape functions), with either global or local supports. Polynomials and non-compactly supported radial basis functions are examples of approximating functions that are defined over the entire domain of interest. See Mikhlin (1971) for a discussion of the use of polynomials and Buhmann (2000) and Powell (1992) for a discussion of the use of radial basis functions. Another type of approximating function is related to interpolation and data fitting procedures. For a survey of various approaches we refer to Atluri and Shen (2002), Dierckx (1995), Franke (1978, 1979), Gordon and Wixon (1978), Lancaster and Salkauskas (1981, 1986), McLain (1974), and Shepard (1968). Typical finite element approximating functions and spline functions have local supports. Babuška, Caloz and Osborn (1994) identified and analysed shape functions that are effective for the approximation of solutions of elliptic equations with rough coefficient; the idea in this paper was extended and developed in Babuška and Melenk (1997). The approximating functions used in Babuška *et al.* (1994) and Babuška and Melenk

(1997) can be characterized as solutions of particular homogeneous differential equations. In one dimension, L-splines – a generalization of splines that satisfy a differential equation – are used as approximating functions: see, *e.g.*, Varga (1971). Principles for the selection of shape function were addressed in Babuška, Banerjee and Osborn (2001, 2002*b*).

We note that in the engineering literature many names are used for methods that differ only in their implementation or in the shape functions employed: see, *e.g.*, De and Bathe (2001) and Sukumar, Moes, Moran and Belytschko (2000), among others. For a survey of results on meshless methods we refer to Babuška, Banerjee and Osborn (2002*a*), Belytschko, Krongauz, Organ, Fleming and Krysl (1996), Duarte (1995), Griebel and Schweitzer, eds (2002*a*), Li and Liu (2002), Liu (2002) and Schweitzer (200x).

One of the major problems of meshless methods is the imposition of boundary conditions, especially Dirichlet boundary conditions. It is well known that, if the underlying problem is a Dirichlet BVP, the essential boundary condition is addressed with a method such as the penalty method or the Lagrange multiplier method. On the other hand, the boundary condition of a Neumann problem is natural, and does not need to be explicitly imposed in the variational formulation. In both situations, a simple uniform mesh on a rectangle containing the domain can be used; the mesh need not conform to the boundary and a mesh generator is not needed. These ideas are classical and have been extensively analysed (for example, see Babuška and Aziz (1972)). This way of imposing boundary conditions can be used in the context of meshless methods, and this approach was also mentioned in Li and Liu (2002). The boundary of the domain does come into play in the construction of the stiffness matrix, but a mesh generator is not needed. This approach was generalized and used together with the ideas in Babuška *et al.* (1994), Babuška and Melenk (1997) and Strouboulis, Babuška and Copps (2001*a*) in solving problems with very complex geometries: see, *e.g.*, Strouboulis, Copps and Babuška (2001*b*).

We finally mention a meshless method – the *generalized finite element method* (GFEM) – which attempts to achieve simultaneously the two goals of variationally formulated meshless methods. With this method we begin with a *partition of unity.* Construction of a partition of unity is a relatively simple task. It can be done by various means. One is to use a simple mesh, for example a uniform mesh, and use the associated hat functions as the partition of unity. We could also use ideas from various interpolation procedures, for instance, the Shepard method. It is essential that the construction can, but is not required to, utilize the geometry of the domain. The partition of unity on the domain is obtained by restriction. The partition of unity functions typically have compact supports with small diameters.

Then we multiply the partition of unity functions by functions that are defined separately and independently on the supports of the partition of unity functions. In this way we create shape functions that belong to $H^1(\Omega)$, and can be used in the variational method. We thereby obtain a large flexibility in the construction of shape functions, and the associated trial spaces. This flexibility can be used to construct approximations that utilize the available information, the character of a singularity, or a boundary layer, *e.g.*, on the approximated function (solution). Hence the method achieves the goals mentioned above.

We do face three serious difficulties in the implementation of the GFEM. First there is the problem of numerical integrations when the areas over which we integrate are not simple triangles, simplices, *etc.*, as with the usual FEM. We note, however, that the process is completely parallelizable. A second difficulty is the treatment of essential boundary conditions. The third issue concerns the system of linear equations. It may be singular, and thus certain classical methods, such as multigrid, may not be applicable. These difficulties can be, and have been, overcome in some implementations, so it is clear that the GFEM shows a definite advantage over the classical FEM in certain situations. We mention problems with complex geometry, crack propagation, and analysis of multi-site local damage.

Of course, any new method should be compared with previously developed methods, and the class of problems for which the new method is superior should be identified. Theoretical and practical experience (see Babuška *et al.* (2002*a*), Li and Liu (2002) and Strouboulis *et al.* (2001*b*)) is progressing in this direction. Meshless methods in various forms, *e.g.*, within the framework of collocation or variational methods, are now the subject of many papers and (engineering) books, which mainly focus on practical aspects without serious theoretical analysis.

This paper focuses on ideas and theoretical results. Some are adjustments of old ideas and results. Some results are based on papers that are submitted or in the final stage of preparation. Although we focus on the theory, we have attempted to address theoretical issues that illuminate practical issues. We will show that the results presented here are natural generalizations of the classical FEM, which is a special case of some of the methods presented here. This paper addresses only problems related to linear PDEs.

Various relevant and typical references are provided. The reference list is not comprehensive, but together with the citations in the references provide, in our opinion, a very reasonable description of the current state of the art in meshless methods.

1.3. The scope of this paper

The short Section 2 defines the model problem, a linear elliptic boundary value problem. Section 3.1 presents approximation results when the particles are uniformly distributed. The presented results were obtained using the Fourier transform. Section 3.2 presents an alternative proof of the approximation results that can be generalized to the case of non-uniformly distributed particles. Section 3.3 discusses approximation for arbitrarily distributed particles. Section 4 discusses the construction of shape functions, and presents some results on interpolation and on the asymptotic form of the error. Section 5 addresses the question of superconvergence. Section 6 discusses the generalized finite element method. Section 7 discusses the application of the approximation results developed in Section 3, and discusses the treatment of Dirichlet boundary conditions. Section 8 explains some implementational aspects. Section 9 reports some numerical examples obtained by the GFEM, when the domain is very complex. Finally, Section 10 presents additional results and challenges.

2. The model problem

For concreteness and simplicity we will address the weak solution of the model problem

$$-\Delta u + u = f(x), \quad \text{on } \Omega \subset R^n \tag{2.1}$$

and

$$\frac{\partial u}{\partial n} = 0, \quad \text{on } \partial\Omega, \tag{2.2}$$

or

$$u = 0, \quad \text{on } \partial\Omega, \tag{2.3}$$

where $f \in L_2(\Omega)$ is given. We will assume that Ω is a Lipschitz domain; additional assumptions on $\partial\Omega$ will be given as needed.

The weak solution $u_0 \in H^1(\Omega)$ ($H^1_0(\Omega)$, respectively) satisfies

$$B(u_0, v) = \mathcal{F}(v), \quad \text{for all } v \in H^1(\Omega) \quad (v \in H^1_0(\Omega), \text{respectively}), \tag{2.4}$$

where

$$B(u, v) \equiv \int_\Omega (\nabla u \cdot \nabla v + uv)\, dx \text{ and } \mathcal{F}(v) \equiv \int_\Omega fv\, dx. \tag{2.5}$$

The energy norm of u_0 is defined by

$$\|u_0\|_E \equiv B(u_0, u_0)^{1/2} = \|u_0\|_{H^1(\Omega)}. \tag{2.6}$$

We will write H instead of $H^1(\Omega)$ or $H^1_0(\Omega)$ if no misunderstanding can occur.

Let $S \subset H$ be a finite-dimensional subspace, called the approximation space. Then the Galerkin approximation, $u_S \in S$, to u_0 is determined by

$$\tilde{B}(u_S, v) = \mathcal{F}(v), \quad \text{for all } v \in S, \tag{2.7}$$

where $\tilde{B}$ is either B or a perturbation of B. If $\tilde{B} = B$, it is immediate that

$$\|u_0 - u_S\|_{H^1(\Omega)} = \inf_{\chi \in S} \|u_0 - \chi\|_{H^1(\Omega)}. \tag{2.8}$$

Hence, the main problem is the approximation of u_0 by functions in S.

Remark 1. The *finite element method* (FEM) is the Galerkin method where S is the span of functions with small supports. For the history of the FEM, see Babuška (1994) and the references therein.

Remark 2. The classical Ritz method uses spaces of polynomials on the entire domain Ω for the approximation spaces: see, *e.g.*, Mikhlin (1971).

As mentioned above, the finite element method uses basis functions with small supports, for example, 'hill' functions. The theory of approximation with general hill functions with translation-invariant supports was developed in Babuška (1970) using the Fourier transform. The results in Babuška (1970) were applied to the numerical solution of PDEs in Babuška (1971). A very similar theory, also based on the Fourier transform, was later developed in Strang (1971) and Strang and Fix (1973); see also Li and Liu (1996). Later, hill functions were, in another context, called *particle functions* (see Gingold and Monaghan (1977)). In the 1990s, hill functions began to be used in the framework of *meshless methods.* For a broad survey of meshless methods see Li and Liu (2002). A survey of the approximation properties of radial hill functions, previously referred to as radial basis functions, is given in Buhmann (2000).

In this paper we will survey basic meshless approximation results and their use in the framework of Galerkin methods.

3. Approximation by local functions in $\mathbb{R}^n$: the h-version analysis

As mentioned in Section 2, we are interested in the approximation of functions by particle shape functions. We first consider uniformly distributed particles, and then general (non-uniformly distributed) particles.

3.1. Uniformly distributed particles and associated particle shape functions

Let $\mathbb{Z}$ be the integer lattice, and, for $j = (j_1, \ldots, j_n) \in \mathbb{Z}^n$ and $0 < h \leq 1$, let

$$x_j^h = (j_1 h, \ldots, j_n h) = hj.$$

The points x_j^h are called *uniformly distributed particles.* When considering such families of particles, we often construct associated shape functions, called particle shape functions, as follows. Let $\phi \in H^q(\mathbb{R}^n)$, for some $0 \leq q$, be a function with compact support; let $\eta \equiv$ supp ϕ, and suppose

$$\eta \subset B_\rho = \{x \in \mathbb{R}^n : \|x\|^2 = x_1^2 + \cdots + x_n^2 < \rho\}.$$

We assume that $0 \in \mathring{\eta}$ ($\mathring{\eta}$ is the interior of η). Then define

$$\phi_j^h(x) = \phi_j^h(x_1, \ldots, x_n) = \phi\left(\frac{x - jh}{h}\right) = \phi\left(\frac{x_1 - j_1 h}{h}, \ldots, \frac{x_n - j_n h}{h}\right), \tag{3.1}$$

for $j \in \mathbb{Z}^n$ and $0 < h \leq 1$. Clearly,

$$\eta_j^h \equiv \text{ supp } \phi_j^h = \left\{x : \frac{x - jh}{h} \in \eta\right\} \subset B_{\rho h}^j = \{x : \|x - x_j^h\| < \rho h\},$$

and $x_j^h \in \mathring{\eta}_j^h$. Particles and particle shape functions defined in this way will be called *translation-invariant*, since they satisfy

$$x_{j+l}^h = x_j^h + x_l^h \text{ and } \phi_{j+l}^h(x) = \phi_j^h(x - x_l^h).$$

They are a special case of general (non-uniformly distributed) particles, which will be introduced in Section 3.3. We refer to h as the size of the particle, and the function ϕ is called the *basic shape function.* In this section we will be interested in the approximation properties of

$$V_h^{k,q} \equiv \left\{v = v(x) = \sum_{j \in \mathbb{Z}^n} w_j^h \phi_j^h(x) : w_j^h \in \mathbb{R}\right\}, \tag{3.2}$$

which is the linear span of the associated shape functions, as $h \to 0$. The parameter k in $V_h^{k,q}$ is related to a property of the $\{\phi_j^h(x)\}$, which will be discussed later. We will refer to $V_h^{k,q}$ as the particle space in $\mathbb{R}^n$. The $\{w_j^h\}$ are called *weights.* Specifically, given $u \in H^{k+1}(\mathbb{R}^n)$, we are interested in estimating

$$\inf_{\chi \in V_h^{k,q}} \|u - \chi\|_{H^s(\mathbb{R}^n)}, \tag{3.3}$$

for $0 \leq s \leq \min\{q, k+1\}$. We are especially interested in the maximum μ such that

$$\inf_{\chi \in V_h^{k,q}} \|u - \chi\|_{H^s(\mathbb{R}^n)} \leq C(k,q) h^\mu \|u\|_{H^{k+1}(\mathbb{R}^n)}, \tag{3.4}$$

for $0 \leq s \leq \min\{q, k+1\}$, where the constant $C = C(k,q)$ depends on k, q, but is independent of h (C also depends on ϕ).

Because we are assuming the particles are uniformly distributed, and hence the particles and shape functions are translation-invariant, estimates of the form (3.4) can be obtained via the Fourier transform. This was done in Babuška (1970), Strang (1971) and Strang and Fix (1973). We will cite one of the results in the last paper.

Let

$$\hat{\phi}(\xi) = \hat{\phi}(\xi_1, \ldots, \xi_n) \equiv \int_{\mathbb{R}^n} \phi(x) \mathrm{e}^{-\mathrm{i}x\cdot\xi}\, \mathrm{d}x$$

denote the Fourier transform of $\phi(x)$. We note that $\hat{\phi}(\xi) \in C^\infty(\mathbb{R}^n)$ since $\phi(x)$ has compact support. We use the usual multi-index notation: $\alpha = (\alpha_1, \ldots, \alpha_n)$, with $\alpha_i \geq 0$; $|\alpha| = \alpha_1 + \cdots + \alpha_n$; $x^\alpha = x_1^{\alpha_1} \cdots x_n^{\alpha_n}$; and

$$D^\alpha \hat{\phi} = \frac{\partial^{|\alpha|} \hat{\phi}}{\partial \xi_1^{\alpha_1} \cdots \partial \xi_n^{\alpha_n}}.$$

Theorem 3.1. (Strang and Fix 1973) Suppose $\phi \in H^q(\mathbb{R}^n)$ has compact support, where the smoothness index $q \geq 0$ is an integer. Then the following three conditions are are equivalent:

(1)

$$\hat{\phi}(0) \neq 0 \tag{3.5}$$

and

$$D^\alpha \hat{\phi}(2\pi j) = 0, \quad \text{for } 0 \neq j \in \mathbb{Z}^n \text{ and } |\alpha| \leq k, \tag{3.6}$$

where k is a nonnegative integer.

(2) For $|\alpha| \leq k$,

$$\sum_{j \in \mathbb{Z}^n} j^\alpha \phi(x - j) = \lambda x^\alpha + q^\alpha(x), \quad \text{for all } x \in \mathbb{R}^n, \tag{3.7}$$

where $\lambda \neq 0$ and $q^\alpha(x)$ is a polynomial with degree $< |\alpha|$.

The equality in (3.7) is for almost all $x \in \mathbb{R}^n$. The function of the right-hand side of (3.7) is, of course, continuous. If the function on the left-hand side is continuous, which will be the case if $q > n/2$, then (3.7) will hold for all $x \in \mathbb{R}^n$.

(3) For each $u \in H^{k+1}(\mathbb{R}^n)$ there are weights $w_j^h \in \mathbb{R}$, for $j \in \mathbb{Z}^n$ and $0 < h$, such that

$$\begin{aligned} &\left\| u - \sum_{j \in \mathbb{Z}^n} w_j^h \phi_j^h \right\|_{H^s(\mathbb{R}^n)} \\ &\quad \leq C h^{k+1-s} \|u\|_{H^{k+1}(\mathbb{R}^n)}, \quad \text{for } 0 \leq s \leq \min\{q, k+1\}, \end{aligned} \tag{3.8}$$

and

$$h^n \sum_{j \in \mathbb{Z}^n} (w_j^h)^2 \leq K^2 \|u\|^2_{H^0(\mathbb{R}^n)}. \tag{3.9}$$

Here C and K may depend on q, k, and s, but are independent of u and h. The exponent $k + 1 - s$ is the best possible if k is the largest integer for which (3.7) holds.

If (3.7) holds, the basic shape function ϕ is called *quasi-reproducing of order k*. If (3.7) holds with $\lambda = 1$ and $q^\alpha(x) = 0$, ϕ is called *reproducing of order k*. If ϕ is quasi-reproducing of order k (respectively, reproducing of order k), then the corresponding particle shape functions ϕ_i^h are also called quasi-reproducing of order k (respectively, reproducing of order k). The parameter k in $V_h^{k,q}$, defined in (3.2), is the quasi-reproducing order of the basic shape function ϕ.

Remark 3. If one were to define the notion of quasi-reproducing basic shape function ϕ of order k by the formula

$$\sum_{j\in\mathbb{Z}^n} j^\alpha\phi(x-j) = \lambda_\alpha x^\alpha + q^\alpha(x), \quad \text{for all } x \in \mathbb{R}^n, \quad \text{for } |\alpha| \le k, \tag{3.10}$$

where $\lambda_\alpha \ne 0$, it might appear that this would lead to a larger class of functions. This, however, is not the case; it is easily shown that if ϕ satisfies (3.10), then $\lambda_\alpha = \lambda$, for $|\alpha| \le k$.

In one dimension we can prove more.

Theorem 3.2. (Strang and Fix 1973) Suppose $\phi \in H^q(\mathbb{R})$ (in one dimension) has compact support and satisfies condition (1) of Theorem 3.1, that is, it satisfies (3.5) and (3.6). Then

$$\hat{\phi}(\xi) = Z(\xi)\left(\frac{\sin(\xi/2)}{\xi/2}\right)^{k+1}, \tag{3.11}$$

where $Z(\xi)$ is an entire function.

Proof. Because ϕ has compact support, $\hat{\phi}(\xi)$ is an entire function and, because of (3.5) and (3.6), $\hat{\phi}(0) \ne 0$, and $\hat{\phi}(\xi)$ has zeros of at least order k at $2\pi j$, $0 \ne j \in \mathbb{Z}$. Let

$$\hat{\sigma}_k(\xi) = \left(\frac{\sin(\xi/2)}{\xi/2}\right)^{k+1}. \tag{3.12}$$

The function $\hat{\sigma}_k(\xi)$ is entire with only zeros of order $k+1$ at $2\pi j$, for $0 \ne j \in Z$. Hence

$$Z(\xi) = \hat{\phi}(\xi)/\hat{\sigma}_k(\xi)$$

is entire, and

$$\hat{\phi}(\xi) = Z(\xi)\left(\frac{\sin(\xi/2)}{\xi/2}\right)^{k+1}, \tag{3.13}$$

as desired. □

Theorem 3.3. (Babuška 1970) Suppose $\phi \in H^q(\mathbb{R})$ (in one dimension) has compact support and satisfies condition (1) of Theorem 3.1, that is, it satisfies (3.5) and (3.6). Then, for any $\epsilon > 0$,

$$\operatorname{supp}\phi \not\subset \left[-\frac{(k+1)}{2}+\epsilon, \frac{(k+1)}{2}-\epsilon\right]. \tag{3.14}$$

Proof. Suppose, on the contrary, that

$$\operatorname{supp}\phi \subset [-(k+1)/2+\epsilon, (k+1)/2-\epsilon], \quad \text{for some } \epsilon > 0. \tag{3.15}$$

We will show that this assumption leads to a contradiction.

The function $\hat{\phi}(\xi)$ is entire and, with $\xi = \xi_1 + i\xi_2$, (3.15) implies

$$|\hat{\phi}(\xi)| \leq C\mathrm{e}^{(\frac{(k+1)}{2}-\epsilon)|\xi_2|}. \tag{3.16}$$

This estimate follows directly from the definition of the Fourier transform and assumption (3.15). Using elementary properties of the sine function, we find that

$$\left|\left(\frac{\sin(\xi/2)}{\xi/2}\right)^{k+1}\right| \geq C\frac{\mathrm{e}^{\frac{k+1}{2}|\xi_2|}}{|\xi_2|^{k+1}}, \quad \text{for } |\xi_2| \text{ large.} \tag{3.17}$$

Using (3.5), (3.6), (3.16), and (3.17), we have

$$|Z(\xi)| = \left|\frac{\hat{\phi}(\xi)}{\left(\frac{\sin(\xi/2)}{\xi/2}\right)^{k+1}}\right| \leq C_0 + C_{k+1}|\xi|^{k+1}, \quad \text{for all } \xi \in \mathcal{C}, \tag{3.18}$$

where $Z(\xi)$ is as in (3.11). Since $Z(\xi)$ is entire, estimate (3.18) implies (via a generalization of Liouville's theorem for entire functions) that $Z(\xi)$ is a polynomial of degree $\leq k+1$. Next, we use (3.11) and (3.16) to get

$$\left|Z(\xi)\left(\frac{\sin(\xi/2)}{\xi/2}\right)^{k+1}\right| = |\hat{\phi}(\xi)| \leq C\mathrm{e}^{(\frac{k+1}{2}-\epsilon)|\xi_2|}. \tag{3.19}$$

Combining this estimate with the lower bound in (3.17), we have

$$|Z(\xi)| \leq C|\xi|^{k+1}\mathrm{e}^{-\epsilon|\xi_2|}, \quad \text{for } |\xi_2| \text{ large.} \tag{3.20}$$

This implies $Z(\xi) = 0$. Thus, (3.11) implies $\hat{\phi}(\xi) = 0$, which contradicts (3.5). Thus (3.15) is false, which proves (3.14). □

The case of uniformly distributed particles is very special, but we have considered it, and cited Theorem 3.1 from Strang and Fix (1973) because the result provides necessary and sufficient conditions on the basic shape function ϕ for the validity of the approximability result (3.8) and (3.9), leading to the optimal value for μ in (3.4).

Remark 4. Condition (2) of Theorem 3.1 implies that

$$\sum_{j\in\mathbb{Z}^n} \phi(x-j) = b. \tag{3.21}$$

Hence the functions $\phi(x-j)/b, j \in \mathbb{Z}^n$, form a *partition of unity*. Thus sets $\mathring{\eta}_j^h$ form an open cover of $\mathbb{R}^n$.

Remark 5. Condition (2), *i.e.*, (3.7), of Theorem 3.1 is the definition of the notion of quasi-reproducing of order k, and from Theorem 3.1 we see that this notion is equivalent to condition (1), *i.e.*, to (3.5) and (3.6). It is of interest to have a condition similar to (3.5) and (3.6) that is equivalent to the related notion of reproducing of order k, which can be stated as

$$\sum_{j\in\mathbb{Z}^n} p(j)\phi(x-j) = p(x), \quad \text{for all polynomials } p(x) \text{ of degree} \leq k.$$

It can be shown that ϕ is reproducing of order k if and only if

(a) $\hat{\phi}(0) = 1$,

(b) $D^\alpha\hat{\phi}(0) = 0$, for $1 \leq |\alpha| \leq k$,

(c) $D^\alpha\hat{\phi}(2\pi j) = 0$, for $0 \neq j \in \mathbb{Z}^n$ and $|\alpha| \leq k$.

The proof follows the argument of the proof of Theorem 3.1 in Strang and Fix (1973).

Remark 6. The B-spline of order k, denoted by $\sigma_k(x)$, is the $(k+1)$-fold convolution of the characteristic function of the cube $(-1/2, 1/2)^n$. The support of $\sigma_k(x)$ is the closed cube $[-(k+1)/2, (k+1)/2]^n$, and the Fourier transform of $\sigma_k(x)$ is given by

$$\hat{\sigma}_k(\xi) = \prod_{i=1}^{n}\left(\frac{\sin \xi_i/2}{\xi_i/2}\right)^{k+1}.$$

Here $\hat{\sigma}_k(\xi)$ satisfies (3.5) and (3.6), and thus $\sigma_k(x)$ is quasi-reproducing of order k. We note, however, that $\sigma_k(x)$ is not reproducing of order k for $k > 1$, since $\hat{\sigma}_k(\xi)$ does not satisfy condition (b) in Remark 2 for $k > 1$.

In one dimension ($n = 1$), we can say more. If $\phi(x)$ satisfies (3.5) and (3.6), then from Theorem 3.2, $\hat{\phi}(\xi)$ is the product of $\hat{\sigma}_k(\xi)$ and a suitable entire function $Z(\xi)$. Using the Paley–Wiener theorem (see Rudin (1991)), it can be shown that $Z(\xi)$ is the Fourier transform of a distribution ψ with compact support: $Z(\xi) = \hat{\psi}(\xi)$. We can thus express (3.11) as

$$\hat{\phi}(\xi) = \hat{\psi}(\xi)\hat{\sigma}_k(\xi).$$

Thus any $\phi(x)$ that satisfies (3.5) and (3.6), which may or may not be piecewise polynomial, can be constructed via the convolution of the B-spline with a distribution with compact support. If $n > 1$, no such divisor $\hat{\phi}/\hat{\sigma}_k$ exists in general.

Remark 7. Theorem 3.3 has an especially simple interpretation for ϕ satisfying (3.5) and (3.6), and whose support is a symmetric interval about 0;

namely, $\operatorname{supp}\phi \supseteq \left[-\frac{k+1}{2}, \frac{k+1}{2}\right]$, and hence grows with k. As mentioned above, the support of $\sigma_k(x)$ is $\left[-\frac{k+1}{2}, \frac{k+1}{2}\right]$, and hence $\sigma_k(x)$ has minimal support.

Remark 8. The size of the support η of ϕ, *i.e.*, the diameter of the smallest closed ball containing η, plays an important role when particle shape functions are used in a Galerkin method to approximate the solution of a variationally posed boundary value problem. The Galerkin method leads to a linear algebraic system, where the (stiffness) matrix is banded, and it is well known that such linear systems can be solved efficiently by the elimination method when the bandwidth is small. The size of the bandwidth is directly proportional to the size of η. Thus it is desirable to use a ϕ whose support is as small as possible, of course, without sacrificing the accuracy of the computed solution of the differential equation. In one dimension, as mentioned above, the basic shape function $\sigma_k(x)$ has the minimal support $[-(k+1)/2, (k+1)/2]$, which increases with k.

Remark 9. Suppose $\phi(x) \in H^q(\mathbb{R}^n)$ has compact support and is quasi-reproducing of order k. For $s \leq q$, let

$$\phi^*(x) = \left(1 + \sum_{|\beta| \leq s} \alpha_\beta D^\beta\right)\phi(x).$$

Then $\phi^*(x) \in H^{q-s}(\mathbb{R}^n)$, ϕ^* is quasi-reproducing of order k, and $\operatorname{supp}\phi^* = \operatorname{supp}\phi$.

Remark 10. Suppose $\phi \in H^q(\mathbb{R}^n)$ is quasi-reproducing of order k. It is possible to construct another shape function ϕ^*, in terms of ϕ, that is reproducing of order k, is in $H^q(\mathbb{R}^n)$, but will have a larger support than ϕ. For example, ϕ^* can be constructed as a linear combination of translates of ϕ.

Remark 11. For the sake of simplicity, suppose $k = 2$ or 3 in this remark. Consider the function $\phi(x), x \in \mathbb{R}$, whose Fourier transform is given by

$$\hat{\phi}_k(\xi) = \hat{\sigma}_k(\xi)\left(1 - \frac{\hat{\sigma}_k''(0)}{2}\xi^2\right), \tag{3.22}$$

where $\sigma_k(x)$ is the B-spline of order k (recall $\sigma_k(x)$ is quasi-reproducing of order k and its support is $[-(k+1)/2, (k+1)/2)$. Here $\hat{\phi}_k(\xi)$ satisfies conditions (a), (b), (c) of Remark 2, and hence $\phi_k(x)$ is reproducing of order k. From (3.22) it is clear that

$$\phi_k(x) = \left[\sigma_k * \left(\delta_0 - \frac{\hat{\sigma}_k''(0)}{2}\delta_0''\right)\right](x),$$

where δ_0 is the Dirac distribution, and hence $\operatorname{supp}\phi_k = \operatorname{supp}\sigma_k$. From the expression for $\hat{\sigma}_k(\xi)$, we easily see that $\sigma_k \in H^{k+1/2-\epsilon}(\mathbb{R})$. Using the expression for $\hat{\phi}_k(\xi)$ (*i.e.*, (3.22)), we get $\phi_k \in H^{k-3/2-\epsilon}(\mathbb{R})$, but $\phi_k \notin H^{k-1}(\mathbb{R})$. Thus, for $k = 2$, we have $\phi_2 \in H^{1/2-\epsilon}(\mathbb{R})$, but $\phi_2 \notin H^1(\mathbb{R})$. And for $k = 3$, we see that $\phi_3 \in H^{3/2-\epsilon}(\mathbb{R})$, and hence $\phi_3 \in H^1(\mathbb{R})$. For approximating solutions of second-order differential equations, it is important that the shape functions are in H^1. This dichotomy between the cases $k = 2$ (even) and $k = 3$ (odd), in fact, holds for all k. Thus, for k even, there is no $\phi_k \in H^1(\mathbb{R})$ that is reproducing of order k with support $\left[-\frac{k+1}{2}, \frac{k+1}{2}\right]$. This latter result can be seen as follows. From Remark 3 we know that ϕ_k is the convolution of σ_k and a distribution with compact suport. If we require $\operatorname{supp} \phi_k = \operatorname{supp} \sigma_k$, then this distribution must be supported at the origin, and hence its Fourier transform must be a polynomial. We have examined the case when the polynomial is $1 - \frac{\hat{\sigma}_k''(0)}{2}\xi^2$; the general situation is similar.

Remark 12. The weights in (3.8) depend on u, but they are not unique. We note that the functions ϕ_j^h may be linearly dependent. Taking $q = k+1$ allows application of Theorem 3.1 for all $s \leq k+1$. Taking $q > k+1$, *i.e.*, assuming extra smoothness on the particle shape functions, does not change the estimate. The approximability of the classical finite element shape functions (the hat functions) can be analysed with Theorem 3.1 with $q = k = 1$.

Remark 13. The space $V_h^{k,q}$ is a S^{t,k^*}-*regular system* (this notion will be introduced in Section 3.2), with $k^* = q$ and $t = k+1$. S^{t,k^*}-regular systems are analysed in Babuška and Aziz (1972). They have many important properties, some of which will be used in the following sections.

3.2. Alternative proof for uniformly distributed particles and particle shape functions

In this section we first give an alternative proof that condition (2) of Theorem 3.1 implies estimate (3.8), again for uniformly distributed particles and associated shape functions. This alternative proof does not use the Fourier transform, and it can be naturally generalized to non-uniformly distributed particles.

We review our notation before stating the theorem. Recall that

$$x_j^h = jh, \quad \text{for } j \in \mathbb{Z}^n \text{ and } 0 < h,$$

are the particles, and $\phi \in H^q(\mathbb{R}^n)$, with $q \geq 0$, is the basic shape function. Also $\eta = \operatorname{supp} \phi \subset B_\rho$, and $0 \in \mathring{\eta}$. Then the particle shape functions, $\phi_j^h(x)$, are defined by

$$\phi_j^h(x) = \phi\left(\frac{x - jh}{h}\right);$$

it is immediate that

$$\eta_j^h = \text{ supp } \phi_j^h \subset B_{\rho h}^j,$$

and $x_j^h \in \mathring{\eta}_j^h$.

Theorem 3.4. Suppose $\phi \in H^q(\mathbb{R}^n)$, with *smoothness index* $q \geq 0$, has compact support $\eta \subset B_\rho$, and suppose the $\phi_j^h(x)$ are defined in (3.1). Suppose $k = 0, 1, 2, \ldots$ and suppose, for $|\alpha| \leq k$,

$$\sum_{j \in \mathbb{Z}^n} j^\alpha \phi(x - j) = \lambda x^\alpha + q^\alpha(x); \tag{3.23}$$

here $\lambda \neq 0$, and $q^\alpha(x)$ is a polynomial of degree $< |\alpha|$, *i.e.*, suppose ϕ is quasi-reproducing of order k. Suppose u satisfies

$$\sum_{j \in \mathbb{Z}^n} \|u\|^2_{H^{r_j+1}(B^j_{\overline{\rho}h})} < \infty, \quad \text{where } 0 \leq r_j \leq k, \tag{3.24}$$

where $\overline{\rho} \geq 1$ is sufficiently large and independent of h. Then there exist weights w_l^h such that

$$\left\| u - \sum_{l \in \mathbb{Z}^n} w_l^h \phi_l^h \right\|^2_{H^s(\mathbb{R}^n)} \leq C \sum_{j \in \mathbb{Z}^n} h^{2(r_j+1-s)} \|u\|^2_{H^{r_j+1}(B^j_{\overline{\rho}h})},$$

$$\text{for } 0 \leq s \leq \min_{j \in \mathbb{Z}_n} \{q, r_j + 1\}, \tag{3.25}$$

where C is independent of u and h. If $u \in H^{k'+1}(\mathbb{R}^n)$, where $0 \leq k' \leq k$, then

$$\left\| u - \sum_{l \in \mathbb{Z}^n} w_l^h \phi_l^h \right\|_{H^s(\mathbb{R}^n)} \leq C h^{k'+1-s} \|u\|_{H^{k'+1}(\mathbb{R}^n)}, \quad \text{for } 0 \leq s \leq \min\{q, k'+1\}. \tag{3.26}$$

Proof. The proof is in several steps.

1. Suppose ϕ satisfies (3.23), and write $q^\alpha(x) = \sum_{|\gamma| \leq |\alpha|-1} d_{\gamma\alpha} x^\gamma$. Then

$$\begin{aligned}
\sum_{j \in \mathbb{Z}^n} (x_j^h)^\alpha \phi_j^h(x) &= \sum_{j \in \mathbb{Z}^n} (jh)^\alpha \phi_j^h(x) \\
&= h^{|\alpha|} \sum_{j \in \mathbb{Z}^n} j^\alpha \phi\left(\frac{x}{h} - j\right) \\
&= h^{|\alpha|} \left\{ \lambda \left(\frac{x}{h}\right)^\alpha + q^\alpha\left(\frac{x}{h}\right) \right\}
\end{aligned}$$

$$= \lambda x^\alpha + h^{|\alpha|} \sum_{|\gamma| \leq |\alpha|-1} d_{\gamma\alpha} \left(\frac{x}{h}\right)^\gamma$$

$$= \lambda x^\alpha + \sum_{|\gamma| \leq |\alpha|-1} h^{|\alpha|-|\gamma|} d_{\gamma\alpha} x^\gamma, \quad \text{for } |\alpha| \leq k. \tag{3.27}$$

Equations (3.23) and (3.27) are in fact equivalent: (3.27) could be viewed as a scaled version of (3.23). For any $p \in \mathcal{P}^k$, there is a uniquely determined $w = w_{p,h} \in \mathcal{P}^k$ satisfying

$$p(x) = \sum_{j \in \mathbb{Z}^n} w_{p,h}(x_j^h) \phi_j^h(x), \quad \text{for all } x \in \mathbb{R}^n. \tag{3.28}$$

We first prove the existence of $w_{p,h}$, and begin by considering the monic polynomials $p_\alpha = x^\alpha$. Suppose $|\alpha| = 0$. Then from (3.27) we have

$$1 = \sum_{j \in \mathbb{Z}^n} \frac{1}{\lambda} \phi_j^h(x) = \sum_{j \in \mathbb{Z}^n} w_{\{1\},h}(x_j^h)\, \phi_j^h(x),$$

where $w_{\{1\},h}(x) = 1/\lambda$. Next suppose $|\alpha| = 1$. Using (3.27) again we have

$$x^\alpha = \sum_{j \in \mathbb{Z}^n} \frac{1}{\lambda} (x_j^h)^\alpha \phi_j^h(x) - \frac{h\, d_{0\alpha}}{\lambda} = \sum_{j \in \mathbb{Z}^n} w_{\{x^\alpha\},h}(x_j^h) \phi_j^h(x)$$

where

$$w_{\{x^\alpha\},h}(x) = \frac{x^\alpha}{\lambda} - \frac{h\, d_{0\alpha}}{\lambda^2}.$$

Proceeding in this way, by induction, we get $w_{\{x^\alpha\},h}(x)$ for $|\alpha| \leq k$, where $w_{\{x^\alpha\},h}(x)$ is of the form

$$w_{\{x^\alpha\},h}(x) = e_{\alpha\alpha} x^\alpha + \sum_{|\beta| \leq |\alpha|-1} e_{\alpha\beta} h^{|\alpha|-|\beta|} x^\beta, \tag{3.29}$$

where $e_{\alpha\alpha} = \lambda^{-1}$ and $e_{\alpha\beta}$ are expressions in $d_{\gamma\alpha}$, $|\gamma| < |\alpha|$. For $p(x) = \sum_{|\alpha| \leq k} c_\alpha x^\alpha$, we let $w_{p,h}(x) = \sum_{|\alpha| \leq k} c_\alpha w_{\{x^\alpha\},h}(x)$. It is immediate that

$$p(x) = \sum_{j \in \mathbb{Z}^n} w_{p,h}(x_j^h) \phi_j^h(x),$$

which establishes the existence of $w_{p,h}(x)$. We can show that

$$w_{p,h}(x) = \sum_{|\beta| \leq k} \left[\sum_{|\beta|+1 \leq |\alpha| \leq k,\, \alpha = \beta} c_\alpha d_{\alpha\beta} h^{|\alpha|-|\beta|} \right] x^\beta. \tag{3.30}$$

To prove uniqueness, suppose $w_{p,h}(x) = 0$. We will show that $p(x) = \sum_{|\alpha| \leq k} c_\alpha x^\alpha = 0$. Since $w_{p,h}(x) = 0$, it is clear from (3.30) that the coefficient of x^β is zero for $|\beta| \leq k$, from which we can deduce that $c_\alpha = 0$, $|\alpha| \leq k$, and thus $p(x) = 0$. It will be convenient to write $w_{p,h}(x) = \mathcal{A}^h p$. Then $\mathcal{A}^h : \mathcal{P}^k \to \mathcal{P}^k$ is a bijection satisfying

$$p(x) = \sum_{j \in \mathbb{Z}^n} (\mathcal{A}^h p)(x_j^h)\phi_j^h(x), \quad \text{for all } x \in \mathbb{R}^n, \quad \text{for any } p \in \mathcal{P}^k. \tag{3.31}$$

We define $\mathcal{A} = \mathcal{A}^h$ when $h = 1$. We note that $\mathcal{A}$ satisfies (3.31) with $h = 1$. We also have

$$[(\mathcal{A}^h)^{-1} w](x) = \sum_{j \in \mathbb{Z}^n} w(x_j^h)\phi_j^h, \quad \text{for all } x \in \mathbb{R}^n, \quad \text{for any } w \in \mathcal{P}^k. \tag{3.32}$$

It is also clear from the construction that $\mathcal{A}^h : \mathcal{P}^i \to \mathcal{P}^i$, for $i \leq k$.

2. Define the cells ω_j and ω_j^h:

$$\omega_j = \left\{ x : \|x - j\|_\infty \equiv \max_{i=1,\dots,n} |x_i - j_i| < \rho \right\}$$

and

$$\omega_j^h = \left\{ x : \|x - x_j^h\|_\infty \equiv \max_{i=1,\dots,n} |x_i - x_{j_i}^h| < \rho h \right\}.$$

The families $\{\omega_j\}_{j \in \mathbb{Z}^n}$ and $\{\omega_j^h\}_{j \in \mathbb{Z}^n}$ are open covers of $\mathbb{R}^n$ provided $\rho > 1/2$. Let

$$A_j^h = \{ l \in \mathbb{Z}^n : \eta_l^h \cap \omega_j^h \neq \emptyset \},$$

and define

$$\Omega_j^h = \cup_{l \in A_j^h} \omega_l^h.$$

It is immediate that one can select M and $\bar{\rho}$ such that

$$\text{card } A_j^h \leq M \tag{3.33}$$

and

$$\Omega_j^h \subset B_{\bar{\rho} h}^j. \tag{3.34}$$

The constants M and $\bar{\rho}$ are independent of j and h, but do depend on ϕ; specifically on ρ.

For any $l \in \mathbb{Z}^n$, since $u \in H^{r_l+1}(B_{\bar{\rho}h}^l)$, it is well known (Bramble and Hilbert 1970, Bramble and Hilbert 1971, Ciarlet 1980) that there is a polynomial $p^{l,h} = p_k^{l,h}$ of degree $\leq k$ such that

$$\|u - p^{l,h}\|_{H^s(B_{\bar{\rho}h}^l)} \leq C h^{r_l+1-s} \|u\|_{H^{r_l+1}(B_{\bar{\rho}h}^l)}, \quad \text{for } 0 \leq s \leq r_l + 1 \leq k+1, \tag{3.35}$$

where C is independent of u, h, and l, but does depend on k ($p^{l,h}$ can, in fact, be chosen such that its degree $\leq r_l$). Define the weights

$$w_l^h = (\mathcal{A}^h p^{l,h})(x_l^h). \tag{3.36}$$

Let j be fixed. We will work with the polynomial $p^{j,h}$, which satisfies (3.35) with $l = j$, as well as the polynomial $p^{l,h}$. Using (3.36), we find

$$\begin{aligned}
\left\| u - \sum_{l\in\mathbb{Z}^n} w_l^h \phi_l^h \right\|_{H^s(\omega_j^h)}
&\le \left\| u - \sum_{l\in A_j^h} w_l^h \phi_l^h \right\|_{H^s(\omega_j^h)} \\
&\le \left\| u - \sum_{l\in A_j^h} (\mathcal{A}^h p^{j,h})(x_l^h)\phi_l^h \right\|_{H^s(\omega_j^h)} \\
&\quad + \sum_{l\in A_j^h} |(\mathcal{A}^h p^{j,h})(x_l^h) - (\mathcal{A}^h p^{l,h})(x_l^h)|\, \|\phi_l^h\|_{H^s(\omega_j^h)}. \qquad (3.37)
\end{aligned}$$

We now estimate the two terms on the right side of (3.37).

3. From (3.31) and the definition of A_j^h, we have

$$p(x) = \sum_{l\in\mathbb{Z}^n} (\mathcal{A}^h p)(x_l^h)\phi_l^h(x) = \sum_{l\in A_j^h} (\mathcal{A}^h p)(x_l^h)\phi_l^h(x), \quad \text{for } x \in \omega_j^h,$$

for any $p \in \mathcal{P}^k$. Using this formula and (3.35) with $l = j$, we obtain the estimate

$$\begin{aligned}
\left\| u - \sum_{l\in A_j^h} (\mathcal{A}^h p^{j,h})(x_l^h)\phi_l^h \right\|_{H^s(\omega_j^h)} &= \|u - p^{j,h}\|_{H^s(\omega_j^h)} \\
&\le Ch^{r_j+1-s}\|u\|_{H^{r_j+1}(B^j_{\bar{\rho}h})}, \qquad (3.38)
\end{aligned}$$

for the first term of (3.37).

A scaling argument shows that

$$\|\phi_l^h\|_{H^s(\omega_j^h)} \le h^{-s+n/2}\|\phi\|_{H^s(\mathbb{R}^n)}.$$

Thus

$$\begin{aligned}
&\sum_{l\in A_j^h} |\mathcal{A}^h p^{j,h}(x_l^h) - \mathcal{A}^h p^{l,h}(x_l^h)|\, \|\phi_l^h\|_{H^s(\omega_j^h)} \\
&\qquad \le Ch^{-s+n/2} \sum_{l\in A_j^h} |\mathcal{A}^h p^{j,h}(x_l^h) - \mathcal{A}^h p^{l,h}(x_l^h)|. \qquad (3.39)
\end{aligned}$$

It remains to estimate the right-hand side of this inequality.

For $l \in A_j^h$, $\omega_l^h \subset \Omega_j^h$, and hence, using (3.34), $\omega_l^h \subset B^j_{\bar{\rho}h}$. Also $\omega_l^h \subset B^l_{\bar{\rho}h}$.

Thus, using (3.35) with $s = 0$, we have

$$\begin{aligned}\|p^{j,h} - p^{l,h}\|_{H^0(\omega_l^h)} &\leq \|p^{j,h} - u\|_{H^0(\omega_l^h)} + \|u - p^{l,h}\|_{H^0(\omega_l^h)} \\ &\leq \|p^{j,h} - u\|_{H^0(B^j_{\bar{\rho}h})} + \|u - p^{l,h}\|_{H^0(B^l_{\bar{\rho}h})} \\ &\leq Ch^{r_j+1}\|u\|_{H^{r_j+1}(B^j_{\bar{\rho}h})} \\ &\quad + Ch^{r_l+1}\|u\|_{H^{r_l+1}(B^l_{\bar{\rho}h})}. \end{aligned} \tag{3.40}$$

It is easily shown that there is a constant C such that

$$\|w\|_{L^\infty(\omega_l^h)} \leq Ch^{-n/2}\|w\|_{H^0(\omega_l^h)}, \quad \text{for any } w \in \mathcal{P}^k; \tag{3.41}$$

C is independent of w, h, and l. Applying (3.41) to $w = \mathcal{A}^h p^{j,h} - \mathcal{A}^h p^{l,h}$, we have

$$\begin{aligned}&|(\mathcal{A}^h p^{j,h})(x_l^h) - (\mathcal{A}^h p^{l,h})(x_l^h)| \\ &\quad \leq \|\mathcal{A}^h p^{j,h} - \mathcal{A}^h p^{l,h}\|_{L^\infty(\omega_l^h)} \\ &\quad \leq Ch^{-n/2}\|\mathcal{A}^h p^{j,h} - \mathcal{A}^h p^{l,h}\|_{H^0(\omega_l^h)}. \end{aligned} \tag{3.42}$$

For any $p \in \mathcal{P}^k$, we write $p(x) = \tilde{p}(\frac{x - x_l^h}{h})$, where $\tilde{p} \in \mathcal{P}^k$. Using (3.31) with $h = 1$ (recall that $\mathcal{A} = \mathcal{A}^h$ for $h = 1$), we see that

$$\tilde{p}(x) = \sum_{i \in \mathbb{Z}^n} (\mathcal{A}\tilde{p})(i)\phi(x - i),$$

and therefore

$$\begin{aligned} p(x) &= \tilde{p}\left(\frac{x - x_l^h}{h}\right) \\ &= \sum_{i \in \mathbb{Z}^n} (\mathcal{A}\tilde{p})(i)\phi\left(\frac{x - x_{i+l}^h}{h}\right) \\ &= \sum_{i \in \mathbb{Z}^n} (\mathcal{A}\tilde{p})(i)\phi_{i+l}^h(x) \\ &= \sum_{i \in \mathbb{Z}^n} (\mathcal{A}\tilde{p})(i - l)\phi_i^h(x) \\ &= \sum_{i \in \mathbb{Z}^n} (\mathcal{A}\tilde{p})\left(\frac{x_i^h - x_l^h}{h}\right)\phi_i^h(x). \end{aligned}$$

Comparing the above expression with (3.31) and using the uniqueness of the representation (3.28), we obtain

$$(\mathcal{A}^h p)(x) = (\mathcal{A}\tilde{p})\left(\frac{x - x_l^h}{h}\right). \tag{3.43}$$

We further note that $\|\mathcal{A}\tilde{p}\|_{H^0(\omega_0)}$ is a norm on $\tilde{p}$, and since all norms are equivalent on a finite-dimensional space, we have

$$\|\mathcal{A}\tilde{p}\|_{H^0(\omega_0)} \leq C\|\tilde{p}\|_{H^0(\omega_0)}. \tag{3.44}$$

Therefore, from (3.43) and (3.44), and using the transformation $y = (x - x_l^h)/h$, we have

$$\begin{aligned}
\|\mathcal{A}^h p\|^2_{H^0(\omega_l^h)} = \int_{\omega_l^h} |(\mathcal{A}^h p(x)|^2 \,\mathrm{d}x &= \int_{\omega_0} |(\mathcal{A}\tilde{p}(y)|^2 \,\mathrm{d}y \\
&\leq C \int_{\omega_0} |\tilde{p}(y)|^2 \,\mathrm{d}y \\
&= C \int_{\omega_l^h} |\tilde{p}((x - x_l^h)/h)|^2 \,\mathrm{d}x \\
&= C \int_{\omega_l^h} |p(x)|^2 \,\mathrm{d}x \\
&= C\|p\|^2_{H^0(\omega_l^h)}, \quad \text{for } p \in \mathcal{P}^k,
\end{aligned} \tag{3.45}$$

with C independent of p, l, and h. Combining (3.42), (3.45) with $p^{l,h} - p^{j,h}$ and (3.40) yields

$$\begin{aligned}
&|(\mathcal{A}^h p^{j,h})(x_l^h) - (\mathcal{A}^h p^{l,h})(x_l^h)| \\
&\qquad \leq Ch^{-n/2}(h^{r_j+1}\|u\|_{H^{r_j+1}(B^j_{\bar{\rho}h})} + h^{r_l+1}\|u\|_{H^{r_l+1}(B^l_{\bar{\rho}h})}),
\end{aligned}$$

and hence, using (3.33), we have the estimate

$$\begin{aligned}
&\sum_{l\in A_j^h} |(\mathcal{A}^h p^{j,h})(x_l^h) - (\mathcal{A}^h p^{l,h})(x_l^h)| \\
&\qquad \leq Ch^{-n/2}\{Mh^{r_j+1}\|u\|_{H^{r_j+1}(B^j_{\bar{\rho}h})} \\
&\qquad\qquad + \sum_{l\in A_j^h} h^{r_l+1}\|u\|_{H^{r_l+1}(B^l_{\bar{\rho}h})}\}
\end{aligned} \tag{3.46}$$

for the right side of (3.37). Now we combine (3.37), (3.38), (3.39), and (3.46) to obtain

$$\left\| u - \sum_{l\in\mathbb{Z}^n} w_l^h \phi_l^h \right\|_{H^s(\omega_j^h)} \leq C \sum_{l\in A_j^h} h^{r_l+1-s}\|u\|_{H^{r_l+1}(B^l_{\bar{\rho}h})}. \tag{3.47}$$

4. Finally, we estimate $\|u - \sum_{l\in\mathbb{Z}^n} w_l^h \phi_l^h\|_{H^s(\mathbb{R}^n)}$. Using (3.47), which is valid for all $j \in \mathbb{Z}^n$, and (3.33) we obtain

$$\begin{aligned}\left\|u - \sum_{l\in\mathbb{Z}^n} w_l^h \phi_l^h\right\|^2_{H^s(\mathbb{R}^n)} &\le \sum_{j\in\mathbb{Z}^n} \left\|u - \sum_{l\in\mathbb{Z}^n} w_l^h \phi_l^h\right\|^2_{H^s(\omega_j^h)} \\ &\le C \sum_{j\in\mathbb{Z}^n} \sum_{l\in\mathcal{A}_j^h} h^{2(r_l+1-s)} \|u\|^2_{H^{r_l+1}(B^l_{\bar\rho h})} \\ &\le C \sum_{j\in\mathbb{Z}^n} h^{2(r_j+1-s)} \|u\|^2_{H^{r_j+1}(B^j_{\bar\rho h})},\end{aligned} \tag{3.48}$$

where C (which depends on M) is independent of u and h. This proves (3.25).

Suppose $u \in H^{k'+1}(\mathbb{R}^n)$, where $0 \le k' \le k$. Then taking $r_j = k'$ in (3.48), and using the fact that the overlap in $\{B^j_{\bar\rho h}\}_{j\in\mathbb{Z}^n}$ is bounded independently of h, we get

$$\begin{aligned}\left\|u - \sum_{l\in\mathbb{Z}^n} w_l^h \phi_l^h\right\|_{H^s(\mathbb{R}^n)} &\le Ch^{k'+1-s} \left(\sum_{j\in\mathbb{Z}^n} \|u\|^2_{H^{k'+1}(B^j_{\bar\rho h})} \right)^{1/2} \\ &\le Ch^{k'+1-s} \|u\|_{H^{k'+1}(\mathbb{R}^n)},\end{aligned} \tag{3.49}$$

where C is independent of u and h, which is (3.26). □

Remark 14. Estimates (3.25) and (3.26) have been established provided $\bar\rho$ is sufficiently large; specifically, provided (3.33) holds. As pointed out in connection with (3.34), $\bar\rho$ depends on ρ. Note that the constants C in (3.25) and (3.26) depend on $\bar\rho$.

So far in this section, we have considered functions u defined on $\mathbb{R}^n$, and have presented a result on the approximation of u by particle shape functions. We now consider functions u defined on a bounded domain Ω in $\mathbb{R}^n$ with Lipschitz-continuous boundary. We will show that $V_h^{k,q}$, defined in (3.2), when restricted to Ω, provides accurate approximation to u.

We first recall the well-known extension result (Stein 1970) that there is a bounded extension operator $E : L_2(\Omega) \to L_2(\mathbb{R}^n)$, *i.e.*, an operator E satisfying $Eu|_\Omega = u$ for all $u \in L_2(\Omega)$, such that if $u \in H^m(\Omega)$ then $Eu \in H^m(\mathbb{R}^n)$ and

$$\|Eu\|_{H^m(\mathbb{R}^n)} \le C_m \|u\|_{H^m(\Omega)}, \quad \text{for all } u \in H^m(\Omega),\ m = 0,\ 1, \ldots. \tag{3.50}$$

Here C_m is independent of u but depends on m.

We define the subset $\mathbb{Z}^n_\Omega$ of $\mathbb{Z}^n$, which will be used in the next result, by

$$\mathbb{Z}^n_\Omega = \{j \in \mathbb{Z}^n : \mathring{\eta}_j^h \cap \Omega \ne \emptyset\}, \tag{3.51}$$

where $\eta_j^h = \operatorname{supp} \phi_j^h$.

Theorem 3.5. Suppose $\phi \in H^q(\mathbb{R}^n)$, with smoothness index $q \geq 0$, has compact support $\eta \subset B_\rho$, and is quasi-reproducing of order k. Suppose $u \in H^{k'+1}(\Omega)$, where $0 \leq k' \leq k$. Then there are weights w_j^h such that

$$\left\| u - \sum_{l \in \mathbb{Z}_\Omega^n} w_l^h \phi_l^h \right\|_{H^s(\Omega)} \leq C h^{k'+1-s} \|u\|_{H^{k'+1}(\Omega)}, \quad 0 \leq s \leq \min(q, k'+1), \tag{3.52}$$

where C is independent of u and h.

Proof. Suppose $u \in H^{k'+1}(\Omega)$, and let $\bar{u} = Eu$, where E is the extension operator mentioned above. Applying (3.26) of Theorem 3.4 to $\bar{u}$, there are weights w_l^h such that

$$\left\| \bar{u} - \sum_{l \in \mathbb{Z}^n} w_l^h \phi_l^h \right\|_{H^s(\mathbb{R}^n)} \leq C h^{k'+1-s} \|\bar{u}\|_{H^{k'+1}(\mathbb{R}^n)}.$$

Therefore, from (3.50) with $m = k' + 1$, we have

$$\begin{aligned} \left\| u - \sum_{l \in \mathbb{Z}^n} w_l^h \phi_l^h \right\|_{H^s(\Omega)} &= \left\| \bar{u} - \sum_{l \in \mathbb{Z}^n} w_l^h \phi_l^h \right\|_{H^s(\Omega)} \\ &\leq \left\| \bar{u} - \sum_{l \in \mathbb{Z}^n} w_l^h \phi_l^h \right\|_{H^s(\mathbb{R}^n)} \\ &\leq C h^{k'+1-s} \|\bar{u}\|_{H^{k'+1}(\mathbb{R}^n)} \\ &\leq C h^{k'+1-s} \|u\|_{H^{k'+1}(\Omega)}. \end{aligned} \tag{3.53}$$

From the definition of $\mathbb{Z}_\Omega^h$ in (3.51), it is clear that

$$\left\| u - \sum_{l \in \mathbb{Z}^n} w_l^h \phi_l^h \right\|_{H^s(\Omega)} = \left\| u - \sum_{l \in \mathbb{Z}_\Omega^n} w_l^h \phi_l^h \right\|_{H^s(\Omega)},$$

and therefore from (3.53) we get the desired result. □

By examining the approximation of u, obtained in Theorem 3.5, namely

$$\sum_{l \in \mathbb{Z}_\Omega^n} w_l^h \phi_l^h |_\Omega,$$

we see that the sum involves only those ls for which

$$\operatorname{supp} \phi_l^h \cap \Omega \neq \emptyset,$$

that is, only those particles x_l^h such that $\operatorname{dist}(x_l^h, \Omega) < \rho h$. So the approximation involves particle shape functions corresponding to the particles inside Ω, as well as some particles lying outside Ω. We will denote the span of these shape functions by

$$V_{\Omega,h}^{k,q} = \text{span}\{\phi_j^h|_\Omega : \text{supp } \phi_j^h \cap \Omega \neq \emptyset\}. \tag{3.54}$$

Thus the functions in $V_{\Omega,h}^{k,q}$ are the functions in $V_h^{k,q}$ restricted to Ω.

(t, k^*)-regular systems

We now introduce (t, k^*)-regular systems of functions (*cf.* Babuška and Aziz (1972)). Let $\Omega \subseteq \mathbb{R}^n$, and suppose $S_h(\Omega)$, $0 < h \leq 1$, is a one-parameter family of linear spaces of functions on Ω. For $0 \leq k^* \leq t$, $S_h(\Omega)$ will be called a (t, k^*)-*regular system* and will be denoted by $S_h^{t,k^*}(\Omega)$ if

(1)

$$(1) S_h^{t,k^*}(\Omega) \subset H^{k^*}(\Omega), \quad \text{for } 0 < h \leq 1; \tag{3.55}$$

(2) For every $u \in H^l(\Omega)$, with $0 \leq l$, there is a function $g_h \in S_h^{t,k^*}$ such that

$$\|u - g_h\|_{H^s(\Omega)} \leq Ch^\mu \|u\|_{H^l(\Omega)}, \quad \text{for } 0 \leq s \leq \min\{l, k^*\}, \tag{3.56}$$

where $\mu = \min\{t - s, l - s\}$. The constant C is independent of u and h.

We now introduce two additional notions.

(LA) A (t, k^*)-regular system $S_h^{t,k^*}(\Omega)$ is said to satisfy a *local assumption* if, for $u \in H^l(\Omega)$, with support S, the function $g_h \in S_h^{t,k^*}(\Omega)$ in (3.56) can be chosen so that the support S_h of g_h has the property that

$$S_h \subset S^{\lambda h} \equiv \{x \in \Omega : d(x, S) \leq \lambda h\},$$

where $d(x, S)$ is the distance from x to S, and λ is independent of h.

(IA) We say that $S_h^{t,k^*}(\Omega)$ satisfies an *inverse assumption* (*cf.* Babuška and Aziz (1972)) if there is an $0 \leq \epsilon \leq k^*$ such that

$$\|g\|_{H^{k^*}(\Omega)} \leq Ch^{-(k^*-r)} \|g\|_{H^r(\Omega)},$$
$$\text{for all } k^* - \epsilon \leq r \leq k^* \text{ and all } g \in S_h^{t,k^*}(\Omega),$$

where C is independent of h and g (it may depend on k^* and ϵ).

A (t, k^*)-regular system is referred to as a (t, k)-regular system in classical literature (Babuška and Aziz 1972). We have used k^* instead of k in this paper for notational clarity.

The approximation space $V_{\Omega,h}^{k,q}$, defined in (3.54), is a (t, k^*)-regular system: more precisely, we have the following result.

Theorem 3.6. Suppose $0 \leq q < k + 1$, and suppose $\phi \in H^q(\mathbb{R}^n)$ has compact support and is quasi-reproducing of order k. Then $V_{\Omega,h}^{k,q}$ is a $(k + 1, q)$-regular system.

Proof. We show that $V_{\Omega,h}^{k,q}$ is a (t,k^*)-regular system with $t = k+1$ and $k^* = q$. Since $\phi \in H^q(\mathbb{R}^n)$, it is clear that $V_{\Omega,h}^{k,q} \subset H^q(\Omega)$ and thus (3.55) is immediate with $k^* = q$. Next we show that (3.56) is satisfied. Suppose $u \in H^l(\mathbb{R}^n)$ with $l \geq 0$. If $l = 0$, (3.56) is trivial. So, suppose $1 \leq l$. Applying Theorem 3.5, specifically (3.52) with $k' = \min(k+1, l) - 1$, we get

$$\left\| u - \sum_{j \in \mathbb{Z}_\Omega^n} w_j^h \phi_j^h \right\|_{H^s(\Omega)} \leq Ch^{\min(l-s,k+1-s)} \|u\|_{H^{\min(l,k+1)}(\Omega)} \leq Ch^\mu \|u\|_{H^l(\mathbb{R}^n)},$$

for $0 \leq s \leq \min\{q, \min\{l, k+1\}\} = \min\{l, q\}$ (since $q < k+1$), where $\mu = \min\{k+1-s, l-s\}$. This is (3.56), with $g_h = \sum_{l \in \mathbb{Z}_\Omega^n} w_l^h \phi_l^h |_\Omega$, $t = k+1$, and $k^* = q$. □

We now show that $V_{\Omega,h}^{k,q}$ satisfies the local assumption, LA.

Theorem 3.7. Suppose $\phi \in H^q(\mathbb{R}^n)$, where $0 \leq q \leq k+1$, has compact support and is quasi-reproducing of order k. Then $V_{\Omega,h}^{k,q}$ satisfies the local assumption, LA.

Proof. Suppose $u \in H^l(\Omega)$ such that supp $u = S \subset \Omega$. Consider the approximation of u, obtained in Theorem 3.5, namely

$$g_h = \sum_{j \in \mathbb{Z}_\Omega^n} w_j^h \phi_j^h. \tag{3.57}$$

A careful study of the proofs of Theorems 3.5 and 3.4, and considering the zero extension of u outside Ω, reveals that, for $j \in \mathbb{Z}_\Omega^n$,

$$w_j^h = 0, \quad \text{if and only if } B_{\bar{\rho}h}^j \cap S = \emptyset.$$

Now, for $j \in \mathbb{Z}_\Omega^n$ such that $w_j^h \neq 0$, we know that $\eta_j^h = \text{supp } \phi_j^h \subset B_{\rho h}^j$. Therefore, $S_h = \text{supp } g_h = \{x \in \mathbb{R}^n : d(x, S) \leq (\bar{\rho} + \rho)h\}$, and so we can take $\lambda = (\bar{\rho} + \rho)$ in the definition of LA. For small h, we have $S_h \subset \Omega$. Hence $V_{\Omega,h}^{k,q}$ satisfies the local assumption, LA. □

Remark 15. The particle space $V_h^{k,q}$ is $(k+1, q)$-regular and satisfies the local assumption, LA, for $\Omega = \mathbb{R}^n$.

We note that the particle spaces $V_{\Omega,h}^{k,q}$ and $V_h^{k,q}$ will also satisfy the inverse assumption, IA, if additional conditions are imposed on the shape functions $\{\phi_j^h\}$. We will formulate these conditions in Section 3.3 in the context of non-uniformly distributed particles; the corresponding conditions on the shape functions associated with uniformly distributed particles can then be obtained as a special case.

3.3. Approximation by particle shape functions associated with arbitrary (non-uniformly distributed) particles in $\mathbb{R}^n$: the h-version

In this section we will generalize the major part of Theorem 3.4.

Suppose $\{X^\nu\}_{\nu\in N}$ is a family of countable subsets of $\mathbb{R}^n$; the family is indexed by the parameter ν, which varies over the index set N. The points in X^ν are called *particles*, and will be denoted by $\underline{x}$, to distinguish them from general points in $\mathbb{R}^n$. If it is necessary to emphasize that $\underline{x} \in X^\nu$, we will write $\underline{x} = \underline{x}^\nu$. With each $\underline{x}^\nu \in X^\nu$ we associate

- $h^\nu_{\underline{x}^\nu} = h^\nu_{\underline{x}}$, a positive number;
- $\omega^\nu_{\underline{x}^\nu} = \omega^\nu_{\underline{x}}$, a bounded domain in $\mathbb{R}^n$;
- $\phi^\nu_{\underline{x}^\nu} = \phi^\nu_{\underline{x}}$, a function in $H^q(\mathbb{R}^n)$, with $q \geq 0$ and with $\eta^\nu_{\underline{x}^\nu} = \eta^\nu_{\underline{x}} \equiv \operatorname{supp} \phi^\nu_{\underline{x}^\nu}$ assumed compact.

The numbers $h^\nu_{\underline{x}^\nu} = h^\nu_{\underline{x}}$ will be referred to as the *sizes* of the particles $\underline{x}$, and the functions $\phi^\nu_{\underline{x}^\nu}$ are called the *particle shape functions.* For a given $\nu \in N$, let

$$\mathcal{M}^\nu = \left\{X^\nu, \{h^\nu_{\underline{x}}, \omega^\nu_{\underline{x}}, \phi^\nu_{\underline{x}}\}_{\underline{x}\in X^\nu}\right\}.$$

$\mathcal{M}^\nu$ will be referred to as a *particle-shape function system* – and $\{\mathcal{M}^\nu\}_{\nu\in N}$ as a family of particle-shape function systems. This nomenclature is similar to that used in the FEM when we speak of a triangulation and a family of triangulations.

Regarding the particle-shape function system, we make several assumptions.

A1. For each ν,

$$\bigcup_{\underline{x}\in X^\nu} \omega^\nu_{\underline{x}} = \mathbb{R}^n,$$

i.e., for each ν, $\{\omega^\nu_{\underline{x}}\}_{\underline{x}\in X^\nu}$ is an open cover of $\mathbb{R}^n$.

A2. For $\underline{x} \in X^\nu$, let

$$S^\nu_{\underline{x}} \equiv \{\underline{y} \in X^\nu : \omega^\nu_{\underline{x}} \cap \omega^\nu_{\underline{y}} \neq \emptyset\}.$$

There is a constant $\kappa < \infty$, which may depend on $\{\mathcal{M}^\nu\}_{\nu\in N}$, but neither on ν nor on $\underline{x} \in X^\nu$, such that

$$\operatorname{card} S^\nu_{\underline{x}} \leq \kappa, \quad \text{for all } \underline{x} \in X^\nu \text{ and all } \nu \in N.$$

A3. For all $\underline{x} \in X^\nu$, and $\nu \in N$, $\underline{x} \in \mathring{\eta}^\nu_{\underline{x}}$ and $\mathring{\eta}^\nu_{\underline{x}} \subset \omega^\nu_{\underline{x}}$.

A4. For $\underline{x} \in X^\nu$, let

$$\Omega^\nu_{\underline{x}} = \bigcup_{\underline{y}\in Q^\nu_{\underline{x}}} \omega^\nu_{\underline{y}},$$

where

$$Q^{\nu}_{\underline{x}} \equiv \{\underline{y} \in X^{\nu} : \eta^{\nu}_{\underline{y}} \cap \omega^{\nu}_{\underline{x}} \neq \emptyset\}.$$

There is a $0 < \bar{\rho} < \infty$, which may depend on $\{\mathcal{M}^{\nu}\}_{\nu \in N}$ but is independent of $\underline{x}$ and ν, such that

$$\Omega^{\nu}_{\underline{x}} \subset B^{\underline{x}}_{\bar{\rho} h^{\nu}_{\underline{x}}}, \quad \text{for all } \underline{x} \in X^{\nu} \text{ and } \nu \in N,$$

where $B^{\underline{x}}_{\bar{\rho} h^{\nu}_{\underline{x}}}$ is the ball of radius $\bar{\rho} h^{\nu}_{\underline{x}}$ centred at $\underline{x}$, namely,

$$B^{\underline{x}}_{\bar{\rho} h^{\nu}_{\underline{x}}} = \{x \in \mathbb{R}^n : \|x - \underline{x}\| \leq \bar{\rho} h^{\nu}_{\underline{x}}\}.$$

A5. For each $\underline{x} \in X^{\nu}$, there is a one-to-one mapping $\mathcal{A}^{\nu}_{\underline{x}} : \mathcal{P}^k \to \mathcal{P}^k$ such that

$$\sum_{\underline{y} \in Q^{\nu}_{\underline{x}}} (\mathcal{A}^{\nu}_{\underline{x}} p)(\underline{y}) \phi^{\nu}_{\underline{y}}(x) = p(x), \quad \text{for } x \in \omega^{\nu}_{\underline{x}}, \text{ and any } p \in \mathcal{P}^k, \tag{3.58}$$

and

$$\|\mathcal{A}_{\underline{x}} p\|_{L^2(\omega^{\nu}_{\underline{y}})} \leq C \|p\|_{L^2(\omega^{\nu}_{\underline{y}})}, \quad \text{for all } p \in \mathcal{P}^k, \ \underline{y} \in Q^{\nu}_{\underline{x}}, \text{ and } \underline{x} \in X^{\nu}.$$

A6. For any $0 \leq s \leq q$,

$$\|\phi^{\nu}_{\underline{y}}\|_{H^s(\omega^{\nu}_{\underline{x}})} \leq C (h^{\nu}_{\underline{y}})^{-s+n/2}, \quad \text{for all } \underline{y} \in Q^{\nu}_{\underline{x}}.$$

The constant C may depend on $\{\mathcal{M}^{\nu}\}_{\nu \in N}$, but is independent of $\underline{y}, \underline{x}$, and ν.

A7. There is a constant C such that

$$\|w\|_{L^{\infty}(\omega^{\nu}_{\underline{y}})} \leq C (h^{\nu}_{\underline{y}})^{-n/2} \|w\|_{L^2(\omega^{\nu}_{\underline{y}})}, \quad \text{for any } w \in \mathcal{P}^k,$$

where C is independent of $\underline{y}$ and ν.

Remark 16. From the definitions of $Q^{\nu}_{\underline{x}}$ and $S^{\nu}_{\underline{x}}$, and assumption A3, it is clear that $Q^{\nu}_{\underline{x}} \subset S^{\nu}_{\underline{x}}$. Hence from assumption A2, it is immediate that

$$\text{card } Q^{\nu}_{\underline{x}} \leq \kappa. \tag{3.59}$$

We could, of course, have stated (3.59) as an assumption, but have chosen to state card $S^{\nu}_{\underline{x}} \leq \kappa$ as an assumption because, generally, $S^{\nu}_{\underline{x}}$ is easier to work with than $Q^{\nu}_{\underline{x}}$. We also note that assumptions A1–A4 imply that, for any $x \in \mathbb{R}^n$,

$$\text{card } \{\underline{x} \in X^{\nu} : x \in \mathring{\eta}^{\nu}_{\underline{x}}\} \leq \kappa, \quad \text{for all } \nu \in N. \tag{3.60}$$

Remark 17. We note that the left-hand side of (3.58) in A5 is defined for all $x \in \mathbb{R}^n$, but the equality is required to hold only for $x \in \omega^{\nu}_{\underline{x}}$.

Remark 18. Note that assumptions A1–A7 can be thought of as assumptions on $\mathcal{M}^\nu$, for each $\nu \in N$; they are assumptions on $\{\mathcal{M}^\nu\}_{\nu\in N}$ in that they are assumptions on $\mathcal{M}^\nu$ for each ν *and* that the constants in the assumptions do not depend on ν.

Remark 19. Assumption A5 effectively defines the notion of quasi-reproducing shape functions $\phi^\nu_{\underline{x}}$ of order k. Note that the condition is local: the operator $\mathcal{A}^\nu_{\underline{x}}$ depends on $\underline{x}$, the sum is taken only over $\underline{y} \in Q^\nu_{\underline{x}}$, and the equation holds only for $x \in \omega^\nu_{\underline{x}}$. The shape functions $\phi^\nu_{\underline{x}}$ are said to be reproducing of order k if

$$\sum_{\underline{y}\in X^\nu} p(\underline{y})\phi^\nu_{\underline{y}}(x) = p(x), \quad \text{for } x \in \mathbb{R}^n \text{ and any } p \in \mathcal{P}^k.$$

If the shape functions $\phi^\nu_{\underline{x}}$ are reproducing of order k, then it is immediate that they satisfy A5 with $\mathcal{A}^\nu_{\underline{x}}$ equal to the identity mapping for each $\underline{x}$.

Remark 20. Assumption A5 implies

$$\bigcup_{\underline{x}\in X^\nu} \mathring{\eta}^\nu_{\underline{x}} = \mathbb{R}^n, \quad \text{for each } \nu.$$

Remark 21. Consider uniformly distributed particles, x^h_j, and associated particle shape functions, ϕ^h_j, as defined in Section 3.2. Then with $\underline{x}^\nu = x^h_j$, $h^\nu_{\underline{x}} = h$, $\phi^\nu_{\underline{x}} = \phi^h_j$, and $\omega^\nu_{\underline{x}} = \omega^h_j$, as defined in the proof of Theorem 3.4, the associated particle-shape function system satisfies assumptions A1–A7. Note that $\mathcal{A}^\nu_{\underline{x}} = \mathcal{A}^h_{x^h_j} = \mathcal{A}^h$ satisfies A5.

Suppose $\{\mathcal{M}^\nu\}_{\nu\in N}$ is a family of particle-shape function systems, satisfying A1–A7. Define

$$\mathbb{V}^{k,q}_\nu = \text{span }\{\phi^\nu_{\underline{x}} : \underline{x} \in X^\nu\}, \quad \text{for each } \nu \in N. \tag{3.61}$$

The next theorem gives an approximation error estimate when a function u, defined on $\mathbb{R}^n$, is approximated by an appropriate function in $\mathbb{V}^{k,q}_\nu$, $\nu \in N$.

Theorem 3.8. Suppose the family $\{\mathcal{M}^\nu\}_{\nu\in N}$ of particle-shape function systems satisfies A1–A7, and $h^\nu_{\underline{x}} \le 1$ for $\underline{x} \in X^\nu$, $\nu \in N$. Suppose

$$\sum_{\underline{x}\in X^\nu} \|u\|^2_{H^{r^\nu_{\underline{x}}+1}(B^{\underline{x}}_{\bar{\rho}h^\nu_{\underline{x}}})} < \infty, \quad \text{where } r^\nu_{\underline{x}} \le k, \quad \text{for all } \underline{x} \in X^\nu \text{ and } \nu \in N, \tag{3.62}$$

where $\overline{\rho}$ is introduced in A4. Further, suppose that operators $\mathcal{A}^\nu_{\underline{x}}$, introduced in A5, satisfy

$$\|(\mathcal{A}^\nu_{\underline{x}} - \mathcal{A}^\nu_{\underline{y}})p\|_{H^0(\omega^\nu_{\underline{y}})} \le C(h^\nu_{\underline{x}})^{r^\nu_{\underline{x}}+1}\|p\|_{H^0(\omega^\nu_{\underline{y}})}, \quad \text{for all } p \in \mathcal{P}^k, \ \underline{y} \in Q^\nu_{\underline{x}}, \tag{3.63}$$

for all $\underline{x} \in X^\nu$, $\nu \in N$, where C is independent of $\underline{x}$ and ν. Then there are weights $w^\nu_{\underline{y}} \in \mathbb{R}$, for $\underline{y} \in X^\nu$ and for all $\nu \in N$, such that

$$\left\| u - \sum_{\underline{y}\in X^\nu} w^\nu_{\underline{y}} \phi^\nu_{\underline{y}} \right\|_{H^s(\mathbb{R}^n)} \leq C \left(\sum_{\underline{y}\in X^\nu} (h^\nu_{\underline{y}})^{2(r^\nu_{\underline{y}}+1-s)} \|u\|^2_{H^{r^\nu_{\underline{y}}+1}(B^{\underline{y}}_{\bar{\rho}h^\nu_{\underline{y}}})} \right)^{1/2}, \tag{3.64}$$

for $0 \leq s \leq \inf\{q, r^\nu_{\underline{y}} + 1 : \underline{y} \in X^\nu, \nu \in N\}$. The constant C depends on the constants in Assumptions A1–A7 and on (3.63), but neither on u nor on ν.

Note. If $\{\phi^\nu_{\underline{x}}\}$ is reproducing of order k, then (3.63) is trivially satisfied (*cf.* Remark 19). Since shape functions that are reproducing of order k are mainly used in practice, we have not included (3.63) in the set of basic assumptions (A1–A7).

Proof. The proof of this result is analogous to the proof of Theorem 3.4.

1. The sets $\omega^\nu_{\underline{x}}$ play the role of the sets ω^h_j in the proof of Theorem 3.4. The sets $Q^\nu_{\underline{x}}$, $\Omega^\nu_{\underline{x}}$, $B^{\underline{x}}_{\bar{\rho}h^\nu_{\underline{x}}}$, and the mapping $\mathcal{A}^\nu_{\underline{x}}$ play the roles of the sets $A^h_j, \Omega^h_j, B^j_{\bar{\rho}h}$, and the mapping $\mathcal{A}^h_j$, respectively, in the proof of Theorem 3.4. Assumptions A1–A7 state the properties of these sets and the mappings we will need in this proof.

2. For any $\underline{y} \in X^\nu$, since $u \in H^{r^\nu_{\underline{y}}+1}(B^{\underline{y}}_{\bar{\rho}h^\nu_{\underline{y}}})$, it is well known that there is a polynomial $p^{\underline{y},\nu} = p^{\underline{y},\nu}_k$ of degree $\leq k$, such that

$$\|u - p^{\underline{y},\nu}_k\|_{H^s(B^{\underline{y}}_{\bar{\rho}h^\nu_{\underline{y}}})} \leq C (h^\nu_{\underline{y}})^{r^\nu_{\underline{y}}+1-s} \|u\|_{H^{r^\nu_{\underline{y}}+1}(B^{\underline{y}}_{\bar{\rho}h^\nu_{\underline{y}}})}, \tag{3.65}$$

for $0 \leq s \leq r^\nu_{\underline{y}} + 1 \leq k + 1$, where C is independent of u, ν and $\underline{y}$, but does depend on k ($p^{\underline{y},\nu}_k$ can, in fact, be selected so that its degree $\leq r^\nu_{\underline{y}}$). Define the weights

$$w^\nu_{\underline{y}} = (\mathcal{A}^\nu_{\underline{y}} p^{\underline{y},\nu})(\underline{y}), \tag{3.66}$$

where $\mathcal{A}^\nu_{\underline{y}}$ is the operator introduced in assumption A5.

Let $\underline{x} \in X^\nu$ be fixed. We will work with the polynomial $p^{\underline{x},\nu}$, which satisfies (3.65) with $\underline{y} = \underline{x}$, as well as the polynomial $p^{\underline{y},\nu}$. Using (3.66) we find

$$\begin{aligned}
\Big\| u - \sum_{\underline{y}\in X^\nu} w^\nu_{\underline{y}} \phi^\nu_{\underline{y}} \Big\|_{H^s(\omega^\nu_{\underline{x}})}
&\le \Big\| u - \sum_{\underline{y}\in Q^\nu_{\underline{x}}} w^\nu_{\underline{y}} \phi^\nu_{\underline{y}} \Big\|_{H^s(\omega^\nu_{\underline{x}})} \\
&\le \Big\| u - \sum_{\underline{y}\in Q^\nu_{\underline{x}}} (\mathcal{A}^\nu_{\underline{x}} p^{\underline{x},\nu})(\underline{y})\, \phi^\nu_{\underline{y}} \Big\|_{H^s(\omega^\nu_{\underline{x}})} \\
&\quad + \sum_{\underline{y}\in Q^\nu_{\underline{x}}} \big|(\mathcal{A}^\nu_{\underline{x}} p^{\underline{x},\nu})(\underline{y}) - (\mathcal{A}^\nu_{\underline{y}} p^{\underline{y},\nu})(\underline{y})\big|\, \|\phi^\nu_{\underline{y}}\|_{H^s(\omega^\nu_{\underline{x}})}. \qquad (3.67)
\end{aligned}$$

We now estimate the two terms on the right-hand side of (3.67).

3. From assumption A5, we know that

$$\sum_{\underline{y}\in Q^\nu_{\underline{x}}} (\mathcal{A}^\nu_{\underline{x}} p)(\underline{y}) \phi^\nu_{\underline{y}}(x) = p(x), \quad \text{for } x \in \omega^\nu_{\underline{x}}, \text{ and any } p \in \mathcal{P}^k.$$

Using this formula and (3.65) with $\underline{y} = \underline{x}$, we obtain the estimate

$$\begin{aligned}
\Big\| u - \sum_{\underline{y}\in Q^\nu_{\underline{x}}} (\mathcal{A}^\nu_{\underline{x}} p^{\underline{x},\nu})(\underline{y})\, \phi^\nu_{\underline{y}} \Big\|_{H^s(\omega^\nu_{\underline{x}})} &\le \|u - p^{\underline{x},\nu}\|_{H^s(\omega^\nu_{\underline{x}})} \\
&\le C (h^\nu_{\underline{x}})^{r^\nu_{\underline{x}}+1-s} \|u\|_{H^{r^\nu_{\underline{x}}+1}(B^{\underline{x}}_{\bar\rho h^\nu_{\underline{x}}})} \qquad (3.68)
\end{aligned}$$

for the first term.

Using assumption A6, we have

$$\|\phi^\nu_{\underline{y}}\|_{H^s(\omega^\nu_{\underline{z}})} \le C (h^\nu_{\underline{y}})^{-s+n/2}, \quad \text{for all } \underline{y} \in Q^\nu_{\underline{x}}.$$

Thus

$$\begin{aligned}
&\sum_{\underline{y}\in Q^\nu_{\underline{x}}} \big|(\mathcal{A}^\nu_{\underline{x}} p^{\underline{x},\nu})(\underline{y}) - (\mathcal{A}^\nu_{\underline{y}} p^{\underline{y},\nu})(\underline{y})\big|\, \|\phi^\nu_{\underline{y}}\|_{H^s(\omega^\nu_{\underline{x}})} \\
&\quad \le \sum_{\underline{y}\in Q^\nu_{\underline{x}}} (h^\nu_{\underline{y}})^{-s+n/2} \big|(\mathcal{A}^\nu_{\underline{x}} p^{\underline{x},\nu})(\underline{y}) - (\mathcal{A}^\nu_{\underline{y}} p^{\underline{y},\nu})(\underline{y})\big|. \qquad (3.69)
\end{aligned}$$

It remains to estimate the right-hand side of this inequality.

For $\underline{y} \in Q^\nu_{\underline{x}}$, we have $\omega^\nu_{\underline{y}} \subset \Omega^\nu_{\underline{x}}$, and hence $\omega^\nu_{\underline{y}} \subset B^{\underline{x}}_{\bar\rho h^\nu_{\underline{x}}}$, using assumption A4.

Also $\omega_{\underline{y}}^{\nu} \subset B_{\bar{\rho}h_{\underline{y}}^{\nu}}^{\underline{y}}$. Thus, using (3.65) with $s = 0$, we have

$$\begin{aligned}\|p^{\underline{x},\nu} - p^{\underline{y},\nu}\|_{H^0(\omega_{\underline{y}}^{\nu})} &\\ \leq \|p^{\underline{x},\nu} - u\|_{H^0(\omega_{\underline{y}}^{\nu})} + \|u - p^{\underline{y},\nu}\|_{H^0(\omega_{\underline{y}}^{\nu})} &\\ \leq (h_{\underline{x}}^{\nu})^{r_{\underline{x}}^{\nu}+1}\|u\|_{H^{r_{\underline{x}}^{\nu}+1}(B_{\bar{\rho}h_{\underline{x}}^{\nu}}^{\underline{x}})} + (h_{\underline{y}}^{\nu})^{r_{\underline{y}}^{\nu}+1}\|u\|_{H^{r_{\underline{y}}^{\nu}+1}(B_{\bar{\rho}h_{\underline{y}}^{\nu}}^{\underline{y}})}. &\qquad (3.70)\end{aligned}$$

Now, using assumption A7, we have

$$\begin{aligned}\big|(\mathcal{A}_{\underline{x}}^{\nu} p^{\underline{x},\nu})(\underline{y}) - (\mathcal{A}_{\underline{y}}^{\nu} p^{\underline{y},\nu})(\underline{y})\big| &\\ \leq \big|[(\mathcal{A}_{\underline{x}}^{\nu} - \mathcal{A}_{\underline{y}}^{\nu})p^{\underline{x},\nu}](\underline{y})\big| + \big|[\mathcal{A}_{\underline{y}}^{\nu}(p^{\underline{x},\nu} - p^{\underline{y},\nu})](\underline{y})\big| &\\ \leq \|(\mathcal{A}_{\underline{x}}^{\nu} - \mathcal{A}_{\underline{y}}^{\nu})p^{\underline{x},\nu}\|_{L^\infty(\omega_{\underline{y}}^{\nu})} + \|\mathcal{A}_{\underline{y}}^{\nu}(p^{\underline{x},\nu} - p^{\underline{y},\nu})\|_{L^\infty(\omega_{\underline{y}}^{\nu})} &\\ \leq C(h_{\underline{y}}^{\nu})^{-n/2}\big\{\|(\mathcal{A}_{\underline{x}}^{\nu} - \mathcal{A}_{\underline{y}}^{\nu})p^{\underline{x},\nu}\|_{H^0(\omega_{\underline{y}}^{\nu})} &\\ + \|\mathcal{A}_{\underline{y}}^{\nu}(p^{\underline{x},\nu} - p^{\underline{y},\nu})\|_{H^0(\omega_{\underline{y}}^{\nu})}\big\}. &\qquad (3.71)\end{aligned}$$

Also, using assumption A5 and (3.70), we obtain

$$\begin{aligned}\|\mathcal{A}_{\underline{y}}^{\nu}(p^{\underline{x},\nu} - p^{\underline{y},\nu})\|_{H^0(\omega_{\underline{y}}^{\nu})} &\\ \leq C\|p^{\underline{x},\nu} - p^{\underline{y},\nu}\|_{H^0(\omega_{\underline{y}}^{\nu})} &\\ \leq C\big\{(h_{\underline{x}}^{\nu})^{r_{\underline{x}}^{\nu}+1}\|u\|_{H^{r_{\underline{x}}^{\nu}+1}(B_{\bar{\rho}h_{\underline{x}}^{\nu}}^{\underline{x}})} + (h_{\underline{y}}^{\nu})^{r_{\underline{y}}^{\nu}+1}\|u\|_{H^{r_{\underline{y}}^{\nu}+1}(B_{\bar{\rho}h_{\underline{y}}^{\nu}}^{\underline{y}})}\big\}. &\qquad (3.72)\end{aligned}$$

Moreover, from (3.63), we have

$$\|(\mathcal{A}_{\underline{x}}^{\nu} - \mathcal{A}_{\underline{y}}^{\nu})p^{\underline{x},\nu}\|_{H^0(\omega_{\underline{y}}^{\nu})} \leq (h_{\underline{x}}^{\nu})^{r_{\underline{x}}^{\nu}+1}\|p^{\underline{x},\nu}\|_{H^0(\omega_{\underline{y}}^{\nu})}, \qquad (3.73)$$

and from (3.65), with $\underline{y} = \underline{x}$, and recalling that $h_{\underline{x}}^{\nu} \leq 1$, we get

$$\begin{aligned}\|p^{\underline{x},\nu}\|_{H^0(\omega_{\underline{y}}^{\nu})} &\leq \|p^{\underline{x},\nu} - u\|_{H^0(\omega_{\underline{y}}^{\nu})} + \|u\|_{H^0(\omega_{\underline{y}}^{\nu})} \\ &\leq C(h_{\underline{x}}^{\nu})^{r_{\underline{x}}^{\nu}+1}\|u\|_{H^{r_{\underline{x}}^{\nu}+1}(B_{\bar{\rho}h_{\underline{x}}^{\nu}}^{\underline{x}})} + \|u\|_{H^0(\omega_{\underline{y}}^{\nu})} \\ &\leq C\|u\|_{H^{r_{\underline{x}}^{\nu}+1}(B_{\bar{\rho}h_{\underline{x}}^{\nu}}^{\underline{x}})}. \qquad (3.74)\end{aligned}$$

Combining (3.71)–(3.74), we have

$$\begin{aligned}\big|(\mathcal{A}_{\underline{x}}^{\nu} p^{\underline{x},\nu})(\underline{y}) - (\mathcal{A}_{\underline{y}}^{\nu} p^{\underline{y},\nu})(\underline{y})\big| &\qquad (3.75)\\ \leq C(h_{\underline{y}}^{\nu})^{-n/2}\big\{(h_{\underline{x}}^{\nu})^{r_{\underline{x}}^{\nu}+1}\|u\|_{H^{r_{\underline{x}}^{\nu}+1}(B_{\bar{\rho}h_{\underline{x}}^{\nu}}^{\underline{x}})} + (h_{\underline{y}}^{\nu})^{r_{\underline{y}}^{\nu}+1}\|u\|_{H^{r_{\underline{y}}^{\nu}+1}(B_{\bar{\rho}h_{\underline{y}}^{\nu}}^{\underline{y}})}\big\}.\end{aligned}$$

Then we combine (3.69), (3.75), (3.59), and assumption A2 (see (3.59)), to obtain

$$\sum_{\underline{y}\in Q^\nu_{\underline{x}}} \left|(\mathcal{A}^\nu_{\underline{x}} p^{\underline{x},\nu})(\underline{y}) - (\mathcal{A}^\nu_{\underline{y}} p^{\underline{y},\nu})(\underline{y})\right| \|\phi^\nu_{\underline{y}}\|_{H^s(\omega^\omega_{\underline{x}})}$$
$$\leq C\Big\{\kappa (h^\nu_{\underline{x}})^{r^\nu_{\underline{x}}+1} \|u\|_{H^{r^\nu_{\underline{x}}+1}(B^{\underline{x}}_{\bar{\rho}h^\nu_{\underline{x}}})} + \sum_{\underline{y}\in Q^\nu_{\underline{x}}} (h^\nu_{\underline{y}})^{r^\nu_{\underline{y}}+1} \|u\|_{H^{r^\nu_{\underline{y}}+1}(B^{\underline{y}}_{\bar{\rho}h^\nu_{\underline{y}}})}\Big\}$$
$$\leq C \sum_{\underline{y}\in Q^\nu_{\underline{x}}} (h^\nu_{\underline{y}})^{r^\nu_{\underline{y}}+1} \|u\|_{H^{r^\nu_{\underline{y}}+1}(B^{\underline{y}}_{\bar{\rho}h^\nu_{\underline{y}}})}, \tag{3.76}$$

which is an estimate for the second term in (3.67). Thus, from (3.67), (3.68), and (3.76), we have

$$\left\| u - \sum_{\underline{y}\in Q^\nu_{\underline{x}}} w^\nu_{\underline{y}} \phi^\nu_{\underline{y}} \right\|_{H^s(\omega^\nu_{\underline{x}})} \leq C \sum_{\underline{y}\in Q^\nu_{\underline{x}}} (h^\nu_{\underline{y}})^{r^\nu_{\underline{y}}+1} \|u\|_{H^{r^\nu_{\underline{y}}+1}(B^{\underline{y}}_{\bar{\rho}h^\nu_{\underline{y}}})}. \tag{3.77}$$

4. It remains to estimate $\|u - \sum_{\underline{y}\in X^\nu} w^\nu_{\underline{y}} \phi^\nu_{\underline{y}}\|_{H^s(\mathbb{R}^n)}$. Using (3.77), which is valid for all $\underline{x} \in X^\nu$, and assumptions A1, A2, A4, we obtain

$$\left\| u - \sum_{\underline{y}\in X^\nu} w^\nu_{\underline{y}} \phi^\nu_{\underline{y}} \right\|^2_{H^s(\mathbb{R}^n)} \leq \sum_{\underline{x}\in X^\nu} \|u - \sum_{\underline{y}\in X^\nu} w^\nu_{\underline{y}} \phi^\nu_{\underline{y}}\|^2_{H^s(\omega^\nu_{\underline{x}})}$$
$$\leq C \sum_{\underline{x}\in X^\nu} \sum_{\underline{y}\in Q^\nu_{\underline{x}}} (h^\nu_{\underline{y}})^{2(r^\nu_{\underline{y}}+1-s)} \|u\|^2_{H^{r_{\underline{y}}+1}(B^{\underline{y}}_{\bar{\rho}h^\nu_{\underline{y}}})}$$
$$\leq C \sum_{\underline{y}\in X^\nu} (h^\nu_{\underline{y}})^{2(r^\nu_{\underline{y}}+1-s)} \|u\|^2_{H^{r_{\underline{y}}+1}(B^{\underline{y}}_{\bar{\rho}h^\nu_{\underline{y}}})}, \tag{3.78}$$

which is (3.64). □

It will be useful to state estimate (3.64) in Theorem 3.8 in certain alternative ways. Given a family of particle-shape function systems $\{\mathcal{M}^\nu\}_{\nu\in N}$ satisfying A1–A7, define

$$h^\nu = \sup_{\underline{x}\in X^\nu} h^\nu_{\underline{x}}, \quad \text{for each } \nu. \tag{3.79}$$

With this definition, from (3.64) we have

$$\left\| u - \sum_{\underline{y}\in X^\nu} w^\nu_{\underline{y}} \phi^\nu_{\underline{y}} \right\|_{H^s(\mathbb{R}^n)} \leq C \left(\sum_{\underline{y}\in X^\nu} (h^\nu)^{2(r^\nu_{\underline{y}}+1-s)} \|u\|_{H^{r^\nu_{\underline{y}}+1}(B^\nu_{\bar{\rho}h^\nu_{\underline{y}}})} \right)^{1/2}. \tag{3.80}$$

Now, if $r^\nu_{\underline{x}^\nu} = k'$, where $0 \leq k' \leq k$, for all $\underline{y} \in X^\nu$ and ν, then (3.80) leads to the following result.

Theorem 3.9. Suppose the family $\{\mathcal{M}^\nu\}_{\nu\in N}$ of particle-shape function systems satisfies A1–A7, (3.63), and in addition, suppose $h^\nu \leq 1$, for all ν. Suppose $\|u\|_{H^{k'+1}(\mathbb{R}^n)} < \infty$, where $0 \leq k' \leq k$. Then there are weights $w_{\underline{y}}^\nu \in \mathbb{R}$ such that

$$\left\| u - \sum_{\underline{y}\in X^\nu} w_{\underline{y}}^\nu \phi_{\underline{y}}^\nu \right\|_{H^s(\mathbb{R}^n)} \leq C(h^\nu)^{k'+1-s}\|u\|_{H^{k'+1}(\mathbb{R}^n)}, \tag{3.81}$$

for $0 \leq s \leq \min(q, k'+1)$, where C is independent of u and ν.

We note that if $h^{\nu_1} < h^{\nu_2}$, $\nu_1, \nu_2 \in N$, then we would view $\mathcal{M}^{\nu_1}$ as a 'refinement' of $\mathcal{M}^{\nu_2}$.

There is yet another way to state the estimate (3.81). Let $0 < h \leq 1$, and suppose $\{\mathcal{M}^\nu\}_{\nu\in N}$, a family of particle-shape function systems satisfying A1–A7, (3.63), and in addition,

$$h^\nu = \sup_{\underline{x}^\nu \in X^\nu} h_{\underline{x}^\nu}^\nu \leq h, \quad \text{for each } \nu. \tag{3.82}$$

We can now think of $\nu = \nu(h)$ as determined by h, although, of course, many particle-shape function systems satisfy (3.82). We can, in fact, think of having a one-to-one correspondence between ν and h. Thus we can regard h as the parameter and write a family of particle-shape function systems as

$$\{\mathcal{M}^h\}_{0<h\leq 1} = \{X^h, \{h_{\underline{x}}^h, \omega_{\underline{x}}^h, \phi_{\underline{x}}^h\}_{\underline{x}\in X^h}\}_{0<h\leq 1}$$

instead of $\{\mathcal{M}^\nu\}_{\nu\in N}$. With this understanding that $\nu = \nu(h)$, the estimate (3.81) can be written as

$$\left\| u - \sum_{\underline{y}\in X^h} w_{\underline{y}}^h \phi_{\underline{y}}^h \right\|_{H^s(\mathbb{R}^n)} \leq Ch^{k+1-s}\|u\|_{H^{k+1}(\mathbb{R}^n)}. \tag{3.83}$$

We are naturally interested in having $h \downarrow 0$, and hence in considering $\nu(h)$s for which $h^{\nu(h)} \downarrow 0$. More specifically, we will often consider a sequence $h_m \downarrow 0$, and the corresponding sequence of particle systems $\mathcal{M}^{\nu_1}, \mathcal{M}^{\nu_2}, \ldots,$ where $\nu_l = \nu(h_l)$.

We remark that the estimate (3.64) is stronger than (3.81) and (3.83), in that (3.64) uses $h_{\underline{x}^\nu}^\nu$ instead of the larger h^ν, and (3.64) allows a more general regularity assumption on the function u. The viewpoint outlined in this paragraph is similar to the usual view of meshes in the FEM.

For a given family $\{\mathcal{M}^\nu\}_{\nu\in N}$ of particle-shape function systems, we defined the space $\mathbb{V}_\nu^{k,q}$ in (3.61). With h, $0 < h \leq 1$, as the parameter, *i.e.*, for a given family $\mathcal{M}^h$, $0 < h \leq 1$, we will use the space

$$\mathbb{V}_h^{k,q} \equiv \mathbb{V}_{\nu(h)}^{k,q} = \text{span}\ \{\phi_{\underline{x}}^h : \underline{x} \in X^h\}. \tag{3.84}$$

So far, we have discussed the approximation of a function u defined on $\mathbb{R}^n$, by particle shape functions. We now consider u defined on Ω, where Ω is a bounded domain, with Lipschitz-continuous boundary, in $\mathbb{R}^n$. We now show that functions in $\mathbb{V}^{k,q}_{\Omega,h}$, defined by

$$\mathbb{V}^{k,q}_{\Omega,h} = \operatorname{span}\{\phi^h_{\underline{x}}|_\Omega : \phi^h_{\underline{x}} \in \mathbb{V}^{k,q}_h, \quad \text{for } \underline{x} \in A^h_\Omega\}, \tag{3.85}$$

where

$$A^h_\Omega = \{\underline{x} \in X^h : \mathring{\eta}^h_{\underline{x}} \cap \Omega \neq \emptyset\},$$

provide accurate approximation of functions u defined on Ω.

Theorem 3.10. Suppose $\mathcal{M}^h$, $0 < h \leq 1$, is a family of particle-shape function systems satisfying A1–A7 and (3.63). Let $\Omega \subset \mathbb{R}^n$ be a bounded domain with Lipschitz-continuous boundary, and suppose $u \in H^{k'+1}(\Omega)$, where $0 \leq k' \leq k$. Then there are weights $w^h_{\underline{y}} \in \mathbb{R}$ such that

$$\left\| u - \sum_{\underline{y}\in A^h_\Omega} w^h_{\underline{y}} \phi^h_{\underline{y}} \right\|_{H^s(\Omega)} \leq C h^{k'+1-s} \|u\|_{H^{k'+1}(\Omega)}, \tag{3.86}$$

for $0 \leq s \leq \min(q, k'+1)$, where the constant C is independent of u and h.

The proof of this theorem is based on using (3.83) on the extension $\bar{u} = Eu$, and is similar to the proof of Theorem 3.5. We omit the proof of this theorem. We note that the approximation $\sum_{\underline{y}\in A^h_\Omega} w^h_{\underline{y}} \phi^h_{\underline{y}}$, obtained in Theorem 3.10, is such that

$$\sum_{\underline{y}\in A^h_\Omega} w^h_{\underline{y}} \phi^h_{\underline{y}} \Big|_\Omega \in \mathbb{V}^{k,q}_{\Omega,h}.$$

In Section 3.2, we reviewed the notion of (t, k^*)-regular system $S_h(\Omega)$. In the next theorem, we show that $\mathbb{V}^{k,q}_{\Omega,h}$ is a $(k+1, q)$-system.

Theorem 3.11. Suppose $\mathcal{M}^h$, $0 < h \leq 1$, is a family of particle-shape function systems satisfying A1–A7 and (3.63). Let $\Omega \subset \mathbb{R}^n$ be a bounded domain with Lipschitz-continuous boundary. Then $\mathbb{V}^{k,q}_{\Omega,h}$ is a $(k+1, q)$-regular system, where k is the order of quasi-reproducing shape functions in $\mathcal{M}^h$.

The proof of this theorem is similar to the proof of Theorem 3.6, and will be omitted.

Remark 22. The space $\mathbb{V}^{k,q}_{\Omega,h}$ satisfies the local assumption, LA.

Quasi-uniform particle-shape function system
We will call a family of particle-shape function systems $\{\mathcal{M}^h\}_{0<h\le 1}$ quasi-uniform if there is a β, $1 < \beta < \infty$, such that

$$\beta^{-1} \le \frac{h^h_{\underline{x}}}{h^h_{\underline{y}}} \le \beta, \quad \text{for all } \underline{x}, \underline{y} \in X^h \ \text{ and } \ 0 < h \le 1. \tag{3.87}$$

We note that (3.87) is equivalent to

$$\beta^{-1} \le \frac{h}{h^h_{\underline{y}}} \le \beta, \quad \text{for all } \underline{y} \in X^h \ \text{ and } \ 0 < h \le 1, \tag{3.88}$$

where h is defined by (3.82).

Remark 23. We can also define *uniform* particle-shape function system by imposing the condition

$$h^h_{\underline{x}} = h^h_{\underline{y}}, \quad \text{for all } \underline{x}, \underline{y} \in X^h \ \text{ and } \ 0 < h \le 1.$$

We note that the system with uniformly distributed particles and the associated shape functions as defined in Section 3.1 is uniform. But uniform particle-shape function systems may have particles that are not uniformly distributed.

Consider a family of particle-shape function systems $\{\mathcal{M}^h\}_{0<h\le 1}$ satisfying assumptions A1–A7. Let $\Omega \subset \mathbb{R}^n$ be a bounded domain, and suppose $\mathcal{M}^h$ satisfies the following additional assumptions:

- $\mathcal{M}^h$ is quasi-uniform, *i.e.*, (3.88) holds;
- for all $x \in A^h_\Omega$, there is a $\beta > 0$ such that, for $0 \le s \le q$,

$$\beta^{-1} h^{\frac{n}{2}-s} \le \|\phi^h_{\underline{y}}\|_{H^s(\omega^h_{\underline{x}} \cap \Omega)} \le \beta h^{\frac{n}{2}-s}, \quad \text{for all } \underline{y} \in Q^h_{\underline{x}}, \tag{3.89}$$

  where $Q^h_{\underline{x}} = \{\underline{y} \in X^h : \eta^h_{\underline{y}} \cap \omega^h_{\underline{x}} \ne \emptyset\}$ (*cf.* A4);
- for all $w_{\underline{y}} \in \mathbb{R}$, for $\underline{y} \in Q^h_{\underline{x}}$, and $\underline{x} \in A^h_\Omega$, there exists $C > 0$, independent of $\underline{x}$, such that

$$h^{-s} \left[\sum_{\underline{y} \in Q^h_{\underline{x}}} |w_{\underline{y}}|^2 h^n \right]^{1/2} \le C \left\| \sum_{\underline{y} \in Q^h_{\underline{x}}} w_{\underline{y}} \phi^h_{\underline{y}} \right\|_{H^s(\omega^h_{\underline{x}} \cap \Omega)}, \quad \text{for } 0 \le s \le q. \tag{3.90}$$

Then the particle space $\mathbb{V}^{k,q}_{\Omega,h}$ satisfies the inverse assumption IA, introduced in Section 3.2. To see this, consider $\underline{x} \in A^h_\Omega$. Then, using (3.89) and (3.90), we have

$$
\begin{aligned}
\left\| \sum_{\underline{y}\in Q^h_{\underline{x}}} w_{\underline{y}}\phi^h_{\underline{y}} \right\|_{H^q(\omega^h_{\underline{x}}\cap\Omega)} &\le \sum_{\underline{y}\in Q^h_{\underline{x}}} |w_{\underline{y}}|\, \|\phi^h_{\underline{y}}\|_{H^q(\omega^h_{\underline{x}}\cap\Omega)} \\
&\le Ch^{\frac{n}{2}-q}\left(\sum_{\underline{y}\in Q^h_{\underline{x}}} |w_{\underline{y}}|^2\right)^{1/2} \\
&= Ch^{s-q}h^{-s}\left(\sum_{\underline{y}\in Q^h_{\underline{x}}} |w_{\underline{y}}|^2 h^n\right)^{1/2} \\
&\le Ch^{s-q}\left\| \sum_{\underline{y}\in Q^h_{\underline{x}}} w_{\underline{y}}\phi^h_{\underline{y}} \right\|_{H^s(\omega^h_{\underline{x}}\cap\Omega)},
\end{aligned}
\tag{3.91}
$$

where C depends on κ (*cf.* A2). Thus

$$
\begin{aligned}
\left\| \sum_{\underline{y}\in A^h_\Omega} w_{\underline{y}}\phi^h_{\underline{y}} \right\|^2_{H^q(\Omega)} &\le \sum_{\underline{x}\in A^h_\Omega} \left\| \sum_{\underline{y}\in Q^h_{\underline{x}}} w_{\underline{y}}\phi^h_{\underline{y}} \right\|^2_{H^q(\omega^h_{\underline{x}}\cap\Omega)} \\
&\le Ch^{2(s-q)} \sum_{\underline{x}\in A^h_\Omega} \left\| \sum_{\underline{y}\in Q^h_{\underline{x}}} w_{\underline{y}}\phi^h_{\underline{y}} \right\|^2_{H^s(\omega^h_{\underline{x}}\cap\Omega)} \\
&\le Ch^{2(s-q)} \left\| \sum_{\underline{y}\in A^h_\Omega} w_{\underline{y}}\phi^h_{\underline{y}} \right\|^2_{H^s(\omega^h_{\underline{x}}\cap\Omega)}.
\end{aligned}
$$

Since any element g of $\mathbb{V}^{k,q}_{\Omega,h}$ is of the form $\sum_{\underline{y}\in A^h_\Omega} w_{\underline{y}}\phi^h_{\underline{y}}|_\Omega$, we have shown that $\mathbb{V}^{k,q}_{\Omega,h}$ satisfies the inverse assumption, IA. We summarize the above discussion in the following theorem.

Theorem 3.12. Suppose $\mathcal{M}^h$, $0 < h \le 1$, is a family of quasi-uniform particle-shape function systems satisfying A1–A7, (3.89), and (3.90). Let $\Omega \subset \mathbb{R}^n$ be a bounded domain with Lipschitz-continuous boundary. Then $\mathbb{V}^{k,q}_{\Omega,h}$ satisfies the inverse assumption, IA.

Remark 24. We can show that the particle space $\mathbb{V}^{k,q}_h$ also satisfies the inverse assumption IA, if $\mathcal{M}^h$, satisfying A1–A7, also satisfies (3.89) and (3.90) with $\Omega = \mathbb{R}^n$.

So far we have addressed the approximation properties of the spaces $\mathbb{V}_h^{k,q}$ and $\mathbb{V}_{\Omega,h}^{k,q}$, *i.e.*, approximation by the functions of the form

$$\sum_{\underline{y}\in X^h} w_{\underline{y}}^h \phi_{\underline{y}}^h \quad \text{and} \quad \sum_{\underline{y}\in A_\Omega^h} w_{\underline{y}}^h \phi_{\underline{y}}^h.$$

We will now present the approximation properties of the space

$$\mathbb{W}_{\Omega,h}^{k',q} = \left\{ v\big|_\Omega : v = \sum_{\underline{x}\in A_\Omega^h} \phi_{\underline{x}}^h \psi_{\underline{x}}^h, \psi_{\underline{x}}^h \in \mathcal{P}^{k'}(\mathring{\eta}_{\underline{x}}^h) \right\},$$

where $\{\phi_{\underline{x}}^h\}$ form a partition of unity. The space $\mathbb{W}_{\Omega,h}^{k',q}$ was used in Taylor, Zienkiewicz and Onate (1998), and is a special case of the space V^ν considered in Section 6. The approximation properties of the space $\mathbb{W}_{\Omega,h}^{k',q}$ are similar to the approximation properties of $\mathbb{V}_{\Omega,h}^{k,q}$.

Theorem 3.13. Suppose $\mathcal{M}^h, 0 < h \leq 1$, is a family of particle-shape function system that satisfy A1–A7 with $k = 0$ and $\mathcal{A}_{\underline{x}}^h = I$, which implies that the shape functions $\{\phi_{\underline{x}}^h\}$ form a partition of unity. Let $\Omega \in \mathbb{R}^n$ be a bounded domain with Lipschitz-continuous boundary and suppose $u \in H^{k'+1}(\Omega)$, where $0 \leq k'$. Then there exist $\psi_{\underline{x}}^h \in \mathcal{P}^{k'}(\mathring{\eta}_{\underline{x}}^h)$ such that

$$\left\| u - \sum_{\underline{x}\in A_\Omega^h} \phi_{\underline{x}}^h \psi_{\underline{x}}^h \right\|_{H^1(\Omega)} \leq C h^{k'} \|u\|_{H^{k'+1}(\Omega)},$$

where C is independent of u and h.

The proof of this result can be obtained directly from Theorems 6.1–6.3, and we will comment on the proof in Section 6.

Remark 25. In Theorem 3.13, we assumed that $\{\phi_{\underline{x}}^h\}$ are reproducing of order $k = 0$ (*i.e.*, they form a partition of unity) because, as indicated in Section 4, such shape functions are easier to construct. The general case where $\{\phi_{\underline{x}}^h\}$ are assumed to be quasi-reproducing of order k will be addressed in a forthcoming paper. In this situation, for $0 \leq s \leq q$, we expect the error estimate to be

$$\left\| u - \sum_{\underline{x}\in A_\Omega^h} \phi_{\underline{x}}^h \psi_{\underline{x}}^h \right\|_{H^s(\Omega)} \leq C h^{k'+k+1-s} \|u\|_{H^{k'+k+1}(\Omega)}.$$

Remark 26. Note that in the situation addressed in Theorem 3.13, we associate with each particle $\underline{x} \in X^h$, multiple shape functions

$$\phi_{\underline{x}}^h \psi_{\underline{x},j}^h, j = 1, 2, \ldots, N,$$

where $\{\psi^h_{\underline{x},j}\}_{j=1}^N$ is a basis for $\mathcal{P}^{k'}(\mathring{\eta}^h_{\underline{x}})$.This is in contrast to the situation discussed earlier in this section, where with each particle we associated only the single shape function $\phi^h_{\underline{x}}$. In a forthcoming paper, we will address, in a more general context, the use of multiple shape functions associated with a single particle.

4. Construction and selection of particle shape functions

In Section 3, we presented an abstract description of particle-shape function systems with respect to uniform as well as non-uniform distribution of particles. We showed that if these particle-shape function systems satisfy certain properties (assumptions A1–A7 and (3.63)), they will have good approximation properties. In this section we will present an example of a particular particle-shape function system, where the shape functions are reproducing of order k, and show that under certain conditions they satisfy assumptions A1–A7, and hence have good approximation properties. We note that (3.63) is trivially satisfied. This example will also show that a wide variety of particle shape functions can be constructed. Therefore it is important to address the issue of selecting an appropriate class of shape functions that would yield efficient approximation of the solution of a particular problem, or a class of problems. We also present an interpolation result that will indicate a procedure for choosing a class of shape functions, among a given collection of such classes. Such shape functions will yield the smallest value of the usual Sobolev norm interpolation error, when the interpolated function is included in a higher-order Sobolev space.

4.1. An example of a class of particle shape functions

Several particle shape functions have been developed over the past decade. SPH shape functions (Gingold and Monaghan 1977) were introduced in the context of fluid dynamics, whereas Shepard functions (Shepard 1968) and MLS shape functions (Lancaster and Salkauskas 1981) were introduced in the context of data fitting with respect to irregularly distributed particles in higher dimensions. In the 1990s, RKP (reproducing kernel particle) shape functions were introduced (Liu, Jun and Zhang 1995) in the context of approximation of solutions of partial differential equations. In this paper, we describe the construction of RKP shape functions for non-uniform as well as uniform distribution of particles, and relate them to the abstract setting given in Section 3. Specifically, we will show that the resulting particle-shape function system satisfies assumptions A1–A7.

Non-uniformly distributed particles

For $\nu \in N$, N an index set, let $X^\nu = \{x^\nu_i\}_{i\in\mathbb{Z}}$ where $x^\nu_i \in \mathbb{R}^n$. With each $x^\nu_i \in X^\nu$ we associate a positive number h^ν_i. We consider a fixed ν and often suppress the superscript ν, for example, we write x_i and h_i instead of x_i^ν and h_i^ν, respectively. We will comment on ν when appropriate.

Let $w(x) \geq 0$ be a continuous function with compact support, specifically,

$$\eta \equiv \text{ supp } w(x) = \overline{B}_R(0), \quad R > 0. \tag{4.1}$$

The function $w(x)$ is called a weight function (or window function).

The commonly used weight functions in one dimension are as follows.

(a) Gaussian:

$$w(x) = \begin{cases} \dfrac{\mathrm{e}^{\delta(x/R)^2} - \mathrm{e}^{\delta}}{1 - \mathrm{e}^{\delta}}, & |x| \leq R \\ 0, & |x| \geq R, \end{cases} \tag{4.2}$$

where $\delta > 0$.

(b) cubic spline:

$$w(x) = \begin{cases} \frac{2}{3} - 4(x/R)^2 + 4(|x|/R)^3, & |x| \leq R/2 \\ \frac{4}{3} - 4(|x|/R) + 4(x/R)^2 - \frac{4}{3}(|x|/R)^3, & R/2 \leq |x| \leq R \\ 0, & |x| > R. \end{cases} \tag{4.3}$$

(c) conical:

$$w(x) = \begin{cases} [1 - (x/R)^2]^l, & |x| \leq R \\ 0, & |x| > R, \end{cases} \tag{4.4}$$

where $l = 1, 2 \ldots$.

We note that one may consider nonsymmetric versions of some of these weight functions, as was done in Armentano and Duran (2001).

In $\mathbb{R}^n$, $w(x)$ can be constructed from a one-dimensional weight function $w(x)$ (symmetric) as $w(x) = w(\|x\|)$, where $\|x\|$ is the Euclidean length of x. Further, $w(x)$ can also be constructed via $w(x) = \prod_{j=1}^n w({}_jx)$, where $x = ({}_1x, {}_2x, \ldots, {}_nx) \in \mathbb{R}^n$. Consequently, η will be an n-cube. However, we will assume η given by (4.1) in this section.

For each j, we define

$$w_j(x) = w\left(\frac{x - x_j}{h_j}\right). \tag{4.5}$$

Clearly,

$$\eta_j \equiv \text{ supp } w_j(x) = \overline{B}_{Rh_j}(x_j). \tag{4.6}$$

Let

$$Q_i = \{x_j : \mathring{\eta}_i \cap \mathring{\eta}_j \neq \emptyset\}, \tag{4.7}$$

and assume that

$$\cup_{j\in\mathbb{Z}} \mathring{\eta}_j = \mathbb{R}^n, \tag{4.8}$$

$$\text{card } Q_i \le \kappa, \quad \text{for all } i \in \mathbb{Z}, \tag{4.9}$$

where κ is independent of i and ν.

For a given integer k, $k \ge 0$, the RKP shape function $\phi_j(x)$, associated with the particle x_j, is defined by

$$\phi_j(x) = w_j(x) \sum_{|\alpha|\le k} (x - x_j)^\alpha b_\alpha(x), \tag{4.10}$$

where $b_\alpha(x)$ are chosen so that

$$\sum_{j\in\mathbb{Z}} p(x_j)\phi_j(x) = p(x), \quad \text{for } x \in \mathbb{R}^n \text{ and } p \in \mathcal{P}^k(\mathbb{R}^n), \tag{4.11}$$

so that $\{\phi_j(x)\}_{j\in\mathbb{Z}}$ are reproducing of order k. This gives rise to a linear system in $b_\alpha(x)$, namely

$$\sum_{|\alpha|\le k} m_{\alpha+\beta}(x) b_\alpha(x) = \delta_{|\beta|,0}, \quad \text{for } |\beta| \le k, \tag{4.12}$$

where $\delta_{|\beta|,|\alpha|}$ is the Kronecker delta, and

$$m_\alpha(x) = \sum_{j\in\mathbb{Z}} w_j(x)(x - x_j)^\alpha.$$

It is clear from (4.6) and (4.10) that

$$\text{supp } \phi_j(x) = \text{ supp } w_j(x) = \eta_j. \tag{4.13}$$

We now briefly describe the derivation of (4.12). For a fixed $y \in \mathbb{R}^n$, consider

$$p_\beta(x) = (y - x)^\beta, \quad 0 \le |\beta| \le k.$$

Using $p(x) = p_\beta(x)$ in (4.11), we get

$$\sum_{j\in\mathbb{Z}} (y - x_j)^\beta \phi_j(x) = (y - x)^\beta,$$

and letting $y = x$ in the above equality, we have

$$\sum_{j\in\mathbb{Z}} (x - x_j)^\beta \phi_j(x) = \delta_{|\beta|,0}, \quad 0 \le |\beta| \le k. \tag{4.14}$$

Thus (4.11) implies (4.14); one can also show that (4.14) implies (4.11).

Now using (4.10) in (4.14), we get

$$\begin{aligned}\delta_{|\beta|,0} &= \sum_{j\in\mathbb{Z}}(x-x_j)^\beta\phi_j(x)\\ &= \sum_{j\in\mathbb{Z}}(x-x_j)^\beta w_j(x)\sum_{|\alpha|\le k}(x-x_j)^\alpha b_\alpha(x)\\ &= \sum_{|\alpha|\le k} b_\alpha(x)\sum_{j\in\mathbb{Z}} w_j(x)(x-x_j)^{\alpha+\beta}\\ &= \sum_{|\alpha|\le k} m_{\alpha+\beta}(x)b_\alpha(x), \end{aligned}\tag{4.15}$$

which is (4.12).

We now consider the unique solvability of (4.11). For $k=0$, the linear system (4.12) is $m_0(x)b_0(x) = \left[\sum_{i\in\mathbb{Z}} w_i(x)\right]b_0(x) = 1$. Assuming $\sum_{i\in\mathbb{Z}} w_i(x) \neq 0$, $x\in\mathbb{R}^n$, we have $b_0(x) = 1/m_0(x)$. Therefore from (4.10), we have

$$\phi_j(x) = \frac{w_j(x)}{\sum_{i\in\mathbb{Z}} w_i(x)}, \quad j\in\mathbb{Z}.$$

This expression for $\{\phi_j(x)\}$ gives another verification that they form a partition of unity. These shape functions, introduced in Shepard (1968), are called Shepard functions.

The unique solvability of (4.12), for $k\geq 1$, depends on the weight functions and on the distribution of the particles $\{x_i\}$ in $\mathbb{R}^n$. The required distribution of particles is in turn related to the interpolation problem in $\mathbb{R}^n$. It was shown in Han and Meng (2001) that a necessary condition for unique solvability of (4.12) is that, for $x\in\mathbb{R}^n$,

$$\operatorname{card} A(x) \geq \dim \mathcal{P}^k, \tag{4.16}$$

where

$$A(x) = \{x_l : x\in \mathring{\eta}_l\}. \tag{4.17}$$

For $k=1$, Han and Meng (2001) showed that the linear system (4.12) is nonsingular if the following conditions are satisfied.

(a) There are constants $C_1, C_2 > 0$, independent of ν, and $h>0$, such that

$$C_1 \leq \frac{h_i}{h} \leq C_2, \quad \text{for all } i\in\mathbb{Z}. \tag{4.18}$$

(b) There are constants $\tilde{C}_1, \tilde{C}_2 > 0$, independent of ν, such that, for any $x\in\mathbb{R}^n$, there are $(n+1)$ particles $x_{i_l}\in A(x), l=0,\ldots,n$, such that

$$\min_{0\leq l\leq n} w\left(\frac{x-x_{i_l}}{h}\right) \geq \tilde{C}_1 > 0 \tag{4.19}$$

and

$$\text{Volume } K(x_{i_0}, x_{i_1}, \ldots, x_{i_n}) \geq \tilde{C}_2 h^n, \tag{4.20}$$

where $K(x_{i_0}, x_{i_1}, \ldots, x_{i_n})$ is the simplex with vertices $x_{i_0}, x_{i_1}, \ldots, x_{i_n}$.

We will now cast RKP shape functions, discussed above, in the framework of a particle-shape function system, introduced in Section 3.3. We started with a collection of particles $X^\nu = \{x_j^\nu\}_{j\in\mathbb{Z}}$, where $x_j^\nu \in \mathbb{R}^n$, and positive numbers h_j^ν. Corresponding to each particle $x_j^\nu \in X^\nu$, we associated, in (4.10), the RKP shape function, $\phi_j^\nu = \phi_j$ with compact support $\eta_j^\nu = \eta_j = \overline{B}_{Rh_j}(x_j)$, where the parameter R was related to the weight function $w(x)$. It was shown in Han and Meng (2001) that if $w(x) \in C^q(\mathbb{R}^n)$, then $\phi_j \in C^q(\mathbb{R}^n)$, and thus $\phi_j \in H^q(\mathbb{R}^n)$; here we assume $q = 1$. We recall that the conditions (4.8), (4.8), (4.16), (4.18)–(4.20) were required for the construction of shape functions, ϕ_j, $j \in \mathbb{Z}$. We let $\omega_j^\nu = \omega_j \equiv \mathring{\eta}_j$; certainly each ω_j^ν is a bounded domain. We now show that the family of particle-shape function systems $\{\mathcal{M}^\nu\}_{\nu\in N}$, where

$$\mathcal{M}^\nu = \{X^\nu, \{h_i^\nu, \omega_i^\nu, \phi_i^\nu\}\},$$

with these choices for ϕ_i^ν and ω_i^ν, satisfies assumptions A1–A7 in Section 3.3. We will continue to use the notation introduced earlier in this section, and suppress ν; the statements of A1–A7 using this notation should be clear.

- Since $\omega_i = \mathring{\eta}_i$ for $i \in \mathbb{Z}$, assumption A1 follows from (4.8). We also see that the sets $S_{\underline{x}}^\nu \equiv S_i$ and $Q_{\underline{x}}^\nu \equiv Q_i$, introduced in assumptions A2 and A4 are the same. Thus A2 follows from (4.9).
- Assumption A3 is immediate from the definition ω_i.
- Since $\omega_j = \mathring{\eta}_j$, the set $\Omega_{\underline{x}}^\nu$, introduced in assumption A4, is given by $\Omega_{\underline{x}}^\nu = \Omega_i = \cup_{x_j\in Q_i}\mathring{\eta}_j$. Since each η_j is a ball of radius Rh_j, it is easily seen, using (4.18), that assumption A4 is satisfied with $\bar{\rho} = 3RC_2/C_1$.
- RKP shape functions, $\{\phi_j\}$, considered here, are reproducing of order $k = 1$, *i.e.*, they satisfy (4.11) with $k = 1$. Thus A5 is satisfied with $\mathcal{A}_{\underline{x}}^\nu = \mathcal{A}_i = I$ (identity), for all $i \in \mathbb{Z}$, with $k = 1$.
- Han and Meng (2001) showed that, if the weight function $w(x) \in C^q$, then

$$\|\phi_i\|_{W^{s,\infty}(\mathring{\eta}_i)} \leq C(h_i)^{-s}, \quad \text{for } 0 \leq s \leq q \text{ and } i \in \mathbb{Z}.$$

Thus using a scaling argument and this estimate, we obtain

$$\|\phi_i\|_{H^s(\mathring{\eta}_i)} \leq C(h_i)^{-s+n/2}, \quad \text{for } 0 \leq s \leq q \text{ and } i \in \mathbb{Z}, \tag{4.21}$$

where h and h_i satisfy (4.18). Recall that we assumed $q = 1$. Now let $x_j \in Q_i$. Then

$$\|\phi_j\|_{H^s(\mathring{\eta}_i)} = \|\phi_j\|_{H^s(\mathring{\eta}_i\cap\mathring{\eta}_j)} \leq \|\phi_j\|_{H^s(\mathring{\eta}_j)},$$

and combining this with (4.21), we get, for $0 \le s \le 1$,

$$\|\phi_j\|_{H^s(\mathring{\eta}_i)} \le C(h_i)^{-s+n/2}, \quad \text{for all } x_j \in Q_i \text{ and } i \in \mathbb{Z},$$

which is assumption A6 with $q = 1$.

- A scaling argument shows that assumption A7 is satisfied.

We remark that (3.60), together with condition (b) (following (4.18)), establishes a lower bound of κ, namely $(n+1) \le \kappa$.

We have thus shown that assumptions A1–A7, with $k = 1$ and $q = 1$, are satisfied by RKP particle-shape function systems provided (4.8), (4.9), (4.16), (4.18)–(4.20) are satisfied. Thus we can apply Theorem 3.8 to obtain an approximation error estimate for RKP particle-shape function systems. Note that the condition (3.63) in Theorem 3.8 is trivially satisfied with $\mathcal{A}^{\nu}_{\underline{x}} = I$ for all $\underline{x} \in X^{\nu}$. We remark that an interpolation error estimate, under the assumptions (4.8), (4.9), (4.16), (4.18)–(4.20), was also obtained in Han and Meng (2001).

We note that A1–A7 only guarantee good approximability of the shape functions; they do not provide a recipe to construct particle shape functions that are quasi-reproducing or reproducing of order k. In fact the availability of such particle shape functions is assumed in A5. Further assumptions may be needed to construct such shape functions; for example (4.16), (4.18)–(4.20) were needed to construct RKP particle shape functions. Therefore, there should be enough restrictions on the particle distributions and the supports of shape functions for it to be possible to construct these shape functions satisfying A1–A7, thereby ensuring good approximation properties.

Uniformly distributed particles

We consider the uniformly distributed particles $x^h_j = jh$, $j \in \mathbb{Z}^n$ as in Section 3.2. This is a special case of the non-uniformly distributed particles considered in the first part of this section. For each x^h_j, we define $w^h_j(x) = w(\frac{x - x^h_i}{h})$, where $w(x) \ge 0$ is a continuous weight function with compact support $\eta = \overline{B}_R(0)$. Clearly, $\eta^h_j \equiv \text{supp } w^h_j(x) = \overline{B}_{Rh}(x^h_j)$. It can be easily shown that if $R = 3\sqrt{n}/2$ (in fact, we need only $R > \sqrt{n}$), then (4.8), (4.9) with $\kappa = (4R+1)^n$, (4.18) with $C_1 = C_2 = 1$, and (4.20) with $\tilde{C}_2 = 1/2$ are satisfied. If $w(x) = w(r)$, with $r = \|x\|$, is monotonically decreasing in r, then it can also be shown easily that (4.19) is satisfied with $\tilde{C}_1 = w(\sqrt{n})$. Therefore, RKP shape functions $\phi^h_i(x)$, associated with x^h_i, for all $i \in \mathbb{Z}^n$, can be constructed using the procedure described in (4.10), (4.11) and (4.12) for $k = 1$, namely

$$\phi^h_j(x) = w^h_j(x) \sum_{|\alpha| \le k} (x - x^h_j)^{\alpha} b^h_{\alpha}(x), \tag{4.22}$$

where $\{b_\alpha^h(x)\}_{|\alpha|\le k}$ is the solution of

$$\sum_{|\alpha|\le k} m_{\alpha+\beta}^h(x) b_\alpha^h(x) = \delta_{|\beta|,0}, \quad |\beta| \le k, \tag{4.23}$$

with $k = 1$, and

$$m_\alpha^h(x) = \sum_{j\in\mathbb{Z}^n} w_j^h(x)(x - x_j^h)^\alpha. \tag{4.24}$$

The shape functions $\{\phi_j^h\}$ satisfy

$$\sum_{j\in\mathbb{Z}^n} p(x_j^h)\phi_j^h(x) = p(x), \quad \text{for all } x \in \mathbb{R}^n \text{ and } p \in \mathcal{P}^k(\mathbb{R}^n). \tag{4.25}$$

As with the non-uniformly distributed particles, we consider the family of particle-shape function systems

$$\mathcal{M}^h = \{X^h, \{h_{\underline{x}}^h, \omega_{\underline{x}}^h, \phi_{\underline{x}}^h\}_{\underline{x}\in X^h}\}, \quad 0 < h \le 1,$$

for RKP shape functions with respect to uniformly distributed particles, by letting $X^h = \{\underline{x} = x_j^h : j \in \mathbb{Z}^n\}$ and using $h_{\underline{x}}^h = h$, $\omega_{\underline{x}}^h = \eta_j^h$ and $\phi_{\underline{x}}^h = \phi_j^h$. Note that here we used the parameter h instead of ν. We have shown above that conditions (4.8), (4.9), (4.18)–(4.20) are satisfied, with $w(x) = w(r)$, a monotonically decreasing weight function in r, and $R = 3\sqrt{n}/2$. Therefore, based on the discussion on RKP particle-shape function systems for non-uniformly distributed particles, it clear that $\{\mathcal{M}^h\}_{0<h\le 1}$ satisfies assumptions A1–A7 with $k = 1$, ensuring good approximation properties of the RKP shape functions.

We recall that in Section 3.1, the particle shape function $\phi_i^h(x)$ was defined in (3.1) by scaling and translating the basic shape function $\phi(x)$ for uniformly distributed particles, *i.e.*, they were translation-invariant. We will show that the RKP shape functions $\{\phi_i^h\}_{i\in\mathbb{Z}^n}$, constructed via (4.22) and (4.23), also satisfy (3.1) with $\phi(x) = \phi_0^1(x)$ (*i.e.*, with $i = 0$ and $h = 1$). We assume that the linear system (4.23) is invertible for $k \ge 1$.

From (4.22) and (4.23) with $i = 0$ and $h = 1$, we have

$$\phi(x) = w(x) \sum_{|\alpha|\le k} x^\alpha b_\alpha^1(x), \tag{4.26}$$

where $b_\alpha^1(x)$ are the solutions of

$$\sum_{|\alpha|\le k} m_{\alpha+\beta}^1(x) b_\alpha^1(x) = \delta_{|\beta|,0}, \quad \text{for } |\beta| \le k, \tag{4.27}$$

and

$$m_\alpha^1(x) = \sum_{j\in\mathbb{Z}^n} w(x - j)(x - j)^\alpha. \tag{4.28}$$

We replace x by $\frac{x-x_i^h}{h}$ in (4.27) and (4.28) to get

$$\sum_{|\alpha|\le k} m^1_{\alpha+\beta}\left(\frac{x-x_i^h}{h}\right) b_\alpha\left(\frac{x-x_i^h}{h}\right) = \delta_{|\beta|,0}, \quad \text{for } |\beta| \le k, \tag{4.29}$$

where

$$\begin{aligned} m^1_\alpha\left(\frac{x-x_i^h}{h}\right) &= \sum_{j\in\mathbb{Z}^n} w\left(\frac{x-x_i^h}{h} - j\right)\left(\frac{x-x_i^h}{h} - j\right)^\alpha \\ &= \sum_{j\in\mathbb{Z}^n} w\left(\frac{x-x_{i+j}^h}{h}\right)\left(\frac{x-x_{i+j}^h}{h}\right)^\alpha \\ &= \frac{1}{h^{|\alpha|}} \sum_{j\in\mathbb{Z}^n} w_j^h(x)(x-x_j^h)^\alpha \\ &= \frac{1}{h^{|\alpha|}} m_\alpha^h(x). \end{aligned} \tag{4.30}$$

Using (4.30) in (4.29), we get

$$\sum_{|\alpha|\le k} \frac{1}{h^{|\alpha+\beta|}} m^h_{\alpha+\beta}(x) b_\alpha\left(\frac{x-x_i^h}{h}\right) = \delta_{|\beta|,0},$$

and therefore

$$\sum_{|\alpha|\le k} m^h_{\alpha+\beta}(x) \frac{1}{h^{|\alpha|}} b_\alpha\left(\frac{x-x_i^h}{h}\right) = h^{|\beta|}\delta_{|\beta|,0} = \delta_{|\beta|,0}, \quad \text{for all } |\beta| \le k. \tag{4.31}$$

Since the $\{b_\alpha^h(x)\}$ is the unique solution for (4.23), it is clear from (4.31) that

$$b_\alpha^h(x) = \frac{1}{h^{|\alpha|}} b_\alpha\left(\frac{x-x_i^h}{h}\right),$$

and thus from (4.26) we have

$$\begin{aligned} \phi\left(\frac{x-x_i^h}{h}\right) &= w\left(\frac{x-x_i^h}{h}\right) \sum_{|\alpha|\le k} \left(\frac{x-x_i^h}{h}\right)^\alpha b_\alpha\left(\frac{x-x_i^h}{h}\right) \\ &= w_i^h(x) \sum_{|\alpha|\le k} (x-x_i^h)^\alpha \frac{1}{h^{|\alpha|}} b_\alpha\left(\frac{x-x_i^h}{h}\right) \\ &= w_i^h(x) \sum_{|\alpha|\le k} (x-x_i^h)^\alpha b_\alpha^h(x) \\ &= \phi_i^h(x). \end{aligned}$$

Thus, for uniformly distributed particles, RKP shape functions satisfy (3.1), *i.e.*, they are translation-invariant.

Remark 27. To approximate functions defined on a bounded domain Ω, we use the restrictions of RKP shape functions on Ω, as described in Section 3.3 (*cf.* (3.85) and Theorem 3.10). We note that the RKP shape functions corresponding to the particles near the boundary of Ω, as defined here, are different from the RKP shape functions defined in Han and Meng (2001) and Liu *et al.* (1995). But they are the same for particles inside Ω, sufficiently away from the boundary $\partial\Omega$. They are also the same when $\Omega = \mathbb{R}^n$.

Remark 28. RKP shape functions are not available analytically in simple forms. Their values at a point $x \in \mathbb{R}^n$ are computed via (4.10), which involves the solution of the linear system (4.12) for each x. Thus computation of RKP shape functions, which are reproducing of order k, is not easy, especially for higher k. Moreover, as a consequence of (4.16), it is necessary that the supports of these shape functions be large so that the linear system (4.12) is invertible. We note that it is much easier to construct the shape functions for the space $\mathbb{W}^{l,q}_{\Omega,h}$, introduced near the end of Section 3.3.

4.2. Interpolation and selection

In this section, we will address the interpolation of a function in terms of particle shape functions, and will propose a procedure to select a shape function that will yield efficient approximation. We consider uniformly distributed particles $\{x_j^h\}$ in $\mathbb{R}^n$, and the associated particle shape functions $\{\phi_j^h\}$, defined in (3.1), where $\phi \in H^q(\mathbb{R}^n)$ with $q \geq 1$ has compact support; supp $\phi \subset B_R(0)$. We have seen that $\{\phi_j^h\}$ is translation-invariant, $\operatorname{supp}\phi_j^h \in B_{Rh}(x_j^h)$, and in addition

$$\|\phi_j^h\|_{H^1(\mathbb{R}^n)} \leq h^{n/2-1}\|\phi\|_{H^1(\mathbb{R}^n)}. \tag{4.32}$$

We assume that $\{\phi_j^h\}$ is reproducing of order k, *i.e.*, (4.25) holds.

Let Ω be a bounded domain in $\mathbb{R}^n$. We will consider a smooth function $u(x)$ defined in Ω and study the error $u - \tilde{\mathcal{I}}_h u$, where $\tilde{\mathcal{I}}_h u$ is the 'interpolant' of u in terms of ϕ_j^h. The results in this subsection are from Babuška, Banerjee and Osborn (200x), and we refer to this paper for some of the details that we do not present here.

We now define the 'interpolant' $\tilde{\mathcal{I}}_h u$ of a function u. For any $x \in \mathbb{R}^n$, let

$$A_x^h = \{k \in \mathbb{Z}^n : x \in \mathring{\eta}_k^h\}.$$

Here A_x^h is called the influence set for the point x. Then $(\tilde{\mathcal{I}}_h u)(x)$ is defined as

$$(\tilde{\mathcal{I}}_h u)(x) = \sum_{j \in A_x^h} u(x_j^h)\phi_j^h(x). \tag{4.33}$$

In (4.33), we, of course, assume that $u(x_j^h)$ is defined for all $j \in A_x^h$. If $p \in \mathcal{P}^k$, then from (4.25) we have

$$\sum_{j \in A_x^h} p(x_j^h)\phi_j^h(x) = \sum_{j \in \mathbb{Z}^n} p(x_j^h)\phi_j^h(x) = p(x), \quad \text{for all } x \in \mathbb{R}^n, \tag{4.34}$$

i.e., $\tilde{\mathcal{I}}_h p = p$. Now let $u \in H^s(\Omega)$ with $s > n/2$. For some $x \in \Omega$, the particles x_j^h for $j \in A_x^h$ may be outside Ω, and $u(x_j^h)$ may not be defined. To define $\tilde{\mathcal{I}}_h u(x)$ for $u \in H^s(\Omega)$ and for all $x \in \Omega$, we need an extension $\bar{u}$ of u in a ball B_{R_0} containing Ω such that $\operatorname{dist}(\partial\Omega, \partial B_{R_0}) > \rho h$, and $\bar{u} \in H^s(B_{R_0})$. Then,

$$(\tilde{\mathcal{I}}_h u)(x) \equiv (\tilde{\mathcal{I}}_h \bar{u})(x) = \sum_{j \in A_x^h} \bar{u}(x_j^h)\phi_j^h(x), \quad \text{for } x \in \Omega, \tag{4.35}$$

is well defined. For an extension $\bar{u}$, we may use $\bar{u} = Eu$, where Eu was defined in (3.50). Thus, $(\tilde{\mathcal{I}}_h u)(x)$ for $x \in \Omega$ will depend on few values of $\bar{u}(x_j^h)$, where the particle x_j^h is just outside Ω. We remark that $\tilde{\mathcal{I}}_h u$ is not an interpolant of u in the usual sense, since, generally, $\phi_j^h(x_i^h) \neq \delta_{ij}$, and hence $(\tilde{\mathcal{I}}_h u)(x_j^h) \neq u(x_j^h)$.

We define the function

$$\xi_\alpha^h(x) = x^\alpha - \sum_{i \in A_x^h} (x_i^h)^\alpha \phi_i^h(x), \ |\alpha| = k+1, \quad \text{for } x \in \mathbb{R}^n. \tag{4.36}$$

We will also use

$$\xi_\alpha(x) \equiv \xi_\alpha^1(x) = x^\alpha - \sum_{i \in A_x^1} i^\alpha \phi(x-i), \ |\alpha| = k+1, \quad \text{for } x \in \mathbb{R}^n, \tag{4.37}$$

where A_x^1 is A_x^h with $h = 1$. These functions will play an important role in the analysis presented in this subsection, as well as in Section 5. We note that $\xi_\alpha(x)$ is the error between the polynomial x^α, with $|\alpha| = k+1$, and its interpolant when $h = 1$. In one dimension, we will write these functions as $\xi_{k+1}^h(x)$ and $\xi_{k+1}(x)$ respectively.

We begin with certain results about these functions. We first present some notation that will be used in these results. Let I_j^h be the *cell* centred at x_j^h, defined by

$$I_j^h = \{x : \|x - x_j^h\|_\infty \equiv \max_{i=1,\dots,n} |x_i - x_{j_i}| \le h/2\}.$$

For each I_j^h, we define

$$A_j^h = \{k \in \mathbb{Z}^n : \mathring{\eta}_k^h \cap I_j^h \neq \emptyset\},$$

and

$$B_j^h = \{\cup_{k \in A_j^h} B_{Rh}(x_k^h)\} \cup I_j^h.$$

We note that the cardinality of A_j^h is finite, and is bounded independently of j and h. Further, there exists $\bar{R} > 0$, independent of j and h, such that $B_j^h \subset \tilde{B}_j^h \equiv B_{\bar{R}h}(x_j^h)$ and $\cup_{j \in \mathbb{Z}^n} \tilde{B}_j^h = \mathbb{R}^n$.

Lemma 4.1. $\xi_\alpha^h(x)$, with $|\alpha| = k+1$, is periodic, *i.e.*,

$$\xi_\alpha^h(x + x_j^h) = \xi_\alpha^h(x), \quad \text{for any } x_j^h. \tag{4.38}$$

Proof. We first note that

$$(x + x_j^h)^\alpha = x^\alpha + p(x; x_j^h), \tag{4.39}$$

where $p(x; x_j^h)$ is a polynomial in x of degree $\leq k$ with coefficients that depend on x_j^h. Now using (4.39), with $x = x_i^h$, and the fact that the $\{\phi_i^h\}$ is translation-invariant and reproducing of order k, we get

$$\begin{aligned}
\sum_{i \in \mathbb{Z}^n} (x_i^h)^\alpha \phi_i^h(x + x_j^h) &= \sum_{i \in \mathbb{Z}^n} (x_i^h)^\alpha \phi_{i-j}^h(x) \\
&= \sum_{i \in \mathbb{Z}^n} (x_{i+j}^h)^\alpha \phi_i^h(x) \\
&= \sum_{i \in \mathbb{Z}^n} (x_i^h + x_j^h)^\alpha \phi_i^h(x) \\
&= \sum_{i \in \mathbb{Z}^n} (x_i^h)^\alpha \phi_i^h(x) + \sum_{i \in \mathbb{Z}^n} p(x_i^h; x_j^h) \phi_i^h(x) \\
&= \sum_{i \in \mathbb{Z}^n} (x_i^h)^\alpha \phi_i^h(x) + p(x, x_j^h).
\end{aligned} \tag{4.40}$$

From (4.36), (4.39) and (4.40), we get

$$\xi_\alpha^h(x + x_j^h) = x^\alpha - \sum_{i \in \mathbb{Z}^n} (x_i^h)^\alpha \phi_i^h(x) = \xi_\alpha^h(x),$$

which is the desired result. □

Lemma 4.2. Let $\alpha = \alpha(i)$, $i = 1, \ldots, M_k$ be an enumeration of the multi-indices α with $|\alpha(i)| = k+1$. Let I_j^h be the cell centred at the particle x_j^h. Then, for $d_\alpha \in \mathbb{R}$, we have

$$\left\| \sum_{|\alpha|=k+1} \frac{1}{\alpha!} d_\alpha \xi_\alpha^h(x) \right\|_{H^1(I_j^h)}^2 = h^{2k+n} \mathbb{V}^T (A + h^2 B) \mathbb{V}, \tag{4.41}$$

where $\mathbb{V} = [d_{\alpha(1)}, d_{\alpha(2)}, \ldots, d_{\alpha(M_k)}]^T$ and A, B are $M_k \times M_k$ matrices given by

$$A_{lm} = \int_I \frac{1}{\alpha(l)!\alpha(m)!} \nabla\xi_{\alpha(l)} \cdot \nabla\xi_{\alpha(m)}\, \mathrm{d}x, \tag{4.42}$$

$$B_{lm} = \int_I \frac{1}{\alpha(l)!\alpha(m)!} \xi_{\alpha(l)}\xi_{\alpha(m)}\, \mathrm{d}x, \tag{4.43}$$

respectively, and $I = [-1/2, 1/2]^n$.

Note. The matrices A and B are independent of I_j^h.

Proof. A simple scaling argument, used with (3.1), shows that

$$\xi_\alpha\left(\frac{x}{h}\right) = h^{-(k+1)}\xi_\alpha^h(x).$$

Now, using the periodicity of $\xi_\alpha^h(x)$, a standard scaling argument, and this identity, we have

$$\begin{aligned}
\left\| \sum_{|\alpha|=k+1} \frac{1}{\alpha!} d_\alpha \nabla\xi_\alpha^h(x) \right\|^2_{H^0(I_j^h)} &= \left\| \sum_{|\alpha|=k+1} \frac{1}{\alpha!} d_\alpha \nabla\xi_\alpha^h(x) \right\|^2_{H^0(I_0^h)} \\
&= h^{2(k+1)} \left\| \sum_{|\alpha|=k+1} \frac{1}{\alpha!} d_\alpha \nabla\left[\xi_\alpha\left(\frac{x}{h}\right)\right] \right\|^2_{H^0(I_0^h)} \\
&= h^{2(k+1)} h^{n-2} \left\| \sum_{|\alpha|=k+1} \frac{1}{\alpha!} d_\alpha \nabla\xi_\alpha(y) \right\|^2_{H^0(I)} \\
&= h^{2k+n}\mathbb{V}^T A\mathbb{V}.
\end{aligned} \tag{4.44}$$

Using a similar argument, we have

$$\left\| \sum_{|\alpha|=k+1} \frac{1}{\alpha!} d_\alpha \xi_\alpha^h(x) \right\|^2_{H^0(I_j^h)} = h^{2k+2+n}\mathbb{V}^T B\mathbb{V}.$$

Combining this identity with (4.44), we get

$$\left\| \sum_{|\alpha|=k+1} \frac{1}{\alpha!} d_\alpha \xi_\alpha^h(x) \right\|^2_{H^1(I_j^h)} = h^{2k+n}\mathbb{V}^T (A + h^2 B)\mathbb{V},$$

which is the desired result. □

Lemma 4.3. Let I_j^h be the cell centred at x_j^h, and consider the corresponding set $\tilde{B}_j^h$. Suppose $u \in H^{k+2+q}(\tilde{B}_j^h)$ with $q > \frac{n}{2}$ when $n \geq 2$, and $q = 0$ when $n = 1$. Then,

(a) for any $\delta > 0$,

$$\begin{aligned} &\|u - \tilde{\mathcal{I}}_h u\|^2_{H^1(I_j^h)} \\ &\leq (1+\delta^2)\left\| \sum_{|\alpha|=k+1} \frac{1}{\alpha!}(D^\alpha u)(x_j^h)\xi_\alpha^h(x)\right\|^2_{H^1(I_j^h)} \\ &\quad + \left(1 + \frac{1}{\delta^2}\right) C h^{2k+2} \sum_{|\alpha|=k+2} \|D^\alpha u\|^2_{H^q(\tilde{B}_j^h)}. \end{aligned} \tag{4.45}$$

and

(b) for any $\delta > 0$,

$$\begin{aligned} &\left\| \sum_{|\alpha|=k+1} \frac{1}{\alpha!}(D^\alpha u)(x_j^h)\xi_\alpha^h(x)\right\|^2_{H^1(I_j^h)} \\ &\leq (1+\delta^2)\|u - \tilde{\mathcal{I}}_h u\|^2_{H^1(I_j^h)} \\ &\quad + \left(1 + \frac{1}{\delta^2}\right) C h^{2k+2} \sum_{|\alpha|=k+2} \|D^\alpha u\|^2_{H^q(\tilde{B}_j^h)}. \end{aligned} \tag{4.46}$$

The proof of this result is based on Taylor's theorem, a bound on the remainder in Taylor's theorem, and a bound on the interpolant of the same remainder. We do not include the proof here, and refer to Babuška *et al.* (200x).

We will now study the interpolation error $u - \tilde{\mathcal{I}}_h u$, where u is a smooth function in Ω. An interpolation error estimate for RKP shape functions, namely $\|u - \tilde{\mathcal{I}}_h u\|_{W^{l,q}(\Omega)} \approx O(h^{k+1-l})$ for certain values of l, was proved in Han and Meng (2001); the same estimate for $q = 2$ was proved in Liu, Li and Belytschko (1997). A similar order of convergence in the $H^{1,\infty}$ norm was also obtained for MLS shape functions in Armentano (2002) and Armentano and Duran (2001). We note that the definitions of $\tilde{\mathcal{I}}_h u$ for the RKP shape functions and MLS shape functions, presented in these papers, are slightly different from our definition as given in (4.35). From the proof of the our next result, we will obtain an estimate of $\|u - \tilde{\mathcal{I}}_h u\|_{H^1(\Omega)}$ where the shape functions are reproducing of order k. Moreover, this theorem gives some information on the size of $h^k\|u - \tilde{\mathcal{I}}_h u\|_{H^1(\Omega)}$, which facilitates the selection of 'good' shape functions, which will be discussed later.

We now present the main result of this section. We define certain sets which will be used in this result:

$$\bar{\mathcal{A}}^h = \{k \in \mathbb{Z}^n : \Omega \cap \mathring{I}_k^h \neq \emptyset\}, \quad \bar{\Omega}_h = \cup_{j\in\bar{\mathcal{A}}^h} I_j^h,$$
$$\underline{\mathcal{A}}^h = \{k \in \mathbb{Z}^n : I_k^h \subset \Omega\}, \quad \underline{\Omega}_h = \cup_{j\in\underline{\mathcal{A}}^h} I_j^h,$$
$$\underline{B}^h = \{\cup_{j\in\underline{\mathcal{A}}^h} \tilde{B}_j^h\} \cup \Omega, \quad \bar{B}^h = \cup_{j\in\bar{\mathcal{A}}^h} \tilde{B}_j^h.$$

It is clear that $\underline{\Omega}_h \subset \Omega \subset \bar{\Omega}_h$, and $|\Omega - \underline{\Omega}_h| \to 0$, $|\bar{\Omega}_h - \Omega| \to 0$ as $h \to 0$. Also $\Omega \subset \underline{B}^h \subset \bar{B}^h$, and $|\underline{B}^h - \Omega| \to 0$, $|\bar{B}^h - \Omega| \to 0$ as $h \to 0$.

Theorem 4.4. (Babuška *et al.* (200x)) Let $\bar{\lambda}$ be the largest eigenvalue of the matrix A given in (4.42). Suppose $q > \frac{n}{2}$ when $n \geq 2$, and $q = 0$ when $n = 1$. Then we have

$$\sup_{u\in H^{k+2+q}(\Omega)} \lim_{h\to 0} \frac{\|u - \tilde{\mathcal{I}}_h u\|^2_{H^1(\Omega)}}{h^{2k} Q_h(u)} = \bar{\lambda}, \tag{4.47}$$

where

$$Q_h(u) = |u|^2_{H^{k+1}(\Omega)} + h \sum_{|\alpha|=k+2} \|D^\alpha u\|^2_{H^q(\Omega)}. \tag{4.48}$$

Note. In (4.47), we consider $u \in H^{k+2+q}(\Omega)$ such that $u \notin \mathcal{P}^k$.

Proof. We will first prove that, for $u \in H^{k+2+q}(\Omega)$,

$$\lim_{h\to 0} \frac{\|u - \tilde{\mathcal{I}}_h u\|^2_{H^1(\Omega)}}{h^{2k} Q_h(u)} = \frac{\int_\Omega V^T(x) A V(x)\,\mathrm{d}x}{|u|^2_{H^{k+1}(\Omega)}}, \tag{4.49}$$

where

$$V^T(x) = [D^{\alpha(1)}u(x), D^{\alpha(2)}u(x), \ldots, D^{\alpha(M_k)}u(x)],$$

and $\alpha(i)$, $1 \leq i \leq M_k$, are the multi-indices with $|\alpha(i)| = k + 1$.

Let $u \in H^{k+2+q}(\Omega)$, and suppose $\bar{u}$ is an extension of u, as discussed before. Since, $\Omega \subset \bar{\Omega}_h$, we have

$$\|u - \tilde{\mathcal{I}}_h u\|^2_{H^1(\Omega)} \leq \|\bar{u} - \tilde{\mathcal{I}}_h \bar{u}\|^2_{H^1(\bar{\Omega}_h)} = \sum_{j\in\bar{\mathcal{A}}^h} \|\bar{u} - \tilde{\mathcal{I}}_h \bar{u}\|^2_{H^1(I_j^h)}.$$

Therefore, using (4.45), (4.41), and recalling that $\bar{B}^h = \cup_{j\in\bar{\mathcal{A}}^h} \tilde{B}_j^h$, we get for any $\delta > 0$,

$$\|u - \tilde{\mathcal{I}}_h u\|^2_{H^1(\Omega)} \leq (1 + \delta^2) \sum_{j\in\bar{\mathcal{A}}^h} \left\| \sum_{|\alpha|=k+1} \frac{1}{\alpha!} (D^\alpha \bar{u})(x_j^h) \xi_\alpha^h(x) \right\|^2_{H^1(I_j^h)}$$
$$+ \left(1 + \frac{1}{\delta^2}\right) C h^{2k+2} \sum_{j\in\bar{\mathcal{A}}^h} \sum_{|\alpha|=k+2} \|D^\alpha \bar{u}\|^2_{H^q(\tilde{B}_j^h)}$$

$$\leq (1+\delta^2)h^{2k} \sum_{j\in\bar{A}^h} h^n V_j^T (A+h^2B)V_j$$
$$+ \left(1+\frac{1}{\delta^2}\right) Ch^{2k+2} \sum_{|\alpha|=k+2} \|D^\alpha \bar{u}\|^2_{H^q(\bar{B}^h)}, \tag{4.50}$$

where

$$V_j^T = [D^{\alpha(1)}\bar{u}(x_j^h), D^{\alpha(2)}\bar{u}(x_j^h), \ldots, D^{\alpha(M_k)}\bar{u}(x_j^h)].$$

Therefore, dividing (4.50) by $h^{2k}Q_h(u)$, where $Q_h(u)$ is defined in (4.48), we get

$$\frac{\|u-\tilde{\mathcal{I}}_h u\|^2_{H^1(\Omega)}}{h^{2k}Q_h(u)} \leq (1+\delta^2)\frac{\sum_{j\in\bar{A}^h} h^n V_j^T(A+h^2B)V_j}{Q_h(u)}$$
$$+ \left(1+\frac{1}{\delta^2}\right) Ch^2 \frac{\sum_{|\alpha|=k+2} \|D^\alpha\bar{u}\|^2_{H^q(\bar{B}^h)}}{Q_h(u)}. \tag{4.51}$$

A typical term of the quadratic form $V_j^T(A+h^2B)V_j$ is

$$D^{\alpha(i)}\bar{u}(x_j^h)(A_{il}+h^2B_{il})D^{\alpha(l)}\bar{u}(x_j^h).$$

Since

$$\lim_{h\to 0} \sum_{j\in\bar{A}^h} h^n D^{\alpha(i)}\bar{u}(x_j^h)A_{il}D^{\alpha(l)}\bar{u}(x_j^h) = \int_\Omega D^{\alpha(i)}u(x)A_{il}D^{\alpha(l)}u(x)\,\mathrm{d}x$$

and

$$\lim_{h\to 0} h^2 \sum_{j\in\bar{A}^h} h^n D^{\alpha(i)}\bar{u}(x_j^h)B_{il}D^{\alpha(l)}\bar{u}(x_j^h) = 0,$$

we have

$$\lim_{h\to 0} \sum_{j\in\bar{A}^h} h^n V_j^T(A+h^2B)V_j = \int_\Omega V^T(x)AV(x)\,\mathrm{d}x. \tag{4.52}$$

Since $|\bar{B}^h - \Omega| \to 0$ as $h \to 0$, we have

$$\lim_{h\to 0} \sum_{|\alpha|=k+2} \|D^\alpha\bar{u}\|^2_{H^q(\bar{B}^h)} = \sum_{|\alpha|=k+2} \|D^\alpha u\|^2_{H^q(\Omega)}. \tag{4.53}$$

Also $\lim_{h\to 0} Q_h(u) = |u|_{H^{k+1}(\Omega)}$. Thus, for any $\delta > 0$, using (4.52) and (4.53) in (4.51), we get

$$\limsup_{h\to 0} \frac{\|u-\tilde{\mathcal{I}}_h u\|^2_{H^1(\Omega)}}{h^{2k}Q_h(u)} \leq (1+\delta^2)\frac{\int_\Omega V^T(x)AV(x)\,\mathrm{d}x}{|u|^2_{H^{k+1}(\Omega)}},$$

and, since $\delta > 0$ is arbitrary, we have

$$\limsup_{h\to 0} \frac{\|u - \tilde{\mathcal{I}}_h u\|^2_{H^1(\Omega)}}{h^{2k} Q_h(u)} \le \frac{\int_\Omega V^T(x) A V(x)\, dx}{|u|^2_{H^{k+1}(\Omega)}}. \tag{4.54}$$

Following the argument leading to (4.54), but using $\underline{\mathcal{A}}^h$, $\underline{\mathcal{B}}^h$, and (4.46) instead of $\bar{\mathcal{A}}^h$, $\bar{\mathcal{B}}^h$, and (4.45), respectively, we can also show that

$$\frac{\int_\Omega V^T(x) A V(x)\, dx}{|u|^2_{H^{k+1}(\Omega)}} \le \liminf_{h\to 0} \frac{\|u - \tilde{\mathcal{I}}_h u\|^2_{H^1(\Omega)}}{h^{2k} Q_h(u)}. \tag{4.55}$$

Combining (4.54) and (4.55), we see that

$$\lim_{h\to 0} \frac{\|u - \tilde{\mathcal{I}}_h u\|^2_{H^1(\Omega)}}{h^{2k} Q_h(u)}$$

exists, and

$$\lim_{h\to 0} \frac{\|u - \tilde{\mathcal{I}}_h u\|^2_{H^1(\Omega)}}{h^{2k} Q_h(u)} = \frac{\int_\Omega V^T(x) A V(x)\, dx}{|u|^2_{H^{k+1}(\Omega)}},$$

which is (4.49).

Since $\bar{\lambda}$ is the largest eigenvalue of the matrix A, from the usual variational characterization of eigenvalues we have

$$\int_\Omega V^T(x) A V(x)\, dx \le \bar{\lambda} \int_\Omega \sum_{i=1}^{M_k} |D^{\alpha(i)} u(x)|^2\, dx = \bar{\lambda} |u|^2_{H^{k+1}(\Omega)}.$$

Thus, from (4.49) we get

$$\lim_{h\to 0} \frac{\|u - \tilde{\mathcal{I}}_h u\|^2_{H^1(\Omega)}}{h^{2k} Q_h(u)} \le \bar{\lambda}, \quad \text{for any } u \in H^{k+2+q}(\Omega).$$

Hence

$$\sup_{u \in H^{p+2+q}(\Omega)} \lim_{h\to 0} \frac{\|u - \tilde{\mathcal{I}}_h u\|^2_{H^1(\Omega)}}{h^{2k} Q_h(u)} \le \bar{\lambda}. \tag{4.56}$$

Let $\bar{v} = [v_1, v_2, \ldots, v_{M_k}]^T$ be an eigenvector of A corresponding to $\bar{\lambda}$. Then it is easily seen that there is a $u \in \mathcal{P}^{k+1}$ such that the vector $V(x) = \bar{v}$. For this particular u, we have

$$\frac{\int_\Omega V^T(x) A V(x)\, dx}{|u|^2_{H^{k+1}(\Omega)}} = \bar{\lambda}.$$

Hence, from (4.56) we conclude that

$$\sup_{u \in H^{k+2+q}(\Omega)} \lim_{h \to 0} \frac{\|u - \tilde{\mathcal{I}}_h u\|^2_{H^1(\Omega)}}{h^{2k} Q_h(u)} = \bar{\lambda},$$

which is the desired result. □

Remark 29. We know from (4.35) that the interpolant of a smooth function depends on its extension to $\mathbb{R}^n$. But it is clear from the proof of Theorem 4.4 that (4.47) is valid for any extension satisfying (3.50).

Remark 30. We note that the same result holds for the H^1-seminorm of the interpolation error, *i.e.*, for $q > \frac{n}{2}$ when $n \geq 2$, and $q = 0$ when $n = 1$, we have

$$\sup_{u \in H^{k+2+q}(\Omega)} \lim_{h \to 0} \frac{|u - \tilde{\mathcal{I}}_h u|^2_{H^1(\Omega)}}{h^{2k}[|u|^2_{H^{k+1}(\Omega)} + h \sum_{|\alpha|=k+2} \|D^\alpha u\|^2_{H^q(\Omega)}]} = \bar{\lambda}.$$

Remark 31. From (4.51) in the proof of Theorem 4.4, we can obtain an interpolation error estimate,

$$\|u - \tilde{\mathcal{I}}_h u\|_{H^1(\Omega)} \leq C h^k \|u\|_{H^{k+2+q}(\Omega)},$$

where C may depend on Ω, but is independent of u and h. We note, however, that this is not the optimal error estimate. For an outline of the proof, see Babuška *et al.* (200x).

We have seen in Remark 31 that, if the particle shape functions are reproducing of order k, then for a smooth function u,

$$\|u - \tilde{\mathcal{I}}_h u\|_{H^1(\Omega)} \approx O(h^k),$$

where $\tilde{\mathcal{I}}_h u$ is the interpolation of u as defined in (4.35). There are many classes of shape functions that have these properties. We have seen in Section 4.1 that translation-invariant RKP shape functions depend on the weight function $w(x)$, and different choices of $w(x)$ will generate different classes of such shape functions.

We will assess the approximability of a family $\{\phi_j^h\}$ of shape functions by the size of $\bar{\lambda}$, the largest eigenvalue of the matrix A defined in (4.42). We note that $\bar{\lambda}$ is computable, and depends only on the basic shape function $\phi(x)$. We emphasize that $\bar{\lambda}$ does not depend on u or on h. From (4.47), we know that

$$\frac{\|u - \tilde{\mathcal{I}}_h u\|_{H^1(\Omega)}}{h^p \sqrt{Q_h(u)}} \lessapprox \sqrt{\bar{\lambda}}, \quad \text{for small } h.$$

Thus we see that $\bar{\lambda}$ is a useful measure of the approximability of the family $\{\phi_j^h\}$, determined from the basic shape function $\phi(x)$.

We will illustrate our selection scheme in one dimension, and will rank the shape functions according to to their approximability. In one dimension,

$$\bar{\lambda} = \left(\frac{|\xi_{k+1}|_{H^1(0,1)}}{(k+1)!} \right)^2.$$

In the rest of this paper we will suppress $H^1(0,1)$ in $|\xi_{k+1}|_{H^1(0,1)}$, and instead write $|\xi_{k+1}|_1$.

We considered four different classes of RKP shape functions, reproducing of order 1, corresponding to four different weight functions $w(x)$. These weight functions were given by (4.2) with $\delta = 2$, (4.3), and (4.4) with $l = 2, 4$. We then computed $|\xi_{k+1}|_1$ for each of these four classes of shape functions for $R = 1.7$; we obtained

$$|\xi_{k+1}|_1 = \begin{cases} 0.237, & \text{for } w(x) \text{ in (4.4)}, \ l = 2, \\ 0.203, & \text{for } w(x) \text{ in (4.2)}, \ \delta = 2, \\ 0.095, & \text{for } w(x) \text{ in (4.3)}, \\ 0.029, & \text{for } w(x) \text{ in (4.4)}, \ l = 4. \end{cases}$$

We choose the RKP shape functions corresponding to $w(x)$ given in (4.4) with $l = 4$, since these shape functions yield the smallest value of $|\xi_{k+1}|_1$. We note that the value of $|\xi_{k+1}|_1$ depends strongly on R, and the shape function corresponding to $w(x)$ given in (4.4) with $l = 4$ may not be our choice for other values of R. We refer to Babuška *et al.* (200x) for further discussion on this issue.

To validate our criterion for selection of the shape functions, we have considered the function $u(x) = x^4$ on the interval $\Omega = (0,1)$ and computed the error $|u - \tilde{\mathcal{I}}_h u|_{H^1(\Omega)}$. $\tilde{\mathcal{I}}_h u$ is the interpolant of u with respect to the four classes of RKP shape functions described in the last paragraph, with $h = 1/n$, $n = 40, 50, \ldots, 100$. We note that the definition of $\tilde{\mathcal{I}}_h u$ requires the values of $u(x)$ in a small neighbourhood of Ω, and we have extended $u = x^4$ outside Ω by itself.

We summarize the results in Table 4.1. We note that columns 2–5 correspond to different classes of RKP shape functions constructed using different weight functions $w(x)$; for example, the column headed 'Cubic spline' refers to the cubic spline weight function given by (4.3). It is clear that the error $|u - \tilde{\mathcal{I}}_h u|_{H^1(\Omega)}$ can be ranked according to the size of $|\xi_2|_1$ for the four choices of $\omega(x)$ considered here with $R = 1.7$; the error and $|\xi_2|_1$ are both minimal when $w(x)$ is the conical weight function with $l = 4$.

This selection scheme is based on (4.47), and we know from Remark 29 that (4.47) is valid for any extension. We refer to Babuška *et al.* (200x) for an experimental illustration of this fact. We remark that this selection scheme is also valid for the projection error, which will be indicated by our results in the next section.

Table 4.1. The H^1-seminorm of the error, $|u - \tilde{\mathcal{I}}_h u|_{H^1(\Omega)}$, where $\tilde{\mathcal{I}}_h u$ is the interpolant of $u(x) = x^4$ using RKP shape functions that are reproducing of order 1, corresponding to different weight functions $w(x)$. The radius of support of $\omega(x)$ is $R = 1.7$.

n	$\lvert u - \tilde{\mathcal{I}}_h u\rvert_{H^1(\Omega)}$			
	Conical: $l = 2$	Gauss: $\delta = 2$	Cubic spline	Conical: $l = 4$
40	1.607e-2	1.376e-2	6.435e-3	2.283e-3
50	1.281e-2	1.096e-2	5.130e-3	1.730e-3
60	1.066e-2	9.112e-3	4.267e-3	1.396e-3
70	9.126e-3	7.800e-3	3.653e-3	1.172e-3
80	7.980e-3	6.819e-3	3.194e-3	1.012e-3
90	7.090e-3	6.058e-3	2.838e-3	8.908e-4
100	6.379e-3	5.449e-3	2.553e-3	7.962e-4

5. Superconvergence of the gradient of the solution in L_2

Superconvergence is an important feature of finite element methods, which allows an accurate approximation of the derivatives of the solution of the underlying BVP. In this section, we will discuss the idea of superconvergence when particle shape functions are used to approximate the solution of a BVP. We will consider uniformly distributed particles and the associated particle shape functions, which were developed in Sections 3.1 and 3.2. For uniformly distributed particles, a careful analysis in one dimension can be easily generalized to higher dimensions. Thus, in this section, we present the results in one dimension, thereby avoiding some technical details arising in the higher-dimensional analysis.

We will use the notation introduced in Section 3.1, but restricted to one dimension, *i.e.*, for $h > 0$, we consider $x_j^h = jh$, $j \in \mathbb{Z}$, and the corresponding shape function ϕ_j^h defined in (3.1). We assume that the shape functions are reproducing of order k. We use the following notation:

$$I_j^h = (x_j^h, x_{j+1}^h), \quad A_j^h = \{m \in \mathbb{Z} : \eta_m^h \cap I_j^h \neq \emptyset\};$$
$$I_j = (j, j+1), \qquad A_j \equiv A_j^1 \ \text{ (with } h = 1).$$

We assume that

$$\operatorname{card}(A_j) \leq \kappa,$$

or equivalently,

$$\operatorname{card}(A_j^h) \leq \kappa,$$

where κ is independent of j and h. We assume that the basic shape function $\phi(x)$ is such that, for any $v(x) = \sum_{i\in\mathbb{Z}} c_i\phi_i(x)$ for $x \in I_0$, there exist positive constants C_1, C_2, independent of v, but possibly depending on κ, such that

$$C_1 \sum_{j\in A_0} c_i^2 \leq \int_{I_0} v^2\,\mathrm{d}x \leq C_2 \sum_{j\in A_0} c_i^2. \tag{5.1}$$

This implies that the functions $\{\phi_i(x)\}_{i\in A_0}$ are linearly independent in I_0, that is,

$$\sum_{j\in A_0} c_j\phi_j(x) = 0, \quad x \in I_0 \text{ implies } c_j = 0, j \in A_0.$$

Throughout this section, we use C, C_1, C_2 as generic constants, which will have different values in different places.

Consider $\Omega = (-c, d) \subset \mathbb{R}$. Let $u_0 \in H^1(\Omega)$ be the solution of the Neumann problem

$$B(u_0, v) = \mathcal{F}(v), \quad \text{for all } v \in H^1(\Omega), \tag{5.2}$$

where

$$B(u, v) = \int_\Omega (u'v' + uv)\,\mathrm{d}x \quad \text{and} \quad \mathcal{F}(v) = \int_\Omega fv\,\mathrm{d}x$$

as in (2.4) and (2.5). We will often use the notation $B^F(u, v)$ to denote the above bilinear form, where the Ω is replaced by another domain F.

Let $u_h \in V_{\Omega,h}^{k,q}$ be the solution of

$$B(u_h, v) = \mathcal{F}(v), \quad \text{for all } v \in V_{\Omega,h}^{k,q}, \tag{5.3}$$

where $V_{\Omega,h}^{k,q}$ was defined in (3.54). It is clear from (5.2) and (5.3) that

$$B(u_0 - u_h, v) = 0, \quad \text{for all } v \in V_{\Omega,h}^{k,q}, \tag{5.4}$$

and we easily have

$$\|u_h\|_{H^1(\Omega)} \leq \|u_0\|_{H^1(\Omega)}. \tag{5.5}$$

Recall that the functions in $V_{\Omega,h}^{k,q}$ are restrictions of the functions in $S_h \equiv V_h^{k,q}$ on Ω (*cf.* (3.2) and (3.54)). Thus (5.4) is true when $V_{\Omega,h}^{k,q}$ is replaced by S_h.

We assume that, for any $\rho > 0$,

$$\|u_0 - u_h\|_{L_2(B_\rho(0))} \leq Ch^{k+1}\|u_0\|_{H^{k+1}(\Omega)}\rho^{\frac{1}{2}}, \tag{5.6}$$

and there are positive constants C_1, C_2, independent of u, h, and ρ, such that

$$C_1 h^k \rho^{\frac{1}{2}} \leq \frac{\|u_0' - u_h'\|_{L_2(B_\rho(0))}}{\|u_0\|_{H^{k+1}(\Omega)}} \leq C_2 h^k \rho^{\frac{1}{2}}, \tag{5.7}$$

where $B_\rho(0) = \{x : |x| < \rho\}$ and $B_\rho(0) \subset \Omega$. We will write $B_\rho \equiv B_\rho(0)$ throughout this section.

The main goal of this section is to investigate the error $u'(x) - u_h'(x)$ in a neighbourhood of $x = 0$, *i.e.*, for $x \in B_H \subset\subset \Omega$ and $H = h^\gamma$, $\gamma < 1$, where γ will be chosen later. We will prove the following result.

Theorem 5.1. Suppose u_0 and u_h satisfy (5.6) and (5.7), and let $e_h = u_0 - u_h$. Moreover, assume that $u_0 \in W_\infty^{k+2}(B_{2H})$. Then, for sufficiently small h, there exists $\epsilon^* > 0$ such that

$$\frac{\|e_h' - T(u_0)\xi_{k+1}^{h}{}'\|_{L_2(B_H)}}{\|e_h'\|_{L_2(B_H)}} \leq Ch^{\epsilon^*},$$

where

$$T(u_0) = \frac{u_0^{(k+1)}(0)}{(k+1)!} \quad \text{and} \quad \xi_{k+1}^h(x) = h^{k+1}\xi_{k+1}\left(\frac{x}{h}\right);$$

ξ_{k+1} is defined in (4.37).

Remark 32. Theorem 5.1 is a superconvergence result. It shows that

$$\|e_h' - T(u_0)\xi_{k+1}^{h}{}'\|_{L_2(B_H)} \ll \|e_h'\|_{L_2(B_H)}.$$

This allows us, for example, to analyse the effectiveness of an error estimator, as was done in Babuška, Strouboulis, Upadhyay and Gangaraj (1996) and Babuška and Strouboulis (2001).

Since the all the results in this paper have been presented in terms of L_2-based norms (*i.e.*, in terms of the usual Sobolev norms), we also present this result in terms of L_2-based norms. Superconvergence in L_∞ will be addressed in a forthcoming paper. Assuming superconvergence in L_∞, the superconvergence points and superconvergence recoveries in the case of particle shape functions can be obtained as in Babuška and Strouboulis (2001). At the end of this section, we will see an example where the superconvergence points are distributed in a different way to that of the classical FEM.

Remark 33. The essential aspects of superconvergence analysis in the classical FEM are interior estimates, developed in Nitsche and Schatz (1974), Schatz and Wahlbin (1995) and Wahlbin (1995). This analysis strongly utilizes the polynomial character of the shape functions. Here, in the case of particle shape functions, we had to develop another approach to the analysis of superconvergence, which is based on weighted Sobolev spaces. The main idea of the proof of our superconvergence result is to show that *locally* the approximation error is asymptotically the same as the error in the interpolation of a polynomial of degree $k+1$ by particle shape functions. The analysis is technical; we present the main idea of this analysis in this section.

Remark 34. Assumptions (5.6) and (5.7) are directly related to the control of pollution, as in the FEM. The assumption $u_0 \in W_\infty^{k+1}(B_\rho)$ is analogous to the assumption in the FEM (see Babuška *et al.* (1996) and Babuška and Strouboulis (2001)).

To prove Theorem 5.1, we will first develop certain ideas and establish several technical results. To that end, for given parameters $H = h^\gamma$, with $\gamma < 1$, and $\alpha \geq 1$, we define the function $g(x)$ by

$$g(x) = \begin{cases} 1, & -H \leq x \leq H, \\ \mathrm{e}^{-\alpha(x-H)}, & x > H, \\ \mathrm{e}^{\alpha(H+x)}, & x < -H, \end{cases} \tag{5.8}$$

where α is such that $\alpha h < 1$, and will be chosen later. We note that a proper choice of γ and α is crucial for the analysis presented in this section. Often, we will use $g \equiv g(x)$, $g_i \equiv g(x_i^h)$ and $g_{i+\frac{1}{2}} \equiv g(x_i^h + h/2)$.

Generalized interpolant and certain norm estimates

We first introduce the idea of a generalized interpolant of a function u, which is different to the $\tilde{\mathcal{I}}_h u$ defined in Section 4.2. Let $\tilde{I}_0 \equiv I_{-1} \cup I_0 \cup \{0\} = (-1, 1)$ and $\tilde{A}_0 \equiv A_{-1} \cup A_0$. Then, from (5.1), it is clear that there are positive constants C_1, C_2, independent of $v = \sum_{i \in \mathbb{Z}} c_i \phi_i(x)$, but possibly depending on κ, such that

$$C_1 \sum_{j \in \tilde{A}_0} c_i^2 \leq \int_{\tilde{I}_0} v^2 \, \mathrm{d}x \leq C_2 \sum_{j \in \tilde{A}_0} c_i^2, \tag{5.9}$$

which implies that $\{\phi_i(x)\}_{i \in \tilde{A}_0}$ are also linearly independent in $\tilde{I}_0$. We define $\psi_0(x) = \sum_{i \in \tilde{A}_0} a_i \phi_i(x)$ with supp $\psi_0 = \overline{\tilde{I}_0}$ (closure of $\tilde{I}_0$), such that

$$\begin{aligned} &\int_{\tilde{I}_0} \psi_0(x) \phi_0(x) \, \mathrm{d}x = 1, \\ &\int_{\tilde{I}_0} \psi_0(x) \phi_j(x) \, \mathrm{d}x = 0, \quad \text{for all } j \in \tilde{A}_0, \quad j \neq 0. \end{aligned} \tag{5.10}$$

Using (5.9), we can show that

$$\|\psi_0\|_{L_2(\tilde{I}_0)} \leq C. \tag{5.11}$$

We also note that, since $\{\phi_i(x)\}_{i \in \tilde{A}_0}$ form a partition unity on $\tilde{I}_0$, from (5.10) we have

$$\int_{\tilde{I}_0} \psi_0(x) \, \mathrm{d}x = \int_{\tilde{I}_0} \psi_0(x) \sum_{i \in \tilde{A}_0} \phi_i(x) \, \mathrm{d}x = \int_{\tilde{I}_0} \psi_0(x) \phi_0(x) \, \mathrm{d}x = 1. \tag{5.12}$$

Let $\psi_i^h(x) = \psi_0(\frac{x}{h} - i)$. Then supp $\psi_i^h = \overline{\tilde{I}_i^h}$, where $\tilde{I}_i^h = (x_{i-1}^h, x_{i+1}^h)$. Note that $\cup_{i\in Z}\tilde{I}_i^h = \mathbb{R}$. Now, for a given $v \in L_2(\mathbb{R})$, we define the *generalized interpolant* of v as

$$\tilde{\mathcal{I}}_h^* v(x) = \sum_{i\in\mathbb{Z}} \Psi_i^h(v)\phi_i^h(x), \tag{5.13}$$

where

$$\Psi_i^h(v) = \frac{1}{h}\int_{\tilde{I}_i^h} \psi_i^h(x)v(x)\,\mathrm{d}x. \tag{5.14}$$

We note that $\tilde{\mathcal{I}}_h^* v(x)$ depends on the $v(y)$ for $y \in \cup_{i\in A^h(x)}\tilde{I}_i^h$, where $A^h(x) = \{l \in \mathbb{Z} : x \in \eta_l^h\}$. We also define

$$\tilde{A}_i^h = \{m \in \mathbb{Z} : \eta_m^h \cap \tilde{I}_i^h \neq \emptyset\}.$$

Lemma 5.2. Suppose $v(x) = \sum_{i\in\mathbb{Z}} c_i^h \phi_i^h(x)$. Then

$$c_i^h = \Psi_i^h(v), \quad \text{and} \tag{5.15}$$

$$\tilde{\mathcal{I}}_h^* v(x) = v(x). \tag{5.16}$$

Proof. From (5.10) and the definition of $\Psi_i^h(v)$ in (5.14), $i \in \mathbb{Z}$, we have

$$\begin{aligned}
\Psi_i^h(v) &= \frac{1}{h}\int_{\tilde{I}_i^h} \psi_i^h(x)v(x)\,\mathrm{d}x \\
&= \frac{1}{h}\int_{\tilde{I}_i^h} \psi_0\Big(\frac{x}{h} - i\Big) \sum_{j\in\tilde{A}_i^h} c_j^h \phi\Big(\frac{x}{h} - j\Big)\,\mathrm{d}x \\
&= \int_{\tilde{I}_0} \psi_0(y) \sum_{j\in\tilde{A}_i^h} c_j^h \phi_{j-i}(y)\,\mathrm{d}y \\
&= c_i^h,
\end{aligned}$$

which is (5.15). Now using (5.15) in (5.13), we get (5.16). □

Remark 35. We note that if v is a local linear combination of the $\{\phi_i^h\}$, *i.e.*, in a bounded open interval, then $\tilde{\mathcal{I}}_h^* v = v$ only in the interior of that open interval. More precisely, $\tilde{\mathcal{I}}_h^* v = v$ in an interval I if v is a linear combination of the $\{\phi_i\}$ in $\cup_{x\in I}\cup_{i\in A^h(x)}\tilde{I}_i^h$.

We will use the following result later.

Lemma 5.3. Let Ω be a bounded interval, and suppose $u \in L_2(\Omega)$. Then

$$\|\tilde{\mathcal{I}}_h^* E u\|_{H^1(\mathbb{R})} \le C h^{-1} \|u\|_{L_2(\Omega)},$$

where E is the extension operator satisfying (3.50).

Proof. We first note that the extension Eu of u satisfies $\|Eu\|_{L_2(\mathbb{R})} \leq C\|u\|_{L_2(\Omega)}$. Now, from (5.13) and (4.32),

$$\|\tilde{\mathcal{I}}_h^* Eu\|_{H^1(\tilde{I}_i^h)}^2 \leq C \sum_{j\in\tilde{A}_i^h} |\Psi_j^h(Eu)|^2 \; \|\phi_j^h\|_{H^1(\mathring{\eta}_j)}^2$$
$$\leq Ch^{-1} \sum_{j\in\tilde{A}_i^h} |\Psi_j^h(Eu)|^2, \tag{5.17}$$

where C depends on κ; and using the Schwartz inequality on (5.14) with $v = Eu$, and a scaling argument, we get

$$|\Psi_j^h(Eu)|^2 \leq \frac{1}{h^2}\left(\int_{\tilde{I}_j^h} \psi_j^h(x) Eu(x)\,\mathrm{d}x\right)^2$$
$$\leq \frac{1}{h^2}\left[\int_{\tilde{I}_j^h} (\psi_j^h)^2\,\mathrm{d}x\right]\left[\int_{\tilde{I}_j^h} (Eu)^2\,\mathrm{d}x\right]$$
$$\leq \frac{1}{h}\|\psi_0\|_{L_2(\tilde{I}_0)}^2 \left[\int_{\tilde{I}_j^h} (Eu)^2\,\mathrm{d}x\right]. \tag{5.18}$$

Thus, from (5.17), (5.18), and the fact that $\|Eu\|_{L_2(\mathbb{R})} \leq C\|u\|_{L_2(\Omega)}$, we have

$$\|\tilde{\mathcal{I}}_h^* Eu\|_{H^1(\mathbb{R})}^2 \leq Ch^{-2}\|u\|_{L_2(\Omega)}^2,$$

which is the desired result. □

Remark 36. We can also show that

$$\|\tilde{\mathcal{I}}_h^* Eu\|_{L_2(\mathbb{R})} \leq C\|u\|_{L_2(\Omega)},$$

using the same arguments as in the proof of Lemma 5.3.

Consider the function $v(x) = \sum_{i\in A_j^h} c_i^h \phi_i^h(x)$ on I_j^h. Then, using scaling, translation, and (5.1), we have

$$C_1 h \sum_{j\in A_j^h} (c_i^h)^2 \leq \int_{I_j^h} v^2\,\mathrm{d}x \leq C_2 h \sum_{j\in A_j^h} (c_i^h)^2, \tag{5.19}$$

where C_1, C_2 are positive constants, independent of h and j, but possibly depending on κ. Using (5.19), we can show that if $v(x) = \sum_{i\in\mathbb{Z}} c_i^h \phi_i^h(x) = 0$ in L_2, then $c_i^h = 0$, for all $i \in \mathbb{Z}$, *i.e.*, $\{\phi_i^h\}$ are linearly independent.

We will now prove certain lower bounds for $\int_{I_j^h} gv^2\,\mathrm{d}x$ and $\int_{I_j^h} gv'^2\,\mathrm{d}x$, where $g(x)$ has been defined before. We first prove the following inequality.

Lemma 5.4. Let i_0, i_1 be integers such that $i_0 < i_1$, and suppose $\{c_i\}_{i=i_0}^{i_1}$ are real numbers. Then there exists a positive constant C, depending only on $i_1 - i_0$, such that, for any k, $i_0 \le k \le i_1$, we have

$$\sum_{i=i_0}^{i_1} g_{i+\frac{1}{2}}(c_i - c_k)^2 \le C \sum_{i=i_0}^{i_1-1} g_{i+\frac{1}{2}}(c_{i+1} - c_i)^2. \tag{5.20}$$

Proof. Suppose the integers i_0, i_1 are such that $H < i_0 h < i_1 h$, where $H = h^\gamma$, $\gamma < 1$. Then

$$\sum_{i=i_0}^{i_1} g_{i+\frac{1}{2}}(c_i - c_k)^2 = \sum_{i=i_0}^{k-1} g_{i+\frac{1}{2}}(c_i - c_k)^2 + \sum_{i=k+1}^{i_1} g_{i+\frac{1}{2}}(c_i - c_k)^2. \tag{5.21}$$

We first note that

$$\begin{aligned}\sum_{i=k+1}^{i_1} & g_{i+\frac{1}{2}}(c_i - c_k)^2 \\ &\le C \sum_{i=k+1}^{i_1} \sum_{j=k}^{i-1} g_{i+\frac{1}{2}}(c_{j+1} - c_j)^2 \\ &= C \sum_{i=k+1}^{i_1} \sum_{j=k}^{i-1} \left(1 + \frac{g_{i+\frac{1}{2}} - g_{j+\frac{1}{2}}}{g_{j+\frac{1}{2}}}\right) g_{j+\frac{1}{2}}(c_{j+1} - c_j)^2. \end{aligned} \tag{5.22}$$

But from the definition of $g(x)$ in (5.8), we have

$$\left(1 + \frac{g_{i+\frac{1}{2}} - g_{j+\frac{1}{2}}}{g_{j+\frac{1}{2}}}\right) \le \mathrm{e}^{-\alpha(i-j)h} \le \mathrm{e}^{\alpha(i_1-i_0)h} \le C, \tag{5.23}$$

and using this in (5.22), we get

$$\begin{aligned}\sum_{i=k+1}^{i_1} g_{i+\frac{1}{2}}(c_i - c_k)^2 &\le C \sum_{i=k+1}^{i_1} \sum_{j=k}^{i-1} g_{j+\frac{1}{2}}(c_{j+1} - c_j)^2 \\ &\le C \sum_{j=k}^{i_1-1} g_{j+\frac{1}{2}}(c_{j+1} - c_j)^2, \end{aligned} \tag{5.24}$$

where C depends on $(i_1 - i_0)$.

Using similar arguments we can show that

$$\sum_{i=i_0}^{k-1} g_{i+\frac{1}{2}}(c_k - c_i)^2 \le C(k - 1 - i_0) \sum_{j=i_0}^{k-1} g_{j+\frac{1}{2}}(c_{j+1} - c_j)^2, \tag{5.25}$$

where C depends on $(i_1 - i_0)$. Therefore, combining (5.21), (5.24), and

(5.25), we have

$$\sum_{i=i_0}^{i_1} g_{i+\frac{1}{2}}(c_i - c_k)^2 \le C \sum_{i=i_0}^{i_1} g_{j+\frac{1}{2}}(c_{j+1} - c_j)^2, \tag{5.26}$$

where C depends on $(i_i - i_0)$. Using similar arguments, we can prove (5.26) for all integers i_0, i_1 such that $i_0 < i_1$. □

Lemma 5.5. Suppose $v(x) = \sum_{i\in\mathbb{Z}} c_i^h \phi_i^h(x)$. Then:

(a) there are positive constants C_1, C_2, independent of v, h and j, but possibly depending on κ, such that

$$C_1 h \sum_{i\in A_j^h} g_i (c_i^h)^2 \le \int_{I_j^h} g v^2 \,\mathrm{d}x \le C_2 h \sum_{i\in A_j^h} g_i (c_i^h)^2; \tag{5.27}$$

(b) there is a positive constant C, independent of v and h, such that

$$\frac{1}{h} \sum_{i\in\mathbb{Z}^n} g_{i+\frac{1}{2}} (c_{i+1}^h - c_i^h)^2 \le C \int_{\mathbb{R}} g v'^2 \,\mathrm{d}x. \tag{5.28}$$

Proof. **(a)** Consider $j \in \mathbb{Z}$ and the corresponding A_j^h such that, for $i \in A_j^h$, $H < x_i^h$. Let $g_M = \max_{i\in A_j^h} g(x_i^h)$ and $g_m = \min_{i\in A_j^h} g(x_i^h)$. Then, it is easy to check that $\frac{g_M}{g_m} \le C$, where C depends κ. Now, using (5.19), we have

$$\begin{aligned} h \sum_{i\in A_j^h} g_i (c_i^h)^2 &\le g_M h \sum_{i\in A_j^h} (c_i^h)^2 \\ &\le \frac{g_M}{C_1 g_m} \int_{I_j^h} g v^2 \,\mathrm{d}x \\ &\le C \int_{I_j^h} g v^2 \,\mathrm{d}x. \end{aligned} \tag{5.29}$$

Using a similar argument, we get

$$\int_{I_j^h} g v^2 \,\mathrm{d}x \le C \sum_{i\in A_j^h} g_i (c_i^h)^2.$$

Combining the above with (5.29) gives the required result. Using similar arguments, we can prove (5.27) for any $j \in \mathbb{Z}$. □

(b) Let $u = \sum_{i\in\mathbb{Z}} c_i \phi_i(x)$. Then, from (5.14) and (5.15) with $h = 1$, we have

$$c_i = \Psi_i^1(u) = \int_{i-1}^{i+1} \psi_i^1(x) u(x) \,\mathrm{d}x, \tag{5.30}$$

and therefore

$$c_{i+1} - c_i = \int_i^{i+2} \psi^1_{i+1}(x)u(x)\,\mathrm{d}x - \int_{i-1}^{i+1} \psi^1_i(x)u(x)\,\mathrm{d}x$$
$$= \int_{i-1}^{i+2} (\psi^1_{i+1}(x) - \psi^1_i(x))u(x)\,\mathrm{d}x. \tag{5.31}$$

Let $F(x) = \int_{i-1}^{x}[\psi^1_{i+1}(t) - \psi^1_i(t)]\,\mathrm{d}t$. Using translation and (5.12), it is easily seen that

$$\int_{i-1}^{i+1} \psi^1_i(t)\,\mathrm{d}t = \int_i^{i+2} \psi^1_{i+1}(t)\,\mathrm{d}t = 1,$$

and therefore $F(i-1) = F(i+2) = 0$. Also, using the Schwartz inequality and (5.11), we can show that

$$\int_{i-1}^{i+2} F^2\,\mathrm{d}x \le C.$$

Now, using the above bound, integrating (5.31) by parts, and using the Schwartz inequality, we get

$$(c_{i+1} - c_i)^2 = \left(\int_{i-1}^{i+2} Fu'\,\mathrm{d}x\right)^2 \le C\int_{i-1}^{i+2} u'^2\,\mathrm{d}x. \tag{5.32}$$

Let $v = \sum_{i\in\mathbb{Z}} c^h_i\phi^h_i(x)$. Then, by a standard scaling argument, we have

$$\int_{x^h_{i-1}}^{x^h_{i+2}} (v'(x))^2\,\mathrm{d}x = \frac{1}{h}\int_{i-1}^{i+1} (u'(y))^2\,\mathrm{d}y, \tag{5.33}$$

where $u(y) = \sum_{i\in\mathbb{Z}} c^h_i\phi_i(y)$. Therefore, from (5.32) and (5.33), we have

$$\frac{1}{h}(c^h_{i+1} - c^h_i)^2 \le C\int_{x^h_{i-1}}^{x^h_{i+2}} v'^2\,\mathrm{d}x. \tag{5.34}$$

From the definition of $g(x)$, we can show that

$$\left(1 + \frac{g_{i+1/2} - g(x)}{g(x)}\right) \le C, \quad \text{for } x \in (x^h_{i-1}, x^h_{i+2}).$$

Therefore,

$$\frac{1}{h}g_{i+\frac{1}{2}}(c^h_{i+1} - c^h_i)^2 \le C\int_{x^h_{i-1}}^{x^h_{i+2}} gv'^2\left(1 + \frac{g_{i+\frac{1}{2}} - g}{g}\right)\mathrm{d}x$$
$$\le C\int_{x^h_{i-1}}^{x^h_{i+2}} gv'^2\,\mathrm{d}x,$$

and hence

$$\frac{1}{h}\sum_{i\in\mathbb{Z}} g_{i+\frac{1}{2}}(c_{i+1}^h - c_i^h)^2 \le C\sum_{i\in\mathbb{Z}}\int_{x_{i-1}^h}^{x_{i+2}^h} gv'^2\,\mathrm{d}x \le C\int_{\mathbb{R}} gv'^2\,\mathrm{d}x,$$

which is the required result. □

Remark 37. We note that it is possible to show that

$$\int_{\mathbb{R}} gv'^2\,\mathrm{d}x \le C\frac{1}{h}\sum_{i\in\mathbb{Z}^n} g_{i+\frac{1}{2}}(c_{i+1}^h - c_i^h)^2,$$

and together with (5.28) we see that $\frac{1}{h}\sum_{i\in\mathbb{Z}^n} g_{i+\frac{1}{2}}(c_{i+1}^h - c_i^h)^2$ is equivalent to $|v|^2_{H^1(\mathbb{R})}$. The proof of this fact is easier than the proof of (5.28), and we do not provide the proof here.

A perturbed bilinear form $B_\Theta(u,v)$ and related results
For a given $\Theta \ge 1$, we now consider the bilinear form

$$B_\Theta^{\mathbb{R}}(u,v) \equiv B^{\mathbb{R}}(u,v) + \Theta D^{\mathbb{R}}(u,v),$$

where

$$B^{\mathbb{R}}(u,v) = \int_{\mathbb{R}} (u'v' + uv)\,\mathrm{d}x \quad \text{and} \quad D^{\mathbb{R}}(u,v) = \int_{\mathbb{R}} uv\,\mathrm{d}x.$$

We will write $B_\Theta(u,v) \equiv B_\Theta^{\mathbb{R}}(u,v)$, but will use $B_\Theta^F(u,v)$ when the domain of integration is F instead of $\mathbb{R}$. Also, we will use $D^F(u,v)$, where $\mathbb{R}$ is replaced by a domain F in the definition of $D^{\mathbb{R}}(u,v)$.

Let $H^1_{g,\Theta}$ and $H^1_{g^{-1},\Theta}$ be Hilbert spaces defined as

$$H^1_{g,\Theta} = \left\{u : \|u\|^2_{1,g,\Theta} \equiv \int_{\mathbb{R}} gu'^2\,\mathrm{d}x + (1+\Theta)\int_{\mathbb{R}} gu^2\,\mathrm{d}x < \infty\right\},$$

$$H^1_{g^{-1},\Theta} = \left\{u : \|u\|^2_{1,g^{-1},\Theta} \equiv \int_{\mathbb{R}} g^{-1}u'^2\,\mathrm{d}x + (1+\Theta)\int_{\mathbb{R}} g^{-1}u^2\,\mathrm{d}x < \infty\right\}.$$

We will choose Θ later. The choice of Θ, along with the choices of γ and α, mentioned before, is important for the main result of this section. We assume that $\alpha^2/\bar{\Theta} < 1$ where $\bar{\Theta} = 1 + \Theta$.

We will often suppress Θ in $\|u\|_{1,g,\Theta}$ and $\|u\|_{1,g^{-1},\Theta}$ and instead write $\|u\|_{1,g}$ and $\|u\|_{1,g^{-1}}$ respectively. We will also use the fact that $|g'/g| \le \alpha$, which is obvious from the definition of $g(x)$.

Remark 38. The space $H^1_{g,\Theta}$ is directed towards obtaining interior estimates of e_h', *i.e.*, e_h' is locally characterized through the use of the space $H_{g,\Theta}$.

We now consider $B_\Theta(\cdot,\cdot) : H^1_{g,\Theta} \times H^1_{g^{-1},\Theta} \to \mathbb{R}$.

Lemma 5.6. The bilinear form $B_\Theta(\cdot,\cdot)$ is bounded on $H^1_{g,\Theta} \times H^1_{g^{-1},\Theta}$, that is,

$$B_\Theta(u,v) \le C\|u\|_{1,g}\|v\|_{1,g^{-1}}, \quad \text{for all } u \in H^1_{g,\Theta}, \quad v \in H^1_{g^{-1},\Theta}.$$

Proof. Let $u \in H^1_{g,\Theta}$ and $v \in H^1_{g^{-1},\Theta}$. Then

$$\begin{aligned}
&B_\Theta(u,v)\\
&= \int_{\mathbb{R}} [u'v' + (1+\Theta)uv]\,\mathrm{d}x\\
&= \int_{\mathbb{R}} [g^{1/2}u'g^{-1/2}v' + (1+\Theta)^{1/2}g^{1/2}u(1+\Theta)^{1/2}g^{-1/2}v]\,\mathrm{d}x\\
&\le C\left[\int_{\mathbb{R}} (gu'^2 + (1+\Theta)gu^2)\,\mathrm{d}x\right]^{1/2}\left[\int_{\mathbb{R}} (g^{-1}v'^2 + (1+\Theta)g^{-1}v^2)\,\mathrm{d}x\right]^{1/2}\\
&= C\|u\|_{1,g}\|v\|_{1,g^{-1}}.
\end{aligned}$$

□

Lemma 5.7. Suppose $\alpha^2/\bar{\Theta} < 1$. Then there is a constant $C > 0$, which depends on $\alpha^2/\bar{\Theta}$, such that

$$\inf_{u\in H^1_{g,\Theta}} \sup_{v\in H^1_{g^{-1},\Theta}} \frac{B_\Theta(u,v)}{\|u\|_{1,g}\|v\|_{1,g^{-1}}} \ge C > 0.$$

Proof. Suppose $u \in H^1_{g,\Theta}$. We consider $v = gu$. Now,

$$\begin{aligned}
B_\Theta(u,v) &= \int_{\mathbb{R}} [u'v' + \bar{\Theta}uv]\,\mathrm{d}x\\
&= \int_{\mathbb{R}} [u'(gu' + g'u) + \bar{\Theta}gu^2]\,\mathrm{d}x\\
&= \int_{\mathbb{R}} [gu'^2 + \bar{\Theta}gu^2]\,\mathrm{d}x + \int_{\mathbb{R}} uu'g'\,\mathrm{d}x. \qquad (5.35)
\end{aligned}$$

Now, for $\epsilon > 0$,

$$\begin{aligned}
\left|\int_{\mathbb{R}} uu'g'\,\mathrm{d}x\right| &= \left|\int_{\mathbb{R}} guu'\left(\frac{g'}{g}\right)\mathrm{d}x\right|\\
&\le \alpha \int_{\mathbb{R}} |g^{1/2}ug^{1/2}u'|\,\mathrm{d}x\\
&\le \alpha\left[\epsilon\int_{\mathbb{R}} gu'^2\,\mathrm{d}x + \frac{1}{\epsilon}\int_{\mathbb{R}} gu^2\,\mathrm{d}x\right],
\end{aligned}$$

and therefore, from (5.35), we get

$$B_\Theta(u,v) \geq \int_{\mathbb{R}} (gu'^2 + \bar{\Theta} g u^2)\,\mathrm{d}x - \alpha\left[\epsilon \int_{\mathbb{R}} gu'^2\,\mathrm{d}x + \frac{1}{\epsilon}\int_{\mathbb{R}} gu^2\,\mathrm{d}x\right]$$
$$= (1-\alpha\epsilon)\int_{\mathbb{R}} gu'^2\,\mathrm{d}x + \left(1 - \frac{\alpha}{\epsilon\bar{\Theta}}\right)\int_{\mathbb{R}} \bar{\Theta} g u^2\,\mathrm{d}x. \tag{5.36}$$

We choose ϵ such that $\alpha\epsilon < 1$ and $\alpha/\epsilon\bar{\Theta} < 1$, and therefore, from (5.36), we have

$$B_\Theta(u,v) \geq C_1 \|u\|_{1,g}^2, \tag{5.37}$$

where

$$C_1 = \min\left[(1-\alpha\epsilon), \left(1 - \frac{\alpha}{\epsilon\bar{\Theta}}\right)\right] > 0. \tag{5.38}$$

We next show that $\|v\|_{1,g^{-1}} \leq C_2\|u\|_{1,g}$. First note that

$$\int_{\mathbb{R}} g^{-1} v'^2\,\mathrm{d}x = \int_{\mathbb{R}} g^{-1}(gu' + g'u)^2\,\mathrm{d}x$$
$$= \int_{\mathbb{R}} gu'^2\,\mathrm{d}x + \int_{\mathbb{R}} g^{-1} g'^2 u^2\,\mathrm{d}x + 2\int_{\mathbb{R}} g'uu'\,\mathrm{d}x. \tag{5.39}$$

Now,

$$\int_{\mathbb{R}} g^{-1} g'^2 u^2\,\mathrm{d}x = \int_{\mathbb{R}} g\left(\frac{g'}{g}\right)^2 u^2\,\mathrm{d}x \leq \alpha^2 \int_{\mathbb{R}} gu^2\,\mathrm{d}x, \tag{5.40}$$

and

$$2\int_{\mathbb{R}} g'uu'\,\mathrm{d}x = 2\int_{\mathbb{R}} g\left(\frac{g'}{g}\right)uu'\,\mathrm{d}x \leq 2\int_{\mathbb{R}} |\alpha g^{1/2} u g^{1/2} u'|\,\mathrm{d}x$$
$$\leq \int_{\mathbb{R}} (gu'^2 + \alpha^2 g u^2)\,\mathrm{d}x. \tag{5.41}$$

Therefore, using (5.40) and (5.41) in (5.39), we get

$$\int_{\mathbb{R}} g^{-1} v'^2\,\mathrm{d}x \leq 2\int_{\mathbb{R}} gu'^2\,\mathrm{d}x + \frac{2\alpha^2}{\bar{\Theta}}\int_{\mathbb{R}} \bar{\Theta} g u^2\,\mathrm{d}x. \tag{5.42}$$

Thus, combining

$$\bar{\Theta}\int_{\mathbb{R}} g^{-1} v^2\,\mathrm{d}x = \bar{\Theta}\int_{\mathbb{R}} g^{-1} g^2 u^2\,\mathrm{d}x = \bar{\Theta}\int_{\mathbb{R}} gu^2\,\mathrm{d}x$$

with (5.42), we get

$$\|v\|_{1,g^{-1}}^2 \leq 2\int_{\mathbb{R}} gu'^2\,\mathrm{d}x + \left(1 + \frac{2\alpha^2}{\bar{\Theta}}\right)\int_{\mathbb{R}} \bar{\Theta} g u^2\,\mathrm{d}x. \tag{5.43}$$

Since $\alpha^2/\bar{\Theta} < 1$, from (5.43) we have

$$\|v\|_{1,g^{-1}}^2 \leq 3\|u\|_{1,g}^2. \tag{5.44}$$

Thus $v \in H^1_{g^{-1},\Theta}$, and combining (5.37) and (5.44), we get

$$\inf_{u \in H^1_{g,\Theta}} \sup_{v \in H^1_{g^{-1},\Theta}} \frac{B_\Theta(u,v)}{\|u\|_{1,g}\|v\|_{1,g^{-1}}} \geq C > 0,$$

where

$$C = \frac{\min[(1-\alpha\epsilon),(1-\frac{\alpha}{\epsilon\Theta})]}{\sqrt{3}}. \qquad \square$$

We now prove the inf-sup condition on $S_h \times S_h$. In the proof, we will use the function $d_i(x), x \in I^h_k$ and $i \in A^h_k$, to denote the following similar functions:

$$\frac{g_i - g(x)}{\sqrt{g_i\, g(x)}}, \qquad \frac{g_{i+\frac{1}{2}} - g(x)}{\sqrt{g_{i+\frac{1}{2}}\, g(x)}}, \qquad \frac{g_{i+\frac{1}{2}} - g_{l_k+\frac{1}{2}}}{\sqrt{g_{l_k+\frac{1}{2}}\, g(x)}},$$

where $l_k \in A^h_k$. It is easily seen from the definition of $g(x)$ that

$$|d_i(x)| \leq C\alpha h. \tag{5.45}$$

Lemma 5.8. Suppose $\alpha^2/\bar{\Theta} < C_1$ and $\alpha h < C_2$, where C_1, C_2 are sufficiently small. Then there is a constant $C > 0$, independent of u, v, and h, but possibly depending on κ and $\alpha^2/\bar{\Theta}$, such that, for sufficiently small h,

$$\inf_{u \in S_h} \sup_{v \in S_h} \frac{B_\Theta(u,v)}{\|u\|_{1,g}\|v\|_{1,g^{-1}}} \geq C > 0. \tag{5.46}$$

Proof. Let $u = \sum_{i\in\mathbb{Z}} c^h_i \phi^h_i$ in S_h such that $\|u\|_{1,g} \leq \infty$. Then, for $x \in I^h_k$, we have $u = \sum_{i \in A^h_k} c^h_i \phi^h_i$. Since $\sum_{i\in A^h_k} {\phi^h_i}'(x) = 0$ for $x \in I^h_k$, we have

$$u'(x) = \sum_{i\in A^h_k} c^h_i {\phi^h_i}'(x) = \sum_{i\in A^h_k} (c^h_i - c^h_{l_k}){\phi^h_i}'(x), \quad x \in I^h_k,$$

where $l_k \in A^h_k$ is a fixed integer for given k.

We now choose $v = \sum_{i\in\mathbb{Z}} c^h_i g_{i+\frac{1}{2}} \phi^h_i$ in S_h and, as before, for $x \in I^h_k$,

$$\begin{aligned} v'(x) &= \sum_{i\in A^h_k} c^h_i g_{i+\frac{1}{2}} {\phi^h_i}'(x) \\ &= \sum_{i\in A^h_k} (c^h_i g_{i+\frac{1}{2}} - c^h_{l_k} g_{l_k+\frac{1}{2}}){\phi^h_i}'(x) \\ &= \sum_{i\in A^h_k} (c^h_i - c^h_{l_k}) g_{i+\frac{1}{2}} {\phi^h_i}'(x) + c^h_{l_k} \sum_{i\in A^h_k} (g_{i+\frac{1}{2}} - g_{l_k+\frac{1}{2}}){\phi^h_i}'(x). \end{aligned}$$

Now,

$$\int_{\mathbb{R}} u'v'\,dx = \int_{\mathbb{R}} gu'^2\,dx + \int_{\mathbb{R}} u'(v' - gu')\,dx. \tag{5.47}$$

For $\epsilon > 0$, we have

$$\left|\int_{\mathbb{R}} u'(v'-gu')\,\mathrm{d}x\right| = \left|\int_{\mathbb{R}} g^{1/2}u'\frac{(v'-gu')}{g^{1/2}}\,\mathrm{d}x\right| \\ \leq \epsilon\int_{\mathbb{R}} gu'^2\,\mathrm{d}x + \frac{1}{\epsilon}\int_{\mathbb{R}}\frac{(v'-gu')^2}{g}\,\mathrm{d}x. \tag{5.48}$$

Now, from the definition of v' and u',

$$\int_{I_k^h}\frac{1}{g}(v'-gu')^2\,\mathrm{d}x = \int_{I_k^h}\left[\sum_{i\in A_k^h}(c_i^h - c_{l_k}^h)\frac{g_{i+\frac{1}{2}}-g}{g^{1/2}}\phi_i^{h'} \right. \\ \left. + c_{l_k}^h\sum_{i\in A_k^h}\frac{g_{i+\frac{1}{2}}-g_{l_k+\frac{1}{2}}}{g^{1/2}}\phi_i^{h'}\right]^2\,\mathrm{d}x \\ \leq C\int_{I_k^h}\left[\sum_{i\in A_k^h}(c_i^h - c_{l_k}^h)\frac{g_{i+\frac{1}{2}}-g}{g^{1/2}}\phi_i^{h'}\right]^2\,\mathrm{d}x \\ + C\int_{I_k^h}(c_{l_k}^h)^2\left[\sum_{i\in A_k^h}\frac{g_{i+\frac{1}{2}}-g_{l_k+\frac{1}{2}}}{g^{1/2}}\phi_i^{h'}\right]^2\,\mathrm{d}x. \tag{5.49}$$

The first term of the right-hand side of the above inequality, employing (5.45) and (5.20), gives

$$\int_{I_k^h}\left[\sum_{i\in A_k^h}(c_i^h - c_{l_k}^h)\frac{g_{i+\frac{1}{2}}-g}{g^{1/2}}\phi_i^{h'}\right]^2\,\mathrm{d}x \\ = \int_{I_k^h}\left[\sum_{i\in A_k^h}(c_i^h - c_{l_k}^h)(g_{i+\frac{1}{2}})^{1/2}\frac{g_{i+\frac{1}{2}}-g}{(g_{i+\frac{1}{2}})^{1/2}g^{1/2}}\phi_i^{h'}\right]^2\,\mathrm{d}x \\ \leq C\int_{I_k^h}\sum_{i\in A_k^h}(c_i^h - c_{l_k}^h)^2 g_{i+\frac{1}{2}}d_i^2(\phi_i^{h'})^2\,\mathrm{d}x \\ \leq C\alpha^2h^2\sum_{i\in A_k^h}(c_i^h - c_{l_k}^h)^2 g_{i+\frac{1}{2}}\int_{I_k^h}(\phi_i^{h'})^2\,\mathrm{d}x \\ \leq C\alpha^2h^2\frac{1}{h}\sum_{i\in A_k^h}(c_i^h - c_{l_k}^h)^2 g_{i+\frac{1}{2}} \\ \leq C\alpha^2h^2\frac{1}{h}\sum_{i,(i+1)\in A_k^h}(c_{i+1}^h - c_i^h)^2 g_{i+\frac{1}{2}}, \tag{5.50}$$

where C is independent of α, h, but depends on κ.

The second term of the right-hand side of (5.49), employing (5.45), gives

$$
\begin{aligned}
&(c_{l_k}^h)^2 \int_{I_k^h} \left[\sum_{i\in A_k^h} \frac{g_{i+\frac{1}{2}} - g_{l_k+\frac{1}{2}}}{g^{1/2}} \phi_i^{h'} \right]^2 \mathrm{d}x \\
&= (c_{l_k}^h)^2 g_{l_k+\frac{1}{2}} \int_{I_k^h} \left[\sum_{i\in A_k^h} \frac{g_{i+\frac{1}{2}} - g_{l_k+\frac{1}{2}}}{(g_{l_k+\frac{1}{2}})^{1/2} g^{1/2}} \phi_i^{h'} \right]^2 \mathrm{d}x \\
&\le C (c_{l_k}^h)^2 g_{l_k+\frac{1}{2}} \sum_{i\in A_k^h} \int_{I_k^h} d_i^2 (\phi_i^{h'})^2 \,\mathrm{d}x \\
&\le C\alpha^2 h (c_{l_k}^h)^2 g_{l_k}, \qquad (5.51)
\end{aligned}
$$

where C depends on κ, but is independent of α, h. Therefore, from (5.49), (5.50), and (5.51) we have

$$
\int_{I_k^h} \frac{1}{g}(v' - gu')^2 \,\mathrm{d}x \le C\alpha^2 h^2 \frac{1}{h} \sum_{i,(i+1)\in A_k^h} (c_{i+1}^h - c_i^h)^2 g_{i+\frac{1}{2}} + C\alpha^2 h (c_{l_k}^h)^2 g_{l_k}.
$$

Now summing the above inequality over $k \in \mathbb{Z}$, and using (5.27) and (5.28), we get

$$
\begin{aligned}
\int_{\mathbb{R}} \frac{1}{g}(v' - gu')^2 \,\mathrm{d}x &= \sum_{k\in\mathbb{Z}} \int_{I_k^h} \frac{1}{g}(v' - gu')^2 \,\mathrm{d}x \\
&\le C\alpha^2 h^2 \frac{1}{h} \sum_{k\in\mathbb{Z}} \sum_{i,(i+1)\in A_k^h} (c_{i+1}^h - c_i^h)^2 g_{i+\frac{1}{2}} \\
&\quad + C\alpha^2 h \sum_{k\in\mathbb{Z}} \sum_{i\in A_k^h} (c_i^h)^2 g_i \\
&\le C\alpha^2 h^2 \frac{1}{h} \sum_{i\in\mathbb{Z}} (c_{i+1}^h - c_i^h)^2 g_{i+\frac{1}{2}} \\
&\quad + C\alpha^2 \sum_{k\in\mathbb{Z}} \int_{I_k^h} g u^2 \,\mathrm{d}x \\
&\le C\alpha^2 h^2 \int_{\mathbb{R}} g u'^2 \,\mathrm{d}x + C\alpha^2 \int_{\mathbb{R}} g u^2 \,\mathrm{d}x. \qquad (5.52)
\end{aligned}
$$

Then, from (5.48) and (5.52) we have

$$\begin{aligned}\int_{\mathbb{R}} u'(v' - gu')\,\mathrm{d}x &\le \epsilon \int_{\mathbb{R}} gu'^2\,\mathrm{d}x \\ &\quad + \frac{1}{\epsilon}\left[C\alpha^2 h^2 \int_{\mathbb{R}} gu'^2\,\mathrm{d}x + C\alpha^2 \int_{\mathbb{R}} gu^2\,\mathrm{d}x\right] \\ &= \left(\epsilon + \frac{C\alpha^2 h^2}{\epsilon}\right)\int_{\mathbb{R}} gu'^2\,\mathrm{d}x + \frac{C\alpha^2}{\epsilon\bar{\Theta}}\int_{\mathbb{R}} \bar{\Theta} gu^2\,\mathrm{d}x. \qquad (5.53)\end{aligned}$$

We next consider

$$\bar{\Theta}\int_{\mathbb{R}} uv\,\mathrm{d}x = \bar{\Theta}\int_{\mathbb{R}} gu^2\,\mathrm{d}x + \bar{\Theta}\int_{\mathbb{R}} u(v - gu)\,\mathrm{d}x. \qquad (5.54)$$

For $\epsilon_1 > 0$, we have

$$\begin{aligned}\left|\int_{\mathbb{R}} u(v - gu)\,\mathrm{d}x\right| &= \left|\int_{\mathbb{R}} g^{1/2} u \frac{v - gu}{g^{1/2}}\,\mathrm{d}x\right| \\ &\le \epsilon_1 \int_{\mathbb{R}} gu^2\,\mathrm{d}x + \frac{1}{\epsilon_1}\int_{\mathbb{R}} \frac{(v - gu)^2}{g}\,\mathrm{d}x. \qquad (5.55)\end{aligned}$$

Now,

$$\begin{aligned}\int_{I_k^h} \frac{(v - gu)^2}{g}\,\mathrm{d}x &= \int_{I_k^h} \frac{1}{g}\left[\sum_{i\in A_k^h} c_i^h (g_i - g)\phi_i^h\right]^2 \mathrm{d}x \\ &= \int_{I_k^h} \left[\sum_{i\in A_k^h} c_i^h {g_i}^{1/2} \frac{(g_i - g)}{{g_i}^{1/2} g^{1/2}}\phi_i^h\right]^2 \mathrm{d}x \\ &\le C\int_{I_k^h} \sum_{i\in A_k^h} (c_i^h)^2 g_i d_i^2 {\phi_i^h}^2\,\mathrm{d}x \\ &\le C\alpha^2 h^2 h \sum_{i\in A_k^h} (c_i^h)^2 g_i.\end{aligned}$$

Therefore, using (5.27), we get

$$\begin{aligned}\int_{\mathbb{R}} \frac{(v - gu)^2}{g}\,\mathrm{d}x &= \sum_{k\in\mathbb{Z}} \int_{I_k^h} \frac{(v - gu)^2}{g}\,\mathrm{d}x \\ &\le \sum_{k\in\mathbb{Z}} C\alpha^2 h^2 h \sum_{i\in A_k^h} (c_i^h)^2 g_i \\ &\le C\alpha^2 h^2 \int_{\mathbb{R}} gu^2\,\mathrm{d}x. \qquad (5.56)\end{aligned}$$

Thus, from (5.55), (5.56), we have

$$\bar{\Theta}\left|\int_{\mathbb{R}} u(v-gu)\,\mathrm{d}x\right| \le \left(\epsilon_1 + \frac{C\alpha^2 h^2}{\epsilon_1}\right)\int_{\mathbb{R}} \bar{\Theta} g u^2\,\mathrm{d}x, \tag{5.57}$$

and combining (5.47), (5.53), (5.54), and (5.57), we get

$$\begin{aligned}
|B_\Theta(u,v)| &\ge \int_{\mathbb{R}} g u'^2\,\mathrm{d}x + \bar{\Theta}\int_{\mathbb{R}} g u^2\,\mathrm{d}x \\
&\quad - \left|\int_{\mathbb{R}} u'(v'-gu')\,\mathrm{d}x\right| - \bar{\Theta}\left|\int_{\mathbb{R}} u(v-gu)\,\mathrm{d}x\right| \\
&\ge \left(1-\epsilon-\frac{C\alpha^2h^2}{\epsilon}\right)\int_{\mathbb{R}} g u'^2\,\mathrm{d}x \\
&\quad + \left(1-\epsilon_1-\frac{C\alpha^2h^2}{\epsilon_1}-\frac{C\alpha^2}{\epsilon\bar{\Theta}}\right)\int_{\mathbb{R}} \bar{\Theta} g u^2\,\mathrm{d}x.
\end{aligned}$$

Now we can choose ϵ and ϵ_1, for sufficiently small h, such that

$$|B_\Theta(u,v)| \ge C_1\|u\|^2_{1,g}, \tag{5.58}$$

where $C_1 > 0$, since $\alpha^2/\bar{\Theta} \ll 1$, $\alpha h \ll 1$ by assumption.

We now show that $\|v\|_{1,g^{-1}} \le C\|u\|_{1,g}$. From the definition of v', we have

$$\begin{aligned}
\int_{I_k^h} g^{-1} v'^2\,\mathrm{d}x &= \int_{I_k^h} g^{-1}\Bigg[\sum_{i\in A_k^h}(c_i^h - c_{l_k}^h) g_{i+\frac{1}{2}}\phi_i^{h\prime} \\
&\qquad + c_{l_k}\sum_{i\in A_k^h}(g_{i+\frac{1}{2}} - g_{l_k+\frac{1}{2}})\phi_i^{h\prime}\Bigg]^2\,\mathrm{d}x \\
&\le \int_{I_k^h}\left[\sum_{i\in A_k^h}(c_i^h - c_{l_k}^h)\frac{g_{i+\frac{1}{2}}}{g^{1/2}}\phi_i^{h\prime}\right]^2\,\mathrm{d}x \\
&\quad + c_{l_k}^2\int_{I_k^h}\left[\sum_{i\in A_k^h}\frac{(g_{i+\frac{1}{2}} - g_{l_k+\frac{1}{2}})}{g^{1/2}}\phi_i^{h\prime}\right]^2\,\mathrm{d}x.
\end{aligned} \tag{5.59}$$

Now,

$$\begin{aligned}
&\int_{I_k^h}\left[\sum_{i\in A_k^h}(c_i^h - c_{l_k}^h)\frac{g_{i+\frac{1}{2}}}{g^{1/2}}\phi_i^{h\prime}\right]^2\,\mathrm{d}x \\
&= \int_{I_k^h}\left[\sum_{i\in A_k^h}(c_i^h - c_{l_k}^h)g^{1/2}\phi_i^{h\prime} + \sum_{i\in A_k^h}(c_i^h - c_{l_k}^h)g_{i+\frac{1}{2}}^{1/2}\left(\frac{g_{i+\frac{1}{2}} - g}{g_{i+\frac{1}{2}}^{1/2}g^{1/2}}\right)\phi_i^{h\prime}\right]^2\,\mathrm{d}x
\end{aligned}$$

$$\leq C\int_{I_k^h} g\left[\sum_{i\in A_k^h}(c_i^h-c_{l_k}^h)\phi_i^{h\prime}\right]^2 \mathrm{d}x + C\int_{I_k^h}\sum_{i\in A_k^h}(c_i^h-c_{l_k}^h)^2 g_{i+\frac{1}{2}} d_i^2(\phi_i^{h\prime})^2\,\mathrm{d}x$$

$$\leq C\int_{I_k^h} gu'^2\,\mathrm{d}x + C\int_{I_k^h}\sum_{i\in A_k^h}(c_i^h-c_{l_k}^h)^2 g_{i+\frac{1}{2}} d_i^2(\phi_i^{h\prime})^2\,\mathrm{d}x. \tag{5.60}$$

Also using (5.45) and (5.20), we have

$$\sum_{i\in A_k^h}(c_i^h-c_{l_k}^h)^2 g_{i+\frac{1}{2}}\int_{I_k^h} d_i^2(\phi_i^{h\prime})^2\,\mathrm{d}x \leq C\alpha^2h^2\frac{1}{h}\sum_{i\in A_k^h}(c_i^h-c_{l_k}^h)^2 g_{i+\frac{1}{2}} \tag{5.61}$$

$$\leq C\alpha^2h^2\frac{1}{h}\sum_{i,i+1\in A_k^h}(c_{i+1}^h-c_i^h)^2 g_{i+\frac{1}{2}}.$$

Therefore, using (5.60), (5.61) and (5.51) in (5.59), we get

$$\int_{I_k^h} g^{-1}v'^2\,\mathrm{d}x$$

$$\leq C\int_{I_k^h} gu'^2\,\mathrm{d}x + C\alpha^2h(c_{l_k}^h)^2 g_{l_k} + C\alpha^2h^2\frac{1}{h}\sum_{i,i+1\in A_k^h}(c_{i+1}^h-c_i^h)^2 g_{i+\frac{1}{2}}$$

$$\leq C\int_{I_k^h} gu'^2\,\mathrm{d}x + C\alpha^2h\sum_{i\in A_k^h}(c_i^h)^2 g_i + C\alpha^2h^2\frac{1}{h}\sum_{i,i+1\in A_k^h}(c_{i+1}^h-c_i^h)^2 g_{i+\frac{1}{2}}$$

$$\leq C\int_{I_k^h} gu'^2\,\mathrm{d}x + C\alpha^2\int_{I_k^h} gu^2\,\mathrm{d}x + C\alpha^2h^2\frac{1}{h}\sum_{i,i+1\in A_k^h}(c_{i+1}^h-c_i^h)^2 g_{i+\frac{1}{2}}.$$

Now, summing the above inequality for all k and using (5.20), we get

$$\int_{\mathbb{R}} g^{-1}v'^2\,\mathrm{d}x \leq C(1+\alpha^2h^2)\int_{\mathbb{R}} gu'^2\,\mathrm{d}x + C\alpha^2\int_{\mathbb{R}} gu^2\,\mathrm{d}x. \tag{5.62}$$

Again,

$$\int_{\mathbb{R}} g^{-1}v^2\,\mathrm{d}x = \int_{\mathbb{R}} g^{-1}\left(\sum_{i\in\mathbb{Z}} c_i^h g_i\phi_i^h\right)^2 \mathrm{d}x = \int_{\mathbb{R}}\left(\sum_{i\in\mathbb{Z}} c_i^h\frac{g_i}{g^{1/2}}\phi_i^h\right)^2 \mathrm{d}x. \tag{5.63}$$

Now using (5.45), we get

$$\int_{I_k^h}\left(\sum_{i\in A_k^h} c_i^h\frac{g_i}{g^{1/2}}\phi_i^h\right)^2 \mathrm{d}x$$

$$= \int_{I_k^h}\left[\sum_{i\in A_k^h} c_i^h g^{1/2}\phi_i^h + \sum_{i\in A_k^h} c_i^h\frac{g_i-g}{g^{1/2}}\phi_i^h\right]^2 \mathrm{d}x$$

$$\begin{aligned}
&\le C\int_{I_k^h}\left(\sum_{i\in A_k^h} c_i^h g^{1/2}\phi_i^h\right)^2 \mathrm{d}x + C\int_{I_k^h}\sum_{i\in A_k^h}(c_i^h)^2 g_i\left(\frac{g_i-g}{g^{1/2}g_i^{1/2}}\right)^2 \phi_i^{h\,2}\,\mathrm{d}x\\
&\le C\int_{I_k^h} gu^2\,\mathrm{d}x + C\sum_{i\in A_k^h}(c_i^h)^2 g_i\int_{I_k^h} d_i^2\phi_i^{h\,2}\,\mathrm{d}x\\
&\le C\int_{I_k^h} gu^2\,\mathrm{d}x + C\alpha^2h^2h\sum_{i\in A_k^h}(c_i^h)^2 g_i\\
&\le C\int_{I_k^h} gu^2\,\mathrm{d}x + C\alpha^2h^2C\int_{I_k^h} gu^2\,\mathrm{d}x\\
&\le C(1+\alpha^2h^2)\int_{I_k^h} gu^2\,\mathrm{d}x,
\end{aligned}$$

and therefore, from (5.63) and the above inequality,

$$\begin{aligned}
\int_{\mathbb{R}} g^{-1}v^2\,\mathrm{d}x &\le C\sum_{k\in\mathbb{Z}}\int_{I_k^h}\left(\sum_{i\in A_k^h} c_i^h\frac{g_i}{g^{1/2}}\phi_i^h\right)^2 \mathrm{d}x\\
&\le C(1+\alpha^2h^2)\sum_{k\in\mathbb{Z}}\int_{I_k^h} gu^2\,\mathrm{d}x\\
&= C(1+\alpha^2h^2)\int_{\mathbb{R}} gu^2\,\mathrm{d}x.
\end{aligned}$$

Thus, combining (5.62) and the above inequality, we have

$$\begin{aligned}
\|v\|_{1,g^{-1}}^2 &= \int_{\mathbb{R}} g^{-1}v'^2\,\mathrm{d}x + \bar{\Theta}\int_{\mathbb{R}} g^{-1}v^2\,\mathrm{d}x\\
&\le C(1+\alpha^2h^2)\int_{\mathbb{R}} gu'^2\,\mathrm{d}x + C\alpha^2\int_{\mathbb{R}} gu^2\,\mathrm{d}x\\
&\quad + C(1+\alpha^2h^2)\int_{\mathbb{R}}\bar{\Theta}gu^2\,\mathrm{d}x\\
&\le C(1+\alpha^2h^2)\int_{\mathbb{R}} gu'^2\,\mathrm{d}x\\
&\quad + \left[\frac{C\alpha^2}{\bar{\Theta}} + C(1+\alpha^2h^2)\right]\int_{\mathbb{R}}\bar{\Theta}gu^2\,\mathrm{d}x\\
&\le C\left(1+\alpha^2h^2+\frac{\alpha^2}{\bar{\Theta}}\right)\|u\|_{1,g}^2 \le C_2\|u\|_{1,g}^2. \qquad (5.64)
\end{aligned}$$

Finally, combining (5.58) and (5.64) we get the desired result. □

Projection with respect to $B_\Theta(u,v)$
Suppose $u \in H^1_{g,\Theta}$ and let $P_\Theta u$ be the projection of u onto S_h defined by

$$B_\Theta(P_\Theta u, v) = B_\Theta(u,v), \quad \text{for all } v \in S_h.$$

The projection $P_\Theta u$ exists (see Babuška and Aziz (1972)), and it is clear from Lemmas 5.8 and 5.6 that

$$\|P_\Theta u\|_{1,g} \le C \sup_{v\in S_h} \frac{B_\Theta(u,v)}{\|v\|_{1,g^{-1}}} \le C\|u\|_{1,g}. \tag{5.65}$$

We first note that, for fixed h, α, and Θ, the polynomials belong to the space $H^1_{g,\Theta}$. Moreover, for fixed h, α, and Θ, we can also show, using (5.27) and Remark 37 (page 66), that $\tilde{\mathcal{I}}_h(x^{k+1}) \in H^1_{g,\Theta}$, where $\tilde{\mathcal{I}}_h(x^{k+1})$ is the interpolant of x^{k+1}, as defined in (4.35).

We now present some simple facts about polynomials and periodic functions.

Lemma 5.9. Let the shape functions $\{\phi_i^h\}_{i\in Z}$ be reproducing of order k. Then

(a) $$P_\Theta x^i = x^i, \quad 0 \le i \le k, \tag{5.66}$$

(b) $$P_\Theta \tilde{\mathcal{I}}_h(x^{k+1}) = \tilde{\mathcal{I}}_h(x^{k+1}), \tag{5.67}$$

where $\tilde{\mathcal{I}}_h(x^{k+1})$ is the interpolant of x^{k+1} as defined in Section 4.

The proofs of these facts are immediate.

Lemma 5.10. Suppose $f \in H^1_{g,\Theta}$ is periodic, *i.e.*, $f(x + x_k^h) = f(x)$ for all k. Then $P_\Theta f$ is also periodic.

Proof. Let $\tilde{f}(x) = f(x + x_k^h)$. Then $[P_\Theta \tilde{f}](x) = [P_\Theta f](x + x_k^h)$. Now $f(x) = \tilde{f}(x)$ since f is periodic, and thus, from the uniqueness of the projection P_Θ, we have $[P_\Theta f](x + x_k^h) = [P_\Theta f](x)$, *i.e.*, $P_\Theta f$ is periodic. □

Remark 39. We note that, if

$$v(x) = \sum_{i\in\mathbb{Z}} c_i^h \phi_i^h(x)$$

is a periodic function, *i.e.*, $v(x + x_k^h) = v(x)$ for any k, then v is a constant. This could be shown as follows. Since $v(x + x_k^h) = v(x)$, we have

$$v(x + x_k^h) = \sum_{i\in\mathbb{Z}} c_i^h \phi_i^h(x + x_k^h) = \sum_{i\in\mathbb{Z}} c_{i+k}^h \phi_i^h(x) = \sum_{i\in\mathbb{Z}} c_i^h \phi_i^h(x) = v(x),$$

which implies that

$$\sum_{i\in\mathbb{Z}} [c_{i+k}^h - c_i^h]\phi_i^h(x) = 0, \quad \text{for all } x \in \mathbb{R}.$$

Using (5.19), we can show that $\{\phi_i^h\}_{i\in\mathbb{Z}}$ is linearly independent in $\mathbb{R}$. Thus we deduce that $c_{i+k}^h = c_i^h = C$ (constant), for all $i \in \mathbb{Z}$. Recalling that $\{\phi_i^h\}_{i\in\mathbb{Z}}$ forms a partition of unity, we get $v(x) = C\sum_{i\in\mathbb{Z}} \phi_i^h(x) = C$.

We now define

$$\xi_{k+1}^{\Theta}(x) \equiv x^{k+1} - P_{\Theta}x^{k+1}, \tag{5.68}$$

which, together with the next lemma, will play a central role in the final result of this section.

Lemma 5.11. Let $\xi_{k+1}^{\Theta}(x)$ be as defined in (5.68) and consider

$$\xi_{k+1}^h(x) = x^{k+1} - \sum_{i\in\mathbb{Z}} (x_i^h)^{k+1}\phi_i^h(x)$$

as defined in (4.36). Then

$$\xi_{k+1}^{\Theta}{}'(x) = \xi_{k+1}^h{}'(x). \tag{5.69}$$

Proof. We first note, from the definition of $\tilde{\mathcal{I}}_h x^{k+1}$, that $\xi_{k+1}^h(x) = x^{k+1} - \tilde{\mathcal{I}}_h x^{k+1}$. Now, using (5.67), we have

$$\begin{aligned}
\xi_{k+1}^{\Theta} &= x^{k+1} - P_{\Theta}x^{k+1} \\
&= x^{k+1} - \tilde{\mathcal{I}}_h x^{k+1} + \tilde{\mathcal{I}}_h x^{k+1} - P_{\Theta}x^{k+1} \\
&= \xi_{k+1}^h - P_{\Theta}[x^{k+1} - \tilde{\mathcal{I}}_h x^{k+1}] \\
&= \xi_{k+1}^h - P_{\Theta}[\xi_{k+1}^h].
\end{aligned} \tag{5.70}$$

But we know from Lemma 4.1 that $\xi_{k+1}^h(x)$ is periodic, and therefore from Lemma 5.10 and Remark 39 we infer that $P_{\Theta}[\xi_{k+1}^h]$ is a constant. Thus, from (5.70), we get

$$\xi_{k+1}^{\Theta}{}'(x) = \xi_{k+1}^h{}'(x),$$

which is the desired result. □

Proof of Theorem 5.1. The proof will be given in several steps.

1. Let E be the extension operator satisfying (3.50). Then, for $x \in B_H \equiv B_H(0)$, we have

$$[u_0 - u_h](x) = [u_0 - P_{\Theta}(Eu_0) - \{Eu_h - P_{\Theta}(Eu_h)\} + \{P_{\Theta}(Eu_0) - P_{\Theta}(Eu_h)\}](x),$$

and therefore

$$(u_0' - u_h{}')(x) = \{u_0 - P_{\Theta}(Eu_0)\}'(x) - \delta_h'(x) + \rho_h'(x), \tag{5.71}$$

where

$$\delta_h = Eu_h - P_{\Theta}(Eu_h), \tag{5.72}$$

$$\rho_h = P_{\Theta}(Eu_0) - P_{\Theta}(Eu_h). \tag{5.73}$$

Since $u_0 = Eu_0$ in $B_H(0)$, from Taylor's theorem we have

$$Eu_0(x) = \sum_{j=0}^{k} \frac{u_0^{(j)}(0)}{j!} x^j + \frac{u_0^{(k+1)}(0)}{(k+1)!} x^{k+1} + R_{k+1}(Eu_0)(x), \tag{5.74}$$

where $R_{k+1}(Eu_0)(x)$ is the remainder given by

$$R_{k+1}(Eu_0)(x) = \frac{1}{(k+1)!} \int_0^x (x-t)^{k+1} (Eu_0)^{(k+2)}(t)\, \mathrm{d}t. \tag{5.75}$$

Since P_Θ is a linear operator, we have

$$P_\Theta(Eu_0)(x) = \sum_{j=0}^{k} \frac{u_0^{(j)}(0)}{j!} P_\Theta x^j + \frac{u_0^{(k+1)}(0)}{(k+1)!} P_\Theta x^{k+1} + P_\Theta R_{k+1}(Eu_0)(x). \tag{5.76}$$

We know from (5.66) that $P_\Theta x^j = x^j, 0 \leq j \leq k$. Therefore, by first subtracting (5.76) from (5.74), then differentiating the identity, and finally using (5.69), we have

$$\begin{aligned}
&\{Eu_0 - P_\Theta(Eu_0)\}'(x) \\
&= \frac{u_0^{(k+1)}(0)}{(k+1)!} \{x^{k+1} - P_\Theta x^{k+1}\}'(x) + [R_{k+1}(Eu_0)]'(x) - [P_\Theta R_{k+1}(Eu_0)]'(x) \\
&= \frac{u_0^{(k+1)}(0)}{(k+1)!} {\xi_{k+1}^\Theta}'(x) + [R_{k+1}(Eu_0)]'(x) - [P_\Theta R_{k+1}(Eu_0)]'(x) \\
&= \frac{u_0^{(k+1)}(0)}{(k+1)!} {\xi_{k+1}^h}'(x) + [R_{k+1}(Eu_0)]'(x) - [P_\Theta R_{k+1}(Eu_0)]'(x).
\end{aligned} \tag{5.77}$$

Thus from (5.71), (5.77), and using $e_h(x) \equiv [u_0 - u_h](x)$, we get for $x \in B_H$,

$$\begin{aligned}
{e_h}'(x) &- \frac{u_0^{(k+1)}(0)}{(k+1)!} {\xi_{k+1}^\Theta}'(x) \\
&= [R_{k+1}(Eu_0)]'(x) - [P_\Theta R_{k+1}(Eu_0)]'(x) - \delta_h' + \rho_h'.
\end{aligned} \tag{5.78}$$

2. From (5.75), we have

$$[R_{k+1}(Eu_0)]'(x) = \frac{1}{k!} \int_0^x (x-t)^k (Eu_0)^{(k+2)}(t)\, \mathrm{d}t,$$

and since $\|u_0\|_{W_\infty^{k+2}(B_{2H})} \leq C$, we have, for $x \in B_{2H}$,

$$\begin{aligned}
\int_{B_H} |[R_{k+1}(Eu_0)]'|^2\, \mathrm{d}x &\leq \int_{B_{2H}} g|[R_{k+1}(Eu_0)]'|^2\, \mathrm{d}x \\
&\leq CH^{2k+2} H |u_0|^2_{W_\infty^{k+2}(B_{2H})}.
\end{aligned} \tag{5.79}$$

Similarly, again from (5.75), we get

$$\int_{B_H} |R_{k+1}(Eu_0)|^2 \, dx \le \int_{B_{2H}} g|R_{k+1}(Eu_0)|^2 \, dx$$
$$\le CH^{2k+4} H |u_0|^2_{W_\infty^{k+2}(B_{2H})}. \tag{5.80}$$

3. It can be shown from the definition of $g(x)$ that, for $0 \le j \le k+1$,

$$\int_{2H}^{\infty} g x^{2j} \, dx = \int_{2H}^{\infty} e^{-\alpha(x-H)} x^{2j} \, dx \le C e^{-\alpha H}, \tag{5.81}$$

where C depends on $k+1$. Now, from (5.65) we get

$$\int_{B_H} |[P_\Theta R_{k+1}(Eu_0)]'|^2 \, dx \le \|P_\Theta R_{k+1}(Eu_0)\|^2_{1,g} \le C\|R_{k+1}(Eu_0)\|^2_{1,g}. \tag{5.82}$$

We note that, from (5.74), we have

$$[R_{k+1}(Eu_0)]'(x) = (Eu_0)'(x) - \sum_{j=0}^{k} \frac{u_0^{(j+1)}(0)}{(j+1)!} x^j.$$

Therefore, using (5.81) and the fact that

$$\int_{2H}^{\infty} g|(Eu_0)'|^2 \, dx \le e^{-\alpha H} \int_{2H}^{\infty} |(Eu_0)'|^2 \, dx \le e^{-\alpha H} |Eu_0|^2_{H^1(\mathbb{R})},$$

we have

$$\int_{2H}^{\infty} g|[R_{k+1}(Eu_0)]'|^2 \, dx$$
$$\le C \int_{2H}^{\infty} g|(Eu_0)'|^2 \, dx + C \sum_{j=0}^{k} \left(\frac{u_0^{(j+1)}(0)}{(j+1)!} \right)^2 \int_{2H}^{\infty} g x^{2j} \, dx$$
$$\le C e^{-\alpha H} \{ |Eu_0|^2_{H^1(\mathbb{R})} + C\|u_0\|^2_{W_\infty^{k+1}(B_{2H})} \}$$
$$\le C e^{-\alpha H} \{ \|u_0\|^2_{H^1(\Omega)} + C\|u_0\|^2_{W_\infty^{k+2}(B_{2H})} \}, \tag{5.83}$$

where C depends on k. Similarly, we can show that

$$\int_{-\infty}^{-2H} g|[R_{k+1}(Eu_0)]'|^2 \, dx \le C e^{-\alpha H} \{ \|u_0\|^2_{H^1(\Omega)} + C\|u_0\|^2_{W_\infty^{k+2}(B_{2H})} \},$$

which together with (5.83) imply that

$$\int_{\mathbb{R}-B_{2H}} g|[R_{k+1}(Eu_0)]'|^2 \, dx \le C e^{-\alpha H} \{ \|u_0\|^2_{H^1(\Omega)} + C\|u_0\|^2_{W_\infty^{k+2}(B_{2H})} \}. \tag{5.84}$$

Using similar arguments, we can show that

$$\int_{\mathbb{R}-B_{2H}} g|R_{k+1}(Eu_0)|^2\,\mathrm{d}x \le C\mathrm{e}^{-\alpha H}\{\|u_0\|^2_{L_2(\Omega)} + C\|u_0\|^2_{W^{k+2}_\infty(B_{2H})}\}. \quad (5.85)$$

Now, combining (5.79), (5.80), (5.82), (5.84), and (5.85) we get

$$\begin{aligned}
&\int_{B_H} |[P_\Theta R_{k+1}(Eu_0)]'|^2\,\mathrm{d}x \\
&\quad\le C\|R_{k+1}(Eu_0)\|^2_{1,g} \\
&\quad= C\int_{\mathbb{R}} g|[R_{k+1}(Eu_0)]'|^2\,\mathrm{d}x + C\bar{\Theta}\int_{\mathbb{R}} g|R_{k+1}(Eu_0)|^2\,\mathrm{d}x \\
&\quad\le C(1+\bar{\Theta}H^2)H^{2k+2}H|u_0|^2_{W^{k+2}_\infty(B_{2H})} \\
&\qquad+ C(1+\bar{\Theta})\mathrm{e}^{-\alpha H}\{\|u_0\|^2_{H^1(\Omega)} + C\|u_0\|^2_{W^{k+2}_\infty(B_{2H})}\}. \qquad (5.86)
\end{aligned}$$

4. We first note from (5.72) that

$$\int_{B_H} {\delta_h'}^2\,\mathrm{d}x \le \|\delta_h\|^2_{1,g} = \|Eu_h - P_\Theta(Eu_h)\|^2_{1,g}. \quad (5.87)$$

Let $P_\Theta(Eu_h) = \tilde{\mathcal{I}}^*_h Eu_h + \mathcal{E}$. Then $\mathcal{E} \in S_h$. Now from Lemma 5.6 and the definition of P_Θ, we have, for all $v \in S_h$,

$$\begin{aligned}
B_\Theta(\mathcal{E}, v) &= B_\Theta(P_\Theta(Eu_h) - \tilde{\mathcal{I}}^*_h Eu_h, v) \\
&= B_\Theta(Eu_h - \tilde{\mathcal{I}}^*_h Eu_h, v) \\
&\le C\|Eu_h - \tilde{\mathcal{I}}^*_h Eu_h\|_{1,g}\|v\|_{1,g^{-1}},
\end{aligned}$$

and hence from Lemma 5.7 we get

$$\|E\|_{1,g} \le C \sup_{v\in S_h} \frac{B_\Theta(E,v)}{\|v\|_{1,g^{-1}}} \le C\|Eu_h - \tilde{\mathcal{I}}^*_h Eu_h\|_{1,g}.$$

Thus,

$$\begin{aligned}
\|Eu_h - P_\Theta(Eu_h)\|_{1,g} &\le \|Eu_h - \tilde{\mathcal{I}}^*_h Eu_h\|_{1,g} + \|\mathcal{E}\|_{1,g} \\
&\le C\|Eu_h - \tilde{\mathcal{I}}^*_h Eu_h\|_{1,g}. \qquad (5.88)
\end{aligned}$$

We now estimate the right-hand side of the above inequality. We first note that $Eu_h(x) = u_h(x)$ for $x \in \Omega$. Consider $\underline{\Omega} \subset \Omega$ such that (see Remark 35 on page 61)

$$B_{2H} \subset \underline{\Omega} \quad \text{and} \quad \tilde{\mathcal{I}}^*_h Eu_h|_{\underline{\Omega}} = Eu_h|_{\underline{\Omega}} = u_h|_{\underline{\Omega}}.$$

Therefore, from Lemma 5.3 and using (5.5),

$$
\begin{aligned}
\int_{\mathbb{R}} g[(Eu_h - \tilde{\mathcal{I}}_h^* Eu_h)']^2 \,\mathrm{d}x &= \int_{\mathbb{R}-\Omega} g[(Eu_h - \tilde{\mathcal{I}}_h^* Eu_h)']^2 \,\mathrm{d}x \\
&\le \mathrm{e}^{-\alpha H}[|Eu_h|^2_{H^1(\mathbb{R})} + |\tilde{\mathcal{I}}_h^* Eu_h|^2_{H^1(\mathbb{R})}] \\
&\le C\mathrm{e}^{-\alpha H}[|Eu_h|^2_{H^1(\mathbb{R})} + \frac{1}{h^2}\|Eu_h\|^2_{L_2(\mathbb{R})}] \\
&\le \frac{C}{h^2}\mathrm{e}^{-\alpha H}\|u_h\|^2_{H^1(\Omega)} \\
&\le \frac{C}{h^2}\mathrm{e}^{-\alpha H}\|u_0\|^2_{H^1(\Omega)}.
\end{aligned} \tag{5.89}
$$

Similarly, we can show using Remark 36 (on page 62) that

$$
\int_{\mathbb{R}} g[Eu_h - \tilde{\mathcal{I}}_h^* Eu_h]^2 \,\mathrm{d}x \le C\mathrm{e}^{-\alpha H}\|u_0\|^2_{L_2(\Omega)},
$$

and thus combining this estimate with (5.89) we get

$$
\|Eu_h - \tilde{\mathcal{I}}_h^* Eu_h\|^2_{1,g} \le \frac{C\bar{\Theta}}{h^2}\mathrm{e}^{-\alpha H}\|u_0\|^2_{H^1(\Omega)}.
$$

Now, from (5.87), (5.88), and the above estimate, we get

$$
\int_{B_H} {\delta_h'}^2 \,\mathrm{d}x \le \frac{C\bar{\Theta}}{h^2}\mathrm{e}^{-\alpha H}\|u_0\|^2_{H^1(\Omega)}. \tag{5.90}
$$

5. We first note from (5.73) that

$$
\int_{B_H(0)} {\rho_h'}^2 \,\mathrm{d}x \le \|\rho_h\|^2_{1,g} = \|P_\Theta(Eu_0) - P_\Theta(Eu_h)\|^2_{1,g}. \tag{5.91}
$$

Now, using (5.4), we have, for all $v \in S_h$,

$$
\begin{aligned}
&B_\Theta(\rho_h, v) \\
&\quad = B_\Theta(P_\Theta Eu_0 - P_\Theta Eu_h, v) \\
&\quad = B_\Theta(Eu_0 - Eu_h, v) \\
&\quad = B^\Omega(Eu_0 - Eu_h, v) + B^{\mathbb{R}-\Omega}(Eu_0 - Eu_h, v) + \Theta D^{\mathbb{R}}(Eu_0 - Eu_h, v) \\
&\quad = B^\Omega(u_0 - u_h, v) + B^{\mathbb{R}-\Omega}(Eu_0 - Eu_h, v) + \Theta D^{\mathbb{R}}(Eu_0 - Eu_h, v) \\
&\quad = B^{\mathbb{R}-\Omega}(Eu_0 - Eu_h, v) + \Theta D^{\mathbb{R}-\Omega}(Eu_0 - Eu_h, v) + \Theta D^\Omega(Eu_0 - Eu_h, v) \\
&\quad = B_\Theta^{\mathbb{R}-\Omega}(Eu_0 - Eu_h, v) + \Theta D^\Omega(u_0 - u_h, v).
\end{aligned} \tag{5.92}
$$

Also, for $v \in S_h$,

$$\begin{aligned}
&B_{\Theta}^{\mathbb{R}-\Omega}(Eu_0 - Eu_h, v) \\
&\quad = \int_{\mathbb{R}-\Omega} [(Eu_0 - Eu_h)'v' + \bar{\Theta}(Eu_0 - Eu_h)v] \,\mathrm{d}x \\
&\quad \le C \| Eu_0 - Eu_h \|_{1,g,\mathbb{R}-\Omega} \| v \|_{1,g^{-1},\mathbb{R}-\Omega} \\
&\quad \le C \| Eu_0 - Eu_h \|_{1,g,\mathbb{R}-\Omega} \|v\|_{1,g^{-1}},
\end{aligned} \tag{5.93}$$

where

$$\begin{aligned}
\| v \|_{1,g^{-1},\mathbb{R}-\Omega}^2 &= \int_{\mathbb{R}-\Omega} g^{-1} v'^2 \,\mathrm{d}x + \bar{\Theta} \int_{\mathbb{R}-\Omega} g^{-1} v^2 \,\mathrm{d}x; \\
\| Eu_0 - Eu_h \|_{1,g,\mathbb{R}-\Omega}^2 &= \int_{\mathbb{R}-\Omega} g (Eu_0 - Eu_h)'^2 \,\mathrm{d}x \\
&\quad + \bar{\Theta} \int_{\mathbb{R}-\Omega} g (Eu_0 - Eu_h)^2 \,\mathrm{d}x.
\end{aligned}$$

From the definition of $g(x)$, we can show that

$$\begin{aligned}
\| Eu_0 - Eu_h \|_{1,g,\mathbb{R}-\Omega}^2 &\le \mathrm{e}^{-\alpha H} \bar{\Theta} \| Eu_0 - Eu_h \|_{H^1(\mathbb{R}-\Omega)}^2 \\
&\le C\mathrm{e}^{-\alpha H} \bar{\Theta} \| u_0 - u_h \|_{H^1(\Omega)}^2 \\
&\le C\mathrm{e}^{-\alpha H} \bar{\Theta} \| u_0 \|_{H^1(\Omega)}^2.
\end{aligned} \tag{5.94}$$

Now, using the definition of $g(x)$ and (5.6) with $R = 2H$, we get

$$\begin{aligned}
\int_{\Omega} g(u_0 - u_h)^2 \,\mathrm{d}x &= \int_{B_{2H}} g(u_0 - u_h)^2 \,\mathrm{d}x + \int_{\Omega - B_{2H}} g(u_0 - u_h)^2 \,\mathrm{d}x \\
&\le \| u_0 - u_h \|_{L_2(B_{2H})}^2 + \mathrm{e}^{-\alpha H} \| u_0 - u_h \|_{L_2(\Omega)}^2 \\
&\le C h^{2k+2} H \| u_0 \|_{H^{k+1}(\Omega)}^2 + \mathrm{e}^{-\alpha H} \| u_0 \|_{H^1(\Omega)}^2,
\end{aligned}$$

and therefore

$$\begin{aligned}
&\frac{\Theta}{\|v\|_{1,g^{-1}}} D^{\Omega}(u_0 - u_h, v) \\
&\quad = \frac{\Theta}{\|v\|_{1,g^{-1}}} \int_{\Omega} (u_0 - u_h) v \,\mathrm{d}x \\
&\quad \le \frac{\Theta}{\|v\|_{1,g^{-1}}} \left(\int_{\Omega} g(u_0 - u_h)^2 \,\mathrm{d}x \right)^{1/2} \left(\int_{\Omega} g^{-1} v^2 \,\mathrm{d}x \right)^{1/2} \\
&\quad \le \Theta^{\frac{1}{2}} C h^{k+1} H^{\frac{1}{2}} \| u_0 \|_{H^{k+1}(\Omega)} + \Theta^{\frac{1}{2}} \mathrm{e}^{-\alpha H/2} \| u_0 \|_{H^1(\Omega)}.
\end{aligned} \tag{5.95}$$

From the inf-sup condition (5.46) and using (5.92) and (5.93), we have

$$\begin{aligned}\|\rho_h\|_{1,g} &\leq C \sup_{v\in S_h} \frac{B_\Theta(\rho_h, v)}{\|v\|_{1,g^{-1}}}\\ &\leq C|\!|\!|Eu_0 - Eu_h|\!|\!|_{1,g,\mathbb{R}-\Omega} + \sup_{v\in S_h} \frac{\Theta}{\|v\|_{1,g^{-1}}} D^\Omega(u_0 - u_h, v),\end{aligned}$$

and thus, using (5.91), (5.94) and (5.95), we have

$$\begin{aligned}\int_{B_H} {\rho_h'}^2 \,\mathrm{d}x &\leq \|\rho_h\|^2_{1,g}\\ &\leq C\mathrm{e}^{-\alpha H}\bar{\Theta}\|u_0\|^2_{H^1(\Omega)} + C\Theta h^{2k+2} H\|u_0\|^2_{H^{k+1}(\Omega)}. \end{aligned}\tag{5.96}$$

6. We first note from (5.69) that ${\xi^\Theta_{k+1}}'(x) = {\xi^h_{k+1}}'(x)$, where ξ^h_{k+1} is defined in (4.36). Let $T(u_0) \equiv \frac{u_0^{(k+1)}(0)}{(k+1)!}$. Then, from (5.78), we have

$$\begin{aligned}&{e_h}'(x) - T(u_0){\xi^h_{k+1}}'(x)\\ &\qquad = [R_{k+1}(Eu_0)]'(x) - [P_\Theta R_{k+1}(Eu_0)]'(x) - \delta_h' + \rho_h',\end{aligned}$$

and therefore, from (5.79), (5.86), (5.90), and (5.96), we have

$$\begin{aligned}&\int_{B_H} \left({e_h}' - T(u_0){\xi^h_{k+1}}'\right)^2 \mathrm{d}x\\ &\quad\leq C\int_{B_H} |[R_{k+1}(Eu_0)]'|^2 \,\mathrm{d}x + C\int_{B_H} |[P_\Theta R_{k+1}(Eu_0)]'|^2 \,\mathrm{d}x\\ &\qquad + C\int_{B_H} {\delta_h'}^2 \,\mathrm{d}x + C\int_{B_H} \rho_h^2 \,\mathrm{d}x\\ &\quad\leq CH^{2k+2}H|u_0|^2_{W^{k+2}_\infty(B_{2H})} + C(1+\bar{\Theta}H^2)H^{2k+2}H|u_0|^2_{W^{k+2}_\infty(B_{2H})}\\ &\qquad + C(1+\bar{\Theta})\mathrm{e}^{-\alpha H}\{\|u_0\|^2_{H^1(\Omega)} + C\|u_0\|^2_{W^{k+2}_\infty(B_{2H})}\} + \frac{C\bar{\Theta}}{h^2}\mathrm{e}^{-\alpha H}\|u_0\|^2_{H^1(\Omega)}\\ &\qquad + C\mathrm{e}^{-\alpha H}\bar{\Theta}\|u_0\|^2_{H^1(\Omega)} + C\Theta h^{2k+2}H\|u_0\|^2_{H^{k+1}(\Omega)}\\ &\quad\leq C\Big[H^{2k+2}H + (1+\bar{\Theta}H^2)H^{2k+2}H + (1+\bar{\Theta})\mathrm{e}^{-\alpha H}\\ &\qquad + \frac{\bar{\Theta}}{h^2}\mathrm{e}^{-\alpha H} + \Theta h^{2k+2}H\Big]M(u_0),\end{aligned}\tag{5.97}$$

where

$$M(u_0) = \|u_0\|^2_{H^{k+1}(\Omega)} + \|u_0\|^2_{W^{k+2}_\infty(B_{2H})}.$$

We will now choose α, Θ, and H, where $H = h^\gamma$ and $\gamma < 1$. First we choose γ such that

$$H^{k+2} = h^{k+1}, \tag{5.98}$$

which implies that

$$\begin{aligned} h^{\gamma(k+2)} &= h^{k+1}, \quad \text{or} \\ \gamma(k+2) &= k+1, \quad \text{or} \\ \gamma &= \frac{k+1}{k+2} < 1. \end{aligned}$$

Let $\epsilon > 0$, which depends on γ, be such that $\epsilon^* \equiv 1 - \gamma - \epsilon > 0$. We will now choose α such that

$$\mathrm{e}^{-\alpha H} \le h^{2k+2} h^2 h^{2\gamma+2\epsilon} H = h^{2k+4+3\gamma+2\epsilon}. \tag{5.99}$$

This implies that

$$\alpha \ge C_1 (\ln h^{-1}) h^{-\gamma},$$

where $C_1 = 2k + 4 + 3\gamma + 2\epsilon$. Since $h^{-\epsilon} > \ln h^{-1}$ for small h, we take

$$\alpha \equiv C_1 h^{-(\gamma+\epsilon)}. \tag{5.100}$$

We now choose

$$\bar{\Theta} \equiv (C_2)^2 h^{-2(\gamma+\epsilon)}, \quad C_2 > C_1. \tag{5.101}$$

We note from (5.100) that $\alpha h = C_1 h^{1-\gamma-\epsilon} = C_1 h^{\epsilon^*} < 1$ for small h, and $\lim_{h\to 0} \alpha h = 0$. Thus αh can be made sufficiently small; this was one of the assumptions in Lemma 5.8. Also, by choosing C_2 sufficiently large in (5.101), we can make $\frac{\alpha^2}{\bar{\Theta}} = (C_1/C_2)^2 \ll 1$, *i.e.*, sufficiently small, which was another assumption in Lemma 5.8. Thus the conclusion of Lemma 5.8 is true for the choices of α and $\bar{\Theta}$ given in (5.100) and (5.101), respectively.

Now, for these choices of γ, α, and $\bar{\Theta}$, we have

$$\bar{\Theta} h^{2k+2} = \bar{\Theta} h^{2(\gamma+\epsilon)} h^{2k} h^{2(1-\gamma-\epsilon)} = C_2^2 h^{2k+2\epsilon^*}. \tag{5.102}$$

Using (5.98) and (5.102) we have

$$\bar{\Theta} H^2 H^{2k+2} = \bar{\Theta} H^{2k+4} = \bar{\Theta} h^{2k+2} = C_2^2 h^{2k+2\epsilon^*}. \tag{5.103}$$

Also, from (5.99), we get

$$\bar{\Theta} \mathrm{e}^{-\alpha H} \le h^{2k+4} H \bar{\Theta} h^{2\gamma+2\epsilon} \le C_2^2 h^{2k+2} H, \tag{5.104}$$

and

$$\frac{\bar{\Theta}}{h^2} \mathrm{e}^{-\alpha H} \le h^{2k+2} H \bar{\Theta} h^{2\gamma+2\epsilon} = C_2^2 h^{2k+2} H. \tag{5.105}$$

Thus, using (5.98) and (5.102)–(5.105) in (5.97), we obtain

$$\| {e_h}' - T(u_0) {\xi_{k+1}^h}' \|_{L_2(B_H)} \le C h^{k+\epsilon^*} H^{1/2} M(u_0)^{\frac{1}{2}}, \tag{5.106}$$

and hence, using (5.7) with $\rho = H$, we have

$$\frac{\|e_h{}' - T(u_0)\xi^h_{k+1}{}'\|_{L_2(B_H)}}{\|e_h{}'\|_{L_2(B_H)}} \le Ch^{\epsilon^*},$$

where $M(u_0)^{\frac{1}{2}}/\|u_0\|_{H^{k+1}(\Omega)} \le C$, which is the desired result. □

Remark 40. The balancing of various terms in step 6 of the proof of Theorem 5.1 is similar to the balancing used in the proof of superconvergence of FEM solutions (see Babuška and Strouboulis (2001) and Babuška *et al.* (1996)).

Remark 41. Assuming that our superconvergence result is valid in L_∞, *i.e.*, assuming that for $x \in B_H$ there exists $\epsilon^* > 0$, such that

$$e_h'(x) = A(u_0)h^k \xi_{k+1}{}'\left(\frac{x}{h}\right) + O(h^{k+\epsilon^*}),$$

we see that the zeros of $\xi_{k+1}{}'(\frac{x}{h})$ are the superconvergence points. In Figure 5.1, we have presented the plot of $\xi_{k+1}{}'(y)$ for the RKP shape functions,

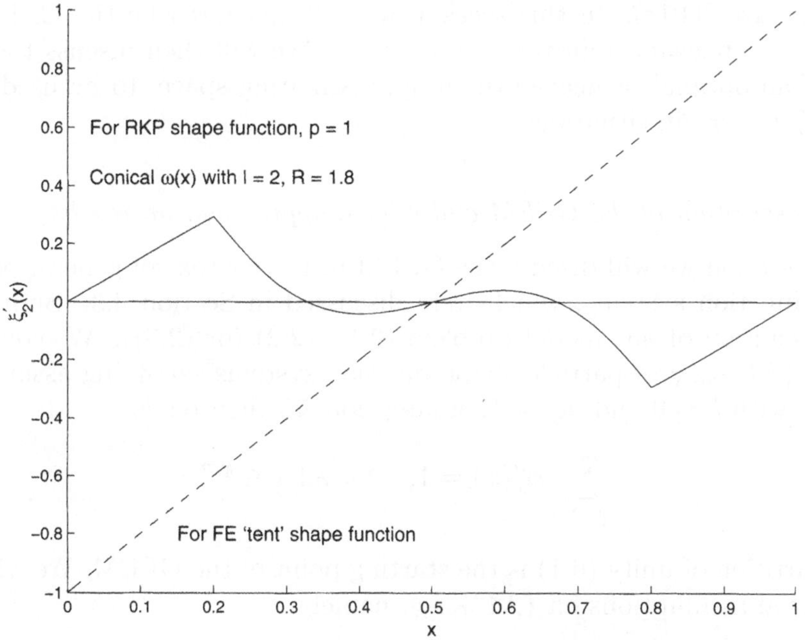

Figure 5.1. The plot of $\xi_2{}'(y)$, $0 \le y \le 1$ for (a) RKP shape functions, reproducing of order $k = 1$, corresponding to the conical weight function with $l = 2$, $R = 1.8$, (b) standard 'tent' functions used in the FEM.

reproducing of order $k = 1$, with respect to the weight function $w(x)$ given by (4.4) with $l = 2$ in one dimension. We have also included the plot of $\xi_{k+1}'(y)$, $k = 1$ (the dashed curve) for the standard tent functions that are used as shape functions in the FEM. We note that $\xi_2'(y)$ for the tent function has only one zero, whereas $\xi_2'(y)$ has five zeros. Thus the superconvergence points for the RKP shape function could be distributed quite differently to the corresponding points for standard tent functions in the FEM.

6. The generalized finite element method

The idea of the *generalized finite element method* (GFEM) was first introduced in Babuška *et al.* (1994) to address elliptic problems with rough coefficients. This idea was later extended, and called the *partition of unity method* (PUM), in Babuška and Melenk (1997) and Melenk and Babuška (1996). In the current literature, the PUM is referred to as the particle-partition of unity method (Griebel and Schweitzer 2002*a*, 2002*b*, 2002*c*), the method of finite spheres (De and Bathe 2001), the cloud method (Oden, Duarte and Zienkiewicz 1998), the eXtended finite element method (Daux, Moes, Dolbow, Sukumar and Belytschko 2000), and the GFEM (Strouboulis *et al.* 2001*a*, 2001*b*). In this section, we will first describe the GFEM and present the relevant approximation results. We will then discuss the selection of an optimal or near-optimal approximating space, to be used in the GFEM, in certain situations.

6.1. Description of the GFEM and related approximation results

In this section we will discuss the GFEM in the context of general particle-shape function systems, which were discussed in Section 3.3. Suppose u_0 is the solution of our model problem (2.1), (2.2) (or (2.3)). We consider a family $\{\mathcal{M}^\nu\}_{\nu \in N}$ of particle-shape function systems satisfying assumptions A1–A7 with $k = 0$ and $\mathcal{A}^\nu_{\underline{x}} = I$; assumption A5 then reads

$$\sum_{\underline{x} \in X^\nu} \phi^\nu_{\underline{x}}(x) = 1, \quad \text{for all } x \in \mathbb{R}^n. \tag{6.1}$$

The partition of unity (6.1) is the starting point of the GFEM. We will need additional assumptions on $\{\mathcal{M}^\nu\}_{\nu \in N}$, namely,

$$\|\phi^\nu_{\underline{x}}\|_{L^\infty(\mathbb{R}^n)} \leq C_1 \tag{6.2}$$

and

$$\|\nabla \phi^\nu_{\underline{x}}\|_{L^\infty(\mathbb{R}^n)} \leq \frac{C_2}{\operatorname{diam}(\eta^\nu_{\underline{x}})}, \tag{6.3}$$

for all $\underline{x} \in X^\nu$, and all $\nu \in N$. In (6.3), we implicitly assume that $q > (n/2) + 1$. We also assume that there is a constant C such that

$$\mathrm{diam}(\eta^\nu_{\underline{x}}) \le C, \quad \text{for all } \underline{x} \in X^\nu \text{ and } \nu.$$

For each $\underline{x} \in X^\nu$, we assume that we have a finite-dimensional space $V^\nu_{\underline{x}}$ of functions with good approximation properties. We refer to $V^\nu_{\underline{x}}$ as *local approximating spaces.* We define a set of particles A^ν_Ω, namely,

$$A^\nu_\Omega = \{\underline{x} \in X^\nu : \mathring{\eta}^\nu_{\underline{x}} \cap \Omega \ne \emptyset\}, \tag{6.4}$$

for each $\nu \in N$. From (6.1) we have

$$\sum_{\underline{x} \in A^\nu_\Omega} \phi^\nu_{\underline{x}}(x) = 1, \quad \text{for all } x \in \Omega. \tag{6.5}$$

For an approximating space on Ω, we then consider

$$V^\nu = \left\{ v\big|_\Omega : v = \sum_{\underline{x} \in A^\nu_\Omega} \phi^\nu_{\underline{x}} \psi^\nu_{\underline{x}}, \text{ where } \psi^\nu_{\underline{x}} \in V^\nu_{\underline{x}} \right\}. \tag{6.6}$$

The GFEM is the Galerkin method (2.7) with $\tilde{B} = B$ and $S = V^\nu$, and we will denote the approximate solution u_S, obtained from the GFEM, by u_{GFEM}. When GFEM is used to approximate the solution u_0 of the Neumann problem, $V^\nu_{\underline{x}}$ can be any finite-dimensional subspace of $H^1(\mathring{\eta}^\nu_{\underline{x}})$. But when the GFEM is used to approximate the solution u_0 of the Dirichlet problem, with the boundary condition (2.3), the functions in $V^\nu_{\underline{x}}$ are required to satisfy $v|_{\mathring{\eta}^\nu_{\underline{x}} \cap \partial\Omega} = 0$, for particles $\underline{x}$ for which $|\mathring{\eta}^\nu_{\underline{x}} \cap \Omega| > 0$. Thus the approximating space $V^\nu \subset H^1_0(\Omega)$.

Our next theorem states an approximation result for V^ν. We will follow the ideas presented in Babuška *et al.* (2002*a*, 1994), Babuška and Melenk (1997), Melenk and Babuška (1996), and Strouboulis *et al.* (2001*a*, 2001*b*).

Theorem 6.1. Suppose $u \in H^1(\Omega)$ and suppose, for all $\underline{x} \in A^\nu_\Omega$, there exists $\psi^\nu_{\underline{x}} \in V^\nu_{\underline{x}}$ such that

$$\|u - \psi_{\underline{x}}\|_{L_2(\eta^\nu_{\underline{x}} \cap \Omega)} \le \epsilon_1(\underline{x}), \tag{6.7}$$

$$\|\nabla(u - \psi_{\underline{x}})\|_{L_2(\eta^\nu_{\underline{x}} \cap \Omega)} \le \epsilon_2(\underline{x}). \tag{6.8}$$

Then the function

$$u_{ap} = \sum_{\underline{x} \in A^\nu_\Omega} \phi^\nu_{\underline{x}} \psi^\nu_{\underline{x}} \in V^\nu \tag{6.9}$$

satisfies

$$\|u - u_{ap}\|_{L_2(\Omega)} \le \kappa^{1/2} C_1 \left(\sum_{\underline{x} \in A^\nu_\Omega} \epsilon_1^2(\underline{x}) \right)^{1/2} \tag{6.10}$$

and

$$\|\nabla(u - u_{ap})\|_{L_2(\Omega)} \le (2\kappa)^{1/2} \left(\sum_{\underline{x} \in A_\Omega^\nu} \left(\frac{C_2}{\operatorname{diam}(\eta_{\underline{x}}^\nu)} \right)^2 \epsilon_1^2(\underline{x}) + C_1^2 \epsilon_2^2(\underline{x}) \right)^{1/2}, \tag{6.11}$$

where C_1 and C_2 are given in (6.2) and (6.3).

Proof. We will prove only (6.11), since (6.10) can be proved similarly. Since $\phi_{\underline{x}}^\nu$, for $\underline{x} \in A_\Omega^\nu$, form a partition of unity for Ω (see (6.5)), we have

$$\begin{aligned} &\|\nabla(u - u_{ap})\|_{L_2(\Omega)}^2 \\ &\quad = \|\nabla \sum_{\underline{x} \in A_\Omega^\nu} \phi_{\underline{x}}^\nu (u - \psi_{\underline{x}})\|_{L_2(\Omega)}^2 \\ &\quad \le 2\| \sum_{\underline{x} \in A_\Omega^\nu} (u - \psi_{\underline{x}}) \nabla \phi_{\underline{x}}^\nu \|_{L_2(\Omega)}^2 + 2\| \sum_{\underline{x} \in A_\Omega^\nu} \phi_{\underline{x}}^\nu \nabla (u - \psi_{\underline{x}})\|_{L_2(\Omega)}^2. \end{aligned} \tag{6.12}$$

For any $x \in \Omega$, the sums $\sum_{\underline{x} \in A_\Omega^\nu} (u - \psi_{\underline{x}}) \nabla \phi_{\underline{x}}^\nu$ and $\sum_{\underline{x} \in A_\Omega^\nu} \phi_{\underline{x}}^\nu \nabla (u - \psi_{\underline{x}})$ have at most κ nonzero terms (see Remark 16 and (3.60)). Therefore,

$$\left| \sum_{\underline{x} \in A_\Omega^\nu} (u - \psi_{\underline{x}}) \nabla \phi_{\underline{x}}^\nu \right|^2 \le \kappa \sum_{\underline{x} \in A_\Omega^\nu} |(u - \psi_{\underline{x}}) \nabla \phi_{\underline{x}}^\nu|^2,$$

and

$$\left| \sum_{\underline{x} \in A_\Omega^\nu} \phi_{\underline{x}}^\nu \nabla (u - \psi_{\underline{x}}) \right|^2 \le \kappa \sum_{\underline{x} \in A_\Omega^\nu} |\phi_{\underline{x}}^\nu \nabla (u - \psi_{\underline{x}})|^2.$$

Hence, from (6.12), (6.7), (6.8), recalling that $\operatorname{supp}(\phi_{\underline{x}}^\nu) = \eta_{\underline{x}}^\nu$, we have

$$\begin{aligned} &\|\nabla(u - u_{ap})\|_{L_2(\Omega)}^2 \\ &\quad \le 2\kappa \sum_{\underline{x} \in A_\Omega^\nu} \|(u - \psi_{\underline{x}}) \nabla \phi_{\underline{x}}^\nu\|_{L_2(\Omega)}^2 + 2\kappa \sum_{\underline{x} \in A_\Omega^\nu} \|\phi_{\underline{x}}^\nu \nabla (u - \psi_{\underline{x}})\|_{L_2(\Omega)}^2 \\ &\quad = 2\kappa \sum_{\underline{x} \in A_\Omega^\nu} \|(u - \psi_{\underline{x}}) \nabla \phi_{\underline{x}}^\nu\|_{L_2(\Omega \cap \eta_{\underline{x}}^\nu)}^2 + 2\kappa \sum_{\underline{x} \in A_\Omega^\nu} \|\phi_{\underline{x}}^\nu \nabla (u - \psi_{\underline{x}})\|_{L_2(\Omega \cap \eta_{\underline{x}}^\nu)}^2 \\ &\quad \le 2\kappa \sum_{\underline{x} \in A_\Omega^\nu} \left(\left(\frac{C_2}{\operatorname{diam}(\eta_{\underline{x}}^\nu)} \right)^2 \epsilon_1^2(\underline{x}) + C_1^2 \epsilon_2^2(\underline{x}) \right), \end{aligned}$$

which is the desired result. □

Remark 42. We note that $\epsilon_1(\underline{x})$, $\epsilon_2(\underline{x})$ in (6.7), (6.8) depend on the parameter ν.

We will now show that both terms of estimate (6.11) are of the same order with additional assumptions on V^ν. These additional assumptions depend on the boundary conditions of the approximated function.

Theorem 6.2. Suppose $u_0 \in H^1(\Omega)$ is the solution of the Neumann problem (2.1), (2.2), and suppose there exists $\psi_{\underline{x}}^\nu \in V_{\underline{x}}^\nu$, $\underline{x} \in A_\Omega^\nu$, such that (6.7) and (6.8) are satisfied. Moreover, assume that, for $\underline{x} \in A_\Omega^h$, the space $V_{\underline{x}}^\nu$ contains constant functions and that

$$\inf_{\lambda\in\mathbb{R}} \|v-\lambda\|_{L_2(\eta_{\underline{x}}^\nu\cap\Omega)} \leq C\,(\mathrm{diam}(\eta_{\underline{x}}^\nu))\,\|\nabla v\|_{L_2(\eta_{\underline{x}}^\nu\cap\Omega)}, \quad \text{for all } v \in H^1(\eta_{\underline{x}}^\nu\cap\Omega), \tag{6.13}$$

where C is independent of $\underline{x} \in X^\nu$ and ν. Then there exists $\tilde{\psi}_{\underline{x}}^\nu \in V_{\underline{x}}^\nu$ so that the corresponding function,

$$\tilde{u}_{ap} = \sum_{\underline{x}\in A_\Omega^\nu} \phi_{\underline{x}}^\nu \tilde{\psi}_{\underline{x}}^\nu \in V^\nu,$$

satisfies

$$\|u_0 - \tilde{u}_{ap}\|_{H^1(\Omega)} \leq C\left(\sum_{\underline{x}\in A_\Omega^h} \epsilon_2^2(\underline{x})\right)^{1/2}, \tag{6.14}$$

where C is independent of u_0 and ν.

Proof. Let $\psi_{\underline{x}}^\nu \in V_{\underline{x}}^\nu$, $\underline{x} \in A_\Omega^\nu$, satisfy (6.7) and (6.8). Define $\tilde{\psi}_{\underline{x}}^\nu = \psi_{\underline{x}}^\nu + r_{\underline{x}}^\nu$, where $r_{\underline{x}}^\nu \in \mathbb{R}$ satisfies

$$\|u_0 - \tilde{\psi}_{\underline{x}}^\nu\|_{L_2(\eta_{\underline{x}}^\nu\cap\Omega)} = \inf_{\lambda\in\mathbb{R}} \|u_0 - \psi_{\underline{x}}^\nu - \lambda\|_{L_2(\eta_{\underline{x}}^\nu\cap\Omega)}. \tag{6.15}$$

Since $V_{\underline{x}}^\nu$ contains constant functions, it is clear that $\tilde{\psi}_{\underline{x}} \in V_{\underline{x}}$. Also, from (6.15), (6.13) with $v = u_0 - \psi_{\underline{x}}^\nu$, and (6.8), we have

$$\begin{aligned}\|u - \tilde{\psi}_{\underline{x}}\|_{L_2(\eta_{\underline{x}}^\nu\cap\Omega)} &\leq C\,\mathrm{diam}(\eta_{\underline{x}}^\nu)\,\|\nabla(u-\psi_{\underline{x}}^\nu)\|_{L_2(\eta_{\underline{x}}^\nu\cap\Omega)}\\ &\leq C\,\mathrm{diam}(\eta_{\underline{x}}^\nu)\,\epsilon_2(\underline{x}).\end{aligned} \tag{6.16}$$

Let $\tilde{u}_{ap} = \sum_{\underline{x}\in A_\Omega^\nu} \phi_{\underline{x}}^\nu \tilde{\psi}_{\underline{x}}$. Recall that $\phi_{\underline{x}}^\nu$, $\underline{x} \in A_\Omega^\nu$, is a partition of unity for Ω. Then, following the arguments in the proof of Theorem 6.1 and using (3.60), (6.2), we can show that

$$\begin{aligned}\|u - \tilde{u}_{ap}\|_{L_2(\Omega)}^2 &= \left\|\sum_{\underline{x}\in A_\Omega^h} \phi_{\underline{x}}^\nu (u - \tilde{\psi}_{\underline{x}})\right\|_{L_2(\Omega)}^2\\ &\leq \kappa \sum_{\underline{x}\in A_\Omega^h} \|\phi_{\underline{x}}^\nu (u - \tilde{\psi}_{\underline{x}})\|_{L_2(\Omega)}^2\end{aligned}$$

$$= \kappa \sum_{\underline{x} \in A_\Omega^h} \|\phi_{\underline{x}}^\nu (u - \tilde{\psi}_{\underline{x}})\|^2_{L_2(\Omega \cap \eta_{\underline{x}}^\nu)}$$

$$\leq C \sum_{\underline{x} \in A_\Omega^h} \|(u - \tilde{\psi}_{\underline{x}})\|^2_{L_2(\Omega \cap \eta_{\underline{x}}^\nu)}, \qquad (6.17)$$

and using (6.16) in this inequality we get

$$\|u - \tilde{u}_{ap}\|^2_{L_2(\Omega)} \leq C \sum_{\underline{x} \in A_\Omega^h} (\mathrm{diam}(\eta_{\underline{x}}^\nu))^2 \epsilon_2^2(\underline{x}). \qquad (6.18)$$

Again, following the arguments in the proof of Theorem (6.1), and using (6.2), (6.3), we can show that

$$\|\nabla(u - \tilde{u}_{ap})\|^2_{L_2(\Omega)}$$

$$\leq 2\kappa \sum_{\underline{x} \in A_\Omega^\nu} \|(u - \tilde{\psi}_{\underline{x}}^\nu) \nabla \phi_{\underline{x}}^\nu\|^2_{L_2(\Omega \cap \eta_{\underline{x}}^\nu)} + 2\kappa \sum_{\underline{x} \in A_\Omega^\nu} \|\phi_{\underline{x}}^\nu \nabla(u - \tilde{\psi}_{\underline{x}}^\nu)\|^2_{L_2(\Omega \cap \eta_{\underline{x}}^\nu)}$$

$$\leq C \sum_{\underline{x} \in A_\Omega^h} \frac{1}{(\mathrm{diam}(\eta_{\underline{x}}^\nu))^2} \|u - \tilde{\psi}_{\underline{x}}^\nu\|^2_{L_2(\Omega \cap \eta_{\underline{x}}^\nu)}$$

$$+ C \sum_{\underline{x} \in A_\Omega^h} \|\nabla(u - \tilde{\psi}_{\underline{x}}^\nu)\|^2_{L_2(\Omega \cap \eta_{\underline{x}}^\nu)}. \qquad (6.19)$$

By first noting that $\nabla(u - \tilde{\psi}_{\underline{x}}^\nu) = \nabla(u - \psi_{\underline{x}}^\nu)$, and then using (6.16) and (6.8) in the above inequality, we get

$$\|\nabla(u - \tilde{u}_{ap})\|^2_{L_2(\Omega)} \leq C \sum_{\underline{x} \in A_\Omega^\nu} \epsilon_2^2(\underline{x}). \qquad (6.20)$$

Combining this with (6.18) we get (6.14), where we used that $\mathrm{diam}(\eta_{\underline{x}}^\nu) \leq C$ for all $\underline{x} \in X^\nu$ and ν. □

Theorem 6.3. Suppose $u_0 \in H_0^1(\Omega)$ is the solution of the Dirichlet problem (2.1), (2.3), and suppose $V_{\underline{x}}^\nu$, $\underline{x} \in A_\Omega^\nu$, satisfy the following assumptions.

(a) For all $\underline{x} \in A_\Omega^h$ such that $\eta_{\underline{x}}^\nu \cap \partial\Omega = \emptyset$, $V_{\underline{x}}^\nu$ contains constant functions, and (6.7), (6.8), and (6.13) hold.

(b) For all $\underline{x} \in A_\Omega^h$ such that $|\eta_{\underline{x}}^\nu \cap \partial\Omega| > 0$, functions $v \in V_{\underline{x}}^\nu$ satisfy $v\big|_{\eta_{\underline{x}}^\nu \cap \partial\Omega} = 0$, and there is a constant C, independent of $\underline{x}$ and ν, such that

$$\|v\|_{L_2(\eta_{\underline{x}}^\nu \cap \Omega)} \leq C\, (\mathrm{diam}(\eta_{\underline{x}}^\nu))\, \|\nabla v\|_{L_2(\eta_{\underline{x}}^\nu \cap \Omega)}, \qquad (6.21)$$

for all $v \in H^1(\eta_{\underline{x}}^\nu \cap \Omega)$ satisfying $v = 0$ on $\partial\Omega$. Moreover (6.7) and (6.8) hold for u satisfying $u\big|_{\eta_{\underline{x}}^\nu \cap \partial\Omega} = 0$.

Then there exists $\tilde{\psi}^\nu_{\underline{x}} \in V^\nu_{\underline{x}}$ so that the corresponding function,

$$\tilde{u}_{ap} = \sum_{\underline{x}\in A^\nu_\Omega} \phi^\nu_{\underline{x}} \tilde{\psi}^\nu_{\underline{x}} \in V^\nu,$$

satisfies

$$\|u_0 - \tilde{u}_{ap}\|_{H^1(\Omega)} \le C \left(\sum_{\underline{x}\in A^h_\Omega} \epsilon_2^2(\underline{x}) \right)^{1/2}, \tag{6.22}$$

where C is independent of u_0 and ν.

Proof. We first divide the set A^ν_ω into two disjoint sets, namely,

$$A^\nu_{\Omega,I} = \{\underline{x} \in A_\Omega : \eta^\nu_{\underline{x}} \cap \partial\Omega = \emptyset\}, \quad \text{and}$$
$$A^\nu_{\Omega,B} = \{\underline{x} \in A_\Omega : \eta^\nu_{\underline{x}} \cap \partial\Omega \neq \emptyset\}.$$

Let $\psi^\nu_{\underline{x}} \in V^\nu_{\underline{x}}$, $\underline{x} \in A^\nu_\Omega$, satisfy (6.7) and (6.8). Define $\tilde{\psi}^\nu_{\underline{x}}$, for $\underline{x} \in A^\nu_{\Omega,I}$, as in the proof of Theorem 6.2. We know from assumption (a) that, for $\underline{x} \in A^\nu_{\Omega,I}$, (6.13) holds and $V^\nu_{\underline{x}}$ contains constant functions. Therefore, following the argument leading to (6.16) in Theorem 6.2, we get

$$\|u_0 - \tilde{\psi}_{\underline{x}}\|_{L_2(\eta^\nu_{\underline{x}}\cap\Omega)} \le C \operatorname{diam}(\eta^\nu_{\underline{x}})\, \epsilon_2(\underline{x}), \quad \underline{x} \in A^\nu_{\Omega,I}. \tag{6.23}$$

For $\underline{x} \in A^\nu_{\Omega,B}$, we set $\tilde{\psi}^\nu_{\underline{x}} = \psi^\nu_{\underline{x}}$. Now, $u_0\big|_{\eta^\nu_{\underline{x}}\cap\partial\Omega} = 0$, and from assumption (b), we know that $\psi^\nu_{\underline{x}}\big|_{\eta^\nu_{\underline{x}}\cap\partial\Omega} = 0$ for $\underline{x} \in A^\nu_{\Omega,B}$. Thus, using (6.21), with $v = u_0 - \psi^\nu_{\underline{x}}$, and (6.8), we have

$$\begin{aligned}\|u_0 - \tilde{\psi}_{\underline{x}}\|_{L_2(\eta^\nu_{\underline{x}}\cap\Omega)} &= \|u - \psi_{\underline{x}}\|_{L_2(\eta^\nu_{\underline{x}}\cap\Omega)} \\ &\le C \operatorname{diam}(\eta^\nu_{\underline{x}})\, \epsilon_2(\underline{x}), \quad \underline{x} \in A^\nu_{\Omega,B}.\end{aligned} \tag{6.24}$$

Following the same steps that lead to (6.17) in the proof of Theorem 6.2, and using (6.23) and (6.24), we get

$$\begin{aligned}\|u_0 - u_{ap}\|^2_{L_2(\Omega)} &\le C \sum_{\underline{x}\in A^\nu_\Omega} \|u_0 - \tilde{\psi}^\nu_{\underline{x}}\|^2_{L_2(\eta^\nu_{\underline{x}}\cap\Omega)} \\ &= C \sum_{\underline{x}\in A^\nu_{\Omega,I}} \|u_0 - \tilde{\psi}^\nu_{\underline{x}}\|^2_{L_2(\eta^\nu_{\underline{x}}\cap\Omega)} + C \sum_{\underline{x}\in A^\nu_{\Omega,B}} \|u_0 - \tilde{\psi}^\nu_{\underline{x}}\|^2_{L_2(\eta^\nu_{\underline{x}}\cap\Omega)} \\ &\le C \sum_{\underline{x}\in A^\nu_\Omega} (\operatorname{diam}(\eta^\nu_{\underline{x}}))^2 \epsilon_2^2(\underline{x}).\end{aligned} \tag{6.25}$$

Similarly, following the steps leading to (6.20) in the proof of Theorem 6.2, we get

$$\|\nabla(u - \tilde{u}_{ap})\|^2_{L_2(\Omega)} \le C \sum_{\underline{x}\in A^\nu_\Omega} \epsilon_2^2(\underline{x}), \tag{6.26}$$

and combining this with (6.25), we get (6.22), where we used the assumption that $\operatorname{diam}(\eta_{\underline{x}}^{\nu}) \leq C$ for all $\underline{x} \in X^{\nu}$ and ν. □

Remark 43. It is clear from (6.14) and (2.8) that, if u_0 is the solution of (2.1), (2.2), then

$$\|u_0 - u_{\mathrm{GFEM}}\|_{H^1(\Omega)} \leq C \left(\sum_{\underline{x} \in X^{\nu}} \epsilon_2^2(\underline{x}) \right)^{1/2},$$

provided the local approximation spaces $V_{\underline{x}}^{\nu}$ contain constant functions, and (6.13) holds. The above estimate is also true if u_0 is the solution of (2.1), (2.3) provided conditions (a) and (b) of Theorem 6.3 are satisfied. We note that, in the latter case, *i.e.*, when u_0 satisfies the Dirichlet boundary condition, $u_0|_{\partial\Omega} = 0$, the space $V_{\underline{x}}^{\nu}$, corresponding to a particle $\underline{x}$ such that $\eta_{\underline{x}}^{\nu}$ intersects $\partial\Omega$, does not need to include constant functions, but the functions in $V_{\underline{x}}^{\nu}$ have to satisfy the Dirichlet boundary condition on $\eta_{\underline{x}}^{\nu} \cap \partial\Omega$.

Remark 44. Conditions (a), (b) in Theorem 6.3, and (6.13) are known as the *uniform Poincaré property*. These conditions put restrictions on the shapes of the $\{\eta_{\underline{x}}^{\nu}\}$. For a detailed discussion of this property, see Babuška and Melenk (1997).

Remark 45. The constant C_2 in (6.3) is related to the ratio of the radius of the largest ball contained in $\eta_{\underline{x}}^{\nu}$ to the radius of the smallest ball that contains $\eta_{\underline{x}}^{\nu}$. A similar condition is also assumed in the classical FEM. If this ratio is uniformly bounded for all $\underline{x} \in A_{\Omega}^{\nu}$ and ν, then (6.13) holds.

Remark 46. In practical computations, one can easily construct particle-shape function systems (with $k = 0$), such that conditions (6.2), (6.3), (6.13), and conditions (a), (b) of Theorem 6.3 are satisfied.

Remark 47. We observed that a partition of unity is the starting point for the construction of approximating space for the GFEM. It is important to emphasize that the construction of partition unity for $k = 0$ is simple: for instance, it could be constructed by Shepard's approach, as discussed in Section 4.

Remark 48. We have assumed that our particle-shape function system satisfies A1–A7 with $k = 0$ (and hence it reproduces polynomials of degree 0), and we have seen that the quality of the approximation in Theorems 6.1–6.3 depends entirely on the approximability properties of the spaces $V_{\underline{x}}^{\nu}$, as quantified by $\epsilon_1(\underline{x})$ and $\epsilon_2(\underline{x})$. If we used a particle-shape function system that reproduced polynomials of degree 1 ($k = 1$), then the space V^{ν} defined in 6.6 would be enlarged, and its approximability would be improved, possibly only marginally, but this improvement would not be directly visible from (6.11) (or (6.14) or (6.18)). Note that Theorems 6.1–6.3 are directed towards the use of nonpolynomial approximating functions, where the rate of convergence cannot be easily defined.

To clarify this point, suppose that for $\phi_{\underline{x}}^{\nu}$ we use the usual FE hat functions of degree 1, and $V_{\underline{x}}^{\nu}$ is the space of constants. Then the GFEM is the classical FEM, with the usual rate of convergence of $O(h)$. However, (6.11) (or (6.14) or (6.18)) does not establish this rate. As a second example, let $V_{\underline{x}}^{\nu}$ be the space of linear polynomials. Then the GFEM is a FE method, but not a usual one. The method has the rate of convergence $O(h^2)$, but (6.11) (or (6.14) or (6.18)) only establishes $O(h)$.

Remark 49. The space $\mathbb{W}_{\Omega,h}^{k',q}$, introduced near the end of Section 3.3, is a special case of the space V^{ν}, where we take $V_{\underline{x}}^{\nu}$ to be $\mathcal{P}^{k'}(\mathring{\eta}_{\underline{x}}^{h})$. The proof of Theorem 3.13 is obtained directly from Theorems 6.1–6.3 by considering $V_{\underline{x}}^{\nu} = \mathcal{P}^{k'}(\mathring{\eta}_{\underline{x}}^{h})$ and applying a standard polynomial approximation result.

The estimates in Theorems 6.2 and 6.3 are quite general, and allow us to employ available information on the approximated function u. Convergence of the approximation can be obtained by considering $\nu_i \in N$, $i = 1, 2, \ldots,$ such that $h^{\nu_i} \downarrow 0$, where h^{ν} is defined in (3.79). This is reminiscent of the h-version of the FEM. Convergence of the approximation can also be attained by keeping ν fixed, and selecting a sequence of spaces $V_{\underline{x}}^{\nu,i}$, $i = 1, 2, \ldots,$ so that they are complete in $H^1(\mathring{\eta}_{\underline{x}}^{\nu})$ or in a space $\mathcal{W}(\mathring{\eta}_{\underline{x}}^{\nu}) \subset H^1(\mathring{\eta}_{\underline{x}}^{\nu})$ that is known to include the approximated function u_0. This is a generalization of the p-version of the FEM.

6.2. *Selection of $V_{\bar{x}}$ and 'handbook' problems*

We saw in Section 6.1 that it is important to select spaces $V_{\underline{x}}^{\nu}$ with good local approximation properties. Principles for selecting shape functions that take advantage of available information on the approximated function were formulated in Babuška, Banerjee and Osborn (2001, 2002*b*). We will use these ideas to discuss the selection of the space $V_{\underline{x}}^{\nu}$. In this section we will suppress ν in our notation.

Let $H_1(\eta_{\underline{x}})$ and $H_2(\eta_{\underline{x}})$ be two Hilbert spaces, and suppose $H_2(\eta_{\underline{x}}) \subset H_1(\eta_{\underline{x}})$. Then

$$d_n(H_2, H_1) = \inf_{\substack{S_n \subset H_1 \\ \dim S_n = n}} \sup_{\substack{u \in H_2 \\ \|u\|_{H_2} \le 1}} \inf_{\chi \in S_n} \|u - \chi\|_{H_1}$$

is called the n-width of the H_2-unit ball in H_1. Let $V_{\underline{x}}^{(n)}$ be an n-dimensional subspace of H_1, and let

$$\Psi(V_{\underline{x}}^{(n)}, H_2, H_1) = \sup_{\substack{u \in H_2 \\ \|u\|_{H_2} \le 1}} \inf_{\chi \in V_{\underline{x}}^{(n)}} \|u - \chi\|_{H_1},$$

which is called the *sup-inf*. We will write $\Psi(V_{\underline{x}}^{(n)})$ for $\Psi(V_{\underline{x}}^{(n)}, H_2, H_1)$ if the spaces H_1, H_2 are evident from context. It is clear that

$$d_n(H_2, H_1) = \inf_{\substack{V_{\underline{x}}^{(n)} \subset H_1 \\ \dim V_{\underline{x}}^{(n)} = n}} \Psi(V_{\underline{x}}^{(n)}, H_2, H_1).$$

If an n-dimensional subspace ${}^0V_{\underline{x}}^{(n)}$ satisfies

$$\Psi({}^0V_{\underline{x}}^{(n)}, H_2, H_1) \leq C d_n(H_2, H_1),$$

where $C > 1$ is a constant, independent of n, then we will refer to ${}^0V_{\underline{x}}^{(n)}$ as a *nearly optimal subspace* relative to H_1 and H_2. An n-dimensional subspace ${}^0\bar{V}_{\underline{x}}^{(n)}$ that satisfies

$$\Psi({}^0\bar{V}_{\underline{x}}^{(n)}, H_2, H_1) = d_n(H_2, H_1),$$

is referred to as an *optimal subspace* relative to H_1 and H_2. An optimal subspace ${}^0\bar{V}_{\underline{x}}^{(n)}$ leads to the minimal error that can be achieved with an n-dimensional space, namely, $d_n(H_2, H_1)$; a nearly optimal subspace leads to essentially the same error, $d_n(H_2, H_1)$.

Suppose we are interested in using the GFEM to approximate the solution u_0 of the Dirichlet problem,

$$\begin{cases} \Delta u_0 = 0, & \text{in } \Omega, \\ u_0 = g, & \text{on } \partial\Omega, \end{cases}$$

where Ω is a bounded domain in $\mathbb{R}^2$. Then, for each $\underline{x} \in X^\nu$, we seek a finite-dimensional space $V_{\underline{x}}$ that contains a good approximation $\psi_{\underline{x}}$ to u_0 on $\eta_{\underline{x}}$ (*cf.* (6.7), (6.8)). This will be done by taking advantage of the available information on $u_0\big|_{\eta_{\underline{x}}\cap\Omega}$, namely that $u_0\big|_{\eta_{\underline{x}}\cap\Omega}$ is harmonic. We now illustrate this procedure.

We suppose that $\eta_{\underline{x}}$ is a disk in $\mathbb{R}^2$ and, for the sake of simplicity, suppose $\eta_{\underline{x}}$ is the unit disk. Let $H_1 = \{u \in H^1(\mathring{\eta}_{\underline{x}}) : u \text{ is harmonic in } \eta_{\underline{x}}\}$. For the space H_2, we use $\mathcal{W}(\mathring{\eta}_{\underline{x}})$, a (regularity) space known to contain u_0. More precisely, we suppose $\mathcal{W}(\mathring{\eta}_{\underline{x}})$ is a linear manifold in $\{u \in H^1(\mathring{\eta}_{\underline{x}}) : u \text{ is harmonic}\}$ and that $|||u|||$ is a norm on $\mathcal{W}(\mathring{\eta}_{\underline{x}})$ that is rotationally invariant and satisfies $\|u\|_{H^1(\mathring{\eta}_{\underline{x}})} \leq |||u|||$, for all $u \in \mathcal{W}(\mathring{\eta}_{\underline{x}})$. Moreover, we assume $\mathcal{W}(\mathring{\eta}_{\underline{x}})$ is complete with respect to $|||\cdot|||$, *i.e.*, $\{\mathcal{W}(\mathring{\eta}_{\underline{x}}), |||\cdot|||\}$ is a Hilbert space. We note that $\mathcal{W}(\mathring{\eta}_{\underline{x}})$ could be any higher-order (isotropic) Sobolev space.

It is well known that any $u \in H_1$ is characterized by its trace on the boundary $I = \partial\eta_{\underline{x}}$; these traces will be in

$$S = \{u : \mathbb{R} \to \mathbb{R} : u \text{ is } 2\pi\text{-periodic}, u \in H^{1/2}(I)\}.$$

Any $u \in S$ can be expanded in its Fourier series

$$u(\theta) = a_0 + \sum_{k=1}^{\infty}(a_k \cos k\theta + b_k \sin k\theta). \tag{6.27}$$

It is immediate that

$$|u|^2_{H^{1/2}(I)} = a_0^2 + \sum_{k=1}^{\infty}(a_k^2 + b_k^2)k,$$

where $|u|_{H^{1/2}(I)}$ is a Sobolev norm of order 1/2 on I, and the series in (6.27) converges in $H^{1/2}(I)$-norm. So we have a one-to-one correspondence between $u(r,\theta) \in H_1$ ((r,θ) are polar coordinates) and $u(\theta) \in S$, which we express by writing $u(r,\theta) \sim u(\theta)$. We easily find that

$$\|u\|^2_{H^1(\mathring{\eta}_{\underline{x}})} = |u|^2_{H^{1/2}(I)} = a_0^2 + \sum_{j=1}^{\infty}(a_j^2 + b_j^2)j. \tag{6.28}$$

Thus we identify the space H_1 with $H^{1/2}(I)$.

Since $|\!|\!|u|\!|\!|$ is rotationally invariant, the corresponding norm on $u(\theta)$ will be translation-invariant, and we can thus show that

$$|\!|\!|u|\!|\!|^2 = a_0^2 + \sum_{j=1}^{\infty}(a_j^2 + b_j^2)j\beta_j, \tag{6.29}$$

where, since $\|u\|_{H^1(\mathring{\eta}_{\underline{x}})} \le |\!|\!|u|\!|\!|$, we have $\beta_j \ge 1$. If we now define

$$H^{\beta}(I) = \{u \in S : |u|_{\beta} < \infty\},$$

where

$$|u|^2_{\beta} = a_0^2 + \sum_{k=1}^{\infty}(a_k^2 + b_k^2)k\beta_k, \tag{6.30}$$

then we see that $u(r,\theta) \in H_2$ if and only if $u(\theta) \in H^{\beta}(I)$ and $|\!|\!|u|\!|\!| = |u|_{\beta}$. We thus identify the space H_2 with $H^{\beta}(I)$.

We will now find an optimal subspace ${}^0V^{(n)}_{\underline{x}}$ relative to H_1 and H_2. We will exploit the correspondence $u(r,\theta) \sim u(\theta)$, and find ${}^0V^{(n)}_{\underline{x}}$ by first identifying an optimal subspace relative to $\bar{H}_1 = H^{1/2}(I)$ and $\bar{H}_2 = H^{\beta}(I)$.

Let $M_n = \{m_1, m_2, \ldots, m_n\}$ be a set of n positive integers, and consider

$$V^{M_n} = \left\{ u \in H^{1/2}(I) : u = a_0 + \sum_{k \in M_n}(a_k \cos k\theta + b_k \sin k\theta) \right\}. \tag{6.31}$$

Clearly, V^{M_n} is a $(2n+1)$-dimensional space.

Lemma 6.4. Let $\bar{H}_1 = H^{1/2}(I)$, $\bar{H}_2 = H^{\beta}(I)$, where $\beta = (\beta_1, \beta_2, \dots)$, $\beta_k \geq 1$, and let V^{M_n} be as defined in (6.31). Then

$$\Psi(V^{M_n}, \bar{H}_2, \bar{H}_1) = (\gamma(V^{M_n}))^{-\frac{1}{2}}, \tag{6.32}$$

where

$$\gamma(V^{M_n}) = \inf_{i \notin M_n} \beta_i.$$

Proof. Consider $u \in \bar{H}_2$ given by

$$u = a_0 + \sum_{k=1}^{\infty} (a_k \cos k\theta + b_k \sin k\theta).$$

Then, from (6.31) we get

$$\inf_{\chi \in V^{M_n}} |u - \chi|^2_{\bar{H}_1} = \sum_{k \in N - M_n} (a_k^2 + b_k^2) k,$$

where N is the set of all positive integers. Therefore, from (6.30) and the definition of $\gamma(V^{M_n})$ we have

$$\begin{aligned}
\inf_{\chi \in V^{M_n}} \frac{|u - \chi|^2_{\bar{H}_1}}{|u|^2_{\bar{H}_2}} &= \frac{\sum_{k \in N - M_n} (a_k^2 + b_k^2) k}{a_0^2 + \sum_{k \in N} (a_k^2 + b_k^2) k \beta_k} \\
&\leq \frac{\sum_{k \in N - M_n} (a_k^2 + b_k^2) k}{\sum_{k \in N - M_n} (a_k^2 + b_k^2) k \beta_k} \\
&\leq \frac{1}{\gamma(V^{M_n})}.
\end{aligned}$$

Thus,

$$\sup_{u \in \bar{H}_2} \inf_{\chi \in V^{M_n}} \frac{|u - \chi|^2_{\bar{H}_1}}{|u|^2_{\bar{H}_2}} \leq \frac{1}{\gamma(V^{M_n})}. \tag{6.33}$$

Let $\epsilon > 0$ be arbitrary. Then there is an $m_0 \notin M_n$, $m_0 \geq 1$, such that

$$\beta_{m_0} \leq \gamma(V^{M_n}) + \epsilon. \tag{6.34}$$

Consider $u_{m_0} = \cos m_0 \theta$. Clearly, $u_{m_0} \notin V^{M_n}$, and therefore, from (6.27),

$$\inf_{\chi \in V^{M_n}} |u_{m_0} - \chi|^2_{\bar{H}_1} = |u_{m_0}|^2_{\bar{H}_1} = m_0.$$

Also, from (6.30), we have $|u_{m_0}|^2_{\bar{H}_2} = m_0 \beta_{m_0}$. Therefore, using (6.34), we get

$$\sup_{u \in \bar{H}_2} \inf_{\chi \in V^{M_n}} \frac{|u - \chi|^2_{\bar{H}_1}}{|u|^2_{\bar{H}_2}} \geq \inf_{\chi \in V^{M_n}} \frac{|u_{m_0} - \chi|^2_{\bar{H}_1}}{|u_{m_0}|^2_{\bar{H}_2}} = \frac{1}{\beta_{m_0}} \geq \frac{1}{\gamma(V^{M_n}) + \epsilon}.$$

From this estimate and (6.33), we have

$$\frac{1}{\gamma(V^{M_n}) + \epsilon} \leq \sup_{u \in \bar{H}_2} \inf_{\chi \in V^{M_n}} \frac{|u - \chi|^2_{\bar{H}_1}}{|u|^2_{\bar{H}_2}} \leq \frac{1}{\gamma(V^{M_n})}.$$

Since ϵ is arbitrary, we get (6.32). □

Lemma 6.5. Let $\bar{H}_1 = H^{1/2}(I)$ and $\bar{H}_2 = H^{\beta}(I)$, where $\beta = (\beta_1, \beta_2, \dots)$, $\beta_k \geq 1$. Then

$$d_{2n}(H_2, H_1) = (\gamma_n^*)^{-\frac{1}{2}},$$

where

$$\gamma_n^* = \sup_{m_1, m_2, \dots, m_n} \inf_{i \notin M_n} \beta_i.$$

The proof of this theorem follows immediately from Lemma 6.4.

Theorem 6.6. Suppose $H_1 = \{u \in H^1(\mathring{\eta}_{\underline{x}}) : u \text{ is harmonic}\}$ and $H_2 = \mathcal{W}(\mathring{\eta}_{\underline{x}})$ with the norm $\|u\|_{\beta} = |u|_{\beta}$, given in (6.30), with $\beta_j \geq 1$. Suppose in addition that the sequence β_j is non-decreasing. Then the space

$${}^0V_{\underline{x}}^{(2n+1)} = \operatorname{span}\{r^j \cos j\theta, r^j \sin j\theta\}_{j=0}^n,$$

i.e., the span of first $(2n+1)$ harmonic polynomials, is optimal relative to H_1 and any H_2 (*i.e.*, any of the spaces H_2 we are considering).

Proof. Using the correspondence $u(r, \theta) \sim u(\theta)$, we can study the optimality of a finite-dimensional subspace relative to H_1 and H_2, by studying the optimality of a subspace relative to $\bar{H}_1$ and $\bar{H}_2$. The result follows directly from Lemma 6.5. □

Remark 50. Obviously the condition on β in Theorem 6.6 holds for any (isotropic) Sobolev space.

Remark 51. Let us return to the solution of the Dirichlet problem mentioned above. Suppose $\eta_{\underline{x}}$ is far from the boundary of Ω. Then, on $\eta_{\underline{x}}$, the character of the solution u_0 is approximately the same in any direction. Thus it is appropriate to embed u_0 in a space with a rotationally invariant norm – a usual (isotropic) Sobolev space, for example. Further, we have learned that, on $\eta_{\underline{x}}$, u_0 is well approximated by harmonic polynomials. The situation is, however, somewhat different when $\eta_{\underline{x}}$ is near the boundary. Then u_0 would be strongly influenced by the boundary values $g(x)$. Hence some other shape functions, constructed, for example, by the 'handbook' approach (see below), which themselves reflected these boundary values, would be 'best'.

Thus the optimal shape functions are the solution of the Laplace equation. This approach could also be used in other situations. Vekua (1967) defines and studies analogues of harmonic polynomials for differential equations with analytic coefficients. Babuška and Melenk (1997) and Laghrouche and Bettess (2000) construct and use special shape functions for the Helmholtz equation, $-\Delta u - ku = 0$. Special shape functions for problems in composite materials were used in Strouboulis, Zhang and Babuška (200x).

In this section, we saw an example of choosing an optimal local approximating space $V_{\underline{x}}$, which turned out to be the span of first $(2n+1)$ harmonic polynomials. In other problems, different local approximating spaces, consisting of optimal or near-optimal approximating functions, are recommended. These optimal or near-optimal approximating functions are solutions of other boundary value problems (posed on $\eta_{\underline{x}} \cap \Omega$). Such locally posed problems are called *handbook problems* and their solutions, which may be available analytically or computed numerically, are called *handbook functions*. This nomenclature is reminiscent of the solved problems and their solutions (via formulae, tables *etc.*) which are used in engineering (Tada, Paris and Irwin 1973). This idea is also used in commercial codes (Szabo, Babuška and Actis 1998).

One of the main advantages of the GFEM is that only simple meshes are used, which need not reflect the boundary, *e.g.*, uniform finite element meshes. Also, in each $\eta_{\underline{x}}$, one can use a space $V_{\underline{x}}$ of arbitrary dimension (depending on $\underline{x}$). $V_{\underline{x}}$ could be space of polynomials or any other space of functions depending on the local properties of the approximated function.

Choosing $V_{\underline{x}}$ to be the space of polynomials of low degree p (and using $\{\phi_{\underline{x}}\}$ that are reproducing of order k), we obtain the h-version of the FEM. All other classical versions of the FEM – the p and h-p versions – are special cases of the GFEM.

The GFEM, with special shape functions, was effectively used to solve differential equations with rough coefficients and, more generally, in problems with micro-structures: see Babuška *et al.* (1994). Babuška and Osborn (2000) showed that, for differential equations with rough coefficients, the classical FEM can converge arbitrarily slowly. With the GFEM, in contrast, with appropriately chosen shape functions, an exponential rate of convergence can be achieved: see Matache, Babuška and Schwab (2000). The GFEM is a very powerful approach for solving problems with microstructures: see Strouboulis *et al.* (2001*b*) and Section 9. The GFEM can also be advantageously used in linear and nonlinear crack propagation problems (Moes, Dolbow and Belytschko 1999, Wells and Sluis 2001, Wells, de Borst and Sluis 2002), and problems with boundary layers (Duarte and Babuška 2002).

7. Solutions of elliptic boundary value problems

In this section we will discuss the approximate solution of the model problem (2.1)–(2.2) (or (2.3)), introduced in Section 2, by a meshless method. We will address the Neumann boundary condition (2.2) and the Dirichlet boundary condition (2.3) separately. These problems have the variational formulation (2.4).

For $0 < h \leq 1$, we consider a family of particle-shape function systems

$$\{\mathcal{M}^h\}_{0<h\leq 1} = \{X^h, \{h_{\underline{x}}^h, \omega_{\underline{x}}^h, \phi_{\underline{x}}^h\}_{\underline{x}\in X^h}\}_{0<h\leq 1},$$

satisfying assumptions A1–A7 in Section 3.3 and (3.63). Recall that (3.63) is trivially satisfied if the shape functions are reproducing of order k. The family $\{\mathcal{M}^h\}_{0<h\leq 1}$ was introduced in Section 3.3; recall that

$$\sup_{\underline{x}\in X^h} h_{\underline{x}}^h \leq h.$$

We will be interested in assessing the approximation error as $h \downarrow 0$.

Let u_0 be the solution of (2.4), where Ω is a bounded domain with Lipschitz-continuous boundary. In this section, we will sometimes assume that the boundary of Ω is smooth. We will use the space $\mathbb{V}_{\Omega,h}^{k,q}$, defined in (3.85), to approximate u_0. It was shown in Theorem 3.11 that $\mathbb{V}_{\Omega,h}^{k,q}$ is $(k+1,q)$-regular. Moreover, $\mathbb{V}_{\Omega,h}^{k,q}$ satisfies the local assumption LA. Recall that k is the order of the quasi-reproducing shape functions considered in $\{\mathcal{M}^h\}$ and q is the smoothness index of these shape functions. The parameters k and q are in assumptions A1–A7 and we assume that $q \leq k+1$. We also recall that $\mathbb{V}_{\Omega,h}^{k,q}$ does not involve all the particles in X^h; it only involves particles in the set

$$A_\Omega^h = \{\underline{x} \in X^h : \mathring{\eta}_{\underline{x}}^h \cap \Omega \neq \emptyset\}. \tag{7.1}$$

Various classes of shape functions can be used for $\phi_{\underline{x}}^h$ in the system $\{\mathcal{M}^h\}$. In Section 4, one such class of shape functions, namely RKP shape functions, were discussed, and references related to other classes of shape functions used in practice were provided.

We note that it is possible to construct particle-shape function systems $\mathcal{M}^h$, satisfying A1–A7, such that the set of particles $X^h \subset \Omega$ and the corresponding $\mathbb{V}_{\Omega,h}^{k,q}$ have the desired approximation properties. We do not consider such $\mathbb{V}_{\Omega,h}^{k,q}$ in this section, and we will further remark on this issue in the next subsection.

Let $u_S = u_h \in \mathbb{V}_{\Omega,h}^{k,q}$ be the approximate solution defined by (2.7) with $S = \mathbb{V}_{\Omega,h}^{k,q}$. Since $\mathbb{V}_{\Omega,h}^{k,q}$ is $(k+1,q)$-regular, we note that $\mathbb{V}_{\Omega,h}^{k,q} \subset H = H^1(\Omega)$ provided $q \geq 1$. Thus u_h is the solution of

$$\begin{cases} u_h \in \mathbb{V}_{\Omega,h}^{k,q} \\ \tilde{B}(u_0, v) = \displaystyle\int_\Omega fv\,\mathrm{d}x, \quad \text{for all } v \in \mathbb{V}_{\Omega,h}^{k,q}, \end{cases} \tag{7.2}$$

where the bilinear form $\tilde{B}$ is either B, given in (2.5), or a perturbation of B. Clearly, u_h is the solution of a Galerkin method. This Galerkin method is a *meshless method* since the construction of the test and the trial space, *i.e.*, $\mathbb{V}^{k,q}_{\Omega,h}$, does not require a mesh. As we remarked in Section 1, avoiding mesh generation is one of the main features and advantages of meshless methods.

In this section, we will consider u_h as an approximation of u_0 and primarily study the error $u_0 - u_h$. We set some notation that will be used in this study in the following sections. We define

$$E^h_{\underline{x}} \equiv \mathring{\eta}^h_{\underline{x}} \cap \partial\Omega, \quad \underline{x} \in A^h_\Omega, \tag{7.3}$$

and

$$A^h_{\partial\Omega} = \{\underline{x} \in A_\Omega : E^h_{\underline{x}} \neq \emptyset\}. \tag{7.4}$$

Thus $A^h_{\partial\Omega}$ is the set of particles $\{\underline{x}\}$ such that $\mathring{\eta}^h_{\underline{x}}$ has non-empty intersection with $\partial\Omega$.

7.1. A meshless method for Neumann boundary conditions

In this section we will address the approximation of solution u_0 of (2.1) and (2.2) by the meshless method. The analysis presented here is based on the ideas and results in Babuška (1971) and Babuška and Aziz (1972). See also the references listed in these articles.

The solution u_0 of (2.1), (2.2) can be variationally characterized by (2.4), which is

$$\begin{cases} u_0 \in H^1(\Omega) \\ B(u_0, v) = \int_\Omega f v \, dx, \quad \text{for all } v \in H^1(\Omega). \end{cases} \tag{7.5}$$

We wish to approximate u_0 by u_h, the solution of (7.2) with $\tilde{B} = B$. For an error estimate, from (2.8) we have

$$\|u_0 - u_h\|_{H^1(\Omega)} \leq \inf_{\chi \in \mathbb{V}^{k,q}_{\Omega,h}} \|u_0 - \chi\|_{H^1(\Omega)}.$$

Suppose $u_0 \in H^l(\Omega)$. Then, since $\mathbb{V}^{k,q}_{\Omega,h}$ is $(k+1, q)$-regular and $q \geq 1$, we have

$$\|u_0 - u_h\|_{H^1(\Omega)} \leq Ch^\mu \|u_0\|_{H^l(\Omega)},$$

where $\mu = \min(k, l-1)$. We summarize this in the following theorem.

Theorem 7.1. Suppose $u_0 \in H^l(\Omega)$, with $l \geq 1$, is the solution of (7.5), where $\partial\Omega$ is Lipschitz-continuous. Let $u_h \in \mathbb{V}^{k,q}_{\Omega,h}$, with $q \geq 1$, be the approximate solution given by (7.2) with $\tilde{B} = B$. Then

$$\|u_0 - u_h\|_{H^1(\Omega)} \le h^\mu \|u_0\|_{H^l(\Omega)}, \tag{7.6}$$

where

$$\mu = \min(k, l-1). \tag{7.7}$$

We note that the computation of u_h, in Theorem 7.1, depends on the definition of $\mathbb{V}^{k,q}_{\Omega,h}$ and involves particles that are also outside Ω. In the literature, especially in the engineering literature, (t, k^*)-regular particle spaces are constructed using particles inside Ω, but the support of some of the corresponding particle shape functions could be partly outside Ω. The apparent reason for such construction is that the approximate solution is viewed as an interpolant with respect to data inside Ω, and hence only the particles that are inside Ω are considered. This is certainly not necessary.

The construction of the approximation space S (in (2.7)) using particles only inside Ω may sometimes lead to better conditioning of the underlying linear system. On the other hand, such construction is more expensive and the approximations could show boundary layer behaviour (Babuška *et al.* 200x).

7.2. Meshless methods for Dirichlet boundary conditions

In this section we consider the approximation of the solution u_0 of the Dirichlet boundary value problem (2.1) and (2.3) by meshless methods. The variational characterization of u_0 is given by

$$\begin{cases} u_0 \in H^1_0(\Omega) \\ B(u_0, v) = \displaystyle\int_\Omega f v \,\mathrm{d}x, \quad \text{for all } v \in H^1_0(\Omega). \end{cases} \tag{7.8}$$

The Galerkin method (2.7) to approximate u_0 would require that the approximating space S be a subspace of $H = H^1_0(\Omega)$ and thus that the approximating functions satisfy the essential homogeneous Dirichlet boundary condition. Unlike shape functions used in the FEM, the particle shape functions $\phi^h_{\underline{x}}$ (we consider h as the parameter), considered in Section 3.3, *do not* in general satisfy the so-called 'Kronecker delta' property, *i.e.*, $\phi^h_{\underline{x}}(\underline{y}) \neq \delta_{\underline{x},\underline{y}}$, $\underline{x}, \underline{y} \in X^h$. This is also true for translation-invariant particle shape functions discussed in Section 3.2 (see Section 4.2). Thus it is difficult to construct a subspace $S \subset \mathbb{V}^{k,q}_{\Omega,h}$ such that S could be used in (2.7) as the approximation space and the functions in S satisfy the Dirichlet boundary condition.

In the literature, several meshless methods have been proposed to approximate the solutions of Dirichlet boundary value problems. They are meshless methods in the sense that they use $\mathbb{V}^{k,q}_{\Omega,h}$ as the approximating space. These methods are:

(1) the penalty method,
(2) the Lagrange multiplier method,
(3) the Nitsche and related methods,
(4) the collocation method,
(5) combination of meshless and finite element methods,
(6) the characteristic function method.

In this section we will describe these methods. We note that the GFEM, discussed in Section 6, uses an approximating space different from $\mathbb{V}^{k,q}_{\Omega,h}$, and can also be used to approximate the solution of a Dirichlet boundary value problem.

We will assume that the boundary $\partial\Omega$ of Ω is sufficiently smooth. The smoothness assumption on the boundary simplifies the arguments presented here, but various results could be obtained when the boundary is not smooth.

The penalty method

The main idea of the penalty method is to use a perturbed variational principle. For $\sigma > 0$, we consider the bilinear form

$$\tilde{B}(u,v) \equiv B_\sigma(u,v) \equiv B(u,v) + h^{-\sigma} D(u,v), \tag{7.9}$$

where

$$B(u,v) = \int_\Omega (\nabla u \cdot \nabla v + uv)\,\mathrm{d}x, \tag{7.10}$$

$$D(u,v) = \int_{\partial\Omega} uv\,\mathrm{d}x. \tag{7.11}$$

We note that (7.10) is the bilinear form given in (2.5). We consider the solution $u_h = u_{\sigma,h} \in \mathbb{V}^{k,q}_{\Omega,h}$ of (7.2), namely

$$B_\sigma(u_{\sigma,h}, v) = \int_\Omega fv\,\mathrm{d}x, \quad \text{for all } v \in \mathbb{V}^{k,q}_{\Omega,h}. \tag{7.12}$$

We note that $u_{\sigma,h}$ is u_S, where u_S is defined in (2.7). For $v \in H^1(\Omega)$, let

$$Q_\sigma(v) = B(v,v) + h^{-\sigma} D(v,v) - 2\int_\Omega fv\,\mathrm{d}x. \tag{7.13}$$

It is well known that

$$Q_\sigma(u_{\sigma,h}) = \min_{v \in \mathbb{V}^{k,q}_{\Omega,h}} Q_\sigma(v). \tag{7.14}$$

We now present a convergence result for the penalty method.

Theorem 7.2. Suppose $u_0 \in H^l(\Omega) \cap H_0^1(\Omega)$, $l > 3/2$, is the solution of (7.8). Let $u_{\sigma,h} \in \mathbb{V}_{\Omega,h}^{k,q}$ be the solution of (7.12). Then, for any $0 < \epsilon < \min(l - 3/2, 1/2)$, we have

$$\|u_0 - u_{\sigma,h}\|_{H^1(\Omega)} \leq C(\epsilon) h^{\mu} \|u_0\|_{H^l(\Omega)}, \tag{7.15}$$

where

$$\mu = \min\left(k,\, l-1,\, \frac{\sigma}{2},\, k+\frac{1}{2}-\frac{\sigma}{2}-\epsilon,\, l-\frac{1}{2}-\frac{\sigma}{2}-\epsilon\right), \tag{7.16}$$

and $C(\epsilon)$ depends on ϵ, but is independent of h and u_0.

Proof. For any $v \in H^1(\Omega)$, we define

$$R_\sigma(v) = B(u_0 - v, u_0 - v) + h^{-\sigma} D\left(\frac{\partial u_0}{\partial n} h^\sigma + v,\, \frac{\partial u_0}{\partial n} h^\sigma + v\right). \tag{7.17}$$

Then, from Green's theorem,

$$\begin{aligned} R_\sigma(v) &= B(u_0, u_0) + B(v, v) - 2B(u_0, v) \\ &\quad + h^\sigma D\left(\frac{\partial u_0}{\partial n}, \frac{\partial u_0}{\partial n}\right) + h^{-\sigma} D(v, v) + 2D\left(\frac{\partial u_0}{\partial n}, v\right) \\ &= B(u_0, u_0) + h^\sigma D\left(\frac{\partial u_0}{\partial n}, \frac{\partial u_0}{\partial n}\right) \\ &\quad + B(v, v) + h^{-\sigma} D(v, v) - 2\int_\Omega f v \,\mathrm{d}x \\ &= B(u_0, u_0) + h^\sigma D\left(\frac{\partial u_0}{\partial n}, \frac{\partial u_0}{\partial n}\right) + Q_\sigma(v), \quad \text{for all } v \in H^1(\Omega), \end{aligned}$$

where $Q_\sigma(v)$ is given by (7.13). Therefore,

$$\min_{v \in \mathbb{V}_{\Omega,h}^{k,q}} R_\sigma(v) = B(u_0, u_0) + h^\sigma D\left(\frac{\partial u_0}{\partial n}, \frac{\partial u_0}{\partial n}\right) + \min_{v \in \mathbb{V}_{\Omega,h}^{k,q}} Q_\sigma(v),$$

and thus from (7.14) we get

$$R_\sigma(u_{\sigma,h}) = \min_{v \in \mathbb{V}_{\Omega,h}^{k,q}} R_\sigma(v).$$

Hence, from (7.17) and the above relation, we have

$$\begin{aligned} \|u_0 - u_{\sigma,h}\|_{H^1(\Omega)}^2 &= B(u_0 - u_{\sigma,h}, u_0 - u_{\sigma,h}) \\ &\leq R_\sigma(u_{\sigma,h}) \\ &\leq R_\sigma(v), \quad \text{for all } v \in \mathbb{V}_{\Omega,h}^{k,q}. \end{aligned} \tag{7.18}$$

Since $\mathbb{V}_{\Omega,h}^{k,q}$ is $(k+1, q)$-regular with $q \geq 1$, there is a $g_h \in \mathbb{V}_{\Omega,h}^{k,q}$ such that

$$\|u_0 - g_h\|_{H^s(\Omega)} \leq C h^{\mu} \|u_0\|_{H^l(\Omega)}, \tag{7.19}$$

where $\mu = \min(k+1-s, l-s)$ and $0 \le s \le 1$. Now, from (7.17) with $v = g_h$ and using the Schwartz inequality, we have

$$R_\sigma(g_h) \le C\left(\|u_0 - g_h\|^2_{H^1(\Omega)} + h^\sigma \int_{\partial\Omega} \left(\frac{\partial u_0}{\partial n}\right)^2 \mathrm{d}s + h^{-\sigma} \int_{\partial\Omega} g_h^2 \,\mathrm{d}s \right). \quad (7.20)$$

We will estimate the right-hand side of the above inequality. We first note that $u_0 = 0$ on $\partial\Omega$. Let $0 < \epsilon < \min(l - \frac{3}{2}, \frac{1}{2})$. Then, using a trace inequality and (7.19) with $s = (1/2) + \epsilon$, we get

$$\begin{aligned} \|g_h\|^2_{L_2(\partial\Omega)} = \|u_0 - g_h\|^2_{L_2(\partial\Omega)} &\le C(\epsilon)\|u_0 - g_h\|^2_{H^{\frac{1}{2}+\epsilon}(\Omega)} \\ &\le C(\epsilon) h^{2\mu_1} \|u_0\|^2_{H^l(\Omega)}, \end{aligned} \quad (7.21)$$

where $\mu_1 = \min(k + \frac{1}{2} - \epsilon, l - \frac{1}{2} - \epsilon)$. Also from a trace inequality, we have

$$\left\|\frac{\partial u_0}{\partial n}\right\|^2_{L_2(\partial\Omega)} \le C(\epsilon)\|u_0\|^2_{H^{\frac{3}{2}+\epsilon}(\Omega)}. \quad (7.22)$$

Now using (7.21), (7.22), and (7.19), with $s = 1$, in (7.20), we get

$$\begin{aligned} R_\sigma(g_h) &\le C(\epsilon)\left(h^{2\min(k,l-1)} + h^\sigma + h^{2\mu_1 - \sigma}\right)\|u_0\|^2_{H^l(\Omega)} \\ &\le C(\epsilon) h^{2\mu}\|u_0\|^2_{H^l(\Omega)}, \end{aligned} \quad (7.23)$$

where $\mu = \min(k, l-1, \frac{\sigma}{2}, k + \frac{1}{2} - \frac{\sigma}{2} - \epsilon, l - \frac{1}{2} - \frac{\sigma}{2} - \epsilon)$. Finally, combining (7.18) and (7.23) we get the desired result. □

Remark 52. If we consider $\mathbb{V}^{k,q}_{\Omega,h}$ in Theorem 7.2 such that $k+1 \ge l > 3/2$, then with $\sigma = l - \frac{1}{2} - \epsilon$ it is easy to see that (7.15) holds with

$$\mu = \frac{1}{2}\left(l - \frac{1}{2} - \epsilon\right).$$

Estimate (7.15) can be improved. We present the following result, based on the analysis in Babuška (1970, 1971, 1972, 1973*b*), without proof.

Theorem 7.3. Suppose $u_0 \in H^l(\Omega) \cap H^1_0(\Omega)$ is the solution of (7.8). Let $u_{\alpha,h} \in \mathbb{V}^{k,q}_{\Omega,h}$ be the solution of (7.12). If $k + 1 \ge l \ge 2$, then, for any $\epsilon > 0$, we have

$$\|u_0 - u_{\alpha,h}\|_{H^1(\Omega)} \le C(\epsilon) h^{\mu-\epsilon}\|u_0\|_{H^l(\Omega)}, \quad (7.24)$$

where $C(\epsilon)$ is independent of u_0 and h but depends on ϵ, and μ is given by

$$\mu = \min\left(\sigma,\ l + \sigma - 2,\ l + \frac{\sigma}{2} - \frac{3}{2},\ \frac{k+1-\kappa}{k}(l-1)\right), \quad (7.25)$$

where

$$\kappa = \max\left(1, \frac{1+\sigma}{2}\right). \quad (7.26)$$

Remark 53. If $\sigma \geq 1$ in Theorem 7.3, then $l + \sigma - 2 \geq l + \frac{\sigma}{2} - \frac{3}{2}$, and therefore μ in (7.24) is given by

$$\mu = \min\left(\sigma, l + \frac{\sigma}{2} - \frac{3}{2}, \frac{k+1-\kappa}{k}(l-1)\right), \tag{7.27}$$

where κ is as in (7.26).

Example. Consider $l = 2$ and $\sigma = 1$ in Theorem 7.3. Then, from (7.26), we have $\kappa = 1$, and (7.27) yields $\mu = \min(1, 1, 1) = 1$. This is the optimal rate of convergence. For higher values of l and $k + 1 \geq l$, there is a loss in the rate of convergence and we get a sub-optimal rate of convergence, *i.e.*, $\mu < \min(l - 1, k)$.

Thus, from Theorem 7.3 we conclude the following.

- It is advantageous to use $\mathbb{V}^{k,q}_{\Omega,h}$ with higher values of k, since it leads to higher accuracy. For example, if $l = 4$ and $\sigma = 3$, then $\kappa = 2$ and (7.27) yields $\mu = 3(\frac{k-1}{k})$. Thus higher values of k will increase accuracy. But higher values of k reduce the sparsity of the resulting linear system.
- Too small or too large a value of σ may decrease the accuracy significantly. For example, if $\sigma = 2k + 1$ then $\kappa = k + 1$, and (7.27) yields $\mu = 0$.

The use of penalty methods was recently suggested in the literature, *e.g.*, Atluri and Shen (2002), Liu (2002) and Zhu and Atluri (1998), without any theoretical analysis. An empirical penalty value of σ, unrelated to l or k, was suggested in Zhu and Atluri (1998).

The Lagrange multiplier method

The theory of Lagrange multiplier methods, in the context of finite element methods, was developed in Babuška (1973*a*) (see also Babuška and Aziz (1972)). This theory can also be extended to meshless methods.

It is known (*cf.* Babuška and Aziz (1972)) that the effectiveness of this method depends on a delicate relationship between the approximating space $S^{t,k^*}_h(\Omega)$ and the space of Lagrange multipliers $S^{t,k^*}_{h_1}(\partial\Omega)$, where both $S^{t,k^*}_h(\Omega)$ and $S^{t,k^*}_{h_1}(\partial\Omega)$ satisfy an inverse assumption. In the context of meshless methods, we let the approximating space to be the particle space $\mathbb{V}^{k,q}_{\Omega,h}$. We know that $\mathbb{V}^{k,q}_{\Omega,h}$ is $(k + 1, q)$-regular, and satisfies the inverse assumption, IA, under additional hypotheses given at the end of Section 3.3. The space of Lagrange multipliers $S^{k+1,q}_{h_1}(\partial\Omega)$ has the same (t, k^*)-regularity as the approximating space, and the functions in $S^{k+1,q}_{h_1}(\partial\Omega)$ are defined only on $\partial\Omega$ with respect to particles on $\partial\Omega$. Thus, $\partial\Omega$ must contain enough particles. We note that the functions in $S^{k+1,q}_{h_1}(\partial\Omega)$ are not restrictions of functions in

$\mathbb{V}^{k,q}_{\Omega,h}$ on $\partial\Omega$. Then, following the analysis in Babuška (1973*a*) and Babuška and Aziz (1972), one can show that, if the size of the supports of the basis elements of $S^{k+1,q}_{h_1}(\partial\Omega)$ is of the same order as $|E^h_{\underline{x}}|$, $\underline{x} \in A^h_{\partial\Omega}$ ($E^h_{\underline{x}}$ and $A^h_{\partial\Omega}$ defined in (7.3) and (7.4), respectively), then the approximate solution obtained from the Lagrange multiplier method converges. If the size of the supports of the basis elements of $S^{k+1,q}_{h_1}(\partial\Omega)$ is smaller than $\eta^h_{\underline{x}}$, $\underline{x} \in A^h_{\partial\Omega}$, then the method is unstable. This relationship was further analysed in Pitkäranta (1979, 1980).

The Lagrange multiplier technique leads to the optimal rate of convergence in comparison to the penalty method, where the optimal rate of convergence is usually not attained. But, as we mentioned before, the sufficient conditions for convergence are quite delicate.

Recently, the Lagrange multiplier technique was applied in the context of meshless methods without any theoretical analysis, in Belytschko, Lu and Gu (1994), Liu (2002), Lu, Belytschko and Gu (1994) and Mukherjee and Mukherjee (1997).

The Nitsche and related methods

Because of the delicate nature of Lagrange multiplier methods, there has been some interest in looking for other methods to deal with the issue of the imposition of Dirichlet boundary conditions, and to avoid complications that are present in Lagrange multiplier methods. To that end, certain methods were proposed in Barbara and Hughes (1991) and Stenberg (1995). But a similar method was proposed much earlier by Nitsche (1970–1971). We will discuss the Nitsche method, following the presentation in Stenberg (1995). We will still assume that $\partial\Omega$ is smooth.

To approximate the solution u_0 of (7.8) by the Nitsche method, we consider the particle space $\mathbb{V}^{k,q}_{\Omega,h}$ with $q \geq 2$. We also assume that:

- card $(A^h_{\partial\Omega}) \leq \kappa$ and
$$C_1 h \leq h_{E^h_{\underline{x}}} \leq C_2 h, \quad \underline{x} \in A^h_{\partial\Omega}, \tag{7.28}$$
where $h_{E^h_{\underline{x}}} \equiv |E^h_{\underline{x}}|$, for $\underline{x} \in A^h_{\partial\Omega}$ ($E^h_{\underline{x}}$ and $A^h_{\partial\Omega}$ are defined in (7.3) and (7.4), respectively);
- there exists $0 < \mathcal{K} < \infty$, $\mathcal{K} = \mathcal{K}(X^h)$, such that
$$\left\| \frac{\partial v}{\partial n} \right\|_{-\frac{1}{2},h} \leq \mathcal{K}[B(v,v)]^{1/2}, \quad \text{for all } v \in \mathbb{V}^{k,q}_{\Omega,h}, \tag{7.29}$$
where $B(u,v)$ was defined in (7.10) and
$$\left\| \frac{\partial v}{\partial n} \right\|^2_{-\frac{1}{2},h} = \sum_{\underline{x} \in A^h_{\partial\Omega}} h_{E^h_{\underline{x}}} \left\| \frac{\partial v}{\partial n} \right\|^2_{H^0(E^h_{\underline{x}})}. \tag{7.30}$$

We define the bilinear form $\tilde{B}(u,v) = B_\gamma(u,v)$, where

$$B_\gamma(u,v) = B(u,v) - D\Big(\frac{\partial u}{\partial n}, v\Big) - D\Big(\frac{\partial v}{\partial n}, u\Big) + \gamma \sum_{\underline{x}\in A^h_{\partial\Omega}} h_{E^h_{\underline{x}}}{}^{-1} \int_{E^h_{\underline{x}}} uv \, \mathrm{d}s,$$

with $\gamma > 0$; $B(u,v)$, $D(u,v)$ are as defined in (7.10), (7.11), respectively. The approximate solution $u_{h,\gamma} \in \mathbb{V}^{k,q}_{\Omega,h}$, obtained from the Nitsche method, is given by

$$B_\gamma(u_{h,\gamma}, v) = \int_\Omega f v \, \mathrm{d}x, \quad \text{for all } v \in \mathbb{V}^{k,q}_{\Omega,h}. \tag{7.31}$$

For $u \in H^2(\Omega)$, we define the norm

$$\|\!|u|\!\|^2 = B(u,u) + \left\| \frac{\partial u}{\partial n} \right\|^2_{-\frac{1}{2},h} + \|u\|^2_{\frac{1}{2},h},$$

where

$$\|u\|^2_{\frac{1}{2},h} = \sum_{\underline{x}\in A^h_{\partial\Omega}} h_{E^h_{\underline{x}}}{}^{-1} \|u\|^2_{H^0(E^h_{\underline{x}})},$$

and $\|\frac{\partial u}{\partial n}\|^2_{-\frac{1}{2},h}$ is given by (7.30) with v replaced by u. We first note that, from the Schwartz inequality, we have

$$\begin{aligned}
\sum_{\underline{x}\in A^h_{\partial\Omega}} h^{-1}_{E^h_{\underline{x}}} \int_{E^h_{\underline{x}}} uv \, \mathrm{d}s &\le \sum_{\underline{x}\in A^h_{\partial\Omega}} \left(\int_{E^h_{\underline{x}}} h^{-1}_{E^h_{\underline{x}}} u^2 \, ds \right)^{1/2} \left(\int_{E^h_{\underline{x}}} h^{-1}_{E^h_{\underline{x}}} v^2 \, \mathrm{d}s \right)^{1/2} \\
&\le \left[\sum_{\underline{x}\in A^h_{\partial\Omega}} h^{-1}_{E^h_{\underline{x}}} \|u\|^2_{H^0(E^h_{\underline{x}})} \right]^{1/2} \left[\sum_{\underline{x}\in A^h_{\partial\Omega}} h^{-1}_{E^h_{\underline{x}}} \|v\|^2_{H^0(E^h_{\underline{x}})} \right]^{1/2} \\
&= \|u\|_{\frac{1}{2},h} \|v\|_{\frac{1}{2},h}.
\end{aligned} \tag{7.32}$$

Also,

$$\begin{aligned}
D\Big(u, \frac{\partial v}{\partial n}\Big) &\le \sum_{\underline{x}\in A^h_{\partial\Omega}} \int_{E^h_{\underline{x}}} \left| h^{-1/2}_{E^h_{\underline{x}}} u \, h^{1/2}_{E^h_{\underline{x}}} \frac{\partial v}{\partial n} \right| \mathrm{d}s \\
&\le \sum_{\underline{x}\in A^h_{\partial\Omega}} h^{-1/2}_{E^h_{\underline{x}}} \|u\|_{H^0(E^h_{\underline{x}})} h^{1/2}_{E^h_{\underline{x}}} \left\| \frac{\partial v}{\partial n} \right\|_{H^0(E^h_{\underline{x}})} \\
&\le \|u\|_{\frac{1}{2},h} \left\| \frac{\partial v}{\partial n} \right\|_{-\frac{1}{2},h}.
\end{aligned} \tag{7.33}$$

Using the same arguments used to obtain (7.33), we get

$$D\left(\frac{\partial u}{\partial n}, v\right) \leq \left\|\frac{\partial u}{\partial n}\right\|_{-\frac{1}{2},h} \|v\|_{\frac{1}{2},h}. \tag{7.34}$$

Now, using (7.32)–(7.34), it can easily be shown that

$$B_\gamma(u,v) \leq (1+\gamma)\, |\!|\!| u |\!|\!|\, |\!|\!| v |\!|\!|. \tag{7.35}$$

We now show that, for a proper value of γ, the bilinear form $B_\gamma(u,v)$ is coercive.

Lemma 7.4. Suppose $\mathcal{K}^2 < \gamma$, where $\mathcal{K}$ satisfies (7.29). Then,

$$B_\gamma(v,v) \geq C_1^* |\!|\!| v |\!|\!|^2, \quad \text{for all } v \in \mathbb{V}^{k,q}_{\Omega,h}, \tag{7.36}$$

where $C^* = C^*(X^h) > 0$.

Proof. Let $v \in \mathbb{V}^{k,q}_{\Omega,h}$, and choose any $\epsilon > 0$. From the definition of $B_\gamma(u,v)$ and (7.33) with $u = v$, we have

$$\begin{aligned}
B_\gamma(v,v) &= B(v,v) - 2D\left(v, \frac{\partial v}{\partial n}\right) + \gamma\|v\|^2_{\frac{1}{2},h} \\
&\geq B(v,v) - 2\|v\|_{\frac{1}{2},h}\left\|\frac{\partial v}{\partial n}\right\|_{-\frac{1}{2},h} + \gamma\|v\|^2_{\frac{1}{2},h} \\
&\geq B(v,v) - \epsilon\|v\|^2_{\frac{1}{2},h} - \frac{1}{\epsilon}\left\|\frac{\partial v}{\partial n}\right\|^2_{-\frac{1}{2},h} + \gamma\|v\|^2_{\frac{1}{2},h} \\
&= B(v,v) - \frac{1}{\epsilon}\left\|\frac{\partial v}{\partial n}\right\|^2_{-\frac{1}{2},h} + (\gamma-\epsilon)\|v\|^2_{\frac{1}{2},h} \\
&\geq \left(1 - \frac{\mathcal{K}^2}{\epsilon}\right)B(v,v) + (\gamma-\epsilon)\|v\|^2_{\frac{1}{2},h}.
\end{aligned}$$

Therefore, considering $\epsilon = \frac{1}{2}(\mathcal{K}^2 + \gamma)$ in the above inequality, we get

$$B_\gamma(v,v) \geq C_1[B(v,v) + \|v\|^2_{\frac{1}{2},h}], \tag{7.37}$$

where $C_1 = \min\left(\frac{\gamma-\mathcal{K}^2}{\gamma+\mathcal{K}^2}, \frac{\gamma-\mathcal{K}^2}{2}\right)$. Now, from the definition of $|\!|\!| \cdot |\!|\!|$ and using (7.29), we get

$$\begin{aligned}
|\!|\!| v |\!|\!|^2 &\leq B(v,v) + \mathcal{K}^2 B(v,v) + \|v\|^2_{\frac{1}{2},h} \\
&\leq (1+\mathcal{K}^2)[B(v,v) + \|v\|^2_{\frac{1}{2},h}].
\end{aligned}$$

Thus, combining the above inequality with (7.37), we get

$$B_\gamma(v,v) \geq C^* |\!|\!| v |\!|\!|^2,$$

where $C^* = \frac{1}{1+\mathcal{K}^2}\min(\frac{\gamma-\mathcal{K}^2}{\gamma+\mathcal{K}^2}, \frac{\gamma-\mathcal{K}^2}{2})$, which is (7.36). □

We now present the following result.

Theorem 7.5. Suppose $u_0 \in H^l(\Omega)$, for $l \geq 2$, is the solution of (7.8). Let $u_{h,\gamma} \in \mathbb{V}^{k,q}_{\Omega,h}$, with $q \geq 2$, be the solution of (7.31), where $\mathbb{V}^{k,q}_{\Omega,h}$ satisfies (7.28) and (7.29). Then

$$\|u_0 - u_{h,\gamma}\|_{H^1(\Omega)} \leq \frac{C(1+\gamma)}{C^*(X^\nu)} h^\mu \|u_0\|_{H^l(\Omega)}, \quad \mu = \min(k, l-1), \tag{7.38}$$

where $C^*(X^\nu)$ is as in (7.36).

Proof. It is easy to see that

$$B_\gamma(u_0, v) = \int_\Omega f v \, \mathrm{d}x, \quad \text{for all } v \in H^1(\Omega),$$

and therefore,

$$B_\gamma(u_0 - u_{h,\gamma}, v) = 0, \quad \text{for all } v \in \mathbb{V}^{k,q}_{\Omega,h}. \tag{7.39}$$

Now, for any $g_h \in \mathbb{V}^{k,q}_{\Omega,h}$, using (7.36), (7.39) and (7.35), we have

$$\begin{aligned} |\!|\!| g_h - u_{h,\gamma} |\!|\!|^2 &\leq \frac{1}{C^*}\hat{B}_\gamma(g_h - \hat{u}_{h,\gamma}, g_h - u_{h,\gamma}) \\ &\leq \frac{1}{C^*}\hat{B}_\gamma(g_h - u_0, g_h - u_{h,\gamma}) \\ &\leq \frac{(1+\gamma)}{C^*} |\!|\!| u_0 - g_h |\!|\!| \, |\!|\!| g_h - u_{h,\gamma} |\!|\!|, \end{aligned}$$

and hence

$$|\!|\!| g_h - u_{h,\gamma} |\!|\!| \leq \frac{(1+\gamma)}{C^*} |\!|\!| u_0 - g_h |\!|\!|.$$

Therefore,

$$\begin{aligned} |\!|\!| u_0 - u_{h,\gamma} |\!|\!| &\leq |\!|\!| u_0 - g_h |\!|\!| + |\!|\!| g_h - u_{h,\gamma} |\!|\!| \\ &\leq C |\!|\!| u_0 - g_h |\!|\!|, \quad \text{for all } g \in \mathbb{V}^{k,q}_{\Omega,h}. \end{aligned} \tag{7.40}$$

Now, using (7.28) and a trace inequality, we have

$$\begin{aligned} \|u_0 - g_h\|^2_{\frac{1}{2},h} &= \sum_{\underline{x} \in A^h_{\partial\Omega}} h^{-1}_{E^h_{\underline{x}}} \|u_0 - g_h\|^2_{H^0(E^h_{\underline{x}})} \\ &\leq C h^{-1} \|u_0 - g_h\|^2_{H^0(\partial\Omega)} \end{aligned}$$

$$\leq Ch^{-1}\Big(\frac{1}{h}\|u_0 - g_h\|^2_{H^0(\Omega)} + h\|u_0 - g_h\|^2_{H^1(\Omega)}\Big)$$

$$= C\big(h^{-2}\|u_0 - g_h\|^2_{H^0(\Omega)} + \|u_0 - g_h\|^2_{H^1(\Omega)}\big), \tag{7.41}$$

where C depends on κ. Also, using a similar argument, we have

$$\left\|\frac{\partial(u_0 - g_h)}{\partial n}\right\|^2_{-\frac{1}{2},h} \leq C\big(h^2\|u_0 - g_h\|^2_{H^2(\Omega)} + \|u_0 - g_h\|^2_{H^1(\Omega)}\big). \tag{7.42}$$

where C depends on κ. Thus from (7.41) and (7.42) we get

$$\begin{aligned}|\!|\!| u_0 - g_h |\!|\!|^2 &= \|u_0 - g_h\|^2_{H^1(\Omega)} + \left\|\frac{\partial(u_0 - g_h)}{\partial n}\right\|^2_{-\frac{1}{2},h} + \|u_0 - g_h\|^2_{\frac{1}{2},h}\\ &\leq Ch^{-2}\sum_{j=0}^{2} h^{2j}\|u_0 - g_h\|^2_{H^j(\Omega)},\end{aligned}$$

and hence, from the definition of $|\!|\!| \cdot |\!|\!|$ and (7.40), we have

$$\begin{aligned}\|u_0 - \hat{u}_{h,\gamma}\|_{H^1(\Omega)} &\leq |\!|\!| u_0 - \hat{u}_{h,\gamma} |\!|\!| \\ &\leq Ch^{-1}\sum_{j=0}^{2} h^j\|u_0 - g_h\|_{H^j(\Omega)}, \quad \text{for all } g_h \in \mathbb{V}^{k,q}_{\Omega,h}.\end{aligned} \tag{7.43}$$

Finally, we choose $g \in \mathbb{V}^{k,q}_{\Omega,h}$ such that

$$\|u_0 - g\|_{H^s(\Omega)} \leq Ch^{\mu_1 - s}\|u_0\|_{H^l(\Omega)}, \quad 0 \leq s \leq 2,$$

where $\mu_1 = \min(k+1, l)$. Using this in (7.43) we get the desired result. □

We now discuss situations where the assumption required to prove Theorem 7.5 is valid. The major problem is to estimate $\mathcal{K}(X^h)$ given in (7.29). We would like to have $\mathcal{K}(X^h) \leq C$, uniformly for all $0 < h \leq 1$. If the supports $\eta^h_{\underline{x}}$ of the particle function $\phi^h_{\underline{x}}$ are 'reasonable', *e.g.*, in $\mathbb{R}^2$ or in $\mathbb{R}^3$, then it is easy to see that the necessary condition for $\mathcal{K}(X^h) \leq C$ is that $h_{E^h_{\underline{x}}} \geq \alpha|\eta^h_{\underline{x}}|$ for $\underline{x} \in A^h_{\partial\Omega}$. This can be enforced by properly selecting the set of particles X^h. This aspect can also affect the design of adaptive meshless (Nitsche) methods. Since the estimates of these constants are difficult to estimate accurately, we may select larger values of γ in (7.31) so that Theorem 7.5 is valid.

The Nitsche method presented here is superior to both the penalty method and Lagrange multiplier methods. The Nitsche method, in the framework of meshless methods, was addressed in Babuška *et al.* (2002*a*) and implemented in Schweitzer (200x) and Griebel and Schweitzer (2002*e*).

The collocation method
The collocation method, in the framework of meshless methods, was recently proposed in Atluri and Shen (2002), Zhu and Atluri (1998) and Hagen (1996). The method consists of adding constraint equation, at certain points of the boundary $\partial\Omega$, to the stiffness matrix. No analysis was presented to address the convergence of the approximate solution obtained from this method. Collocation using radial basis functions was analysed in Franke and Schaback (1998).

Combination of meshless methods and the finite element method
This method was proposed, *e.g.*, in Krongauz and Belytschko (1996). The main idea in this approach is to use classical finite elements (which could also be interpreted as particle functions) in a neighbourhood of the boundary $\partial\Omega$, and to select other particle functions such that their supports do not intersect $\partial\Omega$.

The characteristic function method
The method was proposed in connection to the Ritz method when the approximating functions were global polynomials (see Mikhlin (1971) and Kantorovich and Krylov (1958)). If a domain Ω has a smooth boundary $\partial\Omega$, there exists a smooth function Φ such that

$$\begin{aligned} \Phi(x) &> 0, \quad x \in \Omega, \\ \Phi(x) &= 0, \quad x \in \partial\Omega, \\ \text{and } |\nabla\Phi(x)| &\geq \alpha > 0, \quad x \in \partial\Omega. \end{aligned}$$

Let $S_h^\Phi = \{u : u = \Phi v,\ v \in \mathbb{V}_{\Omega,h}^{k,q}\}$. Then it is obvious that $S_h^\Phi \subset H_0^1(\Omega)$. We approximate the solution u_0 of (2.1) and (2.3) by $u_h \in S_h^\Phi$, where u_h is the solution u_S of (2.7) with $S = S_h^\Phi$.

For $u_0 \in H^l(\Omega) \cap H_0^1(\Omega)$, $l \geq 2$, we define $w_0 = \frac{u_0}{\Phi}$. Then, using Hardy's inequality (Theorem 329 of Hardy, Littlewood and Polya (1952)), one can show that $w_0 \in H^{l-1}(\Omega)$. Using this result, we obtain the following theorem.

Theorem 7.6. Suppose $u_0 \in H^l(\Omega) \cap H_0^1(\Omega)$, and suppose $l \geq 2$. Then there exists $w_h \in \mathbb{V}_{\Omega,h}^{k,q}$ such that $g_h = \Phi w_h$ satisfies

$$\|u_0 - g_h\|_{H^1(\Omega)} \leq Ch^\mu \|u_0\|_{H^l(\Omega)}, \quad \mu = \min(k, l-2). \tag{7.44}$$

Proof. Recall that $\mathbb{V}_{\Omega,h}^{k,q}$ is $(k+1, q)$-regular with $q \geq 1$. Then there exists $w_h \in \mathbb{V}_{\Omega,h}^{k,q}$ such that

$$\|w_0 - w_h\|_{H^1(\Omega)} \leq Ch^\mu \|w_0\|_{H^{l-1}(\Omega)} \leq Ch^\mu \|u_0\|_{H^l(\Omega)}, \tag{7.45}$$

where $\mu = \min(k, l-2)$. Now, from the definition of w_0, we have

$$u_0 - \Phi w_h = u_0 - \Phi w_0 + \Phi(w_0 - w_h) = \Phi(w_0 - w_h),$$

and hence, using (7.45), we have

$$\|u_0 - \Phi w_h\|_{H^1(\Omega)} \le C\|w_0 - w_h\|_{H^1(\Omega)} \le Ch^\mu \|u_0\|_{H^l(\Omega)}, \quad \mu = \min(k, l-2),$$

which is the desired result. □

Remark 54. It is clear from (2.8) and (7.44) that, for $l \ge 2$,

$$\|u_0 - u_h\|_{H^1(\Omega)} \le Ch^\mu \|u_0\|_{H^l(\Omega)}, \quad \mu = \min(k, l-2).$$

We further note that this order of convergence cannot, in general, be improved.

The generalized finite element method

We note that all the methods described so far primarily use $\mathbb{V}^{k,q}_{\Omega,h}$ as the approximating space. The GFEM, on the other hand, uses different approximating spaces, as we have seen in Section 6. The use of these approximating spaces makes the GFEM extremely flexible.

We recall that in the GFEM we start with a partition of unity with respect to a simple mesh that need not conform to the boundary of the domain. This partition unity could be the particle shape functions defined in Section 3. Then 'handbook' functions are used as local approximating spaces. The Dirichlet boundary condition can be implemented by choosing the local approximation space $V_{\underline{x}}$, for $\underline{x} \in A^h_{\partial\Omega}$, such that the functions in $V_{\underline{x}}$ satisfy the Dirichlet boundary conditions.

We have presented a few approaches on how to use meshless approximation to approximate solutions of PDEs. To impose Dirichlet boundary conditions on meshless approximation is a challenge, and we looked into some methods that can overcome this difficulty. While discussing these methods, we assumed that the boundary of the domain is smooth, for simplicity. But the results presented here can be generalized to cover nonsmooth boundaries, especially piecewise smooth boundaries.

Some methods were implemented and reported in the literature, but lacked rigorous theoretical analysis. All the methods reported here have certain advantages as well as disadvantages. If the particle space $\mathbb{V}^{k,q}_{\Omega,h}$ is used as an approximating space, then in our opinion the Nitsche method is very promising, because it is relatively robust and easy to implement. But we note that $\mathbb{V}^{k,q}_{\Omega,h}$ is difficult to construct for higher values of k, and the use of $\mathbb{V}^{k,q}_{\Omega,h}$ with lower values of k reduces the accuracy of the method. On the other hand, the GFEM uses a partition of unity (the basis functions of $\mathbb{V}^{k,q}_{\Omega,h}$ with $k = 0$), which is easy to construct, and higher accuracy can be attained by using suitable local approximation spaces.

8. Implementational aspects of meshless methods

In this section, we will briefly discuss the implementational aspects of meshless methods and the GFEM. As for the finite element method, the implementation of meshless methods and the GFEM has four major parts:

(1) construction of particle shape functions,
(2) construction of the stiffness matrix,
(3) solution of the linear system of equations,
(4) *a posteriori* error estimation, adaptivity, and computation of data of interest.

We now discuss these items in turn.

Construction of particle shape functions
In the classical finite element method, we start with a mesh that is related to the domain, and then shape functions are defined with respect to the chosen mesh. In a meshless method, we start with particles $\{\underline{x}\}$. Corresponding to each particle $\underline{x}$, a particle shape function with compact support $\eta_{\underline{x}}$ is constructed, such that the $\{\mathring{\eta}_{\underline{x}}\}$ form an open cover of the domain Ω. The construction of shape functions that are reproducing of order $k = 0$ or 1 is not difficult. For $k = 0$, one may use Shepard's approach (Shepard 1968) as described in Section 4.1. For $k = 1$ and for an appropriate particle distribution, one may first construct a mesh using tetrahedra such that the particles are the nodes of the mesh (*i.e.*, the vertices of the tetrahedra). This procedure is not difficult, as there are efficient codes available for constructing such a mesh. The shape function corresponding to the particle $\underline{x}$ can be taken to be the standard hat functions, whose support is the union of all the simplices with $\underline{x}$ as one of its vertices. We note, however, that, for $k = 1$, smoother shape functions can also be constructed (see Han and Meng (2001)). For $k = 0, 1$, we have to check that $\operatorname{card}(S_{\underline{x}}) \leq \kappa$, κ is independent of $\underline{x}$, where $S_{\underline{x}} = \{\underline{y} : \eta_{\underline{y}} \cap \eta_{\underline{x}} \neq \emptyset\}$. For the Nitsche method, described in Section 7.2, we also have to check that $\mathcal{K}$, defined in (7.29), is bounded. The construction of particle shape functions for $k \geq 2$ is more expensive than for $k = 0, 1$, and it may be more difficult to check assumptions A1–A7 and (3.63), which ensure convergence.

In contrast, the GFEM uses only a partition of unity, and thus particle shape functions with $k = 0, 1$, described in the last paragraph, can be used for this purpose. Also, a simple regular distribution of particles could be used to construct the partition of unity. The space of local shape functions, $V_{\underline{x}}$, could be created analytically or through 'handbook' solutions. Dirichlet boundary conditions are also treated by appropriate selection of $V_{\underline{x}}$, and hence we do not have to use special methods, *e.g.*, the penalty method, the Nitsche method, *etc.*, which simplifies the implementation.

Construction of the stiffness matrix

The construction of the stiffness matrix for a meshless method is laborious and delicate. In fact, this is where we pay the price for avoiding mesh generation. The elements of the resulting stiffness matrix are integrals, which have to be numerically evaluated over various regions. These regions are not simple tetrahedra as in the finite element method, where they naturally come from a mesh. These regions, for a meshless method, are of the form $\eta_{\underline{x}} \cap \eta_{\underline{y}} \cap \Omega$, $\underline{x}, \underline{y} \in X^{\nu}$, and can be extremely complicated. Also, the integrals have to be evaluated accurately, as it is known that inaccurate numerical integration leads to very poor results (see, for example, Chen, Wu, Yoon and You (2001)). A special numerical integration scheme is given in De and Bathe (2001), where the $\{\eta_{\underline{x}}\}$ are balls and the region of integration is the intersection of two balls. The problem of effective integration has also been addressed in Dolbow and Belytschko (1999), Griebel and Schweitzer (2002*b*), Schweitzer (200x) and Strouboulis *et al.* (2001*a*, 2001*b*). The numerical integration poses additional problems in the GFEM when singular functions are included in the local approximating space $V_{\underline{x}}$. Standard integration schemes in this situation yield poor accuracy. This problem in the GFEM was handled in Strouboulis *et al.* (2001*a*) by using adaptive numerical integration. Because of this sensitivity to numerical integration, the use of adaptive integration is preferred in GFEMs.

Thus we see that an accurate and effective numerical integration scheme to approximate the elements of the stiffness matrix is essential for the success of meshless methods. We will remark further on this issue in the next subsection. We note that numerical integration and construction of stiffness matrices in these methods are parallelizable.

Solution of the linear system

The exact stiffness matrix (without numerical integration) obtained from a meshless method could be positive definite with a large condition number. This is caused by using shape functions with large overlap between their supports, which makes the shape functions 'almost' linearly dependent. Moreover, the exact stiffness matrix obtained from the GFEM could be positive semi-definite, as shown in Strouboulis *et al.* (2001*a*). But the underlying linear system obtained from the GFEM is always consistent, *i.e.*, the linear system has non-unique solutions. The lack of unique solvability of the linear system does not imply that the GFEM produces non-unique solutions. In fact, if the vector $\{c_{\underline{x},j}\}_{1 \le j \le n_{\underline{x}}}$, with dim $V_{\underline{x}} = n_{\underline{x}}$, is a solution of this linear system, then the solution

$$u_h = \sum_{\underline{x}} \sum_{j=1}^{n_{\underline{x}}} \phi_{\underline{x}}\, c_{\underline{x},j}\, \psi_{\underline{x}}^j,$$

obtained from the GFEM, where $\psi_{\underline{x}}^j$ is a basis of $V_{\underline{x}}$, is unique.

We have already mentioned the importance of numerical integration in evaluating the elements of the stiffness matrix obtained from a meshless method. We further note that the elements of the load vector is also evaluated by numerical integration. To obtain a consistent linear system (after the use of numerical integration), the numerical integration scheme applied to compute an element of the load vector should be same as the scheme used to compute the corresponding row of the stiffness matrix.

To find the solution of the linear system obtained from a meshless method (or from the GFEM), one can use a special direct solver based on elimination or an iterative solver. Strouboulis *et al.* (2001*a*) used direct solvers, *e.g.*, subroutines MA27 and MA47 of the Harwell Subroutine Library, to solve the sparse positive semi-definite linear system obtained from the GFEM. The use of these solvers was successful even when the nullity of the stiffness matrix was large. It was also shown in Strouboulis *et al.* (2001*a*) that round-off errors did not play a significant role in solving the linear system, *i.e.*, the round-off error was almost the same as when the standard finite element linear system is solved by the elimination method.

An iterative algorithm for solving such linear systems was given in Strouboulis *et al.* (2001*a*). The idea of this algorithm, which has been used in many situations, is to perturb the stiffness matrix by adding a small multiple of the identity matrix. The perturbed matrix, say P, is positive definite and any solver could be used to solve $Px = b$. Using this fact and a few iterations of a simple iterative technique, a solution of the original linear system could be obtained. We refer to Strouboulis *et al.* (2001*a*) for a complete description of the effectiveness of this iterative algorithm.

We have noted before that the linear system obtained from the meshless method is consistent even if the stiffness matrix is positive semi-definite. In this situation, a solver based on conjugate gradient methods can also be used. The convergence in this situation is similar to the convergence of conjugate gradient methods when applied to solve the linear system obtained from the standard finite element method. The multigrid method is not directly applicable to the linear system when the stiffness matrix is positive semi-definite, since the eigenfunction corresponding to the zero eigenvalue of the stiffness matrix is global and oscillatory. The same is also true when the stiffness matrix is 'almost' singular. However, a special version of the multigrid method was proposed in Schweitzer (200x) and Griebel and Schweitzer (2002*c*), when the underlying partition of unity is reproducing of order $k = 0$. For another multigrid method, see Xu and Zikatanov (2002).

A posteriori error estimation, adaptivity and programming

The rigorous theory of *a posteriori* error estimation originated in Babuška and Melenk (1997) and other estimates, based on various averaging, were also used. These estimators can be used as error indicators for adaptivity purposes. For adaptive approaches in meshless methods, we refer readers to Schweitzer (200x) and Belytschko, Liu and Singer (1998*b*).

We finally mention that programming the meshless method is an important issue and it requires specific concepts. For this aspect of the meshless method, we refer to Dolbow and Belytschko (1998), Griebel and Schweitzer (2002*d*), and Strouboulis *et al.* (2001*a*, 2001*b*).

9. Examples

Meshless methods have been applied to linear and nonlinear elliptic problems, as well as to problems related to other differential equations: we refer to Li and Liu (2002). However, it is essential to characterize the types of problems where this method is, or could be, superior to standard methods (Belytschko, Gerlach, Krongauz, Krysl and Dolbow 1998*a*).

In this paper, we address only the application of the meshless method on a class of linear, elliptic problems. As stated in the Introduction, one of the main advantages of meshless methods is that it avoids mesh generation. This is essential when the domain is complex. Another advantage of this method is that it allows the use of various 'special' local shape functions to improve the accuracy.

The generalized finite element method (GFEM) was discussed in detail by Strouboulis *et al.* (2001*b*), and it was shown that the method is effective. Three types of meshes with successive refinements were used in that paper, and we present one of these meshes in Figure 9.1. This is a simple finite element mesh and it *does not* reflect the geometry of the underlying domain. Then, using the linear finite elements as a partition of unity, and special functions for local approximation, an improvement in the rate of convergence was achieved. Detailed numerical data, with comments on various aspects of the method, for instance numerical integration, *etc.*, were presented in

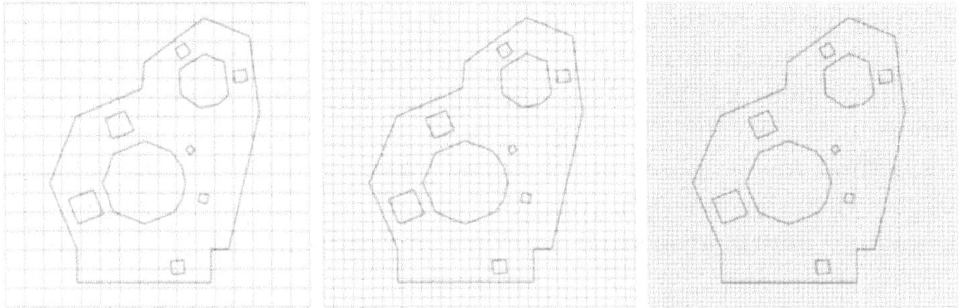

Figure 9.1. A mesh used in Strouboulis *et al.* (2001*b*) for the construction of a partition of unity in the context of the GFEM to solve a problem posed on the domain.

Strouboulis *et al.* (2001*b*). We note, however, that although the domain considered in this example (*i.e.*, the domain in Figure 9.1) was simple, and classical finite element methods with mesh refinement or an adaptive procedure could have been used, the analysis and data presented in Strouboulis *et al.* (2001*b*) clearly show the scope and potential of the GFEM.

As mentioned above, the power of the GFEM lies in handling problems where the underlying domain has complex geometry. Three types of complex domain, shown in Figure 9.2, were analysed by the GFEM in Strouboulis *et al.* (2001*b*). Another complex domain with fibres, analysed in the same paper, is shown in Figure 9.3, where the fibre distribution was taken from Babuška, Andersson, Smith and Levin (1999). To construct finite element meshes for these domains is very complex and nearly impossible. 'Handbook' problems that characterize the local behaviour of the approximated solution (*e.g.*, in the neighbourhood of a crack, fibres, *etc.*) were used to construct special shape functions for these problems.

The GFEM has an advantage in dealing with problems with singularities (in the neighbourhood of geometric edges) in three dimensions. When the basic finite element tetrahedral mesh is used in such problems, it is well known that classical edge refinement is cumbersome. This problem was handled using the GFEM in Duarte, Babuška and Oden (2000), where a refinement by special functions, at positions indicated by an error indicator, was performed. The GFEM was also used to handle difficulties stemming from orthotropic problems in Duarte and Babuška (2002).

There are other types of problem where the GFEM is quite effective. They include multisite problems, where many crack configurations are present, and crack propagation problems, where the geometry of the domain changes. The GFEM could be used in such problems by considering local approximating spaces consisting of functions that are discontinuous over the cracks,

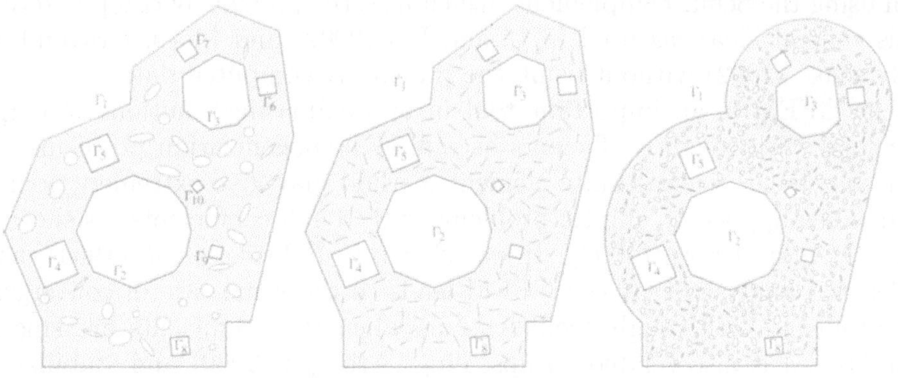

Figure 9.2. The three types of domain analysed by the GFEM in Strouboulis *et al.* (2001*b*).

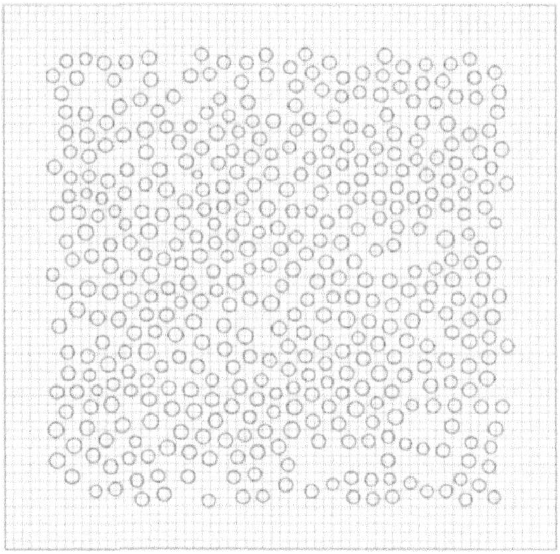

Figure 9.3. The problem of fibre composite type analysed in Strouboulis *et al.* (2001*b*).

and including a singular function with respect to the tip of the crack into the local approximating space. Then the propagation of the crack is computed (using stress intensity factors); the old singular function in the approximating space is replaced by a new singular function, new discontinuous functions are added in the same space, and the process of computing the propagation of crack and changing the local approximation spaces is repeated. In this process, the matrix of the underlying linear system at a particular step can be obtained by augmenting the matrix corresponding to the previous step with new rows and columns, and it is possible to solve the new linear system using the Schur complement, which uses the previously computed data. This general idea was used in Li and Liu (2002) and Moes, Gravouil and Belytschko (2002) without using the previously computed data.

The GFEM is an important tool in approximating solutions of elliptic problems with rough coefficients as well as homogenization problems. We mention that the usual finite element method may give extremely poor results when applied to elliptic problems with rough coefficients, as shown in Babuška and Osborn (2000). It was shown in Babuška *et al.* (1994), using a detailed analysis, that the GFEM leads to the same rate of convergence for problems with rough coefficients as when the coefficients are smooth. The GFEM is also related to upscaling (Arbogast 2000) and stabilization (Hughes 1995).

We emphasize that in this paper we have considered only a small (but important) family of problems. We have shown that the use of meshless methods, particularly the use of the GFEM, on these problems is advantageous in comparison to the standard finite element method. Of course, there are other types of problems, especially nonlinear problems, which we have not addressed in this paper.

10. Future challenges

We have addressed some mathematical problems concerning meshless methods. This method is, in fact, a family of approaches that sometimes differ in implementational details, and are referred to by a variety of names. Particular forms of meshless methods are used for solving important engineering problems in solid and fluid mechanics – linear and nonlinear, stationary and time-dependent problems, with fixed and changing boundary conditions, for differential and integral equations of various types. There is a rapidly growing literature on the subject, which is mainly directed towards engineering and focuses on particular problems, whose mathematical and theoretical aspects are not completely understood. For a description of the state of the art we refer to Atluri and Shen (2002), Babuška *et al.* (2002*a*), Griebel and Schweitzer, eds (2002*a*), Li and Liu (2002), G. R. Liu (2002) and Zhang, Liu, Li, Qian and Hao (2002).

We presented basic approximation results in L_2-based spaces, and discussed their use in the approximate solution of linear elliptic boundary value problems. The approximation theory we developed is applicable to virtually all variationally formulated problems, provided that the stability of the variational method is guaranteed (the inf-sup condition, sometimes called the BB condition, is satisfied). In contrast to coercive problems, proving stability for non-coercive problems is a delicate issue.

Meshless methods, in particular the GFEM, permit the use of non-polynomial and non-smooth shape functions with a certain special character. This feature was successfully utilized in solving partial differential equations with rough coefficients and, more generally, in problems with microstructures. The GFEM can be directly related to upscaling and to stabilization. Special shape functions related to the oscillatory behaviour of the solutions, as in the case of the Helmholtz equation, have been successfully used. The GFEM was used in problems in which the discontinuities propagate, as in linear and nonlinear crack problems. Meshless methods and GFEM can be used for solving higher-order differential equations because it is possible to construct shape functions with high regularity. There are, of course, many remaining issues to be addressed: we mention, in particular, the challenging problems of adaptive selection of shape functions and of proving *a posteriori* error estimates.

Meshless methods and the GFEM are generalizations of the classical FEM, in its h, p and hp versions. Hence all the theoretical and implementational problems for the h, p and hp versions of the FEM have analogues for meshless methods; and there are additional problems, having special features, with meshless methods.

Today, many versions of meshless methods, based on various discretization principles, for example, variational principles, collocation, *etc.*, are used in an *ad hoc* way, with emphasis on constructive results. Thus it is important to create an effective framework in which to study these methods, and to assess their effectiveness.

Acknowledgements

The authors would like to thank Elsevier Science for allowing the use of Figures 9.1, 9.2 and 9.3 in this paper. They were published in the journal *Computer Methods in Applied Mechanics and Engineering.* The authors would also like to thank Professor Weimin Han for reading and commenting on a preliminary version of this paper.

REFERENCES

T. Arbogast (2000), Numerical subgrid upscaling of two-phase flow in porous media, in *Numerical Treatment of Multiphase Flows in Porous Media* (Z. Chen, ed.), Vol. 552 of *Lecture Notes in Physics*, Springer, pp. 35–49.

M. G. Armentano (2002), 'Error estimates in Sobolev spaces for moving least square approximation', *SIAM J. Numer. Anal.* **39**, 38–51.

M. G. Armentano and R. G. Duran (2001), 'Error estimates for moving least square approximation', *Appl. Numer. Math.* **37**, 397–416.

S. N. Atluri and S. Shen (2002), *The Meshless Local Petrov–Galerkin Method*, Tech Science Press.

I. Babuška (1970), 'Approximation by hill functions', *Comment Math. Univ. Carolinae* **11**, 787–811.

I. Babuška (1971), The finite element method for elliptic differential equations, in *Numerical Solution of Partial Differential Equations II, SYNSPADE 1970* (B. Hubbard, ed.), Academic Press, London, pp. 69–106.

I. Babuška (1972), 'Approximation by hill functions II', *Comment Math. Univ. Carolinae* **13**, 1–22.

I. Babuška (1973*a*), 'The finite element method with Lagrange multiplier', *Numer. Math.* **20**, 179–192.

I. Babuška (1973*b*), 'The finite element method with penalty', *Math. Comp.* **27**, 221–228.

I. Babuška (1994), Courant element: Before and after, in *Finite Element Methods: Fifty Years of the Courant Element*, Vol. 164 of *Lecture Notes in Pure and Applied Mathematics*, Marcel Dekker, pp. 37–51.

I. Babuška and A. K. Aziz (1972), Survey lectures on the mathematical foundations of the finite element method, in *Mathematical Foundations of the Finite Element Method with Applications to Partial Differential Equations* (A. K. Aziz, ed.), Academic Press, pp. 3–345.

I. Babuška and J. M. Melenk (1997), 'The partition of unity finite element method', *Internat. J. Numer. Meth. Engr.* **40**, 727–758.

I. Babuška and J. Osborn (2000), 'Can a finite element method perform arbitrarily badly?', *Math. Comp.* **69**, 443–462.

I. Babuška and T. Strouboulis (2001), *The Finite Element Method and its Reliability*, Clarendon Press, Oxford.

I. Babuška, B. Andersson, P. J. Smith and K. Levin (1999), 'Damage analysis of fiber composites, Part 1: Statistical analysis on fiber scale', *Comput. Meth. Appl. Mech. Engr.* **172**, 27–77.

I. Babuška, U. Banerjee and J. Osborn (2001), On principles for the selection of shape functions for the generalized finite element method, Technical Report 01-16, TICAM, University of Texas at Austin.

I. Babuška, U. Banerjee and J. Osborn (2002*a*), Meshless and generalized finite element methods: A survey of major results, in *Meshfree Methods for Partial Differential Equations* (M. Griebel and M. A. Schweitzer, eds), Springer.

I. Babuška, U. Banerjee and J. Osborn (2002*b*), 'On principles for the selection of shape functions for the generalized finite element method', *Comput. Methods Appl. Mech. Engr.* **191**, 5595–5629.

I. Babuška, U. Banerjee and J. Osborn (200x), 'On the approximability and the selection of particle shape functions'. Submitted to *Numer. Math.*

I. Babuška, G. Caloz and J. Osborn (1994), 'Special finite element methods for a class of second order elliptic problems with rough coefficients', *SIAM J. Numer. Anal.* **31**, 945–981.

I. Babuška, T. Strouboulis, C. S. Upadhyay and S. K. Gangaraj (1996), 'Computer-based proof of existence of superconvergence points in the finite element method: Superconvergence of the derivatives in the finite element solution of Laplace's, Poisson's, and elasticity equations', *Numerical Methods for PDEs* **12**, 347–392.

J. C. Barbara and T. J. R. Hughes (1991), 'The finite element method with Lagrange multipliers on the boundary circumventing the Babuška–Brezzi condition', *Comput. Methods Appl. Mech. Engr.* **85**, 109–128.

T. Belytschko, Y. Krongauz, D. Organ, M. Fleming and P. Krysl (1996), 'Meshless methods: An overview and recent developments', *Comput. Methods Appl. Mech. Engr.* **139**, 3–47.

T. Belytschko, C. Gerlach, Y. Krongauz, P. Krysl and J. Dolbow (1998*a*), Why meshfree methods, in *Proc. IV World Congress on Computational Mechanics, Buenos Aires, Argentina.*

T. Belytschko, W. K. Liu and M. Singer (1998*b*), On adaptivity and error criteria for meshfree methods, in *Advances in Adaptive Computational Methods in Mechanics* (P. Ladeveze and J. T. Oden, eds), pp. 217–228.

T. Belytschko, Y. Y. Lu and L. Gu (1994), 'Element-free Galerkin methods', *Internat. J. Numer. Meth. Engr.* **37**, 229–256.

J. H. Bramble and S. R. Hilbert (1970), 'Estimation of linear functionals on Sobolev spaces with application to Fourier transforms and spline interpolation', *SIAM J. Numer. Anal.* **7**, 113–124.

J. H. Bramble and S. R. Hilbert (1971), 'Bounds on a class of linear functionals with applications to Hermite interpolation', *Numer. Math.* **16**, 362–369.

M. D. Buhmann (2000), Radial basis functions, in *Acta Numerica*, Vol. 9, Cambridge University Press, pp. 1–38.

J. S. Chen, C. T. Wu, S. Yoon and Y. You (2001), 'A stabilized conforming nodal integration for Galerkin meshfree methods', *Internat. J. Numer. Meth. Engr.* **50**, 435–466.

P. G. Ciarlet (1980), *Finite Element Methods for Elliptic Problems*, North-Holland.

C. Daux, N. Moes, J. Dolbow, N. Sukumar and T. Belytschko (2000), 'Arbitrary branched and intersecting cracks with eXtended finite element method', *Internat. J. Numer. Meth. Engr.* **48**, 1741–1760.

S. De and K. J. Bathe (2001), 'The method of finite spheres with improved numerical integration', *Computers and Structures* **79**, 2183–2196.

P. Dierckx (1995), *Curve and Surface Fitting with Splines*, Clarendon Press, Oxford.

J. Dolbow and T. Belytschko (1998), 'An introduction to programming the meshless element free Galerkin method', *Arch. Comput. Methods Engr.* **5**, 207–241.

J. Dolbow and T. Belytschko (1999), 'Numerical integration of the Galerkin weak form in meshfree methods', *Comput. Mech.* **23**, 219–230.

C. A. M. Duarte (1995), A review of some meshless methods to solve partial differential equations, Technical Report 95-06, TICAM, University of Texas at Austin.

C. A. M. Duarte and I. Babuška (2002), 'Mesh independent *p*-orthotropic enrichment using generalized finite element method', *Internat. J. Numer. Meth. Engr.* **55**, 1477–1492.

C. A. M. Duarte, I. Babuška and J. T. Oden (2000), 'Generalized finite element method for three dimensional structural problems', *Computers and Structures* **77**, 219–232.

C. Franke and R. Schaback (1998), 'Convergence order estimates of meshless collocation methods using radial basis functions', *Adv. Comp. Math.* **8**, 381–399.

R. Franke (1978), Smooth surface approximation by a local method of interpolation at scattered points, Technical Report NPS-53-78-002, Naval Postgraduate School.

R. Franke (1979), A critical comparison of some methods for interpolation of scattered data, Technical Report NPS-53-79-003, Naval Postgraduate School. Available from NTIS AD-A081 688/4.

R. A. Gingold and J. J. Monaghan (1977), 'Smoothed particle hydrodynamics: Theory and application to non-spherical stars', *Mon. Not. R. Astr. Soc.* **181**, 375–389.

R. A. Gingold and J. J. Monaghan (1982), 'Kernel estimates as a basis for general particle methods in hydrodynamics', *J. Comp. Phys.* **46**, 429–453.

W. J. Gordon and K. Wixon (1978), 'Shepard's method of metric interpolation to bivariate and multivariate data', *Math. Comp.* **32**, 253–264.

M. Griebel and M. A. Schweitzer, eds (2002*a*), *Meshfree Methods for Partial Differential Equations*, Vol. 26 of *Lecture Notes in Computational Science and Engineering*, Springer.

M. Griebel and M. A. Schweitzer (2002*b*), 'A particle-partition of unity method, Part II: Efficient cover construction and reliable integration', *SIAM J. Sci. Comp.* **23**, 1655–1682.

M. Griebel and M. A. Schweitzer (2002*c*), 'A particle-partition of unity method, Part III: A multilevel solver', *SIAM J. Sci. Comp.* **24**, 377–409.

M. Griebel and M. A. Schweitzer (2002*d*), A particle-partition of unity method, Part IV: Parallelization, in *Meshfree Methods for Partial Differential Equations* (M. Griebel and M. A. Schweitzer, eds), Springer.

M. Griebel and M. A. Schweitzer (2002*e*), A particle-partition of unity method, Part V: Boundary conditions, in *Geometric Analysis and Nonlinear Partial Differential Equations* (S. Hildebrand and H. Karcher, eds), Springer.

D. Hagen (1996), 'Element-free Galerkin methods in combination with finite element approaches', *Comput. Methods Appl. Mech. Engr.* **139**, 237–262.

W. Han and X. Meng (2001), 'Error analysis of the reproducing kernel particle method', *Comput. Methods Appl. Mech. Engr.* **190**, 6157–6181.

G. H. Hardy, J. E. Littlewood and G. Polya (1952), *Inequalities*, Cambridge University Press.

T. J. R. Hughes (1995), 'Multiscale phenomena, Green's functions, the Dirichlet to Neumann formulation, subgrid scale models, bubbles and the origins of stabilized methods', *Comput. Methods Appl. Mech. Engr.* **127**, 387–401.

I. V. Kantorovich and V. I. Krylov (1958), *Approximate Methods of Higher Analysis*, Interscience Publishers.

Y. Krongauz and T. Belytschko (1996), 'Enforcement of essential boundary conditions in meshless approximation using finite elements', *Comput. Methods Appl. Mech. Engr.* **131**, 133–145.

O. Laghrouche and P. Bettess (2000), Solving short wave problems using special finite elements: Towards an adaptive approach, in *Mathematics of Finite Elements and Applications X* (J. Whiteman, ed.), Elsevier, pp. 181–195.

P. Lancaster and K. Salkauskas (1981), 'Surfaces generated by moving least squares method', *Math. Comp.* **37**, 141–158.

P. Lancaster and K. Salkauskas (1986), *Curve and Surface Fitting: An Introduction*, Academic Press, London.

S. Li and W. K. Liu (1996), 'Moving least squares reproducing kernel particle method, Part II: Fourier analysis', *Comput. Methods Appl. Mech. Engr.* **139**, 159–194.

S. Li and W. K. Liu (2002), 'Meshfree and particle methods and their application', *Applied Mechanics Review* **55**, 1–34.

G. R. Liu (2002), *Meshless Methods*, CRC Press, Boca Raton.

W. K. Liu, S. Jun and Y. F. Zhang (1995), 'Reproducing kernel particle methods', *Int. J. Numer. Meth. Fluids* **20**, 1081–1106.

W. K. Liu, S. Li and T. Belytschko (1997), 'Moving least square reproducing kernel particle method: Methodology and convergence', *Comput. Methods Appl. Mech. Engr.* **143**, 422–453.

Y. Y. Lu, T. Belytschko and L. Gu (1994), 'A new implementation of the element free Galerkin method', *Comput. Methods Appl. Mech. Engr.* **113**, 397–414.

A. M. Matache, I. Babuška and C. Schwab (2000), 'Generalized *p*-FEM in homogenization', *Numer. Math.* **86**, 319–375.

D. H. McLain (1974), 'Drawing contours from arbitrary data points', *Comp. J.* **17**, 89–97.

J. M. Melenk and I. Babuška (1996), 'The partition of unity finite element method: Theory and application', *Comput. Methods Appl. Mech. Engr.* **139**, 289–314.

S. G. Mikhlin (1971), *Numerical Performance of Variational Methods*, Nordhoff.

A. Moes, J. Dolbow and T. Belytschko (1999), 'A finite element method for crack growth without remeshing', *Internat. J. Numer. Engr.* **46**, 131–150.

N. Moes, A. Gravouil and T. Belytschko (2002), 'Non-planar 3D crack growth by extended finite element and level sets, Part 1: Mechanical model', *Internat. J. Numer. Meth. Engr.* **53**, 2549–2568.

J. J. Monaghan (1982), 'Why particle methods work', *SIAM J. Sci. Stat. Comp.* **3**, 422–433.

J. J. Monaghan (1988), 'An introduction to SPH', *Comp. Phys. Comm.* **48**, 89–96.

Y. K. Mukherjee and S. Mukherjee (1997), 'On boundary conditions in the element-free Galerkin method', *Computational Mechanics* **19**, 264–270.

K. Nanbu (1980), 'Direct simulation scheme derived from the Boltzmann equation', *J. Phys. Soc. Japan* **49**, 20–40.

H. Neunzert and J. Struckmeier (1995), Particle methods for the Boltzmann equation, in *Acta Numerica*, Vol. 4, Cambridge University Press, pp. 417–457.

J. Nitsche (1970–1971), 'Über ein Variationsprinzip zur Lösung von Dirichlet-Problemen bei Verwendung von Teilräumen, die keinen Randbedingungen unterworfen sind', *Abh. Math. Univ. Hamburg* **36**, 9–15.

J. Nitsche and A. H. Schatz (1974), 'Interior estimates for Ritz–Galerkin methods', *Math. Comp.* **28**, 937–958.

J. T. Oden, C. A. Duarte and O. C. Zienkiewicz (1998), 'A new cloud based hp finite element method', *Comput. Methods Appl. Mech. Engr.* **153**, 117–126.

J. Pitkäranta (1979), 'Boundary subspaces for finite element method with Lagrange multipliers', *Numer. Math.* **33**, 273–289.

J. Pitkäranta (1980), 'Local stability conditions for the Babuska method of Lagrange multipliers', *Math. Comp.* **35**, 1113–1129.

M. J. D. Powell (1992), The theory of radial basis function approximation in 1990, in *Advances in Numerical Analysis, II* (W. Light, ed.), Clarendon Press, pp. 105–210.

W. Rudin (1991), *Functional Analysis*, McGraw-Hill.

A. H. Schatz and L. B. Wahlbin (1995), 'Maximum norm estimates for finite element methods, Part II', *Math. Comp.* **64**, 907–928.

M. A. Schweitzer (200x), *A Parallel Multilevel Partition of Unity Method.* Book manuscript.

D. Shepard (1968), A two-dimensional interpolation function for irregularly spaced points, in *Proc. ACM Nat. Conf.*, pp. 517–524.

E. M. Stein (1970), *Singular Integrals and Differentiability Properties of Functions*, Princeton University Press.

R. Stenberg (1995), 'On some techniques for approximating boundary conditions in the finite element method', *J. Comp. Appl. Math.* **63**, 139–148.

G. Strang (1971), The finite element method and approximation theory, in *Numerical Solution of Partial Differential Equations II, SYNSPADE 1970* (B. Hubbard, ed.), Academic Press, London, pp. 547–584.

G. Strang and G. Fix (1973), A Fourier analysis of finite element variational method, in *Constructive Aspects of Functional Analysis*, Edizioni Cremonese, pp. 795–840.

T. Strouboulis, I. Babuška and K. Copps (2001*a*), 'The design and analysis of the generalized finite element method', *Comput. Methods Appl. Mech. Engr.* **181**, 43–69.

T. Strouboulis, K. Copps and I. Babuška (2001*b*), 'The generalized finite element method', *Comput. Methods Appl. Mech. Engr.* **190**, 4081–4193.

T. Strouboulis, L. Zhang and I. Babuška (200x), Generalized finite element method using mesh-based handbooks: Applications in domains with many voids. Manuscript.

N. Sukumar, N. Moes, B. Moran and T. Belytschko (2000), 'Extended finite element method for three dimensional crack modelling', *Internat. J. Numer. Meth. Engr.* **48**, 1549–1570.

B. Szabo, I. Babuška and R. Actis (1998), 'A simpler way to handle complex analysis', *Machine Design.*

H. Tada, P. C. Paris and G. R. Irwin (1973), *The Stress Analysis of Cracks Handbook*, Del Research Corporation, Hellertown, PA.

R. L. Taylor, O. C. Zienkiewicz and E. Onate (1998), 'A hierarchical finite element method based on partition of unity', *Comput. Methods Appl. Mech. Engr.* **152**, 73–84.

R. S. Varga (1971), *Functional Analysis and Approximation Theory in Numerical Analysis*, Regional Conference Series in Applied Mathematics, SIAM, Philadelphia, PA.

I. N. Vekua (1967), *New Methods for Solving Elliptic Equations*, North-Holland.

L. B. Wahlbin (1995), *Superconvergence in Galerkin Finite Element Methods*, Vol. 1605 of *Lecture Notes in Mathematics*, Springer.

G. N. Wells and L. J. Sluis (2001), 'A new method for modelling cohesive finite elements', *Internat. J. Numer. Meth. Engr.* **50**, 2667–2682.

G. N. Wells, R. de Borst and L. J. Sluis (2002), 'A consistent geometrical nonlinear approach for delamination', *Internat. J. Numer. Meth. Engr.* **54**, 1333–1355.

J. Xu and L. T. Zikatanov (2002), On multigrid methods for generalized finite element methods, in *Meshfree Methods for Partial Differential Equations* (M. Griebel and M. A. Schweitzer, eds), Springer, pp. 401–418.

L. T. Zhang, W. K. Liu, S. F. Li, D. Qian and S. Hao (2002), Survey of multiscale meshfree particle methods, in *Meshfree Methods for Partial Differential Equations* (M. Griebel and M. A. Schweitzer, eds), Springer, pp. 441–458.

T. Zhu and S. N. Atluri (1998), 'A modified collocation method and a penalty formulation for enforcing essential boundary conditions', *Computational Mechanics* **21**, 165–178.

Acta Numerica (2003), pp. 127–180
DOI: 10.1017/S0962492902000107

Continuous dependence and error estimation for viscosity methods

Bernardo Cockburn
School of Mathematics,
University of Minnesota,
Minneapolis, Minnesota 55455, USA
E-mail: cockburn@math.umn.edu

In this paper, we review some ideas on continuous dependence results for the entropy solution of hyperbolic scalar conservation laws. They lead to a complete $L^\infty(L^1)$-approximation theory with which error estimates for numerical methods for this type of equation can be obtained. The approach we consider consists in obtaining continuous dependence results for the solutions of parabolic conservation laws and deducing from them the corresponding results for the entropy solution. This is a natural approach, as the entropy solution is nothing but the limit of solutions of parabolic scalar conservation laws as the viscosity coefficient goes to zero.

CONTENTS

1. Introduction

In this paper, we review some of the ideas on continuous dependence and error estimation for numerical methods for the Cauchy problem for the scalar hyperbolic conservation law

$$u_t + \nabla \cdot \mathbf{f}(u) = r, \qquad \text{in } \mathbb{R}^d \times (0,T),$$
$$u(t=0) = u_0, \qquad \text{on } \mathbb{R}^d,$$

that have been entertained during the last three decades.

It is well known that continuous dependence results are essential in order to have a mathematically sound approximation theory upon which the devising and analysis of numerical methods can be based. To illustrate this point, let us assume that we are interested in finding an approximation v, given by the numerical scheme $\mathbb{L}_h(v) = 0$, to u such that

$$\|u - v\| \leq \mathsf{tol},$$

where $\|\cdot\|$ is some norm and tol is a given positive parameter representing our tolerance. Let us show how to use a continuous dependence result for the solution u, with respect to the right-hand side r, to study this problem. If u_i is the solution of the above Cauchy problem with $r = r_i$, $i = 1, 2$, the estimate

$$\|u_1 - u_2\| \leq \Phi(r_1 - r_2)$$

would allow us to compare the solution u with any other function v by simply setting $u_1 = u$, $u_2 = v$ and

$$r_2 = v_t + \nabla \cdot \mathbf{f}(v).$$

Indeed, in this case, we would obtain

$$\|u - v\| \leq \Phi(-R(v)),$$

where

$$R(v) = v_t + \nabla \cdot \mathbf{f}(v) - r$$

is nothing but the residual of v. Thus, to achieve our goal, it is enough to find v such that

$$\Phi(-R(v)) \leq \mathsf{tol}.$$

We see that the study of how to find the approximation v becomes the study of how to minimize the nonlinear functional $\Phi(-R(v))$ in an optimal way with the restriction that the function v is determined by the numerical scheme $\mathbb{L}_h(v) = 0$. This is in fact one of the most important problems in modern computational partial differential equations.

Unfortunately, the main difficulty in dealing with the Cauchy problem for the scalar hyperbolic conservation law is that its solution is not unique, as we will show in Section 2. Given that every physically relevant phenomenon should be modelled by a mathematically well-posed problem, we might then wonder how is it possible that scalar hyperbolic conservation laws are considered at all? The answer is that these equations are obtained from a well-posed problem by formally neglecting terms modelling effects considered to be non-dominant. A typical example occurs when, in the convection–diffusion Cauchy problem,

$$(u_\nu)_t + \nabla \cdot \mathbf{f}(u_\nu) - \nu\,\Delta u_\nu = r, \qquad \text{in } \mathbb{R}^d \times (0,T),$$
$$u_\nu(t=0) = u_0, \qquad \text{on } \mathbb{R}^d,$$

the term $\nu\,\Delta u_\nu$ modelling the viscosity effects is neglected because diffusion is considered to be unimportant. In Sections 2 and 3, we elaborate on this point.

Fortunately, this formal procedure can still be rendered meaningful if we obtain continuous dependence results for the well-posed convection–diffusion problem which do not break down when we let the viscosity coefficient tend to zero: see Sections 4 and 5. Indeed, armed with such results, it is possible to achieve the following.

(i) *Prove* the existence and uniqueness of the function $u = \lim_{\nu \downarrow 0} u_\nu$ which is thus considered to be the physically relevant solution of the scalar hyperbolic conservation law: see Section 5.

(ii) Prove that these continuous dependence results *also* hold for this solution: see Section 5.

(iii) *Characterize* this solution: see Section 7.

(iv) Construct a *complete approximation theory* for this type of solution: see Sections 6 and 8.

In Section 9 we discuss the evolution of some ideas leading to this approach, and in Section 10 we end with some concluding remarks. In the Appendix, we sketch the proofs of the main results.

The above-mentioned approach has recently been proposed for Hamilton–Jacobi equations and might also be used to study strongly degenerate scalar equations and viscosity solutions of nonlinear degenerate parabolic equations. To apply such an approach to hyperbolic systems is certainly a Herculean task, especially in view of the now classical result by Rauch (1986), which states that "for most non-scalar systems of conservation laws in dimension greater than one, one does not have BV estimates of the form

$$\|\nabla u(t)\|_{L^1} \le F(\|\nabla u(0)\|_{L^1})."$$

Here F is a continuous function such that $F(0) = 0$ and F is Lipschitz at 0. On the other hand, in some cases, we might be closer to achieving that goal than would have been expected only a few years ago, given the recent developments in the theory of hyperbolic systems: see Bianchini and Bressan (2001).

Let us mention that the reader interested in the theoretical aspects of hyperbolic problems should consult the books by Bressan (2000), Dafermos (2000), Lax (1972), Liu (1997), Serre (1999, 2000) and Whitham (1974). The reader interested in the numerics for this kind of problem should consult Barth and Deconink (1999), Cockburn and Shu (2001), Cockburn, Karniadakis and Shu (2000), Eymard, Gallouët and Herbin (2000), Godlewski and Raviart (1996), Holden and Risebro (2002), Johnson (1998), Kröner (1997), Kröner, Ohlberger and Rohde (1999), LeVeque (1990), Lucier (2003), Tadmor (1998), Shu (1998), and Toro (1997).

An elegant theory of continuous dependence in a negative-order norm for the physically relevant solutions of the conservation law

$$u_t + (f(u))_x = 0,$$

when f is strictly convex, is reviewed in Tadmor (1998); here we are concerned with general nonlinearities $\mathbf{f}$ in a multi-dimensional setting. Finally, let us point out that, although parts of the material presented in this review can be found in some of the above-mentioned references, in particular in Lucier (2003), the approach proposed here and many of its main results cannot be found in any of them.

2. The main difficulty: the loss of well-posedness

In many instances, nonlinear hyperbolic problems arise from a well-posed problem by neglecting the terms that capture physical phenomena considered to be *non-dominant*; unfortunately, this renders the problem ill-posed. This loss of well-posedness is the main difficulty we face when dealing with nonlinear hyperbolic problems.

In this section, we illustrate this phenomenon in the setting of a simple traffic flow model. First, we display a well-posed parabolic model; the 'driver's awareness of the conditions ahead' giving a parabolic character to the model. We then argue that, if the driver's awareness is negligible, a scalar nonlinear hyperbolic conservation law is obtained, which gives rise to an ill-posed problem. We show that this happens because, although the neglected physical phenomenon can be correctly considered to be unimportant in most parts of the domain, *it is still crucial in small parts of the domain*, namely, near a strong variation of the density of cars. Thus, the removal of this essential physical information has, not surprisingly, disastrous consequences.

2.1. *The model*

If ρ represents the density of cars on a highway and v represents the flow velocity, the conservation of mass is

$$\rho_t + (\rho\, v)_x = 0.$$

Following Whitham (1974), we take the flow velocity to be the following function of the density of cars ρ and of its gradient ρ_x:

$$\rho\, v = f(\rho) - \nu\, \rho_x.$$

It is reasonable to assume that the so-called density flow $f(\rho)$ is of the form $\rho\, V(\rho)$, where $\rho \mapsto V(\rho)$ is a decreasing mapping which, for a given density, say $\rho^\star$, is equal to zero: this corresponds to the situation in which the cars are bumper to bumper. Maybe the simplest function V satisfying these properties is

$$V(\rho) = v_{\max}\left(1 - \frac{\rho}{\rho^\star}\right),$$

where $v_{\max}$ represents the maximum velocity. The term $\nu\, \rho_x$ models the 'awareness of conditions ahead', since, when we perceive a high increase in the density of cars ahead, we try to decelerate to avoid a potentially dangerous situation. With this choice of flow velocity, our conservation law becomes the parabolic equation

$$\rho_t + (f(\rho))_x - \nu\, \rho_{xx} = 0,$$

which, after the change of variables $t := t/T$, $x := x/L$, where $T = L/v_{\max}$, reads

$$\phi_t + (f(\phi))_x - \epsilon\, \phi_{xx} = 0, \tag{2.1}$$

where $\phi = \rho/\rho^\star$ and

$$f(\phi) := \phi\,(1-\phi), \qquad \epsilon = \frac{\nu}{L\, v_{\max}}. \tag{2.2}$$

The dimensionless parameter ϵ measures the ability of the driver to react properly to a high concentration of cars at a distance L when the maximum speed is $v_{\max}$. Clearly, if ϵ is very small, it seems reasonable to formally drop the second-order term from our model and consider instead the scalar nonlinear conservation law

$$\phi_t + (f(\phi))_x = 0. \tag{2.3}$$

Next, we explore the consequences of this formal procedure.

2.2. *Travelling waves*

The simplest way to do that is to consider travelling wave solutions of (2.1) and study their limit as the parameter ϵ tends to zero.

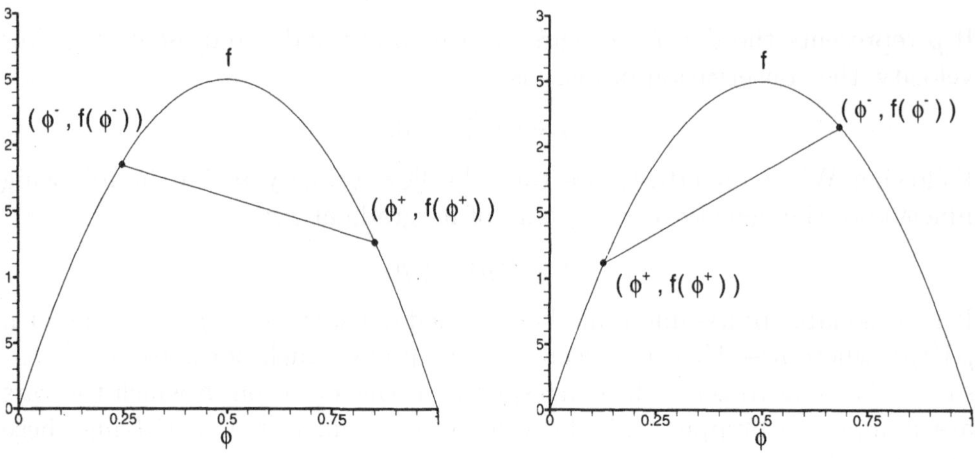

Figure 2.1. The condition for existence of a travelling wave joining ϕ^- to ϕ^+: satisfied (left) and not satisfied (right).

Travelling wave solutions of the parabolic conservation law (2.1) are solutions of the form $\phi(x,t) = \varphi((x-ct)/\epsilon))$ that satisfy the conditions $\lim_{z\to\infty^\pm}\varphi(z) = \phi^\pm$ and $\lim_{z\to\infty^\pm}\varphi'(z) = 0$. Inserting this expression for ϕ in the conservation law (2.1), and integrating once, we get

$$\varphi' = f(\varphi) - \mathcal{L}(\varphi), \tag{2.4}$$

where the function $\mathcal{L}$ and the *velocity* c of the travelling wave are given by

$$\mathcal{L}(\varphi) = f(\phi^+) + c\,(\varphi - \phi^+), \qquad c = \frac{f(\phi^+) - f(\phi^-)}{\phi^+ - \phi^-}. \tag{2.5}$$

Note that $\mathcal{L}$ is nothing but the linear function that coincides with f at $\phi^\pm$.

It is easy to see that there is a solution of the above ordinary differential equation if and only if

$$\begin{aligned} (f(\varphi) - \mathcal{L}(\varphi)) > 0 \quad &\text{for } \phi^+ > \phi > \phi^-, \\ (f(\varphi) - \mathcal{L}(\varphi)) < 0 \quad &\text{for } \phi^+ < \phi < \phi^-, \end{aligned} \tag{2.6}$$

and if the conditions at infinity are satisfied, that is, if

$$\lim_{\varphi\to\phi^\pm} \int_{\varphi(0)}^{\varphi} \frac{\mathrm{d}\theta}{f(\theta) - \mathcal{L}(\theta)} = \pm\infty. \tag{2.7}$$

Note that condition (2.6) states that the graph of f on the interval (ϕ^-, ϕ^+) (resp., (ϕ^+, ϕ^-)) lies *strictly* above (resp., below) the straight line joining the points $(\phi^\pm, f(\phi^\pm))$. Thus, for the concave nonlinearity f given by (2.2), it becomes simply $\phi^- < \phi^+$; see Figure 2.1.

We summarize these findings in the following result.

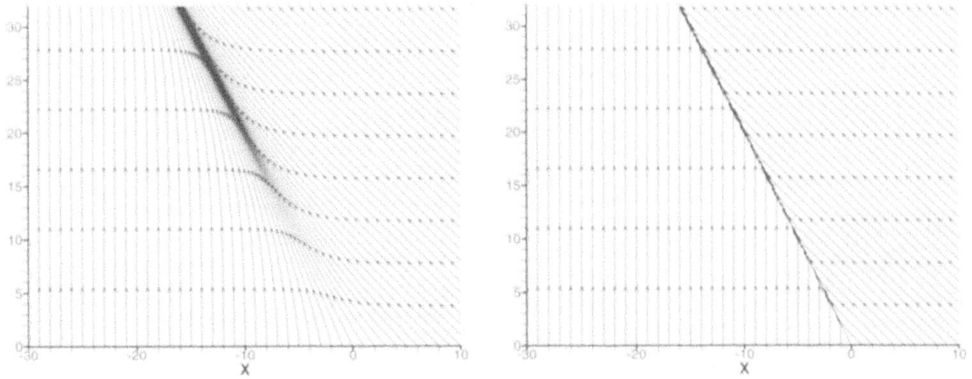

Figure 2.2. The curves $\{(x(t),t) : \dot{x} = f'(\varphi(\frac{x-ct}{\epsilon}))\}$ for $\epsilon = 1$ (left) and $\epsilon = 0.1$ (right) for the case $\phi^- = 1/2$ and $\phi^+ = 1$.

Theorem 2.1. A travelling wave solution of the parabolic conservation law (2.1) exists if and only if conditions (2.6) and (2.7) are satisfied.

Next, we study what happens when we let the coefficient ϵ tend to zero. We begin by noting that the limit of the travelling wave solution is

$$\hat{\phi}(x,t) = \phi_0(x-ct), \qquad \phi_0(x) = \lim_{\epsilon \downarrow 0} \varphi(x/\epsilon) = \begin{cases} \phi^+ & \text{if } x > 0, \\ \phi^- & \text{if } x < 0. \end{cases}$$

We illustrate this passage to the limit in Figure 2.2.

To see what Cauchy problem this limit is a solution of, we proceed as follows. Multiplying the parabolic conservation law (2.1) by an arbitrary function η in $C_0^\infty(\mathbb{R} \times [0,T))$, the set of infinitely differentiable functions of compact support on $\mathbb{R} \times [0,T)$, integrating over the strip $\mathbb{R} \times [0,T)$ and integrating by parts, we obtain

$$\int_0^T \int_{\mathbb{R}} (\phi(x,t)\, \eta(x,t)_t + f(\phi(x,t))\, \eta(x,t)_x)\, \mathrm{d}x\, \mathrm{d}t + \int_{\mathbb{R}} \phi(x,0)\, \eta(x,0)\, \mathrm{d}x + \Psi = 0,$$

where Ψ is defined by

$$\Psi = \epsilon \int_0^T \int_{\mathbb{R}} \phi(x,t)\, \eta(x,t)_{xx}\, \mathrm{d}x\, \mathrm{d}t.$$

Taking $\phi(x,t) = \varphi(\frac{x-ct}{\epsilon})$ and letting ϵ tend to zero, we obtain that $\hat{\phi}$ satisfies

$$\int_0^T \int_{\mathbb{R}} \left(\hat{\phi}(x,t)\, \eta(x,t)_t + f(\hat{\phi}(x,t))\, \eta(x,t)_x \right) \mathrm{d}x\, \mathrm{d}t + \int_{\mathbb{R}} \phi_0(x)\, \eta(x,0)\, \mathrm{d}x = 0,$$

for every function η in $C_0^\infty(\mathbb{R} \times [0,T))$. Any such function $\hat{\phi}$ is said to be a

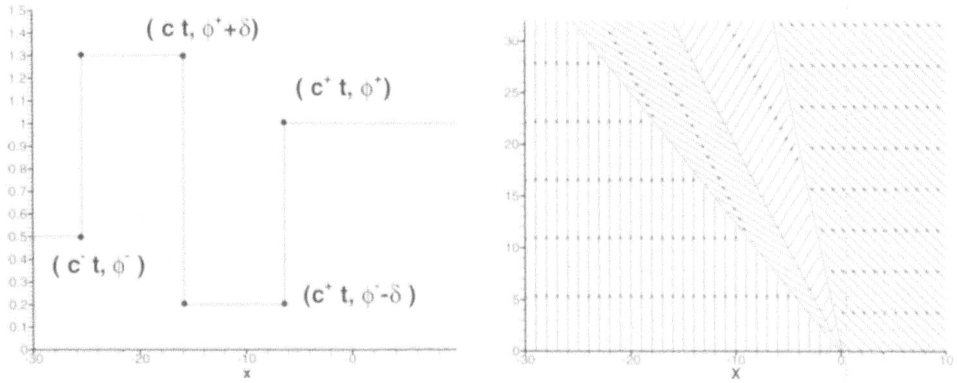

Figure 2.3. The weak solution $\phi^\delta(\cdot, t)$ (left) and the curves $(x(t), t)$ (right) for the case $\phi^- = 1/2$, $\phi^+ = 1$ and $\delta = 0.3$.

weak solution of the hyperbolic conservation law

$$\phi_t + (f(\phi))_x = 0$$

with initial data ϕ_0. This fact seems to indicate that dropping the second-order term $\epsilon\, \phi_{xx}$ from the parabolic conservation law (2.1) is actually mathematically justified. However, it is possible to construct *infinitely many* weak solutions of the hyperbolic conservation law which satisfy the same initial condition satisfied by $\hat{\phi}$.

Indeed, it is easy to verify that, for every value of the nonnegative parameter δ, the following is a weak solution of the hyperbolic conservation law (2.3) with initial condition ϕ_0:

$$\phi^\delta(x,t) = \begin{cases} \phi^- & \text{if } c^- < x/t, \\ \phi^+ + \delta & \text{if } c < x/t < c^-, \\ \phi^- - \delta & \text{if } c^+ < x/t < c, \\ \phi^+ & \text{if } x/t < c^+, \end{cases}$$

where $c = 1 - (\phi^+ + \phi^-)$, $c^- = c - \delta$, $c^+ = c + \delta$; see in Figure 2.3 the above weak solution for $\phi^- = 1/2$, $\phi^+ = 1$ and $\delta = 0.3$.

Note that for $\delta = 0$ we obtain the limit of the travelling wave solutions, $\hat{\phi}$. For $\delta > 0$, however, we obtain weak solutions with three discontinuity curves. Two of them, namely $x/t = c^-$ and $x/t = c^+$, can be obtained as the limit of travelling wave solutions of the parabolic conservation law (2.1), since the conditions for the existence of travelling waves given by Theorem 2.1 are satisfied. This not the case, however, for the discontinuity $x/t = c$. In other words, this discontinuity does not 'remember' anything about the physics captured in the modelling of the awareness of the conditions lying ahead.

Thus, as claimed at the beginning of the section, dropping the second-order term $\epsilon\,\phi_{xx}$ from the parabolic conservation law (2.1) results in the loss of the well-posedness of the Cauchy problem for the nonlinear hyperbolic equation (2.3).

3. Physically relevant solutions

It is thus clear that, since there might be infinitely many weak solutions of the scalar hyperbolic conservation law, it is important to be able to select the *physically relevant* solution. Of course, this notion can only be relative to an already existing model. We thus say that a weak solution of the hyperbolic conservation law (2.3) is physically relevant with respect to the well-posed model

$$(\phi_\omega)_t + (f(\phi_\omega))_x + L(\omega;\phi_\omega) = 0,$$

where $L(0;\cdot) = 0$, if it is the limit of solutions ϕ_ω as ω tends to zero.

The purpose of this section is to show that a weak solution of the hyperbolic conservation law (2.3) can be physically relevant with respect to two different well-posed models simultaneously, and that it can also be physically relevant with respect to one model but not another. We illustrate this phenomenon by using the travelling wave solutions introduced in the previous section.

This implies that, when dealing with hyperbolic conservation laws, we must specify the well-posed problem with respect to which the weak solutions are physically relevant. Only then we are going to be able to recover the well-posedness lost when the operator $L(\omega;\cdot)$ is formally dropped from the equation.

3.1. Two parabolic models

We begin by displaying two examples of models that produce the same physically relevant solution as our original model (2.1). The first is

$$\phi_t + (f(\phi))_x - \epsilon\,(\mathsf{A}(\phi)\phi_x)_x = 0.$$

It is easy to see that the travelling wave solutions must satisfy

$$\varphi' = \frac{f(\varphi) - \mathcal{L}(\varphi)}{\mathsf{A}(\varphi)},$$

and, if $\mathsf{A}(\phi) \geq \mathsf{A}_0 > 0$, Theorem 2.1 holds if condition (2.7) is replaced by

$$\lim_{\varphi\to\phi^\pm} \int_{\varphi(0)}^{\varphi} \frac{\mathsf{A}(\theta)\,\mathrm{d}\theta}{f(\theta) - \mathcal{L}(\theta)} = \pm\infty.$$

As a consequence, the limit of the travelling wave as $\epsilon \downarrow 0$ coincides with the limit obtained with the original model.

The second example is

$$\phi_t + (f(\phi))_x - \epsilon\,(\mathsf{B}(\phi_x))_x = 0,$$

where B is a strictly increasing function. Indeed, in this case, since the travelling wave solutions satisfy

$$\varphi' = \mathsf{B}^{-1}(f(\varphi) - \mathcal{L}(\varphi)),$$

Theorem 2.1 holds if condition (2.7) is replaced by

$$\lim_{\varphi \to \phi^\pm} \int_{\varphi(0)}^{\varphi} \frac{\mathrm{d}\theta}{\mathsf{B}^{-1}(f(\theta) - \mathcal{L}(\theta))} = \pm\infty.$$

Once again, the limit of the travelling wave as $\epsilon \downarrow 0$ coincides with the limit obtained with the original model.

3.2. The modified Korteweg–de Vries–Burgers model

Let us now consider the model

$$\phi_t + (f(\phi))_x - \epsilon\,\phi_{xx} - \omega\phi_{xxx} = 0,$$

where $f(\phi) = \phi^3$. In this case, the travelling wave solutions are solutions of the form $\phi(x,t) = \varphi((x - ct)/\epsilon)$ that satisfy the conditions $\lim_{z\to\infty^\pm} \varphi(z) = \phi^\pm$, $\lim_{z\to\infty^\pm} \varphi'(z) = 0$, and $\lim_{z\to\infty^\pm} \varphi''(z) = 0$. Hence, they are solutions of

$$\varphi' + \mathsf{r}\varphi'' = f(\phi) - \mathcal{L}(\phi),$$

where $\mathsf{r} = \omega/\epsilon^2$, satisfying the above conditions at $\pm\infty$. Equivalently, they are orbits of the dynamical system

$$\frac{\mathrm{d}}{\mathrm{d}t}\begin{pmatrix}\varphi\\ \eta\end{pmatrix} = \begin{pmatrix}\eta\\ \frac{1}{\mathsf{r}}(f(\varphi) - \mathcal{L}(\varphi) - \eta)\end{pmatrix},$$

that connect the equilibrium points $(\phi^\pm, 0)$. We can immediately see that the conditions for these orbits to exist might be different from the conditions for the existence of travelling waves given by Theorem 2.1.

Jacobs, McKinney and Shearer (1995) considered the case $f(\phi) = \phi^3$. They showed that, when $\epsilon > 0$ and $\omega < 0$, the travelling wave solutions exist if and only if the travelling wave solutions of

$$\phi_t + (f(\phi))_x - \epsilon\,\phi_{xx} = 0$$

exist. This implies that a weak solution is physically relevant with respect to both models at the same time. However, this is no longer true if $\omega > 0$. Indeed, it was shown that there are travelling wave solutions depending on the ratio $\mathsf{r} = \epsilon^2/\omega$ that are *not* travelling wave solutions of the original model. Thus a weak solution of the hyperbolic conservation law (2.3) can be physically relevant with respect to one model and not the other. Moreover, since the travelling wave solutions of the model under consideration depend on the ratio r, a weak solution of (2.3) might be physically relevant with respect to the model for one value or but not another.

3.3. Phase transitions in solids

A similar, but more complicated, situation appears in the so-called viscosity–capillarity model for phase transitions in van der Waal fluids proposed by Truskinovsky (1982) and Slemrod (1984). Such a model reads

$$\begin{aligned}\gamma_t &= v_x,\\ v_t &= (\sigma(\gamma))_x + \nu\, v_{xx} - \lambda\, \gamma_{xxx},\end{aligned}$$

where v is the velocity, γ the strain, and σ the stress; the parameter ν is the viscosity and the parameter λ the capillarity. Some solids admit several types of crystals and this is reflected in the fact that σ is not an increasing function of the strain. As a consequence, if we drop the information about the viscosity and capillarity effects, we obtain the system

$$\begin{aligned}\gamma_t &= v_x,\\ v_t &= (\sigma(\gamma))_x,\end{aligned}$$

which is not even hyperbolic, as it changes type as σ' changes sign.

The weak solutions of the above model that are physically relevant with respect to the viscosity–capillarity model has been studied in Abeyaratne and Knowles (1991*a*, 1991*b*). Just as for the modified Korteweg–deVries–Burgers model, they have been shown to depend on the ratio λ/ν^2. This fact has to be taken into consideration when devising numerical schemes that converge to those weak solutions; see Slemrod and Flaherty (1986), Shu (1992), Cockburn and Gau (1996) and Zhong, Hou and LeFloch (1996).

3.4. Compressible fluid flow

Perhaps the most widely known example of the situation under consideration is the case of Navier–Stokes for compressible fluid flow. Using classical tensor notation, they can be written as follows:

$$\begin{aligned}\rho_t + (\rho\, v_j)_{,j} &= 0,\\ (\rho\, v_i)_t + (\rho\, v_i\, v_j - \sigma_{ij})_{,j} &= f_i,\\ (\rho\, e)_t + (\rho\, e\, v_j - \sigma_{ij}\, v_i + q_i)_{,j} &= f_i\, v_i,\end{aligned}$$

where ρ is the density, v the velocity, e the internal energy, and f the external body forces. The viscous stress σ and the heat flux q are given by

$$\begin{aligned}\sigma_{ij} &= (-p + \lambda\, v_{i,i})\, \delta_{ij} + \mu\, (v_{i,j} + v_{j,i}),\\ q_i &= -\kappa\, T_{,i},\end{aligned}$$

where p is the pressure and T the temperature.

From the above equations, the following equation for the entropy s can be obtained:

$$(\rho\, s)_t + (\rho\, s\, v_j)_{,j} = \psi,$$

where

$$\psi = \frac{\lambda}{T}(v_{i,i})^2 + \frac{\mu}{2\,\lambda}\left(v_{i,j} + v_{j,i} - \frac{2}{3}\, v_{k,k}\,\delta_{ij}\right)^2 + \frac{1}{T}(\kappa\, T_{,i})_{,i}. \qquad (3.1)$$

If Ω is any domain on whose border we have $v_i\, n_i = 0$ and $T_{,i}\, n_i = 0$, we get

$$\frac{\mathrm{d}}{\mathrm{d}t}\int_\Omega \rho\, s\,\mathrm{d}x = \int_\Omega \Psi\,\mathrm{d}x,$$

where

$$\Psi = \frac{\lambda}{T}(v_{i,i})^2 + \frac{\mu}{2\,\lambda}\left(v_{i,j} + v_{j,i} - \frac{2}{3}\, v_{k,k}\,\delta_{ij}\right)^2 + \frac{\kappa}{T^2}(T_{,i}\, T_{,i}). \qquad (3.2)$$

By the second law of thermodynamics, the viscosity coefficients λ and μ, and the heat transfer coefficient κ are positive and hence we have the so-called *entropy condition*

$$\frac{\mathrm{d}}{\mathrm{d}t}\int_\Omega \rho\, s\,\mathrm{d}x \geq 0.$$

If convection is the dominant feature of the flow and we formally neglect the effect of the viscosity and that of heat transfer, we obtain the nonlinear first-order system

$$\begin{aligned} \rho_t + (\rho\, v_j)_{,j} &= 0, \\ (\rho\, v_i)_t + (\rho\, v_i\, v_j)_{,j} + p_{,i} &= f_i, \\ (\rho\, e)_t + (\rho\, e\, v_j - p\, v_j)_{,j} &= f_i\, v_i, \end{aligned}$$

usually called the Euler equations. Only the weak solutions of these equations that satisfy the so-called entropy condition are considered to be physically relevant. For this reason, they are called the *entropy solutions.*

3.5. Convection–diffusion equations

It would have been ideal to work with the Navier–Stokes equations of compressible fluid flow and the corresponding Euler equations. Unfortunately, the link between these two equations, in particular concerning continuous dependence results, constitutes the subject of ongoing research.

In fact, such a link exists only for the scalar conservation law. For this reason, in the remainder of the paper, we are going to deal with the weak solutions of the hyperbolic conservation law

$$u_t + \nabla\cdot\mathbf{f}(u) = r,$$

which are physically relevant with respect to the convection–diffusion model

$$(u_\nu)_t + \nabla \cdot \mathbf{f}(u_\nu) - \nu \Delta u_\nu = r.$$

In other words, we consider the weak solutions obtained by the so-called *vanishing viscosity method*, that is, as limits, when the viscosity coefficient ν tends to zero, of the solutions u_ν of the convection–diffusion equation.

Note that since the weak solutions of the scalar hyperbolic conservation law we are considering are obtained by the vanishing viscosity method, it would be quite natural to call them *viscosity* solutions. Moreover, this would emphasize the strong link between these solutions and the so-called *viscosity* solutions of the Hamilton–Jacobi equation

$$u_t + H(\nabla u) = r;$$

see Crandall, Ishii and Lions (1992) and the references therein. These solutions, as we would expect, are obtained by the vanishing viscosity method, that is, as limits, when the viscosity coefficient ν tends to zero, of the solutions u_ν of

$$(u_\nu)_t + H(\nabla u_\nu) - \nu \Delta u_\nu = r.$$

However, the viscosity solutions for scalar hyperbolic problems are usually called *entropy* solutions (Lax 1972), as a mathematical generalization of what happens for the actual entropy solutions of compressible fluid flow.

4. Continuous dependence for parabolic solutions

In this section, we consider the initial value problem

$$u_t + \nabla \cdot \mathbf{f}(u) - \nu \Delta u = r, \qquad \text{in } \mathbb{R}^d \times (0,T), \tag{4.1}$$

$$u(t=0) = u_0, \qquad \text{on } \mathbb{R}^d, \tag{4.2}$$

and estimate the effect on the solution u of changes in the initial data u_0 and the right-hand side r. In other words, we consider the solution of

$$v_t + \nabla \cdot \mathbf{f}(v) - \nu \Delta v = s, \qquad \text{in } \mathbb{R}^d \times (0,T), \tag{4.3}$$

$$v(t=0) = v_0, \qquad \text{on } \mathbb{R}^d, \tag{4.4}$$

and get a simple upper bound for the quantity

$$|(u-v)(t)|_U := \int_{\mathbb{R}^d} U(u(x,t) - v(x,t))\,\mathrm{d}x,$$

where each nonnegative function U is determined to handle the effects of nonlinear convection suitably.

We show that such functions U are associated with L^1-like norms and seminorms and that the upper bound we obtain gives rise to several powerful results, such as *a priori* bounds of the exact solutions; continuous dependence with respect to the initial data, the right-hand side and the nonlinearity $\mathbf{f}$; and to *a posteriori* error estimates. In spite of this, the estimate breaks down as the viscosity coefficient ν tends to zero. Fortunately, a modification of this approach does give a suitable continuous dependence result which, however, will be considered in the next section.

4.1. Looking for suitable functions U

In what follows, we assume that u and v are sufficiently smooth and decay to zero at infinity in such a way that all the formal steps we perform are justified. These properties do hold when the Lipschitz nonlinearity $\mathbf{f}$, the right-hand sides r and s, and the initial data u_0 and v_0 are very smooth and decay to zero at infinity sufficiently fast; see Friedman (1964).

We begin with the following simple continuous dependence result.

Lemma 4.1. Let u be the solution of (4.1) and (4.2), and let v be the solution of (4.3) and (4.4). Let U be any smooth function such that $U(0)=0$. Then

$$|u(T) - v(T)|_U + \Theta_U = |u_0 - v_0|_U + \int_0^T \int_{\mathbb{R}^d} U'(u-v)(r-s)\,\mathrm{d}x\,\mathrm{d}t,$$

where

$$\Theta_U = \int_0^T \int_{\mathbb{R}^d} (Y_{\mathbf{f},U} + Z_{\nu,U})\,\mathrm{d}x\,\mathrm{d}t$$

and

$$\begin{aligned} Y_{\mathbf{f},U} &= -U''(u-v)(\mathbf{f}(u) - \mathbf{f}(v)) \cdot \nabla(u-v), \\ Z_{\nu,U} &= \nu\, U''(u-v)\,|\nabla(u-v)|^2. \end{aligned}$$

Proof. From equations (4.1) and (4.3), we have

$$(u-v)_t + \nabla \cdot (\mathbf{f}(u) - \mathbf{f}(v)) - \nu\Delta(u-v) = r - s.$$

Multiplying by $U'(u-v)$, we obtain

$$\begin{aligned} (U(u-v))_t + \nabla \cdot \big(U'(u-v)\,(\mathbf{f}(u) - \mathbf{f}(v))\big) + Y_{\mathbf{f},U} \\ - \nu\Delta U(u-v) + Z_{\nu,U} = U'(u-v)\,(r-s). \end{aligned}$$

Integrating over $\mathbb{R}^d \times (0,T)$, we obtain the result. This completes the proof. $\square$

Note that if the convection is linear, that is, if $\mathbf{f}(u) = \mathbf{a}\,u$, we have that

$$Y_{\mathbf{f},U} = -U''(u-v)\,(u-v)\,\mathbf{a} \cdot \nabla(u-v), = \nabla \cdot \left(\mathbf{a} \int_0^{u-v} z\,U''(z)\,\mathrm{d}z\right),$$

and so

$$\Theta_U = \int_0^T \int_{\mathbb{R}^d} (Y_{\mathbf{f},U} + Z_{\nu,U})\,\mathrm{d}x\,\mathrm{d}t = \int_0^T \int_{\mathbb{R}^d} Z_{\nu,U}\,\mathrm{d}x\,\mathrm{d}t.$$

This means that Θ_U is a nonnegative functional if U is a convex function, as this would render nonnegative the term $Z_{\nu,U}$. This property of U seems to be necessary to capture the dissipative nature of the Laplacian.

We would like to have Θ_U nonnegative even for a general nonlinearity $\mathbf{f}$. Hence we should find U such that $e\,U''(e) = 0$, as this ensures that $Y_{\mathbf{f},U} = 0$ given that $\mathbf{f}$ is Lipschitz. For such a function U, the functional $|\cdot|_U$ would then be perfectly tailored to handle nonlinear convection easily.

We can actually get such a function as the limit of a sequence of smooth functions $\{ U_\epsilon \}_{\epsilon>0}$ such that $\lim_{\epsilon\to 0} e\,U''_\epsilon(e) = 0$. For example, we can take

$$U_\epsilon(e) = a\,\Upsilon_\epsilon(e) + b\,\Upsilon_\epsilon(-e),$$

where a, b are nonnegative parameters, and

$$\Upsilon_\epsilon(e) = \int_0^{e/\epsilon} (e - \epsilon\, s)\, \mu(s)\, \mathrm{d}s,$$

where μ is a smooth nonnegative function with support in $(0,1)$ and integral equal to one. Then we set

$$U(e) = \lim_{\epsilon\downarrow 0} U_\epsilon(e) \quad \text{and} \quad U'(e) = \lim_{\epsilon\downarrow 0} U'_\epsilon(e).$$

For the special cases

$$U(e) = \begin{cases} \max\{0, e\} & \text{for } a = 1,\ b = 0, \\ \max\{0, -e\} & \text{for } a = 0,\ b = 1, \\ |e| & \text{for } a = 1,\ b = 1, \end{cases} \tag{4.5}$$

Lemma 4.1 gives the following result.

Theorem 4.2. Let u be the solution of (4.1) and (4.2) and let v be the solution of (4.3) and (4.4). Let U be given by (4.5). Then

$$|u(T) - v(T)|_U \leq |u_0 - v_0|_U + \int_0^T |(r - s)(t)|_U\, \mathrm{d}t.$$

Proof. The result follows from Lemma 4.1 and the fact that, for U as in (4.5), we have $\lim_{\epsilon\downarrow 0} \Theta_{U_\nu} \geq 0$, and, for U given by (4.5),

$$\lim_{\epsilon\downarrow 0} \int_0^T \int_{\mathbb{R}^d} U'_\epsilon(u - v)(r - s)\, \mathrm{d}x\, \mathrm{d}t \leq \int_0^T \int_{\mathbb{R}^d} U(r - s)\, \mathrm{d}x\, \mathrm{d}t.$$

This completes the proof. □

4.2. Properties of the exact solutions

As we see next, this powerful theorem gives rise to *a priori* estimates on the exact solutions; continuous dependence results on the initial data, the right-hand side, and the nonlinearity $\mathbf{f}$; and to *a posteriori* error estimates.

To state these results, we use the following notation.

Definition. We write

$$\|v\|_{L^1} = \|v\|_{L^1(\mathbb{R}^d)}, \quad |v|_{TV} = \sup_{\mathbf{w}\in C_0^\infty(\mathbb{R}^d):\|\mathbf{w}\|_{L^\infty(\mathbb{R}^d)}=1} \int_{\mathbb{R}^d} v\,\nabla\cdot\mathbf{w}\,dx,$$

where $\|\mathbf{w}\|_{L^\infty(I)} = \max_{1\le i\le d} \|\mathbf{w}_i\|_{L^\infty(I)}$. Note that, if v is smooth,

$$|w|_{TV} = \sum_{i=1}^{d} \|\partial_{x_i} w\|_{L^1}.$$

We shall also write

$$\|w\|_{L^\infty(B)} = \sup_{t\in[0,T]} \|w(t)\|_B, \quad \|w\|_{L^1(B)} = \int_0^T \|w(t)\|_B\,dt,$$

for $B = L^1$ and $B = TV$. Thus we shall speak of the $L^1(L^1)$ norm, for example.

Proposition 4.3. (*A priori* estimates) Let u be the solution of (4.1) and (4.2). Then:

(i) the range of $u(T)$ is included in $[a(T), b(T)]$, where

$$a(T) = \inf_{x\in\mathbb{R}^d} u_0(x) - \int_0^T \sup_{x\in\mathbb{R}^d} \max\{-r(x,t),0\}\,dt,$$

$$b(T) = \sup_{x\in\mathbb{R}^d} u_0(x) + \int_0^T \sup_{x\in\mathbb{R}^d} \max\{r(x,t),0\}\,dt;$$

(ii) $\|u(T)\|_{TV} \le \|u_0\|_{TV} + \|r\|_{L^1(TV)}$.

See the Appendix for a proof.

Proposition 4.4. (Continuous dependence) Now, let v be the solution of (4.3) and (4.4). Then:

(i) if $s = r$ then $u_0 \ge v_0$ implies $u(T) \ge v(T)$;

(ii) $\|u(T) - v(T)\|_{L^1} \le \|u_0 - v_0\|_{L^1} + \|r - s\|_{L^1(L^1)}$;

(iii) $\|u_t(T)\|_{L^1} \le \|\mathbf{f}\|_{L^\infty(I)}\,|u_0|_{TV} + \nu\,\|\Delta u_0\|_{L^1} + \|r_t\|_{L^1(L^1)}$, where I is the range of u_0.

Let u_i be the solution of (4.1) and (4.2) with $\mathbf{f} = \mathbf{f}_i$, $i = 1, 2$. Then:

(iv) $\|u_1(T) - u_2(T)\|_{L^1} \le \|\mathbf{f}_1' - \mathbf{f}_2'\|_{L^\infty(I)}\,\|u_2\|_{L^1(TV)}$, where I is the range of u_2 on $\mathbb{R}^d \times (0,T)$.

See the Appendix for a proof.

Let us emphasize that this result allows us to extend the notion of solution to initial data, right-hand sides, and nonlinearities $\mathbf{f}$ which need not be extremely smooth. To see this, note that, if u_n is the smooth solution of (4.1) and (4.2) with $u_0 = u_{0,n}$ and $r = r_n$ and the sequence $\{(u_{0,n}, r_n)\}$ converges strongly in $L^1(\mathbb{R}^d) \times L^1(0,T; L^1(\mathbb{R}^d))$ to a limit $(u_0^\star, r^\star)$, then (ii) implies that $\{u_n\}_{n\geq 0}$ is a Cauchy sequence in $\mathcal{C}^0(0,T; L^1(\mathbb{R}^d))$. Its limit, $u^\star$, is thus the physically relevant weak solution of (4.1) and (4.2) with initial data $u_0^\star$ and right-hand side $r^\star$. A similar argument holds when u_n is the smooth solution of (4.1) and (4.2) with $\mathbf{f} = \mathbf{f}_n$.

Proposition 4.5. (*A posteriori* error estimate) Let u be the solution of (4.1) and (4.2) and let v be a smooth function whose residual

$$R(v) = v_t + \nabla \cdot \mathbf{f}(v) - \nu \Delta v - r,$$

belongs to $L^1(0,T; L^1(\mathbb{R}^d))$. Then, for U given by (4.5),

$$|u(T) - v(T)|_U \leq |u_0 - v(0)|_U + \int_0^T |-R(v)(t)|_U \, \mathrm{d}t.$$

Proof. The result follows from Theorem 4.2 by simply taking $s = r + R(v)$. □

This result allows us to estimate the quality of the approximation v to u without knowing the exact solution u; it is thus extremely useful in practical computations. In particular, when $U(e) = |e|$, we have the simple estimate

$$\|u(T) - v(T)\|_{L^1} \leq \|u_0 - v(0)\|_{L^1} + \|R(v)\|_{L^1(L^1)}.$$

Unfortunately, this estimate is of no use as the viscosity coefficient tends to zero.

4.3. Breakdown of the estimate as the viscosity tends to zero

To see this, we begin by pointing out that, by Lemma 4.1, the quantity

$$\|u_0 - v(0)\|_{L^1} + \|R(v)\|_{L^1(L^1)}$$

is in fact an upper bound for the sum

$$\|u(T) - v(T)\|_{L^1} + \lim_{\epsilon \downarrow 0} \Theta_{U_\nu},$$

where $U(e) = |e|$. Next we show that these two terms might be of significantly different sizes when the viscosity parameter ν tends to zero.

Let us consider the one-dimensional case, $d = 1$, and the travelling wave solutions

$$u(x,t) = \varphi((x - ct)/\nu), \qquad v(x,t) = \varphi((x - ct)/\nu').$$

Then we have

$$\lim_{\nu\downarrow 0}\lim_{\nu'\downarrow 0}|u(T)-v(T)|_U = \lim_{\nu'\downarrow 0}\lim_{\nu\downarrow 0}|u(T)-v(T)|_U = 0.$$

On the other hand, we have

$$\begin{aligned}\lim_{\epsilon\downarrow 0}\Theta_{U_\nu} &= \lim_{\epsilon\downarrow 0}\int_0^T\int_{\mathbb{R}}(u_x - v_x)\,(U'_\epsilon(u-v))_x\,\mathrm{d}x\,\mathrm{d}t\\ &= 2\int_0^T \nu|(u_x - v_x)(ct,t)|\,\mathrm{d}t\\ &= 2\,|1-\nu/\nu'|\int_0^T |f(\varphi(0)) - \mathcal{L}(\varphi(0))|\,\mathrm{d}t,\end{aligned}$$

given that

$$\eta\, w_x = \varphi' = f(\varphi) - \mathcal{L}(\varphi),$$

for $w(x,t) = \varphi((x-c\,t)/\eta)$, since φ is a travelling wave solution; see (2.4). This implies

$$\lim_{\nu\downarrow 0}\lim_{\epsilon\downarrow 0}\Theta_{U_\nu} = 2\,T\,|f(\varphi(0)) - \mathcal{L}(\varphi(0))| > 0,$$

and

$$\lim_{\nu'\downarrow 0}\lim_{\epsilon\downarrow 0}\Theta_{U_\nu} = \infty.$$

This computation shows that Theorem 4.2 does not give us any useful information about $\|u(T)-v(T)\|_{L^1}$ when the viscosity coefficients tend to zero, as claimed. It also shows that

$$\lim_{\nu'\downarrow 0}\|R(v)\|_{L^1(L^1)} = \infty,$$

which indicates that measuring the residual $R(v)$ in the $L^1(0,T;L^1(\mathbb{R}^d))$-norm is certainly not a good idea. Next, we show how to overcome this difficulty.

5. Robustness in the viscosity coefficient

In this section, we present a new continuous dependence result which, this time, gives meaningful information even when the viscosity coefficients tend to zero. In fact, we can even use it to prove the existence and uniqueness of the entropy solution of

$$u_t + \nabla\cdot\mathbf{f}(u) = r, \qquad \text{in } \mathbb{R}^d\times(0,T), \tag{5.1}$$

$$u(t=0) = u_0, \qquad \text{on } \mathbb{R}^d. \tag{5.2}$$

Another advantage of this new continuous dependence result is that it also holds for the entropy solution.

In other words, in this section we find continuous dependence result for the parabolic problems which are automatically *inherited* by the entropy solution of the hyperbolic problem.

5.1. Doubling the space variable

The new estimate is obtained by introducing two changes in the previous approach. The first is to take v to be the solution of

$$v_t + \nabla \cdot \mathbf{f}(v) - \nu' \Delta v = s, \qquad \text{in } \mathbb{R}^d \times (0,T), \tag{5.3}$$

$$v(t=0) = v_0, \qquad \text{on } \mathbb{R}^d. \tag{5.4}$$

Note that the viscosity coefficient ν' is not necessarily equal to the viscosity coefficient ν, as occurred in the previous section.

The second is to measure the difference between the right-hand sides, $r-s$, in a much weaker way. To do that, we introduce the auxiliary function

$$\varphi_{\epsilon_x}(z) = \Pi_{i=1}^d \frac{1}{\epsilon_x}\, \omega(z_i/\epsilon_x),$$

where ω is an even, nonnegative smooth function with support $[-1,1]$ and integral equal to one. We are going to use the following numbers associated with ω:

$$\mathsf{c}_1 = \int_{-1}^{1} |z|\,\omega(z)\,\mathrm{d}z \quad \text{and} \quad \mathsf{c}_2 = \int_{-1}^{1} |\omega'(z)|\,\mathrm{d}z.$$

We obtain the following variation of the first continuous dependence result, Theorem 4.2.

Theorem 5.1. Let u be the solution of (4.1) and (4.2) and let v be the solution of (5.3) and (5.4). Let U be given by (4.5). Then,

$$|u(T)-v(T)|_U \le |u_0 - v_0|_U + \inf_{\epsilon_x>0}\left(\mathsf{A}\,\epsilon_x + \frac{(\sqrt{\nu}-\sqrt{\nu'})^2}{\epsilon_x}\,\mathsf{B} + \Lambda_U^{\epsilon_x}(v)\right),$$

where

$$\Lambda_U^{\epsilon_x}(v) = \int_0^T\!\!\int_{\mathbb{R}^d} \sup_{c\in\mathbb{R}} E_U^{\epsilon_x}(c,v;x,t)\,\mathrm{d}x\,\mathrm{d}t,$$

and

$$E_U^{\epsilon_x}(c,v;x,t) = \int_{\mathbb{R}^d} \varphi_{\epsilon_x}(x-y)\,U'(c-v(y,t))\,(r(x,t)-s(y,t))\,\mathrm{d}y.$$

Moreover,

$$\mathsf{A} = \mathsf{c}_1\ \left(\min\{\|u(T)\|_{TV}, \|v(T)\|_{TV}\} + \min\{\|u_0\|_{TV}, \|v_0\|_{TV}\}\right),$$

$$\mathsf{B} = \mathsf{c}_2\ \min\{\|u\|_{L^1(TV)}, \|v\|_{L^1(TV)}\}.$$

This result extends the first continuous dependence result, Theorem 4.2, in two ways. First, it renders explicit the influence of the difference of viscosity coefficients. The second is that it measures the difference of right-hand sides in a weaker way; indeed, note that, in the expression of $E_U^{\epsilon_x}(c, v; x, t)$, the functions $r = r(x,t)$ and $s = s(y,t)$ depend on different space variables. This *doubling of the space variable* is a powerful technique that renders this result possible: see the proof in the Appendix. As a consequence, the estimate does not break down when the viscosity coefficients ν and ν' tend to zero.

5.2. Letting the viscosity tend to zero: the entropy solution

We illustrate this phenomenon by showing how to use Theorem 5.1 to establish the existence and uniqueness of the entropy solution. Then we show that Theorem 5.1 also holds when u is taken to be the entropy solution.

Proposition 5.2. Let u_i be the solution of (4.1) and (4.2) with $\nu = \nu_i$, $i = 1, 2$. Then

$$\|u_1(T) - u_2(T)\|_{L^1} \leq \mathsf{D}\,\sqrt{24\mathsf{c}_1\mathsf{c}_2\,T}\,\left|\sqrt{\nu_1} - \sqrt{\nu_2}\right|,$$

where $\mathsf{D} = \|u_0\|_{TV} + \|r\|_{L^1(TV)}$.

Proof. Since, by (ii) of Proposition 4.3,

$$\max\{\|u_1\|_{L^\infty(TV)}, \|u_2\|_{L^\infty(TV)}\} \leq \mathsf{D},$$

Theorem 5.1 reads

$$\|u_1(T) - u_2(T)\|_U \leq \inf_{\epsilon_x > 0}\left(2\,\mathsf{c}_1\,\mathsf{D}\,\epsilon_x + \frac{\mathsf{c}_2}{\epsilon_x}\,(\sqrt{\nu_1} - \sqrt{\nu_2})^2\,T\,\mathsf{D} + \Lambda_U^{\epsilon_x}(v)\right),$$

and since

$$\Lambda_U^{\epsilon_x}(v) \leq \mathsf{c}_1\epsilon_x\|r\|_{L^1(TV)} \leq \mathsf{c}_1\epsilon_x\mathsf{D},$$

we have

$$\|u_1(T) - u_2(T)\|_{L^1} \leq \inf_{\epsilon_x > 0}\left(3\,\mathsf{c}_1\,\epsilon_x + \mathsf{c}_2\,\frac{\left(\sqrt{\nu_1} - \sqrt{\nu_2}\right)^2}{\epsilon_x}\,T\right)\mathsf{D}.$$

The result follows by minimizing with respect to the parameter ϵ_x. This completes the proof. □

The above result states that the sequence $\{\,u_\nu\,\}_{\nu>0}$ of exact solutions of the parabolic initial value problem (4.1) and (4.2) is a Cauchy sequence in $\mathcal{C}^0(0,T;L^1(\mathbb{R}^d))$. As a consequence, it converges to a unique limit u that belongs to $\mathcal{C}^0(0,T;L^1(\mathbb{R}^d))$, which is in fact merely the entropy solution. Moreover, we immediately have that

$$\|u_\nu(T) - u(T)\|_{L^1} \leq \mathsf{D}\,\sqrt{24\mathsf{c}_1\mathsf{c}_2\,T}\,\sqrt{\nu}.$$

Now, we only have to invoke Theorem 5.1 and Proposition 5.2 to obtain the following result.

Theorem 5.3. Theorem 5.1 holds when u is taken to be the entropy of (5.1) and (5.2).

This proves the claim that the entropy solution inherits the continuous dependence results that hold for the solutions of the parabolic problem. A direct consequence of this result is that Propositions 4.3, 4.4 and 4.5 also hold for entropy solutions.

For simplicity, from now on, we are going to assume that

$$u_0 \in L^\infty(\mathbb{R}^d) \times BV(\mathbb{R}^d), \tag{5.5}$$

$$r \in L^\infty(\mathbb{R}^d \times (0,T)) \cap L^1(0,T; BV(\mathbb{R}^d)), \tag{5.6}$$

$$r_t \in L^1(0,T; L^1(\mathbb{R}^d)), \tag{5.7}$$

$$\mathbf{f} \in \mathcal{C}^1, \tag{5.8}$$

where

$$BV(\mathbb{R}^d) = \{v \in L^1(\mathbb{R}^d) : |v|_{TV} < \infty\}.$$

Note that the above results imply that, in this case, the entropy solution u belongs to the space

$$\mathcal{C}^0(0,T; L^1(\mathbb{R}^d)) \cap L^\infty(0,T; L^\infty(\mathbb{R}^d) \cap BV(\mathbb{R}^d)),$$

and that u_t belongs to

$$L^\infty(0,T; L^1(\mathbb{R}^d)).$$

Note also that, in the above theorem, v is supposed to be the solution of a parabolic equation instead of a function that could have the same regularity the entropy solution has. In the next section, we show how to overcome this deficiency.

6. Approximation theory for entropy solutions

In this section, we present two extensions of Theorem 5.3. In the first, the function v is taken to be in the space of uniformly Lipschitz functions from $[0,T]$ to $L^1(\mathbb{R}^d)$. This estimate, however, depends explicitly on the modulus of continuity in time of both u and v. In the second extension, we partially remove this constraint and obtain an estimate that is completely independent of u, but valid for v as before, or dependent on the modulus of continuity (in time) of u, but valid for functions v that only need to be continuous from the left, as functions from $[0,T]$ to $L^1(\mathbb{R}^d)$.

The results discussed in this section constitute the main tools for studying approximations to entropy solutions of scalar hyperbolic conservation laws.

6.1. Doubling the time variable

To remove the requirement that the function v be the solution of (4.3) and (4.4) in Theorem 5.1 is quite simple. We only have to replace s by

$$v_t + \nabla_y \cdot (v) - \nu' \Delta_y v,$$

in the expression for $E_U^{\epsilon_x}$. Moreover, to further reduce the regularity of v, we can let ν' tend to zero, in which case we get that

$$\begin{aligned} E_U^{\epsilon_x}(c, v; x, t) &= \int_{\mathbb{R}^d} \varphi_{\epsilon_x}(x-y)\, U'(c-v)\,(r - v_t - \nabla_y \cdot \mathbf{f}(v))\,\mathrm{d}y \\ &= \int_{\mathbb{R}^d} \varphi_{\epsilon_x}(x-y)\,\big(U'(c-v)\,(r - v_t) + \nabla_y \cdot \mathbf{G}(c,v)\big)\,\mathrm{d}y \\ &= \int_{\mathbb{R}^d} \varphi_{\epsilon_x}(x-y)\, U'(c-v)\,(r - v_t)\,\mathrm{d}y \\ &\quad - \int_{\mathbb{R}^d} \nabla_y \varphi_{\epsilon_x}(x-y) \cdot \mathbf{G}(c,v)\,\mathrm{d}y, \end{aligned}$$

where $v = v(y,t)$ and $r = r(x,t)$. We immediately see that, as a function of y, the function v can now be discontinuous. This could be achieved because, having doubled the space variable, we were able to integrate by parts. Thus, to be able to do a similar manoeuvre with the time derivative, we only have to double the time variable.

6.2. The first estimate

In what follows, we use the auxiliary functions

$$\varphi_{\epsilon_t}(z) = \frac{1}{\epsilon_t}\,\eta(z/\epsilon_t) \quad \text{and} \quad \Phi_{\epsilon_t}(t) = 2\int_0^t \varphi_{\epsilon_t}(z)\,\mathrm{d}z,$$

where η is an even, nonnegative smooth function with support $[-1,1]$ and integral equal to one. We have the following extension of Theorem 5.1.

Theorem 6.1. Let u be the entropy solution of (5.1) and (5.2) and let v be any bounded function such that

$$v \in L^\infty(0,T; BV(\mathbb{R}^d)) \quad \text{and} \quad v_t \in L^\infty(0,T; L^1(\mathbb{R}^d)).$$

Let U be given by (4.5). Then,

$$|u(T) - v(T)\,|_U \le |u_0 - v_0\,|_U + \inf_{\epsilon_x,\epsilon_t > 0}\big(\mathsf{A}\,\epsilon_x + \mathsf{C}\,\epsilon_t + \Lambda_U^{\epsilon_x,\epsilon_t}(v)\big),$$

where

$$\Lambda_U^{\epsilon_x,\epsilon_t}(v) = \frac{1}{\Phi_{\epsilon_t}(T)} \int_0^T\!\!\int_{\mathbb{R}^d} \sup_{c\in\mathbb{R}} E_U^{\epsilon_x,\epsilon_t}(v, c; x, t)\,\mathrm{d}x\,\mathrm{d}t,$$

and

$$E_U^{\epsilon_x,\epsilon_t}(v,c;x,t) = \int_{\mathbb{R}^d} \varphi_{\epsilon_t,\epsilon_x}\, U(c-v)\,\mathrm{d}y\Big|_{\tau=0}^{\tau=T} - \int_0^T\!\!\int_{\mathbb{R}^d} U(c-v)\,(\varphi_{\epsilon_t,\epsilon_x})_t\,\mathrm{d}y\,\mathrm{d}\tau$$
$$- \int_0^T\!\!\int_{\mathbb{R}^d} \big(\mathbf{G}(c,v)\cdot\nabla_y\varphi_{\epsilon_t,\epsilon_x} - U'(c-v)\,r\,\varphi_{\epsilon_t,\epsilon_x}\big)\,\mathrm{d}y\,\mathrm{d}\tau$$

where $\varphi_{\epsilon_t,\epsilon_x} = \varphi_{\epsilon_t}(t-\tau)\,\varphi_{\epsilon_x}(x-y)$, $r = r(x,t)$, $v = v(y,\tau)$, and

$$\mathbf{G}(c,v) = \int_v^c U'(c-w)\,\mathbf{f}'(w)\,\mathrm{d}w.$$

Moreover,

$$\mathsf{A} = \mathsf{c}_1\,\big(\min\{\|u(T)\|_{TV}, \|v(T)\|_{TV}\} + \min\{\|u_0\|_{TV}, \|v_0\|_{TV}\}\big),$$
$$\mathsf{C} = \|u_t\|_{L^\infty(L^1)} + \|v_t\|_{L^\infty(L^1)}.$$

See the Appendix for a proof.

Note that this result allows us to compare the entropy solution u with functions v with much lower regularity than required in Theorem 5.1. Accordingly, an even weaker measure of the residual of v, $\Lambda_U^{\epsilon_x,\epsilon_t}(v)$, is used.

6.3. The second estimate

In the result we state next, we take the auxiliary function η to be the characteristic function of the interval $[-1,1]$. Let us recall that $\varphi_{\epsilon_t}(z) = \frac{1}{\epsilon_t}\,\eta(z/\epsilon_t)$. We have the following variation of Theorem 6.1.

Theorem 6.2. Let u be the entropy solution of (5.1) and (5.2) and let v be any bounded measurable function, continuous from the left as a function from $[0,T]$ into $L^1(\mathbb{R}^d)$. Let U be given by (4.5). Then,

$$|u(T) - v(T)\,|_U \le 2\,|u_0 - v_0\,|_U + 4 \inf_{\epsilon_x,\epsilon_t>0}\big(\mathsf{A}\,\epsilon_x + \mathsf{C}\,\epsilon_t + \Lambda_U^{\epsilon_x,\epsilon_t}\big),$$

where $\Lambda_U^{\epsilon_x,\epsilon_t}$ is defined in Theorem 6.1 and

$$\mathsf{A} = \mathsf{c}_1\,\big(\min\{\|u(T)\|_{TV}, \|v(T)\|_{TV}\} + \min\{\|u_0\|_{TV}, \|v_0\|_{TV}\}\big),$$
$$\mathsf{C} = 2\,\min\{\|u_t\|_{L^\infty(L^1)}, \|v_t\|_{L^\infty(L^1)}\}.$$

See the Appendix for a proof.

Note that the constants A and C can be bounded by quantities independent of either u or v, as desired. We can see that, if the modulus of continuity in time of v is bounded, the estimate is completely independent of the entropy solution u. If, on the other hand, we made the constants A and C depend only on u, then v does not have to satisfy any continuity restriction. The price we pay for this is innocuous.

In the next two sections, we present applications of the results obtained in this section. First, we obtain a characterization of the entropy solution and then error estimates for the entropy solution and some of its approximations.

7. Characterization of the entropy solution

In this section, we use the *a posteriori* error estimate of Theorem 6.1 to obtain a characterization of the entropy solution.

However, before treating the general case, we illustrate the technique in the one-dimensional case, $d = 1$, when the right-hand side r is equal to zero and the entropy solution is smooth except on a single curve. This result extends the characterization result of weak solutions that are limits of travelling wave solutions of parabolic problems.

7.1. Piecewise smooth entropy solutions

Consider the entropy solution of the problem

$$u_t + (f(u))_x = 0, \qquad \text{in } \mathbb{R} \times (0,T), \tag{7.1}$$
$$u(t=0) = u_0, \qquad \text{on } \mathbb{R}. \tag{7.2}$$

If we assume that v is smooth in two disjoint open sets, namely,

$$\Omega^- = \{(x,t) : x < x(t), t \in (0,T)\},$$
$$\Omega^+ = \{(x,t) : x > x(t), t \in (0,T)\},$$

which are separated by the curve

$$\Gamma = \{(x(t),t) : t \in (0,T)\},$$

then Theorem 6.1 takes a particularly simple form.

To state it, we need to introduce some notation. We let $\mathbf{n}^\pm$ denote the unit normal to Γ outward with respect to $\Omega^\pm$ and set

$$v^\pm(x(t),t) = \lim_{h\downarrow 0} v((x(t),t) - h\,\mathbf{n}^\pm).$$

Finally, we denote the *jump* of the normal component of the vector $\mathbf{q}(v)$ across the discontinuity curve Γ by

$$[\![\mathbf{q}(v)\cdot\mathbf{n}]\!] = \mathbf{q}(v^+)\cdot\mathbf{n}^+ + \mathbf{q}(v^-)\cdot\mathbf{n}^-.$$

We can now state the result.

Corollary 7.1. Let u be the entropy solution of (7.1) and (7.2). Then

$$\|u(T) - v(T)\|_{L^1} \le \|u(0) - v_0\|_{L^1} + \|R(v)\|_{L^1(\mathbb{R}\times(0,T)\setminus\Gamma)} + \|\mathcal{R}(v)\|_{L^1(\Gamma)},$$

where

$$R(v) = v_t + (f(v))_x,$$

and

$$\mathcal{R}(v) = \max\left\{0, \sup_{c\in\mathbb{R}} -[\![(G(c,v), U(c-v))\cdot\mathbf{n}]\!]\right\}.$$

In the above expression, $U(e) = |e|$ and $G(u,c) = \big(f(u) - f(c)\big)\,U'(u-c)$.

See the Appendix for a proof.

A direct consequence of this result is the following characterization of the entropy solution. It is obtained by simply realizing that $v = u$ if each of the three terms of the upper bound for $\|u(T) - v(T)\|_{L^1}$ is equal to zero.

Proposition 7.2. Assume that there is a function v, smooth on the set $\Omega^- \cup \Omega^+$ such that:

(i) $v(t = 0) = u_0$;

(ii) $v_t + (f(v))_x = 0$ in $\Omega^- \cup \Omega^+$;

(iii) $\dfrac{dx}{dt} = \dfrac{f(v^+) - f(v^-)}{v^+ - v^-}$ on Γ;

(iv) for each $t \in (0, T)$, the graph of f on the interval (v^-, v^+) (resp., (v^+, v^-)) does not lie below (resp., above) the straight line joining the points $(v^\pm, f(v^\pm))$.

Then v is the entropy solution (7.1) and (7.2).

See the Appendix for a proof.

This result shows that the entropy solution coincides with the strong solution when it is smooth; this is equivalent to requiring that the residual $R(v)$ be zero on the open set $\Omega^- \cup \Omega^+$. It also shows that if the entropy solution has a curve of discontinuity Γ, conditions (iii) and (iv) characterize the discontinuity jumps as well as the curve Γ itself. In fact, these two properties are equivalent to the condition that the expression $\mathcal{R}(v)$ be identically zero on the discontinuity curve Γ.

Finally, note that, if the entropy solution is a piecewise constant function, that is, if

$$u(x,t) = \begin{cases} u^-, & x < x(t), \\ u^+, & x > x(t), \end{cases}$$

condition (iii) states that Γ is a straight line, and that condition (iv) generalizes the graph condition (2.6) on the discontinuities of entropy solutions that are limits of travelling wave solutions of the parabolic problem.

7.2. The entropy solution

The characterizations of the previous subsection rest on the following characterization of the entropy solution in terms of inequality (7.3). This result is in fact a direct *consequence* of the continuous dependence results obtained for the parabolic solutions.

Theorem 7.3. Assume that the smoothness conditions (5.5), (5.6), (5.7) and (5.8) are satisfied. Then the entropy solution of (5.1) and (5.2) is the only measurable bounded function continuous form the left as a function form $[0,T]$ into $L^1(\mathbb{R}^d)$ such that, for all nonnegative functions φ in $C_0^\infty(\mathbb{R}^d \times [0,T])$,

$$\mathsf{E}(u_0, c; v, \varphi) \le 0, \quad \forall\, c \in \mathbb{R}, \tag{7.3}$$

where

$$\begin{aligned}\mathsf{E}(v_0, c; v, \varphi) = &\int_{\mathbb{R}^d} \varphi(T)\, U(c - v(T))\, \mathrm{d}y - \int_{\mathbb{R}^d} \varphi\, U(c - v_0)\, \mathrm{d}y \\ &- \int_0^T\!\!\int_{\mathbb{R}^d} \big(U(c-v)\,\varphi_t + \mathbf{G}(c,v)\cdot\nabla_y\varphi - U'(c-v)\, r\,\varphi\big)\, \mathrm{d}y\, \mathrm{d}\tau,\end{aligned}$$

where

$$U(e) = |e| \quad \text{and} \quad \mathbf{G}(c,v) = \int_v^c U'(c-w)\, \mathbf{f}'(w)\, \mathrm{d}w.$$

For a proof, see the Appendix. Here, let us simply point out that, for $r = 0$, the link between the continuous dependence Theorems 6.1 and 6.2 is established by the following simple equality:

$$E_U^{\epsilon_x,\epsilon_t}(v, c; x, t) = \mathsf{E}(u_0, c; v, \varphi_{\epsilon_x,\epsilon_t}).$$

We can now show that the entropy solution is also the physically relevant solution with respect to the following model:

$$(u_\omega)_t + \nabla\cdot\mathbf{f}(u_\omega) - \nabla\cdot\mathbb{L}(\omega; u_\omega, \nabla u_\omega) = 0,$$

where $\mathbb{L}(0;\cdot,\cdot) = 0$, provided

$$\mathbb{L}(\cdot;\cdot,p)\cdot p \ge 0 \quad \forall\, p \in \mathbb{R}^d, \tag{7.4}$$

and

$$\lim_{\omega\downarrow 0} \int_0^T\!\!\int_{\mathbb{R}^d} |\mathbb{L}(\omega; u_\omega, \nabla u_\omega)|\, \mathrm{d}x\, \mathrm{d}t = 0. \tag{7.5}$$

Let us show that this is indeed true for the case $r = 0$. Thus, for any nonnegative test function φ, we have

$$\begin{aligned}\mathsf{E}(u_0, c; u_\omega, \varphi) &= -\int_0^T\!\!\int_{\mathbb{R}^d} U'(c-u_\omega)\,\big((u_\omega)_t + \nabla\cdot\mathbf{f}(u_\omega)\big)\,\varphi\, \mathrm{d}x\, \mathrm{d}t \\ &= \mathsf{E}_{\mathsf{small}} + \mathsf{E}_{\mathsf{diss}},\end{aligned}$$

where

$$\begin{aligned}\mathsf{E}_{\mathsf{diss}} &= \int_0^T\!\!\int_{\mathbb{R}^d} U'(c-u_\omega)\, \mathbb{L}(\omega; u_\omega, \nabla u_\omega)\cdot\nabla\varphi\, \mathrm{d}x\, \mathrm{d}t, \\ \mathsf{E}_{\mathsf{small}} &= -\int_0^T\!\!\int_{\mathbb{R}^d} U''(c-u_\omega)\, \mathbb{L}(\omega; u_\omega, \nabla u_\omega)\cdot\nabla u_\omega\, \varphi\, \mathrm{d}x\, \mathrm{d}t.\end{aligned}$$

Since φ is nonnegative, U is convex, and $\mathbb{L}$ satisfies the positivity condition (7.4), we have that

$$\mathsf{E}_{\mathsf{diss}} \leq 0.$$

Moreover, it is easy to see that, by the convergence condition (7.5),

$$\mathsf{E}_{\mathsf{small}} \leq \|\nabla\varphi\|_{L^\infty(L^\infty)} \int_0^T\!\!\int_{\mathbb{R}^d} |\mathbb{L}(\omega; u_\omega, \nabla u_\omega)| \,\mathrm{d}x\,\mathrm{d}t \to 0,$$

as $\omega \downarrow 0$. Of course, we assumed that $u_\omega(t=0) = u_0$. This extends a similar result obtained in Section 3.1 for travelling wave solutions.

7.3. The entropy inequality

The inequality (7.3) is called *entropy inequality* as it is reminiscent of a similar situation for compressible fluid flow. Indeed, the conservation of the entropy $\rho\, s$ given by the compressible Navier–Stokes equations reads

$$(\rho\, s)_t + (\rho\, s\, v_j)_{,j} = \psi,$$

where $\rho\, s\, \mathbf{v}$ is the entropy flux and

$$\psi = \frac{\lambda}{T}(v_{i,i})^2 + \frac{\mu}{2\,\lambda}\left(v_{i,j} + v_{j,i} - \frac{2}{3}\, v_{k,k}\, \delta_{ij}\right)^2 + \frac{1}{T}(\kappa\, T_{,i})_{,i}.$$

If we multiply the conservation law by a nonnegative φ in $\mathcal{C}_0^\infty(\mathbb{R}^d \times [0,T])$ and integrate, we get

$$\begin{aligned}\mathbb{E}(\varphi) &= \int_{\mathbb{R}^d} \varphi(T)\,(\rho\, s)(T)\,\mathrm{d}y - \int_{\mathbb{R}^d} \varphi(0)\,(\rho\, s)(0)\,\mathrm{d}y \\ &\quad - \int_0^T\!\!\int_{\mathbb{R}^d} \big((\rho\, s)\,\varphi_t + \rho\, s\,\mathbf{v}\cdot\nabla_y\varphi\big)\,\mathrm{d}y\,\mathrm{d}\tau \\ &= \int_0^T\!\!\int_{\mathbb{R}^d} \psi\,\varphi\,\mathrm{d}y\,\mathrm{d}\tau\end{aligned}$$

where ψ is given by (3.1). This implies that

$$\mathbb{E}(\varphi) = \mathbb{E}_{\mathsf{small}} + \mathbb{E}_{\mathsf{diss}},$$

where

$$\begin{aligned}\mathbb{E}_{\mathsf{small}} &= -\kappa \int_0^T\!\!\int_{\mathbb{R}^d} \frac{1}{T}\nabla T \cdot \nabla\varphi\,\mathrm{d}y\,\mathrm{d}\tau, \\ \mathbb{E}_{\mathsf{diss}} &= \int_0^T\!\!\int_{\mathbb{R}^d} \Psi\,\varphi\,\mathrm{d}y\,\mathrm{d}\tau,\end{aligned}$$

and Ψ is given by (3.2). By the second law of thermodynamics, $\Psi \geq 0$, and so

$$\mathbb{E}_{\mathsf{diss}} \geq 0.$$

Moreover, if

$$\mathbb{E}_{\mathsf{small}} \to 0,$$

as $\kappa \downarrow 0$, we get the so-called entropy inequality

$$\mathbb{E}(\varphi) \geq 0.$$

Because of this, and since the function $U(c-u)$ has a role analogous to that of the entropy $\rho\, s$, and $G(c,u)$ a role analogous to that of the entropy flux $\rho\, s\, \mathbf{v}$, they are usually referred to in the same manner.

8. Error estimates for the Engquist–Osher scheme

In this section, we apply the *a posteriori* error estimates of Section 6 to obtain an estimate of the distance between the entropy solution of the simple Cauchy problem

$$u_t + \nabla \cdot \mathbf{f}(u) = 0, \qquad \text{in } \mathbb{R} \times (0,T), \tag{8.1}$$

$$u(t=0) = u_0, \qquad \text{on } \mathbb{R}, \tag{8.2}$$

and the approximate solution, u_h, determined by a model monotone numerical scheme, the Engquist–Osher scheme.

To define the scheme, we discretize the time interval $(0,T)$ in intervals $J^n = (t^n, t^{n+1})$ and set $\Delta^n = t^{n+1} - t^n$ and $t^N = T$. We also discretize $\mathbb{R}$ in intervals $I_j = (x_{j-1/2}, x_{j+1/2})$ and set $\Delta_j = x_{j+1/2} - x_{j-1/2}$. The approximate solution u_h is continuous from the left in time and is equal to the value u_i^n on the rectangle $I_i \times J^n$.

These values are determined in the following way. For $n = 0, \ldots, N-1$,

$$\frac{u_j^{n+1} - u_j^n}{\Delta^n} + \frac{\widehat{f}_{j+1/2}^n - \widehat{f}_{j-1/2}^n}{\Delta_j} = 0, \quad \forall\, j \in \mathbb{Z}, \tag{8.3}$$

$$u_j^0 = \frac{1}{\Delta_j} \int_{I_j} u_0(x)\,\mathrm{d}x, \quad \forall\, j \in \mathbb{Z}, \tag{8.4}$$

and the so-called Engquist–Osher numerical flux $\widehat{f}_{j+1/2}^n = \widehat{f}(u_i^n, u_{i+1}^n)$ is given by

$$\widehat{f}(a,b) = f^+(a) + f^-(b), \tag{8.5}$$

where

$$f^+(a) = \int_d^a \max\{0, f'(s)\}\,\mathrm{d}s, \qquad f^-(b) = -\int_d^b \max\{0, -f'(s)\}\,\mathrm{d}s,$$

and d is an arbitrary, but fixed, real value.

We could use the *a posteriori* error estimate of Theorem 6.1 with $v = \tilde{u}_h$ if we take $\tilde{u}_h$ to be the continuous function equal to the value u_i^n at the point (x_i, t^n). We can also apply Theorem 6.2 with $v = u_h$ without any modification. Of course, we can always insert all the information about the approximate solution u_h and then simplify the corresponding upper bound as much as possible. Next, we consider the result of such operations.

To state it, we introduce the *rate of entropy dissipation* (RED) of the numerical scheme. It is a piecewise constant function whose value for $(x,t) \in I_n \times J^n$ is

$$\mathrm{RED}(x,t) = \sup_{c\in\mathbb{R}} \left(\frac{U(c-u_j^{n+1}) - U(c-u_j^n)}{\Delta^n} + \frac{\widehat{G}^n_{j+1/2}(c) - \widehat{G}^n_{j-1/2}(c)}{\Delta_j} \right),$$

where the numerical entropy flux $\widehat{G}^n_{j+1/2}(c) = G^+(c, u_j^n) + G^-(c, u_{j+1}^n)$ is given by

$$G^{\pm}(c,v) = \int_v^c U'(c-s)\,(f^{\pm})'(s)\,\mathrm{d}s.$$

Theorem 8.1. Let u be the entropy solution of (8.1) and (8.2) and let u_h be the approximate solution of the Engquist–Osher scheme (8.3) and (8.4). Let U be given by (4.5). Then,

$$|u(T) - u_h(T)|_U \le \Phi(u_h),$$

where

$$\Phi(u_h) = 2\,|u_0 - u_h(0)|_U + C\,|u_0|_{TV}\Theta_0^{1/2} + \int_0^T\!\!\int_{\mathbb{R}} \mathrm{RED}\,\mathrm{d}x\,\mathrm{d}t,$$

where

$$\Theta_0 = \sum_{n=0}^{N-1}\sum_{j\in\mathbb{Z}} \left(|\widehat{f}^n_{j+1/2} - f(u_j^n)| + |\widehat{f}^n_{j-1/2} - f(u_j^n)| \right) \Delta_j\,\Delta^n.$$

Proof. Then, by Theorem 6.2 we have

$$|u(T) - u_h(T)|_U \le 2\,|u_0 - u_h(0)|_U + 2 \inf_{\epsilon_x,\epsilon_t>0} \left(\mathsf{A}\,\epsilon_x + \mathsf{C}\,\epsilon_t + \Lambda_U^{\epsilon_x,\epsilon_t} \right),$$

where

$$\Lambda_U^{\epsilon_x,\epsilon_t} = \frac{1}{\Phi_{\epsilon_t}(T)} \int_0^T\!\!\int_{\mathbb{R}^d} \sup_{c\in\mathbb{R}} E_U^{\epsilon_x,\epsilon_t}(v,c;x,t)\,\mathrm{d}x\,\mathrm{d}t.$$

A simple computation gives

$$\Lambda_U^{\epsilon_x,\epsilon_t} = \Lambda_{\mathsf{small}} + \Lambda_{\mathsf{diss}},$$

where

$$\Lambda_{\mathsf{small}} = \left(\frac{1}{\epsilon_t} + \frac{1}{\epsilon_x} \right) \Theta_0, \qquad \Lambda_{\mathsf{diss}} = \int_0^T\!\!\int_{\mathbb{R}} \mathrm{RED}\,\mathrm{d}x\,\mathrm{d}t.$$

The result follows after minimizing on ϵ_t and ϵ_x. This completes the proof. □

If we base an adaptive algorithm on the above estimate, it is important to know if given an arbitrary tolerance TOL, it is possible to find a mesh such that the quantity $\Phi(u_h)$ can actually be made smaller than TOL. It is not difficult to see that, in fact, this can always be done.

Indeed, since it is possible to show that

$$|u_0 - u_h(0)|_U \leq |u_0|_{TV}\, h, \qquad \mathrm{RED}_j^n \leq 0,$$
$$\sum_{j\in\mathbb{Z}} \left|f(u_{j+1}^n) - f(u_j^n)\right| \leq \sup_{w\in Rg(u_0)} |f'(w)|\, |u_0|_{TV},$$

provided that

$$\sup_{w\in Rg(u_0)} |f'(w)| \sup_{j\in\mathbb{Z}} \frac{\Delta^n}{\Delta_j} \leq 1,$$

where $Rg(u_0)$ is the convex hull of the range of u_0, we immediately see that we get the *a priori* estimate

$$\Phi(u_h) \leq C\,(\Delta x)^{1/2},$$

for some constant C independent of $\Delta x = \sup_{j\in\mathbb{Z}} \Delta_j$.

This concludes the presentation of the material we wanted to discuss in this paper. It has not followed the historical development of the topic, but rather the short introduction to the subject for graduate students given in Cockburn (1999). No references were provided, in order to focus on the ideas themselves. Next, we compensate for this omission and present a brief overview of the historical development of the topic under consideration.

9. The historical evolution of some ideas

9.1. A priori error estimates I

The definition of the entropy solution (in terms of the entropy inequality) was first introduced in the seminal paper by Kružkov (1970). Therein, Kružkov proved its existence, uniqueness and stability with respect to initial data. He proved the existence of the entropy solution by using the vanishing viscosity method. First, he obtained estimates of moduli of continuity of the solutions, u_ν, of the parabolic problem which are independent of the viscosity coefficient, ν. Then, he used a compactness argument to show that, when ν tends to zero, the functions u_ν converge to a limit that satisfies the entropy inequality. He obtained uniqueness and stability by using the entropy inequality and the doubling of variables technique (which was also introduced in this paper). For our simple model scalar conservation law, the continuous dependence result

$$\|u_1(T) - u_2(T)\|_{L^1} \leq \|u_{0,1}(T) - u_{0,2}(T)\|_{L^1}$$

was obtained, where u_i is the entropy solution of the Cauchy problem with initial data $u_{0,i}$, $i = 1, 2$; see Proposition 4.4 and Theorem 5.3. No other error estimate was obtained.

Error estimates and, in fact, a powerful approximation theory for the entropy solution was developed a few years later by Kuznetsov (1976). The main result is (essentially) what we have labelled as Theorem 6.1. It was applied to obtain the first *a priori* error estimate between the entropy solution u and the parabolic solution u_ν (when $r = 0$), namely,

$$\|u - u_\nu\|_{L^\infty(L^1)} \leq C\,|u_0|_{TV}\,\sqrt{T\,\nu},$$

and the first *a priori* error estimate between the entropy solution and the approximate solution u_h defined by a monotone scheme on uniform Cartesian grids

$$\|u - u_h\|_{L^\infty(L^1)} \leq C\,|u_0|_{TV}\,\sqrt{T\,\Delta x}.$$

(The rate of convergence of $(\Delta x)^{1/2}$ is the best possible rate for monotone schemes when the initial data are functions of bounded variation, as was shown many years later in Şabac (1997).)

Around the same time, the paper by Harten, Hyman and Lax (1976) on monotone schemes appeared, which gave a physical argument indicating that convergence to the entropy solution should always take place. Moreover, it was shown that monotone schemes are necessarily at most first-order accurate, and that second- (and higher)-order schemes might *not* converge to the entropy solution. To illustrate this, we show in Figure 9.1 the entropy solution of the problem

$$u_t + (u^2/2)_x = 0, \qquad u(x, t = 0) = \begin{cases} 1, & x \in (.4, .6), \\ 0, & \text{otherwise}, \end{cases}$$

and the approximation given by a monotone scheme, the Engquist–Osher scheme, and a formally second-order accurate scheme, the Lax–Wendroff method.

A few years later, convergence of monotone schemes (in uniform Cartesian grids) to the entropy solution was proved by Crandall and Majda (1980). A compactness argument was used which could be considered to be the discrete counterpart of the argument used in Kružkov (1970). Apparently, the authors of this result were unaware of the approximation theory introduced in Kuznetsov (1976).

However, this theory has since been used by most researchers working on error estimation for scalar hyperbolic conservation laws. Thus, Sanders (1983) used it to prove that

$$\|u - u_h\|_{L^\infty(L^1)} \leq C\,|u_0|_{TV}\,\sqrt{T\,\Delta x},$$

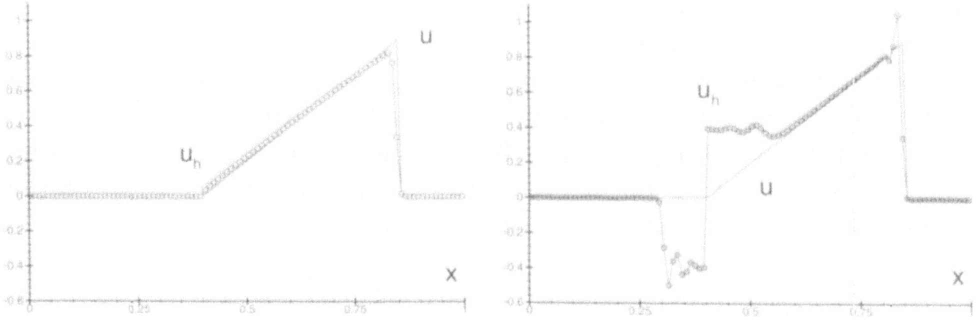

Figure 9.1. The entropy solution, u, and its approximation u_h at time $T = 1/2$: Engquist–Osher scheme (left) and Lax–Wendroff scheme (right).

for u_h determined by monotone schemes defined in non-uniform Cartesian grids. His main contribution was to show that the total variation of the approximation u_h was uniformly bounded in time. This estimate immediately implies the uniform estimate of the modulus of continuity in time, essential for the application of Theorem 6.1.

Then Lucier (1985*a*) obtained the bound

$$\|u - u_h\|_{L^\infty(L^1)} \leq |u_0|_{TV} \left(\Delta x + \frac{2}{\sqrt{3}} \frac{\Delta x \, T^{1/2}}{(\Delta t)^{1/2}} \right),$$

for the numerical schemes of Godunov, Glimm and LeVeque in one space dimension. To prove the estimate for LeVeque's method, he had to prove that, if u_i is the entropy solution with $\mathbf{f} = \mathbf{f}_i$, $i = 1, 2$, then

$$\|u_1 - u_2\|_{L^\infty(L^1)} \leq C\,T\,|u_0|_{TV}\,\|\mathbf{f}_1' - \mathbf{f}_2'\|_{L^\infty(I)};$$

see the similar result for the parabolic solutions in Proposition 4.4 and Theorem 5.3.

Later, error estimates of the form

$$\|u - u_h\|_{L^\infty(L^1)} \leq C\,|u_0|_{TV}\,\sqrt{T}\,(\Delta x)^\delta,$$

where $\delta \in (0, 1/2]$, for the so-called quasi-monotone schemes (which include Petrov–Galerkin methods) were obtained in Cockburn (1989, 1990*a*, 1990*b*) for the case of Cartesian grids. The parameter δ is equal to $1/2$ for monotone schemes and controls the spurious production of entropy per cell typical of high-order accurate schemes. The main approximation result in this papers was the modification of the approximation theory introduced in Kuznetsov (1976) to include what we have called the smooth functions U, and its application to non-monotone schemes defined in several space dimensions.

9.2. A priori error estimates II

In *all* the above papers, an estimate of the modulus of continuity in time was obtained, as required by Theorem 6.1. However, to obtain such an estimate is extremely difficult; indeed, there is no known numerical scheme defined in unstructured grids for which this property has been proved. This is true even for the monotone schemes, the simplest schemes for hyperbolic conservation laws.

To obtain *a priori* error estimates thus became an extremely difficult task, and the main focus in this area shifted to the search for weaker smoothness properties with which convergence, and not error estimates, could be proved. Thus, the theory of convergence of measure-valued functions (DiPerna 1985) was used. (See the monograph by Málek, Nečas, Rokyta and Ružička (1996) on weak and measure-valued solutions for a treatment of the initial boundary-value problem.) This was first done by Szepessy (1989, 1991), where the streamline diffusion method with shock-capturing terms was proved to converge to the entropy solution; a refinement of this approach was obtained a few years later by Jaffré, Johnson and Szepessy (1995). This approach was also used by Coquel and LeFloch (1991) to prove convergence of finite difference methods, and by Cockburn, Coquel and LeFloch (1995), Kröner, Noelle and Rokyta (1995) and Nöelle (1995) to prove finite volume methods defined in unstructured triangulations.

Since the main difficulty in using the approximation theory introduced by Kuznetsov (1976) was the estimate of the modulus of continuity in time of the approximate solution (the constant C in Theorem 6.1), a switch to Theorem 6.2 would solve the problem since the corresponding C is independent of such a modulus. This was noticed by Cockburn, Coquel and LeFloch (1994), Kröner and Rokyta (1994) and Vila (1994), where it was shown that

$$\|u - u_h\|_{L^\infty(L^1)} \leq C\,|u_0|_{TV}^{1/2}\,\|u_0\|_{L^2}^{1/2}\,(\Delta x)^{1/4}, \tag{9.1}$$

where u_h is given by monotone methods defined in general triangulations; see also Nöelle (1996). This approach was further developed in Cockburn and Gremaud (1996*a*), where a similar estimate was proved for the shock-capturing discontinuous Galerkin method and the estimate

$$\|u - u_h\|_{L^\infty(L^1)} \leq C\,|u_0|_{TV}^{1/2}\,\|u_0\|_{L^2}^{3/4}\,(\Delta x)^{1/8}$$

was shown for the shock-capturing streamline diffusion method. The lack of optimality in the rate of convergence is due to the fact that the total variation of the approximate solution cannot be proved to be uniformly bounded in time. This lack of uniform boundedness of the total variation has a greater impact on the streamline diffusion method, because of its use of continuous approximations. In fact, only a weaker estimate of the form

$$\left(\int_0^T \int_{\mathbb{R}^d} |\nabla u|^2 \,\mathrm{d}x\right)^{1/2} \leq \|u_0\|_{L^2(\mathbb{R}^d)} \,(\Delta x)^{-1/2}$$

can be proved. The discrete version of this estimate relies on the continuity of the L^2 projection into finite-dimensional spaces of discontinuous functions: see Cockburn (1991).

Finally, let us point out that, for each of the above two methods, the approximate solution is a piecewise polynomial of degree $k \geq 0$. The fact that the *a priori* estimates do not reflect the influence of the polynomial degree is certainly a drawback.

Extensions of these results to the conservation law

$$u_t + \nabla \cdot (\mathbf{v}\, f(u)) = 0,$$

where $\mathbf{v}$ is a divergence-free function, were obtained in Eymard, Gallouët, Ghilani and Herbin (1998), where the estimate (9.1) was proved. Extensions to the conservation law

$$u_t + \nabla \cdot \mathbf{F}(x, t, u) = 0$$

where $\nabla_x \cdot \mathbf{F} = 0$, were carried out in Chainais-Hillairet (1999). More importantly, the error $\|u(T) - u_h(T)\|_{L^1(\Omega)}$ was bounded in terms of the behaviour of u_h in the domain of dependence associated with Ω; this constitutes the first *local* error estimate for this kind of problem. See Chainais-Hillairet and Champier (2001) for extensions to the case in which the right-hand side is not equal to zero.

A formalization of the techniques used in the above-mentioned papers was obtained in Bouchut and Perthame (1998). In particular, the following fundamental fact was identified. If, for $U(e) = |e|$, we have that

$$U'(v - c)\, R(v) \leq \nabla \cdot H_c,$$

then

$$\|u(T) - v(T)\|_{L^1} \leq \|u_0 - v_0\|_{L^1} + C\, \|H\|_{L^1(L^1)}^{1/2},$$

if $|H_c| \leq |H|$ for all $c \in \mathbb{R}$. Let us also mention that, using this technique, the following result was obtained. If u_i is the entropy solution with $r = 0$, $\mathbf{f} = \mathbf{f}_i$, $i = 1, 2$, with $\mathbf{f}_1(0) = \mathbf{f}_2(0)$, then

$$\|u_1 - u_2\|_{L^\infty(L^1)} \leq C\, \left(T\, |u_0|_{L^1}\, |u_0|_{TV}\, Q\right)^{1/2},$$

where

$$Q = \sup_{\xi \in \mathbb{R}\setminus\{0\}} \frac{|\mathbf{f}_1(\xi) - \mathbf{f}_2(\xi)|}{|\xi|}.$$

Compare this with the continuous dependence result (iv) of Proposition 4.4.

9.3. A priori error estimates III

Lucier (1986) considered and analysed moving-mesh methods in one space dimension. The methods used approximate solutions u_h which were piecewise constant or piecewise linear between points called *nodes*; these nodes were then evolved in a suitable way. For strictly convex, smooth nonlinearities f, and N nodes, it was shown that

$$\|u(T) - u_h(T)\|_{L^1} \leq C\, N^{-1},$$

for piecewise constant approximations (Dafermos' method) and

$$\|u(T) - u_h(T)\|_{L^1} \leq C\, N^{-2},$$

for piecewise linear approximations. These are remarkable results, not only because the rate of convergence is not the classical rate of order $N^{-1/2}$, but because they give information about the approximability of the entropy solution by functions that are piecewise polynomials between nodes. From these results, regularity properties of the exact solution can be *deduced.*

This idea was developed in the papers by Lucier (1988) and DeVore and Lucier (1990, 1996). In particular, it was shown that, if the initial data u_0 belongs to the Besov space $B^{\alpha}_{\sigma}(L^{\sigma})$, where $\sigma = 1/(1+\alpha)$ and $\alpha > 0$, then the same is true for the entropy solution $u(\cdot, t)$ for $t > 0$. It was also shown that this property does not hold for any Besov space $B^{\alpha}_{\sigma}(L^{\sigma})$ with $\alpha > 1$. Note that, since $\sigma < 1$, the space L^{σ} is not a locally convex space.

9.4. A priori error estimates IV

Note that Theorem 6.2 gives an estimate of the form

$$\|u - u_h\|_{L^\infty(L^1)} \leq \Phi(u_h).$$

This means that, in order to obtain an *a priori* error for the above estimate, we are bound to obtain regularity properties of the approximate solution u_h, and this is extremely difficult, as we have seen. An ideal way out of this unfortunate predicament would be to be able to interchange the roles of u and u_h in Theorem 6.2, as we would then obtain an estimate of the form

$$\|u_h - u\|_{L^\infty(L^1)} \leq \Phi_h(u).$$

In this way, the estimate would be totally independent of the regularity properties of u_h.

The main idea is to realize that, while $\Phi(u_h)$ is a measure of the *residual* of u_h, the functional $\Phi_h(u)$ should be nothing but a measure of the *truncation error* of u. Thus, instead of basing the approximation theory on the entropy inequality for u, it would be enough to base it on the discrete entropy entropy inequality for the approximate solution u_h. That this can actually be carried out was proved in the papers of Cockburn and Gremaud (1996*b*, 1997) and

Cockburn, Gremaud and Yang (1998), when u_h is defined by monotone schemes. In particular, it was proved that

$$\|u - u_h\|_{L^\infty(L^1)} \leq C\,|u_0|_{TV}\,\sqrt{T\,\Delta x},$$

for the Lax–Friedrichs method (monotone scheme) in a uniform grid of triangles without using any regularity properties of the approximate solution. This result cannot be proved with any other technique.

The application of this approach to other numerical schemes remains to be carried out.

9.5. *A posteriori error estimates and adaptivity*

We thus see that the theory of *a priori* error estimates for entropy solutions, which should be based on estimates of the *truncation error*, was slowed down because it was based on Theorems 6.1 and 6.2, which are based on estimates of the *residual* instead. For this reason, they are, so to speak, *natural a posteriori* error estimates. Conversely, the use of these *a posteriori* error estimates as a stepping stone to obtain *a priori* error estimates also obscured the fact that they could be used as the basis of adaptive algorithms.

These estimates were for the first time recognized as such in Cockburn and Gremaud (1996*a*), where *a posteriori* error estimates for the shock-capturing streamline diffusion and the discontinuous Galerkin methods of arbitrary order were obtained; however, they still remain to be numerically tested. On the other hand, this lack of clarity did not prevent the introduction of the first stable adaptive algorithm for scalar hyperbolic conservation laws in Lucier (1985*b*). Numerical evidence was shown which indicated that, if an error of size ϵ is required, a computational complexity of order ϵ^{-3} was required instead of the complexity of standard monotone schemes of order ϵ^{-4}. See also Lucier and Overbeek (1987).

Later, adaptive algorithms were implemented by Kröner and Ohlberger (2000) for a monotone scheme, and then by Gosse and Makridakis (2000) for monotone second-order finite-difference schemes. The implementation by Kröner and Ohlberger (2000) used a local version of Theorem 6.1 obtained in Chainais-Hillairet (1999).

The only study of the ratio of the upper bound of the error to the error itself, usually called the *effectivity index*, has been done in Cockburn and Gau (1995) for the Engquist–Osher scheme in one space dimension. There, the *a posteriori* error estimate of Corollary 7.1 was obtained and then applied to a continuous approximate solution u_h obtained by a bilinear interpolation of the values u_j^n. The effectivity index was shown to remain close to one for smooth entropy solutions and for entropy solutions with discontinuities and linear convection. The case of nonlinear convection and entropy solutions with discontinuities was not treated therein; it remains to be studied.

9.6. Continuous dependence for parabolic problems

The technique used in Kuznetsov (1976) to obtain error estimates for the entropy solution was thought to be impossible to extend to the parabolic case. However, this extension was actually carried out by Cockburn and Gripengerg (1999), who obtained continuous dependence of the solution of the scalar degenerate parabolic equation

$$u_t + \nabla \cdot \mathbf{f}(u) - \Delta\varphi(u) = 0$$

on the nonlinearities $\mathbf{f}$ and φ. The main continuous dependence result for parabolic solutions, Theorem 5.1, which is the basis for the approximation theory for entropy solutions, Theorems 6.1 and 6.2, was proposed in Cockburn (1999), and is based on the technique introduced in Cockburn and Gripengerg (1999). A simple exposition of this approach can be found in the proof of Lemma A1 in the Appendix.

In particular, Cockburn and Gripengerg (1999) proved that, if u_i is the (semi-group) solution of the above equation with $\varphi = \varphi_i$, and $u_i(t=0) = u_0$, for $i = 1, 2$, then

$$\|u_1 - u_2\|_{L^\infty(L^1)} \leq C \left\| \sqrt{\varphi_1'} - \sqrt{\varphi_2'} \right\|_{L^\infty(I)},$$

where I is the convex hull of the range of u_0. We note that Benilan and Crandall (1981) studied the dependence with respect to φ in the case $\mathbf{f} = \mathbf{0}$, but their results are not written in terms of explicit estimates. This means that a complete approximation theory for entropy solutions of degenerate parabolic equations can be obtained in exactly the same way as for the entropy solution of the scalar hyperbolic conservation law.

On the other hand, an approximation theory for degenerate parabolic equations can also be constructed from the characterization of their entropy solutions obtained in Carrillo (1999) (see also the extension to more general boundary conditions in Mascia, Porretta and Terracina (2002)), just as the approximation theory in Kuznetsov (1976) was obtained from the characterization of the entropy solution for scalar hyperbolic conservation laws given in Kružkov (1970). Using this approach, comparisons between the entropy solution of

$$(u_1)_t + \nabla \cdot (\mathbf{v} f(u_1)) - \Delta\varphi(u_1) = 0$$

and that of

$$(u_2)_t + \nabla \cdot (\mathbf{v} f(u_2)) - \Delta(\varphi(u_2) + \epsilon\, u_2) = 0$$

have recently been obtained. Indeed, Evje and Karlsen (2002) proved that

$$\|u_1 - u_2\|_{L^\infty(L^1)} \leq C\, \epsilon^{1/2},$$

and Eymard, Gallouët and Herbin (2002*a*) proved that

$$\|u_1 - u_2\|_{L^\infty(L^1(\Omega))} \leq C\, \epsilon^{1/5},$$

where Ω is a bounded domain. Also using this approach, *a posteriori* error estimates for the equation

$$u_t + \nabla \cdot (\mathbf{v} f(u)) - \Delta\varphi(u) = \lambda u$$

have been obtained (Ohlberger 2001) and adaptivity strategies devised and numerically tested.

So far, there are no *a priori* error estimates for numerical schemes for degenerate parabolic equations. The first convergence result for monotone schemes for strongly degenerate equations in one space dimension was obtained by Evje and Karlsen (2000*b*); see also Evje and Karlsen (2000*a*). Another convergence result for finite volume methods in several space dimensions has been obtained by Eymard, Gallouët, Herbin and Michel (2002*b*).

9.7. Hamilton–Jacobi equations

The theory of entropy solutions of scalar hyperbolic conservation laws runs parallel, in many instances, to the theory of viscosity solutions of Hamilton–Jacobi equations. For example, the counterpart of the paper by Kružkov (1970) about the entropy solution could be considered to be the papers of Crandall and Lions (1983) and Crandall, Evans and Lions (1984) on the characterization of the viscosity solution.

The error estimate obtained by Kuznetsov (1976) between the entropy solution and the approximation given by a monotone scheme, namely

$$\|u - u_h\|_{L^\infty(L^1)} \leq C\,|u_0|_{TV}\,\sqrt{T\,\Delta x},$$

corresponds to the error estimate obtained in Crandall and Lions (1984) between the viscosity solution and the approximation given by a monotone scheme, namely,

$$\|u - u_h\|_{L^\infty(L^\infty)} \leq C\,|u_0|_{W^{1,\infty}}\,\sqrt{T\,\Delta x}.$$

The continuous dependence results on the entropy solution of

$$u_t + \nabla \cdot \mathbf{f}(u) - \Delta\varphi(u) = 0$$

on the nonlinearities in Cockburn and Gripengerg (1999) has as counterpart the continuous dependence results on the viscosity solution of

$$u_t + F(u, D_x u, D_x^2 u) = 0,$$

with respect to the nonlinearity F in Cockburn, Gripengerg and Londen (2001) and Gripenberg (2002); see also Jakobsen and Karlsen (2002).

The counterpart of the *a posteriori* error estimate in Theorem 6.1 is contained in Albert, Cockburn, French and Peterson (2002*a*) for the steady state Hamilton–Jacobi equation

$$u + H(\nabla u) = f,$$

in a periodic setting where a careful numerical study of the effectivity index is carried out. In particular, for monotone schemes and strictly convex Hamiltonians, it is shown that the effectivity index increases like $|\ln \Delta x|$, and not like $(\Delta x)^{-}1/2$ as would be expected if the general theory of monotone schemes is used.

For the transient case, see Albert, Cockburn, French and Peterson (2002*b*).

The counterpart of the material presented in this paper for entropy solutions is contained in the paper by Cockburn and Qian (2002), where the steady state case was considered.

Finally, the theory of continuous dependence in negative-order norms for the physically relevant solutions of the conservation law

$$u_t + (f(u))_x = 0,$$

when f is strictly convex, reviewed in Tadmor (1998) has its counterpart in the theory of continuous dependence in L^1 of viscosity solutions for convex Hamiltonians introduced in Lin and Tadmor (2001).

A result for the Bellman's equation

$$F(D_t u, D_x^2 u, D_x u, u, x, t) = 0$$

that does not have any counterpart in the theory of strongly degenerate convection–diffusion equations is the estimate of the rate of convergence of monotone schemes obtained in Krylov (2000) for fairly general nonlinearities F. It reads

$$-C_\star\,(\Delta x)^{\delta_\star} \leq u - u_h \leq C^\star\,(\Delta x)^{\delta^\star},$$

where $C_\star$ and $C^\star$ and positive constants and $\delta_\star$ and $\delta_\star$ only depend on the smoothness of the coefficients and the function F. When F is Lipschitz, we obtain the best estimate, namely

$$\delta_\star = \frac{1}{3} \quad \text{and} \quad \delta^\star = \frac{1}{21}.$$

10. Concluding remarks and open problems

In this paper, we have given an overview of a theory of continuous dependence and error estimation for the entropy solution of scalar hyperbolic conservation laws. We have stressed the idea that it is essential to obtain continuous dependence results for the original well-posed problem which do not break down when the viscosity coefficient tends to zero. The existence and uniqueness of the entropy solution are direct consequences of these results, as is the fact that the entropy solution inherits the continuous dependence results valid for the parabolic solutions. This procedure is to be contrasted with the traditional approach of using compactness arguments to obtain the existence of an entropy solution, obtain the so-called entropy inequality and only then obtain continuous dependence results. How to extend this approach to more complicated scalar hyperbolic conservation laws, to the corresponding initial boundary value problem, and then to hyperbolic systems remains a challenging open problem.

The other very important open problem is that of studying the minimization of the *nonlinear*, *nonsmooth* functional

$$\Phi(-R(v)),$$

where v is an approximation of the entropy solution u. The devising of iterative adaptive algorithms that guarantee a decrease of the error by a given factor per iteration is certainly the main problem to solve. Strongly related to this issue is the study of the relation between the upper bound of the error and the error itself,

$$\frac{\Phi(-R(v))}{\|u - v\|_{L^\infty(L^1)}}.$$

To give an idea of the difficulty of this task, let us recall that this problem has only been recently been solved for finite element approximations of linear strongly elliptic problems in Morin, Nochetto and Siebert (2000). Moreover, in such a case, the functional to minimize was quadratic and smooth!

Acknowledgements
The author would like to thank Wolfgang Dahmen, Markus Keel, Fernando Reitich and Eitan Tadmor for feedback on the material presented here. The author would also like to thank Kenneth Karlsen for valuable information about degenerate parabolic equations. Finally, the author would like to thank Michael Breuss for a careful reading of the manuscript.

REFERENCES

R. Abeyaratne and J. Knowles (1991*a*), 'Implications of viscosity and strain gradient effects for the kinetics of propagating phase boundaries in solids', *SIAM J. Appl. Math.* **51**, 1205–1221.

R. Abeyaratne and J. Knowles (1991*b*), 'Kinetic relations and the propagation of phase boundaries in solids', *Arch. Rational Mech. Anal.* **114**, 119–154.

S. Albert, B. Cockburn, D. French and T. Peterson (2002*a*), '*A posteriori* error estimates for general numerical methods for Hamilton–Jacobi equations, Part I: The steady state case', *Math. Comp.* **71**, 49–76.

S. Albert, B. Cockburn, D. French and T. Peterson (2002*b*), *A posteriori* error estimates for general numerical methods for Hamilton–Jacobi equations, Part II: The time-dependent case, Vol. III of *Finite Volumes for Complex Applications*, Hermes Penton Science, pp. 17–24.

T. Barth and H. Deconink, eds (1999), *High-Order Methods for Computational Physics*, Vol. 9 of *Lecture Notes in Computational Science and Engineering*, Springer.

P. Benilan and M. Crandall (1981), 'The continuous dependence on φ of solutions of $u_t - \Delta\varphi(u) = 0$', *Indiana Univ. Math. J.* **30**, 161–177.

S. Bianchini and A. Bressan (2001), 'Vanishing viscosity solutions of nonlinear hyperbolic systems', Preprint *SISSA ref. 86/2001/M November.*

F. Bouchut and B. Perthame (1998), 'Kružkov's estimates for scalar conservation laws revisited', *Trans. Amer. Math. Soc.* **350**, 2847–2870.

A. Bressan (2000), *Hyperbolic Systems of Conservation Laws: The One-Dimensional Cauchy Problem*, Vol. 20 of *Oxford Lecture Series in Mathematics and its Applications*, Oxford University Press, Oxford.

J. Carrillo (1999), 'Entropy solutions for nonlinear degenerate problems', *Arch. Rational Mech. Anal.* **147**, 269–361.

C. Chainais-Hillairet (1999), 'Finite volume schemes for a nonlinear hyperbolic equation: Convergence towards the entropy solution and error estimate', *Modél. Math. Anal. Numér.* **33**, 129–156.

C. Chainais-Hillairet and S. Champier (2001), 'Finite volume schemes for nonhomogeneous scalar conservation laws: Error estimate', *Numer. Math.* **88**, 607–639.

B. Cockburn (1989), 'The quasi-monotone schemes for scalar conservation laws, I', *SIAM J. Numer. Anal.* **26**, 1325–1341.

B. Cockburn (1990*a*), 'The quasi-monotone schemes for scalar conservation laws, II', *SIAM J. Numer. Anal.* **27**, 247–258.

B. Cockburn (1990*b*), 'The quasi-monotone schemes for scalar conservation laws, III', *SIAM J. Numer. Anal.* **27**, 259–276.

B. Cockburn (1991), 'On the continuity in $BV(\Omega)$ of the L^2-projection into finite element spaces', *Math. Comp.* **57**, 551–561.

B. Cockburn (1999), A simple introduction to error estimation for nonnlinear hyperbolic conservation laws: Some ideas, techniques, and promising results, in *Proc. 1998 EPSRC Summer School in Numerical Analysis, SSCM*, Vol. 26 of *The Graduate Student's Guide to Numerical Analysis*, Springer, pp. 1–46.

B. Cockburn and H. Gau (1995), '*A posteriori* error estimates for general numerical methods for scalar conservation laws', *Mat. Aplic. e Comp.* **14**, 37–45.

B. Cockburn and H. Gau (1996), 'A model numerical scheme for the propagation of phase transitions in solids', *SIAM J. Sci. Comput.* **17**, 1092–1121.

B. Cockburn and P.-A. Gremaud (1996*a*), 'Error estimates for finite element methods for nonlinear conservation laws', *SIAM J. Numer. Anal.* **33**, 522–554.

B. Cockburn and P.-A. Gremaud (1996*b*), '*A priori* error estimates for numerical methods for scalar conservation laws, Part I: The general approach', *Math. Comp.* **65**, 533–573.

B. Cockburn and P.-A. Gremaud (1997), '*A priori* error estimates for numerical methods for scalar conservation laws, Part II: Flux-splitting monotone schemes on irregular Cartesian grids', *Math. Comp.* **66**, 547–572.

B. Cockburn and G. Gripengerg (1999), 'Continuous dependence on the nonlinearities of solutions of degenerate parabolic equations', *J. Diff. Eqns* **151**, 231–251.

B. Cockburn and J. Qian (2002), Continuous dependence results for Hamilton–Jacobi equations, in *Collected Lectures on the Preservation of Stability under Discretization* (D. Estep and S. Tavener, eds), SIAM, pp. 67–90.

B. Cockburn and C.-W. Shu (2001), 'Runge–Kutta discontinuous Galerkin methods for convection-dominated problems', *J. Sci. Comput.* **16**, 173–261.

B. Cockburn, F. Coquel and P. LeFloch (1994), 'An error estimate for finite volume methods for multidimensional conservation laws', *Math. Comp.* **63**, 77–103.

B. Cockburn, F. Coquel and P. LeFloch (1995), 'Convergence of the finite volume method for multidimensional conservation laws', *SIAM J. Numer. Anal.* **32**, 687–705.

B. Cockburn, P.-A. Gremaud and J. Yang (1998), '*A priori* error estimates for hyperbolic conservation laws, Part III: Multidimensional flux-splitting monotone schemes in non-Cartesian grids', *SIAM J. Numer. Anal.* **35**, 1775–1803.

B. Cockburn, G. Karniadakis and C.-W. Shu, eds (2000), *Discontinuous Galerkin Methods: Theory, Computation and Applications*, Vol. 11 of *Lecture Notes in Computational Science and Engineering*, Springer.

B. Cockburn, G. Gripengerg and S.-O. Londen (2001), 'Continuous dependence on the nonlinearity of viscosity solutions of parabolic equations', *J. Diff. Eqns* **170**, 180–187.

F. Coquel and P. LeFloch (1991), 'Convergence of finite difference schemes for conservation laws in several space dimensions: The corrected antidiffusive flux approach', *Math. Comp.* **57**, 169–210.

M. Crandall and P. Lions (1983), 'Viscosity solutions of Hamilton–Jacobi equations', *Trans. Amer. Math. Soc.* **277**, 1–42.

M. Crandall and P. Lions (1984), 'Two approximations of solutions of Hamilton–Jacobi equations', *Math. Comp.* **43**, 1–19.

M. Crandall and A. Majda (1980), 'Monotone difference approximations for scalar conservation laws', *Math. Comp.* **34**, 1–21.

M. Crandall, L. Evans and P. Lions (1984), 'Some properties of viscosity solutions of Hamilton–Jacobi equations', *Trans. Amer. Math. Soc.* **282**, 478–502.

M. Crandall, H. Ishii and P. Lions (1992), 'User's guide to viscosity solutions of second-order partial differential equations', *Bull. Amer. Math. Soc.* **27**, 1–67.

C. Dafermos (2000), *Hyperbolic Conservation Laws in Continuum Physics*, Vol. 325 of *Grundlehren der Mathematischen Wissenschaften* (*Fundamental Principles of Mathematical Sciences*), Springer, Berlin.

R. DeVore and B. Lucier (1990), 'High order regularity for conservation laws', *Indiana Univ. Math. J.* **39**, 413–430.

R. DeVore and B. Lucier (1996), 'On the size and smoothness of solutions to nonlinear hyperbolic conservation laws', *SIAM J. Math. Anal.* **27**, 684–707.

R. DiPerna (1985), 'Measure-valued solutions to conservation laws', *Arch. Rat. Mech. Anal.* **88**, 223–270.

S. Evje and K. Karlsen (2000*a*), 'Discrete approximations of BV solutions to doubly nonlinear degenerate parabolic equations', *Numer. Math.* **86**, 377–417.

S. Evje and K. Karlsen (2000*b*), 'Monotone difference approximations of BV solutions to degenerate convection–diffusion equations', *SIAM J. Numer. Anal.* **37**, 1838–1860.

S. Evje and K. Karlsen (2002), 'An error estimate for viscous approximate solutions of degenerate parabolic equations', *J. Nonlin. Math. Phys.* **9**, 1–20.

R. Eymard, T. Gallouët, M. Ghilani and R. Herbin (1998), 'Error estimates for the approximate solutions of a nonlinear hyperbolic equation given by finite volume schemes', *IMA J. Numer. Anal.* **18**, 563–594.

R. Eymard, T. Gallouët and R. Herbin (2000), Finite volume methods, in *Handbook of Numerical Analysis, Vol. VII*, North-Holland, Amsterdam, pp. 713–1020.

R. Eymard, T. Gallouët and R. Herbin (2002*a*), 'Error estimate for approximate solutions of a nonlinear convection–diffusion problem', *Adv. Diff. Eqns* **7**, 419–440.

R. Eymard, T. Gallouët, R. Herbin and A. Michel (2002*b*), 'Convergence of a finite volume scheme for nonlinear degenerate parabolic equations', *Numer. Math.* **92**, 41–82.

A. Friedman (1964), *Partial Differential Equations of Parabolic Type*, Prentice-Hall, Englewood Cliffs, NJ.

E. Godlewski and P.-A. Raviart (1996), *Numerical Approximation of Hyperbolic Systems of Conservation Laws*, Vol. 118 of *Applied Mathematical Sciences*, Springer, Paris.

L. Gosse and C. Makridakis (2000), 'Two *a posteriori* error estimates for one-dimensional scalar conservation laws', *SIAM J. Numer. Anal.* **38**, 964–988.

G. Gripenberg (2002), 'Estimates for viscosity solutions of parabolic equations with Dirichlet boundary conditions', *Proc. Amer. Math. Soc.* **130**, 3651–3660.

A. Harten, J. M. Hyman and P. Lax (1976), 'On finite difference approximations and entropy conditions for shocks', *Comm. Pure Appl. Math.* **29**, 297–322.

H. Holden and N. Risebro (2002), *Front Tracking for Hyperbolic Conservation Laws*, Vol. 152 of *Applied Mathematical Sciences*, Springer, New York.

D. Jacobs, B. McKinney and M. Shearer (1995), 'Traveling wave solutions of the modified Korteweg–deVries–Burgers equations', *J. Diff. Eqns* **116**, 448–467.

J. Jaffré, C. Johnson and A. Szepessy (1995), 'Convergence of the discontinuous Galerkin finite element method for hyperbolic conservation laws', *Math. Models Methods Appl. Sci.* **5**, 367–386.

E. Jakobsen and K. Karlsen (2002), 'Continuous dependence estimates for viscosity solutions of fully nonlinear degenerate parabolic equations', *J. Diff. Eqns* **183**, 497–525.

C. Johnson (1998), Adaptive finite element methods for conservation laws, in *Advanced Numerical Approximation of Nonlinear Hyperbolic Equations* (A. Quarteroni, ed.), Vol. 1697 of *Lecture Notes in Mathematics; Subseries Fondazione C.I.M.E., Firenze*, Springer, pp. 269–323.

D. Kröner (1997), *Numerical Schemes for Conservation Laws*, Wiley–Teubner series *Advances in Numerical Mathematics*, Wiley, Chichester.

D. Kröner and M. Ohlberger (2000), '*A posteriori* error estimates for upwind finite volume schemes for nonlinear conservation laws in multidimensions', *Math. Comp.* **69**, 25–39.

D. Kröner and M. Rokyta (1994), 'Convergence of upwind finite volume methods for scalar conservation laws in two dimensions', *SIAM J. Numer. Anal.* **31**, 324–343.

D. Kröner, S. Noelle and M. Rokyta (1995), 'Convergence of higher order upwind finite volume schemes on unstructured grids for scalar conservation laws in several space dimensions', *Numer. Math.* **71**, 527–560.

D. Kröner, M. Ohlberger and C. Rohde, eds (1999), *An Introduction to Recent Developments in Theory and Numerics for Conservation Laws*, Vol. 5 of *Lecture Notes in Computational Science and Engineering*, Springer, Berlin. (Papers from the International School on Theory and Numerics for Conservation Laws held in Freiburg/Littenweiler, October 20–24, 1997.)

S. N. Kružkov (1970), 'First order quasilinear equations in several independent variables', *Math. USSR-Sb* **10**, 217–243.

N. Krylov (2000), 'SPDEs in $L_Q((0,\tau], L_P)$ spaces', *Electron. J. Probab.* **5**, #13.

N. N. Kuznetsov (1976), 'Accuracy of some approximate methods for computing the weak solutions of a first-order quasi-linear equation', *USSR Comp. Math. Math. Phys.* **16**, 105–119.

P. Lax (1972), *Hyperbolic Systems of Conservation Laws and the Mathematical Theory of Shock waves*, Vol. 11 of *CMBS Regional Conference Series in Applied Mathematics*, SIAM, Philadephia.

R. J. LeVeque (1990), *Numerical Methods for Conservation Laws*, Birkhäuser.

C.-T. Lin and E. Tadmor (2001), 'L^1-stability and error estimates for approximate Hamilton–Jacobi solutions', *Numer. Math.* **87**, 701–735.

T.-P. Liu (1997), *Hyperbolic and Viscous Systems of Conservation Laws*, Vol. 18 of *CMBS Regional Conference Series in Applied Mathematics*, SIAM, Philadephia.

B. J. Lucier (1985*a*), 'Error bounds for the methods of Glimm, Godunov and LeVeque', *SIAM J. Numer. Anal.* **22**, 1074–1081.

B. J. Lucier (1985*b*), 'A stable adaptive scheme for hyperbolic conservation laws', *SIAM J. Numer. Anal.* **22**, 180–203.

B. J. Lucier (1986), 'A moving mesh numerical method for hyperbolic conservation laws', *Math. Comp.* **46**, 59–69.

B. J. Lucier (1988), 'Regularity through approximation for scalar conservation laws', *SIAM J. Numer. Anal.* **19**, 763–773.

B. J. Lucier (200x), *Numerical Analysis of Conservation Laws.* In preparation.

B. J. Lucier and R. Overbeek (1987), 'A parallel adaptive numerical scheme for hyperbolic systems of conservation laws', *SIAM J. Sci. Statist. Comput.* **8**, S203–S219.

J. Málek, J. Nečas, M. Rokyta and M. Ružička (1996), *Weak and Measure-Valued Solutions to Evolutionary PDEs*, Vol. 13 of *Applied Mathematics and Mathematical Computation*, Chapman & Hall, London.

C. Mascia, A. Porretta and A. Terracina (2002), 'Non-homogeneous Dirichlet problems for degenerate parabolic–hyperbolic equations', *Arch. Rational Mech. Anal.* **163**, 87–124.

P. Morin, R. Nochetto and K. Siebert (2000), 'Data oscillation and convergence of adaptive FEM', *SIAM J. Numer. Anal.* **38**, 466–488.

S. Nöelle (1996), 'A note on entropy inequalities and error estimates for higher-order accurate finite volume schemes on irregular families of grids', *Math. Comp.* **65**, 1155–1163.

S. Nöelle (1995), 'Convergence of higher order finite volume schemes on irregular grids', *Adv. Comput. Math.* **3**, 197–218.

M. Ohlberger (2001), '*A posteriori* error estimates for vertex centered finite volume approximations of convection–diffusion–reaction equations', *Modél. Math. Anal. Numér.* **35**, 355–387.

J. Rauch (1986), 'BV estimates fail for most quasilinear hyperbolic systems in dimensions greater than one', *Commun. Math. Phys.* **106**, 481–484.

F. Şabac (1997), 'The optimal convergence rate of monotone finite difference methods for hyperbolic conservation laws', *SIAM J. Numer. Anal.* **34**, 2306–2318.

R. Sanders (1983), 'On convergence of monotone finite difference schemes with variable spacing differencing', *Math. Comp.* **40**, 91–106.

D. Serre (1999), *Systems of Conservation Laws, 1: Hyperbolicity, Entropies, Shock Waves*, Cambridge University Press, Cambridge. Translated from the 1996 French original by I. N. Sneddon.

D. Serre (2000), *Systems of Conservation Laws, 2: Geometric Structures, Oscillations, and Initial-Boundary Value Problems*, Cambridge University Press, Cambridge. Translated from the 1996 French original by I. N. Sneddon.

C.-W. Shu (1992), 'A numerical method for systems of conservation laws of mixed type admitting hyperbolic flux splitting', *J. Comput. Phys.* **100**, 424–429.

C.-W. Shu (1998), Essentially non-oscillatory and weighted essentially non-oscillatory schemes for hyperbolic conservation laws, in *Advanced Numerical Approximation of Nonlinear Hyperbolic Equations* (A. Quarteroni, ed.), Vol. 1697 of *Lecture Notes in Mathematics: Subseries Fondazione C.I.M.E., Firenze*, Springer, pp. 325–432.

M. Slemrod (1984), 'Dynamic phase transitions in a van der Waals fluid', *J. Diff. Eqns* **52**, 1–23.

M. Slemrod and J. Flaherty (1986), Numerical integration of a Riemman problem for a van der Waals fluid, in *Phase Transformations* (E. Aifantis and J. Gittus, eds), Elsevier Applied Science, pp. 203–212.

A. Szepessy (1989), 'Convergence of a shock-capturing streamline-diffusion finite element method for scalar conservation laws in two space dimensions', *Math. Comp.* **53**, 527–545.

A. Szepessy (1991), 'Convergence of a streamline-diffusion finite element method for a conservation law with boundary conditions', *RAIRO Modél. Math. Anal. Numér.* **25**, 749–783.

E. Tadmor (1998), Approximate solutions of nonlinear conservation laws, in *Advanced Numerical Approximation of Nonlinear Hyperbolic Equations* (A. Quarteroni, ed.), Vol. 1697 of *Lecture Notes in Mathematics: Subseries Fondazione C.I.M.E., Firenze*, Springer, pp. 1–149.

E. Toro (1997), *Riemann Solvers and Numerical Methods for Fluid Dynamics*, Springer.

L. Truskinovsky (1982), 'Equilibrium phase interfaces', *Dokl. Akad. Nauk. SSSR* **256**, 306–310.

J.-P. Vila (1994), 'Convergence and error estimates for finite volume schemes for general multidimensional scalar conservation laws', *Model. Math. Anal. Numér.* **28**, 267–295.

G. B. Whitham (1974), *Linear and Nonlinear Waves*, Wiley, New York.

X. Zhong, T. Hou and P. LeFloch (1996), 'Computational methods for propagating phase boundaries', *J. Comput. Phys.* **124**, 192–216.

Appendix: Proofs of some results

In this appendix, we sketch the proofs of some of the results in this paper.

Proof of Proposition 4.3 (a priori error estimates for parabolic solutions). Property (i) follows from Theorem 4.2 by first taking $v(x,t) = a(t)$, $s(t) = -\sup_{x\in\mathbb{R}^d} \max\{-r(x,t),0\}$ and $U(e) = \max\{-e,0\}$, and then $v(x,t) = b(t)$, $s(t) = \sup_{x\in\mathbb{R}^d} \max\{r(x,t),0\}$ and $U(e) = \max\{e,0\}$.

Property (ii) is obtained by proving that, for each $i = 1,\ldots,d$, we have

$$\|\partial_{x_i} u(T)\|_{L^1} \le \|\partial_{x_i} u_0\|_{L^1} + \|\partial_{x_i} r\|_{L^1(L^1)},$$

and then summing over i. To prove the above inequality, we take, in Theorem 4.2, $v(x,t) = u(x+h\,e_i,t)$, $s(x,t) = r(x+h\,e_i,t)$, and $U(e) = |e|$, where e_i is the ith canonical vector in $\mathbb{R}^d$. The result follows after dividing by $|h|$ and letting h tend to zero.

This completes the proof. □

Proof of Proposition 4.4 (continuous dependence results for parabolic solutions). Property (i) follows from Theorem 4.2 by taking $s = r$ and $U(e) = \max\{-e,0\}$, and property (ii) by taking $s = r$ and $U(e) = |e|$. To obtain property (iii), we begin by taking $v(t) = u(t+\delta t)$ in (ii), dividing by $\delta t > 0$ and letting it tend to zero, obtaining

$$\|u_t(T)\|_{L^1} \le \|u_t(0)\|_{L^1} + \|r_t\|_{L^1(L^1)}.$$

Then we use the estimate

$$\|u_t(0)\|_{L^1} \le \|\mathbf{f}\|_{L^\infty(I)}\,|u_0|_{TV} + \nu\,\|\Delta u_0\|_{L^1}.$$

Finally, property (iv) follows by taking $u = u_1$, $v = u_2$ and $s = r + \mathbf{f}_1(u_2) - \mathbf{f}_2(u_2)$. □

Proof of Theorem 5.1 (continuous dependence for parabolic equations with the doubling of the space variable technique). One way to introduce the technique of doubling the space variable is to consider working with the convolution $\varphi_{\epsilon_x} * u$, because this regularization inherits the smoothness of the convolution kernel φ_{ϵ_x}; we would then avoid the problems associated with the appearance of strong gradients as the viscosity coefficient becomes smaller and smaller.

Our objective would now be to estimate

$$|\varphi_{\epsilon_x} * u - v|_U,$$

but it is preferable to symmetrize the role of u and v and work instead with the functional

$$\langle u, v\rangle_U^{\epsilon_x} := \int_{\mathbb{R}^d\times\mathbb{R}^d} \varphi_{\epsilon_x}(x-y)\,U(u(x)-v(y))\,\mathrm{d}y\,\mathrm{d}x. \tag{A.1}$$

Indeed, not only do we have that

$$\max\{|\varphi_{\epsilon_x} * u - v|_U,\, |u - \varphi_{\epsilon_x} * v|_U\} \le \langle u, v\rangle_U^{\epsilon_x},$$

for all convex, nonnegative functions U, but also

$$\big||u - v|_U - \langle u, v\rangle_U^{\epsilon_x}\big| \le \langle u, u\rangle_U^{\epsilon_x} \le \mathrm{c}_1\, \epsilon_x\, \|\nabla u\|_{L^1}, \tag{A.2}$$

for U, as in (4.5).

We thus see that an estimate on $\langle u, v\rangle_U^{\epsilon_x}$ immediately implies an estimate on $|u - v|_U$. We also see that the introduction of the convolution naturally leads to the so-called *doubling of the variables* technique, as now $u = u(x, t)$ and $v = v(y, t)$ depend on different space variables.

To prove Theorem 5.1, we begin by obtaining the following extension of Lemma 4.1.

Lemma A1. Let u be the solution of (4.1) and (4.2) and let v be the solution of (5.3) and (5.4). Let U be any smooth function such that $U(0)=0$. Then,

$$\langle u(T), v(T)\rangle_U^{\epsilon_x} + \Theta_U^{\epsilon_x} = \langle u_0, v_0\rangle_U^{\epsilon_x} + V_U^{\epsilon_x} + R_U^{\epsilon_x},$$

where

$$\begin{aligned}
\Theta_U^{\epsilon_x} &= \int_0^T\!\!\int_{\mathbb{R}^d\times\mathbb{R}^d} \varphi_{\epsilon_x}(x-y)\,(Y_{\mathbf{f},U,\epsilon_x} + Z_{\nu,U,\epsilon_x})\,\mathrm{d}x\,\mathrm{d}y\,\mathrm{d}t,\\
V_U^{\epsilon_x} &= (\sqrt{\nu_1} - \sqrt{\nu_2})^2 \int_0^T\!\!\int_{\mathbb{R}^d\times\mathbb{R}^d} U(u-v)\,\Delta_y\varphi_{\epsilon_x}(x-y)\,\mathrm{d}x\,\mathrm{d}y\,\mathrm{d}t,\\
R_U^{\epsilon_x} &= \int_0^T\!\!\int_{\mathbb{R}^d\times\mathbb{R}^d} \varphi_{\epsilon_x}(x-y)\,U'(u-v)\,(r-s)\,\mathrm{d}x\,\mathrm{d}y\,\mathrm{d}t,
\end{aligned}$$

and

$$\begin{aligned}
Y_{\mathbf{f},U} &= \nabla_y v \cdot \int_v^u U''(w-v)(\mathbf{f}'(w) - \mathbf{f}'(v))\,\mathrm{d}w,\\
Z_{\nu,U,\epsilon_x} &= |\sqrt{\nu}\,\nabla_x u - \sqrt{\nu'}\nabla_y v|^2\, U''(u-v).
\end{aligned}$$

In the above expressions, $u = u(x,t)$, $r = r(x,t)$, $v = v(y,t)$ and $s = s(y,t)$.

Proof. Let us begin by pointing out that, since we have doubled the variables, we indicate that we are differentiating with respect to the variable x by the sub-index 'x', and with respect to the variable y by the sub-index 'y'. Thus, we rewrite equations (4.1) and (5.3) as follows:

$$\begin{aligned}
u_t + \nabla_x \cdot \mathbf{f}(u) - \nu\Delta_x u &= r,\\
v_t + \nabla_y \cdot \mathbf{f}(v) - \nu'\Delta_y v &= s.
\end{aligned}$$

Subtracting the second equation from the first and multiplying by $U'(u-v)$, we get

$$(U(u-v))_t + C_{\mathbf{f}} + D_\nu = U'(u-v)\,(r-s),$$

where

$$\begin{aligned} C_{\mathbf{f}} &= U'(u-v)\,(\nabla_x\cdot\mathbf{f}(u) - \nabla_y\cdot\mathbf{f}(v)),\\ D_\nu &= U'(u-v)\,(\nu\Delta_x u - \nu'\Delta_y v). \end{aligned}$$

Next, we rewrite the above expression by using the fact that $u = u(x,t)$ does not depend on y and that $v = v(y,t)$ does not depend on x.

Let us begin with the term associated with convection, $C_{\mathbf{f}}$. Since

$$\begin{aligned} U'(u-v)\,\nabla_x\cdot\mathbf{f}(u) &= \nabla_x\cdot\mathbf{F}(u,v),\\ -U'(u-v)\,\nabla_y\cdot\mathbf{f}(v) &= \nabla_y\cdot\mathbf{G}(u,v), \end{aligned}$$

where

$$\begin{aligned} \mathbf{F}(u,v) &= \int_v^u U'(w-v)\,\mathbf{f}'(w)\,\mathrm{d}w,\\ \mathbf{G}(u,v) &= \int_v^u U'(u-w)\,\mathbf{f}'(w)\,\mathrm{d}w, \end{aligned}$$

we obtain that

$$\begin{aligned} C_{\mathbf{f}} &= \nabla_x\cdot\mathbf{F}(u,v) + \nabla_y\cdot\mathbf{G}(u,v)\\ &= (\nabla_x+\nabla_y)\cdot\mathbf{F}(u,v) + \nabla_y\cdot(\mathbf{G}(u,v)-\mathbf{F}(u,v))\\ &= (\nabla_x+\nabla_y)\cdot\mathbf{F}(u,v) + Y_{\mathbf{f},U}. \end{aligned}$$

Now let us work on the term associated to diffusion, D_ν. Since

$$\begin{aligned} U'(u-v)\,\Delta_x u &= \Delta_x\,U(u-v) - U''(u-v)\,|\nabla_x u|^2,\\ -U'(u-v)\,\Delta_y v &= \Delta_y\,U(u-v) - U''(u-v)\,|\nabla_y v|^2, \end{aligned}$$

we get that

$$D_\nu = (\nu\Delta_x + \nu'\Delta_y)\,U(u-v) + U''(u-v)\,\big(\nu|\nabla_x u|^2 + \nu'|\nabla_y v|^2\big).$$

Completing squares in the last term of the right-hand side by adding and subtracting the expression

$$2\sqrt{\nu\nu'}\,U''(u-v)\,\nabla_x u\cdot\nabla_y v = -2\sqrt{\nu\nu'}\,\nabla_x\cdot\nabla_y U(u-v),$$

we obtain

$$D_\nu = \mathbb{L}_\nu\,U(u-v) + Z_{\nu,U},$$

where

$$\mathbb{L}_\nu = \nu\Delta_x + 2\sqrt{\nu\nu'}\,\nabla_x\cdot\nabla_y + \nu'\Delta_y.$$

As a consequence, we get

$$\begin{aligned} (U(u-v))_t &+ (\nabla_x+\nabla_y)\cdot\mathbf{F}(u,v) + Y_{\mathbf{f},U}\\ &- \mathbb{L}_\nu\,U(u-v) + Z_{\nu,U} = U'(u-v)\,(r-s). \end{aligned}$$

Now we multiply the above equation by the convolution kernel $\varphi_{\epsilon_x}(x-y)$ and integrate over $\mathbb{R}^d \times \mathbb{R}^d$ the on $(0,T)$. We get

$$\langle u(T), v(T)\rangle_U^{\epsilon_x} + \Theta_U^{\epsilon_x} + \widehat{C}_{\mathbf{f}} = \langle u_0, v_0\rangle_U^{\epsilon_x} + \widehat{D}_\nu + R_U^{\epsilon_x},$$

where

$$\widehat{C}_{\mathbf{f}} = \int_0^T\int_{\mathbb{R}^d\times\mathbb{R}^d} \varphi_{\epsilon_x}(x-y)(\nabla_x + \nabla_y)\cdot \mathbf{F}(u,v)\,\mathrm{d}x\,\mathrm{d}y\,\mathrm{d}t,$$

$$\widehat{D}_\nu = \int_0^T\int_{\mathbb{R}^d\times\mathbb{R}^d} \varphi_{\epsilon_x}(x-y)\,\mathbb{L}_\nu U(u-v)\,\mathrm{d}x\,\mathrm{d}y\,\mathrm{d}t.$$

After a simple integration by parts, we obtain that $\widehat{C}_{\mathbf{f}} = 0$ and $\widehat{D}_\nu = V_U^{\epsilon_x}$ since

$$(\nabla_x + \nabla_y)\varphi_{\epsilon_x}(x-y) = 0, \quad \mathbb{L}_\nu\varphi_{\epsilon_x}(x-y) = (\sqrt{\nu} - \sqrt{\nu'})^2\,\Delta_y\varphi_{\epsilon_x}(x-y).$$

This completes the proof. □

We can now prove Theorem 5.1. We have, by (A.2),

$$|u(T) - v(T)\,|_U \le \langle u(T), v(T)\rangle_U^{\epsilon_x} + \mathsf{c}_1\,\epsilon_x\ \min\{\|u(T)\|_{TV}, \|v(T)\|_{TV}\},$$
$$\langle u_0, v_0\rangle_U^{\epsilon_x} \le |u_0 - v_0\,|_U + \mathsf{c}_1\,\epsilon_x\ \min\{\|u_0\|_{TV}, \|v_0\|_{TV}\}.$$

Hence

$$|u(T) - v(T)\,|_U \le |u_0 - v_0\,|_U + \mathsf{c}_1\,\epsilon_x\ \mathsf{A} + \langle u(T), v(T)\rangle_U^{\epsilon_x} - \langle u_0, v_0\rangle_U^{\epsilon_x}.$$

Then, by Lemma A1, with U given by (4.5),

$$|u(T) - v(T)\,|_U \le |u_0 - v_0\,|_U + \mathsf{c}_1\,\epsilon_x\ \mathsf{A} + V_U^{\epsilon_x} + R_U^{\epsilon_x}.$$

Next, let us estimate $V_U^{\epsilon_x}$. Integrating by parts, we get

$$V_U^{\epsilon_x} = (\sqrt{\nu}\ - \sqrt{\nu'})^2 \int_0^T\int_{\mathbb{R}^d\times\mathbb{R}^d} U'(u-v)\,\nabla_y v\cdot\nabla_y\varphi_{\epsilon_x}(x-y)\,\mathrm{d}x\,\mathrm{d}y\,\mathrm{d}t,$$

where $u = u(x,t)$ and $v = v(y,t)$, and since $|U'| \le 1$,

$$V_U^{\epsilon_x} \le (\sqrt{\nu}\ - \sqrt{\nu'})^2\,\frac{\mathsf{c}_2}{\epsilon_x}\,\|v\|_{L^1(TV)}.$$

Similarly, since $\Delta_y\varphi_{\epsilon_x}(x-y) = \Delta_x\varphi_{\epsilon_x}(x-y)$, we obtain

$$V_U^{\epsilon_x} \le (\sqrt{\nu}\ - \sqrt{\nu'})^2\,\frac{\mathsf{c}_2}{\epsilon_x}\,\|u\|_{L^1(TV)}.$$

Finally, we have, for U given by (4.5),

$$R_U^{\epsilon_x} = \int_0^T\int_{\mathbb{R}^d\times\mathbb{R}^d} \varphi_{\epsilon_x}(x-y)\,U'(u-v)\,(r(x,t) - s(y,t))\,\mathrm{d}y\,\mathrm{d}x\,\mathrm{d}t \le \Lambda_U^{\epsilon_x}(v).$$

This completes the proof. □

Proof of Theorem 6.1 (continuous dependence with the doubling of the space and time variables). The only difference between the technique of doubling the space variable and doubling the time variable is associated with the treatment of the boundaries of the time domain, $(0,T)$. We display this difference on the model equations

$$u_t = F(t) \quad \forall\, t \in (0,T), \qquad u(0) = u_0,$$
$$v_\tau = G(\tau) \quad \forall\, \tau \in (0,T), \qquad v(0) = v_0.$$

Lemma A2. For U given by (4.5), we have

$$|u(T) - v(T)|_U \le |u_0 - v_0|_U + \mathsf{D}\,\epsilon_t + \Theta_U^{\epsilon_t},$$

where

$$\mathsf{D} = |u_t|_{L^\infty(0,T)} + |v_t|_{L^\infty(0,T)},$$

$$\Theta_U^{\epsilon_t} = \frac{1}{\Phi_{\epsilon_t}(T)} \int_0^T \int_0^T \varphi_{\epsilon_t}(t-\tau)\, U'(u(t) - v(\tau))\,(F(t) - G(\tau))\,\mathrm{d}\tau\,\mathrm{d}t.$$

Proof. Let U be any smooth function. Then

$$(\partial_t + \partial_\tau)\, U(u(t) - v(\tau)) = U'(u(t) - v(\tau))\,(F(t) - G(\tau)).$$

Multiplying by $\varphi_{\epsilon_t}(t-\tau)$ and using the fact that $(\partial_t + \partial_\tau)\,\varphi_{\epsilon_t}(t-\tau) = 0$, we get

$$D_u + D_v = \Phi_{\epsilon_t}(T)\,\Theta_U^{\epsilon_t},$$

where

$$D_u = \int_0^T \varphi_{\epsilon_t}(t-\tau)\, U(u(t) - v(\tau))\,\mathrm{d}\tau \Big|_{t=0}^{t=T},$$

$$D_v = \int_0^T \varphi_{\epsilon_t}(t-\tau)\, U(u(t) - v(\tau))\,\mathrm{d}t \Big|_{\tau=0}^{\tau=T}.$$

Now, note that, for U given by (4.5), the triangle inequality gives

$$\int_0^T \varphi_{\epsilon_t}(T-\tau)\, U(u(T) - v(\tau))\,\mathrm{d}\tau \ge \frac{1}{2}\,\Phi_{\epsilon_t}(T)\, U(u(T) - v(T)) - \frac{\epsilon_t}{2}\,\Phi_{\epsilon_t}(T)\,|v_t|_{L^\infty(0,T)}.$$

In a similar manner, the following inequalities are obtained:

$$D_u \ge \frac{1}{2}\,\Phi_{\epsilon_t}(T)\,\big(U(u(T) - v(T)) - U(u_0 - v_0)\big) - \Phi_{\epsilon_t}(T)\,|v_t|_{L^\infty(0,T)},$$
$$D_v \ge \frac{1}{2}\,\Phi_{\epsilon_t}(T)\,\big(U(u(T) - v(T)) - U(u_0 - v_0)\big) - \Phi_{\epsilon_t}(T)\,|u_t|_{L^\infty(0,T)}.$$

This completes the proof. □

Using the above lemma, we prove Theorem 6.1. If v is the solution of (5.3) and (5.4), then it is not difficult to combine Lemma A2 with the proof of Theorem 5.1, to obtain

$$|u(T) - v(T)|_U \le |u_0 - v_0|_U + \inf_{\epsilon_x,\epsilon_t>0}\left(\mathsf{A}\,\epsilon_x + \frac{(\sqrt{\nu} - \sqrt{\nu'})^2}{\epsilon_x}\mathsf{B} + \mathsf{C}\,\epsilon_t + \Lambda_U^{\epsilon_x,\epsilon_t}(v)\right),$$

where

$$\Lambda_U^{\epsilon_x,\epsilon_t}(v) = \frac{1}{\Phi_{\epsilon_t}(T)}\int_0^T\int_{\mathbb{R}^d} \mathcal{E}_U^{\epsilon_x,\epsilon_t}(v,u;x,t)\,\mathrm{d}x\,\mathrm{d}t,$$

and

$$\mathcal{E}_U^{\epsilon_x,\epsilon_t}(v,u;x,t) = \int_0^T\int_{\mathbb{R}^d} \varphi_{\epsilon_t,\epsilon_x}\,U'(u-v)\,(r-s)\,\mathrm{d}y\,\mathrm{d}\tau.$$

Now, note that we can write

$$U'(u-v)\,(r-s) = U'(u-v)\,r + (U(u-v))_\tau + \nabla_y\cdot\mathbf{G}(u,v) + \nu'\,U'(u-v)\,\Delta_y v,$$

and so

$$\begin{aligned}
\mathcal{E}_U^{\epsilon_x,\epsilon_t}(v,u;x,t) &= \int_{\mathbb{R}^d}\varphi_{\epsilon_t,\epsilon_x}U(u-v)\,\mathrm{d}y\Bigg|_{\tau=0}^{\tau=T} - \int_0^T\int_{\mathbb{R}^d} U(u-v)\,(\varphi_{\epsilon_t,\epsilon_x})_t\,\mathrm{d}y\,\mathrm{d}\tau\\
&\quad - \int_0^T\int_{\mathbb{R}^d}\big(\mathbf{G}(u,v)\cdot\nabla_y\varphi_{\epsilon_t,\epsilon_x} - U'(u-v)\,r\,\varphi_{\epsilon_t,\epsilon_x}\big)\,\mathrm{d}y\,\mathrm{d}\tau\\
&\quad + \nu'\int_0^T\int_{\mathbb{R}^d} U'(u-v)\,\Delta_y v\;\varphi_{\epsilon_t,\epsilon_x}\,\mathrm{d}y\,\mathrm{d}\tau\\
&\le \sup_{c\in\mathbb{R}} E_U^{\epsilon_x,\epsilon_t}(v,c;x,t) + \nu'\int_0^T\int_{\mathbb{R}^d} U'(u-v)\,\Delta_y v\;\varphi_{\epsilon_t,\epsilon_x}\,\mathrm{d}y\,\mathrm{d}\tau.
\end{aligned}$$

If we let ν' and then ν tend to zero, we obtain the result for smooth v. Now the result follows from a classical density argument. This completes the proof. □

Proof of Theorem 6.2 (continuous dependence with a variation of the doubling of the time-variable technique).

Lemma A3. For U given by (4.5), we have

$$|u(T) - v(T)|_U \le 2\,|u_0 - v_0|_U + 4\,\mathsf{D}\,\epsilon_t + 4\sup_{z\in[0,\epsilon_t]\cup\{T\}}\Theta_U^{\epsilon_t}(z),$$

where

$$\mathsf{D} = 2\,\min\{|u_t|_{L^\infty(0,T)}, |v_t|_{L^\infty(0,T)}\},$$

$$\Theta_U^{\epsilon_t}(z) = \frac{1}{\Phi_{\epsilon_t}(z)}\int_0^z\int_0^z \varphi_{\epsilon_t}(t-\tau)\,U'(u(t)-v(\tau))\,(F(t)-G(\tau))\,\mathrm{d}\tau\,\mathrm{d}t.$$

Proof. We proceed exactly as in the proof of Lemma A2, to get

$$D_u + D_v = \Phi_{\epsilon_t}(T)\,\Theta_U^{\epsilon_t}.$$

Next, we estimate D_u in a different way. Setting $e(t) = U(u(t) - v(t))$, we get

$$\begin{aligned}\int_0^T \varphi_{\epsilon_t}(T-\tau)\,U(u(T)-v(\tau))\,\mathrm{d}\tau &\geq \int_0^T \varphi_{\epsilon_t}(T-\tau)\,e(\tau)\,\mathrm{d}\tau\\ &\quad - \frac{\epsilon_t}{2}\Phi_{\epsilon_t}(T)\,|u_t|_{L^\infty(0,T)},\\ \int_0^T \varphi_{\epsilon_t}(0-\tau)\,U(u_0-v(\tau))\,\mathrm{d}\tau &\leq \int_0^T \varphi_{\epsilon_t}(\tau)\,e(\tau)\,\mathrm{d}\tau\\ &\quad + \frac{\epsilon_t}{2}\,\Phi_{\epsilon_t}(T)\,|u_t|_{L^\infty(0,T)}.\end{aligned}$$

As a consequence, we obtain

$$\begin{aligned}D_u &\geq \int_0^T (\varphi_{\epsilon_t}(T-\tau) - \varphi_{\epsilon_t}(\tau))\,e(\tau)\,\mathrm{d}\tau - \epsilon_t\,\Phi_{\epsilon_t}(T)\,|u_t|_{L^\infty(0,T)},\\ D_v &\geq \frac{1}{2}\,\Phi_{\epsilon_t}(T)\,(e(T)) - e(0)) - \epsilon_t\,\Phi_{\epsilon_t}(T)\,|u_t|_{L^\infty(0,T)}.\end{aligned}$$

In a similar way, we get

$$\begin{aligned}D_u &\geq \frac{1}{2}\,\Phi_{\epsilon_t}(T)\,(e(T)) - e(0)) - \epsilon_t\,\Phi_{\epsilon_t}(T)\,|v_t|_{L^\infty(0,T)},\\ D_v &\geq \int_0^T (\varphi_{\epsilon_t}(T-\tau) - \varphi_{\epsilon_t}(\tau))\,e(\tau)\,\mathrm{d}\tau - \epsilon_t\,\Phi_{\epsilon_t}(T)\,|v_t|_{L^\infty(0,T)}.\end{aligned}$$

As a consequence,

$$e(T) + \frac{2}{\Phi_{\epsilon_t}(T)}\int_0^T \varphi_{\epsilon_t}(T-\tau)\,e(\tau)\,\mathrm{d}\tau \leq \zeta(T) + \frac{2}{\Phi_{\epsilon_t}(T)}\int_0^T \varphi_{\epsilon_t}(\tau)\,e(\tau)\,\mathrm{d}\tau,$$

where

$$\zeta(T) = e(0) + 2\,\epsilon_t\,\mathsf{D} + 2\,\Theta_U^{\epsilon_t}(T).$$

Thus, if $T \leq \epsilon_t$, $\varphi_{\epsilon_t}(T-\tau) = \varphi_{\epsilon_t}(\tau)$ for all $\tau \in [0,T]$, and so

$$e(T) \leq \zeta(T).$$

Now, if $T \geq \epsilon_t$,

$$e(T) \leq \zeta(T) + \sup_{\tau\in(0,\epsilon_t)} e(\tau) \leq 2 \sup_{z\in[0,\epsilon_t]\cup\{T\}} \zeta(z),$$

and the result follows. This completes the proof. $\square$

Proof of Corollary 7.1 (*characterization of piecewise smooth entropy solutions*). By Theorem 6.1, with $U(e) = |e|$, we have

$$\|u(T) - v(T)\|_{L^1} \le \|u_0 - v_0\|_{L^1} + \inf_{\epsilon_x, \epsilon_t > 0} \big(\mathsf{A}\,\epsilon_x + \mathsf{C}\,\epsilon_t + \Lambda_U^{\epsilon_x, \epsilon_t}(v)\big),$$

where

$$\Lambda_U^{\epsilon_x,\epsilon_t}(v) = \frac{1}{\Phi_{\epsilon_t}(T)} \int_0^T \int_{\mathbb{R}} \sup_{c \in \mathbb{R}} E_U^{\epsilon_x,\epsilon_t}(v, c; x, t)\, \mathrm{d}x\, \mathrm{d}t,$$

and, in our case, $E_U^{\epsilon_x,\epsilon_t}(v, c; x, t)$ is equal to

$$\int_{\Omega^- \cup \Omega^+} U'(c - v)\, R(v)\, \varphi_{\epsilon_t,\epsilon_x}\, \mathrm{d}y\, \mathrm{d}\tau - \int_\Gamma [\![(G(c, v), U(c - v)) \cdot \mathbf{n}]\!]\, \varphi_{\epsilon_t,\epsilon_x}\, \mathrm{d}\gamma$$
$$\le \int_{\Omega^- \cup \Omega^+} |R(v)|\, \varphi_{\epsilon_t,\epsilon_x}\, \mathrm{d}y\, \mathrm{d}\tau + \int_\Gamma \mathcal{R}(v)\, \varphi_{\epsilon_t,\epsilon_x}\, \mathrm{d}\gamma.$$

Hence

$$\Lambda_U^{\epsilon_x,\epsilon_t}(v) \le \|R(v)\|_{L^1(\mathbb{R}\times(0,T)\setminus\Gamma)} + \|\mathcal{R}(v)\|_{L^1(\Gamma)},$$

and the result follows. This completes the proof. □

Proof of Proposition 7.2 (*characterization of piecewise smooth entropy solutions*). Let v be any function satisfying the conditions (i) and (ii). Then, by Corollary 7.1,

$$\|u(T) - v(T)\|_{L^1} \le \int_0^T \mathcal{R}(v)(x(t), t)\, \mathrm{d}t.$$

Next we show that $\mathcal{R}(v) = 0$ if and only if conditions (iii) and (iv) are satisfied. This implies that $u = v$, as claimed.

So, by definition, $\mathcal{R}(v) = 0$ on Γ if

$$-[\![(G(v, c), U(v - c)) \cdot \mathbf{n}]\!] \le 0$$

for all $c \in \mathbb{R}$ on each point of Γ. From this inequality, and since

$$\mathbf{n}^\pm = \pm\rho\Big(-1, \frac{\mathrm{d}x}{\mathrm{d}t}\Big) \quad \text{where} \quad \rho^{-1} = \Big|\Big(-1, \frac{\mathrm{d}x}{\mathrm{d}t}\Big)\Big|,$$

we get that

$$(f(v^+) - f(v^-)) - \frac{\mathrm{d}x}{\mathrm{d}t}(v^+ - v^-) \le 0 \qquad \text{for } c < a = \min\{v^-, v^+\},$$
$$(f(v^+) - f(v^-)) - \frac{\mathrm{d}x}{\mathrm{d}t}(v^+ - v^-) \ge 0 \qquad \text{for } c > b = \max\{v^-, v^+\}.$$

These two inequalities are equivalent to condition (iii).

Finally, for $c \in (a, b)$ we get

$$\mathrm{sign}(v^+ - v^-)\Big((f(v^+) - 2\,f(c) + f(v^-)) - \frac{\mathrm{d}x_i}{\mathrm{d}t}(v^+ - 2\,c + v^-)\Big) \le 0.$$

After a few simple algebraic manipulations, we obtain the inequality

$$\frac{f(v^+) - f(v^-)}{v^+ - v^-} \leq \frac{f(c) - f(v^-)}{c - v^-} \qquad \forall\, c \in (a, b),$$

which is equivalent to condition (iv). This completes the proof. □

Proof of Theorem 7.3 (characterization of the entropy solution). Let u_ν be the solution of the parabolic problem (4.1) and (4.2). Then

$$\mathsf{E}(u_0, c; u_\nu, \varphi) = -\nu \int_0^T \int_{\mathbb{R}^d} U'(c - u_\nu)\, \Delta\, u_\nu\, \varphi\, \mathrm{d}y\, \mathrm{d}t \leq \nu\, |u_\nu|_{L^1(TV)}\, \|\nabla \varphi\|_{L^\infty},$$

and by Proposition 5.2,

$$\mathsf{E}(u_0, c; u, \varphi) = \lim_{\nu \downarrow 0} \mathsf{E}(u_0, c; u_\nu, \varphi) \leq 0.$$

This shows that the set of functions satisfying the inequality (7.3) is not empty.

Now, let v be a function such that

$$\mathsf{E}(u_0, c; v, \varphi) \leq 0.$$

Then, by Theorem 6.2 we have

$$|u(T) - v(T)|_U \leq 2 \inf_{\epsilon_x, \epsilon_t > 0} \left(\mathsf{A}\, \epsilon_x + \mathsf{C}\, \epsilon_t + \Lambda_U^{\epsilon_x, \epsilon_t} \right),$$

where

$$\Lambda_U^{\epsilon_x, \epsilon_t} = \frac{1}{\Phi_{\epsilon_t}(T)} \int_0^T \int_{\mathbb{R}^d} \sup_{c \in \mathbb{R}} E_U^{\epsilon_x, \epsilon_t}(v, c; x, t)\, \mathrm{d}x\, \mathrm{d}t.$$

Since

$$E_U^{\epsilon_x, \epsilon_t}(v, c; x, t) = E_{\mathsf{small}} + E_{\mathsf{diss}},$$

where

$$E_{\mathsf{small}} = -\int_0^T \int_{\mathbb{R}^d} U'(c - v)\, (r(x, t) - r(y, \tau))\, \varphi_{\epsilon_t, \epsilon_x}\, \mathrm{d}y\, \mathrm{d}\tau,$$

$$E_{\mathsf{diss}} = \mathsf{E}(u_0, c; v, \varphi_{\epsilon_x, \epsilon_t}),$$

we have that

$$E_{\mathsf{small}} \leq \int_0^T \int_{\mathbb{R}^d} |r(x, t) - r(y, \tau)|\, \varphi_{\epsilon_t, \epsilon_x}\, \mathrm{d}y\, \mathrm{d}\tau,$$

$$E_{\mathsf{diss}} \leq 0,$$

and so

$$\Lambda_U^{\epsilon_x, \epsilon_t} \leq \mathsf{c}_1\, \epsilon_x\, |r|_{L^1(TV)} + \epsilon_t\, |r_t|_{L^1(L^1)}.$$

Minimizing over the parameters ϵ_t and ϵ_x, we see that v is in fact the entropy solution u. This completes the proof. □

Acta Numerica (2003), pp. 181–266
DOI: 10.1017/S0962492902000119

Computational high frequency wave propagation

Björn Engquist
PACM, Department of Mathematics,
Princeton University,
Princeton, NJ 08544, USA
E-mail: engquist@math.princeton.edu

Olof Runborg
Department of Numerical Analysis
and Computer Science,
Royal Institute of Technology (KTH),
10044 Stockholm, Sweden
E-mail: olofr@nada.kth.se

Numerical simulation of high frequency acoustic, elastic or electro-magnetic wave propagation is important in many applications. Recently the traditional techniques of ray tracing based on geometrical optics have been augmented by numerical procedures based on partial differential equations. Direct simulations of solutions to the eikonal equation have been used in seismology, and lately approximations of the Liouville or Vlasov equation formulations of geometrical optics have generated impressive results. There are basically two techniques that follow from this latter approach: one is wave front methods and the other moment methods. We shall develop these methods in some detail after a brief review of more traditional algorithms for simulating high frequency wave propagation.

CONTENTS

1. Introduction

The numerical approximation of high frequency wave propagation is important in many applications. Examples are the simulation of seismic, acoustic and optical waves, and microwaves. When the *essential frequencies* in the wave field are relatively high, and thus the wavelengths are short compared to the overall size of the computational domain, direct simulation using the standard wave equations will be very costly, and approximate models for wave propagation must be used. Fortunately, there exist good approximations of many wave equations precisely for very high frequency solutions. Even for linear wave equations these approximations are often nonlinear. It is the goal of this paper to discuss numerical simulations based on such high frequency approximations.

We consider the linear scalar wave equation

$$u_{tt} - c(\boldsymbol{x})^2 \Delta u = 0, \qquad (t, \boldsymbol{x}) \in \mathbb{R}^+ \times \Omega, \qquad \Omega \subset \mathbb{R}^d, \tag{1.1}$$

where $c(\boldsymbol{x})$ is the local speed of wave propagation of the medium. We complement (1.1) with initial or boundary data that generate high frequency solutions. The exact form of the data will not be important here, but a typical example would be $u(t, \boldsymbol{x}) = A(t, \boldsymbol{x}) \exp(\mathrm{i}\omega(c(\boldsymbol{x})t - \boldsymbol{k} \cdot \boldsymbol{x}))$ at $t = 0$ and with $|\boldsymbol{k}|^2 = \sum_{j=1}^{d} k_j^2 = 1$ and the frequency $\omega \gg 1$. With $u(t, \boldsymbol{x}) = \exp(\mathrm{i}\omega t) v(\boldsymbol{x})$, the solution in frequency domain is given by the Helmholtz equation

$$\Delta v + \frac{\omega^2}{c(\boldsymbol{x})^2} v = 0, \qquad \boldsymbol{x} \in \Omega. \tag{1.2}$$

We shall continue with the time domain formulation in the Introduction and later come back to approximations in frequency domain.

In the direct numerical simulation of (1.1) the accuracy of the solution is determined by the number of grid points or elements per wavelength. The computational cost of maintaining constant accuracy grows algebraically with the frequency, and for sufficiently high frequencies a direct numerical simulation of (1.1) is no longer feasible. Numerical methods based on approximations of (1.1) are needed.

In this paper we consider variants of geometrical optics, which are asymptotic approximations obtained when the frequency tends to infinity. These approximations are widely used in applications such as computational electromagnetics, acoustics, optics and geophysics. Instead of the oscillating wave field u, the unknowns in standard geometrical optics equations are the phase ϕ and the amplitude A, neither of which depends on the parameter ω, and typically vary on a much coarser scale than u. Hence they should in principle be easier to compute numerically.

The derivation of the geometrical optics equations in the linear case is classical: see, for instance, the book by Whitham (1974). Formally, the equations follow if we assume a series expansion of the form

$$u(t, \boldsymbol{x}) = e^{i\omega\phi(t,\boldsymbol{x})} \sum_{k=0}^{\infty} A_k(t, \boldsymbol{x})(i\omega)^{-k}. \tag{1.3}$$

Entering this expression into (1.1) and summing terms of the same order in ω to zero, we obtain separate equations for the unknown dependent variables in (1.3). The $\mathcal{O}(\omega^2)$ terms give the equation for the phase function ϕ. It satisfies the Hamilton–Jacobi-type *eikonal equation*

$$\phi_t + c(\boldsymbol{x})\,|\nabla\phi| = 0, \tag{1.4}$$

where $|\cdot|$ denotes the Euclidean norm in $\mathbb{R}^d$, $|\boldsymbol{x}| = \left(\sum_{j=1}^d x_j^2\right)^{1/2}$, for $\boldsymbol{x} = (x_1, \ldots, x_d)^T \in \mathbb{R}^d$. For the $\mathcal{O}(\omega)$ terms we get the *transport equation* for A_0:

$$(A_0)_t + c(\boldsymbol{x})\frac{\nabla\phi \cdot \nabla A_0}{|\nabla\phi|} + \frac{c(\boldsymbol{x})^2\Delta\phi - \phi_{tt}}{2c(\boldsymbol{x})\,|\nabla\phi|}A_0 = 0. \tag{1.5}$$

For large ω we can discard the remaining terms in (1.3).

Some typical wave phenomena, such as diffraction, are lost in the infinite frequency approximation. Moreover, the approximation breaks down at caustics, where the amplitude A_0 is unbounded. For these situations, correction terms can be derived, such as those given by Keller (1962) in his *geometrical theory of diffraction* (GTD), further developed by Kouyoumjian and Pathak (1974), for instance. The geometry of Ω and boundary conditions are accounted for in GTD. A closer study of the solution's asymptotic behaviour close to caustics was made by Ludwig (1966) and Kravtsov (1964), among others. Generalizations of the series expansion (1.3), also valid at caustics and when the solution contains several crossing waves with different phase functions, were studied by Maslov (1965) and Duistermaat (1974), for example. For a rigorous treatment of propagation of singularities in linear partial differential equations, see Hörmander (1983–1985). We will briefly comment on some of these techniques in Section 2.5.

The traditional way to compute travel times of high frequency waves is through *ray tracing*. See Section 2 for a derivation of the ray equations (1.6) and the equations (1.7), (1.8) and (1.9). The travel time of a wave is given directly by the phase function ϕ, and ray tracing corresponds to solving the eikonal equation (1.4) through the method of characteristics, *i.e.*, solving the system of ordinary differential equations (ODEs)

$$\frac{d\boldsymbol{x}}{dt} = \nabla_p H(\boldsymbol{x}, \boldsymbol{p}), \qquad \frac{d\boldsymbol{p}}{dt} = -\nabla_x H(\boldsymbol{x}, \boldsymbol{p}), \tag{1.6}$$
$$H(\boldsymbol{x}, \boldsymbol{p}) = c(\boldsymbol{x})|\boldsymbol{p}|, \qquad \boldsymbol{x}, \boldsymbol{p} \in \mathbb{R}^d,$$

where the momentum variable $\boldsymbol{p}$ is usually called the 'slowness' vector, and

∇_p and ∇_x are the gradients taken with respect to $\boldsymbol{p}$ and $\boldsymbol{x}$, respectively, that is,

$$\nabla_p = \left(\frac{\partial}{\partial p_1}, \ldots, \frac{\partial}{\partial p_d}\right)^T, \qquad \nabla_x = \left(\frac{\partial}{\partial x_1}, \ldots, \frac{\partial}{\partial x_d}\right)^T.$$

There are also ODEs for the amplitude. Suppose the source is a curve $\boldsymbol{x}_0(r)$ in $\mathbb{R}^2$ with $\phi(\boldsymbol{x}_0(r)) \equiv 0$. Let $(\boldsymbol{x}(t,r), \boldsymbol{p}(t,r))$ be the solution of (1.6) with $\boldsymbol{x}(0,r) = \boldsymbol{x}_0(r)$, and $\boldsymbol{p}(0,r) = \nabla\phi(\boldsymbol{x}_0(r))$. Then

$$A_0(\boldsymbol{x}(t,r)) = A_0(\boldsymbol{x}_0(r))\sqrt{\frac{|\boldsymbol{x}_{0r}(r)|c(\boldsymbol{x}(t,r))}{|\boldsymbol{x}_r(t,r)|c(\boldsymbol{x}_0(r))}}. \tag{1.7}$$

The vector $\boldsymbol{x}_r$ is obtained by solving the auxiliary ODEs

$$\frac{\mathrm{d}}{\mathrm{d}t}\begin{pmatrix}\boldsymbol{x}_r\\ \boldsymbol{p}_r\end{pmatrix} = \begin{pmatrix} D^2_{px}H & D^2_{pp}H \\ -D^2_{xx}H & -D^2_{px}H\end{pmatrix}\begin{pmatrix}\boldsymbol{x}_r\\ \boldsymbol{p}_r\end{pmatrix}, \quad \begin{pmatrix}\boldsymbol{x}_r(0,r)\\ \boldsymbol{p}_r(0,r)\end{pmatrix} = \begin{pmatrix}\boldsymbol{x}_{0r}(r)\\ \partial_r\nabla\phi(\boldsymbol{x}_0(r))\end{pmatrix}. \tag{1.8}$$

The initial data in (1.8) represent the local shape of the ray's source, which is an additional piece of information needed to compute the amplitude along rays.

Finally, we can adopt a purely kinetic viewpoint, which will prove to be useful as a basis for some new numerical techniques. The kinetic model is based on the interpretation that rays are trajectories of particles following Hamiltonian dynamics. We introduce the phase space $(t, \boldsymbol{x}, \boldsymbol{p})$, where $\boldsymbol{p}$ is the slowness vector defined above. The evolution of a particle in this space is governed by (1.6). Letting $f(t, \boldsymbol{x}, \boldsymbol{p})$ be a particle density function, it will satisfy the Liouville equation

$$f_t + \nabla_p H \cdot \nabla_x f - \nabla_x H \cdot \nabla_p f = 0. \tag{1.9}$$

In Figure 1.1(b) we see a snapshot of a wave front propagating in a heterogeneous medium. The front can be accurately followed by using equation (1.9), Figure 1.1(c). The faint fronts in the upper part of these figures represent reflections and are not captured by (1.9) but they vanish in the limit as $\omega \to \infty$. Figure 1.2(c) shows that ray tracing may produce diverging rays that fail to cover the domain. With ray tracing it is also difficult to compute the amplitude and to find the minimum travel time in regions where rays cross.

Recently, new computational methods based on partial differential equations (PDEs) have been proposed to avoid some of the drawbacks of ray tracing. Interest was initially focused on solving the eikonal equation (1.4) numerically, and different types of upwind finite difference methods have been used to compute the viscosity solution of (1.4).

One problem with (1.4) is that it cannot produce solutions with multiple phases, corresponding to crossing rays. There is no super-position principle.

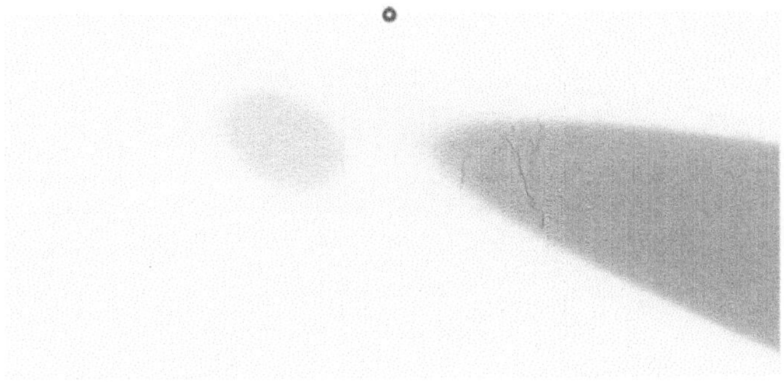

(a) Index of refraction and source point, marked by circle

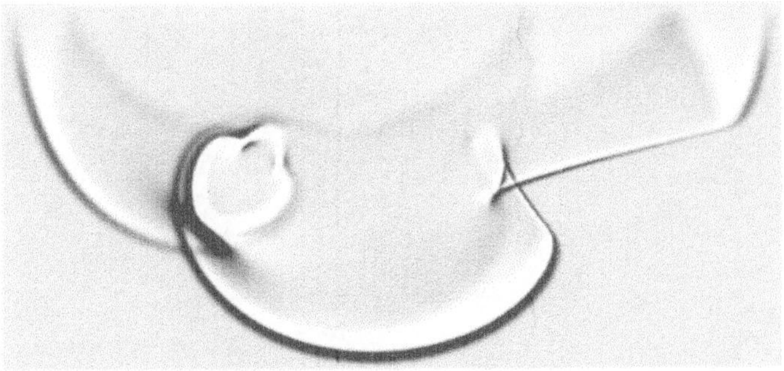

(b) Wave equation solution

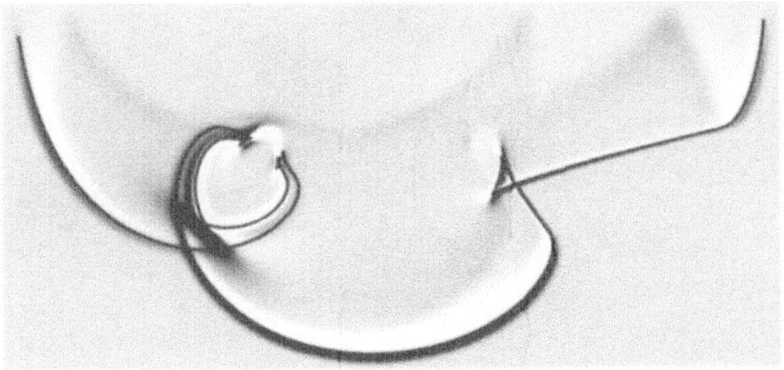

(c) Wave equation solution and wave front

Figure 1.1. Comparison between different techniques for the same problem. A wave propagates from a point source through a hetereogeneous medium. (a) Source and index of refraction of the medium. Dark and light areas represent high and low index of refraction, respectively. (b) Snapshot of a resolved numerical solution of the wave equation, where the solution is represented by grey scale levels. (c) The same solution with a wave front construction solution overlaid.

(a) Eikonal equation solution

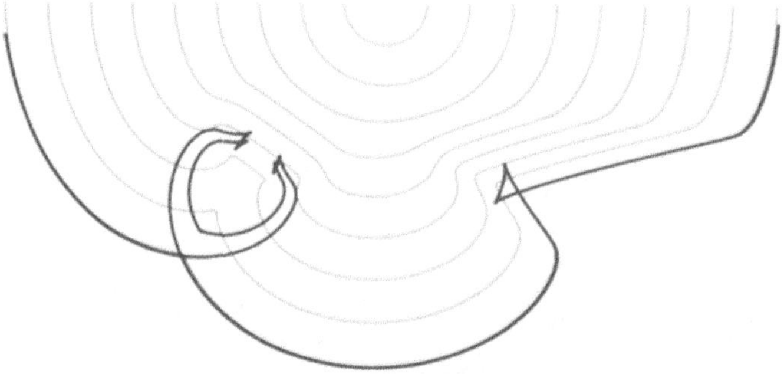

(b) Eikonal equation solution and wave front

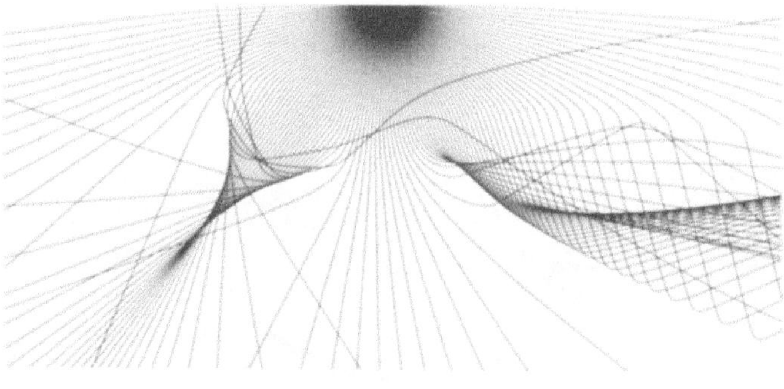

(c) Ray-traced solution

Figure 1.2. Comparison between different techniques for the same problem. (a) Iso curves of a solution to the eikonal equation. (b) The same solution with a wave front construction solution overlaid. (c) A ray-traced solution.

At points where the correct physical solution should have a multivalued phase, the viscosity solution picks out the phase corresponding to the first arriving wave (Crandall and Lions 1983); see Figure 1.2(a). Hence, the eikonal equation only gives the first arrival travel time. A multivalued solution, however, can be constructed by patching together the solutions of several eikonal equations: see Section 3.2. See also Benamou (2003) for another survey of Eulerian methods for geometrical optics.

In this paper we shall mainly focus on numerical techniques based on the Liouville equation (1.9). It has the advantage of the linear superposition property of the ray equations and, like the eikonal equation, the solution is defined by a PDE and can easily be computed on a uniform Eulerian grid.

There is, however, a serious drawback with direct numerical approximation of the Liouville equation. It has the phase included in the set of independent variables, and straightforward simulation would be computationally very costly. We will discuss two ways to remedy this problem and focus the presentation on techniques that have not been surveyed before. In one, special wave front solutions are computed. The other is based on reducing the number of independent variables by introducing equations for moments.

Wave front methods are related to ray tracing. The evolution of a wave front is tracked in the physical or the phase space. We shall mainly follow the presentation in Engquist, Runborg and Tornberg (2002), in which the front evolution is defined directly by the Liouville equation and the tracking is done by the segment projection method. Level set and fast marching methods will also be discussed.

The moment method relies on the closure assumption that only a finite number of rays cross at each point in time and space. Then f in the Liouville equation (1.9) is of a special form, and it can be transformed into a finite system of equations representing the moments of f, set in the reduced space $(t, \boldsymbol{x})$.

Finally, let us mention that simulation of high frequency wave propagation is practically important in many applications. Classical optics will clearly require formulations other than the wave equation as the basis for computations. Visible light has wavelengths of the order of hundreds of nanometers. Thus, in numerical simulation, the number of unknowns would be prohibitively large for simulation over dimensions of metres.

A modern application of geometrical optics is computer visualization. The rendering of images is based on geometrical optics together with different *radiosity* boundary conditions. Usually the wave velocity is constant, resulting in straight rays, and the computational problem is mainly geometric.

For acoustic problems the computational domain is often smaller compared to the wavelength; the wave equation can be directly approached by a numerical method, and no geometrical optics is needed. High frequency techniques become interesting, however, for very large distances, which, for instance, may occur in underwater acoustics.

High frequency approximation techniques also apply to other wave equations, and we shall give two examples of practical importance. The first is Maxwell's equations for electromagnetic waves in a lossless medium, that is,

$$\begin{aligned} \varepsilon(\boldsymbol{x})E_t &= \nabla \times H - J_e(\boldsymbol{x}), \\ \mu(\boldsymbol{x})H_t &= -\nabla \times E \\ \nabla \cdot (\varepsilon(\boldsymbol{x})E) &= \rho(\boldsymbol{x}), \\ \nabla \cdot (\mu(\boldsymbol{x})H) &= 0, \end{aligned} \tag{1.10}$$

where $E(t, \boldsymbol{x})$ and $H(t, \boldsymbol{x})$ are the electric and magnetic fields, respectively, ε and μ are the electric permittivity and magnetic permeability, respectively, J_e is the electric current density, and ρ is the electric charge density. Direct simulation based on (1.10) is common when the wavelength is not too short relative to the size of the computational domain. The geometrical theory of diffraction is the method of choice when the relative wavelength is very short. The latter is, for example, the case in the study of locations of base stations for cell phones in a city.

The other example is elastic wave propagation, given, for example, by

$$\rho(\boldsymbol{x})\boldsymbol{u}_{tt} = \nabla \cdot \boldsymbol{\sigma}(\boldsymbol{x}, \nabla \boldsymbol{u}), \tag{1.11}$$

where $\boldsymbol{u}(t, \boldsymbol{x})$ is the displacement vector, ρ is the density, and $\boldsymbol{\sigma}$ is the stress tensor. Seismic wave propagation is a challenging problem of this type. Both the forward and the inverse problems are of great interest, and may require geometrical optics-type approximations when the relative wavelength is short.

2. Mathematical background

In this chapter we derive the equations that are used in geometrical optics. We thus study the Cauchy problem for the scalar wave equation (1.1):

$$\begin{aligned} u_{tt}(\boldsymbol{x}, t) - c(\boldsymbol{x})^2 \Delta u(\boldsymbol{x}, t) &= 0, & \boldsymbol{x} &\in \mathbb{R}^d,\ t > 0, \\ u(\boldsymbol{x}, 0) &= u_0(\boldsymbol{x}), & \boldsymbol{x} &\in \mathbb{R}^d, \\ u_t(\boldsymbol{x}, 0) &= u_1(\boldsymbol{x}), & \boldsymbol{x} &\in \mathbb{R}^d. \end{aligned} \tag{2.1}$$

Here $c(\boldsymbol{x})$ is the local wave velocity of the medium. We also define the *index of refraction* as $\eta(\boldsymbol{x}) = c_0/c(\boldsymbol{x})$ with the reference velocity c_0 (*e.g.*, the speed of light in a vacuum). For simplicity we will henceforth let $c_0 = 1$. When c is constant, equation (1.1) admits the simple plane wave solution

$$u(t, \boldsymbol{x}) = A e^{i\omega(ct - \boldsymbol{k}\cdot\boldsymbol{x})}, \qquad |\boldsymbol{k}| = 1, \tag{2.2}$$

where $\boldsymbol{k}$ is the wave vector giving the direction of propagation and A is a constant representing the amplitude. Both $\boldsymbol{k}$ and A are determined by appropriate initial data. For more complicated waves and when c is not constant, we need to replace $ct - \boldsymbol{k} \cdot \boldsymbol{x}$ by a general *phase function* ϕ, and also permit the amplitude to depend on time and space. Hence, (1.1) has solutions of the type

$$u(t, \boldsymbol{x}) = A(t, \boldsymbol{x}) \mathrm{e}^{\mathrm{i}\omega\phi(t,\boldsymbol{x})}. \tag{2.3}$$

The level curves of ϕ correspond to the wave fronts of a propagating wave: *cf.* Figure 2.1.

Since (1.1) is linear, the superposition principle is valid and a sum of solutions is itself a solution. The generic solution to (1.1) is, at least locally, described by a finite sum of terms like (2.3), with the amplitudes and phases being smooth functions that depend only mildly on the frequency ω. Typically this setting only breaks down at a small set of points, namely focus points, caustica and discontinuities in $c(\boldsymbol{x})$.

The solutions contain length and time scales that become very small as the frequency increases. In the direct numerical solution of (1.1) a substantial number of grid points per wavelength and dimension is needed to maintain constant accuracy. The work therefore grows algebraically with frequency. For sufficiently high frequencies or short wavelengths, it is unrealistic to compute the wave field directly. Fortunately, this is often the regime for which high frequency asymptotic approximations are quite accurate.

We will assume the geometrical optics approximation that $\omega \to \infty$. This means that, for the moment, we accept the loss of diffraction phenomena in the solution, and that the approximation of the wave amplitude breaks down at caustics. There are three strongly related formulations of geometrical optics, which we will review here. In Section 2.5 we consider some other approximations besides the pure geometrical optics.

2.1. Eikonal equations

Let us now derive Eulerian PDEs for the phase and the amplitude functions that are formally valid in the limit when $\omega \to \infty$. This is motivated by the observation that the phase and amplitudes of (2.3) generically vary on a much larger scale than the solution u itself, and should therefore be easier to compute. In the homogeneous case (2.2), for instance, $\phi = ct - \boldsymbol{k} \cdot \boldsymbol{x}$ stays nonoscillating and bounded independently of ω.

To begin with, we assume that the solution to (1.1) can be described by the asymptotic WKB expansion (Hörmander 1983–1985),

$$u = \mathrm{e}^{\mathrm{i}\omega\phi(t,\boldsymbol{x})} \sum_{k=0}^{\infty} A_k(t, \boldsymbol{x})(\mathrm{i}\omega)^{-k}. \tag{2.4}$$

This form is a slight generalization of (2.3) that also includes a series expansion in powers of $1/\omega$ of the amplitude. We now substitute the expression (2.4) into (1.1) and, following the procedure outlined in the Introduction, equate coefficients of powers of ω to zero. For ω^2, this gives the *eikonal equation*,

$$\phi_t \pm c\,|\nabla\phi| = 0. \tag{2.5}$$

In fact, because of the sign ambiguity, we get two eikonal equations. Without loss of generality we will henceforth consider the one with a plus sign. For ω^1, we get the *transport equation* for the first amplitude term,

$$(A_0)_t + c\frac{\nabla\phi\cdot\nabla A_0}{|\nabla\phi|} + \frac{c^2\Delta\phi - \phi_{tt}}{2c\,|\nabla\phi|}A_0 = 0. \tag{2.6}$$

For higher-order terms of $1/\omega$, we get additional transport equations

$$(A_{k+1})_t + c\frac{\nabla\phi\cdot\nabla A_{k+1}}{|\nabla\phi|} + \frac{c^2\Delta\phi - \phi_{tt}}{2c\,|\nabla\phi|}A_{k+1} + \frac{c^2\Delta A_k - (A_k)_{tt}}{2c\,|\nabla\phi|} = 0 \tag{2.7}$$

for the remaining amplitude terms. When ω is large, only the first term in the expansion (2.4) is significant, and the problem is reduced to computing the phase ϕ and the first amplitude term A_0. Note that, once ϕ is known, the transport equations are linear equations with variable coefficients.

Instead of the time-dependent wave equation (1.1) we can consider the frequency domain problem. Setting $u(t,\boldsymbol{x}) = v(\boldsymbol{x})\exp(\mathrm{i}\omega t)$, with ω fixed, v satisfies the Helmholtz equation

$$c^2\Delta v + \omega^2 v = 0. \tag{2.8}$$

Substituting the series

$$v = \mathrm{e}^{\mathrm{i}\omega\tilde\phi(\boldsymbol{x})}\sum_{k=0}^{\infty}\tilde A_k(\boldsymbol{x})(\mathrm{i}\omega)^{-k} \tag{2.9}$$

into (2.8), we get an alternative, frequency domain, version of the pair (2.5) and (2.6),

$$|\nabla\tilde\phi| = 1/c = \eta, \qquad 2\nabla\tilde\phi\cdot\nabla\tilde A_0 + \Delta\tilde\phi\tilde A_0 = 0. \tag{2.10}$$

With consistent initial and boundary data, $\phi(t,\boldsymbol{x}) = \tilde\phi(\boldsymbol{x}) - t$. We note that, since the family of curves $\{\boldsymbol{x} \mid \phi(t,\boldsymbol{x}) = \tilde\phi(\boldsymbol{x}) - t = 0\}$, parametrized by $t \geq 0$, describes a propagating wave front in (2.9), we often directly interpret the frequency domain phase $\tilde\phi(\boldsymbol{x})$ as the *travel time* of a wave; the difference in phase between two points on the same characteristic signifies the time it takes for a wave to travel between them.

We will drop the zero index in what follows and simply denote A_0 by A. We also drop the tilde for the frequency domain quantities.

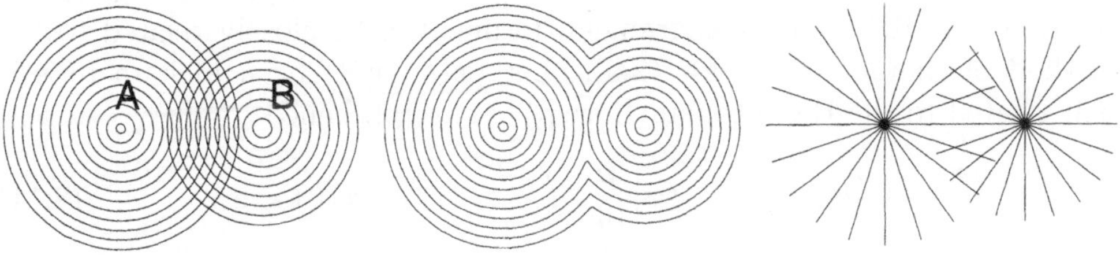

(a) Correct solution (b) Eikonal equation (c) Ray tracing

Figure 2.1. Solution after some time to a homogeneous problem with two point sources, A and B, where the A source began transmitting slightly before the B source. Figure (a) shows the physically correct solution with two superimposed wave fields. Level curves of their phase functions are plotted. Figure (b) shows level curves of ϕ in the viscosity solution of the eikonal equation (2.5). Note that the superposition principle does not hold. Instead, the first arriving wave takes precedence over the second at each point. Figure (c) shows a ray-traced solution.

One problem with the eikonal and transport equations is that they do not accept solutions with multiple phases. There is no superposition principle for the nonlinear eikonal equation: *cf.* Figure 2.1. A finite sum of solutions of the form (2.3), with slowly varying A and ϕ, can in general not be well approximated by the first term in the ansatz (2.4) at high frequencies.

The eikonal equation is a nonlinear Hamilton–Jacobi-type equation with Hamiltonian $H(\boldsymbol{x}, \boldsymbol{p}) = c(\boldsymbol{x})|\boldsymbol{p}|$. As in the case of hyperbolic conservation laws, extra conditions are needed for this type of equation to have a unique solution. These were given in Crandall and Lions (1983) and the solution is known as the *viscosity solution*, which is the analogue of the entropy solution for conservation laws. As can be deduced from the previous paragraph, the viscosity solution does not have to agree with the correct physical solution in all cases. At points where the correct solution should have a multivalued phase, the viscosity solution picks out the phase corresponding to the first arriving wave.

It is well known that solutions of Hamilton–Jacobi equations can develop kinks, *i.e.*, discontinuities in the gradient, just as shocks appear in the solutions of conservation laws. In the case of the eikonal equation, the kinks are located where the physically correct phase solution should become multivalued: *cf.* Figure 2.1. We notice that the transport equation (2.6) has a factor involving $\Delta\phi$, which is not bounded at kinks, and therefore we can expect blow-up of A_0 at these points.

2.2. Ray equations

Another formulation of geometrical optics is *ray tracing*, which gives the solution via ODEs. This Lagrangian formulation is closely related to the method of characteristics for (2.5). Let $(\boldsymbol{x}(t), \boldsymbol{p}(t))$ be a bicharacteristic pair related to the Hamiltonian $H(\boldsymbol{x}, \boldsymbol{p}) = c(\boldsymbol{x})|\boldsymbol{p}|$, hence

$$\frac{\mathrm{d}\boldsymbol{x}}{\mathrm{d}t} = \nabla_p H(\boldsymbol{x}, \boldsymbol{p}) = c(\boldsymbol{x})\frac{\boldsymbol{p}}{|\boldsymbol{p}|}, \qquad \boldsymbol{x}(0) = \boldsymbol{x}_0, \tag{2.11}$$

$$\frac{\mathrm{d}\boldsymbol{p}}{\mathrm{d}t} = -\nabla_x H(\boldsymbol{x}, \boldsymbol{p}) = -|\boldsymbol{p}|\nabla c(\boldsymbol{x}), \qquad \boldsymbol{p}(0) = \boldsymbol{p}_0. \tag{2.12}$$

In d dimensions the bicharacteristics are curves in $2d$-dimensional *phase space* $(\boldsymbol{x}, \boldsymbol{p}) \in \mathbb{R}^{d\times d}$. It follows immediately that H is constant along them, $H(\boldsymbol{x}(t), \boldsymbol{p}(t)) = H(\boldsymbol{x}_0, \boldsymbol{p}_0)$. We are interested in solutions for which $H \equiv 1$. In this case the projections on physical space, $\boldsymbol{x}(t)$, are usually called *rays*, and we can reduce (2.11) and (2.12) to

$$\frac{\mathrm{d}\boldsymbol{x}}{\mathrm{d}t} = \frac{1}{\eta^2}\boldsymbol{p}, \qquad \boldsymbol{x}(0) = \boldsymbol{x}_0, \tag{2.13}$$

$$\frac{\mathrm{d}\boldsymbol{p}}{\mathrm{d}t} = \frac{\nabla\eta}{\eta}, \qquad \boldsymbol{p}(0) = \boldsymbol{p}_0, \quad |\boldsymbol{p}_0| = \eta(\boldsymbol{x}_0). \tag{2.14}$$

Solving (2.13) and (2.14) is called *ray tracing.* It should be noted here that if $\eta =$ const the rays are just straight lines.

We use the frequency domain version of the eikonal equation, (2.10), to explain the significance of the bicharacteristics when the solution ϕ is smooth. It can be written as

$$H(\boldsymbol{x}, \nabla\phi(\boldsymbol{x})) = 1, \tag{2.15}$$

with H as above. By differentiating (2.15) with respect to $\boldsymbol{x}$, we get

$$\nabla_x H(\boldsymbol{x}, \nabla\phi(\boldsymbol{x})) + D^2\phi(\boldsymbol{x})\nabla_p H(\boldsymbol{x}, \nabla\phi(\boldsymbol{x})) = 0.$$

Here D^2 represents the Hessian. Then for any curve $\boldsymbol{y}(t)$ we have the identity

$$\begin{aligned}\frac{\mathrm{d}}{\mathrm{d}t}\nabla\phi(\boldsymbol{y}(t)) &= D^2\phi(\boldsymbol{y}(t))\frac{\mathrm{d}\boldsymbol{y}(t)}{\mathrm{d}t}\\ &= D^2\phi(\boldsymbol{y}(t))\left[\frac{\mathrm{d}\boldsymbol{y}(t)}{\mathrm{d}t} - \nabla_p H(\boldsymbol{y}(t), \nabla\phi(\boldsymbol{y}(t)))\right]\\ &\quad - \nabla_x H(\boldsymbol{y}, \nabla\phi(\boldsymbol{y}(t))).\end{aligned}$$

Taking $\boldsymbol{x}(t)$ to be the curve for which the expression in brackets vanishes, we see that $(\boldsymbol{x}(t), \nabla\phi(\boldsymbol{x}(t)))$ is a bicharacteristic. By the uniqueness of solutions to (2.11), (2.12), we therefore have that $\boldsymbol{p}(t) \equiv \nabla\phi(\boldsymbol{x}(t))$ if we take $\boldsymbol{p}_0 = \nabla\phi(\boldsymbol{x}_0)$. Hence, with this initialization, the rays are therefore always orthogonal to the level curves of ϕ, since $\mathrm{d}\boldsymbol{x}/\mathrm{d}t$ is parallel to $\boldsymbol{p} = \nabla\phi$ by (2.13). Moreover, for our particular H,

$$\frac{\mathrm{d}}{\mathrm{d}t}\phi(\boldsymbol{x}(t)) = \nabla\phi(\boldsymbol{x}(t)) \cdot \frac{\mathrm{d}\boldsymbol{x}(t)}{\mathrm{d}t} = \boldsymbol{p}(t) \cdot \nabla_p H(\boldsymbol{x}(t), \boldsymbol{p}(t)) = H(\boldsymbol{x}(t), \boldsymbol{p}(t)) = 1. \tag{2.16}$$

Thus, as long as ϕ is smooth, the solution to (2.15) along the ray is given by the simple expression

$$\phi(\boldsymbol{x}(t)) = \phi(\boldsymbol{x}_0) + t. \tag{2.17}$$

Since ϕ corresponds to travel time, this also shows that the parametrization t in (2.11) and (2.12) actually corresponds to unscaled time; the ray $\boldsymbol{x}(t)$ traces one point on a propagating wave front at time t. The absolute value of its time derivative $|\mathrm{d}\boldsymbol{x}/\mathrm{d}t|$ is precisely the local speed of propagation $c(\boldsymbol{x})$ by (2.13), and since $\boldsymbol{p}$ is parallel to $\mathrm{d}\boldsymbol{x}/\mathrm{d}t$, while $|\boldsymbol{p}| = H(\boldsymbol{x}, \boldsymbol{p})c(\boldsymbol{x})^{-1} = c(\boldsymbol{x})^{-1}$ by (2.15), the vector $\boldsymbol{p}$ is often called the *slowness* vector.

As was discussed in Section 2.1, the solution of the eikonal equation (2.5) is valid up to the point where discontinuities appear in the gradient of ϕ. This is where the phase should become multivalued but, by the construction, cannot. The bicharacteristics, however, do not have this problem, and we can extend their validity to all t: see Figure 2.1.

The ODEs for the bicharacteristics are sometimes solved using another parametrization than time. Setting $\mathrm{d}t = \eta(\boldsymbol{x}(t))^2\,\mathrm{d}\tau$, we get a simple ODE for $\boldsymbol{x}$,

$$\frac{\mathrm{d}^2\boldsymbol{x}}{\mathrm{d}\tau^2} = \frac{1}{2}\nabla\eta(\boldsymbol{x})^2. \tag{2.18}$$

This can still be interpreted as a Hamiltonian system, with a different H, but in this case an accompanying ODE must be solved to obtain the solution ϕ, *i.e.*, the travel time, along the ray,

$$H = \frac{|\boldsymbol{p}|^2 - \eta^2}{2}, \qquad \frac{\mathrm{d}}{\mathrm{d}\tau}\phi(\boldsymbol{x}(\tau)) = \eta(\boldsymbol{x}(\tau))^2. \tag{2.19}$$

The rays can also be derived from the calculus of variations, using Fermat's principle. By analogy with the least action principle in classical mechanics, it says that the rays between two points are stationary curves of the functional

$$I[\gamma] = \int_\gamma \eta(\boldsymbol{x})\,\mathrm{d}\boldsymbol{x},$$

taken over all curves γ starting and ending at the points in question. The Euler–Lagrange equations for this optimization problem give the same bicharacteristics as (2.11) and (2.12), but the formulation is also well defined for non-differentiable η. The integral represents the length of γ under the measure $\eta\,\mathrm{d}s$ and therefore we often describe the rays as the *shortest optical path* between two points.

In order to compute the amplitude along a ray we also need information about the local shape of the ray's source. Let $(\boldsymbol{x}(t,\boldsymbol{x}_0),\boldsymbol{p}(t,\boldsymbol{x}_0))$ denote the bicharacteristic originating in $\boldsymbol{x}_0$ with $\boldsymbol{p}(0,\boldsymbol{x}_0)=\nabla\phi(\boldsymbol{x}_0)$, hence $\boldsymbol{x}(0,\boldsymbol{x}_0)=\boldsymbol{x}_0$. Let $J(t,\boldsymbol{x}_0)$ be the Jacobian of $\boldsymbol{x}$ with respect to initial data, $J=D_{x_0}\boldsymbol{x}(t,\boldsymbol{x}_0)$. By differentiating (2.13) we get

$$\begin{aligned}\frac{\partial J}{\partial t}=D_{x_0}\frac{\partial \boldsymbol{x}(t,\boldsymbol{x}_0)}{\partial t}&=D_{x_0}c(\boldsymbol{x}(t,\boldsymbol{x}_0))^2\boldsymbol{p}(t,\boldsymbol{x}_0)\\&=D_{x_0}c^2(\boldsymbol{x}(t,\boldsymbol{x}_0))\nabla\phi(\boldsymbol{x}(t,\boldsymbol{x}_0))\\&=\left(D_xc^2\nabla\phi\right)J.\end{aligned}$$

Assume that J is nonsingular and let $J=S\Lambda S^{-1}$ be a Jordan decomposition, so that the diagonal entries of Λ are the eigenvalues $\{\lambda_j\}$ of J. Setting $q=\det J\prod_j\lambda_j$, and using the fact that $\operatorname{tr}(T^{-1}AT)=\operatorname{tr}A$, we have

$$\begin{aligned}\frac{\partial q}{\partial t}&=q\operatorname{tr}\bigl(\Lambda^{-1}\Lambda_t\bigr)\\&=q\operatorname{tr}\bigl(S\Lambda^{-1}S^{-1}S\Lambda_tS^{-1}+(S\Lambda^{-1})S^{-1}S_t(S\Lambda^{-1})^{-1}+(S^{-1})_tS\bigr)\\&=q\operatorname{tr}\bigl(J^{-1}J_t\bigr)\\&=q\operatorname{tr}\bigl(J^{-1}\left(Dc^2\nabla\phi\right)J\bigr)\\&=q\operatorname{tr}\bigl(Dc^2\nabla\phi\bigr)=q\nabla\cdot c^2\nabla\phi.\end{aligned}$$

Therefore differentiation along the ray gives

$$\begin{aligned}\frac{\mathrm{d}}{\mathrm{d}t}\bigl[A^2(\boldsymbol{x}(t,\boldsymbol{x}_0))\eta(\boldsymbol{x}(t,\boldsymbol{x}_0))^2q(t,\boldsymbol{x}_0)\bigr]&=q(\nabla A^2\eta^2)\cdot\frac{\partial\boldsymbol{x}}{\partial t}+qA^2\eta^2\nabla\cdot c^2\nabla\phi\\&=q\nabla\cdot(A^2\nabla\phi)\\&=qA\bigl[2\nabla A\cdot\nabla\phi+\Delta\phi A\bigr]=0,\end{aligned}$$

using (2.10) in the last step. It follows that the amplitude is given by the expression

$$A(\boldsymbol{x}(t,\boldsymbol{x}_0))=A(\boldsymbol{x}_0)\frac{\eta(\boldsymbol{x}_0)}{\eta(\boldsymbol{x}(t,\boldsymbol{x}_0))}\sqrt{\left|\frac{q(0,\boldsymbol{x}_0)}{q(t,\boldsymbol{x}_0)}\right|}. \tag{2.20}$$

For example, an outgoing spherical wave in homogeneous medium with $\eta\equiv 1$ is given by $\boldsymbol{x}(t)=\boldsymbol{x}_0+t\boldsymbol{x}_0/|\boldsymbol{x}_0|$. Then $J=I+t(I/|\boldsymbol{x}_0|-\boldsymbol{x}_0\boldsymbol{x}_0^T/|\boldsymbol{x}_0|^3)$ and $q=\det J=(1+t/|\boldsymbol{x}_0|)^{d-1}=(|\boldsymbol{x}|/|\boldsymbol{x}_0|)^{d-1}$ in d dimensions. Consequently, by (2.20), we get the well-known amplitude decay of such waves, $A\sim|\boldsymbol{x}|^{-(d-1)/2}$.

The determinant q is often called the *geometrical spreading*, since it measures the amplification of an infinitesimal area transported by the rays. It vanishes at caustics, and we see clearly from this expression that the amplitude is unbounded close to these points. (Strictly speaking we have only shown (2.20) as long as J is nonsingular, and then, by continuity, $q(0)$ and

$q(t)$ have the same sign. The expression is, however, also valid after caustic points, with the absolute values placed under the root sign.)

In order to compute A we thus need q, the determinant of $D_{x_0}\boldsymbol{x}$. The elements of this matrix are given by another ODE system. After differentiating (2.11) and (2.12) with respect to $\boldsymbol{x}_0$, we obtain

$$\frac{\mathrm{d}}{\mathrm{d}t}\begin{pmatrix} D_{x_0}\boldsymbol{x} \\ D_{x_0}\boldsymbol{p} \end{pmatrix} = \begin{pmatrix} D^2_{px}H & D^2_{pp}H \\ -D^2_{xx}H & -D^2_{px}H \end{pmatrix}\begin{pmatrix} D_{x_0}\boldsymbol{x} \\ D_{x_0}\boldsymbol{p} \end{pmatrix}, \tag{2.21}$$

with initial data

$$D_{x_0}\boldsymbol{x}(0,\boldsymbol{x}_0) = I, \qquad D_{x_0}\boldsymbol{p}(0,\boldsymbol{x}_0) = D^2\phi(\boldsymbol{x}_0).$$

We note that the system matrix here only depends on $\boldsymbol{x}$ and $\boldsymbol{p}$.

Since we have the constraint $H(\boldsymbol{x},\boldsymbol{p}) = 1$, or $|\boldsymbol{p}| = \eta(\boldsymbol{x})$, the dimension of the phase space $(\boldsymbol{x},\boldsymbol{p})$ can actually be reduced by one. We have not done this reduction in the equations above, and (2.13), (2.14), (2.20) and (2.21) are in this sense all overdetermined. We will here show the reduced equations in two dimensions.

Setting $\boldsymbol{p} = \eta(\cos\theta, \sin\theta)$, we can use θ as a dependent variable in (2.13) and (2.14) instead of $\boldsymbol{p}$. We then get, with $\boldsymbol{x} = (x, y)$,

$$\frac{\mathrm{d}x}{\mathrm{d}t} = c(x,y)\cos\theta, \tag{2.22}$$

$$\frac{\mathrm{d}y}{\mathrm{d}t} = c(x,y)\sin\theta, \tag{2.23}$$

$$\frac{\mathrm{d}\theta}{\mathrm{d}t} = \frac{\partial c}{\partial x}\sin\theta - \frac{\partial c}{\partial y}\cos\theta. \tag{2.24}$$

Suppose the source is a curve $\boldsymbol{x}_0(r)$ in $\mathbb{R}^2$ parametrized by r, and $\phi(\boldsymbol{x}_0(r)) \equiv 0$. Set $\tilde{\boldsymbol{x}}(t,r) := \boldsymbol{x}(t,\boldsymbol{x}_0(r))$ and $\tilde{\boldsymbol{p}}(t,r) := \boldsymbol{p}(r,\boldsymbol{x}_0(r))$. Then $\phi(\tilde{\boldsymbol{x}}(t,r)) = t$ by (2.17) and $\tilde{\boldsymbol{x}}_t \perp \tilde{\boldsymbol{x}}_r$ for all time, since

$$0 = \frac{\partial}{\partial r}\phi(\tilde{\boldsymbol{x}}(t,r)) = \nabla\phi(\tilde{\boldsymbol{x}})\cdot\tilde{\boldsymbol{x}}_r = \boldsymbol{p}\cdot\tilde{\boldsymbol{x}}_r = \eta^2\boldsymbol{x}_t\cdot\tilde{\boldsymbol{x}}_r.$$

We can then introduce the orthogonal matrix $R := [\tilde{\boldsymbol{x}}_r\,\tilde{\boldsymbol{x}}_t]$, with determinant $|\det R| = |\tilde{\boldsymbol{x}}_r||\tilde{\boldsymbol{x}}_t| = |\tilde{\boldsymbol{x}}_r|/\eta(\tilde{\boldsymbol{x}})$. By definition, for $0 \le s \le t$, we have $\boldsymbol{x}(t,\boldsymbol{x}_0) = \boldsymbol{x}(s,\boldsymbol{x}(t-s,\boldsymbol{x}_0))$, and, by differentiating both sides,

$$\boldsymbol{x}_t(t,\boldsymbol{x}_0) = D_{x_0}\boldsymbol{x}(s,\boldsymbol{x}(t-s,\boldsymbol{x}_0))\boldsymbol{x}_t(t-s,\boldsymbol{x}_0).$$

Evaluating at $s = t$ gives

$$\boldsymbol{x}_t(t,\boldsymbol{x}_0) = D_{x_0}\boldsymbol{x}(t,\boldsymbol{x}_0)\boldsymbol{x}_t(0,\boldsymbol{x}_0).$$

Therefore $D_{\boldsymbol{x}_0}\boldsymbol{x}(t,\boldsymbol{x}_0(r))R(0,r) = R(t,r)$ and

$$|q(t,\boldsymbol{x}_0(r))| = |\det D_{x_0}\boldsymbol{x}(t,\boldsymbol{x}_0(r))| = \frac{|\det R(t,r)|}{|\det R(0,r)|} = \frac{|\tilde{\boldsymbol{x}}_r(t,r)|\eta(\boldsymbol{x}_0(r))}{|\partial_r\boldsymbol{x}_0(r)|\eta(\tilde{\boldsymbol{x}}(t,r))},$$

so that

$$A(\boldsymbol{x}(t,r)) = A(\boldsymbol{x}_0(r))\sqrt{\frac{|\partial_r \boldsymbol{x}_0(r)|\eta(\boldsymbol{x}_0(r))}{|\tilde{\boldsymbol{x}}_r(t,r)|\eta(\tilde{\boldsymbol{x}}(t,r))}}. \tag{2.25}$$

We then only need to compute $\tilde{\boldsymbol{x}}_r$ to get the amplitude, which reduces (2.21) to

$$\frac{\mathrm{d}}{\mathrm{d}t}\begin{pmatrix}\tilde{\boldsymbol{x}}_r\\ \tilde{\boldsymbol{p}}_r\end{pmatrix} = \begin{pmatrix} D^2_{px}H & D^2_{pp}H \\ -D^2_{xx}H & -D^2_{px}H\end{pmatrix}\begin{pmatrix}\tilde{\boldsymbol{x}}_r\\ \tilde{\boldsymbol{p}}_r\end{pmatrix}. \tag{2.26}$$

2.3. Kinetic equations

Finally, we can adopt a purely kinetic viewpoint. This is based on the interpretation that rays are trajectories of particles following the Hamiltonian dynamics of (2.11) and (2.12). We introduce the phase space $(t, \boldsymbol{x}, \boldsymbol{p})$, where $\boldsymbol{p}$ is the slowness vector defined above in Section 2.2, and we let $f(t, \boldsymbol{x}, \boldsymbol{p})$ be a particle ('photon') density function. It will satisfy the Liouville equation,

$$f_t + \nabla_p H \cdot \nabla_x f - \nabla_x H \cdot \nabla_p f = 0, \tag{2.27}$$

or, with $H(\boldsymbol{x}, \boldsymbol{p}) = c(\boldsymbol{x})|\boldsymbol{p}|$,

$$f_t + \frac{c(\boldsymbol{x})}{|\boldsymbol{p}|}\boldsymbol{p}\cdot\nabla_x f + \frac{|\boldsymbol{p}|}{\eta^2}\nabla_x \eta \cdot \nabla_p f = 0. \tag{2.28}$$

We are only interested in solutions to (2.11) and (2.12) for which $H \equiv 1$, meaning that f only has support on the sphere $|\boldsymbol{p}| = \eta(\boldsymbol{x})$ in phase space. Because of this we can simplify (2.28) to the Vlasov-type equation

$$f_t + \frac{1}{\eta^2}\boldsymbol{p}\cdot\nabla_x f + \frac{1}{\eta}\nabla_x \eta \cdot \nabla_p f = 0, \tag{2.29}$$

with initial data $f_0(\boldsymbol{x}, \boldsymbol{p})$ vanishing whenever $|\boldsymbol{p}| \neq \eta$. We note that, if $\eta \equiv 1$, the equation (2.29) is just a free transport equation with solution $f(t, \boldsymbol{x}, \boldsymbol{p}) = f_0(\boldsymbol{x} - t\boldsymbol{p}, \boldsymbol{p})$ which corresponds to straight line ray solutions of (2.13) and (2.14).

The Wigner transform provides a direct link between the density function f in (2.29) and the solution to the scalar wave equation (1.1) and Helmholtz equation (1.2). It is an important tool in the study of high frequency, homogenization and random medium limits of these and many other equations, such as the Schrödinger equation (Lions and Paul 1993, Gérard, Markowich, Mauser and Poupaud 1997, Ryzhik, Papanicolaou and Keller 1996, Benamou, Castella, Katsaounis and Perthame 2002). The Wigner transform of $u^\varepsilon(\boldsymbol{x})$ is defined by

$$f^\varepsilon(t, \boldsymbol{x}, \boldsymbol{p}) := \int_{\mathbb{R}^d} \exp(-\mathrm{i}\boldsymbol{y}\cdot\boldsymbol{p})u^\varepsilon(t, \boldsymbol{x} + \varepsilon\boldsymbol{y}/2)\overline{u^\varepsilon(t, \boldsymbol{x} - \varepsilon\boldsymbol{y}/2)}\,\mathrm{d}y.$$

If $\{u^\varepsilon\}$ is bounded in $L^2(\mathbb{R}^d)$ (for instance), then a subsequence of $\{f^\varepsilon\}$

converges weakly in $\mathcal{S}'(\mathbb{R}^d)$, the space of tempered distributions (Lions and Paul 1993). The limit is a locally bounded nonnegative measure, called the Wigner measure or semiclassical measure, which in our case agrees with the density function f in (2.29) above. An important property of the Wigner transform is that, when u^ε is a simple wave,

$$u^\varepsilon(t,\boldsymbol{x}) = A(t,\boldsymbol{x})\mathrm{e}^{\mathrm{i}\phi(t,\boldsymbol{x})/\varepsilon}, \tag{2.30}$$

then $f^\varepsilon \to f$ weakly in $\mathcal{S}'$, and the Wigner measure f represents a 'particle' in phase space of the form

$$f(t,\boldsymbol{x},\boldsymbol{p}) = A^2(t,\boldsymbol{x})\delta(\boldsymbol{p} - \nabla\phi(t,\boldsymbol{x})). \tag{2.31}$$

Even though f^ε is not linear in u^ε, a sum of simple wave solutions to (1.1) of the type (2.30) converges to a sum of 'particle' solutions to (2.29) of the type (2.31): see, *e.g.*, Jin and Li (200x) and Sparber, Mauser and Markowich (2003). Some other references dealing with the rigorous study of the convergence $f^\varepsilon \to f$, and proving that the limiting Wigner measure f satisfies a transport equation such as (2.29), are Castella, Perthame and Runborg (2002), Miller (2000) and Bal, Papanicolaou and Ryzhik (2002). We can also derive (2.29) directly from the wave equation (1.1) using so-called H-measures (Tartar 1990) or microlocal defect measures (Gérard 1991).

From (2.31) it follows that the amplitude at a point $\boldsymbol{x}$ is given as the integral of f over the phase variable,

$$A^2(t,\boldsymbol{x}) = \int_{\mathbb{R}^d} f(t,\boldsymbol{x},\boldsymbol{p})\,\mathrm{d}p.$$

Equation (2.29) can in fact be further reduced by drawing on the constraint $|\boldsymbol{p}| = \eta(\boldsymbol{x})$. Let us use polar coordinates for $\boldsymbol{p}$ in two dimensions, setting $\boldsymbol{p} = r(\cos\theta, \sin\theta)$. We then make the substitution

$$f(t,\boldsymbol{x},r,\theta) = \frac{1}{\eta(\boldsymbol{x})}\delta(r - \eta(\boldsymbol{x}))\tilde{f}(t,\boldsymbol{x},\theta),$$

and integrate (2.28) over all positive r. This gives a similar transport equation for $\tilde{f}$,

$$\tilde{f}_t + \frac{1}{\eta}\cos\theta\tilde{f}_x + \frac{1}{\eta}\sin\theta\tilde{f}_y + \frac{1}{\eta^2}(\eta_y\cos\theta - \eta_x\sin\theta)\tilde{f}_\theta = 0, \tag{2.32}$$

with $\boldsymbol{x} = (x,y)$. Also, with this scaling, that the integral over all phases gives the amplitude

$$\int_0^{2\pi} \tilde{f}(t,\boldsymbol{x},\theta)\,\mathrm{d}\theta = \int_0^{2\pi}\int_0^\infty f(t,\boldsymbol{x},r,\theta)r\,\mathrm{d}r\,\mathrm{d}\theta = \int_{\mathbb{R}^2} f(t,\boldsymbol{x},\boldsymbol{p})\,\mathrm{d}p = A^2(t,\boldsymbol{x}).$$

2.4. Boundary conditions

When a wave hits a sharp interface between two materials there will in general be one reflected and one transmitted wave. The interface is modelled by a rapid variation in the index of refraction η. For simplicity we assume that $\eta(\boldsymbol{x})$, with $\boldsymbol{x} = (x, y)$, only depends on x in two dimensions, and that $\eta = \eta_L$ to the left and $\eta = \eta_R$ to the right of the interface: *cf.* Figure 2.2. In the high frequency limit the solution depends on the limiting ratio between the width δ of the interface and the wavelength $\lambda = 2\pi c/\omega$ of the incident wave.

If $\lambda \ll \delta$ as $\lambda, \delta \to 0$, the geometrical optics equations are also valid at the interface, and if $\boldsymbol{p} = (p_x, p_y) = \eta(\cos\theta, \sin\theta)$ then

$$\frac{\mathrm{d}p_y}{\mathrm{d}t} = 0,$$

by (2.14), so that $\eta \sin\theta$ is constant along the ray. In the limit of a sharp interface this is Snell's law of refraction, usually written in the form

$$\eta_L \sin\theta_{\mathrm{inc}} = \eta_R \sin\theta_{\mathrm{tr}}. \tag{2.33}$$

Similarly, for a plane wave ($A_y = \phi_{yy} = 0$) hitting the interface, the transport equation (2.10) gives

$$(A^2\phi_x)_x = 0,$$

and since $\phi_x = \eta\cos\theta$, we get the corresponding law for the amplitudes of the incident and transmitted waves

$$\eta_L A_{\mathrm{inc}}^2 \cos\theta_{\mathrm{inc}} = \eta_R A_{\mathrm{tr}}^2 \cos\theta_{\mathrm{tr}}. \tag{2.34}$$

In this scaling limit there is no reflected wave.

The most common situation, however, is when $\lambda \gg \delta$ as $\lambda, \delta \to 0$. In this case boundary conditions must be derived directly from the wave equation before passing to the high frequency limit. They follow from assuming that the incident, reflected and transmitted waves have smooth phase functions.

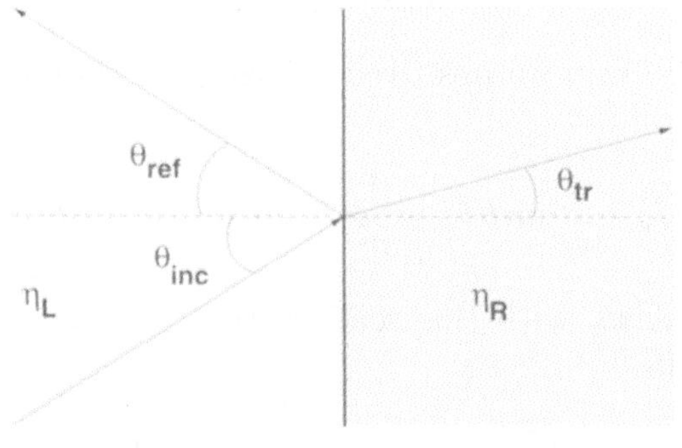

Figure 2.2. Reflection and transmission of a ray at a sharp interface when $\eta_L < \eta_R$.

Continuity of the solution across the interface gives Snell's law (2.33) and the reflection law

$$\theta_{\text{ref}} = \theta_{\text{inc}}.$$

There are also other interface conditions for the solution that depend on the type of wave equation, physics for the problem in question, and the local shape of the interface. Those give expressions for A_{ref} and A_{tr} in terms of A_{inc} as well as the corresponding quantities in (2.21) related to the geometrical spreading, often represented by the wave front's principal radii of curvature. In the case of the scalar wave equation (1.1), with a plane incident wave and a planar interface, continuity of the solution's normal derivative at the interface implies

$$A_{\text{ref}} = \frac{\eta_L \cos\theta_{\text{inc}} - \eta_R \cos\theta_{\text{tr}}}{\eta_L \cos\theta_{\text{inc}} + \eta_R \cos\theta_{\text{tr}}} A_{\text{inc}}, \qquad A_{\text{tr}} = \frac{2\eta_L \cos\theta_{\text{inc}}}{\eta_L \cos\theta_{\text{inc}} + \eta_R \cos\theta_{\text{tr}}} A_{\text{inc}}.$$

For the systems of wave equations (1.10) and (1.11), the interface and boundary conditions typically couple the amplitudes of different components, and one incident wave may generate several transmitted and reflected waves.

2.5. Other models

In this section we shall comment on a few different techniques that are related to our main focus on geometrical optics. These techniques handle high frequencies more efficiently than direct numerical approximation of the wave equation (1.1), and they include some phenomena that are not described by geometrical optics.

Paraxial approximations

Paraxial wave equation approximations are used to study waves propagating in a preferred direction. These approximations allow for numerical approximations of moderately higher frequencies than for regular wave equation methods even if the wave field is approximated directly without introducing phase or amplitude functions. A simple derivation follows from introducing a moving coordinate frame. Assume c to be constant and the waves propagating mainly in the positive x-direction,

$$\tilde{x} = x - ct, \qquad \frac{\partial}{\partial x} = \frac{\partial}{\partial \tilde{x}}, \tag{2.35}$$

$$\tilde{y} = y, \qquad \frac{\partial}{\partial y} = \frac{\partial}{\partial \tilde{y}}, \tag{2.36}$$

$$\tilde{t} = t, \qquad \frac{\partial}{\partial t} = \frac{\partial}{\partial \tilde{t}} - c\frac{\partial}{\partial \tilde{x}}. \tag{2.37}$$

This implies

$$u_{\tilde{t}\tilde{t}} - 2cu_{\tilde{x}\tilde{t}} = c^2 u_{\tilde{y}\tilde{y}},$$

and for waves moving essentially in the positive x-direction, $u_{\tilde{t}\tilde{t}}$ is small and set to zero. Dropping the tilde, the paraxial equation takes the form

$$u_{xt} = -\frac{c}{2} u_{yy}. \tag{2.38}$$

Equation (2.38) is well posed as an evolution equation both in the x- and the t-direction. The slow variation of u with respect to t reduces the computational complexity in numerical approximations.

Higher-order paraxial approximations can be derived from the dispersion relation of the wave equation or from the calculus of pseudo-differential operators. Paraxial approximations are, for example, used in underwater acoustics, in the inverse migration technique in seismology (Claerbout 1976) and as absorbing boundary conditions (Engquist and Majda 1977).

In geometrical optics a paraxial approximation often signifies another simplification, which can be made when there is one preferred coordinate direction. This means that all rays propagate in one direction and do not turn back. The slowness vector component in this direction is always positive. In those cases, time can be replaced by the preferred coordinate direction in the evolution equations, which reduces the dimension of the problem by one in the time-dependent case. Note that the expression is a misnomer in this case, since there is no approximation involved if the assumptions hold.

In two dimensions, $\boldsymbol{x} = (x, y)$, suppose that the x-axis can be used as evolution direction. Time is thus not explicitly needed in the calculation and θ, y and ϕ can be computed as a function of x directly. The phase ϕ (which is also the travel time) must be computed by a separate ODE. Dividing (2.23), (2.24) and (2.16) by (2.22), we get

$$\frac{\mathrm{d}}{\mathrm{d}x}\begin{pmatrix} y \\ \theta \end{pmatrix} = \begin{pmatrix} \tan\theta \\ \eta^{-1}(\eta_y - \eta_x \tan\theta) \end{pmatrix} =: \boldsymbol{u}(y, \theta), \tag{2.39}$$

$$\frac{\mathrm{d}\phi}{\mathrm{d}x} = \frac{\eta}{\cos\theta}. \tag{2.40}$$

These equations are valid as long as there are no turning rays, by which we mean that there is a constant C such that $|\theta| \leq C < \pi/2$.

Let $(y(x,r), \theta(x,r))$ be the ray originating at $x = x_0$, $y = y_0(r)$ and $\theta = \theta_0(r)$, where r is some parametrization of the initial data. Then the amplitude is reduced to

$$A(x, y(x,r)) = A(x_0, y_0(r))\sqrt{\frac{\eta(x_0, y_0(r))|\partial_r y_0(r)|}{\eta(x, y(x,r))|y_r(x,r)|}}. \tag{2.41}$$

Here y_r can be computed through the ODEs

$$\frac{\mathrm{d}}{\mathrm{d}x}\begin{pmatrix} y_r \\ \theta_r \end{pmatrix} = \begin{pmatrix} u_y & u_\theta \\ v_y & v_\theta \end{pmatrix}\begin{pmatrix} y_r \\ \theta_r \end{pmatrix}, \qquad \begin{pmatrix} y_r(0) \\ \theta_r(0) \end{pmatrix} = \frac{\mathrm{d}}{\mathrm{d}r}\begin{pmatrix} y_0(r) \\ \theta_0(r) \end{pmatrix}, \tag{2.42}$$

where $\boldsymbol{u} = (u, v)$ was defined above in (2.39) and $y_r(0, r) = \partial_r y_0(r)$, $\theta_r(0, r) = \partial_r \theta_0(r)$.

The eikonal equation (2.10) can be similarly reduced. In two dimensions we can evolve the equation in the x-direction, giving

$$\phi_x - \sqrt{\eta^2 - \phi_y^2} = 0,$$

which is now a one-dimensional evolution equation, valid as long as $|\phi_y| < \eta$. By the simple modification

$$\phi_x - \sqrt{\max(\eta^2 - \phi_y^2,\ \eta^2 \cos^2 \theta^*)} = 0, \tag{2.43}$$

with $\theta^* < \pi/2$, the equation is also well defined for problems with turning rays. This *paraxial eikonal equation* ignores rays with a propagation angle larger than θ^*, and its solution represents the first arrival time among the remaining rays (Gray and May 1994). See also Symes and Qian (2003) for a rigorous statement and proof of this.

Geometrical theory of diffraction

The geometrical theory of diffraction (GTD) can be seen as a generalization of geometrical optics. It was pioneered by J. Keller in the 1960s (Keller 1962), and provides a systematic technique for adding diffraction effects to the geometrical optics approximation.

Standard geometrical optics excludes diffraction phenomena, which may be too crude an approximation for a scattering problem at moderate frequencies. The derivation of (2.5) and (2.6) in Section 2.1 does not take into account the effects of geometry and boundary conditions, which often gives rise to geometrical optics solution that are discontinuous: see Figure 2.3. In this case the series expansion (2.4) is not adequate. Extra terms must be added to the expansion to match the solution to the boundary conditions. One typical such expansion is

$$u = \mathrm{e}^{\mathrm{i}\omega\phi} \sum_{k=0}^{\infty} A_k (\mathrm{i}\omega)^{-k} + \mathrm{e}^{\mathrm{i}\omega\phi_{\mathrm{d}}} \sum_{k=0}^{\infty} B_k (\mathrm{i}\omega)^{-k-1/2}, \tag{2.44}$$

which is similar to the standard geometrical optics ansatz (2.4), only that a new diffracted wave scaled by $\sqrt{\omega}$ has been added (index d). For high frequencies, the term B_0 is also retained, together with A_0. The *local* geometry of the boundary determines the first B_k coefficients. More elaborate expansions must sometimes be used, such as those given by the *uniform theory of diffraction* (UTD) (Kouyoumjian and Pathak 1974).

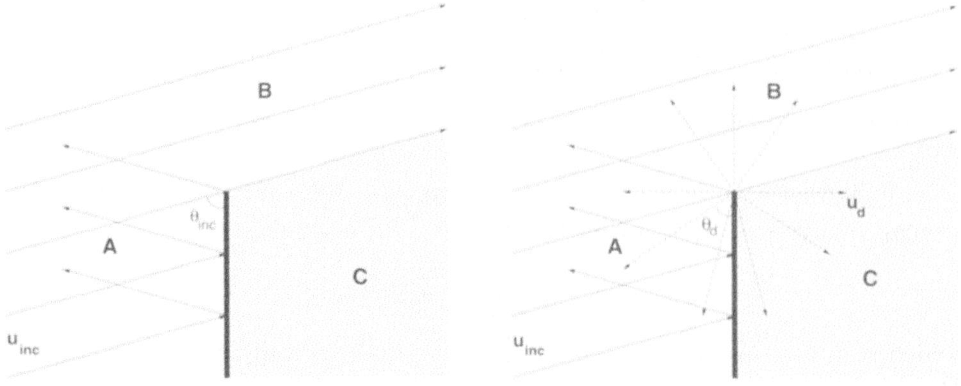

(a) GO (b) GO + GTD

Figure 2.3. A typical geometrical optics solution in two dimensions and in a constant medium ($c \equiv 1$) around a perfectly reflecting halfplane (a), and the same problem augmented with diffracted waves given by GTD (b). In the geometrical optics case, region A contains two phases (incident and reflected), region B one phase (incident), and region C is in shadow, with no phases and hence a zero solution. On the boundaries between the regions the solution is discontinuous.

In general, diffracted rays are induced by rays that form discontinuities in the standard geometrical optics solution. Those rays produce an infinite set of diffracted rays that obey the usual geometrical optics equations. The main computational task, even for GTD, is thus based on the standard GO approximation, which is the central topic of this article. In Figure 2.3 the incident ray hitting the tip of the halfplane splits into a reflected ray that divides regions A and B, and another one that continues past the tip, dividing regions B and C. This ray gives rise to infinitely many diffracted rays shooting out in all directions from the tip of the wedge, which thus acts as an (anisotropic) point source.

The amplitude of each diffracted ray is proportional to the amplitude of the inducing ray and a diffraction coefficient D ($\sim B_0$). The coefficient D depends on the directions of the inducing and diffracted rays, on the frequency and on the local boundary geometry and index of refraction. In a two-dimensional homogeneous medium, the diffraction coefficient D for a halfplane is

$$D(\theta_{\mathrm{d}}, \theta_{\mathrm{inc}}, \omega) = \frac{\mathrm{e}^{\mathrm{i}\pi/4}}{2\sqrt{2\pi\omega}} \left(\frac{1}{\cos\frac{\theta_{\mathrm{d}}-\theta_{\mathrm{inc}}}{2}} \pm \frac{1}{\cos\frac{\theta_{\mathrm{d}}+\theta_{\mathrm{inc}}}{2}} \right), \tag{2.45}$$

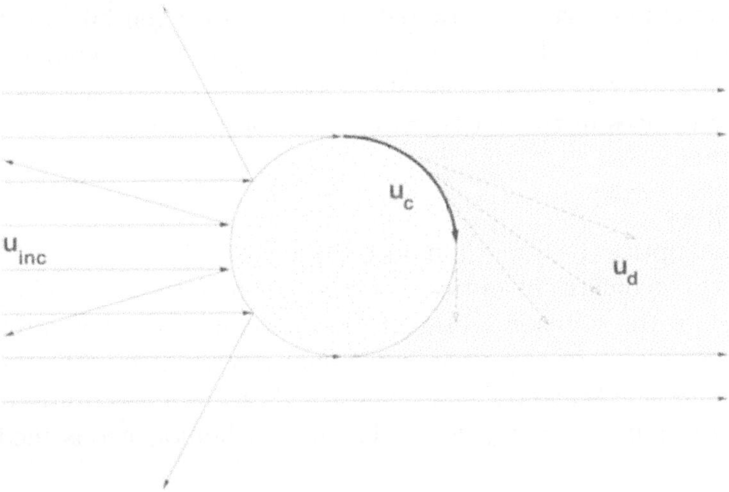

Figure 2.4. Diffraction by a smooth cylinder. The incident field u_{inc} induces a creeping ray u_{c} at the north (and south) pole of the cylinder. As the creeping ray propagates along the surface, it continuously emits surface-diffracted rays u_{d} with exponentially decreasing initial amplitude.

with the definition of the angles as in Figure 2.3(a) and Figure 2.3(b). The expression for the diffracted wave is then

$$u_{\mathrm{d}} = \frac{u_{\mathrm{inc}}}{\sqrt{r}} D(\theta_{\mathrm{d}}, \theta_{\mathrm{inc}}, \omega) \mathrm{e}^{-\mathrm{i}\omega r}, \tag{2.46}$$

where r is the distance to the tip of the halfplane.

It is important to note that diffraction coefficients only depend on the local geometry of the boundary. Relatively few types of coefficients are therefore sufficient for a systematic use of GTD. Diffraction coefficients have been computed for many different canonical geometries, such as wedges, slits and apertures, different wave equations, in particular Maxwell equations, and different materials and boundary conditions.

Another type of diffraction is generated even from smooth scatterers. When an incident field hits a smooth body such that some rays are tangent to the body surface, there will be a shadow zone behind it. The geometrical optics solution will again be discontinuous, and the curve (point in 2D) dividing the shadow part and the illuminated part of the body, will act as a source for surface rays, or *creeping rays*, that propagate along geodesics on the scatterer surface, if the surrounding medium is homogeneous, $\eta \equiv 1$. The creeping ray carries an amplitude proportional to the amplitude of the inducing ray. The amplitude decays exponentially along the creeping ray's trajectory. In three dimensions, the amplitude also changes through geometrical spreading on the surface. At each point on a convex surface, the creeping ray emits surface-diffracted rays in the tangential direction, with its current amplitude. Those rays then follow the usual geometrical optics laws. See Figure 2.4 for an example. Other well-known surface waves are the Rayleigh waves in the elastic wave equation (1.11).

Physical optics

The physical optics (PO) method, also known as Kirchhoff's approximation, combines the geometrical optics (GO) solution with a boundary integral formulation of the solution to the Helmholtz equation. It is often used for scattering problems in, *e.g.*, computational electromagnetics. Let Ω be a perfectly reflecting scatterer in $\mathbb{R}^3$ and divide the solution into an incident and scattered part, $u = u_{\rm inc} + u_{\rm s}$. Then, in a homogeneous medium with $c \equiv 1$,

$$\Delta u_{\rm s} + \omega^2 u_{\rm s} = 0, \qquad \boldsymbol{x} \in \mathbb{R}^3 \setminus \overline{\Omega}, \tag{2.47}$$

$$u_{\rm s} = -u_{\rm inc}, \qquad \boldsymbol{x} \in \partial\Omega, \tag{2.48}$$

together with an outgoing radiation condition. The solution outside Ω is given by the integral

$$u_{\rm s}(\boldsymbol{x}) = -\oint_{\partial\Omega} u_{\rm inc}(\boldsymbol{x}')\frac{\partial G(\boldsymbol{x},\boldsymbol{x}')}{\partial n} + G(\boldsymbol{x},\boldsymbol{x}')\frac{\partial u_{\rm s}(\boldsymbol{x}')}{\partial n}\,{\rm d}x', \tag{2.49}$$

where G is the free space Green's function in three dimensions:

$$G(\boldsymbol{x},\boldsymbol{x}') = \frac{{\rm e}^{{\rm i}\omega|\boldsymbol{x}-\boldsymbol{x}'|}}{4\pi|\boldsymbol{x}-\boldsymbol{x}'|}. \tag{2.50}$$

The unknown in this, exact, expression for the solution is $\partial u_{\rm s}/\partial n$ on the boundary of Ω. In physical optics, this unknown is simply replaced by the geometrical optics solution. For example, if the incident field is a plane wave $u_{\rm inc} = \exp(-{\rm i}\omega\boldsymbol{k}\cdot\boldsymbol{x})$, with $|\boldsymbol{k}| = 1$, then we would use $\partial u_{\rm s}/\partial n = -{\rm i}\omega\boldsymbol{k}\cdot\hat{\boldsymbol{n}}u_{\rm inc}$ for $\boldsymbol{x} \in \partial\Omega$, where $\hat{\boldsymbol{n}}$ is the normal of $\partial\Omega$ at $\boldsymbol{x}$. In the so-called *physical theory of diffraction* (PTD), the GTD extension of the geometrical optics solution is used.

Expression (2.49) gives a rigorous solution to Helmholtz in free space. The PO approximation is made at the boundary. PO can be regarded as a high frequency approximation in the sense that the accuracy increases with frequency. The computational cost is lower than the direct solution of the boundary integral formulation of the wave equation. Unlike GO, however, the cost is typically not frequency-independent, but grows drastically with frequency. Note also that PO is not self-consistent for finite frequencies. The resulting $\partial u_{\rm s}(\boldsymbol{x} \in \partial\Omega)/\partial n$ is not equivalent to the applied GO solution. Iterative schemes to obtain this consistency can be used.

3. Overview of numerical methods

High frequency wave propagation is well approximated by asymptotic formulations like geometrical optics and the geometrical theory of diffraction. These formulations can be the basis of computations or they can be used analytically for the understanding of high frequency phenomena. In this section we shall describe different classes of computational techniques, based on the three different mathematical models for geometrical optics discussed in Section 2 above: see Figure 3.1.

$$u_{tt} - c(\boldsymbol{x})^2 \Delta u = 0$$

Rays	Kinetic	Eikonal
$\frac{\mathrm{d}^2 \boldsymbol{x}}{\mathrm{d}\tau} = \frac{1}{2}\nabla\eta^2$	$f_t + \frac{1}{\eta^2}\boldsymbol{p}\cdot\nabla_x f$	$\|\nabla\phi\| = \eta$
	$+\frac{1}{\eta}\nabla_x\eta\cdot\nabla_p f = 0$	
Ray tracing	Phase space methods	Hamilton–Jacobi methods

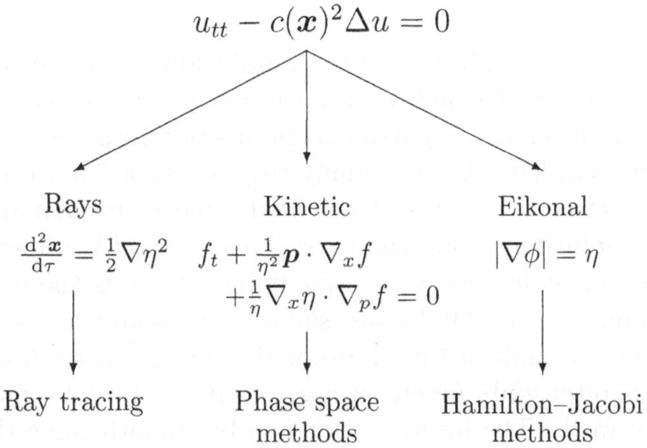

Figure 3.1. Mathematical models and numerical methods.

3.1. Ray tracing

The ray equations derived in Section 2.2 are the basis of ray tracing. The ray $\boldsymbol{x}(t)$ and slowness vector $\boldsymbol{p}(t) = \nabla\phi(\boldsymbol{x}(t))$ are governed by the ODE system (2.13) and (2.14). This system can be augmented by another ODE system for the amplitude, (2.21). Solving those ODEs is called ray tracing and it can be regarded as the method of characteristics applied to the eikonal equation. Some general references on ray tracing are Červený, Molotkov and Psencik (1977), Julian and Gubbins (1977), Langan, Lerche and Cutler (1985) and Thurber and Ellsworth (1980).

Ray tracing is typically not used to solve the complete Cauchy problem, with arbitrary initial and boundary data. Rather, the interest is to find the travel time of a wave from one source point to all points in a domain, or to a limited set of receiver points, together with the corresponding amplitudes in those points. The initial data are thus a single point source. In applications the same information is often needed for many source points, such as all points on a curve. The procedure is then repeated for each source point.

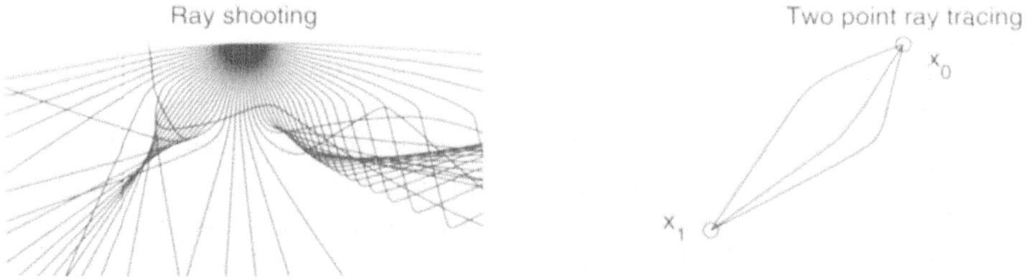

Figure 3.2. Ray shooting and two-point ray tracing. In this case there are three solutions to the two-point ray tracing problem.

Ray tracing gives the phase and the amplitude along the rays, and there is no *a priori* control of which points the ray passes through. One way of obtaining the solution at the particular points of interest is to use *ray shooting*: see Figure 3.2, left. A great many rays are shot from the source point in different directions. The result at the desired receiver points is interpolated from the solutions along the rays. This method is preferred when the travel time is sought for many receiver points, such as the grid points of a discretized domain. The ODEs are solved with standard numerical methods, for instance second- or fourth-order Runge–Kutta methods. The index of refraction is often only given on a grid, and it must be interpolated for the method to work. The interpolation can be smooth, such that the gradient of η in (2.14) exists everywhere, but simple piecewise constant or linear interpolation is also used. The rays are then straight lines or circular arcs within the grid cells, and they can be propagated exactly without an ODE solver. Snell's law of refraction is used at cell boundaries. Interpolating the ray solutions to a uniform grid from a large number of rays is difficult, in particular in shadow zones where few rays penetrate, and in regions where many families of rays cross: *cf.* Figure 3.2.

Another strategy to obtain the solution at a particular point is *two-point ray tracing*, also known as *ray bending*: see Figure 3.2, right. It is often used when there is only a limited number of receiver points. In this setting the ODEs are regarded as a nonlinear elliptic boundary value problem. From (2.18) we get

$$\begin{aligned} \frac{\mathrm{d}^2\boldsymbol{x}}{\mathrm{d}\tau^2} &= \frac{1}{2}\nabla\eta(\boldsymbol{x}(\tau))^2, \qquad (3.1)\\ \boldsymbol{x}(0) &= \boldsymbol{x}_0,\\ \boldsymbol{x}(\tau^*) &= \boldsymbol{x}_1, \end{aligned}$$

where $\boldsymbol{x}_0$ is the source point and $\boldsymbol{x}_1$ is the point of interest: see, *e.g.*, Pereyra, Lee and Keller (1980). Note that τ^*, the parameter value at the end point $\boldsymbol{x}_1$, is an additional unknown that must be determined together with the solution. The equation (3.1) can be solved by a standard shooting method. It can also be discretized and turned into a nonlinear system of equations that can be solved with, for instance, variants of Newton's method. Initial data for the iterative solver can be difficult to find, in particular if there are multiple solutions (arrivals). Also, for two-point ray tracing the index of refraction must be interpolated.

In most problems in computational electromagnetics (CEM) the medium is piecewise homogeneous. This simplifies the calculations, since the solution of (2.13) and (2.14) is trivial given the solution at the boundaries and on the interfaces between media. Rays are straight lines satisfying the reflection law and Snell's law at interfaces. Ray tracing then reduces to the geometrical problem of finding points where rays are reflected and refracted. In the electromagnetic community, ray shooting is often referred to as *shooting and bouncing rays* (SBR), and two-point ray tracing as *ray tracing*: see, *e.g.*, (Ling, Chou and Lee 1989).

Note that the source and receiver points may be at infinity, corresponding to incident and scattered plane waves. For instance, a common problem in CEM is to compute the radar cross section (RCS) of an object. In this case both the source and receiver points are typically at infinity.

3.2. Hamilton–Jacobi methods

To avoid the problem of diverging rays, several PDE-based methods have been proposed for the eikonal and transport equations (2.5), (2.6) and (2.10). When the solution is sought in a domain, this is also computationally a more efficient and robust approach. The equations are solved directly, using numerical methods for PDEs, on a uniform Eulerian grid to control the resolution.

Viscosity solutions

The eikonal equation is a Hamilton–Jacobi-type equation and it has a unique viscosity solution which represents the first arrival travel time (Crandall and Lions 1983). This is also the solution to which monotone numerical finite difference schemes converge, and computing it was the starting point for a number of PDE-based methods. Vidale (1988) and van Trier and Symes (1991) used upwind methods to compute the viscosity solution of the frequency domain eikonal equation

$$|\nabla\phi| = \eta. \tag{3.2}$$

Upwind methods are stable, monotone methods that give good resolution of the kinks usually appearing in a viscosity solution. Importantly, the methods of Vidale (1988) and van Trier and Symes (1991) are explicit: computing the solution at a new grid point only involves previously computed solutions at adjacent grid points. The methods make one sweep over the computational domain, finding the solution at one grid point after another, following an imagined expanding 'grid wave front', propagating out from the source: see Figure 3.3. To ensure causality and to obtain the correct viscosity solution from an explicit scheme, the grid points must be updated in a certain order. Those early methods used a grid wave front with fixed shape (rectangular in Vidale (1988) and circular in van Trier and Symes (1991)) and fail in this respect when there are rays in the exact solution that run parallel to the grid wave front, much in the same way as a paraxial approximation fails when there are turning rays. To avoid failure, the grid wave front could systematically be advanced from the grid point that has the smallest current solution value (minimum travel time). This ensures causality and guarantees a correct result, which was recognized by Qin, Luo, Olsen, Cai and Schuster (1992). The method presented by Qin *et al.* (1992) included simple sorting of the points on the grid wave front according to solution value. The sorting was improved in Cao and Greenhalgh (1994), where an efficient heap sort algorithm was proposed to maintain the right ordering of the points on the grid wave front, as it is advanced. The method of Cao and Greenhalgh (1994) bears a close resemblance to the *fast marching method* (Tsitsiklis 1995, Sethian 1996, Sethian 1999). This is an upwind-based method for efficient evaluation of distances or generalized distance functions such as the phase ϕ in (3.2). It also uses a heap sort algorithm allowing for computationally efficient choices of marching directions. Those methods can be seen as versions of Dijkstra's algorithm

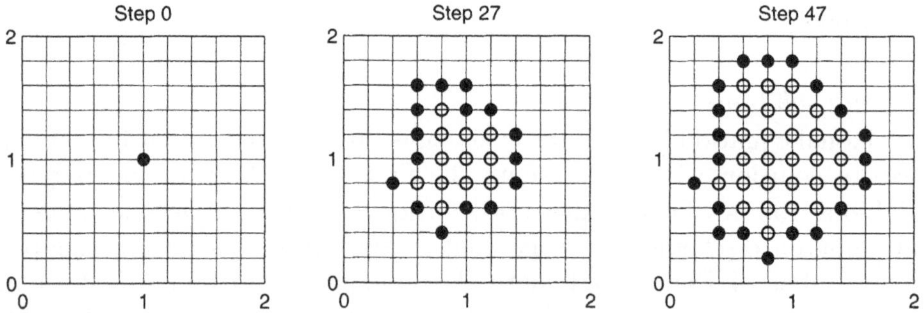

Figure 3.3. An explicit solver for the frequency domain eikonal equation (3.2). Starting from one source point (left), the grid points are updated one at a time in a certain order (middle, right). The outermost points (filled circles) constitute the grid wave front, which propagates outwards from the source, leaving behind it points where the solution is already established (circles). Note that the grid wave front is not necessarily close to an actual wave front.

for finding the shortest path in a network, adapted to a grid-based setting. The overall computational complexity for solving a problem with N grid points is $\mathcal{O}(N \log N)$. Another recent fast method is the *group marching method* (Kim 2000), whose complexity is merely $\mathcal{O}(N)$.

In parallel, high-resolution methods of ENO and WENO type, which had for some time been used in the numerical analysis of nonlinear conservation laws, were adapted to Hamilton–Jacobi equations (Osher and Shu 1991). Those methods were used for the time-dependent eikonal equation (2.5) in Fatemi, Engquist and Osher (1995). Constructing higher-order schemes for methods that use an expanding grid wave front is difficult if the shape of the front changes, as in the fast marching method. For paraxial approximations and methods with fixed-shape grid fronts, the high-resolution methods can be applied directly to obtain higher-order schemes. *Post sweeping* is a technique for avoiding the failures that are associated with turning rays in these methods. The problem at hand is solved in several 'sweeps', using different preferred directions. For each sweep, at each grid point, the smallest of the new and the previously computed solution value is selected (Schneider, Ranzinger, Balch and Kruse 1992, Kim and Cook 1999, Tsai, Cheng, Osher and Zhao 2003).

Multivalued solutions

The eikonal and transport equations only describe one unique wave (phase) at a time. There is no superposition principle in the nonlinear eikonal equation. At points where the correct solution should have a multivalued phase, the viscosity solution picks out the phase corresponding to the first arriving wave. When later arriving waves are also of interest, the viscosity solution is not enough. In inverse seismic problems, for instance, it is recognized that first arrival travel times are often not sufficient to give a good migration image (Geoltrain and Brac 1993). This is a particular problem in complicated inhomogeneous media, where caustics that generate new phases appear in the interior of the computational domain for any type of source. The problem is related to the fact that the first arrival wave is not always the most energetic one (*cf.* the example in Figures 1.1 and 1.2).

One way to obtain more than the first arrival solution is to geometrically decompose the computational domain, and solve the the eikonal solution, with appropriate boundary conditions, in each of the subdomains. The viscosity solutions thus obtained can be pieced together to reconstruct a larger part of the full multibranch solution.

A simple decomposition strategy can be based on detecting kinks in the viscosity solution. The kinks appear where two different branches of the full solution meet: *cf.* Section 2.1. Fatemi *et al.* (1995) made an attempt to compute multivalued travel times with this approach. A second phase, corresponding to the second arrival time, was calculated using two separate

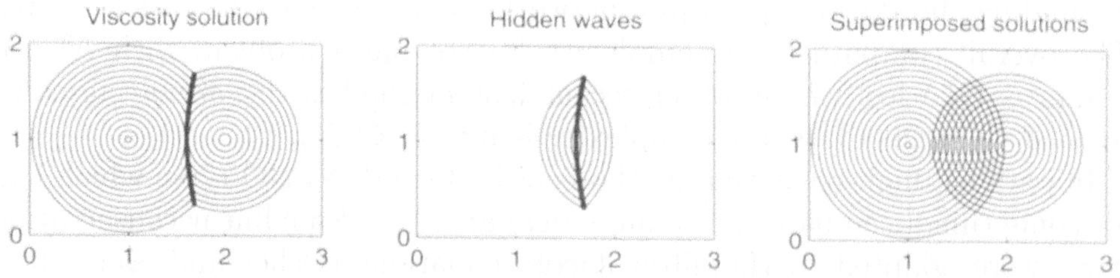

Figure 3.4. Geometrical decomposition by detecting kinks. Bold lines indicate location of the (first) viscosity solution kink. The middle figure shows second viscosity solution where the first solution (left) was applied as boundary condition at the kink.

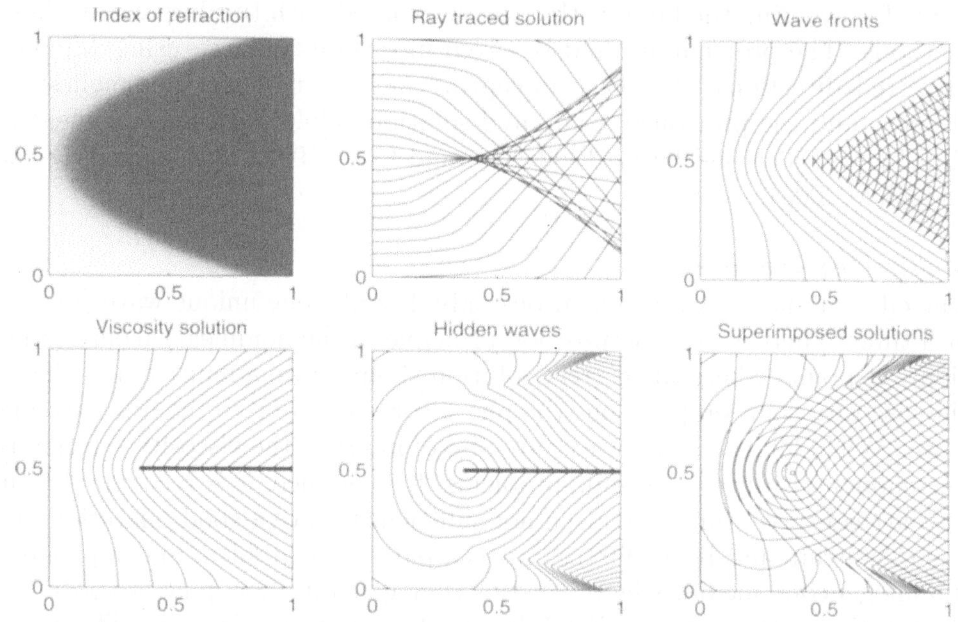

Figure 3.5. Geometrical decomposition by detecting kinks in a problem with a caustic. The top row shows the index of refraction and exact solution. The bottom row shows computed solutions. Bold lines indicate location of the (first) viscosity solution kink.

viscosity solutions of the eikonal equation, with boundary conditions for the second phase given at the location of the kink that had appeared in the first viscosity solution: see Figure 3.4. The same technique was also used at geometric reflecting boundaries. In principle, the same procedure could be repeated, using kinks in the second solution as boundary data for a third phase, and so on.

It is difficult, however, to find a robust way of detecting a kink and to distinguish it from rapid, but smooth, gradient shifts at strong refractions. For more complicated problems, such as the one shown in Figure 3.5, there are difficulties even if the kink could be detected perfectly. In this example, there is no obvious way to find boundary data for the third phase using the singularities in the second viscosity solution (bottom row, middle figure). Moreover, only the part of the second solution lying to the right of the caustic curve that develops (see the ray-traced solution), corresponds to a physical wave. The rest of the solution should be disregarded, including the kinks near the top and bottom right corners.

Another, more *ad hoc*, way of dividing the domain is used in the *big ray tracing* method: see Figure 3.6. It was introduced by Benamou (1996), and extended for use with unstructured grids by Abgrall and Benamou (1999). A limited number of rays are shot from the source point in different directions. The domains bounded by two successive rays are the 'big rays'. In each

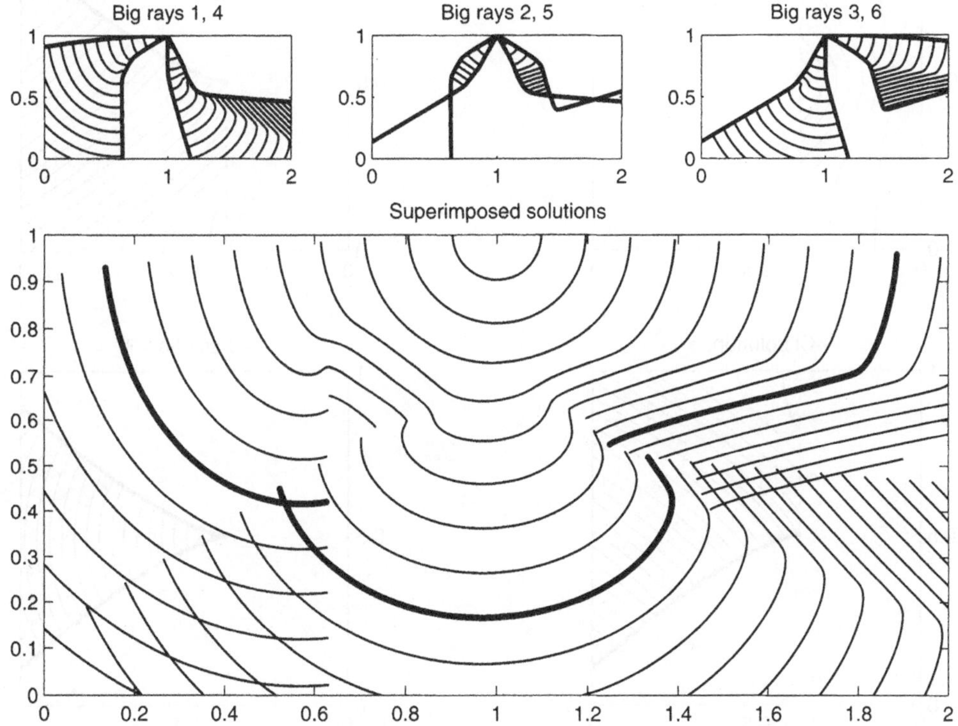

Figure 3.6. Big ray tracing using six big rays for the problem in Figure 1.1. The top row shows viscosity solutions in each ray. The bottom figure shows solutions superimposed; the bold line corresponds to points on the same wave front that was indicated in Figure 1.2(b).

big ray, the viscosity solution is computed. Since the big rays may overlap (*e.g.*, big rays 1 and 3 in Figure 3.6), multivalued solutions can be obtained, although in general the method will not capture all phases. In the presence of caustics the basic method is not so reliable, and it needs to be modified. Then there is, for instance, no guarantee that it includes the viscosity solution among its branches: *cf.* the example in Figure 3.6.

Benamou (1999) introduced a more natural decomposition of the computational domain, which ensures that all phases in the multibranch solution are captured. In his method, the domain is cut along caustic curves. The caustics are detected by solving an accompanying PDE that enables a continuous monitoring of the geometrical spreading. The geometrical spreading

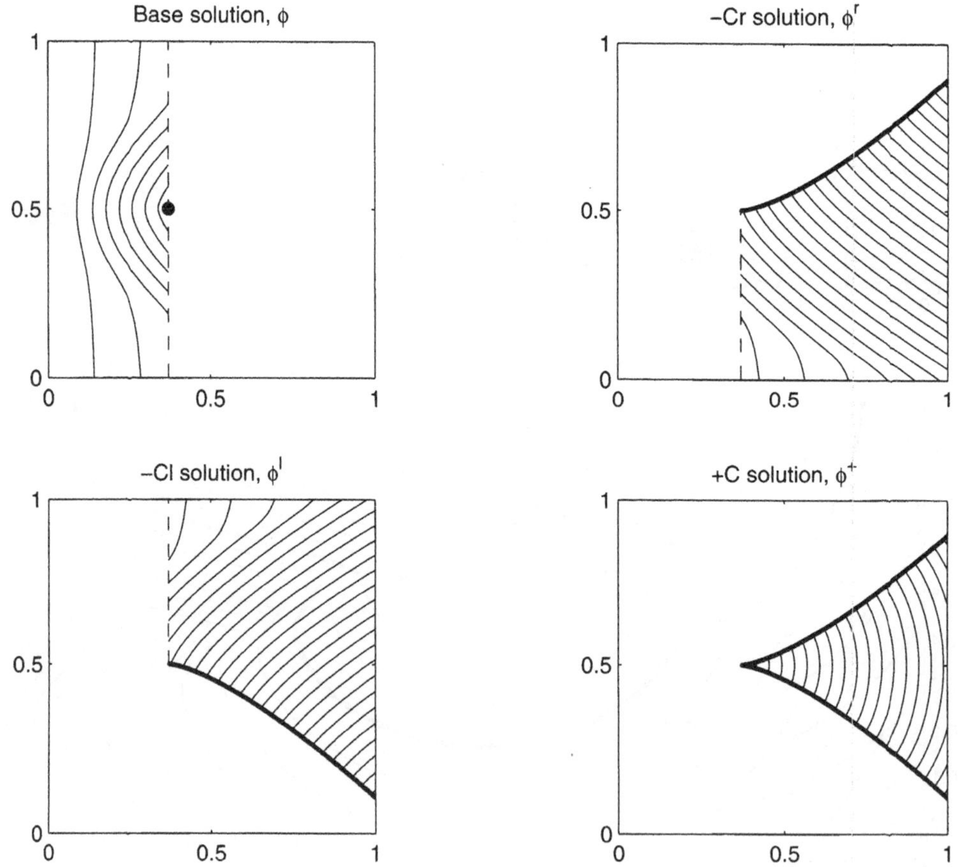

Figure 3.7. Direct computation of multivalued solutions for the problem in Figure 3.5. The computational domain is cut along caustic curves. The filled circle in the top left figure indicates point (x^*, y^*), where the caustic is first detected. Bold lines indicate the caustic curve.

vanishes at caustics, which can therefore be found numerically by checking sign changes in the computed geometrical spreading.

Figure 3.7 exemplifies the method for the problem described in Figure 3.5. In this case the paraxial approximation

$$\phi_x - \sqrt{\eta^2 - {\phi_y}^2} = 0,$$
$$\phi(0,y) = \phi_0(y),$$

is used. The accompanying PDE is the Eulerian version of (2.42). Let the initial data satisfy $\partial_y\phi_0(y) = \eta(0,y)\cos\theta_0(y)$, and, in the notation of Section 2.5 on paraxial approximations (see page 200), set $\delta(x, y(x,r)) = (y_r(x,r), \theta_r(x,r))^T$, with $y_0(r) = r$ and $\theta_0(r)$ as above. Then, by (2.39) and the chain rule, $\delta(x,y)$ satisfies

$$\delta_x + (\tan\theta)\delta_y = \begin{pmatrix} u_y & u_\theta \\ v_y & v_\theta \end{pmatrix}\delta, \qquad \delta(0,y) = \begin{pmatrix} 1 \\ \frac{\mathrm{d}\theta_0(y)}{\mathrm{d}y} \end{pmatrix},$$

where

$$u = \tan\theta, \qquad v = \eta(x,y)^{-1}(\eta_y(x,y) - \eta_x(x,y)\tan\theta), \qquad \tan\theta = \frac{\phi_y}{\phi_x}.$$

The geometrical spreading is the first component of $\delta = (\delta_1, \delta_2)$, and a sign change in $\delta_1 \sim y_r$ indicates a caustic point. When such a point is discovered (call it (x^*, y^*)), the solution is split into three separate branches: $-Cr$, $-C\ell$ and $+C$. The first two are the outer branches, and the last one is the middle branch: see Figure 3.7. Let variables related to the $-Cr$, $-C\ell$ and $+C$ branches be superscripted by r, ℓ and $+$, respectively. For $-Cr$, define the domain

$$\Omega^r(\delta^r) = \{(x,y) \in (x^*, \infty) \times \mathbb{R} \mid \delta_1^r(x,y) > 0\}.$$

This represents the domain below the top caustic curve. The viscosity solution in the $-Cr$ branch is given by the free boundary problem

$$\begin{aligned} \phi_x^r - \sqrt{\eta^2 - {\phi_y^r}^2} &= 0, & (x,y) &\in \Omega^r(\delta^r), \\ \delta_x^r + (\tan\theta^r)\delta_y^r &= \begin{pmatrix} u_y & u_\theta \\ v_y & v_\theta \end{pmatrix}\delta^r, & (x,y) &\in \Omega^r(\delta^r), \\ \phi^r(x^*, y) &= \phi(x^*, y), & y &\le y^*, \\ \delta^r(x^*, y) &= \delta(x^*, y), & y &\le y^*, \end{aligned}$$

where we note that the domain Ω^r depends on the solution. The $-C\ell$ branch is treated similarly. For the middle branch, a Dirichlet problem, coupled by boundary conditions to the other two systems, is solved:

$$\begin{aligned} \phi_x^+ - \sqrt{\eta^2 - \phi_y^{+2}} &= 0, & (x,y) &\in \Omega^\ell(\delta^\ell) \cap \Omega^r(\delta^r), \\ \phi^+(x,y) &= \phi^\ell(x,y), & (x,y) &\in \partial\Omega^\ell(\delta^\ell), \\ \phi^+(x,y) &= \phi^r(x,y), & (x,y) &\in \partial\Omega^r(\delta^r). \end{aligned}$$

The three solutions ϕ^r, ϕ^ℓ and ϕ^+ together make up the full multibranch solution. The strategy can be used recursively, detecting new caustics in the three solutions, and decomposing them into new branches if they appear. For fold caustics, the caustic can be traced more accurately by solving an ODE coupled to the eikonal equations (Benamou, Lafitte, Sentis and Solliec 2003, Solliec 2003). Let the top caustic curve be given by $y^{c\ell}(x) = y(x, s(x))$, where $s(x)$ is an unknown function and $y(x,r)$ is as on page 200. Then, since the geometrical spreading $y_r \equiv 0$ at the caustic,

$$\frac{\mathrm{d}y^{c\ell}}{\mathrm{d}x} = y_x + s_x y_r = y_x = \tan\theta = \frac{\phi_y^r(x, y^{c\ell})}{\phi_x^r(x, y^{c\ell})}.$$

See also Benamou and Solliec (2000).

The *slowness matching method* of Symes (Symes 1996, Symes and Qian 2003), is another method for finding multivalued solutions to the eikonal equation. It is based on the travel time map $\tau(\boldsymbol{x}_{\mathrm{src}}, \boldsymbol{x}_{\mathrm{rcv}})$, which gives the travel time of a wave from a source point $\boldsymbol{x}_{\mathrm{src}}$ to a receiver point $\boldsymbol{x}_{\mathrm{rcv}}$. Hence, if $(\boldsymbol{x}(t), \boldsymbol{p}(t))$ is a bicharacteristic going from $\boldsymbol{x}(0) = \boldsymbol{x}_1$ to $\boldsymbol{x}(T) = \boldsymbol{x}_2$, then $\tau(\boldsymbol{x}_1, \boldsymbol{x}_2) = T$. This function may of course be multivalued if there is more than one such bicharacteristic, and we distinguish between values by their associated arrival slowness, $\boldsymbol{p}(T)$. There is, however, always a neighbourhood of $\boldsymbol{x}_{\mathrm{src}}$ for which $\tau(\boldsymbol{x}_{\mathrm{src}}, \cdot\,)$ is smooth and single-valued. We denote this neighbourhood by $\mathcal{N}(\boldsymbol{x}_{\mathrm{src}})$. For fixed $\boldsymbol{x}_{\mathrm{src}}$, the map satisfies the eikonal equation with respect to $\boldsymbol{x}_{\mathrm{rcv}}$ in $\mathcal{N}(\boldsymbol{x}_{\mathrm{src}})$,

$$|\nabla_{x_{\mathrm{rcv}}} \tau(\boldsymbol{x}_{\mathrm{src}}, \boldsymbol{x}_{\mathrm{rcv}})| = \eta(\boldsymbol{x}_{\mathrm{rcv}}), \quad \boldsymbol{x}_{\mathrm{rcv}} \in \mathcal{N}(\boldsymbol{x}_{\mathrm{src}}), \quad \tau(\boldsymbol{x}_{\mathrm{src}}, \boldsymbol{x}_{\mathrm{src}}) = 0. \tag{3.3}$$

The slowness matching method draws on the following observation. Suppose there is a travel time $\tau(\boldsymbol{x}_1, \boldsymbol{x}_2)$ between the points $\boldsymbol{x}_1$ and $\boldsymbol{x}_2$ with arrival slowness $\boldsymbol{p}$. If there is a third point $\boldsymbol{x}_3$ for which $\boldsymbol{x}_2 \in \mathcal{N}(\boldsymbol{x}_3)$ and the *slowness matching condition* holds at $\boldsymbol{x}_2$,

$$\boldsymbol{p} + \nabla_{x_{\mathrm{rcv}}} \tau(\boldsymbol{x}_3, \boldsymbol{x}_2) = 0, \tag{3.4}$$

then the travel times are additive, that is,

$$\tau(\boldsymbol{x}_1, \boldsymbol{x}_3) = \tau(\boldsymbol{x}_1, \boldsymbol{x}_2) + \tau(\boldsymbol{x}_2, \boldsymbol{x}_3) \tag{3.5}$$

is a travel time between $\boldsymbol{x}_1$ and $\boldsymbol{x}_3$ with arrival slowness $\nabla_{x_r}\tau(\boldsymbol{x}_2, \boldsymbol{x}_3)$. Conversely, if a ray from $\boldsymbol{x}_1$ to $\boldsymbol{x}_3$ passes through a point $\boldsymbol{x}_2 \in \mathcal{N}(\boldsymbol{x}_3)$, then (3.4) and (3.5) hold. This follows from the uniqueness of solutions to the ray equations (2.13) and (2.14), and the fact that $\boldsymbol{p}(t) = \nabla\phi(\boldsymbol{x}(t))$ when ϕ is a smooth solution to the eikonal equation and $(\boldsymbol{x}(t), \boldsymbol{p}(t))$ is a bicharacteristic.

The method divides the domain into M layers of width Δx, with layer boundaries at constant x-coordinates, $x_n = n\Delta x$, $n = 0, \ldots, M$. At each point on layer boundary n, data are stored about the travel times $\tau_n(y)$ and arrival slownesses $\boldsymbol{p}_n(y)$ of rays crossing the boundary: see Figure 3.8, left. These functions will typically be multivalued. In order to compute the corresponding data for layer boundary $n+1$, the travel time map is used to piece together the multivalued solution via the slowness matching principle (3.4). For each point (x_{n+1}, y) on boundary $n+1$, find all points $(x_n, \tilde{y})$ on boundary n with an arrival slowness $\boldsymbol{p}_n(\tilde{y})$ such that

$$\boldsymbol{p}_n(\tilde{y}) + \nabla_{x_r}\tau(x_{n+1}, y;\, x_n, \tilde{y}) = 0. \tag{3.6}$$

Then, for each $\tilde{y}$ satisfying this condition, $\tau_{n+1}(y) = \tau_n(\tilde{y}) + \tau(x_n, \tilde{y};\, x_{n+1}, y)$ is a travel time at (x_{n+1}, y) with arrival slowness

$$\boldsymbol{p}_{n+1}(y) = \nabla_{x_r}\tau(x_n, \tilde{y};\, x_{n+1}, y).$$

See the example in Figure 3.8, right.

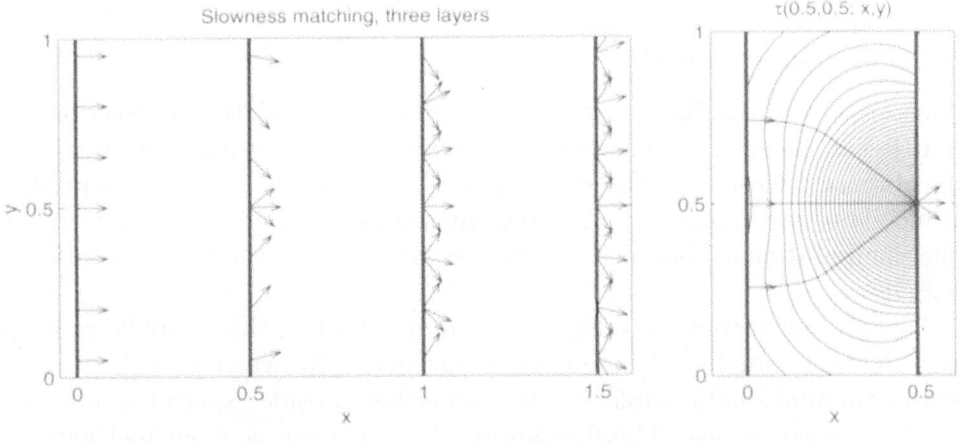

Figure 3.8. Slowness matching method. The left frame shows results for problem in Figure 3.5 with $\Delta x = 0.5$. Arrows indicate the arrival slownesses $\boldsymbol{p}_n$, bold lines indicate layer boundaries. The right frame shows the slowness matching condition and its three solutions at the point $(0.5, 0.5)$ on the second layer boundary. Iso curves of travel time map $\tau(0.5, 0.5;\, x, y)$, and rays associated to the three solutions are shown.

Numerically, the y-coordinate is discretized with a uniform grid, $\{y_j\}$. For each grid point (x_n, y_j), a list is maintained that contains one or more associated travel times and arrival slownesses. The travel time map τ is computed by solving the paraxial eikonal equation (2.43) with $\boldsymbol{x}_{\rm src} = (x_{n+1}, y_j)$ for all j. The slowness matching involves interpolating the travel time map on a regular grid and root finding.

Under the paraxial approximation the solution $\tilde{y}$ to (3.6) satisfies $|y - \tilde{y}| \leq C\,\Delta x$. Therefore, when Δx is small enough, $(x_n, \tilde{y}) \in \mathcal{N}(x_{n+1}, y)$ and the travel time map $\tau(x_{n+1}, y;\,\cdot)$ is smooth, single-valued around $(x_n, \tilde{y})$. Multivalued solutions can still be obtained, however, since there may be multiple solutions to the slowness matching condition (3.6): *cf.* Figure 3.8, right frame. In fact, all branches of the complete multivalued solution will be found in this way, if Δx is small enough and the paraxial approximation holds.

The cost of the slowness matching method is quite high when it is used as described above. Let N_x and N_y be the number of grid points in the x- and y-coordinate directions, respectively. Computing the travel time map τ involves $\mathcal{O}(N_x^2 N_y / M)$ operations, and it is the dominating cost when $M \ll N_x / \sqrt{N_y}$. It should be compared with the $\mathcal{O}(N_x N_y)$ cost of computing the viscosity solution with the paraxial approximation. The travel time map τ can, however, be re-used when the solution is sought for multiple sources, a case for which the slowness matching method is competitive. In this respect it falls in the same category as the fast phase space method of Fomel and Sethian (2002), described below in Section 4.5.

3.3. Phase space methods

This class of methods is based on the kinetic formulation in Section 2.3, and it can be seen as a compromise between ray tracing and Hamilton–Jacobi-based methods. The techniques try to keep the linear superposition principle of ray tracing and the regular representation of the solution over the computational domain that can be achieved by the approximation of a PDE.

We mentioned that the computational drawback of the Liouville equation was the large number of independent variables. To overcome this difficulty with computational complexity, we can either consider special solutions or modify the equations. The first approach leads to wave front methods and the second to moment-based methods.

In wave front methods, an interface representing a wave front is evolved following the kinetic formulation. There are different ways of representing interfaces, leading to different techniques. Lagrangian front tracking has been used, and it is closest to traditional ray tracing. Eulerian methods are based on the segment projection method, the level set method or the fast marching method for interface evolution.

For the other approach, with moment-based methods, new equations with fewer unknowns are derived from the kinetic formulation. A finite number of nonlinear partial differential equations for the moments of the kinetic density function f in (2.29) is obtained using a closure assumption that allows for a limited superposition principle.

We will discuss these methods in more detail below in Section 4 and Section 5, respectively.

3.4. Dynamic surface extension

The method of *dynamic surface extension* was introduced by Steinhoff and collaborators in Steinhoff, Wenren, Underhill and Puskas (1995) and Steinhoff, Fan and Wang (2000), and further refined by Ruuth, Merriman and Osher (2000). There are a few variants of the method, but the dependent variables in this technique are essentially the coordinates of the closest point on a wave front from a given $\boldsymbol{x}$-coordinate. This clever choice of representation and a time-stepping scheme following the rules of geometrical optics allow for linear superposition in an Eulerian representation. The present forms of the method have higher complexity than ray tracing, and certain cases will not be correctly described. One such example is a wave front given by a collapsing circle with two parallel tangent lines. At the time when the circle and the tangent lines have been reduced to one line, the information of the circle is lost and cannot be recovered. The method is quite straightforward and is being further developed.

We may illustrate the capability of the dynamic surface extension method to handle crossing wave fronts by the following simple one-dimensional algorithm. Let $X(x_j, t_0)$ be the location of the front which is closest to the grid point $x_j = j\Delta x$, at the initial time t_0, and let $|c(X)|$ be the velocity at X, with $c > 0$ if the front propagates in the positive x-direction and otherwise negative. The algorithm consists of two steps. In the first step, the location of the fronts are updated,

$$\tilde{X}(x_j, t_{n+1}) = X(x_j, t_n) + \Delta t c(X(x_j, t_n)), \qquad t_n = t_0 + n\Delta t,$$

and in the second step, the fronts are assigned to the appropriate grid points,

$$X(x_j, t_{n+1}) = \tilde{X}(x_{j+\ell}, t_{n+1}),$$

where $\ell \in \{-1, 0, 1\}$ is chosen such that

$$|\tilde{X}(x_{j+\ell}, t_{n+1}) - x_j|$$

is minimal.

It is easy to see how fronts are allowed to cross each other. For example, let $|c| = 1$, $\Delta x = 1$, and let a front at $X = 1/4$ be moving in the positive x-direction and one at $X = 3/4$ be moving in the negative x-direction.

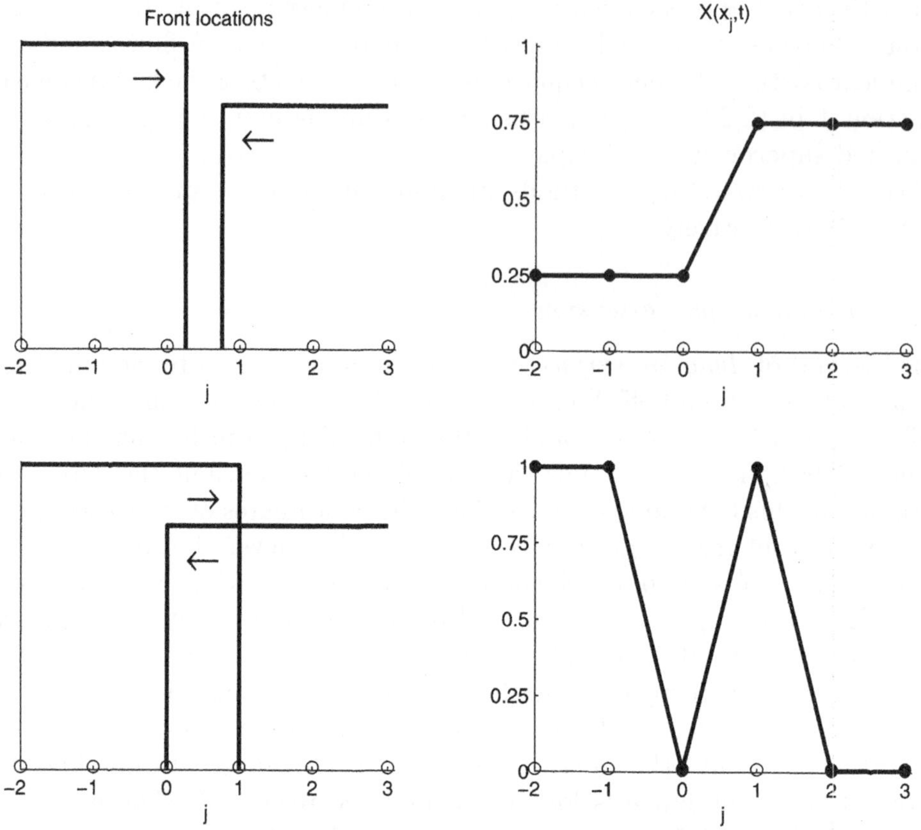

Figure 3.9. Dynamic surface extension. The top row shows the initial front locations and values of $X(x_j, t_0)$. The bottom row shows the state after one step of the algorithm with $\Delta t = 0.75$.

This example is illustrated in Figure 3.9. The algorithm will then give, for $1/2 < \Delta t < 1$,

$$X(x_j, t_0) = \begin{cases} 1/4, & j \leq 0, \\ 3/4, & j > 0, \end{cases} \qquad \tilde{X}(x_j, t_1) = \begin{cases} 1/4 + \Delta t, & j \leq 0, \\ 3/4 - \Delta t, & j > 0, \end{cases}$$

$$X(x_j, t_1) = \begin{cases} 1/4 + \Delta t, & j \leq -1, \\ 3/4 - \Delta t, & j = 0, \\ 1/4 + \Delta t, & j = 1, \\ 3/4 - \Delta t, & j > 1. \end{cases}$$

The grid points at $j = 0, 1$ have registered the new front location, and this information will spread to the other j-values at later times. The extension to multi-dimensional problems also requires an interpolation step, but unfortunately not all cases of front propagation are well represented.

4. Wave front methods

Wave front methods are related to standard ray tracing, but instead of computing a sequence of individual rays a wave front is evolved in physical or phase space. This can be based on the ODE formulation (2.11) and (2.12) or the PDE Liouville equation (2.29).

The propagation of a wave front in the xy-plane is given by the velocity $c(\boldsymbol{x})$ in its normal direction $\hat{\boldsymbol{n}}$. The velocity $\boldsymbol{u} = (u, v)$ of the wave front in the xy-plane is thus

$$(u, v) = c(\boldsymbol{x})\hat{\boldsymbol{n}} = c(\boldsymbol{x})\,(\cos\theta, \sin\theta), \tag{4.1}$$

where θ is the angle between the normal vector and the x-axis. At caustic and focus points, the normal direction is not defined, and front tracking methods based on (4.1) break down.

The tracing of the wave fronts in phase space facilitates problems including the formation of caustics, as is seen in the following simple example. In Figure 4.1, left frame, an initial circular wave front is given in the xy-plane. This frame also displays the phase plane curve γ in $\mathbb{R}^3$ together with its $x\theta$- and $y\theta$-projections. Let the circular wave front contract with time in

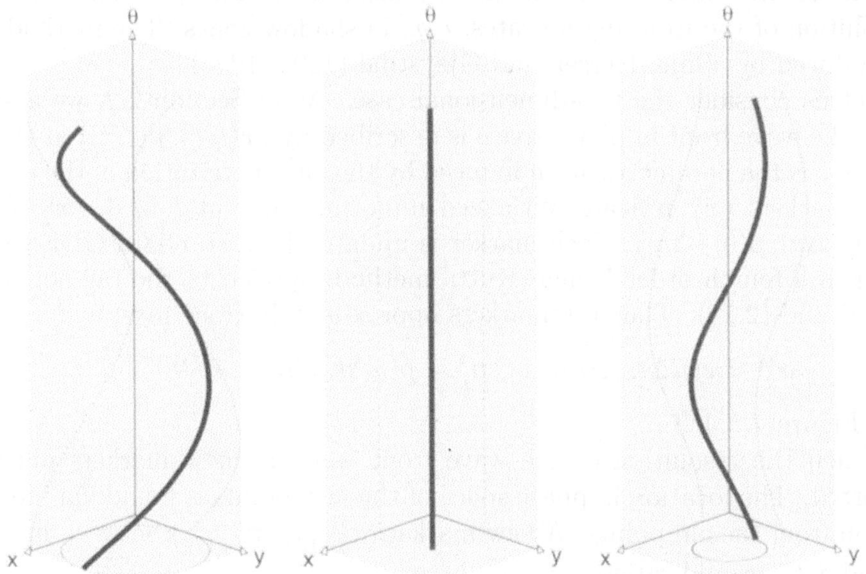

Figure 4.1. Phase plane curve γ: thick line, with projections onto xy-, $x\theta$- and $y\theta$-planes: dotted lines. The left frame shows the initial circular wave front at $t = 0$. The middle frame shows the focus at $t = 1$. The right frame shows the wave front after the focus at $t = 1.5$.

a constant medium, $c(\boldsymbol{x}) \equiv 1$, and be focused to a point $(x, y) = (1,\ 1)$ at time $t = 1$. Although degenerate in the xy-plane, the representations of γ at $t = 1$ in the $x\theta$- and $y\theta$-planes are smooth and the evolution, as well as the computation of amplitudes, can easily be continued to $t > 1$.

For the PDE-based wave front methods in phase space, the evolution of the front is given by (2.29) and the front is represented by some interface propagation technique. We shall here discuss the application of the segment projection method (Engquist *et al.* 2002, Tornberg and Engquist 2003), and also briefly outline level set techniques (Osher and Sethian 1988) and methods based on fast marching (Fomel and Sethian 2002). The segment projection method uses an explicit representation of the wave front, while the level set and fast marching methods use implicit representations: see below. These classes of techniques are based on Eulerian grids and thus there is no need for redistribution of marker points.

4.1. Wave front construction

Wave front construction is a front tracking method in which Lagrangian markers on the phase space wave front are propagated according to the ray equations (2.13) and (2.14). To maintain an accurate description of the front, new markers are adaptively inserted by interpolation when the resolution of the front deteriorates, *e.g.*, in shadow zones. The method was introduced by Vinje, Iversen and Gjøystdal (1992, 1993).

Let us consider the two-dimensional case. As in Section 2.2, we assume that the wave front in phase space is described by $(\boldsymbol{x}(t, r), \boldsymbol{p}(t, r))$ at time t, where r is the parametrization induced by the parametrization of the source. The markers $(\boldsymbol{x}_j^n, \boldsymbol{p}_j^n)$ are initialized uniformly in r at $t = 0$, $(\boldsymbol{x}_j^0, \boldsymbol{p}_j^0) = (\boldsymbol{x}(0, j\Delta r), \boldsymbol{p}(0, j\Delta r))$. Each marker is updated by a standard ODE-solver, such as a fourth order Runge–Kutta method, applied to the ray equations (2.13) and (2.14). Thus the markers approximately trace rays, and

$$\boldsymbol{x}_j^n \approx \boldsymbol{x}(n\Delta t, j\Delta r), \qquad \boldsymbol{p}_j^n \approx \boldsymbol{p}(n\Delta t, j\Delta r), \qquad \forall n > 0, j.$$

See Figure 4.2, left.

When the resolution of the wave front worsens, new markers must be inserted. The location in phase space of the new points is found via interpolation from the old points. A new marker $(\boldsymbol{x}_{j+1/2}^n, \boldsymbol{p}_{j+1/2}^n)$ between markers j and $j + 1$ would satisfy

$$\boldsymbol{x}_{j+1/2}^n \approx \boldsymbol{x}(n\Delta t, j\Delta r + \Delta r/2), \qquad \boldsymbol{p}_{j+1/2}^n \approx \boldsymbol{p}(n\Delta t, j\Delta r + \Delta r/2).$$

See Figure 4.2, middle. When deciding on whether to add new markers, it is not sufficient only to look at the distance in physical space between the old markers, because it degenerates at caustics and focus points. The distance in the phase variable should also be taken into account (Sun 1992). A useful

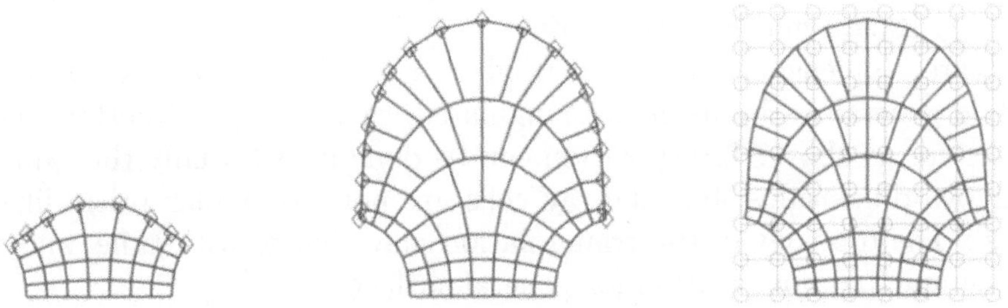

Figure 4.2. Wave front construction. Markers ($\diamond$) on the wave front are propagated as ordinary rays (left). The grid approximates the wave front in physical space, $\boldsymbol{x}(t,r)$ at constant t- and r-values. When the markers move too wide apart to accurately describe the front, new markers are inserted via interpolation (middle). The travel times and possibly amplitudes on the wave front are interpolated onto a regular grid as the front propagates (right).

criterion is to add a new marker between markers j and $j+1$ if

$$|\boldsymbol{x}_{j+1}^n - \boldsymbol{x}_j^n| \geq \mathrm{TOL} \qquad \text{or} \qquad |\boldsymbol{p}_{j+1}^n - \boldsymbol{p}_j^n| \geq \mathrm{TOL}.$$

for some tolerance TOL. This criterion ensures that the phase wave front remains fairly uniformly sampled. Lambaré, Lucio and Hanyga (1996) introduced another criterion, where more points are added when the curvature of the phase space wave front is large. For each marker, they compute the additional quantities

$$\boldsymbol{X}_j^n \approx \boldsymbol{x}_r(n\Delta t, j\Delta r), \qquad \boldsymbol{P}_j^n \approx \boldsymbol{p}_r(n\Delta t, j\Delta r),$$

via the ODE system (2.26). Based on the fact that

$$\begin{aligned}|\boldsymbol{x}(t,r+\Delta r) - \boldsymbol{x}(t,r) - \Delta r \boldsymbol{x}_r(t,r)| &\approx \frac{1}{2}(\Delta r)^2 |\boldsymbol{x}_{rr}| \geq \frac{1}{2}(\Delta r |\boldsymbol{x}_r|)^2 \kappa(r) \\ &\approx \frac{1}{2}|\boldsymbol{x}(t,r+\Delta r) - \boldsymbol{x}(t,r)|^2 \kappa(r),\end{aligned}$$

where $\kappa(r)$ is the curvature, the criterion for adding a new marker is taken as

$$|\boldsymbol{x}_{j+1}^n - \boldsymbol{x}_j^n - \Delta r \boldsymbol{X}_j^n| \geq \mathrm{TOL} \qquad \text{or} \qquad |\boldsymbol{p}_{j+1}^n - \boldsymbol{p}_j^n - \Delta r \boldsymbol{P}_j^n| \geq \mathrm{TOL}.$$

The computed variables $\boldsymbol{X}_j^n$, which is the geometrical spreading, and $\boldsymbol{P}_j^n$ are also used for computing the amplitude and to simplify high-order interpolation when inserting new markers, and in the grid interpolation below.

Finally, the interesting quantities carried by the markers on the wave front, such as travel time and amplitude, are interpolated down on a regular Cartesian grid: see Figure 4.2, right. The wave front construction covers the physical space by quadrilateral 'ray cells'. The interpolation step involves mapping the grid points to the right ray cells, in order to find the markers and marker positions from which to interpolate. This can be complicated. See, *e.g.*, Bulant and Klimeš (1999).

In three dimensions the wave front is a two-dimensional surface. The method generalizes by using a triangulated wave front, and performing the same steps as above. Interpolation can be done in essentially the same way as in two dimensions, but the ray cells are now triangular prism-like 'ray tubes'. The topology of the triangulation may change with time, and there is no simple parametrization for general surfaces.

4.2. Segment projection method

Let us first consider the segment projection method for general interfaces and then apply the technique to geometrical optics. In order to make the presentation more clear, we shall discuss the two-dimensional case, and this is also the case for which the most general software has been developed (Tornberg and Engquist 2003).

The segment projection method is a computational method for tracking the dynamic evolution of interfaces (Tornberg 2000, Tornberg and Engquist 2000). The basic idea is to represent a curve or surface as a union of segments. Each segment is chosen such that it can be given as a function of the independent variables. The representation is thus analogous to a manifold being defined by an atlas of charts. The motions of the individual segments are given by partial differential equations based on the physics describing the evolution of the interfaces.

The segments representing a curve γ in $\mathbb{R}^2$ are here given by functions $Y_j(x)$ and $X_k(y)$. The domains of the independent variables of these functions are projections of the segments onto the coordinate axis. The coordinates of the points on γ are given by $(x, y) = (x, Y_j(x))$ or $(x, y) = (X_k(y), y)$. For each point on a curve γ, there is at least one segment defining the curve. To make the description complete, information about the connectivity of segments must also be provided. For each segment in one variable there is information regarding which part of the curve has overlap with segments in the other variable, as well as pointers to these segments.

The number of segments needed to describe a curve depends on the shape of the curve. An extremum of a function $Y_j(x)$ defines a separation point for the y-segments, as no segment given as a function of y can continue past this point. Similarly, an extremum of a function $X_k(y)$ defines a separation point for the x-segments. A sketch of a distribution of segments is shown in Figure 4.3. For moving interfaces, $Y_j = Y_j(x, t)$ and $X_k = X_k(y, t)$ are also functions of time.

The segments are moved by equations of motion, and after each numerical advection step, the segment representation is re-initialized. Dynamic creation and elimination of segments are employed to follow the evolution

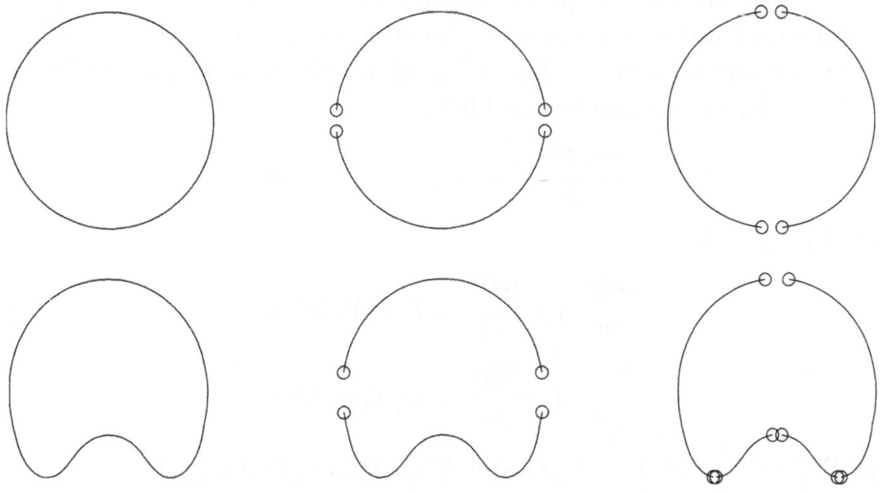

Figure 4.3. Segment structure for circle and deformed circle: curve γ (left), x-segments (middle), y-segments (right).

of the curves. New segments are created if necessary, and segments are removed when they are no longer needed. The connectivity of segments must be kept updated in such a way that the pointers relating segments represent the current configuration. If we assume that the lower deformed circle in Figure 4.3 has evolved from the circle above, a new maximum and two new minima have appeared in the lower x-segment. The number of y-segments should then be increased, as is seen in the figure.

For each segment, the domain of the independent variable must be defined. These segments are numerically given by arrays for Y_j and X_k. The domains of the independent variables, the arrays and information about connectivity between the segments define the structure that represents the curve.

From the definition of an x-segment, an ordered set of numbers is created that contains the start and end points of the segment, together with the extremum points of the segment. The intervals between these points correspond to different segments of the other variable. It is necessary to keep track of the connections between these segments.

Let a velocity field $\boldsymbol{u} = (u, v)^T$ be given, by which the curve should move. The segments $y = Y(x,t)$ and $x = X(y,t)$ are updated according to the partial differential equations

$$\frac{\partial Y}{\partial t} + u\frac{\partial Y}{\partial x} = v, \tag{4.2}$$

$$\frac{\partial X}{\partial t} + v\frac{\partial X}{\partial y} = u. \tag{4.3}$$

Note that there is only one spatial variable present in each of these equations. Quantities that are transported by the velocity field can also be defined as functions on the segments. Let $\mathcal{F}^t(x,y)$ be the flow generated by $\boldsymbol{u}$. If $r(x,y,t)$ evolves according to the ODE

$$\frac{\mathrm{d}r(\mathcal{F}^t,t)}{\mathrm{d}t} = h(\mathcal{F}^t, r(\mathcal{F}^t,t),t),$$

for fixed (x,y), then

$$\frac{\partial R^x}{\partial t} + u\frac{\partial R^x}{\partial x} = h(x,Y,R^x,t), \tag{4.4}$$

$$\frac{\partial R^y}{\partial t} + v\frac{\partial R^y}{\partial y} = h(X,y,R^y,t), \tag{4.5}$$

where $R^x(x,t) = r(x,Y(x,t),t)$ and $R^y(y,t) = r(Y(y,t),y,t)$.

Boundary conditions must be defined for the segments. They are either given in the original problem formulation or interpolated from an overlapping segment in the other coordinate direction. Note that this interpolation is well defined as an interpolation on an irregular mesh from the discrete segment in the other coordinate direction.

After the numerical advection step based on (4.2) and (4.3), we need to review the segment structure. If no new extrema have appeared and no old ones have disappeared, no change needs to be made in the structure of the segments.

Any moving curve γ is represented by overlapping segments. These segments evolve individually and may separate slightly in the overlapping regions due to numerical errors. A re-initialization is applied in every time step to realign the segments. This is done by a weighted interpolation. A segment will typically yield the most accurate description of the curve if its slope is small.

When different parts of γ cross each other, geometric rules for the segment interaction must be given. Examples are the merging of two bubbles in multiphase flow and the reflection of a wave front γ_1 meeting a curve γ_2, representing a perfect reflector.

The advection and re-initialization process for a structure of segments, representing a curve γ, can be summarized as follows.

(1) Advect all x- and y-segments from a velocity field $\boldsymbol{u} = (u,v)$ and evolve associated quantities defined on the segments by numerical approximations of (4.2)–(4.5).

(2) Update the segment structure.

(3) For each segment whose domain of definition has increased, new values need to be defined. These are interpolated from the corresponding segment in the other coordinate direction.

(4) Interpolate the segment between overlapping parts of the x- and y-segments. The new values are assigned using a weight function based on the slopes of the segments.

(5) Rearrange the segment structure from the rules of segment interactions.

These steps are generic and essentially the same for different applications. Common software will thus apply to different problems with only minor modifications, for example, in the advection and the interaction algorithms.

Curves of co-dimension two in $\mathbb{R}^3$ are approximated by their projections onto the two-dimensional coordinate planes. If a curve

$$\gamma(t) : \{X_1(s,t), X_2(s,t), X_3(s,t)\},$$

is parametrized by s and evolves by the velocity field

$$\boldsymbol{u}(\boldsymbol{x},t) = (u_1(\boldsymbol{x},t), u_2(\boldsymbol{x},t), u_3(\boldsymbol{x},t)),$$

we have

$$\frac{\mathrm{d}X_j}{\mathrm{d}t} = u_j(\boldsymbol{x},t), \quad j = 1,2,3.$$

Let the projection of γ onto the $x_j x_k$-plane be represented by a set of segment functions of the type $x_j = X^{jk}(x_k,t)$ and $x_k = X^{kj}(x_j,t)$. The evolution of the segment functions in all three projection planes is then given by the equations

$$\frac{\partial X^{jk}}{\partial t} + u_k \frac{\partial X^{jk}}{\partial x_k} = u_j, \quad j = 1,2,3; \ k = 1,2,3; \ j \neq k. \tag{4.6}$$

The projections in the $x_j x_k$-planes are updated in time following the steps (1)–(5) above. A sixth step is then added with interpolation between the representations in the three coordinate planes. This step is similar to step (4) and is in general needed in order to define u_k and u_j in (4.6). The simulations presented in this paper do not require such interpolations, however.

There is as yet no general software for two-dimensional surfaces in $\mathbb{R}^3$. The principle is analogous to the lower-dimensional case. The surface Σ is represented by functions defined on the three coordinate planes. The functions $x_\ell = X_\ell^{jk}(x_j, x_k, t)$ define the segments as in $\mathbb{R}^2$ and the union of segments defines Σ. Given the velocity field $\boldsymbol{u}(\boldsymbol{x},t)$, the motion of the segments is given by

$$\frac{\partial X_\ell^{jk}}{\partial t} + u_j \frac{\partial X_\ell^{jk}}{\partial x_j} + u_k \frac{\partial X_\ell^{jk}}{\partial x_k} = u_\ell,$$
$$j = 1,2,3, \quad k = 1,2,3, \quad \ell = 1,2,3, \quad j \neq k, \ j \neq \ell, \ k \neq \ell.$$

4.3. Segment projection method for geometrical optics

The segment projection method will be applied to track the evolution in phase space of fronts that are given by geometrical optics. For two space dimensions a curve γ in $\mathbb{R}^3$ is tracked, and for three space dimensions a surface Σ in $\mathbb{R}^5$ is evolved. We shall mainly discuss the two-dimensional case and only give one three-dimensional example on page 230. In the presentation of the method, we may assume that γ or Σ and their projections are functions. The segment projection technique will reduce the general case to a set of segments that are functions of some of the independent variables.

Let the independent variables be x, y and θ. The orthogonal projections of γ onto the xy-, $x\theta$- and $y\theta$-planes are denoted by γ_{xy}, $\gamma_{x\theta}$ and $\gamma_{y\theta}$, respectively. The evolution of $\gamma = \gamma(t)$ will be determined by the two-dimensional segment projection method, as presented in Section 4.2.

From the general equations (4.2) and (4.3) and the velocity field (4.1), we get the Eulerian form of the evolution equations for the x- and y-segments in the xy-plane, respectively:

$$\begin{cases} \dfrac{\partial Y^x}{\partial t} + c(x, Y^x(x,t)) \cos\theta \,\dfrac{\partial Y^x}{\partial x} = c(x, Y^x(x,t)) \sin\theta, \\[2ex] \dfrac{\partial X^y}{\partial t} + c(X^y(y,t), y) \sin\theta \,\dfrac{\partial X^y}{\partial y} = c(X^y(y,t), y) \cos\theta, \end{cases} \tag{4.7}$$

The approach of tracking the front only in the xy-plane, computing θ from the segments, breaks down at caustics. Therefore, the front should be tracked in phase space, and the other two projections are needed. From (2.22), (2.23) and (2.24) we get the velocity field needed to apply (4.2) and (4.3) to the segment equations in the $x\theta$- and $y\theta$-planes. Let the x- and θ-segments in the $x\theta$-plane be denoted by Θ^x and X^θ, and let the y- and θ-segments in the $y\theta$-plane be Θ^y and Y^θ, respectively. The segment equations are

$$\begin{cases} \dfrac{\partial \Theta^x}{\partial t} + c\cos\theta \,\dfrac{\partial \Theta^x}{\partial x} = \alpha, \\[2ex] \dfrac{\partial X^\theta}{\partial t} + \alpha \,\dfrac{\partial X^\theta}{\partial \theta} = c\cos\theta, \end{cases} \qquad \begin{cases} \dfrac{\partial \Theta^y}{\partial t} + c\sin\theta \,\dfrac{\partial \Theta^y}{\partial y} = \alpha, \\[2ex] \dfrac{\partial Y^\theta}{\partial t} + \alpha \,\dfrac{\partial Y^\theta}{\partial \theta} = c\sin\theta, \end{cases} \tag{4.8}$$

$$\alpha = \frac{\partial c(\boldsymbol{x})}{\partial x} \sin\theta - \frac{\partial c(\boldsymbol{x})}{\partial y} \cos\theta. \tag{4.9}$$

The one-dimensional hyperbolic equations above are easily solved by standard numerical methods. Note that the representation of the phase plane curve γ may be degenerate for the projection onto one of the coordinate planes, but there will always be two projections for which γ is well represented.

When η is constant the amplitude on the curve can easily be calculated by post-processing of the results from (4.8). (Below we will compute the amplitude for problems with variable η.) Consider, for instance, an initial curve $(x_0(r), y_0(r))$ with amplitude $A_0(r)$ moving in the normal direction $(\cos\theta_0(r), \sin\theta_0(r))^T$. We let r be the parametrization defined such that $\theta_0(r) = r$. Then by (2.24), since $\alpha \equiv 0$, we will have $\theta(t, r) = r$ also for $t > 0$ and we see that r and θ are therefore the same parametrization for all times. By (2.25), the amplitude at time t is given by

$$A^2(t,\theta) = \frac{A_0^2(\theta)q(\theta,0)}{q(t,\theta)}, \qquad q(t,\theta) = \left((x_\theta(t,\theta))^2 + (y_\theta(t,\theta))^2\right)^{1/2}. \quad (4.10)$$

We note finally that q can be computed from $X^\theta(t,\theta)$ and $Y^\theta(t,\theta)$ in (4.8).

We will also make use of the paraxial approximation, discussed in Section 2.5 (see page 200) in order to reduce two-dimensional problems to one dimension. Time is thus not explicitly needed in the calculation and θ can be computed as a function of x and y, and y as a function of x and θ. From (2.39) we get the velocity field

$$\boldsymbol{u} = (u,\, v)^T = \left(\tan\theta,\ \frac{1}{\eta}(\eta_y - \eta_x \tan\theta)\right)^T$$

for this setting, and the partial differential equations for the segments $\theta = \Theta(x, y)$ and $y = Y(x, \theta)$ are

$$\frac{\partial\Theta}{\partial x} + u\frac{\partial\Theta}{\partial y} = v,$$
$$\frac{\partial Y}{\partial x} + v\frac{\partial Y}{\partial \theta} = u.$$

The travel time T is now a quantity defined on the phase plane curve, and can be computed according to (4.4, 4.5). Let $\mathcal{F}^x(y,\theta) = (\mathcal{F}^x_y, \mathcal{F}^x_\theta)^T$ be the flow generated by $\boldsymbol{u}$. Keeping in mind that T is given by the phase ϕ, we see from (2.40) that $\mathrm{d}T(x, \mathcal{F}^x)/\mathrm{d}x = \eta/\cos\theta$. This yields the following differential equations, defined for the y-segments and θ-segments respectively:

$$\frac{\partial T^y}{\partial x} + u\frac{\partial T^y}{\partial y} = \frac{\eta}{\cos\theta}, \quad (4.12)$$
$$\frac{\partial T^\theta}{\partial x} + v\frac{\partial T^\theta}{\partial \theta} = \frac{\eta}{\cos\theta}.$$

To compute the amplitude we use equations (2.41) and (2.42). Suppose the initial data are given as $y(0, r) = y_0(r)$ and $\theta(0, r) = \theta_0(r)$ with amplitude $A_0(y_0(r), \theta_0(r))$. When Θ is well defined, the amplitude on the segment is given by (2.41):

$$A(x,y) = A_0(\mathcal{F}^{-x}(y,\Theta(x,y)))\sqrt{\frac{\eta(0,\mathcal{F}_y^{-x}(y,\Theta(x,y)))|y_r(0,\mathcal{F}^{-x}(y,\Theta(x,y)))|}{\eta(x,y)|y_r(x,y,\Theta(x,y))|}}.$$

When Y is well defined, the same equality holds after replacing $(y,\Theta(x,y))$ by $(Y(x,\theta),\theta)$. In order to compute $A(x,y)$ we must hence also evolve $\mathcal{F}^{-x}$ and y_r as quantities on the curve. Let J be the Jacobian of $\boldsymbol{u}(x,y,\theta)$ with respect to (y,θ), and set $\boldsymbol{z} = (y_r,\theta_r)^T$. Then, by the definition of $\mathcal{F}^x$ and (2.42),

$$\frac{\mathrm{d}\mathcal{F}^{-x}(x,\mathcal{F}^x)}{\mathrm{d}x} = 0, \qquad \frac{\mathrm{d}\boldsymbol{z}(x,\mathcal{F}^x)}{\mathrm{d}x} = J(x,\mathcal{F}^x)\boldsymbol{z}(x,\mathcal{F}^x), \tag{4.13}$$

for fixed (y,θ). Note that both $\mathcal{F}^{-x}$ and $\boldsymbol{z}$ remain bounded and smooth also at caustics, where the amplitude A becomes infinite. The quantities are then given by the PDEs

$$\frac{\partial F^y}{\partial x} + u\frac{\partial F^y}{\partial y} = 0, \qquad \frac{\partial Z^y}{\partial x} + u\frac{\partial Z^y}{\partial y} = JZ^y,$$

$$\frac{\partial F^\theta}{\partial x} + v\frac{\partial F^\theta}{\partial \theta} = 0, \qquad \frac{\partial Z^\theta}{\partial x} + v\frac{\partial Z^\theta}{\partial \theta} = JZ^\theta.$$

Here, F^y, F^θ give $\mathcal{F}^{-x}$ and Z^y, Z^θ give $\boldsymbol{z}$ on the segments $\Theta(x,y)$ and $Y(x,\theta)$ respectively. Initial data for those equations are $F^y(0,y) = (y,\Theta(0,y))^T$, $F^\theta(0,\theta) = (Y(0,\theta),\theta)^T$ and $Z^y(0,y) = (\partial_r y_0(r)\ ,\partial_r\theta_0(r))^T$ for r given by $y_0(r) = y$ and $Z^\theta(0,\theta) = (\partial_r y_0(r)\ ,\partial_r\theta_0(r))^T$, with $\theta_0(r) = \theta$.

We shall present three computational examples in order to describe different aspects of the method. All of the examples involve caustics and superposition.

Contracting elliptical and ellipsoidal wave front

The initial values in the first example correspond to a one-dimensional elliptical wave front in $\mathbb{R}^2$: see Figure 4.4. The initial motion is contraction and the index of refraction is constant. The projections of the front in phase space onto the $x\theta$- and $y\theta$-planes are smooth even through caustics: see Figure 4.5. The function α defined in (4.9) vanishes, which simplifies the calculations. No differential equation needs to be solved in the xy-plane. The equations (4.8) are sufficient and the xy-location of the wave front is given by (X^θ, Y^θ). With constant index of refraction in these examples, there is no need to perform the interpolation discussed in Section 4.2 as a sixth step in the segment projection process. There is no problem in

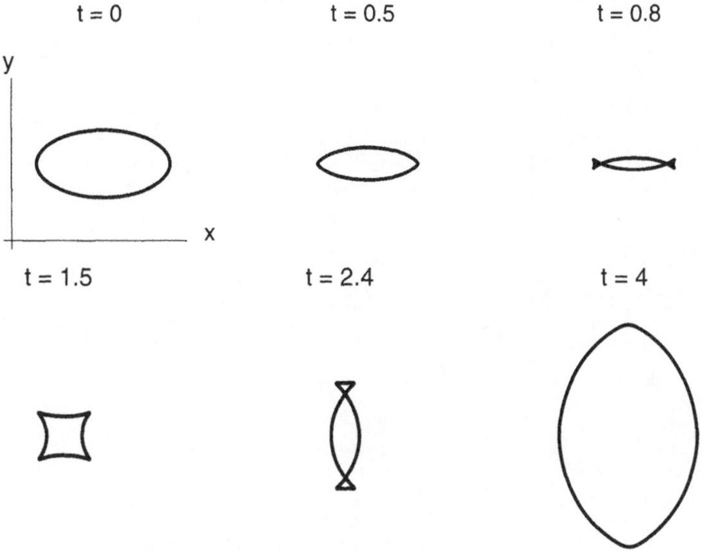

Figure 4.4. Evolution of an initially elliptical wave front in the xy-plane.

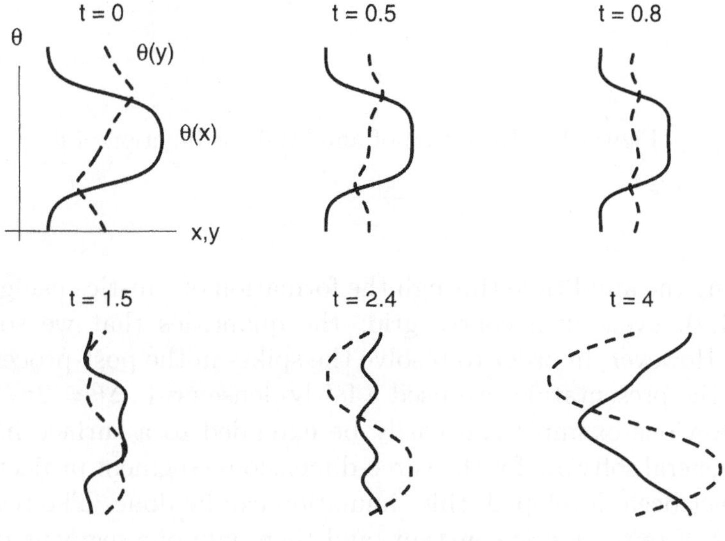

Figure 4.5. Projection of phase plane curves γ: solid lines show $x\theta$-plane, dashed lines show $y\theta$-plane.

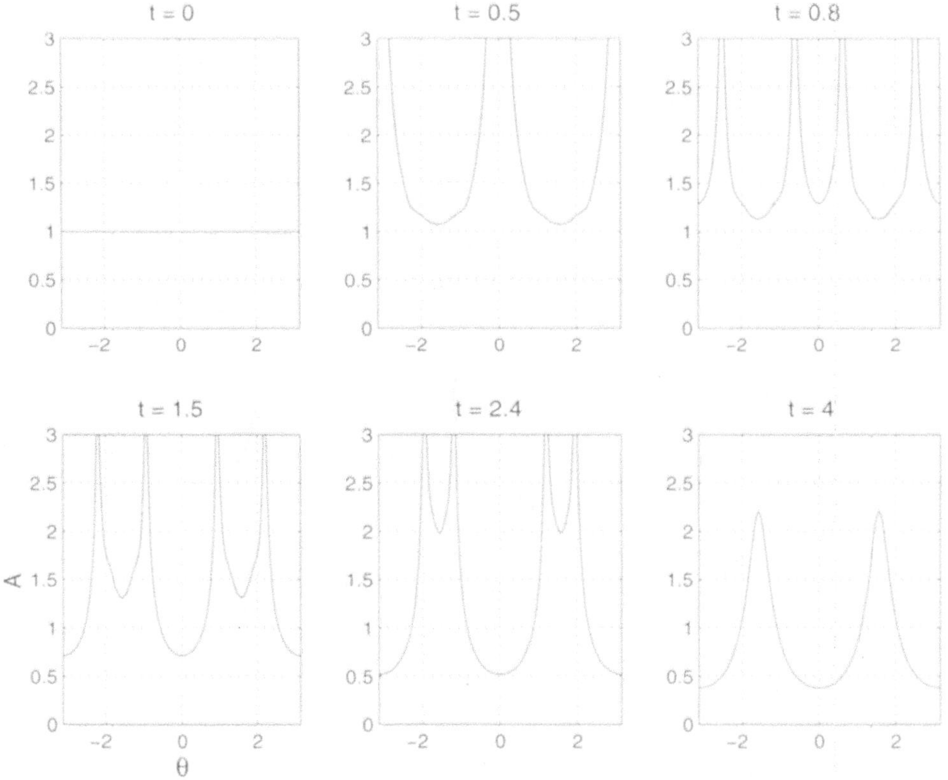

Figure 4.6. Evolution of amplitude as function of θ.

calculating the amplitude through the formation of caustics using (4.10) (see Figure 4.6), even on a coarse grid; the quantities that we solve for are smooth. However, in order to resolve the spikes in the post-processed amplitude for the presentation, we used a fairly dense grid, $\Delta\theta = 2\pi/512 \approx 0.01$.

The previous example can easily be extended to a surface in $\mathbb{R}^3$. Even though general software for the three-dimensional segment projection method has not yet been developed, this simulation can be done. The reason is that the index of refraction is constant, and then only one segment is needed in each of the coordinate planes given in Figure 4.7, which displays the projections of the initial surface in phase space onto the $x\theta_1\theta_2$-, $y\theta_1\theta_2$- and $z\theta_1\theta_2$-spaces. In Figure 4.8 we see the evolution of the wave front in xyz-space at different times. For the general case of variable index of refraction, a larger number of segments could be required. See Tornberg and Engquist (2003) for a simple three-dimensional example with several overlapping segments. The grid resolution used in the computations was $\Delta\theta_1 = \Delta\theta_2 = 2\pi/60 \approx 0.1$.

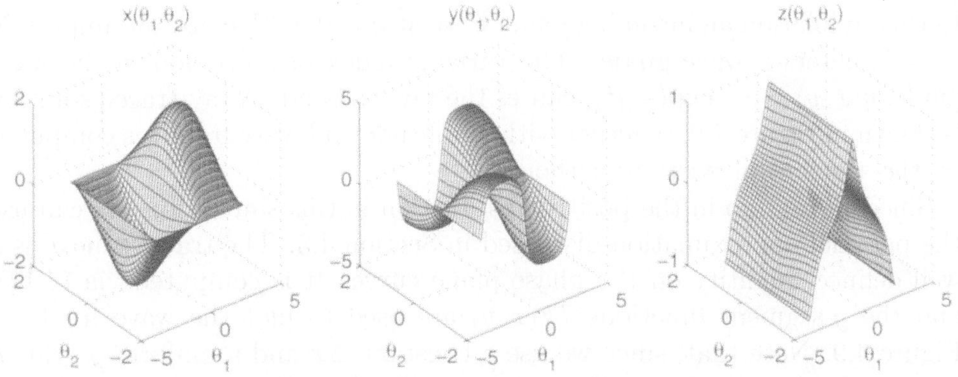

Figure 4.7. Projections of phase space surface Σ onto different coordinate planes at $t = 0$.

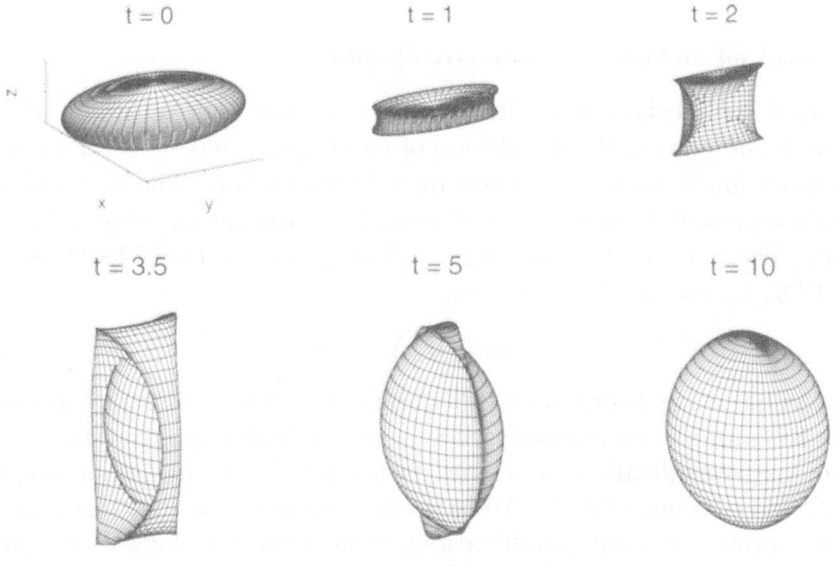

Figure 4.8. Evolution of wave front in xyz-space.

Wave guide

In this simulation an incoming plane wave at $x = 0$, with constant amplitude $A = 1$, enters a wave guide. The variable index of refraction in the wave guide, $\eta(x, y) = 1+\exp(-y^2)$, causes the rays to bend. A ray-traced solution is shown in Figure 4.9, together with amplitude and wave fronts as computed by the segment projection method.

Since all rays go in the positive x-direction in this simulation, we can use the paraxial approximation discussed in Section 4.3. The travel time T is a well-defined quantity on the phase plane curve. It is computed via (4.12), and the y-segment functions $T^y(x, y)$ are used to plot the wave fronts in Figure 4.9. Note that, since we use a constant Δx and a uniform y grid, T is in fact obtained on a uniform $x \times y$ grid. At each point the number of y-segments corresponds to the number of crossing wave fronts.

The phase space curve in the $y\theta$-plane becomes complicated at larger x-values but it is still handled well by the segment projection method: see Figure 4.10. The grid resolution was high ($\Delta y = 7/4096 \approx 0.002$) to get an accurate rendering of the amplitude at the far end of the wave guide.

4.4. Level set methods for geometrical optics

The level set method was introduced by Osher and Sethian (1988) as a general technique for the simulation of moving interfaces. Level set methods for special applications had been introduced earlier. The method uses an implicit representation of an interface in $\mathbb{R}^d$ as the zero level set of a function $\phi(t, \boldsymbol{x})$. The motion of the interface following a velocity field $\boldsymbol{u}(t, \boldsymbol{x})$ is given by a PDE for the level set function ϕ,

$$\phi_t + \boldsymbol{u} \cdot \nabla\phi = 0. \tag{4.14}$$

This technique has been successfully applied to many different types of problems. Examples are multiphase flow, etching, epitaxial growth, image processing and visualization, described in the two books by Osher and Fedkiw (2002) and Sethian (1999). An attractive property is that equation (4.14) can be applied without modifications even if the topology of the interface changes as, for example, when merging occurs in multiphase flow.

For the location of the interface to be well defined, the gradient of ϕ in the direction normal to the interface should be bounded away from zero. In practice, the level set function ϕ is re-initialized at regular time intervals such that it is approximately a signed distance function to the interface.

If $\phi(t, \boldsymbol{x}) = 0$ represents an evolving wave front given by geometrical optics, the velocity is $\boldsymbol{u}(t, \boldsymbol{x}) = c(\boldsymbol{x})\hat{\boldsymbol{n}}(\boldsymbol{x})$, where $\hat{\boldsymbol{n}}$ is the normal vector at the interface, that is,

$$\hat{\boldsymbol{n}}(t, \boldsymbol{x}) = \frac{\nabla\phi}{|\nabla\phi|}.$$

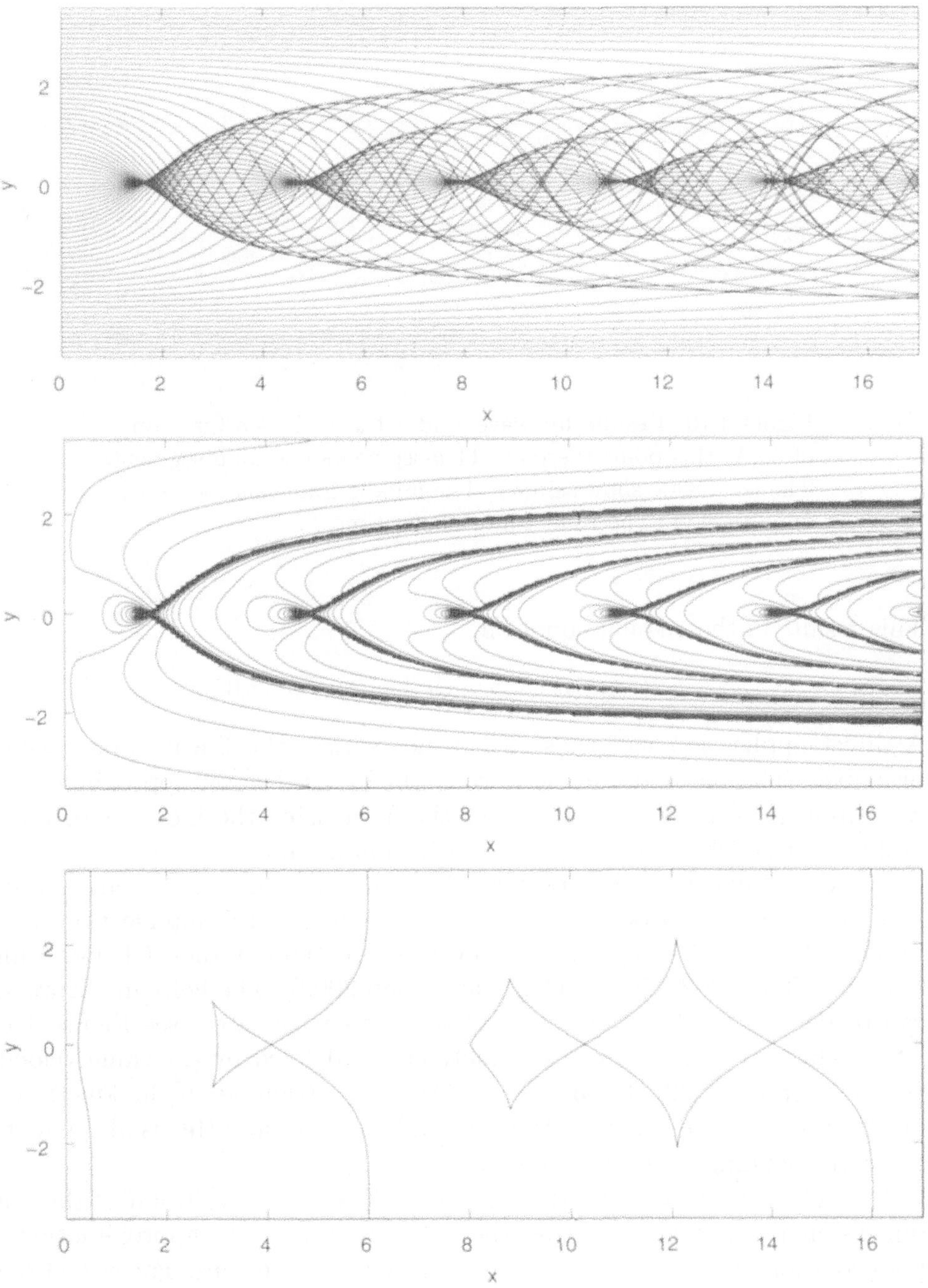

Figure 4.9. Results for the wave guide simulation. The top frame shows rays from initial plane wave. The middle frame shows amplitude, contour lines of $\min(A, 4)$. The bottom frame shows wave fronts in the xy-plane at $T = 0.5$, 6, 16.

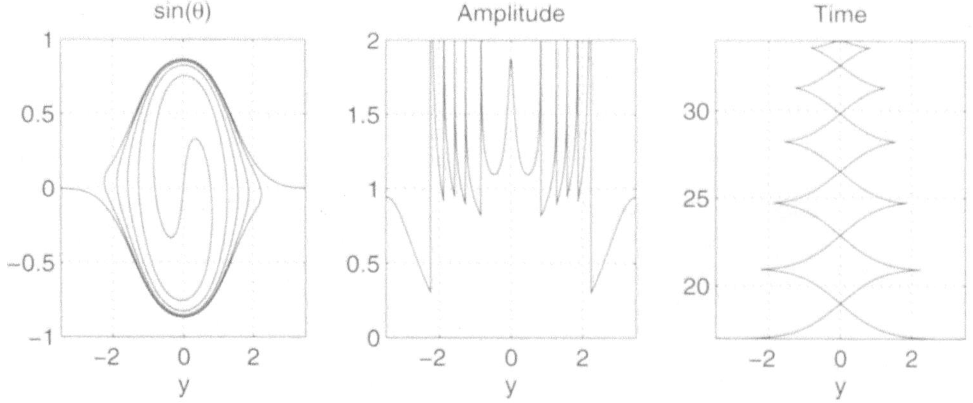

Figure 4.10. Results for wave guide at $x = 17$ as a function of y. At this point there are 11 y-segments and 13 θ-segments. The left frame shows $\sin(\theta)$. The middle frame shows amplitude A. The right frame shows time T.

This results in the eikonal equation,

$$\phi_t + c(\boldsymbol{x})\hat{\boldsymbol{n}} \cdot \nabla\phi = \phi_t + c(\boldsymbol{x})|\nabla\phi| = 0.$$

A direct application will thus clearly not satisfy the linear superposition principle. The method can, however, still be used if we approximate the wave front in phase space and evolve the front using the Liouville equation (2.29), as was done in the segment projection method.

The wave front in the kinetic formulation (2.29) is of higher codimension, and such geometrical objects can be represented by the intersection of interfaces that are given by different level set functions (Osher, Cheng, Kang, Shim and Tsai 2002, Cheng, Osher and Qian 2002). The helix in Figure 4.1 can be defined by the intersection of two regular surfaces: see Figure 4.11. The evolutions of both level set functions are defined by the same velocity vector given by (2.22), (2.23) and (2.24). The advantage of the kinetic formulation is that the superposition principle is valid, and this is also true for the corresponding level set formulation.

A practical problem with this approach is that the evolution of the one-dimensional object representing the wave front requires approximation of the evolution of two level set functions in three dimensions. For wave fronts in $\mathbb{R}^3$, three level set functions in five independent variables and time are required. The computational burden can be reduced by restricting the computation to a small neighbourhood of the wave front. In order to have a well-functioning algorithm, a number of special techniques are useful. Reinitialization of the level set functions ϕ_j, $j = 1, \ldots, d$ in d dimensions should

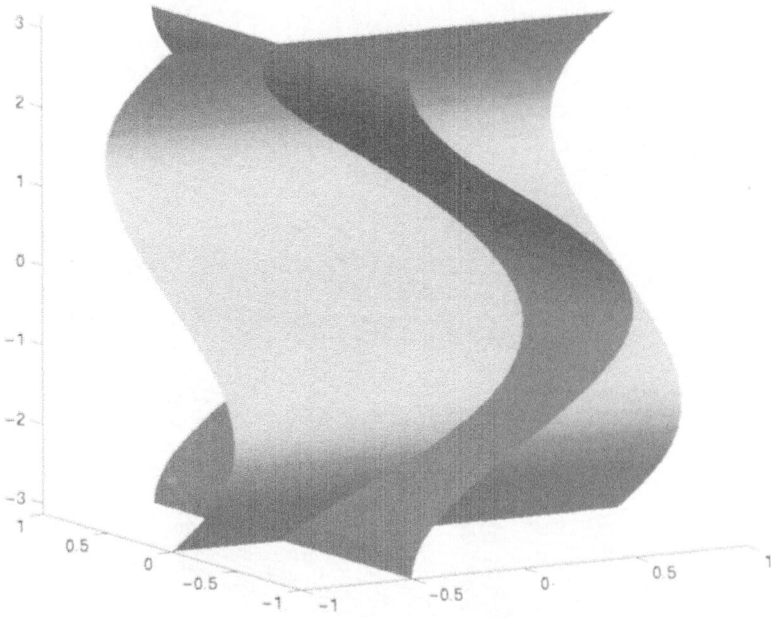

Figure 4.11. Using two level set functions to describe the phase space front. From Osher *et al.* (2002), reproduced with permission.

be performed at regular time intervals, such that

$$|\nabla\phi_j| \approx 1, \qquad \nabla\phi_j \cdot \nabla\phi_k \approx 0, \quad j \neq k,$$

at the interface. In Figure 4.12 we see the result of a level set simulation of a propagating wave front in an inhomogeneous medium simulating high-frequency seismic waves.

4.5. Fast marching in phase space

In Section 3.2 on viscosity solutions (see page 207) the fast marching method was briefly mentioned in the context of viscosity solutions for the frequency domain eikonal equation (2.15). The method can also be applied to a transport equation in phase space, enabling it to capture multivalued solutions. This idea was put forth by Fomel and Sethian (2002).

The transport equation in question (4.15) is an 'escape' equation set in a subdomain Ω of phase space. The unknown, $\hat{\boldsymbol{y}}(\boldsymbol{x}, \boldsymbol{p})$, represents the point on the boundary $\partial\Omega$ (in phase space) where a bicharacteristic originating in $(\boldsymbol{x}, \boldsymbol{p}) \in \Omega$ crosses the boundary. We note that $\hat{\boldsymbol{y}}(\boldsymbol{x}(t), \boldsymbol{p}(t))$ is constant along a bicharacteristic $(\boldsymbol{x}(t), \boldsymbol{p}(t))$. Therefore, after differentiation with respect to t and multiplication by η^2, the chain rule together with the ray

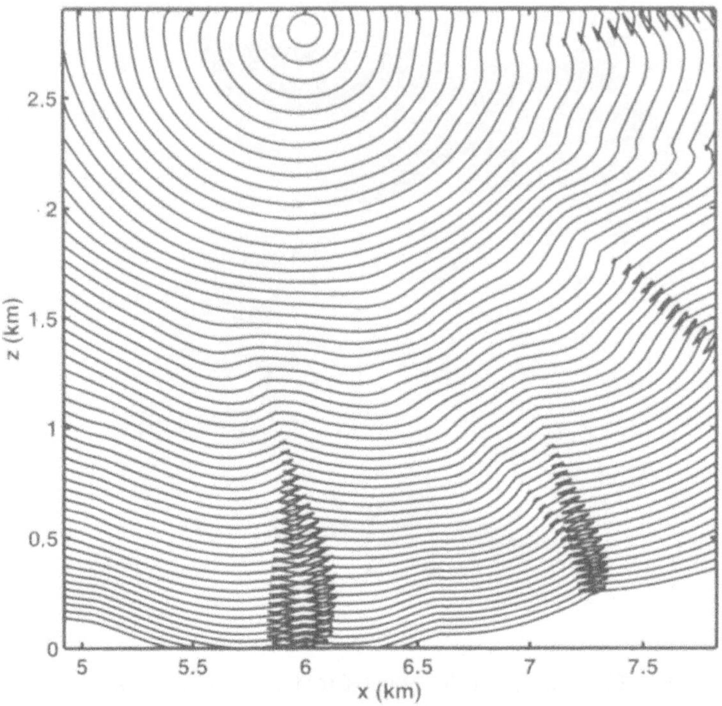

Figure 4.12. Numerical result for Marmousi test problem. From Cheng *et al.* (2002), reproduced with permission.

equations (2.13) and (2.14) give the transport equation

$$\begin{aligned} D_x\hat{\boldsymbol{y}}\,\boldsymbol{p} + \eta D_p\hat{\boldsymbol{y}}\,\nabla\eta &= 0, & (\boldsymbol{x},\boldsymbol{p}) &\in \Omega, \\ \hat{\boldsymbol{y}}(\boldsymbol{x},\boldsymbol{p}) &= (\boldsymbol{x},\boldsymbol{p}), & (\boldsymbol{x},\boldsymbol{p}) &\in \partial\Omega, \end{aligned} \tag{4.15}$$

where $D_x\hat{\boldsymbol{y}}$ and $D_p\hat{\boldsymbol{y}}$ are the Jacobians of $\hat{\boldsymbol{y}}$ with respect to $\boldsymbol{x}$ and $\boldsymbol{p}$, respectively. Note that this is the stationary version of (2.29) with the scalar density function f replaced by the vector $\hat{\boldsymbol{y}}$. There is also an accompanying transport equation for the travel time $T(\boldsymbol{x},\boldsymbol{p})$ from the point $(\boldsymbol{x},\boldsymbol{p})$ to the first boundary crossing. Similarly, since $\partial_t T(\boldsymbol{x}(t),\boldsymbol{p}(t)) = 1$, we get

$$\begin{aligned} \boldsymbol{p}\cdot\nabla_x T + \eta\nabla\eta\cdot\nabla_p T &= \eta^2, & (\boldsymbol{x},\boldsymbol{p}) &\in \Omega, \\ T(\boldsymbol{x},\boldsymbol{p}) &= 0, & (\boldsymbol{x},\boldsymbol{p}) &\in \partial\Omega. \end{aligned} \tag{4.16}$$

Once the solutions to (4.15) and (4.16) have been found, travel times between any two points $\boldsymbol{x}_0$ and $\boldsymbol{x}_1$ can be computed. First, solve

$$\hat{\boldsymbol{y}}(\boldsymbol{x}_0,\boldsymbol{p}_0) = \hat{\boldsymbol{y}}(\boldsymbol{x}_1,\boldsymbol{p}_1) \tag{4.17}$$

for $\boldsymbol{p}_0$ and $\boldsymbol{p}_1$. Then the travel time is $|T(\boldsymbol{x}_0,\boldsymbol{p}_0) - T(\boldsymbol{x}_0,\boldsymbol{p}_1)|$. There may be multiple solutions to (4.17), giving multiple travel times. If $\boldsymbol{x}_1 \in \partial\Omega$, the expression simplifies. Setting $\hat{\boldsymbol{y}} = (\hat{\boldsymbol{x}}, \hat{\boldsymbol{p}})$, we can solve

$$\hat{\boldsymbol{x}}(\boldsymbol{x}_0, \boldsymbol{p}) = \boldsymbol{x}_1$$

for $\boldsymbol{p}$, to get the travel time $T(\boldsymbol{x}_0, \boldsymbol{p})$. To find the travel time at $\boldsymbol{x}_0$ of a wave front that starts at the boundary of Ω in physical space, we instead need to find $\boldsymbol{p}$ such that

$$\hat{\boldsymbol{p}}(\boldsymbol{x}_0, \boldsymbol{p}) = \eta \hat{\boldsymbol{n}}(\hat{\boldsymbol{x}}(\boldsymbol{x}_0, \boldsymbol{p})),$$

where $\hat{\boldsymbol{n}}(\hat{\boldsymbol{x}})$ is the normal of the boundary at $\hat{\boldsymbol{x}}$. Again, the travel time is $T(\boldsymbol{x}_0, \boldsymbol{p})$.

The amplitude can also be obtained directly through post-processing of the solution. Let us consider a point source at $\boldsymbol{x}_0$ in two dimensions. In the notation of Section 2.2, we have

$$A^{-2}(\tilde{\boldsymbol{x}}(t,r)) \sim |\tilde{\boldsymbol{x}}_r(t,r)| \eta(\tilde{\boldsymbol{x}}(t,r)),$$

after assuming that $\boldsymbol{x}_0(r) = \boldsymbol{x}_0 + \varepsilon(\cos r, \sin r)$ and $\varepsilon \to 0$. Set $\boldsymbol{p}_0(r) = \eta(\boldsymbol{x}_0)(\cos r, \sin r)^T$. Then there is a function $t(r)$ such that $\hat{\boldsymbol{x}}(\boldsymbol{x}_0, \boldsymbol{p}_0(r)) = \tilde{\boldsymbol{x}}(t(r), r)$, and, after differentiation with respect to r,

$$D_p \hat{\boldsymbol{x}}\, \boldsymbol{p}_0^{\perp} = \tilde{\boldsymbol{x}}_t t'(r) + \tilde{\boldsymbol{x}}_r.$$

But $\hat{\boldsymbol{p}}(\boldsymbol{x}_0, \boldsymbol{p}_0) \parallel \tilde{\boldsymbol{x}}_t \perp \tilde{\boldsymbol{x}}_r$, and since $|\hat{\boldsymbol{p}}| = \eta(\hat{\boldsymbol{x}})$,

$$A^{-2}(\hat{\boldsymbol{x}}(\boldsymbol{x}_0, \boldsymbol{p}_0)) \sim \hat{\boldsymbol{p}}^{\perp} D_p \hat{\boldsymbol{x}}\, \boldsymbol{p}_0^{\perp}.$$

The method computes the travel time to the boundary and escape location for rays with all possible starting points and starting directions in the domain Ω, at the expense of solving the transport equations (4.15) and (4.16) set in the full phase space. It does so in an Eulerian framework, on a fixed grid. In two dimensions, the phase space is three-dimensional, and the cost for fast marching is $\mathcal{O}(N^3 \log N)$ when each dimension is discretized with N grid points. The corresponding cost for three-dimensional problems is $\mathcal{O}(N^5 \log N)$. This is expensive if only one set of initial data is of interest. In many applications, however, we are interested in solving a problem with the same index of refraction $\eta(\boldsymbol{x})$ for many different initial data (sources). Examples include the inverse problem in geophysics and the computation of bistatic radar cross sections. Then the cost is competitive: *cf.* the slowness matching method in Section 3.2 on multivalued solutions (see page 216).

5. Moment-based methods

In the kinetic formulation of geometrical optics presented in Section 2.3, we interpret rays as particle trajectories governed by the Hamiltonian system (1.6). We let $f(t, \boldsymbol{x}, \boldsymbol{p}) \geq 0$ be the density of particles in phase space.

It satisfies the Liouville equation

$$f_t + \frac{1}{\eta^2}\boldsymbol{p}\cdot\nabla_x f + \frac{1}{\eta}\nabla_x\eta\cdot\nabla_p f = 0. \tag{5.1}$$

Like kinetic equations in general, solving the full equation (5.1) by direct numerical methods would be very expensive, because of the large number of independent variables (six in 3D). Instead we use the classic technique of approximating a kinetic transport equation set in high-dimensional phase space $(t, \boldsymbol{x}, \boldsymbol{p})$, by a finite system of moment equations in the reduced space $(t, \boldsymbol{x})$. See, for instance, Grad (1949) and more recently Levermore (1996). In general the moment equations form a system of conservation laws that gives an approximation of the true solution. The classical example is the compressible Euler approximation of the Boltzmann equation (see remark below). In our setting, the moment system is, however, typically *exact* under the closure assumption that at most N rays cross at any given point in time and space. In fact, this moment system solution is equivalent to N disjoint pairs of eikonal and transport equations (2.5) and (2.6) when the solution is smooth.

Brenier and Corrias (1998) originally proposed this approach for finding multivalued solutions to geometrical optics problems in the one-dimensional homogeneous case. It was subsequently adapted for two-dimensional inhomogeneous problems by Engquist and Runborg (Engquist and Runborg 1996, 1998, Runborg 2000). See also Gosse (2002). More recently, the same technique has been applied to the Schrödinger equation by Jin and Li (200x), Gosse, Jin and Li (200x), and Sparber *et al.* (2003).

In this section we derive and analyse the system of PDEs that follows from the kinetic model in two dimensions, together with the closure assumption that a maximum of N rays passes through any given point in space and time. We consider two different versions of the closure assumption and present some numerical examples.

5.1. Moment equations

We start by defining the moments m_{ij}, with $\boldsymbol{p} = (p_1, p_2)^T$, as

$$m_{ij}(t, \boldsymbol{x}) = \frac{1}{\eta(\boldsymbol{x})^{i+j}} \int_{\mathbb{R}^2} p_1^i p_2^j f(t, \boldsymbol{x}, \boldsymbol{p})\,\mathrm{d}\boldsymbol{p}. \tag{5.2}$$

Next, multiply (5.1) by $\eta^{2-i-j}p_1^i p_2^j$ and integrate over $\mathbb{R}^2$ with respect to $\boldsymbol{p}$, so that

$$\frac{\eta^2}{\eta^{i+j}}\partial_t \int p_1^i p_2^j f\,\mathrm{d}\boldsymbol{p} \\ + \partial_x \frac{1}{\eta^{i+j}} \int p_1^{i+1} p_2^j f\,\mathrm{d}\boldsymbol{p} + \frac{(i+j)\eta_x}{\eta^{i+j+1}} \int p_1^{i+1} p_2^j f\,\mathrm{d}\boldsymbol{p} \tag{5.3}$$

$$+ \partial_y \frac{1}{\eta^{i+j}} \int p_1^i p_2^{j+1} f \, d\boldsymbol{p} + \frac{(i+j)\eta_y}{\eta^{i+j+1}} \int p_1^i p_2^{j+1} f \, d\boldsymbol{p}$$
$$- \frac{\eta_x}{\eta^{i+j-1}} \int i p_1^{i-1} p_2^j f \, d\boldsymbol{p} - \frac{\eta_y}{\eta^{i+j-1}} \int j p_1^i p_2^{j-1} f \, d\boldsymbol{p} = 0,$$

after using integration by parts and the fact that f has compact support in $\boldsymbol{p}$ for the last term. From definition (5.2) we see that formally m_{ij} will satisfy the infinite system of moment equations

$$(\eta^2 m_{ij})_t + (\eta m_{i+1,j})_x + (\eta m_{i,j+1})_y = i\eta_x m_{i-1,j} + j\eta_y m_{i,j-1} - (i+j)(\eta_x m_{i+1,j} + \eta_y m_{i,j+1}), \qquad (5.4)$$

valid for all $i, j \geq 0$. For uniformity in notation we have defined $m_{i,-1} = m_{-1,i} = 0$, $\forall i$.

The system (5.4) is not closed. If truncated at finite i and j, there are more unknowns than equations. To close the system we will make specific assumptions on the form of the density function f. First, in Section 5.2 we consider the case when f is a weighted sum of delta functions in $\boldsymbol{p}$,

$$f(t, \boldsymbol{x}, \boldsymbol{p}) = \sum_{k=1}^{N} g_k \cdot \delta(\boldsymbol{p} - \boldsymbol{p}_k), \qquad \boldsymbol{p}_k = \eta \begin{pmatrix} \cos\theta_k \\ \sin\theta_k \end{pmatrix}, \qquad (5.5)$$

and second, in Section 5.3, the case when f is a sum of Heaviside functions on the sphere in phase space,

$$f(t, \boldsymbol{x}, \boldsymbol{p}) = \frac{1}{\eta}\delta(|\boldsymbol{p}| - \eta) \sum_{k=1}^{N} (-1)^{k+1} H(\theta - \theta_k), \qquad \boldsymbol{p} = |\boldsymbol{p}| \begin{pmatrix} \cos\theta \\ \sin\theta \end{pmatrix}, \quad (5.6)$$

where H is the Heaviside function. Both cases correspond to the assumption of a finite number of rays at each point in time and space. Since f here only depends on a finite number of unknowns ($2N$ for (5.5) and N for (5.6)) the infinite system (5.4) can be reduced to a finite system.

Remark. The derivation above of the multiphase equations for geometrical optics is completely analogous to the derivation of the hydrodynamical limit from a kinetic formulation of gas dynamics. Instead of (5.1), we use the Boltzmann equation

$$f_t + \boldsymbol{p} \cdot \nabla_x f = Q(f, f), \qquad (\boldsymbol{x}, \boldsymbol{p}) \in \mathbb{R}^2 \times \mathbb{R}^2, \qquad (5.7)$$

where $Q(f, f)$ is the collision operator. Moreover, instead of the closure assumptions (5.5) and (5.6), we assume f is a *Maxwellian*, that is,

$$f(t, \boldsymbol{x}, \boldsymbol{p}) = \frac{\rho(t, \boldsymbol{x})}{2\pi T(t, \boldsymbol{x})} \exp\left(-\frac{|\boldsymbol{p} - \boldsymbol{u}(t, \boldsymbol{x})|^2}{2T(t, \boldsymbol{x})} \right), \qquad \boldsymbol{u} = \begin{pmatrix} u \\ v \end{pmatrix}. \qquad (5.8)$$

The lowest moments of Q are zero by its special form, and therefore the first moment equations of (5.7) are the same as those of (5.1) with $\eta \equiv 1$. If we pick the following equations from (5.4),

$$\begin{pmatrix} m_{00} \\ m_{10} \\ m_{01} \\ m_{20}+m_{02} \end{pmatrix}_t + \begin{pmatrix} m_{10} \\ m_{20} \\ m_{11} \\ m_{30}+m_{12} \end{pmatrix}_x + \begin{pmatrix} m_{01} \\ m_{11} \\ m_{02} \\ m_{21}+m_{03} \end{pmatrix}_y = 0, \tag{5.9}$$

and write them in terms of the unknowns $\rho, \boldsymbol{u}, T$ in (5.8) and

$$E \equiv \rho\left(\frac{1}{2}|\boldsymbol{u}|^2 + \frac{3}{2}T\right),$$

we get

$$\begin{pmatrix} \rho \\ \rho u \\ \rho v \\ E \end{pmatrix}_t + \begin{pmatrix} \rho u \\ \rho(u^2+T) \\ \rho uv \\ u(E+\rho T) \end{pmatrix}_x + \begin{pmatrix} \rho v \\ \rho uv \\ \rho(v^2+T) \\ v(E+\rho T) \end{pmatrix}_y = 0, \tag{5.10}$$

the compressible Euler equations for a perfect monoatomic gas.

5.2. Closure with delta functions

To close (5.4) we assume in this section that f can be written as

$$f(t,\boldsymbol{x},\boldsymbol{p}) = \sum_{k=1}^{N} g_k \cdot \delta(\boldsymbol{p}-\boldsymbol{p}_k), \qquad \boldsymbol{p}_k = \eta\begin{pmatrix}\cos\theta_k \\ \sin\theta_k\end{pmatrix}. \tag{5.11}$$

Hence, for fixed values of $\boldsymbol{x}$ and t, the particle density f is nonzero at a maximum of N points, and only when $|\boldsymbol{p}| = \eta(\boldsymbol{x})$. The new variables that we have introduced here are $g_k = g_k(t,\boldsymbol{x})$, which corresponds to the strength (particle density) of ray k, and $\theta_k = \theta_k(t,\boldsymbol{x})$, which is the direction of the same ray. Inserting (5.11) into the definition of the moments (5.2) yields

$$m_{ij} = \sum_{k=1}^{N} g_k \cos^i\theta_k \sin^j\theta_k. \tag{5.12}$$

A system describing N phases needs $2N$ equations, corresponding to the N ray strengths g_k and their directions θ_k. It is not immediately clear which equations to select among the candidates in (5.4). Given the equations for a set of $2N$ moments, we should be able to write the remaining moments in these equations in terms of the leading ones. This is not always possible. For instance, with the choice of m_{20} and m_{02}, for $N = 1$, the quadrant of the angle θ cannot be recovered, and therefore, in general, the sign of the moments cannot be determined. Here we choose the equations for the moments $m_{2\ell-1,0}$ and $m_{0,2\ell-1}$

$$
\begin{aligned}
(\eta^2 m_{2\ell-1,0})_t + (\eta m_{2\ell,0})_x + (\eta m_{2\ell-1,1})_y = \\
(2\ell-1)(\eta_x m_{2\ell-2,0} - \eta_x m_{2\ell,0} - \eta_y m_{2\ell-1,1}), \\
(\eta^2 m_{0,2\ell-1})_t + (\eta m_{1,2\ell-1})_x + (\eta m_{0,2\ell})_y = \\
(2\ell-1)(\eta_y m_{0,2\ell-2} - \eta_x m_{1,2\ell-1} - \eta_y m_{0,2\ell}),
\end{aligned} \tag{5.13}
$$

for $\ell = 1, \ldots, N$, and we collect these moments in a vector,

$$
\boldsymbol{m} = (m_{10}, m_{01}, m_{30}, m_{03}, \ldots, m_{2N-1,0}, m_{0,2N-1})^T. \tag{5.14}
$$

As we will show below, this system of equations for $\boldsymbol{m}$ can be essentially closed, for all N, meaning that, for almost all $\boldsymbol{m}$, we can uniquely determine the remaining moments in (5.13). We introduce new variables,

$$
\boldsymbol{u} = \begin{pmatrix} u_1 \\ u_2 \\ \vdots \\ u_{2N-1} \\ u_{2N} \end{pmatrix} := \begin{pmatrix} g_1 \cos\theta_1 \\ g_1 \sin\theta_1 \\ \vdots \\ g_N \cos\theta_N \\ g_N \sin\theta_N \end{pmatrix}, \tag{5.15}
$$

which have a physical interpretation: the vector (u_{2k-1}, u_{2k}) shows the direction and strength of ray k. The new variables together with (5.12) define a function $\boldsymbol{F}_0$ through the equation

$$
\boldsymbol{F}_0(\boldsymbol{u}) = \boldsymbol{m}. \tag{5.16}
$$

Similarly, they define the functions

$$
\begin{aligned}
\boldsymbol{F}_1(\boldsymbol{u}) &= (m_{20}, m_{11}, \ldots, m_{2N,0}, m_{1,2N-1})^T, \\
\boldsymbol{F}_2(\boldsymbol{u}) &= (m_{11}, m_{02}, \ldots, m_{2N-1,1}, m_{0,2N})^T,
\end{aligned} \tag{5.17}
$$

$$
\boldsymbol{K}(\boldsymbol{u}, \eta_x, \eta_y) = \begin{pmatrix} \eta_x m_{00} - \eta_x m_{2,0} - \eta_y m_{1,1} \\ \eta_y m_{00} - \eta_x m_{1,1} - \eta_y m_{0,2} \\ \vdots \\ (2N-1)(\eta_x m_{2N-2,0} - \eta_x m_{2N,0} - \eta_y m_{2N-1,1}) \\ (2N-1)(\eta_y m_{0,2N-2} - \eta_x m_{1,2N-1} - \eta_y m_{0,2N}) \end{pmatrix}.
$$

These functions permit us to write the equations as a system of nonlinear conservation laws with source terms

$$
\boldsymbol{F}_0(\eta^2 \boldsymbol{u})_t + \boldsymbol{F}_1(\eta \boldsymbol{u})_x + \boldsymbol{F}_2(\eta \boldsymbol{u})_y = \boldsymbol{K}(\boldsymbol{u}, \eta_x, \eta_y). \tag{5.18}
$$

Equivalently, we can write (5.18) as

$$
(\eta^2 \boldsymbol{m})_t + \boldsymbol{F}_1 \circ \boldsymbol{F}_0^{-1}(\eta \boldsymbol{m})_x + \boldsymbol{F}_2 \circ \boldsymbol{F}_0^{-1}(\eta \boldsymbol{m})_y = \boldsymbol{K}(\boldsymbol{F}_0^{-1}(\boldsymbol{m}), \eta_x, \eta_y).
$$

The functions $\boldsymbol{F}_j$ and $\boldsymbol{K}$ are rather complicated nonlinear functions. In the most simple case, $N=1$, the function $\boldsymbol{F}_0$ is the identity, and

$$\boldsymbol{F}_1 = \frac{u_1}{|\boldsymbol{u}|}\begin{pmatrix} u_1 \\ u_2 \end{pmatrix}, \qquad \boldsymbol{F}_2 = \frac{u_2}{|\boldsymbol{u}|}\begin{pmatrix} u_1 \\ u_2 \end{pmatrix}, \qquad \boldsymbol{K} = \frac{\eta_x u_2 - \eta_y u_1}{|\boldsymbol{u}|}\begin{pmatrix} u_2 \\ -u_1 \end{pmatrix}.$$

For $N=2$, let $\boldsymbol{w} = (w_1, w_2)^T$ and

$$\boldsymbol{f}_0 = \begin{pmatrix} w_1 \\ w_2 \\ w_1^3/|\boldsymbol{w}|^2 \\ w_2^3/|\boldsymbol{w}|^2 \end{pmatrix}, \qquad \boldsymbol{f}_1 = \frac{w_1}{|\boldsymbol{w}|}\boldsymbol{f}_0, \qquad \boldsymbol{f}_2 = \frac{w_2}{|\boldsymbol{w}|}\boldsymbol{f}_0,$$

$$\boldsymbol{k} = \frac{\eta_x w_2 - \eta_y w_1}{|\boldsymbol{w}|}\begin{pmatrix} w_2 \\ -w_1 \\ w_1^2 w_2/|\boldsymbol{w}|^2 \\ -w_1 w_2^2/|\boldsymbol{w}|^2 \end{pmatrix}.$$

Then $\boldsymbol{F}_j = \boldsymbol{f}_j(u_1,u_2) + \boldsymbol{f}_j(u_3,u_4)$ for $j = 0,1,2$ and $\boldsymbol{K} = \boldsymbol{k}(u_1,u_2) + \boldsymbol{k}(u_3,u_4)$.

Since the angles θ_k remain unaffected when $\boldsymbol{u}$ is scaled by a constant for all N, the $\boldsymbol{F}_j$ and $\boldsymbol{K}$ are always homogeneous of degree one, $\boldsymbol{F}_j(\alpha\boldsymbol{u}) = \alpha\boldsymbol{F}_j(\boldsymbol{u})$, $\boldsymbol{K}(\alpha\boldsymbol{u},\eta_x,\eta_y) = \alpha\boldsymbol{K}(\boldsymbol{u},\eta_x,\eta_y)$ for all $\alpha \in \mathbb{R}$. Moreover, the source term $\boldsymbol{K}$ always vanishes for constant η.

Properties of the flux functions

In this section we analyse the flux functions and source

$$\boldsymbol{F}_1 \circ \boldsymbol{F}_0^{-1}(\boldsymbol{m}), \qquad \boldsymbol{F}_2 \circ \boldsymbol{F}_0^{-1}(\boldsymbol{m}), \qquad \boldsymbol{K}(\boldsymbol{F}_0^{-1}(\boldsymbol{m}),\eta_x,\eta_y).$$

In order for them to be well defined we must restrict their domain to the case when there are no rays meeting head-on. With this restriction they are also continuous. We have the following result.

Theorem 5.1. Let $\boldsymbol{F}_0$ be the function in (5.16) and let $\boldsymbol{F}_0|U_N$ be its restriction to the domain

$$U_N = \{\boldsymbol{u} \in \mathbb{R}^{2N} \mid 1 + \cos(\theta_k - \theta_\ell) \neq 0, \text{ whenever } g_k g_l > 0, \ \forall k,\ell\},$$

and $M_N = \boldsymbol{F}_0(U_N)$. The composition $m \circ (\boldsymbol{F}_0|U_N)^{-1} : M_N \to \mathbb{R}$ is well defined and continuous for all maps of the form

$$m : U_N \to \mathbb{R}, \qquad m(\boldsymbol{u}) = \sum_{k=1}^{N} g_k h(\theta_k), \tag{5.19}$$

where $h : \mathbb{S} \to \mathbb{R}$ is continuous.

Since $\boldsymbol{F}_1$, $\boldsymbol{F}_2$ and $\boldsymbol{K}$ are all of the form (5.19) we have the following result.

Corollary 5.2. Let $\boldsymbol{F}_j$ and $\boldsymbol{K}$ be the functions in (5.16) and (5.17) and let $\boldsymbol{F}_0|U_N$ and M_N be as in Theorem 5.1. Then the functions

$$\boldsymbol{F}_1 \circ (\boldsymbol{F}_0|U_N)^{-1}(\boldsymbol{m}), \qquad \boldsymbol{F}_2 \circ (\boldsymbol{F}_0|U_N)^{-1}(\boldsymbol{m}), \qquad \boldsymbol{K}((\boldsymbol{F}_0|U_N)^{-1}(\boldsymbol{m}), \eta_x, \eta_y)$$

are well defined and depend continuously on $\boldsymbol{m} \in M_N$.

Remark. If we do not restrict $\boldsymbol{F}_0$ to U_N the result is false. Take, for instance, $\boldsymbol{u} = (-1\ 0\ 1\ 0)^T$ and $\tilde{\boldsymbol{u}} = 2\boldsymbol{u}$ for $N = 2$ so that $\boldsymbol{F}_0(\boldsymbol{u}) = \boldsymbol{F}_0(\tilde{\boldsymbol{u}}) = 0$, but $\boldsymbol{F}_1(\tilde{\boldsymbol{u}}) = 2\boldsymbol{F}_1(\boldsymbol{u}) \neq 0$. Furthermore, with a different choice of moment equations the result does not necessarily hold either. For instance, if instead of (5.14) we use the equations for

$$\boldsymbol{m} = (m_{10}, m_{01}, m_{20}, m_{02})^T$$

when $N = 2$, the functions $\boldsymbol{F}_j$ change, and in general there are two unrelated solutions to $\boldsymbol{F}_0(\boldsymbol{u}) = \boldsymbol{m}$ which $\boldsymbol{F}_1$ does not map to the same point. For example, if $\boldsymbol{u} = (1\ 1\ 0\ -1)^T$ and $\tilde{\boldsymbol{u}} = (1\ -1\ 0\ 1)^T$ then $\boldsymbol{F}_0(\boldsymbol{u}) = \boldsymbol{F}_0(\tilde{\boldsymbol{u}})$, but $\boldsymbol{F}_1(\boldsymbol{u}) = \boldsymbol{F}_1(\tilde{\boldsymbol{u}}) + (0\ \sqrt{2}\ 0\ 0)^T$. The function $\boldsymbol{F}_2 \circ \boldsymbol{F}_0^{-1}$ is ill defined in the same way.

Proof of Theorem 5.1. It will be convenient to work with complex versions of our variables, and we start by introducing the isometry $\mathcal{A} : \mathbb{R}^{2N} \to \mathbb{C}^N$,

$$\mathcal{A}\begin{pmatrix} x_1 \\ x_2 \\ \vdots \\ x_{2N} \end{pmatrix} = \begin{pmatrix} x_1 + \mathrm{i}x_2 \\ \vdots \\ x_{2N-1} + \mathrm{i}x_{2N} \end{pmatrix},$$

identifying $\mathbb{R}^{2N}$ with $\mathbb{C}^N$. We set $\boldsymbol{w} = (w_1, \ldots, w_N)^T := \mathcal{A}\boldsymbol{u}$ and

$$z_k := \cos\theta_k + \mathrm{i}\sin\theta_k, \qquad \boldsymbol{z}_k := \left(z_k, z_k^{-3}, \ldots, z_k^{(2N-1)(-1)^{N+1}}\right)^T, \tag{5.20}$$

so that $w_k = g_k z_k$. Furthermore, define the continuous mapping $Q : \mathbb{C}^N \to \mathbb{C}^N$,

$$\boldsymbol{Q}(\boldsymbol{w}) = \begin{pmatrix} | & | & & | \\ \boldsymbol{z}_1 & \boldsymbol{z}_2 & \ldots & \boldsymbol{z}_N \\ | & | & & | \end{pmatrix} \begin{pmatrix} g_1 \\ \vdots \\ g_N \end{pmatrix}.$$

To relate $\boldsymbol{w}$ to $\boldsymbol{m}$ via this function, we use the trigonometric identity

$$\boldsymbol{z}_k = B\begin{pmatrix} \cos\theta_k + \mathrm{i}\sin\theta_k \\ \vdots \\ \cos^{2N-1}\theta_k + \mathrm{i}\sin^{2N-1}\theta_k \end{pmatrix}, \tag{5.21}$$

where $B = \{b_{k\ell}\} \in \mathbb{R}^{N\times N}$ is a lower-triangular matrix with $b_{k\ell}$ equal to the $(2\ell-1)$th coefficient of the $(2k-1)$th degree Chebyshev polynomial, for $k \leq \ell$. The matrix is nonsingular since $b_{kk} = 4^{k-1} > 0$. From the definition of $\boldsymbol{Q}$ and the identity (5.21) it then follows that

$$\boldsymbol{Q}(\boldsymbol{w}) = \boldsymbol{Q}(\mathcal{A}\boldsymbol{u}) = B\mathcal{A}\boldsymbol{m}, \tag{5.22}$$

where we also recall that $\boldsymbol{F}_0(\boldsymbol{u}) = \boldsymbol{m}$. Before continuing, we show the following lemma.

Lemma 5.3. Let $\{z_k\}$ be N' complex numbers such that $|z_k| = 1$, and let $\{\boldsymbol{z}_k\}$ be the corresponding vectors as defined in (5.20). If $N' \leq 2N$ then $\boldsymbol{z}_k \in C^N$ are linearly independent over $\mathbb{R}$ if and only if

$$z_k^2 \neq z_\ell^2, \qquad k \neq \ell. \tag{5.23}$$

Proof. The necessity is obvious. To show that (5.23) is a sufficient condition, we only need to consider the case $N' = 2N$, since we can always find $2N - N'$ additional z_k such that (5.23) still holds if $N' < 2N$. Suppose therefore that $\{\boldsymbol{z}_k\}_{k=1}^{2N}$ are linearly dependent over $\mathbb{R}$, and that (5.23) is true. Then the real matrix

$$A = \begin{pmatrix} \mathrm{Re}\,(\boldsymbol{z}_1) & \mathrm{Re}\,(\boldsymbol{z}_2) & \cdots & \mathrm{Re}\,(\boldsymbol{z}_{2N}) \\ \mathrm{Im}\,(\boldsymbol{z}_1) & \mathrm{Im}\,(\boldsymbol{z}_2) & \cdots & \mathrm{Im}\,(\boldsymbol{z}_{2N}) \end{pmatrix}, \qquad A \in \mathbb{R}^{2N\times 2N},$$

is singular and we can find a vector $\boldsymbol{\beta} = (\beta_1, \ldots, \beta_{2N})^T \neq 0$ such that $A^T\boldsymbol{\beta} = 0$. Using the fact that $|z_k| = 1$ and $\bar{z}_k = 1/z_k$, this implies

$$P_{\boldsymbol{\beta}}(z_k^2) = 0, \qquad k = 1, \ldots, 2N,$$

where

$$P_{\boldsymbol{\beta}}(z) = \frac{1}{2}\sum_{\ell=1}^{N} \beta_\ell (z^{\ell+N-1} + z^{N-\ell}) + \frac{1}{2i}\sum_{\ell=1}^{N} (-1)^{\ell+1}\beta_{\ell+N}(z^{\ell+N-1} - z^{N-\ell}).$$

But since the degree of $P_{\boldsymbol{\beta}}$ is at most $2N-1$, regardless of $\boldsymbol{\beta}$, it cannot have $2N$ distinct zeros if $\boldsymbol{\beta} \neq 0$. Therefore there must exist k, ℓ such that $z_k^2 = z_\ell^2$, a contradiction. □

Let $\bar{m}(\boldsymbol{w}) := m(\mathcal{A}^{-1}\boldsymbol{w})$ and let $\bar{\boldsymbol{Q}}$ be the restriction of $\boldsymbol{Q}$ to $\mathcal{A}U_N$. We now want to prove that $\bar{m} \circ \bar{\boldsymbol{Q}}^{-1}$ is well defined on $\bar{\boldsymbol{Q}}(\mathcal{A}U_N)$, and we do this by showing that $\bar{\boldsymbol{Q}} \circ \bar{m}^{-1}$ is injective on $\bar{m}(\mathcal{A}U_N)$. Let $\boldsymbol{w}, \tilde{\boldsymbol{w}} \in \mathcal{A}U_N$ be such that $\bar{\boldsymbol{Q}}(\boldsymbol{w}) = \bar{\boldsymbol{Q}}(\tilde{\boldsymbol{w}})$. We need to show that $\bar{m}(\boldsymbol{w}) = \bar{m}(\tilde{\boldsymbol{w}})$ and we use the variables introduced in (5.20). A tilde indicates that a variable relates to $\tilde{\boldsymbol{w}}$. Let N' and $\tilde{N}'$, respectively, be the number of distinct z_k and $\tilde{z}_k$ with $g_k, \tilde{g}_k > 0$. Without loss of generality we order the variables such that $z_{\ell_j} = \cdots = z_{\ell_{j+1}-1}$, with $1 = \ell_1 < \cdots < \ell_{N'+1} = N+1$, and similarly for $\{\tilde{z}_k\}$. With this notation we get

$$\bar{\boldsymbol{Q}}(\boldsymbol{w}) = \sum_{j=1}^{N'} \left(\sum_{k=\ell_j}^{\ell_{j+1}-1} g_k \right) \boldsymbol{z}_{\ell_j} = \sum_{j=1}^{\tilde{N}'} \left(\sum_{k=\tilde{\ell}_j}^{\tilde{\ell}_{j+1}-1} \tilde{g}_k \right) \tilde{\boldsymbol{z}}_{\tilde{\ell}_j} = \bar{\boldsymbol{Q}}(\tilde{\boldsymbol{w}}).$$

The sets of numbers $\{z_{\ell_j}\}_{j=1}^{N'}$ and $\{\tilde{z}_{\tilde{\ell}_j}\}_{j=1}^{\tilde{N}'}$ both satisfy (5.23), because $\boldsymbol{w}, \tilde{\boldsymbol{w}} \in \mathcal{A}U_N$. Therefore, since $N' + \tilde{N}' \leq 2N$, there must exist j and k such that $z_{\ell_j}^2 = \tilde{z}_{\tilde{\ell}_k}^2$ by Lemma 5.3. By induction it follows that $N' = \tilde{N}'$ and, possibly after some reordering,

$$\ell_j = \tilde{\ell}_j, \qquad z_{\ell_j} = s_j \tilde{z}_{\tilde{\ell}_j}, \qquad \sum_{k=\ell_j}^{\ell_{j+1}-1} g_k = s_j \sum_{k=\tilde{\ell}_j}^{\tilde{\ell}_{j+1}-1} \tilde{g}_k, \qquad s_j = \pm 1, \qquad \forall j.$$

But $g_k, \tilde{g}_k$ are positive, and we can conclude that $s_j = 1$ for all j. Thus, $\boldsymbol{w}$ and $\tilde{\boldsymbol{w}}$ are identical up to permutations and to the individual g_k values. We now apply $\bar{m}$ to them:

$$\begin{aligned} \bar{m}(\boldsymbol{w}) &= \sum_{j=1}^{N'} \sum_{k=\ell_j}^{\ell_{j+1}-1} g_k h(z_k) = \sum_{j=1}^{N'} h(z_{\ell_j}) \sum_{k=\ell_j}^{\ell_{j+1}-1} g_k \\ &= \sum_{j=1}^{\tilde{N}'} h(\tilde{z}_{\tilde{\ell}_j}) \sum_{k=\tilde{\ell}_j}^{\tilde{\ell}_{j+1}-1} \tilde{g}_k = \sum_{j=1}^{\tilde{N}'} \sum_{k=\tilde{\ell}_j}^{\tilde{\ell}_{j+1}-1} \tilde{g}_k h(\tilde{z}_k) = \bar{m}(\tilde{\boldsymbol{w}}). \end{aligned}$$

Hence, $\bar{m} \circ \bar{\boldsymbol{Q}}^{-1}$ is well defined on its domain of definition. Now, (5.22) and the fact that $\boldsymbol{F}_0(\boldsymbol{u}) = \boldsymbol{m}$ show that $m \circ (\boldsymbol{F}_0|U_N)^{-1}(\boldsymbol{m}) = \bar{m} \circ \bar{\boldsymbol{Q}}^{-1}(B\mathcal{A}\boldsymbol{m})$, which implies that $m \circ (\boldsymbol{F}_0|U_N)^{-1}$ is well defined on M_N. The continuity follows by approximating U_N by compact sets, and using the following lemma from elementary analysis.

Lemma 5.4. Let K be a compact metric space and suppose $f : K \to X$ and $g : K \to Y$ are continuous functions, and $X = f(K), Y$ are metric spaces. If the composition $f \circ g^{-1} : g(U) \to X$ is injective, then $g \circ f^{-1} : X \to Y$ is continuous. (The function inverses should be interpreted as set functions here.)

Proof. We need to show that, for any open set $U \subset Y$, the set $f \circ g^{-1}(U)$ is open. Since g is continuous and K compact, $g^{-1}(U)^c$ is compact, and consequently $f(g^{-1}(U)^c)$ is also compact by the continuity of f. But $g^{-1}(U^c) = g^{-1}(U)^c$, and hence $f \circ g^{-1}(U^c)$ is compact. Moreover,

$$f \circ g^{-1}(U) \cup f \circ g^{-1}(U^c) = f(g^{-1}(U)) \cup f(g^{-1}(U)^c) = f(K) = X.$$

Since $f \circ g^{-1}$ is injective, $f \circ g^{-1}(U) \cap f \circ g^{-1}(U^c) = \emptyset$, and therefore $f \circ g^{-1}(U) = (f \circ g^{-1}(U^c))^c$, which is open. □

Analysis of the conservation laws

Engquist and Runborg (1996, 1998) showed that the general system (5.18) is nonstrictly hyperbolic for all states $\boldsymbol{u}$ and N. Jin and Li (200x) showed the same for the Schrödinger equation case. The systems are thus not well posed in the strong sense, and they are more sensitive to perturbations than strictly hyperbolic systems. The Jacobian has a Jordan-type degeneracy and there will never be more than N linearly independent eigenvectors for the $2N \times 2N$ system. For a general study of this type of degenerate systems of conservation laws, see Zheng (1998).

A distinguishing feature of the system (5.18) is that it typically has measure solutions of delta function type, even for smooth and compactly supported initial data. These appear when the physically correct solution passes outside the class of solutions that the system (5.18) describes. If initial data dictate a physical solution with M phases for $t > T$, the system (5.18) with $N < M$ phases will have a measure solution for $t > T$; *cf.* Figure 5.3(a, b) and Figure 5.5(a).

For smooth solutions, (5.18) with N phases is equivalent to N pairs of eikonal and transport equations (2.5) and (2.6) if the variables are identified as

$$g_k = A_{0,k}^2, \qquad \begin{pmatrix}\cos\theta_k \\ \sin\theta_k\end{pmatrix} = \frac{\nabla\phi_k}{|\nabla\phi_k|}, \qquad k = 1, \dots, N,$$

(Engquist and Runborg 1996). Note that this is expected from the relationship between equations (2.30) and (2.31) and the remark thereafter. The pair (2.5) and (2.6) form a nonstrictly hyperbolic system, just like (5.18), with the same eigenvalue. Where wave fields meet, the viscosity solution of (2.5) is in general discontinuous. Because of the term $\Delta\phi$ in the source term of (2.6), the first amplitude coefficient A_0 has a concentration of mass at these points. Hence, the two different formulations are also similar in this respect.

There is a close relationship between (5.18) with $N = 1$ and $\eta \equiv 1$,

$$\boldsymbol{u}_t + \boldsymbol{f}(\boldsymbol{u})_x + \boldsymbol{g}(\boldsymbol{u})_y = 0, \qquad \boldsymbol{f}(\boldsymbol{u}) = u_1 \frac{\boldsymbol{u}}{|\boldsymbol{u}|}, \qquad \boldsymbol{g}(\boldsymbol{u}) = u_2 \frac{\boldsymbol{u}}{|\boldsymbol{u}|}, \tag{5.24}$$

and the equations of pressureless gases:

$$\rho_t + (\rho u)_x = 0, \qquad (\rho u)_t + (\rho u^2)_x = 0. \tag{5.25}$$

Indeed, the steady state version of (5.24) is precisely (5.25) if we identify $\rho = g\cos^2\theta$ and $u = \tan\theta$. Moreover, the one-dimensional version of (5.24) corresponds to (5.25), with relativistic effects added if we identify $\rho = g\sin\theta$ and $u = \cos\theta$. We also note that, if we formally let $T \to 0$ in (5.8), we recover (5.11) with $N = 1$ and without the restriction on $|\boldsymbol{p}_1|$. The same formal limit of (5.10) gives the two-dimensional pressureless gas equations.

In the context of non-relativistic pressureless gases, this problem was addressed by Bouchut (1994) and later Brenier and Grenier (Grenier 1995, Brenier and Grenier 1998), and E, Rykov and Sinai (1996), who independently proved global existence of measure solutions to (5.25). The uniqueness question was settled in Bouchut and James (1999). For linear transport equations, related results have been obtained by Bouchut and James (1995) and Poupaud and Rascle (1997). The questions of existence and uniqueness for (5.24) and its one-dimensional version are still open.

The Riemann problem. Since standard numerical schemes are based on solving one-dimensional Riemann problems (LeVeque 1992), we consider this problem for (5.24):

$$\boldsymbol{u}_t + \boldsymbol{f}(\boldsymbol{u})_x = 0, \qquad \boldsymbol{f}(\boldsymbol{u}) = u_1 \frac{\boldsymbol{u}}{|\boldsymbol{u}|}, \qquad \boldsymbol{u}(0, x) = \begin{cases} \boldsymbol{u}_\ell & x < 0, \\ \boldsymbol{u}_r & x > 0. \end{cases} \tag{5.26}$$

At a discontinuity the conservation form gives the Rankine–Hugoniot jump condition,

$$\boldsymbol{f}(\boldsymbol{u}_\ell) - \boldsymbol{f}(\boldsymbol{u}_r) = s(\boldsymbol{u}_\ell - \boldsymbol{u}_r), \tag{5.27}$$

where s represents the propagation speed of the discontinuity. Since $\boldsymbol{f}(\boldsymbol{u}) = \cos\theta \boldsymbol{u}$, the jump condition (5.27) simplifies to

$$\cos\theta_\ell \boldsymbol{u}_\ell - \cos\theta_r \boldsymbol{u}_r = s(\boldsymbol{u}_\ell - \boldsymbol{u}_r).$$

The states to which a given nonzero state $\boldsymbol{u}_\ell$ can connect with a discontinuity, *i.e.*, its Hugoniot locus, is simply $\alpha \boldsymbol{u}_\ell$ for $\alpha \in \mathbb{R}$, with speed of propagation $s = \cos\theta_\ell$ when $\alpha \geq 0$ and $s = \cos\theta_\ell (1+\alpha)/(1-\alpha)$ for $\alpha < 0$. It follows that, unless they are parallel, two nonzero states $\boldsymbol{u}_\ell$ and $\boldsymbol{u}_r$ can only be connected via the intermediate state $\boldsymbol{u}_m = 0$. There will be two types of discontinuity. If $\cos\theta_\ell < \cos\theta_r$, the solution with $\boldsymbol{u}_m = 0$, satisfies the Lax entropy condition (the left discontinuity moves more slowly than the right one). The states' Hugoniot loci and the solution for this type of discontinuity is illustrated in Figure 5.1(a). If $\cos\theta_\ell > \cos\theta_r$, on the other hand, we do not have a solution in the usual weak sense. This situation corresponds to two meeting wave fields. Formally, however, $\boldsymbol{u}_m = t\tilde{\boldsymbol{u}}_m \delta(x - st)$ is a weak solution to the conservation law with these initial data. The conservation form gives a slightly modified jump condition,

$$\cos\theta_\ell \boldsymbol{u}_\ell - \cos\theta_r \boldsymbol{u}_r = \cos\tilde{\theta}_m (\boldsymbol{u}_\ell - \boldsymbol{u}_r) + \tilde{\boldsymbol{u}}_m,$$

with the propagation speed $s = \cos\tilde{\theta}_m$. This construction, a delta function solution to the Riemann problem leading to a modified Rankine–Hugoniot condition, is found also in Zheng (1998) for more general equations.

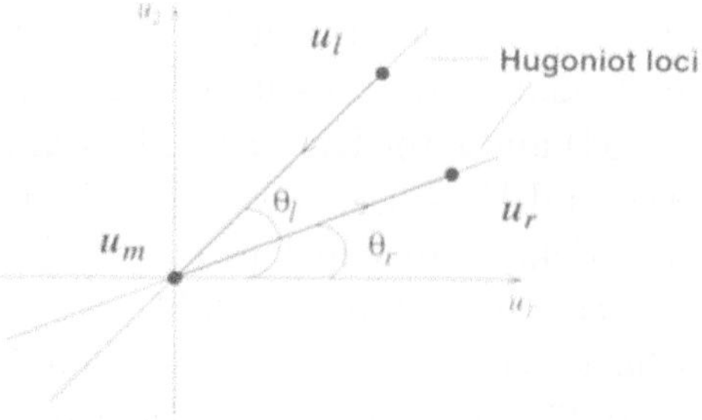

(a) Hugoniot loci of states and solution for contact discontinuity

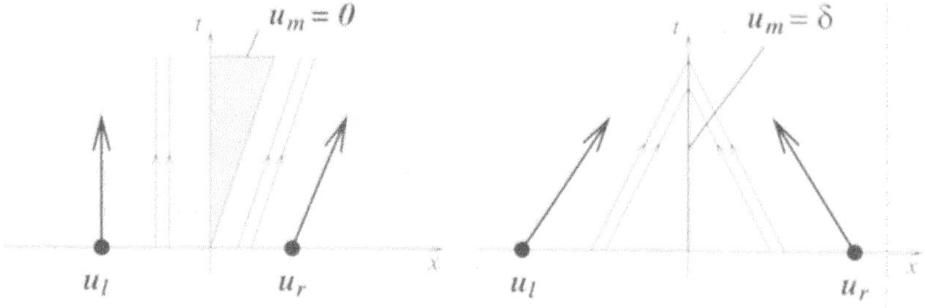

(b) Contact discontinuity (c) Overcompressive shock

Figure 5.1. The Riemann problem, with Hugoniot loci for the left and right states in phase space and the two different types of discontinuity in $(t, \boldsymbol{x})$ space.

It is easily verified that $\boldsymbol{u}$ itself is an eigenvector of the Jacobian of $\boldsymbol{f}$ and that the Jacobian has a double eigenvalue equalling $\cos\theta$. Therefore, the Hugoniot locus will coincide with the integral curves of the system's characteristic fields and, since $\cos\theta$ remains constant along the curves, the fields are linearly degenerate. From this we conclude that the first type of discontinuity is a linear, contact discontinuity; characteristics run parallel to the discontinuity. The linear degeneracy also excludes the possibility of rarefaction wave solutions. The second type of discontinuity will always have two characteristics incident to the discontinuity at each side, because of the double eigenvalue. These discontinuities are thus of overcompressive shock type. The two different discontinuities, plotted in $(t, \boldsymbol{x})$-space, are shown in Figure 5.1(b)–(c).

Entropy. For the analysis of (5.24) it would be useful to find a strictly convex entropy pair for the one-dimensional system. This is, however, not possible since the system is nonstrictly hyperbolic. However, there do exist nonstrictly convex entropy pairs, which can be characterized as follows (Runborg 2000).

Theorem 5.5. Let $U \in C^2$ be convex. There exists a function $F \in C^2$ such that $U(\boldsymbol{u})_t + F(\boldsymbol{u})_x = 0$ for all smooth solutions $\boldsymbol{u} = g(\cos\theta, \sin\theta)$ to

$$\boldsymbol{u}_t + \boldsymbol{f}(\boldsymbol{u})_x = 0, \qquad \boldsymbol{f}(\boldsymbol{u}) = u_1 \frac{\boldsymbol{u}}{|\boldsymbol{u}|}, \tag{5.28}$$

if and only if U is of the form

$$U = gh(\theta) + \text{const}, \qquad h \in C^2(\mathbb{S}), \qquad h + h'' \geq 0.$$

Superposition. The multiple phase systems possess a finite superposition principle in the sense that a sum of N solutions to the single phase system, is a solution to the N-phase system. This follows from a trivial computation if the solutions are smooth. Physical solutions can, however, have discontinuities in g. If we introduce weak solutions, we can show that a sufficient condition for the superposition principle to hold is just that g is bounded and that θ is continuous and has locally bounded variation. A discontinuous θ would typically not be physical, generating a delta shock-type solution, as seen above. We have the following result.

Theorem 5.6. Suppose $\{\boldsymbol{u}_k\}_{k=1}^N$ are N weak solutions to the homogeneous single phase system (5.24) in the sense that $\boldsymbol{u}_k \in L^\infty((0,\infty) \times \mathbb{R}^2)$ and

$$\iint_{t \geq 0} \boldsymbol{u}_k \phi_t + \boldsymbol{f}(\boldsymbol{u}_k)\phi_x + \boldsymbol{g}(\boldsymbol{u}_k)\phi_y \, \mathrm{d}t \, \mathrm{d}\boldsymbol{x} = 0, \qquad \forall \phi \in C_c^1((0,\infty) \times \mathbb{R}^2). \tag{5.29}$$

Moreover, suppose that, for each k and each point in $(0,\infty) \times \mathbb{R}^2$, there is an open neighbourhood on which we can define a continuous function $\theta_k(t, \boldsymbol{x})$ with locally bounded variation such that $\boldsymbol{u}_k = |\boldsymbol{u}_k|(\cos\theta_k, \sin\theta_k)^T$ on that neighbourhood. Then $\boldsymbol{u} = (\boldsymbol{u}_1, \ldots, \boldsymbol{u}_N)^T$ is a weak solution to the homogeneous N-phase system (5.18) in the same sense as (5.29).

Proof. We start by showing that, if $\boldsymbol{v} = (v_1, v_2)^T$ is a weak solution to (5.24) in the sense of Theorem 5.6, then $m_{i,0}$ and $m_{0,i}$, with $i > 1$, are weak solutions in the same sense to the corresponding moment equations, under the given hypotheses. Take $\phi \in C_c^1((0,\infty) \times \mathbb{R}^2)$ and assume without loss of generality that θ is continuous and that $\boldsymbol{v} = g(\cos\theta, \sin\theta)^T$ on supp ϕ. (We can always obtain such a θ after a partition of unity.) Let $M \in C_c^\infty(\mathbb{R}^3)$ be a mollifier with $\int M \, \mathrm{d}t \, \mathrm{d}\boldsymbol{x} = 1$ and set $\theta_\epsilon = \theta \star M_\epsilon$, where $M_\epsilon = M(t/\epsilon, \boldsymbol{x}/\epsilon)/\epsilon^3$. Furthermore, set

$$\psi_s^\epsilon = \phi\left(\cos^{i-1}\theta_\epsilon - \frac{\mathrm{d}\cos^{i-1}\theta_\epsilon}{\mathrm{d}\theta_\epsilon} \sin\theta_\epsilon \cos\theta_\epsilon\right), \qquad \psi_c^\epsilon = \phi \frac{\mathrm{d}\cos^{i-1}\theta_\epsilon}{\mathrm{d}\theta_\epsilon} \cos^2\theta_\epsilon. \tag{5.30}$$

We observe that $\phi_t \cos^i \theta_\epsilon = (\psi_s^\epsilon)_t \cos\theta_\epsilon + (\psi_c^\epsilon)_t \sin\theta_\epsilon$, and similarly for the partial derivatives with respect to x and y. Also, $m_{i,0} = g\cos^i\theta$ on the support of ϕ. This shows that, for all ϵ,

$$\iint_{t\geq 0} m_{i,0}\phi_t \, \mathrm{d}t \, \mathrm{d}\boldsymbol{x} = \iint_{t\geq 0} (m_{i,0} - m_{i,0}^\epsilon)\phi_t + (v_1^\epsilon - v_1)(\psi_s^\epsilon)_t + (v_2^\epsilon - v_2)(\psi_c^\epsilon)_t + v_1(\psi_s^\epsilon)_t + v_2(\psi_c^\epsilon)_t \, \mathrm{d}t \, \mathrm{d}\boldsymbol{x},$$

where the ϵ superscript indicates that a function depends on θ_ϵ instead of θ. The first term of the the right-hand side tends to zero by the dominated convergence theorem. Since $\theta \in \mathrm{BV}_{\mathrm{loc}}$ the expression $\|\phi\partial_t\theta_\epsilon\|_{L^1}$ is bounded independently of ϵ, and therefore

$$\iint_{t\geq 0} |v_1^\epsilon - v_1| \, |(\psi_s^\epsilon)_t| \, \mathrm{d}t \, \mathrm{d}\boldsymbol{x} \leq C \sup_{(t,\boldsymbol{x})\in \operatorname{supp}\phi} |v_1^\epsilon - v_1| \leq C\|\boldsymbol{v}\|_{L^\infty} \sup_{(t,\boldsymbol{x})\in \operatorname{supp}\phi} |\cos\theta_\epsilon - \cos\theta| \to 0,$$

by the continuity of θ. Using the same argument for the remaining terms, we arrive at

$$\iint_{t\geq 0} m_{i,0}\phi_t + m_{i+1,0}\phi_x + m_{i,1}\phi_y \, \mathrm{d}\boldsymbol{x} \, \mathrm{d}t = \iint_{t\geq 0} v_1(\psi_s^\epsilon)_t + \frac{v_1^2}{|\boldsymbol{v}|}(\psi_s^\epsilon)_x + \frac{v_1 v_2}{|\boldsymbol{v}|}(\psi_s^\epsilon)_y \, \mathrm{d}\boldsymbol{x} \, \mathrm{d}t + \iint_{t\geq 0} v_2(\psi_c^\epsilon)_t + \frac{v_2 v_1}{|\boldsymbol{v}|}(\psi_c^\epsilon)_x + \frac{v_2^2}{|\boldsymbol{v}|}(\psi_c^\epsilon)_y \, \mathrm{d}\boldsymbol{x} \, \mathrm{d}t + R^\epsilon, \tag{5.31}$$

where $R^\epsilon \to 0$. But $\psi_c^\epsilon, \psi_s^\epsilon \in C_c^1((0,\infty)\times\mathbb{R}^2)$ and $\boldsymbol{v}$ is a weak solution, so by letting $\epsilon \to 0$ we see that (5.31) in fact equals zero. After replacing $\cos^{i-1}\theta$ with $\sin^{i-1}\theta$ in (5.30) we get the same result for $m_{0,i}$. We can now conclude that, with $m_{i,j}^k = g_k \cos^i\theta_k \sin^j\theta_k$,

$$\sum_{k=1}^{N} \iint_{t\geq 0} m_{2\ell-1,0}^k \phi_t + m_{2\ell,0}^k \phi_x + m_{2\ell-1,1}^k \phi_y \, \mathrm{d}\boldsymbol{x} \, \mathrm{d}t = 0, \qquad \ell = 1, \ldots, N.$$

The same is true for $m_{0,2\ell-1}^k$. But these are just the componentwise statements of

$$\iint_{t\geq 0} \boldsymbol{F}_0(\boldsymbol{u})\phi_t + \boldsymbol{F}_1(\boldsymbol{u})\phi_x + \boldsymbol{F}_2(\boldsymbol{u})\phi_y = 0,$$

and $|\boldsymbol{u}|$ is bounded by $\sum_{k=1}^N |\boldsymbol{u}_k|_\infty$. □

Of course, some of the $\boldsymbol{u}_k$ solutions in Theorem 5.6 can be identically zero, so that, in particular, a weak solution of the single phase system is also a solution of the N-phase system, under the above assumptions.

5.3. *Closure with Heaviside functions*

We will now consider a different way to close (5.4). We discard the amplitude information carried by g_k used in Section 5.2, and only solve for the angles θ_k. In this way we get fewer and less singular equations. The 'correct' values of the unknowns θ_k are, however, not well defined when the physically motivated amplitude is zero. In particular, this is the case at time $t = 0$ for the typical initial value problem with sources given through boundary values (like the problems in Section 5.4). In order to reduce the initialization problem we make the paraxial approximation discussed in Section 2.5, assuming that no rays go in the negative x-direction. This means that data only need to be given on the line $x = 0$. We then consider density functions of the form

$$f(t,\boldsymbol{x},\boldsymbol{p}) = \frac{1}{\eta(\boldsymbol{x})}\delta(|\boldsymbol{p}|-\eta(\boldsymbol{x}))\sum_{k=1}^{N}(-1)^{k+1}H(\theta-\theta_k(t,\boldsymbol{x})), \quad \boldsymbol{p} = |\boldsymbol{p}|\begin{pmatrix}\cos\theta\\ \sin\theta\end{pmatrix}. \tag{5.32}$$

For fixed $(t, \boldsymbol{x})$, the density function f is supported by a set of intervals on the sphere $\{|\boldsymbol{p}| = \eta\}$. The intervals correspond to fans of rays whose edges are given by the unknown angles θ_k. The transport equation (5.1) governs the propagation of all these rays, and in particular the rays at the edges, which will propagate just like ordinary rays as long as f stays of the form (5.32). The values of the N angles θ_k will then coincide with those of a problem with N rays crossing at each point, as long as the assumption (5.32) holds.

The paraxial approximation amounts to the additional assumption that $f(t, \boldsymbol{x}, \boldsymbol{p}) = 0$ when $\boldsymbol{p}\cdot\boldsymbol{e}_x \leq 0$, where $\boldsymbol{e}_x$ is the unit vector in the x-direction, and that the boundary data at $x = 0$ is time-independent. We also adopt the convention that $-\pi/2 < \theta_1 \leq \cdots \leq \theta_N < \pi/2$. The general formula for the moments then follows from (5.32) together with (5.2), namely

$$m_{ij}(t,\boldsymbol{x}) = \sum_{k=1}^{N}(-1)^{k+1}\int_{\theta_k(t,\boldsymbol{x})}^{\pi/2}\cos^i\theta\sin^j\theta\,\mathrm{d}\theta. \tag{5.33}$$

Among the equations in (5.4) we choose the ones for the moments $\{m_{0,\ell}\}$ with $\ell = 0,\ldots,N-1$. By the paraxial approximation, this leads to the steady state equations

$$(\eta m_{1,\ell})_x + (\eta m_{0,\ell+1})_y = \ell(\eta_y m_{0,\ell-1} - \eta_x m_{1,\ell} - \eta_y m_{0,\ell+1}), \qquad \ell = 0,\ldots,N-1. \tag{5.34}$$

Next, we introduce the new variables:

$$\boldsymbol{u} = (u_1,\ldots,u_N)^T, \qquad u_k = \sin\theta_k. \tag{5.35}$$

By evaluating the integrals in (5.33), we then get, for N even,

$$m_{1,\ell} = \sum_{k=1}^{N} \frac{(-1)^k u_k^{\ell+1}}{\ell+1}, \qquad m_{0,\ell} = \sum_{k=1}^{N} (-1)^k R_\ell(u_k),$$

$$R_\ell = \begin{cases} \arcsin(u), & \ell = 0, \\ -\sqrt{1-u^2}, & \ell = 1, \\ \frac{\ell-1}{\ell} R_{\ell-2} - \frac{1}{\ell} u^{\ell-1}\sqrt{1-u^2}, & \ell \geq 2. \end{cases} \tag{5.36}$$

These expressions can in fact also be used to define the moments for odd N (Runborg 2000).

As in Section 5.2 we let $\boldsymbol{m} = (m_{10}, \ldots, m_{1,N-1})^T$. We define the function $\boldsymbol{F}_1$ by $\boldsymbol{F}_1(\boldsymbol{u}) = \boldsymbol{m}$ together with (5.3), and similarly for $\boldsymbol{F}_2$ and $\boldsymbol{K}$. We can then finally write (5.34) as

$$(\eta \boldsymbol{F}_1(\boldsymbol{u}))_x + (\eta \boldsymbol{F}_2(\boldsymbol{u}))_y = \boldsymbol{K}(\boldsymbol{u}, \eta_x, \eta_y), \tag{5.37}$$

or, in terms of $\boldsymbol{m}$,

$$(\eta \boldsymbol{m})_x + (\eta \boldsymbol{F}_2 \circ \boldsymbol{F}_1^{-1}(\boldsymbol{m}))_y = \boldsymbol{K}(\boldsymbol{F}_1^{-1}(\boldsymbol{m}), \eta_x, \eta_y).$$

The functions $\boldsymbol{F}_j$ and $\boldsymbol{K}$ are again rather complicated nonlinear functions. For $N = 1$, the functions are simple:

$$\boldsymbol{F}_1(u_1) = -u_1, \qquad \boldsymbol{F}_2(u_1) = \sqrt{1-u_1^2}, \qquad \boldsymbol{K} = 0.$$

For $N = 2$, let

$$\boldsymbol{f}_1 = \begin{pmatrix} w \\ \frac{1}{2}w^2 \end{pmatrix}, \qquad \boldsymbol{f}_2 = \begin{pmatrix} -\sqrt{1-w^2} \\ \frac{1}{2}\left(\arcsin(w) - w\sqrt{1-w^2}\right) \end{pmatrix},$$

$$\boldsymbol{k} = \begin{pmatrix} 0 \\ \frac{\eta_y}{2}\left(\arcsin(w) + w\sqrt{1-w^2}\right) - \frac{1}{2}\eta_x w^2 \end{pmatrix}.$$

Then $\boldsymbol{F}_j = -\boldsymbol{f}_j(u_1) + \boldsymbol{f}_j(u_2)$ for $j = 1, 2$ and $\boldsymbol{K} = -\boldsymbol{k}(u_1) + \boldsymbol{k}(u_2)$. Finally, for $N = 3$, let

$$\boldsymbol{f}_1 = \begin{pmatrix} w \\ \frac{1}{2}w^2 \\ \frac{1}{3}w^3 \end{pmatrix}, \qquad \boldsymbol{f}_2 = \begin{pmatrix} -\sqrt{1-w^2} \\ \frac{1}{2}\left(\arcsin(w) - w\sqrt{1-w^2}\right) \\ -\frac{1}{3}(2+w^2)\sqrt{1-w^2} \end{pmatrix},$$

$$\boldsymbol{k} = \begin{pmatrix} 0 \\ \eta_y\left(\arcsin(w) + w\sqrt{1-w^2}\right) - \frac{1}{2}\eta_x w^2 \\ -\frac{2}{3}\eta_y(1-w^2)\sqrt{1-w^2} - \frac{2}{3}\eta_x w^3 \end{pmatrix}.$$

Then $\boldsymbol{F}_j = -\boldsymbol{f}_j(u_1) + \boldsymbol{f}_j(u_2) - \boldsymbol{f}_j(u_3)$ for $j = 1, 2$ and $\boldsymbol{K} = -\boldsymbol{k}(u_1) + \boldsymbol{k}(u_2) - \boldsymbol{k}(u_3)$.

If $u_k < u_{k+1}$ for all k, we can compute the gradient of $m_{0,\ell}(\boldsymbol{m})$ explicitly,

$$\nabla_m m_{0,\ell} = V^{-1}\Theta_\ell. \tag{5.38}$$

Here $V = \{v_{k,\ell}\} \in \mathbb{R}^{N\times N}$ is the Vandermonde matrix associated with the points $\boldsymbol{u}$, *i.e.*, $v_{k,\ell} = u_k^{\ell-1}$ (nonsingular by the assumption on $\boldsymbol{u}$), and

$$\Theta_\ell = \left\{u_k^\ell/\sqrt{1-u_k^2}\right\}_{k=1}^N = \{u_k^{\ell-1}\tan\theta_k\}_{k=1}^N \in \mathbb{R}^N.$$

By using (5.38) we get an expression for the Jacobian of $\boldsymbol{F}_2 \circ \boldsymbol{F}_1^{-1}$,

$$\frac{\mathrm{d}\boldsymbol{F}_2 \circ \boldsymbol{F}_1^{-1}}{\mathrm{d}\boldsymbol{m}} = V^T \mathrm{diag}(\{\tan\theta_k\})V^{-T}.$$

We see that this system is strictly hyperbolic as long as $\theta_k \neq \theta_\ell$ for all k, ℓ. See Gosse (2002) for further discussion of the theory for this system and how to couple it with equations for the amplitudes. Here, we simply note that, since $\tan\theta \to \infty$ when $|\theta| \to \pi/2$ or $|u| \to 1$, the Jacobian will blow up at these points. This is expected under the paraxial assumption.

We close this section by establishing the same superposition principle as for the delta equations in Section 5.2.

Theorem 5.7. Suppose $\{u_k\}_{k=1}^M$ are M weak solutions to (5.37) with $N = 1$ in the sense of Theorem 5.6, and $\eta \in C^1$. If u_k are continuous functions with locally bounded variation, then $\boldsymbol{u} = (u_1, \ldots, u_M)^T$ is a weak solution to (5.37) with $N = M$ in the same sense.

Properties of the flux functions

Also in this case, the functions

$$\boldsymbol{F}_2 \circ \boldsymbol{F}_1^{-1}(\boldsymbol{m}) \qquad \text{and} \qquad \boldsymbol{K}(\boldsymbol{F}_1^{-1}(\boldsymbol{m}), \eta_x, \eta_y) \tag{5.39}$$

are well defined and regular on their domains of definition. We consider a slightly more general class of functions than those in (5.39). For a closed interval $I \subset \mathbb{R}$, define the (compact) set of attainable moments, $M_N \subset \mathbb{R}^N$,

$$M_N(I) = \{\boldsymbol{m} \in \mathbb{R}^N \mid \boldsymbol{m} = \boldsymbol{F}_1(\boldsymbol{u}),\ u_1 \leq \cdots \leq u_N,\ u_k \in I\},$$

and introduce the class of mappings from $M_N(I)$ to $\mathbb{R}$ given by

$$J_\psi(\boldsymbol{m}) = \int_I \psi(t) f_m(t)\,\mathrm{d}t, \tag{5.40}$$

where $f_m(t)$ and $\boldsymbol{m}$ are related by

$$\boldsymbol{m} = \boldsymbol{F}_1(\boldsymbol{u}), \qquad f_m(t) = \sum_{k=1}^N (-1)^{k+1} H(t-u_k), \qquad u_1 \leq \cdots \leq u_N, \quad u_k \in I. \tag{5.41}$$

Brenier and Corrias (1998) showed that, if $I = [0, L]$, the mappings given by (5.40) and (5.41) are well defined and continuous on $M_N(I)$ for each $0 < L < \infty$, when ψ has a strictly positive and bounded Nth distributional derivative. These functions were identified as entropies for the moment system in Brenier and Corrias (1998). In general J_ψ is Hölder-continuous, but not continuously differentiable, as seen in the following result, from Runborg (2000).

Theorem 5.8. Let $I = [-L, L]$ for some positive $L < \infty$. The mapping $J_\psi : M_N(I) \to \mathbb{R}$ is well defined by (5.40) and (5.41). If $\psi \in L^p(I)$, with $1 \le p \le \infty$, then

$$J_\psi \in \begin{cases} C^0, & p = 1, \\ C^{0,\frac{p-1}{pN}}, & 0 < p < \infty, \\ C^{0,1/N}, & p = \infty, \end{cases}$$

where $C^{0,\alpha}$ is the set of Hölder-continuous functions with exponent α. If $\psi \in C^M(I)$, then

$$J_\psi \in \begin{cases} C^{M+1}, & N = 1, \\ C^{0,1/\max(N-M,1)}, & N > 1. \end{cases}$$

If $N > 1$ and $\psi \in C^0(I)$, then ∇J_ψ is continuous almost everywhere. It is discontinuous at $\boldsymbol{m} = 0$, unless ψ is a polynomial of degree at most $N - 1$, in which case $J_\psi \in C^\infty$.

When $|u_k| \le L < 1$ then, up to a constant, each element of the flux function $\boldsymbol{F}_2 \circ \boldsymbol{F}_1^{-1}$ is of the form (5.40) and (5.41) with $\psi = u^\ell/\sqrt{1-u^2}$, $\ell = 1, \ldots, N$. The source function $\boldsymbol{K}$ is of a similar form. Hence we have the following corollary.

Corollary 5.9. The flux and source functions (5.39) are well defined and depend Lipschitz-continuously on $\boldsymbol{m} \in M_N[-L, L]$ when $0 < L < 1$. They are not continuously differentiable.

We refer to Runborg (2000, Section 3.1) for the proofs.

5.4. Numerical results

In this section we show results of applying the equations derived in Sections 5.2 and 5.3 to a few different test problems. We consider both homogeneous ($\eta \equiv 1$) and inhomogeneous ($\eta = \eta(\boldsymbol{x})$) media and use closures corresponding to $N = 1, 2, 3$ crossing rays at each point. The equations closed with delta functions, (5.18), are set in two-dimensional space, while the Heaviside equations, (5.37), are reduced to a one-dimensional space by the paraxial approximation. As a shorthand we will refer to the equations as the δ- and the H-equations. For a more complete numerical study, see Runborg (1998, 2000) and Gosse (2002).

As we remarked on page 246, the δ-equations (5.18) are nonstrictly hyperbolic with linearly degenerate fields. This is reflected in their sensitivity to numerical treatment. Even for smooth problems, some standard numerical schemes, such as Godunov, Lax–Friedrichs and Nessyahu–Tadmor with dimensional splitting, converge poorly in L^1 and may fail to converge in L^∞ (Engquist and Runborg 1996, 1998). The standard unsplit Lax–Friedrichs scheme converges well, and Jiang and Tadmor (1998) showed that, with an unsplit, genuinely two-dimensional version of Nessyahu–Tadmor, the expected second-order convergence rate is obtained for smooth problems. This is illustrated by Figure 5.2, Table 5.1 and Table 5.2. It appears that the dimensional splitting aggravates the numerical errors, although for the Godunov scheme James and Gosse (2000) observed that the same type of failure to converge in L^∞ can also occur in the much simpler case of a linear one-dimensional equation with variable coefficients. Kinetic schemes have been recognized to handle nonstrictly hyperbolic problems better. They were used with success for the δ-equations in Jin and Li (200x) and Gosse *et al.* (200x). Also note the importance of treating the source term correctly in heterogeneous media, where it may be very stiff because of large gradients in the index of refraction, for example by using so-called well-balanced schemes (Gosse 2002, Gosse *et al.* 200x).

Another difficulty for the δ-equations is to evaluate the flux functions $\boldsymbol{F}_1 \circ \boldsymbol{F}_0^{-1}$ and $\boldsymbol{F}_2 \circ \boldsymbol{F}_0^{-1}$. In both cases it is necessary to solve a nonlinear system of equations

$$\boldsymbol{F}_0(\boldsymbol{u}) = \boldsymbol{m}, \tag{5.42}$$

for each time step, at each grid point. Solving (5.42) can be difficult. An iterative solver must be used when $N > 2$, which is expensive and requires good initial values. In general, the Jacobian of $\boldsymbol{F}_0$ is singular when two rays are parallel. For iterative methods that use the Jacobian, this is a problem. When $N = 1, 2$ there is an analytical way to invert $\boldsymbol{F}_0$: see Runborg (2000). Furthermore, (5.42) may not have a solution. Although, for the exact solution of the PDE, (5.42) should always be satisfied, truncation errors in the numerical scheme may have perturbed the solution so that $\boldsymbol{m}$ is not in M_N, the range of $\boldsymbol{F}_0$.

The H-equations (5.37) are strictly hyperbolic and numerical schemes are not as sensitive as for the δ-equations. The evaluation of the flux functions is also easier, since it can be reduced to solving polynomial equations of low degree (Runborg 2000). By also accepting complex roots of those polynomial equations, M_N, the domain of definition of the flux function can be continuously extended, avoiding the problem of (5.42) not having a solution.

When the number of physically relevant phases is less than the number of phases supported by the system, we must still give initial data for the nonexistent phases. In the delta case a near-zero value can be given. (It is practical, though, not to use exactly zero since the flux functions have a weak singularity at zero.) Alternatively, the phase can be initialized to the same as another, physically relevant, phase. In the Heaviside case the fictitious phases can obviously not be set to zero. Moreover, setting them to the same as another phase would eliminate them from the equations. For the H-equations with $N = 2$, for instance, $u_1 \equiv u_2$ is a trivial solution. However, Gosse (2002) pointed out that simply initializing $u_{k+1} = u_k + \epsilon$, with a small ϵ, often works well when u_k and u_{k+1} are physically relevant and nonrelevant phases, respectively.

Test problems

One point source. We consider one point source located at $\boldsymbol{s} = (-0.2, 1)$ and compute the solution in the rectangle $[0, 1] \times [0, 2]$. The source is smooth with exact solution $\boldsymbol{u}(t, \boldsymbol{x}) = (\boldsymbol{x} - \boldsymbol{s}) \max(0, t - r)^3/r^2$, $r = \|\boldsymbol{x} - \boldsymbol{s}\|$, which we apply as boundary value at $x = 0$. General results are shown in Figure 5.2(a), where the Lax–Friedrichs method was used to solve the δ-system with $N = 1$ and 40×80 grid points. The difficulties with using the Godunov method for the same problem are highlighted in Figure 5.2(b).

Convergence for the different methods are summarized in Tables 5.1 and 5.2. The numerical error in $u_1 = m_{10}$ is shown measured in the L^1- and L^∞-norms.

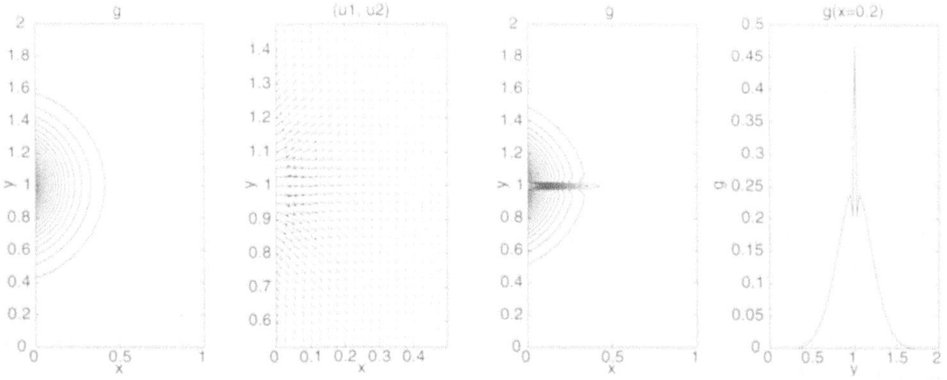

(a) Lax–Friedrichs scheme: contour plot of g (left) and quiver plot of vector field $\boldsymbol{u} = (u_1, u_2)$ (right)

(b) Godunov scheme: contour plot of g (left), and g in a vertical cut at $x = 0.2$ (right)

Figure 5.2. *One point source.* Snapshot of solution of δ-equations with $N = 1$ at time $t = 0.85$ using Lax–Friedrichs and Godunov schemes.

Table 5.1. *One point source.* L^1-norm of the numerical errors for the single-phase δ-equations with different methods. Here n is the number of grid points in the x-direction.

L^1	Lax–Friedrichs unsplit		Lax–Friedrichs split		Godunov split		Nessyahu–Tadmor unsplit		Nessyahu–Tadmor split	
n	error	order	error	order	error	order	error	order	error	order
10	7.78e-3		3.60e-2		1.13e-2		7.98e-3		1.02e-2	
		0.85		0.56		0.80		1.47		1.23
20	4.33e-3		2.44e-2		6.50e-3		2.89e-3		4.35e-3	
		0.92		0.72		0.69		1.74		1.20
40	2.29e-3		1.48e-2		4.04e-3		8.66e-4		1.89e-3	
		0.96		0.82		0.78		1.88		1.03
80	1.18e-3		8.39e-3		2.35e-3		2.35e-4		9.24e-4	
		0.98		0.89		0.85		1.89		0.76
160	5.99e-4		4.53e-3		1.30e-3		6.32e-5		5.45e-4	

Table 5.2. *One point source.* L^∞-norm of the numerical errors for the single-phase δ-equations with different methods. Here n is the number of grid points in the x-direction.

L^∞	Lax–Friedrichs unsplit		Lax–Friedrichs split		Godunov split		Nessyahu–Tadmor unsplit		Nessyahu–Tadmor split	
n	error	order	error	order	error	order	error	order	error	order
10	9.49e-2		1.78e-1		3.04e-1		6.64e-2		8.99e-2	
		1.26		0.85		0.06		1.26		0.91
20	3.97e-2		9.87e-2		2.91e-1		2.76e-2		4.78e-2	
		1.21		0.55		0.02		1.71		1.07
40	1.71e-2		6.73e-2		2.87e-1		8.46e-3		2.28e-2	
		1.15		0.73		0.02		1.70		0.80
80	7.71e-3		4.06e-2		2.83e-1		2.61e-3		1.31e-2	
		1.09		0.85		0.01		1.57		0.91
160	3.63e-3		2.26e-2		2.82e-1		8.75e-4		6.98e-3	

Three point sources. We now consider a problem with three point sources located at coordinates $\boldsymbol{s}_1 = (-0.5, 0.5)$, $\boldsymbol{s}_2 = (-0.5, 1.0)$ and $\boldsymbol{s}_3 = (-0.5, 1.5)$. The exact solutions are

$$\boldsymbol{w}_k(t, \boldsymbol{x}) = A_k(\boldsymbol{x} - \boldsymbol{s}_k)H(t - r_k)/r_k^2,$$
$$r_k = \|\boldsymbol{x} - \boldsymbol{s}_k\|, \quad k = 1, 2, 3,$$

where $A_1 = 1.25$, $A_2 = 0.75$ and $A_3 = 1.0$. The solution is computed in the rectangle $[0, 1] \times [0, 2]$. Figure 5.3 shows the solution at $t = 1.0$ of the δ-equations with $N = 1, 2, 3$. For the $N = 3$ system, the exact solution was given at $x = 0$. For the $N = 2$ system, the first two arriving waves were given at $x = 0$, that is,

$$\boldsymbol{u}_1 = \boldsymbol{w}_2, \qquad \boldsymbol{u}_2 = \begin{cases} \boldsymbol{w}_1 & r_1 < r_3, \\ \boldsymbol{w}_3 & r_1 \geq r_3. \end{cases}$$

Finally, for the $N = 1$ system, the first arriving wave was given at $x = 0$,

$$\boldsymbol{u} = \begin{cases} \boldsymbol{w}_1 & r_1 < r_2, \ r_1 < r_3, \\ \boldsymbol{w}_2 & r_2 \leq r_1, \ r_2 < r_3, \\ \boldsymbol{w}_3 & r_3 \leq r_1, \ r_3 \leq r_2. \end{cases}$$

As expected, the $N = 3$ system is the only one solving this problem correctly. Delta functions appear in the solutions of the $N = 1, 2$ systems, where rays should cross, but cannot, since the systems describe too few phases; *cf.* the analysis of the conservation law in Section 5.2.

Convex lens. In this test problem a plane wave is sent through a smooth convex lens, given by the index of refraction

$$\eta(x, y) = \begin{cases} 1 & d^2 > 1, \\ \left(\frac{4}{3 - \cos(\pi d^2)}\right)^2 & d^2 \leq 1, \end{cases} \qquad d^2 = \left(\frac{x - 0.5}{0.2}\right)^2 + \left(\frac{y - 1}{0.8}\right)^2.$$

We have computed the solution in the square $[0, 2] \times [0, 2]$ for the δ-equations with $N = 1, 2$ and the H-equations with $N = 2, 3$. Figure 5.4 shows the ray angles of the solutions. Here initial data only contain one phase, but the focusing of the lens creates additional phases, which are captured automatically by the multiphase systems.

Wedge. In this test problem a plane wave, injected at $x = 0$ with $\theta(0, y) = 0$ and $g(0, y) = 2$, is refracted by a smooth wedge, modelled by the index of refraction

$$\eta(x, y) = 1.5 - \frac{1}{\pi} \arctan(20((y - 1)^2 - 0.3(x - 0.5))).$$

When it is refracted in the interface a second and third phase appear. A caustic develops around the point $(1, 1)$, fanning out to the right: see Figure 5.5(c).

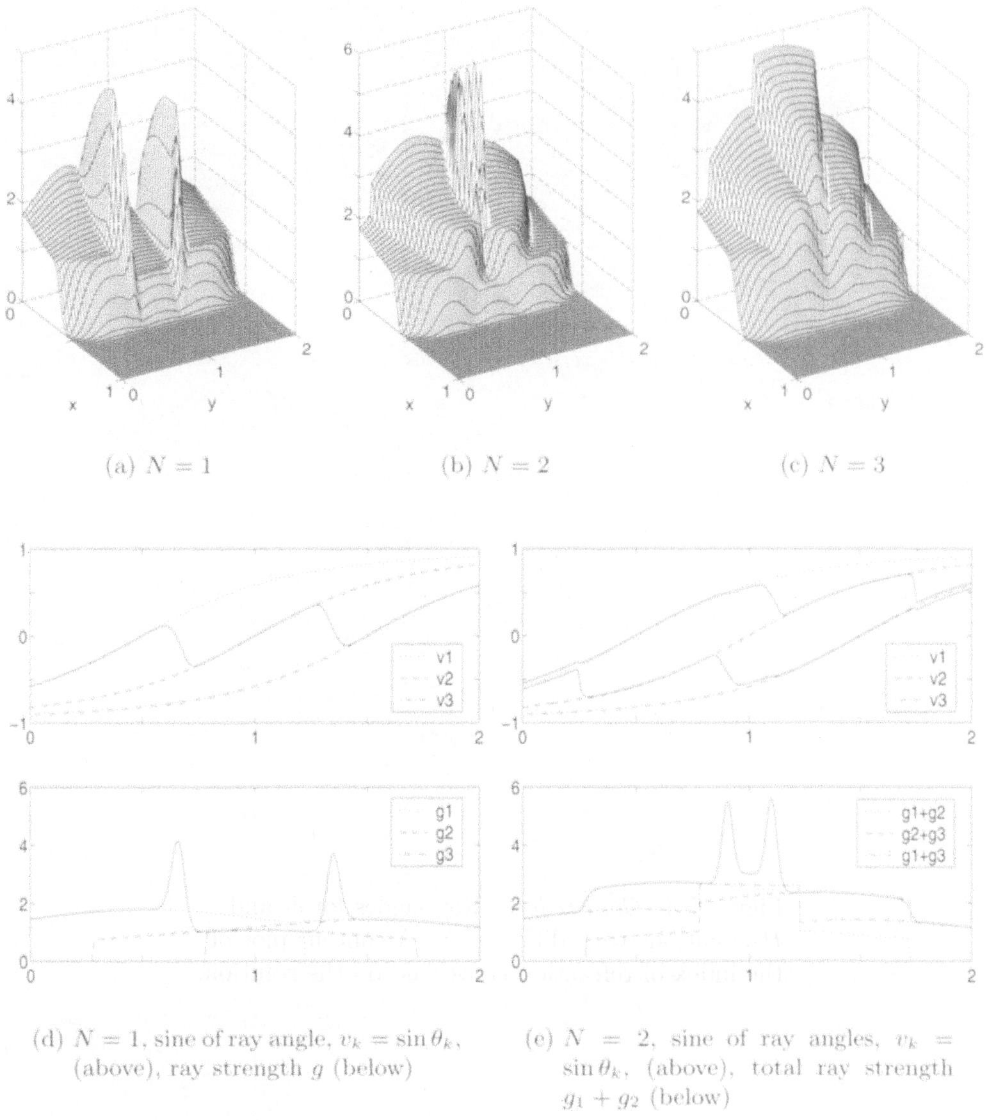

(a) $N = 1$

(b) $N = 2$

(c) $N = 3$

(d) $N = 1$, sine of ray angle, $v_k = \sin\theta_k$, (above), ray strength g (below)

(e) $N = 2$, sine of ray angles, $v_k = \sin\theta_k$, (above), total ray strength $g_1 + g_2$ (below)

Figure 5.3. *Three point sources.* Solution of the δ-equations with $N = 1, 2, 3$. Figures (a), (b), (c) show total ray strength, that is, g, $g_1 + g_2$ and $g_1 + g_2 + g_3$, respectively. Figures (d), (e) show solution in a cut at $x = 0.2$, computed (solid) and exact (dotted, dashed, dash-dotted).

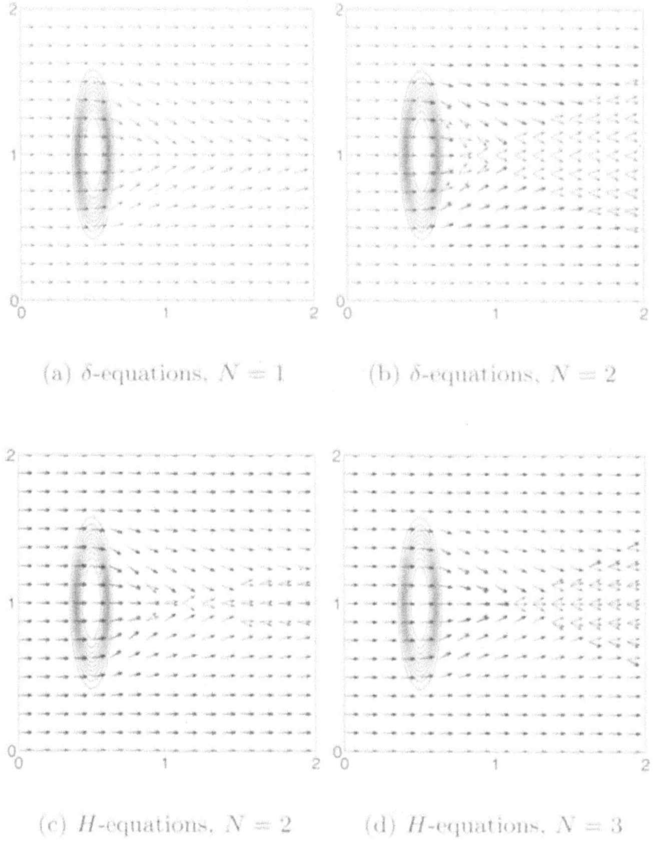

Figure 5.4. *Convex lens.* Ray angles for δ- and H-equations with different N. A contour plot of the index of refraction is overlaid on the solution.

As in the previous problem the δ- and H-equations were solved in the square $[0, 2] \times [0, 2]$. Different aspects of the solutions are shown in Figure 5.5. The δ-equations with $N = 1$ only capture one of the phases, as expected. A delta function appears where rays try to cross. The $N = 2$ system captures both the second phase and the caustic quite well. The H-equations cannot correctly capture the second phase when $N = 2$. However, when $N = 3$ all three phases are captured.

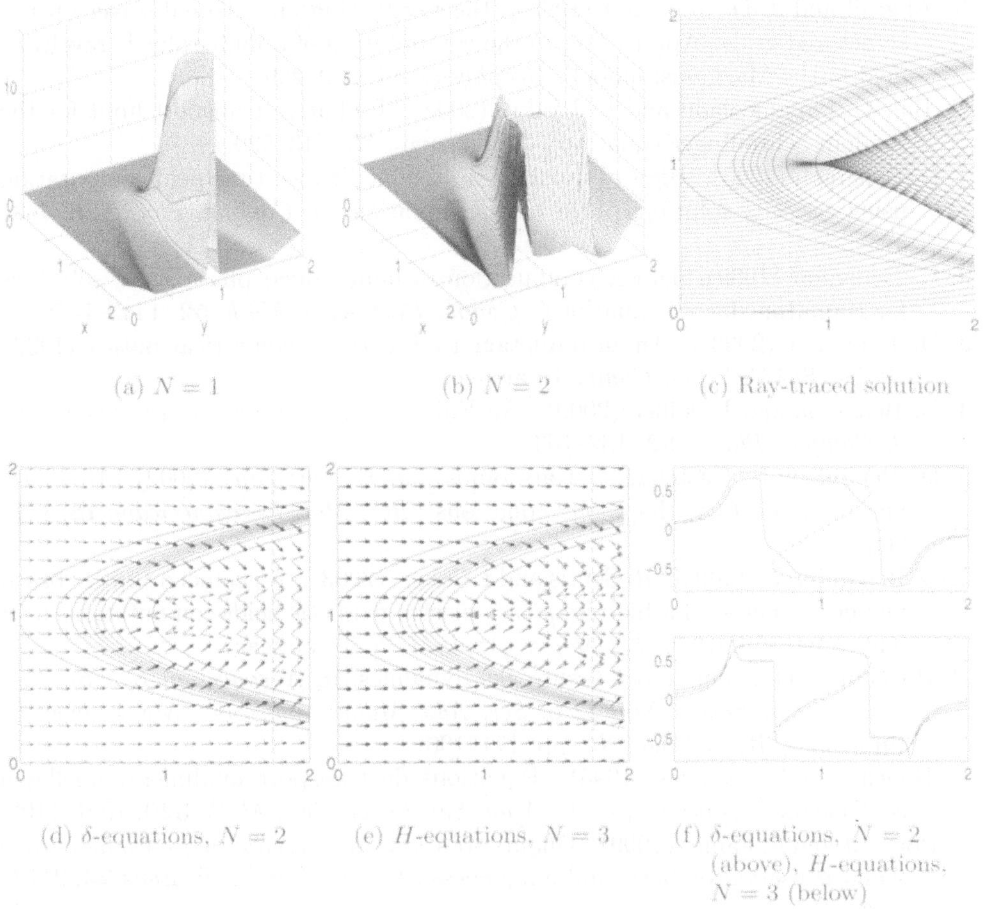

(a) $N = 1$ (b) $N = 2$ (c) Ray-traced solution

(d) δ-equations, $N = 2$ (e) H-equations, $N = 3$ (f) δ-equations, $N = 2$ (above), H-equations, $N = 3$ (below)

Figure 5.5. *Wedge.* Amplitude results, (a), (b), (c), for δ-equations with $N = 1, 2$. Figures (a), (b) show total ray strength, that is, g and $g_1 + g_2$ respectively. Figure (c) shows a ray-traced solution with contour lines of the index of refraction superimposed. Figures (d), (e) show quiver plots of ray angles for δ- and H-equations with $N = 2, 3$. A contour plot of the index of refraction is overlaid on the solution. Figure (f) shows sine of ray angles (solid) in a cut at $x = 1.75$ together with the corresponding values for a ray-traced solution (dashed).

REFERENCES

R. Abgrall and J.-D. Benamou (1999), 'Big ray tracing and eikonal solver on unstructured grids: Application to the computation of a multivalued traveltime field in the Marmousi model', *Geophysics* **64**, 230–239.

G. Bal, G. Papanicolaou and L. Ryzhik (2002), 'Radiative transport limit for the random Schrödinger equation', *Nonlinearity* **15**, 513–529.

J.-D. Benamou (1996), 'Big ray tracing: Multivalued travel time field computation using viscosity solutions of the eikonal equation', *J. Comput. Phys.* **128**, 463–474.

J.-D. Benamou (1999), 'Direct computation of multivalued phase space solutions for Hamilton–Jacobi equations', *Comm. Pure Appl. Math.* **52**, 1443–1475.

J.-D. Benamou (2003), 'An introduction to Eulerian geometrical optics (1992–2002)', *SIAM J. Sci. Comp.* To appear.

J.-D. Benamou and I. Solliec (2000), 'An Eulerian method for capturing caustics', *J. Comput. Phys.* **162**, 132–163.

J.-D. Benamou, F. Castella, T. Katsaounis and B. Perthame (2002), 'High frequency limit of the Helmholtz equations', *Rev. Mat. Iberoamericana* **18**, 187–209.

J.-D. Benamou, O. Lafitte, R. Sentis and I. Solliec (2003), 'A geometric optics based numerical method for high frequency electromagnetic fields computations near fold caustics, Part I', *J. Comput. Appl. Math.* To appear.

F. Bouchut (1994), On zero pressure gas dynamics, in *Advances in Kinetic Theory and Computing*, Vol. 22 of *Ser. Adv. Math. Appl. Sci.*, World Scientific Publishing, River Edge, NJ, pp. 171–190.

F. Bouchut and F. James (1995), 'Equations de transport unidimensionnelles à coefficients discontinus', *C. R. Acad. Sci. Paris Sér. I Math.* **320**, 1097–1102.

F. Bouchut and F. James (1999), 'Duality solutions for pressureless gases, monotone scalar conservation laws, and uniqueness', *Comm. Part. Diff. Eqns* **24**, 2173–2189.

Y. Brenier and L. Corrias (1998), 'A kinetic formulation for multibranch entropy solutions of scalar conservation laws', *Ann. Inst. Henri Poincaré* **15**, 169–190.

Y. Brenier and E. Grenier (1998), 'Sticky particles and scalar conservation laws', *SIAM J. Numer. Anal.* **35**, 2317–2328 (electronic).

P. Bulant and L. Klimeš (1999), 'Interpolation of ray theory traveltimes within ray cells', *Geophys. J. Int.* **139**, 273–282.

S. Cao and S. Greenhalgh (1994), 'Finite-difference solution of the eikonal equation using an efficient, first-arrival, wavefront tracking scheme', *Geophysics* **59**, 632–643.

F. Castella, B. Perthame and O. Runborg (2002), 'High frequency limit of the Helmholtz equation II: Source on a general smooth manifold', *Comm. Part. Diff. Eqns* **27**, 607–651.

V. Červený, I. A. Molotkov and I. Psencik (1977), *Ray Methods in Seismology*, University of Karlova Press.

L.-T. Cheng, S. J. Osher and J. Qian (2002), 'Level set based Eulerian methods for multivalued traveltimes in both isotropic and anisotropic media'. Preprint.

J. Claerbout (1976), *Fundamentals of Geophysical Data Processing*, McGraw-Hill.

M. G. Crandall and P.-L. Lions (1983), 'Viscosity solutions of Hamilton–Jacobi equations', *Trans. Amer. Math. Soc.* **277**, 1–42.

J. J. Duistermaat (1974), 'Oscillatory integrals, Lagrange immersions and unfolding of singularities', *Comm. Pure Appl. Math.* **27**, 207–281.

W. E, Y. G. Rykov and Y. G. Sinai (1996), 'Generalized variational principles, global weak solutions and behavior with random initial data for systems of conservation laws arising in adhesion particle dynamics', *Comm. Math. Phys.* **177**, 349–380.

B. Engquist and A. Majda (1977), 'Absorbing boundary conditions for the numerical simulation of waves', *Math. Comp.* **31**, 629–651.

B. Engquist and O. Runborg (1996), 'Multiphase computations in geometrical optics', *J. Comput. Appl. Math.* **74**, 175–192.

B. Engquist and O. Runborg (1998), Multiphase computations in geometrical optics, in *Hyperbolic Problems: Theory, Numerics, Applications* (M. Fey and R. Jeltsch, eds), Vol. 129 of *Internat. Ser. Numer. Math.*, ETH Zentrum, Zürich, Switzerland.

B. Engquist, O. Runborg and A.-K. Tornberg (2002), 'High frequency wave propagation by the segment projection method', *J. Comput. Phys.* **178**, 373–390.

E. Fatemi, B. Engquist and S. J. Osher (1995), 'Numerical solution of the high frequency asymptotic expansion for the scalar wave equation', *J. Comput. Phys.* **120**, 145–155.

S. Fomel and J. A. Sethian (2002), 'Fast-phase space computation of multiple arrivals', *Proc. Natl. Acad. Sci. USA* **99**, 7329–7334 (electronic).

S. Geoltrain and J. Brac (1993), 'Can we image complex structures with first-arrival traveltime?', *Geophysics* **58**, 564–575.

P. Gérard (1991), 'Microlocal defect measures', *Comm. Part. Diff. Eqns* **16**, 1761–1794.

P. Gérard, P. A. Markowich, N. J. Mauser and F. Poupaud (1997), 'Homogenization limits and Wigner transforms', *Comm. Pure Appl. Math.* **50**, 323–379.

L. Gosse (2002), 'Using K-branch entropy solutions for multivalued geometric optics computations', *J. Comput. Phys.* **180**, 155–182.

L. Gosse, S. Jin and X. Li (200x), 'On two moment systems for computing multiphase semiclassical limits of the Schrödinger equation'. To appear.

H. Grad (1949), 'On the kinetic theory of rarefied gases', *Comm. Pure Appl. Math.* **2**, 331–407.

S. Gray and W. May (1994), 'Kirchhoff migration using eikonal equation traveltimes', *Geophysics* **59**, 810–817.

E. Grenier (1995), 'Existence globale pour le système des gaz sans pression', *CR Acad. Sci. Paris Sér. I Math.* **321**, 171–174.

L. Hörmander (1983–1985), *The Analysis of Linear Partial Differential Operators, I–IV*, Springer, Berlin.

F. James and L. Gosse (2000), 'Numerical approximations of one-dimensional linear conservation equations with discontinuous coefficients', *Math. Comp.* **69**, 987–1015.

G.-S. Jiang and E. Tadmor (1998), 'Nonoscillatory central schemes for multidimensional hyperbolic conservation laws', *SIAM J. Sci. Comput.* **19**, 1892–1917.

S. Jin and X. Li (200x), 'Multi-phase computations of the semiclassical limit of the Schrödinger equation and related problems: Whitham vs. Wigner', *Physica D*. To appear.

B. R. Julian and D. Gubbins (1977), 'Three-dimensional seismic ray tracing', *J. Geophys. Res.* **43**, 95–114.

J. Keller (1962), 'Geometrical theory of diffraction', *J. Opt. Soc. Amer.*

S. Kim (2000), 'An $\mathcal{O}(N)$ level set method for eikonal equations', *SIAM J. Sci. Comput.* **22**, 2178–2193.

S. Kim and R. Cook (1999), '3-D traveltime computation using second-order ENO scheme', *Geophysics* **64**, 1867–1876.

R. G. Kouyoumjian and P. H. Pathak (1974), 'A uniform theory of diffraction for an edge in a perfectly conducting surface', *Proc. IEEE* **62**, 1448–1461.

Y. A. Kravtsov (1964), 'On a modification of the geometrical optics method', *Izv. VUZ Radiofiz.* **7**, 664–673.

G. Lambaré, P. S. Lucio and A. Hanyga (1996), 'Two-dimensional multivalued traveltime and amplitude maps by uniform sampling of ray field', *Geophys. J. Int.* **125**, 584–598.

R. T. Langan, I. Lerche and R. T. Cutler (1985), 'Tracing of rays through heterogeneous media: An accurate and efficient procedure', *Geophysics* **50**, 1456–1465.

R. J. LeVeque (1992), *Numerical Methods for Conservation Laws*, Birkhäuser.

C. D. Levermore (1996), 'Moment closure hierarchies for kinetic theories', *J. Statist. Phys.* **83**, 1021–1065.

H. Ling, R. Chou and S. W. Lee (1989), 'Shooting and bouncing rays: Calculating the RCS of an arbitrarily shaped cavity', *IEEE T. Antenn. Propag.* **37**, 194–205.

P.-L. Lions and T. Paul (1993), 'Sur les mesures de Wigner', *Rev. Mat. Iberoamericana* **9**, 553–618.

D. Ludwig (1966), 'Uniform asymptotic expansions at a caustic', *Comm. Pure Appl. Math.* **19**, 215–250.

V. P. Maslov (1965), *Theory of Perturbations and Asymptotic Methods.*

L. Miller (2000), 'Refraction of high-frequency waves density by sharp interfaces and semiclassical measures at the boundary', *J. Math. Pures Appl. IX* **79**, 227–269.

S. J. Osher and R. P. Fedkiw (2002), *Level Set Methods and Dynamic Implicit Surfaces*, Springer.

S. J. Osher and J. A. Sethian (1988), 'Fronts propagating with curvature-dependent speed: Algorithms based on Hamilton–Jacobi formulations', *J. Comput. Phys.* **79**, 12–49.

S. J. Osher and C.-W. Shu (1991), 'High-order essentially nonoscillatory schemes for Hamilton–Jacobi equations', *SIAM J. Numer. Anal.* **28**, 907–922.

S. J. Osher, L.-T. Cheng, M. Kang, H. Shim and Y.-H. Tsai (2002), 'Geometric optics in a phase-space-based level set and Eulerian framework', *J. Comput. Phys.* **179**, 622–648.

V. Pereyra, W. H. K. Lee and H. B. Keller (1980), 'Solving two-point seismic ray-tracing problems in a heterogeneous medium', *Bull. Seism. Soc. Amer.* **70**, 79–99.

F. Poupaud and M. Rascle (1997), 'Measure solutions to the linear multi-dimensional transport equation with non-smooth coefficients', *Comm. Part. Diff. Eqns* **22**, 337–358.

F. Qin, Y. Luo, K. B. Olsen, W. Cai and G. T. Schuster (1992), 'Finite-difference solution of the eikonal equation along expanding wavefronts', *Geophysics* **57**, 478–487.

O. Runborg (1998), Multiscale and multiphase methods for wave propagation, PhD thesis, NADA, KTH, Stockholm.

O. Runborg (2000), 'Some new results in multiphase geometrical optics', *M2AN Math. Model. Numer. Anal.* **34**, 1203–1231.

S. J. Ruuth, B. Merriman and S. J. Osher (2000), 'A fixed grid method for capturing the motion of self-intersecting wavefronts and related PDEs', *J. Comput. Phys.* **163**, 1–21.

L. Ryzhik, G. Papanicolaou and J. B. Keller (1996), 'Transport equations for elastic and other waves in random media', *Wave Motion* **24**, 327–370.

W. A. Schneider, K. A. Ranzinger, A. H. Balch and C. Kruse (1992), 'A dynamic programming approach to first arrival traveltime computation in media with arbitrary distributed velocities', *Geophysics* **57**, 39–50.

J. A. Sethian (1996), 'A fast marching level set method for monotonically advancing fronts', *Proc. Nat. Acad. Sci. USA* **93**, 1591–1595.

J. A. Sethian (1999), *Level Set Methods and Fast Marching Methods: Evolving Interfaces in Computational Geometry, Fluid Mechanics, Computer Vision, and Materials Science*, 2nd edn, Cambridge University Press, Cambridge.

I. Solliec (2003), Calcul optique multivalué et calcul d'energie électromagnétique en présence d'une caustique de type pli, PhD thesis, Université Pierre et Marie Curie, Paris 6, France.

C. Sparber, N. Mauser and P. A. Markowich (2003), 'Wigner functions vs. WKB-techniques in multivalued geometrical optics', *J. Asympt. Anal.* **33**, 153–187.

J. Steinhoff, M. Fan and L. Wang (2000), 'A new Eulerian method for the computation of propagating short acoustic and electromagnetic pulses', *J. Comput. Phys.* **157**, 683–706.

J. Steinhoff, Y. Wenren, D. Underhill and E. Puskas (1995), Computation of short acoustic pulses, in *Proceedings, 6th International Symposium on CFD, Lake Tahoe NV*.

Y. Sun (1992), Computation of 2D multiple arrival traveltime fields by an interpolative shooting method, in *Soc. Expl. Geophys.*, pp. 1320–1323.

W. W. Symes (1996), A slowness matching finite difference method for traveltimes beyond transmission caustics, Preprint, Department of Computational and Applied Mathematics, Rice University.

W. W. Symes and J. Qian (2003), 'A slowness matching Eulerian method for multivalued solutions of eikonal equations', *SIAM J. Sci. Comp.* To appear.

L. Tartar (1990), 'H-measures, a new approach for studying homogenisation, oscillations and concentration effects in partial differential equations', *Proc. Roy. Soc. Edinburgh Sect. A* **115**, 193–230.

C. H. Thurber and W. L. Ellsworth (1980), 'Rapid solution of ray tracing problems in hetereogeneous media', *Bull. Seism. Soc. Amer.* **70**, 1137–1148.

A.-K. Tornberg (2000), Interface tracking methods with applications to multiphase flows, PhD thesis, NADA, KTH, Stockholm.

A.-K. Tornberg and B. Engquist (2000), Interface tracking in multiphase flows, in *Multifield Problems*, Springer, Berlin, pp. 58–65.

A.-K. Tornberg and B. Engquist (2003), 'The segment projection method for interface tracking', *Comm. Pure Appl. Math.* **56**, 47–79.

Y. R. Tsai, L. T. Cheng, S. Osher and H. K. Zhao (2003), 'Fast sweeping algorithms for a class of Hamilton–Jacobi equations', *SIAM J. Numer. Anal.* To appear.

J. N. Tsitsiklis (1995), 'Efficient algorithms for globally optimal trajectories', *IEEE Trans. Automat. Control* **40**, 1528–1538.

J. van Trier and W. W. Symes (1991), 'Upwind finite-difference calculation of traveltimes', *Geophysics* **56**, 812–821.

J. Vidale (1988), 'Finite-difference calculation of traveltimes', *Bull. Seism. Soc. Amer.* **78**, 2062–2076.

V. Vinje, E. Iversen and H. Gjøystdal (1992), Traveltime and amplitude estimation using wavefront construction, in *Eur. Assoc. Expl. Geophys.*, pp. 504–505.

V. Vinje, E. Iversen and H. Gjøystdal (1993), 'Traveltime and amplitude estimation using wavefront construction', *Geophysics* **58**, 1157–1166.

G. B. Whitham (1974), *Linear and Nonlinear Waves*, Wiley.

Y. Zheng (1998), Systems of conservation laws with incomplete sets of eigenvectors everywhere, in *Advances in Nonlinear Partial Differential Equations and Related Areas*, World Scientific Publishing, River Edge, NJ, pp. 399–426.

Acta Numerica (2003), pp. 267–319
DOI: 10.1017/S0962492902000120

Model reduction methods based on Krylov subspaces

Roland W. Freund
Bell Laboratories, Lucent Technologies,
Room 2C–525,
Murray Hill, NJ 07974-0636, USA
E-mail: freund@research.bell-labs.com

In recent years, reduced-order modelling techniques based on Krylov-subspace iterations, especially the Lanczos algorithm and the Arnoldi process, have become popular tools for tackling the large-scale time-invariant linear dynamical systems that arise in the simulation of electronic circuits. This paper reviews the main ideas of reduced-order modelling techniques based on Krylov subspaces and describes some applications of reduced-order modelling in circuit simulation.

CONTENTS

1. Introduction

Roughly speaking, the problem of model reduction is to replace a given mathematical model of a system or a process by a model that is much 'smaller' than the original model, but still describes, at least 'approximately', certain aspects of the system or process. Clearly, model reduction involves a number of interesting issues. First and foremost is the issue of selecting appropriate approximation schemes that allow the definition of suitable reduced-order models. In addition, it is often important that the reduced-order model preserves certain crucial properties of the original system, such as stability or passivity. Other issues include the characterization of the quality of the models, the extraction of the data from the original model that is needed to actually generate the reduced-order models, and the efficient and numerically stable computation of the models. We refer the reader to Fortuna, Nunnari and Gallo (1992) for a review of general model-reduction techniques, and to the more specialized survey papers by Bultheel and Van Barel (1986), Bultheel and De Moor (2000), Freund (1997, 1999*b*), and Bai (2002), which review methods based on Padé and more general rational approximation, and techniques tailored to applications in VLSI circuit simulation.

In this paper, we discuss Krylov subspace-based reduced-order modelling techniques for large-scale linear dynamical systems, especially those that arise in the simulation of electronic circuits and of microelectromechanical systems.

We begin with a brief description of reduced-order modelling problems in circuit simulation. Electronic circuits are usually modelled as networks whose branches correspond to the circuit elements and whose nodes correspond to the interconnections of the circuit elements. Such networks are characterized by three types of equation. *Kirchhoff's current law* (KCL) states that, for each node of the network, the currents flowing in and out of that node sum up to zero. *Kirchhoff's voltage law* (KVL) states that, for each closed loop of the network, the voltage drops along that loop sum up to zero. *Branch constitutive relations* (BCRs) are equations that characterize the actual circuit elements. For example, the BCR of a linear resistor is Ohm's law. The BCRs are linear equations for simple devices, such as linear resistors, capacitors, and inductors, and they are nonlinear equations for more complex devices, such as diodes and transistors. Furthermore, in general, the BCRs involve time-derivatives of the unknowns, and thus they are *ordinary differential equations* (ODEs). On the other hand, the KCLs and KVLs are linear algebraic equations that only depend on the topology of the circuit.

The KCLs, KVLs, and BCRs can be summarized as a system of first-order, in general nonlinear, *differential-algebraic equations* (DAEs) of the form

$$\frac{\mathrm{d}}{\mathrm{d}t}\, q(\hat{x}, t) + f(\hat{x}, t) = 0, \tag{1.1}$$

together with suitable initial conditions. Here, $\hat{x} = \hat{x}(t)$ is the unknown vector of circuit variables at time t, the vector-valued function $f(\hat{x}, t)$ represents the contributions of nonreactive elements, such as resistors and sources, and the vector-valued function $\frac{\mathrm{d}}{\mathrm{d}t}\, q(\hat{x}, t)$ represents the contributions of reactive elements, such as capacitors and inductors. There are several established methods (Vlach and Singhal 1994), such as sparse tableau, nodal formulation, and modified nodal analysis, for generating a system of equations of the form (1.1) from a so-called *netlist* description of a given circuit. The vector functions $\hat{x}$, f, q, as well as their dimension, depend on the chosen formulation method. The most general method is sparse tableau, which consists of just listing all the KCLs, KVLs, and BCRs. The other formulation methods can be interpreted as starting from sparse tableau and eliminating some of the unknowns by using some of the KCL or KVL equations.

For all the standard formulation methods, the dimension of the system (1.1) is of the order of the number of elements in the circuit. Since today's VLSI circuits can have up to hundreds of millions of circuit elements, systems (1.1) describing such circuits can be of extremely large dimension. Reduced-order modelling allows us to first replace large systems of the form (1.1) by systems of smaller dimension and then tackle these smaller systems by suitable DAE solvers. Ideally, we would like to apply nonlinear reduced-order modelling directly to the nonlinear system (1.1). However, since nonlinear reduction techniques are a lot less developed and less well understood than linear ones, linear reduced-order modelling is almost always employed at present. To that end, we either linearize the system (1.1) or decouple (1.1) into nonlinear and linear subsystems; see, *e.g.*, Freund (1999*b*) and the references given there.

For example, the first case arises in *small-signal analysis*; see, *e.g.*, Freund and Feldmann (1996*b*). Given a *DC operating point*, say $\hat{x}_0$, of the circuit described by (1.1), we linearize the system (1.1) around $\hat{x}_0$. The resulting linearized version of (1.1) is of the following form:

$$E\frac{\mathrm{d}x}{\mathrm{d}t} = Ax + Bu(t), \tag{1.2}$$

$$y(t) = C^T x(t). \tag{1.3}$$

Here, $A = -\mathrm{D}_x f$ is the negative of the Jacobian matrix of f at the DC operating point $\hat{x}_0$, and $E = \mathrm{D}_x q$ is the Jacobian matrix of q at $\hat{x}_0$. Furthermore, $x(t) = \hat{x}(t) - \hat{x}_0$ is the distance of the solution $\hat{x}$ of (1.1) to the DC operating point, $u(t)$ is the vector of excitations applied to the sources of the circuit, and $y(t)$ is the vector of circuit variables of interest. Equations (1.2) and (1.3) represent a *time-invariant linear dynamical system.* Its *state-space dimension*, N, is the length of the vector x of circuit variables. For a circuit with many elements, the system (1.2) and (1.3) is thus of very high dimension. The idea of reduced-order modelling is then to replace the system (1.2) and (1.3) by one of the same form,

$$E_n\frac{\mathrm{d}z}{\mathrm{d}t} = A_n z + B_n u(t),$$

$$y(t) = C_n^T z(t),$$

but of much smaller state-space dimension $n \ll N$.

Time-invariant linear dynamical systems of the form (1.2) and (1.3) also arise when equations describing linear subcircuits of a given circuit are decoupled from the system (1.1) that characterizes the whole circuit; see, *e.g.*, Freund (1999*b*). For example, as discussed in Cheng, Lillis, Lin and Chang (2000), the interconnect and the pin package of VLSI circuits are often modelled as large linear RCL networks consisting only of resistors, capacitors, and inductors. Such linear subcircuits are described by systems of the form (1.2) and (1.3), where $x(t)$ is the vector of circuit variables associated with the subcircuit, and the vectors $u(t)$ and $y(t)$ contain the variables of the connections of the subcircuit to the generally nonlinear remainder of the whole circuit. By replacing, in the nonlinear system (1.1), the linear subsystem (1.2) and (1.3) by a reduced-order model of much smaller state-space dimension, the dimension of (1.1) can be reduced significantly, before a DAE solver is then applied to such a smaller version of (1.1).

The remainder of this paper is organized as follows. In Section 2, we review some basic facts about time-invariant linear dynamical systems. In Section 3, we introduce reduced-order models that are defined via Padé or Padé-type approximation. In Section 4, we discuss the concepts of stability and passivity of linear dynamical systems. In Section 5, we discuss reduced-order modelling approaches based on Lanczos and Lanczos-type methods. In Section 6, we describe the use of the Arnoldi process for reduced-order modelling. In Section 7, we discuss reduced-order modelling of noise-type transfer functions, which arise in circuit-noise computations. Section 8 is concerned with reduced-order modelling of second-order dynamical systems. Finally, in Section 9, we make some concluding remarks.

2. Time-invariant linear dynamical systems

In this section, we review some basic facts about time-invariant linear dynamical systems.

2.1. State-space description

We consider m-input p-output time-invariant linear dynamical systems given by a *state-space description* of the form

$$E\frac{\mathrm{d}x}{\mathrm{d}t} = Ax + Bu(t), \tag{2.1}$$

$$y(t) = C^T x(t) + Du(t), \tag{2.2}$$

together with suitable initial conditions. Here, $A,\ E \in \mathbb{R}^{N\times N}$, $B \in \mathbb{R}^{N\times m}$, $C \in \mathbb{R}^{N\times p}$, and $D \in \mathbb{R}^{p\times m}$ are given matrices, $x(t) \in \mathbb{R}^N$ is the vector of state variables, $u(t) \in \mathbb{R}^m$ is the vector of inputs, $y(t) \in \mathbb{R}^p$ the vector of outputs, N is the state-space dimension, and m and p are the number of inputs and outputs, respectively. Note that systems of the form (1.2) and (1.3) are just a special case of (2.1) and (2.2) with $D = 0$.

The linear system (2.1) and (2.2) is called *regular* if the matrix E in (2.1) is nonsingular, and it is called *singular* or a *descriptor system* if E is singular. In the regular case, the linear system (2.1) and (2.2) can always be rewritten as

$$\frac{\mathrm{d}x}{\mathrm{d}t} = \left(E^{-1}A\right)x + \left(E^{-1}B\right)u(t),$$
$$y(t) = C^T x(t) + Du(t),$$

which is also a system of the form (2.1) and (2.2) with $E = I$. Note that the first equation is just a system of ODEs.

The linear dynamical systems arising in circuit simulation are descriptor systems in general. Therefore, in the following, we allow $E \in \mathbb{R}^{N\times N}$ to be a general, possibly singular, matrix. The only assumption that we make on the matrices A and E in (2.1) is that the matrix pencil $A - sE$ be *regular*, that is, the matrix $A - sE$ is singular only for finitely many values of $s \in \mathbb{C}$.

In the case of singular E, equation (2.1) represents a system of DAEs, rather than ODEs. Solving DAEs is significantly more complex than solving ODEs. Moreover, there are constraints on the possible initial conditions that can be imposed on the solutions of (2.1). For a detailed discussion of DAEs and the structure of their solutions, we refer the reader to Campbell (1980, 1982), Dai (1989), and Verghese, Lévy and Kailath (1981). Here, we only present a brief glimpse of the issues arising in DAEs.

A general descriptor system (2.1) has three different types of *modes*, which are characterized by the eigenstructure of the pencil $A - sE$; see, *e.g.*, Bender and Laub (1987). The finite eigenvalues, $s \in \mathbb{C}$, of the pencil are the *finite dynamic* modes of (2.1). The eigenvectors associated with an infinite eigenvalue $s = \infty$ of the pencil span the space of *nondynamic* solutions of (2.1), and the corresponding eigenvalues $s = \infty$ are the *nondynamic* modes of (2.1). Note that the set of nondynamic solutions of (2.1) is just the null space of E. If $r := \operatorname{rank} E$ is bigger than the degree ρ of the characteristic polynomial $\det(A - sE)$, then the pencil also has $r - \rho$ *impulsive* modes. The impulsive modes correspond to generalized eigenvectors of eigenvalues $s = \infty$ with Jordan blocks of size bigger than 1. A descriptor system is called *impulsive-free* if it has no impulsive modes.

To explain the different types of modes further, we bring the matrices A and E in (2.1) to an appropriate normal form. For any regular pencil $A - sE$, there exist nonsingular matrices P and Q such that

$$P\,(A - sE)\,Q = \begin{bmatrix} A^{(1)} - sI & 0 \\ 0 & I - sJ \end{bmatrix}, \tag{2.3}$$

where the submatrix J is nilpotent. The matrix pencil on the right-hand side of (2.3) is called the *Weierstrass form* of $A - sE$. Assuming that the matrices A and E in (2.1) are already in Weierstrass form, the system (2.1) can be decoupled as follows:

$$\frac{\mathrm{d}x^{(1)}}{\mathrm{d}t} = A^{(1)}x^{(1)} + B^{(1)}u(t), \tag{2.4}$$

$$J\frac{\mathrm{d}x^{(2)}}{\mathrm{d}t} = x^{(2)} + B^{(2)}u(t). \tag{2.5}$$

The first subsystem, (2.4), is just a system of linear ODEs. Thus, for any given initial condition $x^{(1)}(0) = \hat{x}^{(1)}$, there exists a unique solution of (2.4). Moreover, the so-called *free-response* of (2.4), that is, the solutions $x(t)$ for $t \geq 0$ when $u \equiv 0$, consists of combinations of exponential modes at the eigenvalues of the matrix $A^{(1)}$. Note that, in view of (2.3), the eigenvalues of $A^{(1)}$ are just the finite eigenvalues of the pencil $A - sE$ and thus they are the finite dynamic modes. The solutions of the second subsystem, (2.5), however, are of a quite different nature. In particular, the free-response of (2.5) consists of $k_i - 1$ independent impulsive motions for each $k_i \times k_i$ Jordan block of the matrix J; see, *e.g.*, Verghese *et al.* (1981).

For example, consider the case when the nilpotent matrix J in (2.5) is a single $k \times k$ Jordan block, that is,

$$J = \begin{bmatrix} 0 & 1 & 0 & \cdots & 0 \\ 0 & 0 & 1 & \ddots & \vdots \\ \vdots & \ddots & \ddots & \ddots & 0 \\ \vdots & \ddots & \ddots & \ddots & 1 \\ 0 & \cdots & \cdots & 0 & 0 \end{bmatrix} \in \mathbb{R}^{k \times k}.$$

The k components of the free-response $x^{(2)}(t)$ of (2.5) are then given by

$$\begin{aligned} x_1^{(2)}(t) &= -x_2^{(2)}(0-)\delta(t) - x_3^{(2)}(0-)\delta^{(1)}(t) - \cdots - x_k^{(2)}(0-)\delta^{(k-2)}(t), \\ x_2^{(2)}(t) &= -x_3^{(2)}(0-)\delta(t) - x_4^{(2)}(0-)\delta^{(1)}(t) - \cdots - x_k^{(2)}(0-)\delta^{(k-3)}(t), \\ \vdots &= \vdots \\ x_{k-1}^{(2)}(t) &= -x_k^{(2)}(0-)\delta(t), \\ x_k^{(2)}(t) &= 0. \end{aligned}$$

Here, $\delta(t)$ is the delta function and $\delta^{(i)}(t)$ is its ith derivative. Moreover, $x_i^{(2)}(0-)$, $i = 2, 3, \ldots, k$, are the components of the initial conditions that can be imposed on (2.5). Note that there are only $k - 1$ degrees of freedom for the initial condition and that it is not possible to prescribe $x_1^{(2)}(0-)$.

In particular, the free-response of (2.5) corresponding to a 1×1 Jordan block of J is just the zero solution, and there is no degree of freedom for the selection of an initial value corresponding to that block.

Finally, we remark that, in view of (2.3), the eigenvalues of the matrix pencil $A - sE$ corresponding to the subsystem (2.5) are just the infinite eigenvalues of $A - sE$ and thus the nondynamic modes.

2.2. Reduced-order models and transfer functions

The basic idea of reduced-order modelling is to replace a given system by a system of the same type, but with smaller state-space dimension. Thus, a *reduced-order model* of state-space dimension n of a given time-invariant linear dynamical system (2.1) and (2.2) of dimension N is a system of the form

$$E_n \frac{\mathrm{d}z}{\mathrm{d}t} = A_n z + B_n u(t), \tag{2.6}$$

$$y(t) = C_n^T z(t) + D_n u(t), \tag{2.7}$$

where A_n, $E_n \in \mathbb{R}^{n \times n}$, $B_n \in \mathbb{R}^{n \times m}$, $C_n \in \mathbb{R}^{n \times p}$, $D_n \in \mathbb{R}^{p \times m}$, and $n < N$.

The challenge then is to choose the matrices A_n, E_n, B_n, C_n, and D_n in (2.6) and (2.7) such that the reduced-order model in some sense approximates the original system. One possible measure of the approximation quality of reduced-order models is based on the concept of transfer functions.

If we assume zero initial conditions, then, by applying the Laplace transform

$$X(s) = \int_0^\infty x(t)\, \mathrm{e}^{-st}\, \mathrm{d}t$$

to the original system (2.1) and (2.2), we obtain the following algebraic equations:

$$sEX(s) = AX(s) + BU(s),$$
$$Y(s) = C^T X(s) + DU(s).$$

Here, the frequency-domain variables $X(s)$, $U(s)$, and $Y(s)$ are the Laplace transforms of the time-domain variables of $x(t)$, $u(t)$, and $y(t)$, respectively. Note that $s \in \mathbb{C}$. Then, formally eliminating $X(s)$ in the above equations, we arrive at the frequency-domain input-output relation $Y(s) = H(s)U(s)$. Here,

$$H(s) := D + C^T (sE - A)^{-1} B, \quad s \in \mathbb{C}, \tag{2.8}$$

is the so-called *transfer function* of the system (2.1) and (2.2). Note that

$$H : \mathbb{C} \mapsto (\mathbb{C} \cup \infty)^{p \times m},$$

is a $(p \times m)$-matrix-valued rational function.

Similarly, the transfer function, H_n, of the reduced-order model (2.6) and (2.7) is given by

$$H_n(s) := D_n + C_n^T \left(sE_n - A_n\right)^{-1} B_n, \quad s \in \mathbb{C}. \tag{2.9}$$

Note that H_n is also a $(p \times m)$-matrix-valued rational function.

3. Padé and Padé-type models

The concept of transfer functions allows us to define reduced-order models by means of Padé or Padé-type approximation.

3.1. Padé approximants of transfer functions

Let $s_0 \in \mathbb{C}$ be any point such that s_0 is not a pole of the transfer function H given by (2.8). In practice, the point s_0 is chosen such that it is in some sense close to the frequency range of interest. We remark that the frequency range of interest is usually a subset of the imaginary axis in the complex s-plane. Since s_0 is not a pole of H, the function H admits the Taylor expansion

$$H(s) = \mu_0 + \mu_1 \left(s - s_0\right) + \mu_2 \left(s - s_0\right)^2 + \cdots + \mu_j \left(s - s_0\right)^j + \cdots \tag{3.1}$$

about s_0. The coefficients μ_j, $j = 0, 1, \ldots$, in (3.1) are called the *moments* of H about the expansion point s_0. Note that each μ_j is a $p \times m$ matrix.

A reduced-order model (2.6) and (2.7) of state-space dimension n is called an nth *Padé model* (at the expansion point s_0) of the original system (2.1) and (2.2) if the Taylor expansions about s_0 of the transfer function (2.8), H, of the original system and the transfer function (2.9), H_n, of the reduced-order model agree in as many leading terms as possible, that is,

$$H(s) = H_n(s) + \mathcal{O}\left(\left(s - s_0\right)^{q(n)}\right), \tag{3.2}$$

where $q(n)$ is as large as possible. For an introduction to Padé approximation, we refer the reader to Baker, Jr. and Graves-Morris (1996). In Feldmann and Freund (1995*b*) and Freund (1995), it was shown that

$$q(n) \geq \left\lfloor \frac{n}{m} \right\rfloor + \left\lfloor \frac{n}{p} \right\rfloor,$$

with equality in the 'generic' case.

Even though Padé models are defined via the local approximation property (3.2), in practice, they usually are excellent approximations over large frequency ranges. The following single-input single-output example illustrates this statement. The example is a circuit resulting from the so-called PEEC discretization (Ruehli 1974) of an electromagnetic problem. The circuit is an RCL network consisting of 2100 capacitors, 172 inductors, 6990 inductive couplings, and a single resistive source that drives the circuit.

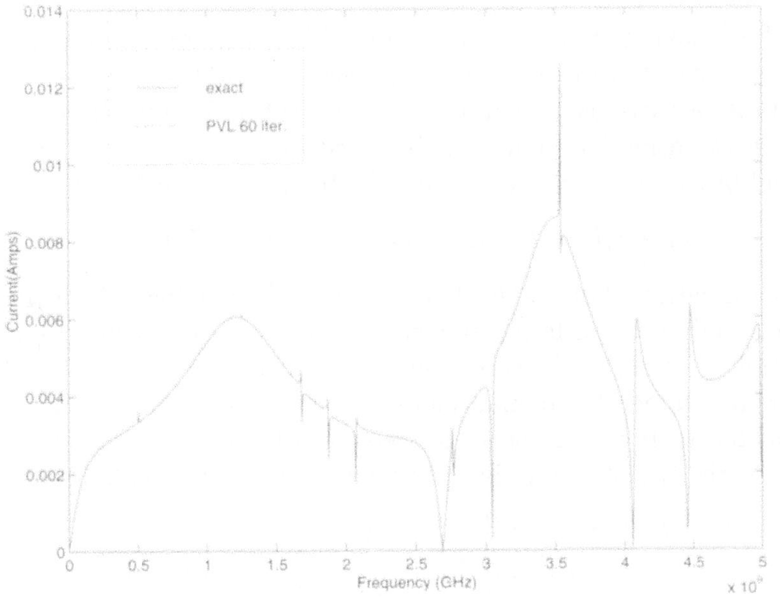

Figure 3.1. The PEEC transfer function, exact and Padé model of dimension $n = 60$.

Modified nodal analysis is used to set up the circuit equations, resulting in a linear dynamical system of dimension $N = 306$. In turns out that a Padé model of dimension $n = 60$ is sufficient to produce an almost exact transfer function in the relevant frequency range $s = 2\pi \mathrm{i}\,\omega$, $0 \le \omega \le 5 \times 10^9$. The corresponding curves for $|H(s)|$ and $|H_{60}(s)|$ are shown in Figure 3.1. The Padé model shown in Figure 3.1 was computed with the PVL method described in Section 5 below.

It is very tempting to compute Padé models by exploiting the definition (3.2) directly. More precisely, we would first explicitly generate the $q(n)$ moment matrices $\mu_0, \mu_1, \ldots, \mu_{q(n)-1}$, and then compute H_n and the system matrices in the reduced-order model (2.6) and (2.7) from these moments. In fact, for the special case $m = p = 1$ of single-input single-output systems, this approach is the *asymptotic waveform evaluation* (AWE) method that was introduced to the circuit simulation community by Pillage and Rohrer (1990). For surveys of AWE and its derivatives, we refer the reader to Chiprout and Nakhla (1994) and Raghavan, Rohrer, Pillage, Lee, Bracken and Alaybeyi (1993). However, computing Padé models directly from the moments is extremely ill-conditioned, and consequently, such an approach is not a viable numerical procedure in general. We discuss these shortcomings of the AWE approach in more detail in Section 3.4 below.

3.2. Reduction to a single matrix

Instead of employing explicit moment matching, the preferred way to compute Padé models is to use Krylov-subspace techniques, such as a suitable Lanczos-type process, as we will describe in Section 5. This becomes possible after the transfer function (2.8) is rewritten in terms of a single matrix M, instead of the two matrices A and E. To this end, let

$$A - s_0 E = F_1 F_2, \quad \text{where} \quad F_1,\, F_2 \in \mathbb{C}^{N \times N}, \tag{3.3}$$

be any factorization of $A - s_0 E$. For example, the matrices $A - s_0 E$ arising in circuit simulation are large, but sparse, and are such that a sparse LU factorization is feasible. In this case, the matrices F_1 and F_2 in (3.3) are the lower and upper triangular factors, possibly with rows and columns permuted due to pivoting, of such a sparse LU factorization of $A - s_0 E$. Using (3.3), the transfer function (2.8) can be rewritten as follows:

$$\begin{aligned} H(s) &= D + C^T \left(sE - A\right)^{-1} B \\ &= D - C^T \left(A - s_0 E - (s - s_0)E\right)^{-1} B \\ &= D - L^T \left(I - (s - s_0)M\right)^{-1} R, \end{aligned} \tag{3.4}$$

where

$$M := F_1^{-1} E F_2^{-1}, \quad R := F_1^{-1} B, \quad \text{and} \quad L := F_2^{-T} C. \tag{3.5}$$

Note that (3.4) only involves one $N \times N$ matrix, namely M, instead of the two $N \times N$ matrices A and E in (2.8). This allows us to apply Krylov-subspace methods to the single matrix M, with the $N \times m$ matrix R and the $N \times p$ matrix L as blocks of right and left starting vectors.

3.3. Padé-type approximants

While Padé models often provide very good approximations in the frequency domain, they also have undesirable properties. In particular, Padé models in general do not preserve stability or passivity of the original system. However, by relaxing the Padé-approximation property (3.2), it is often possible to obtain stable or passive models. More precisely, we call a reduced-order model (2.6) and (2.7) of state-space dimension n an nth *Padé-type model* (at the expansion point s_0) of the original system (2.1) and (2.2) if the Taylor expansions about s_0 of the transfer functions H and H_n of the original system and the reduced-order system agree in a number of leading terms, that is,

$$H(s) = H_n(s) + \mathcal{O}\big((s - s_0)^{q'}\big),$$

where $1 \le q' < q(n)$. Recall that $q(n)$ denotes the optimal approximation order of a true Padé approximant.

Unless $m = p = 1$, the transfer functions H and H_n are matrix-valued, and thus the Padé and Padé-type approximants underlying Padé and Padé-type models are so-called matrix-Padé and matrix-Padé-type approximants in general.

3.4. Explicit moment matching

In this subsection, we restrict ourselves to the single-input single-output case, $m = p = 1$. In this case, in (3.4), both R and L are vectors, and we set $r = R$ and $l = L$. Moreover, we assume that $D = 0$ in (3.4). Thus, (3.4) reduces to the representation

$$H(s) = -l^T \left(I - (s - s_0)M\right)^{-1} r. \tag{3.6}$$

Note that H is a scalar-valued rational function. Correspondingly, the nth Padé approximant H_n defined by (2.9) (with $D_n = 0$) and (3.2) is now also a scalar-valued rational function with numerator and denominator polynomial φ_{n-1} and ψ_n of degree at most $n - 1$ and n, respectively. Instead of (2.9), we represent H_n in terms of these polynomials:

$$H_n(s) = \frac{\varphi_{n-1}(s)}{\psi_n(s)}. \tag{3.7}$$

There are $2n$ free parameters in (3.7), namely the coefficients of the polynomials φ_{n-1} and ψ_n. Except for certain degenerate cases, these parameters can be chosen such that, in (3.2), the first $2n$ moments match:

$$H(s) = H_n(s) + \mathcal{O}\left((s - s_0)^{2n}\right) = \sum_{j=0}^{2n-1} \mu_j \, (s - s_0)^j + \mathcal{O}\left((s - s_0)^{2n}\right).$$

Here, the $\{\mu_j\}$ are the moments defined by the expansion (3.1). Using the representation (3.6) of H, the moments can be expressed as follows:

$$\mu_j = -l^T M^j r, \quad j = 0, 1, 2, \ldots. \tag{3.8}$$

The standard approach to computing H_n is based on the representation (3.7) and on explicit moment generation via (3.8). First, we use (3.8) to compute the leading $2n$ moments,

$$\mu_0, \mu_1, \ldots, \mu_{2n-1}, \tag{3.9}$$

of H, and from these, we then generate the coefficients of the polynomials φ_{n-1} and ψ_n in (3.7) by solving a system of linear equations with a Hankel matrix whose entries are the moments (3.9). This standard approach to computing H_n is employed in the AWE method (Pillage and Rohrer 1990). However, computing Padé approximants using explicit moment computations is inherently numerically unstable, and indeed, in practice, this approach can be employed in a meaningful way only for very moderate values

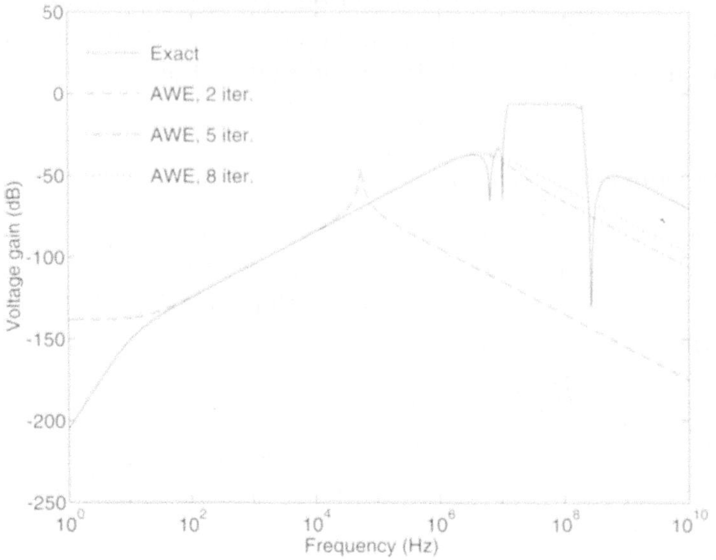

Figure 3.2. Results for simulation of voltage gain with AWE.

of n, such as $n \leq 10$; see Feldmann and Freund (1995a). As we will describe in more detail in Section 5, the numerical problems with AWE can easily be remedied by exploiting the Lanczos–Padé connection (Gragg 1974) and generating the Padé approximant H_n via the classical Lanczos process (Lanczos 1950). This approach was first introduced in Feldmann and Freund (1994) as the *Padé via Lanczos* (PVL) method; see also Gallivan, Grimme and Van Dooren (1994) and Feldmann and Freund (1995a).

While AWE and PVL are mathematically equivalent, their behaviour when run on an actual computer can be vastly different. The reason is that AWE is a numerically unstable algorithm and thus susceptible to round-off errors caused by finite-precision arithmetic. We illustrate the numerical differences between AWE and PVL with a circuit example taken from Feldmann and Freund (1994, 1995a). The circuit simulated here is a voltage filter, where the frequency range of interest is $1 \leq \omega \leq 10^{10}$. This example was first run with AWE, and in Figure 3.2 we show the computed function $|H_n(\mathrm{i}\,\omega)|$, for $n = 2, 5, 8$, together with the exact function $|H(\mathrm{i}\,\omega)|$, each for the frequency range $1 \leq \omega \leq 10^{10}$. Note that H_8 has clearly not yet converged to H. On the other hand, it turned out that the $\{H_n\}$ were hardly changing any more for $n \geq 8$, and in particular, AWE never converged in this example. In Figure 3.3 we show the computed results obtained with PVL for $n = 2, 8, 28$, together with the exact function $|H|$. Note that the results for $n = 8$ (the dotted curves) in Figures 3.1 and 3.2 are vastly different, although they both correspond to the same function H_8. The reason

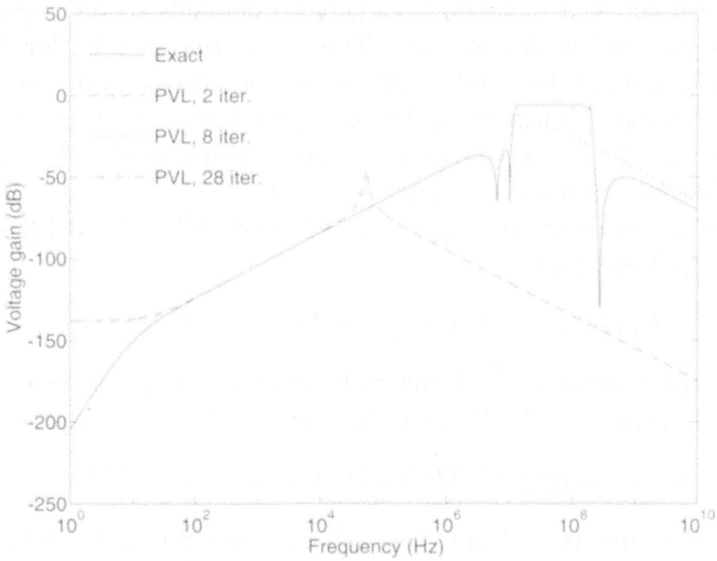

Figure 3.3. Results for simulation of voltage gain with PVL.

for this is that AWE is numerically unstable, while PVL generates H_8 stably. Furthermore, note that PVL converges, with the computed 28th Padé approximant being practically identical to H.

The main reason for the numerical instability of AWE is the explicit generation of the moments (3.9) by means of the formula (3.8). This computation is usually done as follows. We first generate the $2n$ vectors

$$r, M\,r, M^2\,r, \dots, M^{2n-1}\,r, \tag{3.10}$$

and then obtain (3.9) by computing the inner products

$$\mu_j = l^T \cdot (M^j\,r), \quad j = 0, 1, \dots, 2n-1, \tag{3.11}$$

of l with (3.10). An alternative is first to generate the vectors

$$r, M\,r, M^2\,r, \dots, M^{n-1}\,r \quad \text{and} \quad l, M^T\,l, (M^T)^2\,l, \dots, (M^T)^{n-1}\,l, \tag{3.12}$$

and then to obtain (3.9) by computing the inner products

$$\mu_{2j} = \left((M^T)^j\,l\right)^T \cdot (M^j\,r) \quad \text{and} \quad \mu_{2j+1} = \left((M^T)^j\,l\right)^T \cdot \left(M^{j+1}\,r\right) \tag{3.13}$$

for $j = 0, 1, \dots, n-1$. The problem is that the vectors (3.10) quickly converge to an eigenvector corresponding to a dominant eigenvalue of M. As a result, in finite-precision arithmetic, the moments μ_j computed via (3.11), even for rather moderate values of j, contain only information about this dominant eigenvalue. The Padé approximant H_n generated from the moments thus contains only information about part of the spectrum of M.

The transfer function H, however, in general depends on all eigenvalues of M, and not just the dominant ones. This is the reason why, for AWE, the computed H_n usually does not converge to the transfer function H. The alternative approach suffers from the same problem since the two sequences of vectors in (3.12) quickly converge to a right, respectively left, eigenvector corresponding to a dominant eigenvalue of M.

Note that the space spanned by the first set of vectors in (3.12) is just the nth *right Krylov subspace*

$$\mathcal{K}_n(M, r) := \operatorname{span}\{\, r, M\, r, M^2\, r, \dots, M^{n-1}\, r \,\} \tag{3.14}$$

induced by the matrix M and the right starting vector r. Similarly, the second set of vectors in (3.12) spans the nth *left Krylov subspace*

$$\mathcal{K}_n(M^T, l) := \operatorname{span}\{\, l, M^T\, l, (M^T)^2\, l, \dots, (M^T)^{n-1}\, l \,\} \tag{3.15}$$

induced by the matrix M^T and the left starting vector l. While Krylov subspaces are very useful for large-scale matrix computations, the vectors in the definitions (3.14) and (3.15) are in general unsuitable as basis vectors. Indeed, as we just mentioned, they quickly converge, and in particular, they quickly become almost linearly dependent. The remedy is to construct more suitable basis vectors

$$v_1, v_2, \dots, v_n, \dots, \quad \text{and} \quad w_1, w_2, \dots, w_n, \dots, \tag{3.16}$$

such that, for all $n = 1, 2, \dots,$

$$\mathcal{K}_n(M, r) = \operatorname{span}\{\, v_1, v_2, \dots, v_n \,\} \tag{3.17}$$

and

$$\mathcal{K}_n(M^T, l) = \operatorname{span}\{\, w_1, w_2, \dots, w_n \,\}. \tag{3.18}$$

There are two main approaches for constructing basis vectors (3.16), the Lanczos algorithm and the Arnoldi process, which will be discussed in Sections 5 and 6, respectively.

Using the basis vectors (3.16), the explicit moment computations can now easily be avoided. Indeed, instead of the moments (3.9), we now compute so-called *modified moments*

$$w_j^T\, v_j \quad \text{and} \quad w_j^T\, M\, v_j, \quad j = 1, 2, \dots, n. \tag{3.19}$$

In view of (3.17), (3.18), (3.14), (3.15), and (3.13), the modified moments (3.19) contain the very same information as the moments (3.9), and for each $j = 0, 1, \dots, 2n-1$, the jth moment μ_j can be expressed as a suitable linear combination of the numbers (3.19).

4. Stability and passivity

In this section, we discuss the concepts of stability and passivity of linear dynamical systems.

4.1. Stability

An important property of linear dynamical systems is stability. An actual physical system needs to be stable in order to function properly. If a linear dynamical system (2.1) and (2.2) is used as a description of such a physical system, then clearly it should also be stable. Moreover, when a system (2.1) and (2.2) is replaced by a reduced-order model that is then used in a time-domain analysis, the reduced-order model also needs to be stable.

In this subsection, we present a brief discussion of stability of linear descriptor systems. For a more general survey of the various concepts of stability of dynamical systems, we refer the reader to Anderson and Vongpanitlerd (1973) and Willems (1970).

A descriptor system of the form (2.1) and (2.2) is said to be *stable* if its free-response, that is, the solutions $x(t)$, $t \geq 0$, of

$$\begin{aligned} E\frac{\mathrm{d}x}{\mathrm{d}t} &= Ax, \\ x(0) &= x_0, \end{aligned}$$

remain bounded as $t \to \infty$ for any possible initial vector x_0. Recall from the discussion in Section 2.1 that, for singular E, there are certain restrictions on the possible initial vectors x_0.

Stability can easily be characterized in terms of the finite eigenvalues of the matrix pencil $A - sE$; see, *e.g.*, Masubuchi, Kamitane, Ohara and Suda (1997). More precisely, we have the following theorem.

Theorem 4.1. The descriptor system (2.1) and (2.2) is stable if and only if the following two conditions are satisfied:

(i) all finite eigenvalues $s \in \mathbb{C}$ of the matrix pencil $A - sE$ satisfy $\operatorname{Re} s \leq 0$;

(ii) all finite eigenvalues s of $A - sE$ with $\operatorname{Re} s = 0$ are simple.

We stress that, in view of Theorem 4.1, the infinite eigenvalues of the matrix pencil $A - sE$ have no effect on stability. The reason is that these infinite eigenvalues result only in impulsive motions, which go to zero as $t \to \infty$.

Recall that the transfer function H of the descriptor system (2.1) and (2.2) is of the form

$$H(s) = D + C^T \,(sE - A)^{-1} B, \tag{4.1}$$

$$\text{where} \quad A,\, E \in \mathbb{R}^{N \times N},\ B \in \mathbb{R}^{N \times m},\ C \in \mathbb{R}^{N \times m},\ \text{and } D \in \mathbb{R}^{p \times m}. \tag{4.2}$$

Note that any pole of H is necessarily an eigenvalue of the matrix pencil $A - sE$. Hence, it is tempting to determine stability via the poles of H. However, in general, not all eigenvalues of $A - sE$ are poles of H. For example, consider the system

$$\begin{aligned} \frac{\mathrm{d}x}{\mathrm{d}t} &= \begin{bmatrix} 1 & 0 \\ 0 & -1 \end{bmatrix} x + \begin{bmatrix} 0 \\ 1 \end{bmatrix} u(t), \\ y(t) &= \begin{bmatrix} 1 & 1 \end{bmatrix} x(t), \end{aligned}$$

which is taken from Anderson and Vongpanitlerd (1973). The pencil associated with this system is

$$A - sI = \begin{bmatrix} 1-s & 0 \\ 0 & -1-s \end{bmatrix}.$$

Its eigenvalues are ± 1, and hence this system is unstable. The transfer function $H(s) = 1/(s+1)$, however, only has the 'stable' pole -1. Therefore, checking conditions (i) and (ii) of Theorem 4.1 only for the poles of H is, in general, not enough to guarantee stability. In order to infer stability of the system (2.1) and (2.2) from the poles of its transfer function, we need an additional condition, which we formulate next.

Let H be a given $(p \times m)$-matrix-valued rational function. Any representation of H of the form (4.1) with matrices (4.2) is called a *realization* of H. Furthermore, a realization (4.1) of H is said to be *minimal* if the dimension N of the matrices (4.2) is as small as possible. We will also say that the state-space description (2.1) and (2.2) is a minimal realization if its transfer function (4.1) is a minimal realization.

The following theorem is the well-known characterization of minimal realizations in terms of conditions on the matrices (4.2); see, *e.g.*, Verghese *et al.* (1981). We also refer the reader to the related results on controllability, observability, and minimal realizations of descriptor systems given in Chapter 2 of Dai (1989).

Theorem 4.2. Let H be a $(p \times m)$-matrix-valued rational function given by a realization (4.1). Then, (4.1) is a minimal realization of H if and only if the matrices (4.2) satisfy the following five conditions:

(i) $\operatorname{rank} \begin{bmatrix} A - sE & B \end{bmatrix} = N$ for all $s \in \mathbb{C}$ (finite controllability);
(ii) $\operatorname{rank} \begin{bmatrix} E & B \end{bmatrix} = N$ (infinite controllability);
(iii) $\operatorname{rank} \begin{bmatrix} A^T - sE^T & C \end{bmatrix} = N$ for all $s \in \mathbb{C}$ (finite observability);
(iv) $\operatorname{rank} \begin{bmatrix} E^T & C \end{bmatrix} = N$ (infinite observability);
(v) $A \ker(E) \subseteq \operatorname{Im}(E)$ (absence of nondynamic modes).

For descriptor systems given by a minimal realization, stability can indeed be checked via the poles of its transfer function.

Theorem 4.3. Let (2.1) and (2.2) be a minimal realization of a descriptor system, and let H be its transfer function (4.1). Then, the descriptor system (2.1) and (2.2) is stable if and only if all finite poles s_i of H satisfy $\operatorname{Re} s_i \leq 0$ and any pole with $\operatorname{Re} s_i = 0$ is simple.

4.2. Passivity

In circuit simulation, reduced-order modelling is often applied to large passive linear subcircuits, such as RCL networks consisting of only resistors, capacitors, and inductors. When reduced-order models of such subcircuits are used within a simulation of the whole circuit, stability of the overall simulation can only be guaranteed if the reduced-order models preserve the passivity of the original subcircuits; see, *e.g.*, Chirlian (1967), Rohrer and Nosrati (1981), and Lozano, Brogliato, Egeland and Maschke (2000). Therefore, it is important to have techniques to check the passivity of a given reduced-order model.

Roughly speaking, a system is *passive* if it does not generate energy. For descriptor systems of the form (2.1) and (2.2), passivity is equivalent to positive realness of the transfer function. Moreover, such systems can only be passive if they have identical numbers of inputs and outputs. Thus, for the remainder of this subsection, we assume that $m = p$. Then, a system described by (2.1) and (2.2) is passive, that is, it does not generate energy, if and only if its transfer function (4.1) is *positive real*; see, *e.g.*, Anderson and Vongpanitlerd (1973). A precise definition of positive realness is as follows.

Definition 1. An $(m \times m)$-matrix-valued function $H : \mathbb{C} \mapsto (\mathbb{C} \cup \infty)^{m \times m}$ is called *positive real* if the following three conditions are satisfied:

(i) H is analytic in $\mathbb{C}_+ := \{\, s \in \mathbb{C} \mid \operatorname{Re} s > 0 \,\}$;

(ii) $H(\overline{s}) = \overline{H(s)}$ for all $s \in \mathbb{C}$;

(iii) $H(s) + (H(s))^H \succeq 0$ for all $s \in \mathbb{C}_+$.

In Definition 1 and hereafter, the notation $M \succeq 0$ means that the matrix M is Hermitian positive semi-definite. Similarly, $M \preceq 0$ means that M is Hermitian negative semi-definite.

For transfer functions H of the form (4.1), condition (ii) of Definition 1 is always satisfied since the matrices (4.2) are assumed to be real. Furthermore, condition (i) simply means that H cannot have poles in $\mathbb{C}_+$, and this can be checked easily. For the special case $m = 1$ of scalar-valued functions H, condition (iii) states that the real part of $H(s)$ is nonnegative for all s with nonnegative real part. In order to check this condition, it is sufficient to show that the real part of $H(s)$ is nonnegative for all purely imaginary s. This can be done by means of relatively elementary means. For example, in Bai and Freund (2000), a procedure based on eigenvalue computations is proposed. For the general matrix-valued case, $m \geq 1$, however, checking condition (iii) is much more involved. One possibility is to employ a suitable extension of the classical positive real lemma (Anderson 1967, Anderson and Vongpanitlerd 1973, Zhou, Doyle and Glover 1996) that characterizes positive realness of regular linear systems via the solvability of certain linear matrix inequalities (LMIs). Such a version of the positive real lemma for general descriptor systems is stated in Theorem 4.4 below.

We remark that any matrix-valued rational function H has an expansion about $s = \infty$ of the form

$$H(s) = \sum_{j=-\infty}^{j_0} M_j s^j, \tag{4.3}$$

where $j_0 \geq 0$ is an integer. Moreover, the function H has a pole at $s = \infty$ if and only if $j_0 \geq 1$ and $M_{j_0} \neq 0$ in (4.3).

A suitable extension of the classical positive real lemma for regular systems to descriptor systems can now be stated as follows.

Theorem 4.4. (Positive real lemma for descriptor systems) Let H be a real $(m \times m)$-matrix-valued rational function of the form (4.1) with matrices (4.2).

(a) (Sufficient condition.) If the LMIs

$$\begin{bmatrix} A^T X + X^T A & X^T B - C \\ B^T X - C^T & -D - D^T \end{bmatrix} \preceq 0 \quad \text{and} \quad E^T X = X^T E \succeq 0 \tag{4.4}$$

have a solution $X \in \mathbb{R}^{N \times N}$, then H is positive real.

(b) (Necessary condition.) Suppose that (4.1) is a minimal realization of H and that the matrix M_0 in the expansion (4.3) satisfies

$$(D - M_0) + (D - M_0)^T \succeq 0. \tag{4.5}$$

If H is positive real, then there exists a solution $X \in \mathbb{R}^{N \times N}$ of the LMIs (4.4).

A proof of Theorem 4.4 can be found in Freund and Jarre (2000).

The result of Theorem 4.4 allows us to check positive realness by solving semi-definite programming problems of the form (4.4). Note that there are N^2 unknowns in (4.4), namely the entries of the $N \times N$ matrix X. Problems of the form (4.4) can be tackled with interior-point methods; see, *e.g.*, Boyd, El Ghaoui, Feron and Balakrishnan (1994) and Freund and Jarre (2003). However, the computational complexity of these methods grows quickly with N, and thus, these methods are viable only for rather small values of N. On the other hand, it is usually known whether a given system is passive, and the need to numerically check passivity mainly arises for reduced-order models of a given passive model. In this case, the dimension N of the semi-definite programming problem (4.4) is equal to the dimension of the reduced-order model, which is usually small enough for techniques based on Theorem 4.4 to become feasible.

For the special case $E = I$, the result of Theorem 4.4 is just the classical positive real lemma (Anderson 1967, Anderson and Vongpanitlerd 1973, Zhou *et al.* 1996). In this case, (4.4) reduces to the problem of finding a symmetric positive semi-definite matrix $X \in \mathbb{R}^{N\times N}$ such that

$$\begin{bmatrix} A^T X + XA & XB - C \\ B^T X - C^T & -D - D^T \end{bmatrix} \preceq 0.$$

Moreover, if $E = I$, the condition (4.5) is always satisfied, since in this case $M_0 = 0$ and $D + D^T \succeq 0$.

4.3. Linear RCL subcircuits

In circuit simulation, an important special case of passive circuits is linear subcircuits that consist only of resistors, capacitors, and inductors. Such linear RCL subcircuits arise in the modelling of a circuit's interconnect and pin package; see, *e.g.*, Cheng *et al.* (2000), Freund and Feldmann (1997, 1998), Kim, Gopal and Pillage (1994), and Pileggi (1995).

The equations describing linear RCL subcircuits are of the form (2.1) and (2.2) with $D = 0$ and $m = p$. Furthermore, the equations can be formulated such that the matrices A, $E \in \mathbb{R}^{N\times N}$ in (2.1) are symmetric and exhibit a block structure; see Freund and Feldmann (1996*a*, 1998). More precisely, we have

$$A = A^T = \begin{bmatrix} -A_{11} & A_{12} \\ A_{12}^T & 0 \end{bmatrix} \quad \text{and} \quad E = E^T = \begin{bmatrix} E_{11} & 0 \\ 0 & -E_{22} \end{bmatrix}, \tag{4.6}$$

where the submatrices A_{11}, $E_{11} \in \mathbb{R}^{N_1\times N_1}$ and $E_{22} \in \mathbb{R}^{N_2\times N_2}$ are symmetric positive semi-definite, and $N = N_1 + N_2$. Note that, except for the special case $N_2 = 0$, the matrices A and E are indefinite. The special case $N_2 = 0$ arises for RC subcircuits that contain only resistors and capacitors, but no inductors.

If the RCL subcircuit is viewed as an m-terminal component with m inputs and $m = p$ outputs, then the matrices B and C in (2.1) and (2.2) are identical and of the form

$$B = C = \begin{bmatrix} B_1 \\ 0 \end{bmatrix} \quad \text{with} \quad B_1 \in \mathbb{R}^{N_1\times m}. \tag{4.7}$$

In view of (4.6) and (4.7), the transfer function of such an m-terminal RCL subcircuit is given by

$$H(s) = B^T(sE - A)^{-1}B, \quad \text{where} \quad A = A^T, \quad E = E^T. \tag{4.8}$$

We call a transfer function H *symmetric* if it is of the form (4.8) with real matrices A, E, and B.

We will also use the following nonsymmetric formulation of (4.8). Let J be the block matrix

$$J = \begin{bmatrix} I_{N_1} & 0 \\ 0 & -I_{N_2} \end{bmatrix}, \tag{4.9}$$

where I_{N_1} and I_{N_2} are the $N_1 \times N_1$ and $N_2 \times N_2$ identity matrix, respectively.

Note that, by (4.7) and (4.9), we have $B = JB$. Using this relation, as well as (4.6), we can rewrite (4.8) as follows:

$$\begin{gathered} H(s) = B^T \bigl(s\widetilde{E} - \widetilde{A}\bigr)^{-1} B, \\ \text{where} \quad \widetilde{A} = \begin{bmatrix} -A_{11} & A_{12} \\ -A_{12}^T & 0 \end{bmatrix}, \quad \widetilde{E} = \begin{bmatrix} E_{11} & 0 \\ 0 & E_{22} \end{bmatrix}. \end{gathered} \tag{4.10}$$

In this formulation, the matrix $\widetilde{A}$ is no longer symmetric, but now

$$\widetilde{A} + \widetilde{A}^T \preceq 0 \quad \text{and} \quad \widetilde{E} \succeq 0. \tag{4.11}$$

It turns out that the properties (4.11) are the key to ensure positive realness. Indeed, in Freund (2000*b*), we established the following result.

Theorem 4.5. Let $\widetilde{A}$, $\widetilde{E} \in \mathbb{R}^{N \times N}$, and $B \in \mathbb{R}^{N \times m}$. Assume that $\widetilde{A}$ and $\widetilde{E}$ satisfy (4.11), and that the matrix pencil $\widetilde{A} - s\widetilde{E}$ is regular. Then, the $(m \times m)$-matrix-valued function

$$H(s) = B^T \bigl(s\widetilde{E} - \widetilde{A}\bigr)^{-1} B$$

is positive real.

5. Approaches based on Lanczos-type methods

In this section, we discuss the use of Lanczos-type methods for the construction of Padé and Padé-type reduced-order models of time-invariant linear dynamical systems.

5.1. Block Krylov subspaces

We consider general descriptor systems of the form (2.1) and (2.2). As discussed in Section 3.2, the key to using Krylov-subspace techniques for reduced-order modelling of such systems is to first replace the matrix pair A and E by a single matrix M. To this end, let $s_0 \in \mathbb{C}$ be any given point such that the matrix $A - s_0 E$ is nonsingular. Then, with M, R, and L denoting the matrices defined in (3.5), the linear system (2.1) and (2.2) can be rewritten in the following form:

$$M\frac{\mathrm{d}\tilde{x}}{\mathrm{d}t} = (I + s_0 M)\,\tilde{x} + Ru(t), \tag{5.1}$$

$$y(t) = L^T\tilde{x}(t) + Du(t). \tag{5.2}$$

Here, $\tilde{x} := F_2 x$, where F_2 is the matrix from the factorization (3.3). Note that $M \in \mathbb{C}^{N\times N}$, $R \in \mathbb{C}^{N\times m}$, and $L \in \mathbb{C}^{N\times p}$, where N is the state-space dimension of the system, m is the number of inputs, and p is the number of outputs.

The transfer function H of the rewritten system (5.1) and (5.2) is given by (3.4). By expanding (3.4) about s_0, we obtain

$$H(s) = D - \sum_{j=0}^{\infty} L^T M^j R\,(s - s_0)^j. \tag{5.3}$$

Recall from Section 3 that Padé and Padé-type reduced-order models are defined via the leading coefficients of an expansion of H about s_0. In view of (5.3), the jth coefficient of such an expansion can be expressed as follows:

$$-L^T M^j R = -\left((M^{j-i})^T L\right)^T \left(M^i R\right), \quad i = 0, 1, \ldots, j. \tag{5.4}$$

Notice that the factors on the right-hand side of (5.4) are blocks of the *right* and *left block Krylov matrices*

$$\begin{aligned} &\begin{bmatrix} R & MR & M^2R & \cdots & M^iR & \cdots \end{bmatrix} \\ \text{and} \quad &\begin{bmatrix} L & M^TL & \left(M^T\right)^2L & \cdots & \left(M^T\right)^kL & \cdots \end{bmatrix}, \end{aligned} \tag{5.5}$$

respectively. As a result, all the information needed to generate Padé and Padé-type reduced-order models is contained in the block Krylov matrices (5.5). However, simply computing the blocks M^iR and $\left(M^T\right)^iL$ in (5.5) and then generating the leading coefficients of the expansion (5.3) from these blocks is not a viable numerical procedure. The reason is that, in finite-precision arithmetic, as i increases, the blocks M^iR and $\left(M^T\right)^iL$ quickly contain only information about the eigenspaces of the dominant eigenvalue of M. Instead, we need to employ suitable Krylov-subspace methods that generate numerically better basis vectors for the subspaces associated with the block Krylov matrices (5.5).

Next, we give a formal definition of the subspaces induced by (5.5). For the special case $m = p = 1$ of single-input single-output systems, the 'blocks' of the Krylov matrices (5.5) reduce to vectors, and the Krylov subspaces spanned by these vectors are just the standard Krylov subspaces that we introduced in (3.17) and (3.18). For the general case $m,\ p \geq 1$ of multi-input multi-output systems, however, the definition of subspaces induced by (5.5) is more involved. First, note that each block M^iR consists of m column vectors of length N. By scanning these column vectors of the right block Krylov matrix in (5.5) from left to right and by deleting any column that is linearly dependent on columns to its left, we obtain the *deflated* right block Krylov matrix

$$\begin{bmatrix} R_1 & MR_2 & M^2R_3 & \cdots & M^{i_{\max}-1}R_{i_{\max}} \end{bmatrix}. \tag{5.6}$$

This process of detecting and deleting the linearly dependent columns is called *exact deflation*. We remark that the matrix (5.6) is finite, since at most N of the column vectors can be linearly independent. Furthermore, a column $M^i r$ being linearly dependent on columns to its left in (5.5) implies that any column $M^{i'} r$, $i' \geq i$, is linearly dependent on columns to its right. Therefore, in (5.6), for each $i = 1, 2, \ldots, i_{\max}$, the matrix R_i is a submatrix of R_{i-1}, where, for $i = 1$, we set $R_0 = R$.

Let m_i denote the number of columns of R_i. The matrix (5.6) thus has

$$n_{\max}^{(\mathrm{r})} := m_1 + m_2 + \cdots + m_{i_{\max}},$$

columns. For each integer n with $1 \leq n \leq n_{\max}^{(\mathrm{r})}$, we define the nth *right block Krylov subspace* $\mathcal{K}_n(M, R)$ (induced by M and R) as the subspace spanned by the first n columns of the deflated right block Krylov matrix (5.6).

Analogously, by deleting the linearly independent columns of the left block Krylov matrix in (5.5), we obtain a deflated left block Krylov matrix of the form

$$\begin{bmatrix} L_1 & M^T L_2 & \left(M^T\right)^2 L_3 & \cdots & \left(M^T\right)^{i_{\max}-1} L_{k_{\max}} \end{bmatrix}. \tag{5.7}$$

Let $n_{\max}^{(\mathrm{l})}$ be the number of columns of the matrix (5.7). Then, for each integer n with $1 \leq n \leq n_{\max}^{(\mathrm{l})}$, we define the nth *left block Krylov subspace* $\mathcal{K}_n(M^T, L)$ (induced by M^T and L) as the subspace spanned by the first n columns of the deflated left block Krylov matrix (5.7).

For a more detailed discussion of block Krylov subspaces and deflation, we refer the reader to Aliaga, Boley, Freund and Hernández (2000) and Freund (2000*b*).

Next, we discuss reduced-order modelling approaches that employ Lanczos and Lanczos-type methods for the construction of suitable basis vectors for the right and left block Krylov subspaces $\mathcal{K}_n(M, R)$ and $\mathcal{K}_n(M^T, L)$.

5.2. The MPVL algorithm

For the special case $m = p = 1$ of single-input single-output linear dynamical systems, each of the 'blocks' R and L only consists of a single vector, say r and l, and $\mathcal{K}_n(M, r)$ and $\mathcal{K}_n(M^T, l)$ are just the standard nth right and left Krylov subspaces induced by single vectors. The classical Lanczos process (Lanczos 1950) is a well-known procedure for computing two sets of bi-orthogonal basis vectors for $\mathcal{K}_n(M, r)$ and $\mathcal{K}_n(M^T, l)$. Moreover, these vectors are generated by means of three-term recurrences the coefficients of which define a tridiagonal matrix T_n. It turns out that T_n contains all the information that is needed to set up an nth Padé reduced-order model of a given single-input single-output time-invariant linear dynamical system. The associated computational procedure is called the *Padé via Lanczos* (PVL) algorithm (Feldmann and Freund 1994, 1995a).

Here, we describe in some detail an extension of the PVL algorithm to the case of general m-input p-output time-invariant linear dynamical systems. The underlying block Krylov subspace method is the *nonsymmetric band Lanczos algorithm* (Freund 2000a) for constructing two sets of right and left Lanczos vectors,

$$v_1, v_2, \ldots, v_n \quad \text{and} \quad w_1, w_2, \ldots, w_n, \tag{5.8}$$

respectively. These vectors span the nth right and left block Krylov subspaces (induced by M and R, and M^T and L, respectively):

$$\begin{aligned} \operatorname{span}\{\, v_1, v_2, \ldots, v_n \,\} &= \mathcal{K}_n(M, R) \\ \text{and} \quad \operatorname{span}\{\, w_1, w_2, \ldots, w_n \,\} &= \mathcal{K}_n(M^T, L). \end{aligned} \tag{5.9}$$

Moreover, the vectors (5.8) are constructed to be bi-orthogonal:

$$w_j^T v_k = \begin{cases} 0 & \text{if } j \neq k, \\ \delta_j & \text{if } j = k, \end{cases} \qquad \text{for all} \quad j, k = 1, 2, \ldots, n. \tag{5.10}$$

It turns out that the Lanczos vectors (5.8) can be constructed by means of recurrence relations of length at most $m+p+1$. The recurrence coefficients for the first n right Lanczos vectors define an $n \times n$ matrix $T_n^{(\mathrm{pr})}$ that is 'essentially' a band matrix with total bandwidth $m+p+1$. Similarly, the recurrence coefficients for the first n left Lanczos vectors define an $n \times n$ band matrix $\widetilde{T}_n^{(\mathrm{pr})}$ with total bandwidth $m+p+1$. For a more detailed discussion of the structure of $T_n^{(\mathrm{pr})}$ and $\widetilde{T}_n^{(\mathrm{pr})}$, we refer the reader to Aliaga *et al.* (2000) and Freund (2000a).

Algorithm 5.1 below gives a complete description of the numerical procedure that generates the Lanczos vectors (5.8) with properties (5.9) and (5.10). In order to obtain a Padé reduced-order model based on this algorithm, we do not need the Lanczos vectors themselves, but rather the matrix of right recurrence coefficients $T_n^{(\mathrm{pr})}$, the matrices $\rho_n^{(\mathrm{pr})}$ and $\eta_n^{(\mathrm{pr})}$ that contain the recurrence coefficients from processing the starting blocks R and L, respectively, and the diagonal matrix

$$\Delta_n = \operatorname{diag}(\delta_1, \delta_2, \ldots, \delta_n),$$

whose diagonal entries are the δ_j's from (5.10). The following algorithm produces the matrices $T_n^{(\mathrm{pr})}$, $\rho_n^{(\mathrm{pr})}$, $\eta_n^{(\mathrm{pr})}$, and Δ_n as output.

Algorithm 5.1. (Nonsymmetric band Lanczos algorithm)

INPUT: A matrix $M \in \mathbb{C}^{N \times N}$;

A block of m right starting vectors $R = \begin{bmatrix} r_1 & r_2 & \cdots & r_m \end{bmatrix} \in \mathbb{C}^{N \times m}$;

A block of p left starting vectors $L = \begin{bmatrix} l_1 & l_2 & \cdots & l_p \end{bmatrix} \in \mathbb{C}^{N \times p}$.

OUTPUT: The $n \times n$ Lanczos matrix $T_n^{(\mathrm{pr})}$, and the matrices $\rho_n^{(\mathrm{pr})}$, $\eta_n^{(\mathrm{pr})}$, and Δ_n.

(0) For $k = 1, 2, \ldots, m$, set $\hat{v}_k = r_k$.
For $k = 1, 2, \ldots, p$, set $\hat{w}_k = l_k$.
Set $m_{\mathrm{c}} = m$, $p_{\mathrm{c}} = p$, and $\mathcal{I}_v = \mathcal{I}_w = \emptyset$.

For $n = 1, 2, \ldots,$ until convergence or $m_{\mathrm{c}} = 0$ or $p_{\mathrm{c}} = 0$ or $\delta_n = 0$ do:

(1) (If necessary, deflate $\hat{v}_n$.)
Compute $\|\hat{v}_n\|_2$.
Decide if $\hat{v}_n$ should be deflated. If yes, do the following:

(a) Set $\hat{v}_{n-m_{\mathrm{c}}}^{\mathrm{defl}} = \hat{v}_n$ and store this vector. Set $\mathcal{I}_v = \mathcal{I}_v \cup \{\, n - m_{\mathrm{c}} \,\}$.
(b) Set $m_{\mathrm{c}} = m_{\mathrm{c}} - 1$. If $m_{\mathrm{c}} = 0$, set $n = n - 1$ and stop.
(c) For $k = n, n+1, \ldots, n + m_{\mathrm{c}} - 1$, set $\hat{v}_k = \hat{v}_{k+1}$.
(d) Repeat all of step (1).

(2) (If necessary, deflate $\hat{w}_n$.)
Compute $\|\hat{w}_n\|_2$.
Decide if $\hat{w}_n$ should be deflated. If yes, do the following:

(a) Set $\hat{w}_{n-p_{\mathrm{c}}}^{\mathrm{defl}} = \hat{w}_n$ and store this vector. Set $\mathcal{I}_w = \mathcal{I}_w \cup \{\, n - p_{\mathrm{c}} \,\}$.
(b) Set $p_{\mathrm{c}} = p_{\mathrm{c}} - 1$. If $p_{\mathrm{c}} = 0$, set $n = n - 1$ and stop.
(c) For $k = n, n+1, \ldots, n + p_{\mathrm{c}} - 1$, set $\hat{w}_k = \hat{w}_{k+1}$.
(d) Repeat all of step (2).

(3) (Normalize $\hat{v}_n$ and $\hat{w}_n$ to obtain v_n and w_n.)
Set

$$t_{n,n-m_{\mathrm{c}}} = \|\hat{v}_n\|_2, \quad \tilde{t}_{n,n-p_{\mathrm{c}}} = \|\hat{w}_n\|_2,$$
$$v_n = \frac{\hat{v}_n}{t_{n,n-m_{\mathrm{c}}}}, \quad \text{and} \quad w_n = \frac{\hat{w}_n}{\tilde{t}_{n,n-p_{\mathrm{c}}}}.$$

(4) (Compute δ_n and check for possible breakdown.)
Set $\delta_n = w_n^T v_n$. If $\delta_n = 0$, set $n = n - 1$ and stop.

(5) (Orthogonalize the right candidate vectors against w_n.)
For $k = n+1, n+2, \ldots, n+m_c-1$, set

$$t_{n,k-m_c} = \frac{w_n^T \hat{v}_k}{\delta_n} \quad \text{and} \quad \hat{v}_k = \hat{v}_k - v_n\, t_{n,k-m_c}.$$

(6) (Orthogonalize the left candidate vectors against v_n.)
For $k = n+1, n+2, \ldots, n+p_c-1$, set

$$\tilde{t}_{n,k-p_c} = \frac{\hat{w}_k^T v_n}{\delta_n} \quad \text{and} \quad \hat{w}_k = \hat{w}_k - w_n\, \tilde{t}_{n,k-p_c}.$$

(7) (Advance the right block Krylov subspace to get $\hat{v}_{n+m_c}$.)

(a) Set $\hat{v}_{n+m_c} = M\, v_n$.

(b) For $k \in \mathcal{I}_w$ (in ascending order), set

$$\tilde{\sigma} = \left(\hat{w}_k^{\text{defl}}\right)^T v_n, \quad \tilde{t}_{n,k} = \frac{\tilde{\sigma}}{\delta_n},$$

and, if $k > 0$, set

$$t_{k,n} = \frac{\tilde{\sigma}}{\delta_k} \quad \text{and} \quad \hat{v}_{n+m_c} = \hat{v}_{n+m_c} - v_k\, t_{k,n}.$$

(c) Set $k_v = \max\{\,1,\, n-p_c\,\}$.

(d) For $k = k_v, k_v+1, \ldots, n-1$, set

$$t_{k,n} = \tilde{t}_{n,k}\,\frac{\delta_n}{\delta_k} \quad \text{and} \quad \hat{v}_{n+m_c} = \hat{v}_{n+m_c} - v_k\, t_{k,n}.$$

(e) Set

$$t_{n,n} = \frac{w_n^T \hat{v}_{n+m_c}}{\delta_n} \quad \text{and} \quad \hat{v}_{n+m_c} = \hat{v}_{n+m_c} - v_n\, t_{n,n}.$$

(8) (Advance the left block Krylov subspace to get $\hat{w}_{n+p_c}$.)

(a) Set $\hat{w}_{n+p_c} = M^T w_n$.

(b) For $k \in \mathcal{I}_v$ (in ascending order), set

$$\sigma = w_n^T \hat{v}_k^{\text{defl}}, \quad t_{n,k} = \frac{\sigma}{\delta_n},$$

and, if $k > 0$, set

$$\tilde{t}_{k,n} = \frac{\sigma}{\delta_k} \quad \text{and} \quad \hat{w}_{n+p_c} = \hat{w}_{n+p_c} - w_k\, \tilde{t}_{k,n}.$$

(c) Set $k_w = \max\{\,1,\, n-m_c\,\}$.

(d) For $k = k_w, k_w+1, \ldots, n-1$, set

$$\tilde{t}_{k,n} = t_{n,k}\,\frac{\delta_n}{\delta_k} \quad \text{and} \quad \hat{w}_{n+p_c} = \hat{w}_{n+p_c} - w_k\, \tilde{t}_{k,n}.$$

(e) Set

$$\tilde{t}_{n,n} = t_{n,n} \quad \text{and} \quad \hat{w}_{n+p_c} = \hat{w}_{n+p_c} - w_n \, \tilde{t}_{n,n}.$$

(9) Set

$$T_n^{(\mathrm{pr})} = \big[t_{i,k}\big]_{i,k=1,2,\ldots,n},$$

$$\rho_n^{(\mathrm{pr})} = \big[t_{i,k-m}\big]_{i=1,2,\ldots,n;\, k=1,2,\ldots,k_\rho} \quad \text{where} \quad k_\rho = m + \min\{\, 0,\, n - m_c\,\},$$

$$\eta_n^{(\mathrm{pr})} = \big[\tilde{t}_{i,k-p}\big]_{i=1,2,\ldots,n;\, k=1,2,\ldots,k_\eta} \quad \text{where} \quad k_\eta = p + \min\{\, 0,\, n - p_c\,\},$$

$$\Delta_n = \operatorname{diag}\big(\delta_1, \delta_2, \ldots, \delta_n\big).$$

(10) Check if n is large enough. If yes, stop.

Remark 1. When applied to single starting vectors, that is, for the special case $m = p = 1$, Algorithm 5.1 reduces to the classical nonsymmetric Lanczos process (Lanczos 1950).

Remark 2. It can be shown that, at step n of Algorithm 5.1, exact deflation of a vector in the right, respectively left, block Krylov matrix (5.5) occurs if and only if $\hat{v}_n = 0$, respectively $\hat{w}_n = 0$, in step (1), respectively step (2). Therefore, to run Algorithm 5.1 with exact deflation only, we deflate $\hat{v}_n$ if $\|\hat{v}_n\|_2 = 0$ in step (1), and we deflate $\hat{w}_n$ if $\|\hat{w}_n\|_2 = 0$ in step (2). In finite-precision arithmetic, however, so-called *inexact deflation* is employed. This means that, in step (1), $\hat{v}_n$ is deflated if $\|\hat{v}_n\|_2 \le \varepsilon$, and, in step (2), $\hat{w}_n$ is deflated if $\|\hat{w}_n\|_2 \le \varepsilon$, where $\varepsilon = \varepsilon(M) > 0$ is a suitably chosen small constant.

Remark 3. The occurrence of $\delta_n = 0$ in step (4) of Algorithm 5.1 is called a *breakdown*. In finite-precision arithmetic, in step (4) we should also check for *near-breakdowns*, that is, if $\delta_n \approx 0$. In general, it cannot be excluded that breakdowns or near-breakdowns occur, although they are very unlikely. Furthermore, by using so-called *look-ahead* techniques, it is possible to remedy the problem of possible breakdowns or near-breakdowns. For the sake of simplicity, we have stated the band Lanczos algorithm without look-ahead only. A look-ahead version of Algorithm 5.1 is described in Aliaga *et al.* (2000).

The matrices $T_n^{(\mathrm{pr})}$, $\rho_n^{(\mathrm{pr})}$, and $\eta_n^{(\mathrm{pr})}$ produced by Algorithm 5.1 can be viewed as oblique projections of the input data M, R, and L onto the right block Krylov subspace $\mathcal{K}_n(M, R)$ and orthogonally to the left block Krylov subspace $\mathcal{K}_n(M^T, L)$. To give a precise statement of these projection properties, we let

$$V_n := \begin{bmatrix} v_1 & v_2 & \cdots & v_n \end{bmatrix} \quad \text{and} \quad W_n := \begin{bmatrix} w_1 & w_2 & \cdots & w_n \end{bmatrix} \tag{5.11}$$

denote the matrices whose columns are the first n right and left Lanczos vectors, respectively. Then the matrices $T_n^{(\mathrm{pr})}$, $\rho_n^{(\mathrm{pr})}$, $\eta_n^{(\mathrm{pr})}$, and Δ_n generated by Algorithm 5.1 are related to the input data M, R, and L as follows:

$$\begin{aligned} T_n^{(\mathrm{pr})} &= \Delta_n^{-1} W_n^T M V_n, \\ \rho_n^{(\mathrm{pr})} &= \Delta_n^{-1} W_n^T R, \\ \eta_n^{(\mathrm{pr})} &= \Delta_n^{-T} V_n^T L, \\ \Delta_n &= W_n^T V_n. \end{aligned} \tag{5.12}$$

The relations (5.12) can be employed to set up a reduced-order model of dimension n of the linear system (5.1) and (5.2). To this end, we restrict the state vector $\tilde{x}(t)$ in (5.1) and (5.2) to vectors in $\mathcal{K}_n(M, R)$. In view of (5.11), these restricted vectors can be written as

$$\tilde{x}(t) = V_n z(t), \tag{5.13}$$

where $z(t)$ has length n. By applying the oblique projections stated in (5.12) to the linear dynamical system (5.1) and (5.2), and by using (5.13), we obtain the following reduced-order model:

$$T_n^{(\mathrm{pr})} \frac{\mathrm{d}z}{\mathrm{d}t} = \left(I + s_0 T_n^{(\mathrm{pr})}\right) z + \rho_n^{(\mathrm{pr})} u(t), \tag{5.14}$$

$$y(t) = \left(\eta_n^{(\mathrm{pr})}\right)^T \Delta_n z(t) + D u(t). \tag{5.15}$$

Note that the transfer function of this reduced-order model is given by

$$H_n(s) = D - \left(\eta_n^{(\mathrm{pr})}\right)^T \Delta_n \left(I - (s - s_0) T_n^{(\mathrm{pr})}\right)^{-1} \rho_n^{(\mathrm{pr})}. \tag{5.16}$$

The *matrix-Padé via Lanczos* (MPVL) algorithm (Feldmann and Freund 1995*b*, Freund 1995) consists of applying Algorithm 5.1 to the matrices M, R, and L defined in (3.5), and running it for n steps. The matrices $T_n^{(\mathrm{pr})}$, $\rho_n^{(\mathrm{pr})}$, $\eta_n^{(\mathrm{pr})}$, and Δ_n produced by Algorithm 5.1 are then used to set up the reduced-order model (5.14) and (5.15) of the original linear dynamical system (2.1) and (2.2).

It turns out that the reduced-order model (5.14) and (5.15) is indeed a matrix-Padé model of the original system.

Theorem 5.2. (Matrix-Padé model) Suppose that Algorithm 5.1 is run with exact deflation only and that $n \geq \max\{m, p\}$. Then, the reduced-order model (5.14) and (5.15) is a matrix-Padé model of the linear dynamical system (2.1) and (2.2). More precisely, the Taylor expansions about s_0 of the transfer functions H (2.8) and H_n (5.16) agree to as many leading coefficients as possible, that is,

$$H(s) = H_n(s) + \mathcal{O}\left((s - s_0)^{q(n)}\right),$$

where $q(n)$ is as large as possible. In particular,

$$q(n) \geq \left\lfloor \frac{n}{m} \right\rfloor + \left\lfloor \frac{n}{p} \right\rfloor.$$

A proof of Theorem 5.2 is given in Freund (1995). Earlier related results, which required additional assumptions, can be found in de Villemagne and Skelton (1987) and Feldmann and Freund (1995*b*).

5.3. A connection with shifted Krylov-subspace solvers

The representation (3.4) of the transfer function H suggests to employ the machinery of shifted Krylov-subspace methods (Freund 1993) for reduced-order modelling. Indeed, let us define the new variable

$$\sigma(s) := \frac{1}{s - s_0}, \quad s \in \mathbb{C}, \quad s \neq s_0. \tag{5.17}$$

Using (5.17), we can rewrite (3.4) as follows:

$$H(s) = D + \sigma(s)\, L^T (M - \sigma(s)\, I)^{-1} R, \quad s \neq s_0. \tag{5.18}$$

For any $\sigma \in \mathbb{C}$ that is not an eigenvalue of the matrix M, let $X(\sigma)$ denote the unique solution of the block linear system

$$(M - \sigma I) X(\sigma) = R. \tag{5.19}$$

By (5.18) and (5.19), we have

$$H(s) = D + \sigma(s)\, L^T X\,(\sigma(s)) \tag{5.20}$$

for any $s \in \mathbb{C}$ such that $s \neq s_0$ and $\sigma(s)$ is not an eigenvalue of M.

In view of (5.20), we can compute the values $H(s)$ via solution of block linear systems of the form (5.19). Furthermore, (5.19) is a family of shifted systems, that is, the coefficient matrices of (5.19) differ from the fixed matrix M only by scalar multiples of the identity matrix. It is well known that Krylov-subspace methods for the solution of linear equations can exploit this shift structure; see, *e.g.*, Freund (1993) and the references given there. The basic observation is that Krylov subspaces are invariant under additive shifts by scalar multiples of the identity matrix. The underlying Krylov-subspace method thus has to be run only once, and approximate solutions of any shifted system can then be obtained by solving small shifted problems.

Next, we describe one such method, namely a variant of the block *bi-conjugate gradient* (BCG) method (O'Leary 1980), in a little more detail. Our variant of block BCG is based on the band Lanczos method. Recall that, after n steps, Algorithm 5.1 (applied to the matrices M, R, and L) has generated the matrices $T_n^{(\mathrm{pr})}$, $\rho_n^{(\mathrm{pr})}$, and $\eta_n^{(\mathrm{pr})}$ and that these satisfy (5.12). In terms of these matrices, the nth block BCG iterate, $X_n(\sigma)$, for the block system (5.19) can be expressed as follows:

$$X_n(\sigma) = V_n Z_n(\sigma), \tag{5.21}$$

where $Z_n(\sigma)$ is the solution of the shifted block Lanczos system

$$\left(T_n^{(\mathrm{pr})} - \sigma I_n\right) Z_n(\sigma) = \rho_n^{(\mathrm{pr})}. \tag{5.22}$$

Recall that V_n is the matrix of right Lanczos vectors defined in (5.11). Also, note that the coefficient matrix of the system (5.22) is of size $n \times n$. By choosing $\sigma = \sigma(s)$ and inserting the associated block BCG iterate into (5.20), we obtain the approximation

$$H^{(n)}(s) := D + \sigma(s)\, L^T X_n\left(\sigma(s)\right) \approx H(s) \tag{5.23}$$

for the value of the transfer function H at s. Using (5.21), (5.22), (5.12), and (5.17), it readily follows from (5.23) that

$$\begin{aligned} H^{(n)}(s) &:= D + \sigma(s)\left(\eta_n^{(\mathrm{pr})}\right)^T \Delta_n \left(T_n^{(\mathrm{pr})} - \sigma(s)\right)^{-1} \rho_n^{(\mathrm{pr})} \\ &= D - \left(\eta_n^{(\mathrm{pr})}\right)^T \Delta_n \left(I - (s - s_0) T_n^{(\mathrm{pr})}\right)^{-1} \rho_n^{(\mathrm{pr})}. \end{aligned} \tag{5.24}$$

By comparing (5.24) and (5.16), we conclude that

$$H_n(s) = H^{(n)}(s). \tag{5.25}$$

This means that computing approximate values of $H(s)$ via n iterations of the shifted block BCG method is equivalent to matrix-Padé approximation.

Of course, this equivalence no longer holds true when shifted variants of other Krylov-subspace solvers, such as block QMR (Freund and Malhotra 1997), are employed.

5.4. The SyMPVL algorithm

A disadvantage of Padé models is that, in general, they do not preserve the stability and possibly passivity of the original linear dynamical system. In part, these problems can be overcome by means of suitable post-processing techniques, such as the ones described in Bai, Feldmann and Freund (1998), and Bai and Freund (2001*a*). However, the reduced-order models obtained by post-processing of Padé models are necessarily no longer optimal in the sense of Padé approximation. Furthermore, post-processing techniques are not guaranteed to result always in stable and possibly passive reduced-order models.

For special cases, however, Padé models can be shown to be stable and passive. In particular, this is the case for linear dynamical systems describing RC subcircuits, RL subcircuits, and LC subcircuits; see Bai and Freund (2001*b*) and Freund and Feldmann (1996*a*, 1997, 1998).

Next, we describe the SyMPVL algorithm (Freund and Feldmann 1996*a*, 1997, 1998), which is a special version of MPVL tailored to linear RCL subcircuits.

Recall from Section 4.3 that linear RCL subcircuits can be described by linear dynamical systems (2.1) and (2.2) with $D = 0$, symmetric matrices A and E of the form (4.6), and matrices $B = C$ of the form (4.7). Furthermore, the transfer function (4.8), H, is symmetric.

We now assume that the expansion point s_0 for the Padé approximation is chosen to be real and nonnegative, that is, $s_0 \geq 0$. Together with (4.6) it follows that the matrix $A - s_0 E$ is symmetric indefinite, with N_1 nonpositive and N_2 nonnegative eigenvalues. Thus, $A - s_0 E$ admits a factorization of the following form:

$$A - s_0 E = -F_1 J F_1^T, \tag{5.26}$$

where J is the block matrix defined in (4.9). Instead of the general factorization (3.3), we now use (5.26). By (5.26) and (3.5), the matrices M, R, and L are then of the following form:

$$M = F_1^{-1} E F_1^{-T} J, \quad R = F_1^{-1} B, \quad \text{and} \quad L = -J F_1^{-1} C.$$

Since $E = E^T$ and $B = C$, it follows that

$$JM = M^T J \quad \text{and} \quad L = -JR.$$

This means that M is J-symmetric and the left starting block L is (up to its sign) the J-multiple of the right starting block R. These two properties imply that all the right and left Lanczos vectors generated by the band Lanczos Algorithm 5.1 are J-multiples of each other:

$$w_j = J v_j \quad \text{for all} \quad j = 1, 2, \ldots, n.$$

Consequently, Algorithm 5.1 simplifies, in that only the right Lanczos vectors need to be computed. The resulting version of MPVL for computing matrix-Padé models of RCL subcircuits is just the SyMPVL algorithm. The computational costs of SyMPVL are half of that of the general MPVL algorithm.

Let $H_n^{(1)}$ denote the matrix-Padé model generated by SyMPVL after n Lanczos steps. For general RCL subcircuits, however, $H_n^{(1)}$ will not preserve the passivity of the original system.

An additional reduced-order model that is guaranteed to be passive can be obtained as follows, provided that all right Lanczos vectors are stored. Let

$$V_n = \begin{bmatrix} v_1 & v_2 & \cdots & v_n \end{bmatrix}$$

denote the matrix that contains the first n right Lanczos vectors as columns. Then, by projecting the matrices in the representation (4.10) of the transfer function H of the original RCL subcircuit onto the columns of V_n, we obtain the following reduced-order transfer function:

$$H_n^{(2)}(s) = \left(V_n^T B\right)^T \left(s V_n^T \tilde{E} V_n - V_n^T \tilde{A} V_n\right)^{-1} V_n^T B. \tag{5.27}$$

The passivity of the original RCL subcircuit, together with Theorem 4.5, implies that the reduced-order model defined by $H_n^{(2)}$ is indeed passive. Furthermore, in Freund (2000*b*), it is shown that $H_n^{(2)}$ is a matrix-Padé-type approximation of the original transfer function and that, at the expansion point s_0, $H_n^{(2)}$ matches half as many leading coefficients of H as the matrix-Padé approximant $H_n^{(1)}$.

Next, we illustrate the behaviour of SyMPVL with two circuit examples.

5.5. A package model

The first example that arises is the analysis of a 64-pin package model used for an RF integrated circuit. Only eight of the package pins carry signals, the rest being either unused or carrying supply voltages. The package is characterized as a passive linear dynamical system with $m = p = 16$ inputs and outputs, representing 8 exterior and 8 interior terminals. The package model is described by approximately 4000 circuit elements, resistors, capacitors, inductors, and inductive couplings, resulting in a linear dynamical system with a state-space dimension of about 2000.

In Freund and Feldmann (1997), SyMPVL was used to compute a Padé-based reduced-order model of the package, and it was found that a model $H_n^{(1)}$ of order $n = 80$ is sufficient to match the transfer-function components of interest. However, the model $H_n^{(1)}$ has a few poles in the right half of the complex plane, and therefore it is not passive.

In order to obtain a passive reduced-order model, we ran SyMPVL again on the package example, and this time, also generated the projected reduced-order model $H_n^{(2)}$ given by (5.27). The expansion point $s_0 = 5\pi \times 10^9$ was used. Recall that $H_n^{(2)}$ is only a Padé-type approximant and thus less accurate than the Padé approximant $H_n^{(2)}$. Therefore, we now have to go to order $n = 112$ to obtain a projected reduced-order model $H_n^{(2)}$ that matches the transfer-function components of interest. Figures 5.1 and 5.2 show the voltage-to-voltage transfer function between the external terminal of pin no. 1 and the internal terminals of the same pin and the neighbouring pin no. 2, respectively. The plots show results with the projected model $H_n^{(2)}$ and the Padé model $H_n^{(2)}$, both of order $n = 112$, compared with an exact analysis.

In Figure 5.3 we compare the relative error of the projected model $H_{112}^{(2)}$ and the Padé model $H_{112}^{(1)}$ of the same size. Clearly, the Padé model is more

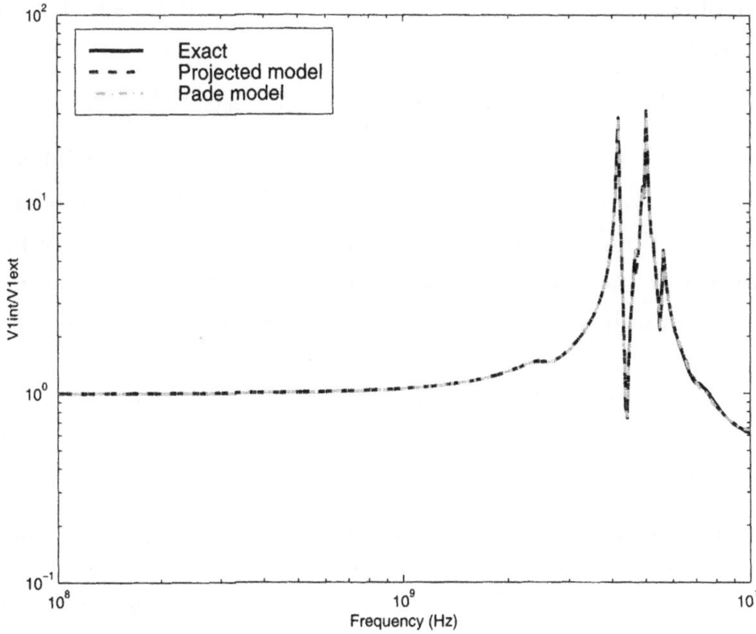

Figure 5.1. Package: pin no. 1 external to pin no. 1 internal, exact, projected model, and Padé model.

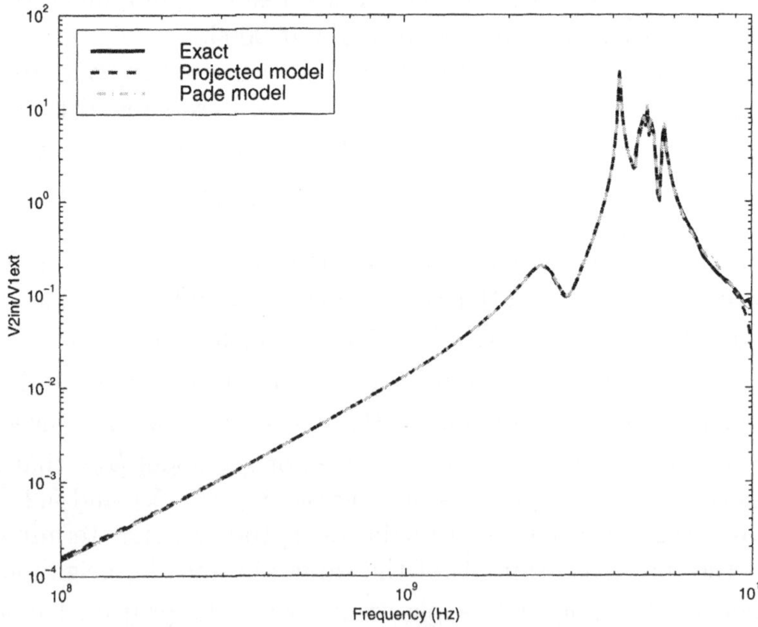

Figure 5.2. Package: pin no 1 external to pin no. 2 internal, exact, projected model, and Padé model.

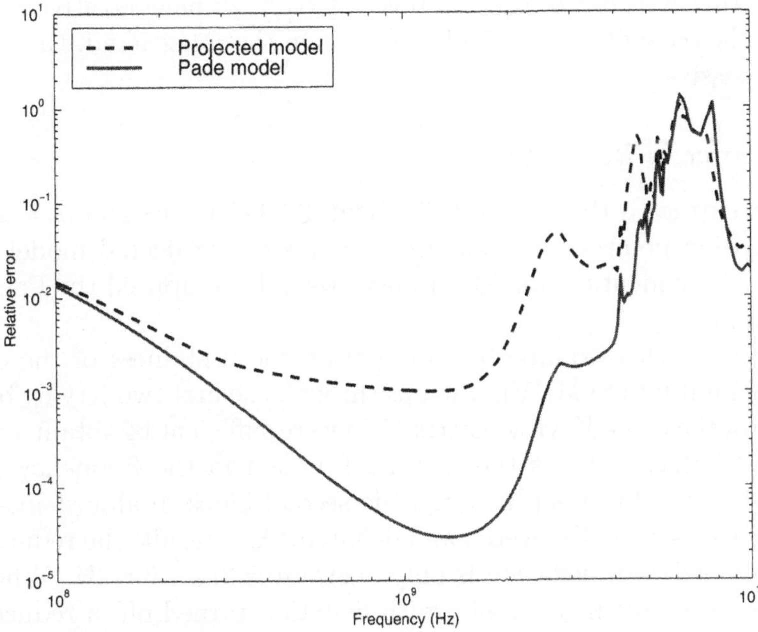

Figure 5.3. Relative error of projected model and Padé model.

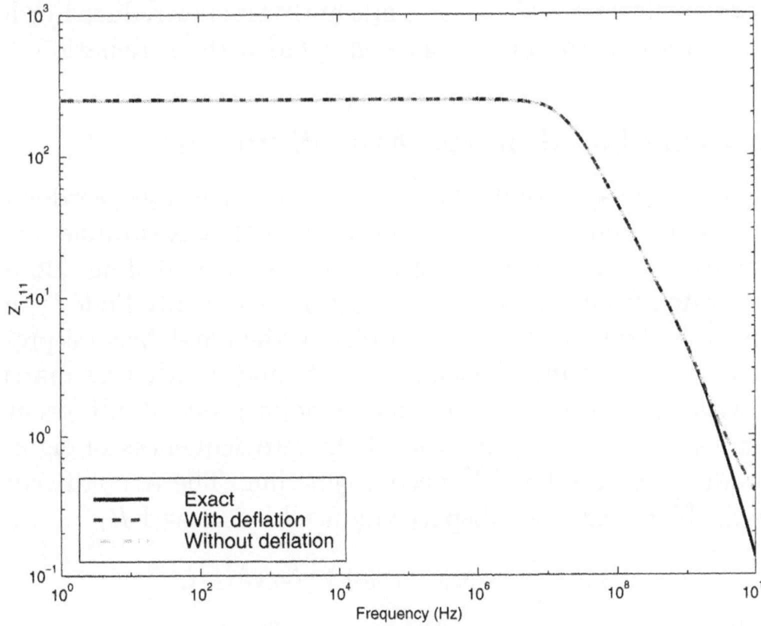

Figure 5.4. Impedance $H_{1,11}$.

accurate. However, out of the 112 poles of $H_{112}^{(1)}$, 22 have positive real parts, violating the passivity of the Padé model. On the other hand, the projected model is passive.

5.6. *An extracted RC circuit*

This is an extracted RC circuit with about 4000 elements and $m = 20$ ports. The expansion point $s_0 = 0$ was used. Since the projected model and the Padé model are identical for RC circuits, we only computed the Padé model via SyMPVL.

The point of this example is to illustrate the usefulness of the deflation procedure built into SyMPVL. Sweeps through the first two Krylov blocks, R and MR, of the block Krylov matrix (5.5) were sufficient to obtain a reduced-order model that matches the transfer function in the frequency range of interest. During the sweep through the second block, 6 almost linearly dependent vectors were discovered and deflated. As a result, the reduced-order model obtained with deflation is only of size $n = 2m - 6 = 34$. When SyMPVL was rerun on this example, with deflation turned off, a reduced-order model of size $n = 40$ was needed to match the transfer function. In Figure 5.4, we show the $H_{1,11}$ component of the reduced-order model obtained with deflation and without deflation, compared to the exact transfer function. Clearly, deflation leads to a significantly smaller reduced-order model that is as accurate as the bigger one generated without deflation.

6. Approaches based on the Arnoldi process

The Arnoldi process (Arnoldi 1951) is another widely used Krylov-subspace method. A band version of the Arnoldi process that is suitable for multiple starting vectors can also be used for reduced-order modelling. However, the models generated from the band Arnoldi process are only Padé-type models.

In contrast to the band Lanczos algorithm, the band Arnoldi process only involves one of the starting blocks, namely R, and it only uses matrix-vector products with M. Moreover, the band Arnoldi process only generates one set of vectors, $v_1, v_2, \dots, v_n$, instead of the two sequences of right and left vectors produced by the band Lanczos algorithm. The Arnoldi vectors span the nth right block Krylov subspace (induced by M and R):

$$\operatorname{span}\{ v_1, v_2, \dots, v_n \} = \mathcal{K}_n(M, R).$$

The Arnoldi vectors are constructed to be orthonormal:

$$V_n^H V_n = I, \quad \text{where} \quad V_n := \begin{bmatrix} v_1 & v_2 & \cdots & v_n \end{bmatrix}.$$

After n iterations, the Arnoldi process has generated the first n Arnoldi vectors, namely the n columns of the matrix V_n, as well as an $n \times n$ matrix

$G_n^{(\mathrm{pr})}$ of recurrence coefficients, and, provided that $n \geq m$, an $n \times m$ matrix $\rho_n^{(\mathrm{pr})}$. The matrices $G_n^{(\mathrm{pr})}$ and $\rho_n^{(\mathrm{pr})}$ are projections of the matrices M and R onto the subspace spanned by the columns of V_n, which is just the block Krylov subspace $\mathcal{K}_n(M, R)$. More precisely, we have

$$G_n^{(\mathrm{pr})} = V_n^H M V_n \quad \text{and} \quad \rho_n^{(\mathrm{pr})} = V_n^H R. \tag{6.1}$$

The band Arnoldi process can be stated as follows.

Algorithm 6.1. (Band Arnoldi process)

INPUT: A matrix $M \in \mathbb{C}^{n \times n}$;
A block of m right starting vectors $R = \begin{bmatrix} r_1 & r_2 & \cdots & r_m \end{bmatrix} \in \mathbb{C}^{n \times m}$.

OUTPUT: The $n \times n$ Arnoldi matrix $G_n^{(\mathrm{pr})}$.
The matrix $V_n = \begin{bmatrix} v_1 & v_2 & \cdots & v_n \end{bmatrix}$ containing the first n Arnoldi vectors, and the matrix $\rho_n^{(\mathrm{pr})}$.

(0) For $k = 1, 2, \ldots, m$, set $\hat{v}_k = r_k$.
Set $m_c = m$ and $\mathcal{I} = \emptyset$.

For $n = 1, 2, \ldots,$ until convergence or $m_c = 0$ do:

(1) (If necessary, deflate $\hat{v}_n$.)
Compute $\|\hat{v}_n\|_2$.
Decide if $\hat{v}_n$ should be deflated. If yes, do the following:

(a) Set $\hat{v}_{n-m_c}^{\mathrm{defl}} = \hat{v}_n$ and store this vector. Set $\mathcal{I} = \mathcal{I} \cup \{\, n - m_c \,\}$.

(b) Set $m_c = m_c - 1$. If $m_c = 0$, set $n = n - 1$ and stop.

(c) For $k = n, n+1, \ldots, n + m_c - 1$, set $\hat{v}_k = \hat{v}_{k+1}$.

(d) Repeat all of step (1).

(2) (Normalize $\hat{v}_n$ to obtain v_n.)
Set

$$g_{n,n-m_c} = \|\hat{v}_n\|_2 \quad \text{and} \quad v_n = \frac{\hat{v}_n}{g_{n,n-m_c}}.$$

(3) (Orthogonalize the candidate vectors against v_n.)
For $k = n+1, n+2, \ldots, n + m_c - 1$, set

$$g_{n,k-m_c} = v_n^H \hat{v}_k \quad \text{and} \quad \hat{v}_k = \hat{v}_k - v_n\, g_{n,k-m_c}.$$

(4) (Advance the block Krylov subspace to get $\hat{v}_{n+m_c}$.)

(a) Set $\hat{v}_{n+m_c} = M\, v_n$.

(b) For $k = 1, 2, \ldots, n$, set

$$g_{k,n} = v_k^H \hat{v}_{n+m_c} \quad \text{and} \quad \hat{v}_{n+m_c} = \hat{v}_{n+m_c} - v_k\, g_{k,n}.$$

(5) (a) For $k \in \mathcal{I}$, set $g_{n,k} = v_n^H \hat{v}_k^{\text{defl}}$.
(b) Set

$$G_n^{(\text{pr})} = \left[g_{i,k}\right]_{i,k=1,2,\ldots,n},$$
$$k_\rho = m + \min\{\, 0,\ n - m_c \,\},$$
$$\rho_n^{(\text{pr})} = \left[g_{i,k-m}\right]_{i=1,2,\ldots,n;\ k=1,2,\ldots,k_\rho}.$$

(6) Check if n is large enough. If yes, stop.

Note that, in contrast to the band Lanczos algorithm, the band Arnoldi process requires the storage of all previously computed Arnoldi vectors.

Like the band Lanczos algorithm, the band Arnoldi process can also be employed for reduced-order modelling. Let M, R, and L be the matrices defined in (3.5). After running Algorithm 6.1 (applied to M and R) for n steps, we have obtained the matrices $G_n^{(\text{pr})}$ and $\rho_n^{(\text{pr})}$, as well as the matrix V_n of Arnoldi vectors. The transfer function H_n of a reduced-order model H_n can now be defined as follows:

$$H_n(s) = \left(V_n^H L\right)^H \left(I - (s - s_0)V_n^H M V_n\right)^{-1} \left(V_n^H R\right).$$

Using the relations (6.1) for $G_n^{(\text{pr})}$ and $\rho_n^{(\text{pr})}$, the formula for H_n reduces to

$$H_n(s) = \left(V_n^H L\right)^H \left(I - (s - s_0)G_n^{(\text{pr})}\right)^{-1} \rho_n^{(\text{pr})}. \tag{6.2}$$

The matrices $G_n^{(\text{pr})}$ and $\rho_n^{(\text{pr})}$ are directly available from Algorithm 6.1. In addition, we also need to compute the matrix

$$\eta_n^{(\text{pr})} = V_n^H L.$$

It turns out that the transfer function (6.2) defines a matrix-Padé-type reduced-order model.

Theorem 6.2. (Matrix-Padé-type model) Suppose that Algorithm 6.1 is run with exact deflation only and that $n \geq m$. Then, the reduced-order model associated with the reduced-order transfer function (6.2) is a matrix-Padé-type model of the linear dynamical system (2.1) and (2.2). More precisely, the Taylor expansions about s_0 of the transfer functions (2.8), H, and (6.2), H_n, agree in at least

$$q'(n) \geq \left\lfloor \frac{n}{m} \right\rfloor$$

leading coefficients:

$$H(s) = H_n(s) + \mathcal{O}\left((s - s_0)^{q'(n)}\right). \tag{6.3}$$

A proof of Theorem 6.2 is given in Freund (2000*b*).

Remark 4. The number $q'(n)$ is the exact number of terms matched in the expansion (6.3) provided that no exact deflations occur in Algorithm 6.1. In the case of exact deflations, the number of matching terms is somewhat higher, but so is the number of matching terms for the matrix-Padé model of Theorem 5.2; see Freund (2000*b*). In particular, the matrix-Padé model is always more accurate than the matrix-Padé-type model obtained from Algorithm 6.1. On the other hand, the band Arnoldi process is certainly simpler than the band Lanczos process. Furthermore, the true orthogonality of the Arnoldi vectors generally results in better numerical behaviour than the bi-orthogonality of the Lanczos vectors.

Remark 5. For the special case of RCL subcircuits, the algorithm PRIMA proposed by Odabasioglu (1996) and Odabasioglu, Celik and Pileggi (1997) can be interpreted as a special case of the Arnoldi reduced-order modelling procedure described here. Furthermore, in Freund (1999*a*) and (2000*b*) it is shown that the reduced-order model produced by PRIMA is mathematically equivalent to the additional passive model produced by SyMPVL. In contrast to PRIMA, however, SyMPVL also produces a true matrix-Padé model, and thus PRIMA does not appear to have any real advantage over – or even be competitive with – SyMPVL.

Remark 6. It is also possible to devise a two-sided Arnoldi procedure and then generate Padé models from it. Such an approach is described in Cullum and Zhang (2002).

7. Circuit-noise computations

In this section, we discuss the use of reduced-order modelling for *circuit-noise computations.* In particular, we show how noise-type transfer functions can be rewritten so that reduced-order modelling techniques for linear dynamical systems can be applied. The material in this section is based on the paper by Feldmann and Freund (1997).

7.1. The problem

Noise in electronic circuits is caused by the stochastical fluctuations in currents and voltages that occur within the devices of the circuit. We refer the reader to Chapter 8 of Davidse (1991) or to van der Ziel (1986) for an introduction to circuit noise and the main noise mechanisms. Noise-analysis algorithms for circuits in DC steady-state have been available for a long time in traditional circuit simulators such as SPICE (Rohrer, Nagel, Meyer and Weber 1971). As we will now describe, simulation techniques based on reduced-order modelling, such as PVL and MPVL, can easily be extended to include noise computations.

Noise in circuit devices is modelled by stochastic processes. In the time domain, a stochastic process is characterized in terms of statistical averages, such as the mean and autocorrelation, and in the frequency domain, it is described by the spectral power density. The main types of noise in integrated circuits are *thermal noise*, *shot noise*, and *flicker noise*. Thermal and shot noise represent *white noise*, that is, their *spectral power densities* do not depend on the frequency ω. Flicker noise is modelled by a stochastic process with a spectral power density that is proportional to $(1/\omega)^\beta$ where β is a constant of about one.

Next, we describe the problem of noise computation for circuits with constant excitation in steady-state (DC). Moreover, we assume that all time-varying circuit elements are independent sources. In this case, the general system of circuit equations (1.1) simplifies to a system of the form

$$\frac{\mathrm{d}}{\mathrm{d}t} q(\hat{x}) + f(\hat{x}) = b_0. \tag{7.1}$$

Here, b_0 denotes the constant excitation vector. Let $\hat{x}_0$ be a DC operating point of the circuit, that is, $\hat{x}_0$ is a constant vector that satisfies $f(\hat{x}_0) = b_0$. Adding noise sources to (7.1) gives

$$\frac{\mathrm{d}}{\mathrm{d}t} q(\hat{x} + x) + f(\hat{x} + x) = b_0 + B\,\nu(t), \tag{7.2}$$

where $\nu(t)$ is a vector stochastic process of length m that describes the noise sources, $B \in \mathbb{R}^{N\times m}$ is the noise-source incidence matrix, and m denotes the number of noise sources. The vector function $x = x(t)$ in (7.2) represents the stochastical deviations of the circuit variables from the DC operating point $\hat{x}_0$ that are caused by the noise sources. By linearizing (7.2) about $\hat{x}_0$ and using the fact that $f(\hat{x}_0) = b_0$, we obtain the following linear system of DAEs:

$$E\frac{\mathrm{d}x}{\mathrm{d}t} = Ax + B\nu(t), \tag{7.3}$$

$$y(t) = C^T x(t). \tag{7.4}$$

Here,

$$A = -\mathrm{D}_x f(\hat{x}_0) \quad \text{and} \quad E = \mathrm{D}_x q(\hat{x}_0) \tag{7.5}$$

that is, A is the negative of the Jacobian matrix of f at the point $\hat{x}_0$ and E is the Jacobian matrix of q at the point $\hat{x}_0$. Furthermore, in (7.4), $y(t)$ is a vector stochastic process of length p describing the stochastical deviations at the outputs of interest due to the noise sources, and $C \in \mathbb{R}^{N\times p}$ is a constant matrix that selects the outputs of interest. Note that (7.3) and (7.4) describe a linear dynamical system of the form (1.2) and (1.3) with m inputs and p outputs. Thus we can use MPVL or, if $m = p = 1$, PVL to generate reduced-order models for (7.3) and (7.4).

For noise computations in the frequency domain, the goal is to compute the $(p \times p)$-cross-spectral power density matrix $S_y(\omega)$ of the vector stochastic process y in (7.4). It turns out that

$$S_y(\omega) = C^T (\mathrm{i}\omega E - A)^{-1} B \, S_\nu(\omega) \, B^T (\mathrm{i}\omega E - A)^{-H} C \tag{7.6}$$

for all $\omega \geq 0$. Here, ω denotes frequency, and $S_\nu(\omega)$ is the given $m \times m$ cross-spectral power density matrix of the noise sources $\nu(t)$ in (7.2). We remark that the diagonal entries of $S_\nu(\omega)$ are the spectral power densities of the noise sources, and that nonzero off-diagonal entries of $S_\nu(\omega)$ occur only if there is coupling between some of the noise sources. Moreover, if all noise sources are white, then S_ν is a constant matrix.

7.2. Reformulation as a transfer function

Clearly, the matrix-valued function (7.6), S_y, does not have the form of a transfer function (2.8). Consequently, the reduced-order modelling techniques we discussed so far cannot be applied directly to S_y. However, for the physical relevant values $\omega \geq 0$ and under some mild assumptions on the form of S_ν, we can rewrite (7.6) as a function of the type (2.8). More precisely, we assume that

$$S_\nu(\omega) = (P(\mathrm{i}\omega))^{-1} \quad \text{for all} \quad \omega \geq 0, \tag{7.7}$$

where

$$P(s) = P_0 + P_1 s + \cdots + P_M s^M, \quad P_i \in \mathbb{C}^{m \times m}, \quad 0 \leq i \leq M, \tag{7.8}$$

is any matrix polynomial of degree M (that is, $P_M \neq 0$). In particular, for the important special case that all noise sources are white, as in the case of thermal and shot noise, we have

$$P(s) = P_0 = S_\nu^{-1} \quad \text{and} \quad M = 0. \tag{7.9}$$

If $S_\nu(\omega)$ does depend on the frequency, as in the case of flicker noise, then the assumption (7.7) is satisfied at least approximately, see Feldmann and Freund (1997).

By inserting (7.7) into (7.6) and setting

$$H(s) := C^T (sE - A)^{-1} B \, (P(s))^{-1} B^T (sE - A)^{-H} C, \quad s \in \mathbb{C}, \tag{7.10}$$

it follows that

$$H(\mathrm{i}\omega) = S_y(\omega) \quad \text{for all} \quad \omega \geq 0. \tag{7.11}$$

The relation (7.11) suggests first generating an approximation H_n to the function H in (7.10) and then using

$$S_y(\omega) \approx H_n(\mathrm{i}\omega) \tag{7.12}$$

as an approximation to S_y. It turns out that the function H can be rewritten as a transfer function of the type (2.8), and thus we can employ MPVL (or PVL if $p = 1$) to obtain H_n as an nth matrix-Padé approximant to H. More precisely, in Feldmann and Freund (1997), it is shown that

$$H(s) = \widetilde{C}^T \left(s\,\widetilde{E} - \widetilde{A}\right)^{-1} \widetilde{C} \quad \text{for all} \quad s \in \mathbb{C}. \tag{7.13}$$

Here, $\widetilde{C} \in \mathbb{C}^{\widetilde{N} \times p}$ and $\widetilde{A},\ \widetilde{E} \in \mathbb{C}^{\widetilde{N} \times \widetilde{N}}$ are matrices given by

$$\widetilde{C} := \begin{bmatrix} C \\ 0_{N \times p} \\ 0_{m \times p} \\ 0_{m \times p} \\ \vdots \\ 0_{m \times p} \end{bmatrix}, \quad \widetilde{A} := \begin{bmatrix} 0 & A^T & 0 & 0 & \cdots & 0 \\ A & 0 & -B & 0 & \cdots & 0 \\ 0 & -B^T & -P_0 & 0 & \ddots & \vdots \\ 0 & 0 & 0 & I & \ddots & 0 \\ \vdots & \ddots & \ddots & \ddots & \ddots & 0 \\ 0 & \cdots & \cdots & 0 & 0 & I \end{bmatrix},$$

$$\widetilde{E} := \begin{bmatrix} 0 & -E^T & 0 & 0 & \cdots & 0 \\ E & 0 & 0 & 0 & \cdots & 0 \\ 0 & 0 & P_1 & P_2 & \cdots & P_M \\ 0 & 0 & I & 0 & \cdots & 0 \\ \vdots & \ddots & \ddots & \ddots & \ddots & 0 \\ 0 & \cdots & \cdots & 0 & I & 0 \end{bmatrix}, \tag{7.14}$$

and $\widetilde{N} := 2 \cdot N + m \cdot M$.

If the matrix polynomial P is linear, that is, $M = 1$ in (7.8), the matrices (7.14) reduce to

$$\widetilde{C} := \begin{bmatrix} C \\ 0 \\ 0 \end{bmatrix}, \quad \widetilde{A} := \begin{bmatrix} 0 & A^T & 0 \\ A & 0 & -B \\ 0 & -B^T & -P_0 \end{bmatrix}, \quad \widetilde{E} := \begin{bmatrix} 0 & -E^T & 0 \\ E & 0 & 0 \\ 0 & 0 & P_1 \end{bmatrix}. \tag{7.15}$$

The important special case (7.9) of white noise is also covered by (7.15) with $P_0 := S_\nu^{-1}$ and $P_1 := 0$. In this case, by eliminating the third block rows and columns in (7.15), the matrices $\widetilde{C}$, $\widetilde{A}$, and $\widetilde{E}$ can be further reduced to

$$\widetilde{C} = \begin{bmatrix} C \\ 0 \end{bmatrix}, \quad \widetilde{A} = \begin{bmatrix} 0 & A^T \\ A & B^T S_\nu B \end{bmatrix}, \quad \widetilde{E} = \begin{bmatrix} 0 & -E^T \\ E & 0 \end{bmatrix}. \tag{7.16}$$

7.3. A PVL simulation

We now present results of a typical simulation with the noise-computation algorithm described in Section 7.2.

The example is a 5th-order Cauer filter that uses ten 741 operational amplifiers as building blocks. The total size of the problem is 463 variables. The noise sources are all white. The circuit has a single input and a single output, and we employ PVL to compute an nth Padé approximant to the

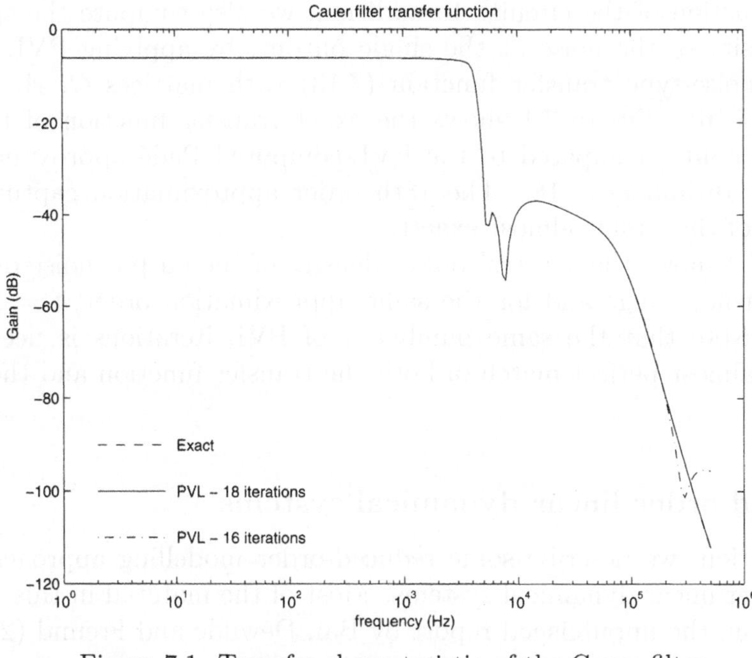

Figure 7.1. Transfer characteristic of the Cauer filter.

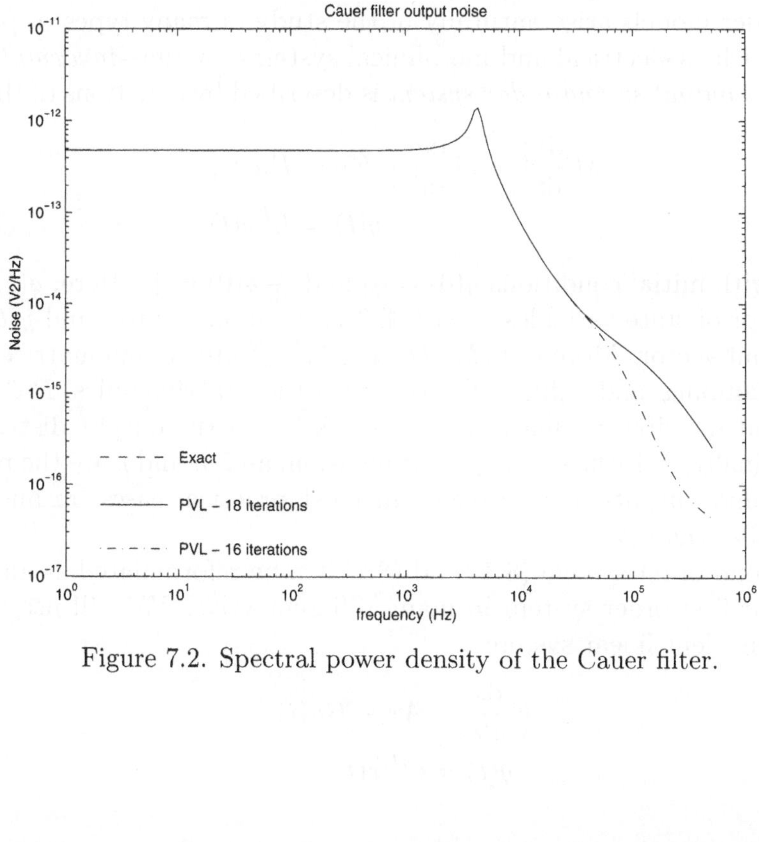

Figure 7.2. Spectral power density of the Cauer filter.

transfer function of the circuit. In addition, we also compute the spectral power density of the noise at the single output, by applying PVL to the rewritten noise-type transfer function (7.13) with matrices C, $\widetilde{A}$, and $\widetilde{E}$ given by (7.16). Figure 7.1 shows the exact transfer function of the 5th order Cauer filter compared to the PVL-computed Padé approximants of order $n = 16$ and $n = 18$. The 18th order approximation captures the behaviour of the circuit almost exactly.

Figure 7.2 shows the spectral power density of the output noise over the same frequency range and for the same approximation order $n = 16$ and $n = 18$. Note that the same number n of PVL iterations is needed to obtain an almost perfect match of both the transfer function and the noise spectrum.

8. Second-order linear dynamical systems

In this section, we describe some reduced-order modelling approaches for second-order linear dynamical systems. Most of the material in this section is taken from the unpublished report by Bai, Dewilde and Freund (2002).

8.1. The problem

Second-order models arise naturally in the study of many types of physical systems, such as electrical and mechanical systems. A *time-invariant multi-input multi-output second-order system* is described by equations of the form

$$M\frac{\mathrm{d}^2q}{\mathrm{d}t^2} + D\frac{\mathrm{d}q}{\mathrm{d}t} + Kq = Pu(t), \tag{8.1}$$

$$y(t) = L^T q(t), \tag{8.2}$$

together with initial conditions $q(0) = q_0$ and $\frac{\mathrm{d}}{\mathrm{d}t}\, q(0) = \dot{q}_0$. Here, $q(t) \in \mathbb{R}^N$ is the vector of state variables, $u(t) \in \mathbb{R}^m$ is the input vector, and $y(t) \in \mathbb{R}^p$ is the output vector. Moreover, M, D, $K \in \mathbb{R}^{N\times N}$ are system matrices, such as mass, damping, and stiffness matrices in structural dynamics, $P \in \mathbb{R}^{N\times m}$ is the input distribution matrix, and $L \in \mathbb{R}^{N\times p}$ is the output distribution matrix. Finally, N is the state-space dimension, and m and p are the number of inputs and outputs, respectively. In most practical cases, m and p are much smaller than N.

The second-order system (8.1) and (8.2) can be reformulated as an equivalent linear first-order system in many different ways. We will use the following equivalent linear system:

$$E\frac{\mathrm{d}x}{\mathrm{d}t} = Ax + Bu(t), \tag{8.3}$$

$$y(t) = C^T x(t), \tag{8.4}$$

where

$$x = \begin{bmatrix} q \\ \frac{\mathrm{d}q}{\mathrm{d}t} \end{bmatrix}, \; A = \begin{bmatrix} -K & 0 \\ 0 & W \end{bmatrix}, \; E = \begin{bmatrix} D & M \\ W & 0 \end{bmatrix}, \; B = \begin{bmatrix} P \\ 0 \end{bmatrix}, \; C = \begin{bmatrix} L \\ 0 \end{bmatrix}.$$

Here, $W \in \mathbb{R}^{N \times N}$ can be any nonsingular matrix. A common choice is the identity matrix, $W = I$. If the matrices M, D, and K are all symmetric and M is nonsingular, as is often the case in structural dynamics, we can choose $W = M$. The resulting matrices A and E in the linearized system (8.3) are then symmetric, and thus preserve the symmetry of the original second-order system.

Assume that, for simplicity, we have zero initial conditions, that is, $q(0) = q_0$, $\frac{\mathrm{d}}{\mathrm{d}t} q(0) = 0$, and $u(0) = 0$ in (8.1) and (8.2). Then, by taking the Laplace transform of (8.1) and (8.2), we obtain the following algebraic system:

$$\begin{aligned} s^2 MQ(s) + DQ(s) + KQ(s) &= PU(s), \\ Y(s) &= L^T Q(s). \end{aligned}$$

Eliminating $Q(s)$ in this system results in the frequency-domain input-output relation $Y(s) = H(s)U(s)$, where

$$H(s) := L^T \left(s^2 M + sD + K\right)^{-1} P$$

is the transfer function. In view of the equivalent linearized system (8.3) and (8.4), the transfer function can also be written as

$$H(s) = C^T (sE - A)^{-1} B.$$

If the matrix K in (8.1) is nonsingular, then $s_0 = 0$ is guaranteed not to be a pole of H. In this case, H can be expanded about $s_0 = 0$ as follows:

$$H(s) = \mu_0 + \mu_1 s + \mu_2 s^2 + \cdots,$$

where the matrices μ_j are the so-called *low-frequency moments.* In terms of the matrices of the linearized system (8.3) and (8.4), the moments are given by

$$\mu_j = -C^T \left(A^{-1}E\right)^j A^{-1} B, \quad j = 0, 1, 2, \ldots.$$

8.2. Frequency-response analysis methods

In this subsection, we describe the use of eigensystem analysis to tackle the second-order system (8.1) and (8.2) directly.

We assume that the input force vector $u(t)$ of (8.1) is time-harmonic:

$$u(t) = \tilde{u}(\omega)\mathrm{e}^{\mathrm{i}\omega t},$$

where ω is the frequency of the system. Correspondingly, we assume that the state variables of the second-order system can be represented as follows:

$$q(t) = \tilde{q}(\omega)\mathrm{e}^{\mathrm{i}\omega t}.$$

The problem of solving the system of second-order differential equations (8.1) then reduces to solving the parametrized linear system of equations

$$(-\omega^2 M + \mathrm{i}\,\omega D + K)\,\tilde{q}(\omega) = P\tilde{u}(\omega) \tag{8.5}$$

for $\tilde{q}(\omega)$. This approach is called the *direct frequency-response analysis method*. For a given frequency ω_0, we can use a linear system solver, either direct or iterative, to obtain the desired vector $\tilde{q}(\omega_0)$.

Alternatively, we can try to reduce the cost of solving the large-scale parametrized linear system of equations (8.5) by first applying an eigensystem analysis. This approach is called the *modal frequency-response analysis* in structural dynamics. The basic idea is to first transfer the coordinates $\tilde{q}(\omega)$ of the state vector $q(t)$ to new coordinates $p(\omega)$ as follows:

$$q(t) \cong W_k p(\omega) \mathrm{e}^{\mathrm{i}\,\omega t}.$$

Here, W_k consists of k selected modal shapes to retain the modes whose resonant frequencies lie within the range of forcing frequencies. More precisely, W_k consists of k selected eigenvectors of the underlying quadratic eigenvalue problem $(\lambda^2 M + \lambda D + K)\,w = 0$. Equation (8.5) is then approximated by

$$(-\omega^2 M W_k + \mathrm{i}\,\omega D W_k + K W_k)\,p(\omega) = P\tilde{u}(\omega).$$

Multiplying this equation from the left by W_k^T, we obtain a $k \times k$ parametrized linear system of equations for $p(\omega)$:

$$(-\omega^2\,(W_k^T M W_k) + \mathrm{i}\,\omega\,(W_k^T D W_k) + (W_k^T K W_k))\,p(\omega) = W_k^T P(\omega).$$

Typically, $k \ll n$. The main question now is how to obtain the desired modal shapes W_k. One possibility is to simply extract W_k from the matrix pair (M, K) by ignoring the contribution of the damping term. This is called the *modal superposition method* in structural dynamics. This approach is applicable under the assumption that the damping term is of a certain form. For example, this is the case for so-called Rayleigh damping $D = \alpha M + \beta K$, where α and β are scalars (Clough and Penzien 1975). In general, however, we may need to solve the full quadratic eigenvalue problem $(\lambda^2 M + \lambda D + K)\,w = 0$ in order to obtain the desired modal shapes W_k. Some of these techniques have been reviewed in the recent survey paper (Tisseur and Meerbergen 2001) on the quadratic eigenvalue problem.

8.3. Reduced-order modelling based on linearization

An obvious approach to constructing reduced-order models of the second-order system (8.1) and (8.2) is to apply any of the model-reduction techniques for linear systems to the linearized system (8.3) and (8.4). In particular, we can employ the Krylov-subspace techniques discussed in Sections 5 and 6.

The resulting approach can be summarized as follows.

1. Linearize the second-order system (8.1) and (8.2) by properly defining the $2N \times 2N$ matrices A and E of the equivalent linear system (8.3) and (8.4). Select an expansion point s_0 'close' to the frequency range of interest and such that the matrix $A - s_0E$ is nonsingular.

2. Apply a suitable Krylov process, such as the nonsymmetric band Lanczos algorithm described in Section 5, to the matrix $M := (A - s_0E)^{-1}E$ and the blocks of right and left starting vectors $R := (A - s_0E)^{-1}B$ and $L := C$ to obtain bi-orthogonal Lanczos basis matrices V_n and W_n for the nth right and left block-Krylov subspaces $\mathcal{K}_n(M, R)$ and $\mathcal{K}_n(M^T, L)$.

3. Approximate the state vector $x(t)$ by $V_n z(t)$, where $z(t)$ is determined by the following linear reduced-order model of the linear system (8.3) and (8.4):

$$E_n \frac{\mathrm{d}z}{\mathrm{d}t} = A_n z + B_n u(t),$$
$$y(t) = C_n^T z(t).$$

Here, $E_n = T_n^{(\mathrm{pr})}$, $A_n = I_n + s_0 T_n^{(\mathrm{pr})}$, $B_n = \rho_n^{(\mathrm{pr})}$, $C_n = \Delta_n^T \eta_n^{(\mathrm{pr})}$, and $T_n^{(\mathrm{pr})}$, $\rho_n^{(\mathrm{pr})}$, $\eta_n^{(\mathrm{pr})}$, and Δ_n are the matrices generated by the nonsymmetric band Lanczos Algorithm 5.1.

In Figure 8.1, we show the results of this approach applied to the linear-drive multi-mode resonator structure described in Clark, Zhou and Pister (1998). The solid lines are the Bode plots of the frequency response of the original second-order system, which is of dimension $N = 63$. The dashed lines are the Bode plots of the frequency response of the reduced-order model of dimension $n = 12$. The relative error between the transfer functions of the original system and the reduced-order model of dimension $n = 12$ is less than 10^{-4} over the frequency range shown in Figure 8.1.

There are a couple of advantages to the linearization approach. First, we can directly employ existing reduced-order modelling techniques developed for linear systems. Second, we can also exploit the structures of the linearized system matrices A and E in a Krylov process to reduce the computational cost. However, the linearization approach also has disadvantages. In particular, it ignores the physical meaning of the original system matrices, and more importantly, the reduced-order models are no longer in a second-order form. For engineering design and control of structural systems, it is often desirable to have reduced-order models that preserve the second-order form: see, *e.g.*, Su and Craig, Jr. (1991).

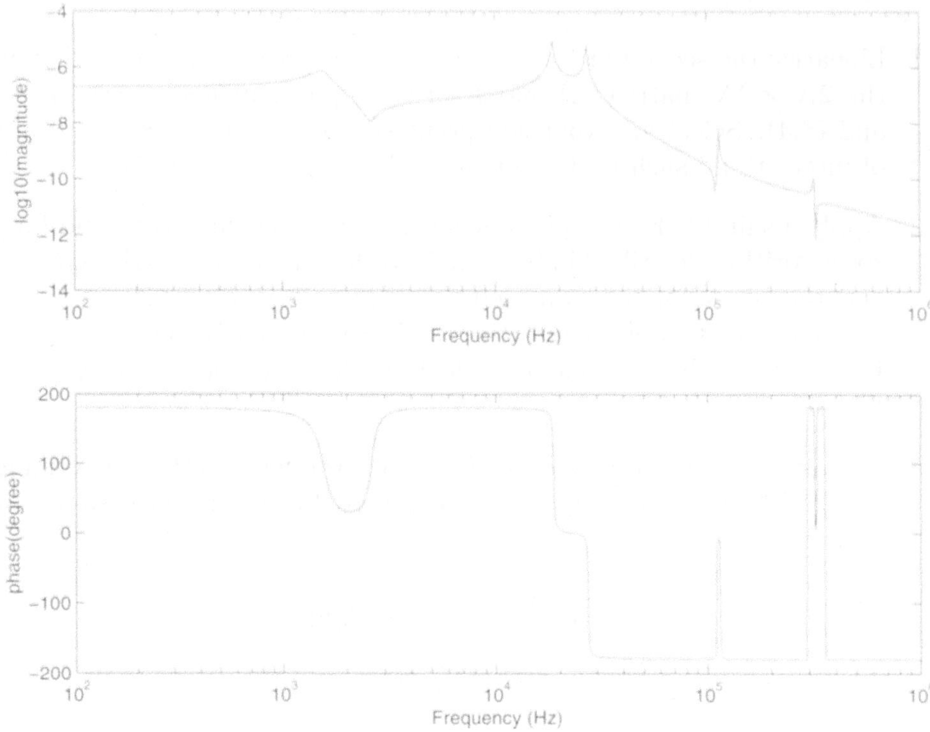

Figure 8.1. Bode plots for the original system and the reduced-order model of dimension $n = 12$.

8.4. Reduced-order modelling based on second-order systems

In this section, we discuss a Krylov-subspace technique that produces a reduced-order model of second-order form. This approach is based on the work by Su and Craig, Jr. (1991).

The key observation is the following. In view of the linearization (8.3) and (8.4) of the second-order system (8.1) and (8.2), the desired Krylov subspace for reduced-order modelling is

$$\operatorname{span}\{\, \widetilde{B},\, (A^{-1}E)\,\widetilde{B},\, (A^{-1}E)^2\,\widetilde{B}, \ldots,\, (A^{-1}E)^{n-1}\,\widetilde{B}\,\}.$$

Here, $\widetilde{B} := -A^{-1}\begin{bmatrix} B & C \end{bmatrix}$. Moreover, we have assumed that the matrix A in (8.3) is nonsingular. Let us set

$$R_j = \begin{bmatrix} R_j^d \\ R_j^v \end{bmatrix} := (-A^{-1}E)^j\,\widetilde{B},$$

where R_j^d is the vector of length N corresponding to the displacement portion of the vector R_j, and R_j^v is the vector of length N corresponding to the velocity portion of the vector R_j: see Su and Craig, Jr. (1991). Then, in view of the structure of the matrices A and E, we have

$$\begin{bmatrix} R_j^d \\ R_j^v \end{bmatrix} = (-A^{-1}E) \begin{bmatrix} R_{j-1}^d \\ R_{j-1}^v \end{bmatrix} = \begin{bmatrix} K^{-1}DR_{j-1}^d + K^{-1}MR_{j-1}^d \\ -R_{j-1}^d \end{bmatrix}.$$

Note that the jth velocity-portion vector R_j^v is the same (up to its sign) as the $(j-1)$st displacement-portion vector R_{j-1}^d. In other words, the second portion R_j^v of R_j is the 'one-step' delay of the first portion R_{j-1}^d of R_j. This suggests that we may simply choose

$$\operatorname{span}\{ R_0^d, R_1^d, R_2^d, \ldots, R_{n-1}^d \} \tag{8.6}$$

as the projection subspace used for reduced-order modelling.

In practice, for numerical stability, we may opt to employ the Arnoldi process to generate an orthonormal basis Q_n of the subspace (8.6). The resulting procedure can be summarized as follows.

Algorithm 8.1. (Algorithm by Su and Craig Jr.)

(0) (Initialization.)
Set $R_0^d = K^{-1}\begin{bmatrix} P & L \end{bmatrix}$, $R_0^v = 0$, $U_0S_0V_0^T = (R_0^d)^TKR_0^d$ (by computing an SVD),
$Q_1^d = R_0^dU_0S_0^{-1/2}$, and $Q_1^v = 0$.

(1) (Arnoldi loop.)
For $j = 1, 2, \ldots, n-1$ do:
Set $R_j^d = K^{-1}\left(DQ_{j-1}^d + MQ_{j-1}^v\right)$ and $R_j^v = -Q_{j-1}^d$.

(2) (Orthogonalization.)
For $i = 1, 2, \ldots, j$ do:
Set $T_i = (Q_i^d)^TKR_j^d$, $R_j^d = R_j^d - Q_i^dT_i$, and $R_j^v = R_j^v - Q_i^vT_i$.

(3) (Normalization.)
Set $U_0S_0V_0^T = (R_j^d)^TKR_j^d$ (by computing an SVD),
$Q_{j+1}^d = R_j^dU_0S_0^{-1/2}$, and $Q_{j+1}^v = R_j^vU_0S_0^{-1/2}$.

An approximation of the state vector $q(t)$ can then be obtained by constraining $q(t)$ to the subspace spanned by the columns of Q_n, that is, $q(t) \approx Q_nz(t)$. Moreover, the reduced-order state vector $z(t)$ is defined as the solution of the following second-order system:

$$M_n\frac{\mathrm{d}^2q}{\mathrm{d}t^2} + D_n\frac{\mathrm{d}q}{\mathrm{d}t} + K_nq = P_nu(t), \tag{8.7}$$

$$y(t) = L_n^Tq(t), \tag{8.8}$$

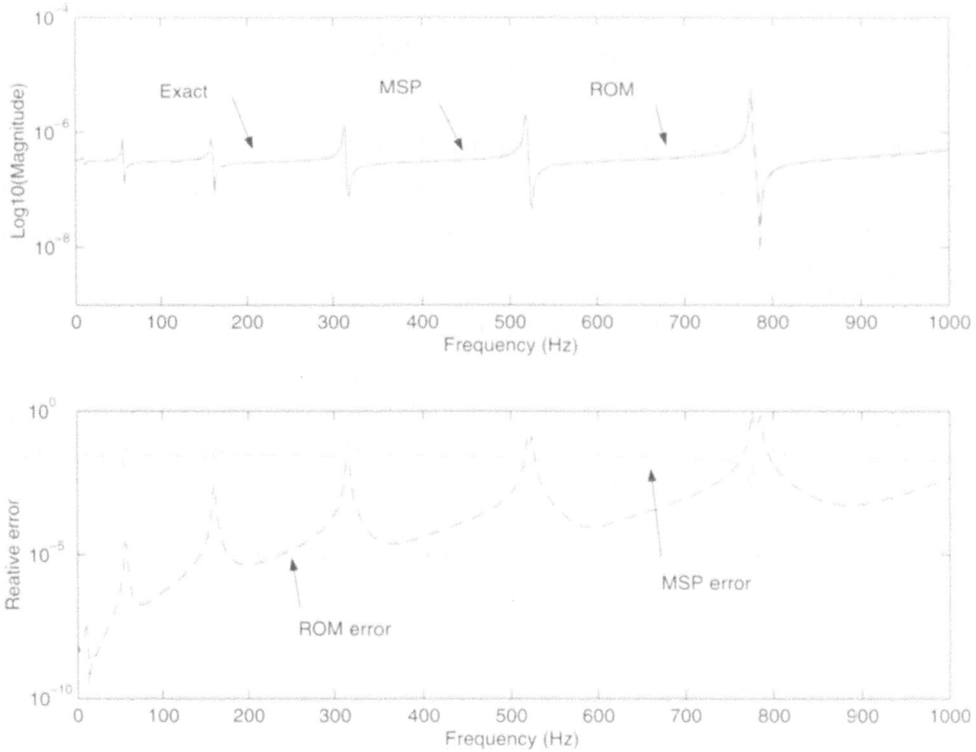

Figure 8.2. Frequency-response analysis (top plot) and relative errors (bottom plot) of a finite-element model of a shaft.

where $M_n := Q_n^T M Q_n$, $D_n := Q_n^T D Q_n$, $K_n := Q_n^T K Q_n$, $P_n := Q_n^T P$, and $L_n := Q_n^T L$. Note that (8.7) and (8.8) describe a reduced-order model in second-order form of the original second-order system (8.1) and (8.2).

Su and Craig, Jr. (1991) describe several advantages of this approach. Here, we present some numerical results of a frequency-response analysis of a second-order system of order $N = 400$, which arises from a finite-element model of a shaft on bearing support with a damper. In the top plot of Figure 8.2, we plot the magnitudes of the transfer function H computed exactly, approximated by the model-superposition (MSP) method, and approximated by the Krylov-subspace technique (ROM). For the MSP method, we used the 80 modal shapes W_{80} from the matrix pencil $\lambda^2 M + K$. The reduced-order model (8.7) and (8.8) is also of dimension $n = 80$. The bottom plot of Figure 8.2 shows the relative errors between the exact transfer function and its approximations based on the MSP method (dash-dotted line) and the ROM method (dashed line). The plots indicate that no accuracy has been lost by the Krylov subspace-based method.

9. Concluding remarks

We have presented a survey of the most common techniques for reduced-order modelling of large-scale linear dynamical systems. By and large, the area of linear reduced-order modelling is fairly well explored, and we have a number of efficient techniques at our disposal. Still, some open problems remain. One such problem is the construction of reduced-order models that preserve stability or passivity and, at the same time, have optimal approximation properties. Particularly in circuit simulation, reduced-order modelling is used to substitute large linear subsystems within the simulation of even larger, generally nonlinear systems. It would be important to better understand the effects of these substitutions on the overall nonlinear simulation.

The systems arising in the simulation of electronic circuits are nonlinear in general, and it would be highly desirable to apply nonlinear reduced-order modelling techniques directly to these nonlinear systems. However, the area of nonlinear reduced-order modelling is in its infancy compared to the state of the art of linear reduced-order modelling. We expect that further progress in model reduction will mainly occur in the area of nonlinear reduced-order modelling.

In this survey, we have focused solely on Krylov subspace-based model-reduction techniques for time-invariant systems. There are of course many other order-reduction approaches that do not fall into this limited category. Methods that we have not treated here include balanced realizations (Moore 1981), Hankel-norm optimal approximations (Glover 1984), order reduction of time-varying systems (Roychowdhury 1999), and proper orthogonal decomposition, which is also known as Karhunen–Loève decomposition (Holmes, Lumley and Berkooz 1996, Glavaški, Marsden and Murray 1998, Rathinam and Petzold 2002).

Acknowledgements

I am grateful to Peter Feldmann, who first introduced me to circuit simulation; many of the results surveyed in this paper are based on joint work with him. I would like to thank Zhaojun Bai for his help with the material on second-order systems and for providing the two numerical examples in Section 8.

REFERENCES

J. I. Aliaga, D. L. Boley, R. W. Freund and V. Hernández (2000), 'A Lanczos-type method for multiple starting vectors', *Math. Comp.* **69**, 1577–1601.

B. D. O. Anderson (1967), 'A system theory criterion for positive real matrices', *SIAM J. Control.* **5**, 171–182.

B. D. O. Anderson and S. Vongpanitlerd (1973), *Network Analysis and Synthesis*, Prentice-Hall, Englewood Cliffs, NJ.

W. E. Arnoldi (1951), 'The principle of minimized iterations in the solution of the matrix eigenvalue problem', *Quart. Appl. Math.* **9**, 17–29.

Z. Bai (2002), 'Krylov subspace techniques for reduced-order modeling of large-scale dynamical systems', *Appl. Numer. Math.* **43**, 9–44.

Z. Bai and R. W. Freund (2000), 'Eigenvalue-based characterization and test for positive realness of scalar transfer functions', *IEEE Trans. Automat. Control* **45**, 2396–2402.

Z. Bai and R. W. Freund (2001*a*), 'A partial Padé-via-Lanczos method for reduced-order modeling', *Linear Algebra Appl.* **332–334**, 139–164.

Z. Bai and R. W. Freund (2001*b*), 'A symmetric band Lanczos process based on coupled recurrences and some applications', *SIAM J. Sci. Comput.* **23**, 542–562.

Z. Bai, P. Feldmann and R. W. Freund (1998), How to make theoretically passive reduced-order models passive in practice, in *Proc. IEEE 1998 Custom Integrated Circuits Conference*, IEEE, Piscataway, NJ, pp. 207–210.

Z. Bai, P. M. Dewilde and R. W. Freund (2002), Reduced-order modeling, Numerical Analysis Manuscript No. 02–4–13, Bell Laboratories, Lucent Technologies, Murray Hill, NJ. Also available online from: `http://cm.bell-labs.com/cs/doc/02`.

G. A. Baker, Jr. and P. Graves-Morris (1996), *Padé Approximants*, 2nd edn, Cambridge University Press, New York.

D. J. Bender and A. J. Laub (1987), 'The linear-quadratic optimal regulator for descriptor systems', *IEEE Trans. Automat. Control* **32**, 672–688.

S. Boyd, L. El Ghaoui, E. Feron and V. Balakrishnan (1994), *Linear Matrix Inequalities in System and Control Theory*, SIAM Publications, Philadelphia, PA.

A. Bultheel and B. De Moor (2000), 'Rational approximation in linear systems and control', *J. Comput. Appl. Math.* **121**, 355–378.

A. Bultheel and M. Van Barel (1986), 'Padé techniques for model reduction in linear system theory: a survey', *J. Comput. Appl. Math.* **14**, 401–438.

S. L. Campbell (1980), *Singular Systems of Differential Equations*, Pitman, London.

S. L. Campbell (1982), *Singular Systems of Differential Equations II*, Pitman, London.

C.-K. Cheng, J. Lillis, S. Lin and N. H. Chang (2000), *Interconnect Analysis and Synthesis*, Wiley, New York.

E. Chiprout and M. S. Nakhla (1994), *Asymptotic Waveform Evaluation*, Kluwer Academic Publishers, Norwell, MA.

P. M. Chirlian (1967), *Integrated and Active Network Analysis and Synthesis*, Prentice-Hall, Englewood Cliffs, NJ.

J. V. Clark, N. Zhou and K. S. J. Pister (1998), MEMS simulation using SUGAR v0.5, in *Proc. Solid-State Sensors and Actuators Workshop*, Hilton Head Island, SC, pp. 191–196.

R. W. Clough and J. Penzien (1975), *Dynamics of Structures*, McGraw-Hill.

J. Cullum and T. Zhang (2002), 'Two-sided Arnoldi and nonsymmetric Lanczos algorithms', *SIAM J. Matrix Anal. Appl.* **24**, 303–319.

L. Dai (1989), *Singular Control Systems*, Vol. 118 of *Lecture Notes in Control and Information Sciences*, Springer, Berlin, Germany.

J. Davidse (1991), *Analog Electronic Circuit Design*, Prentice-Hall, New York.

P. Feldmann and R. W. Freund (1994), Efficient linear circuit analysis by Padé approximation via the Lanczos process, in *Proceedings of EURO-DAC '94 with EURO-VHDL '94*, IEEE Computer Society Press, Los Alamitos, CA, pp. 170–175.

P. Feldmann and R. W. Freund (1995*a*), 'Efficient linear circuit analysis by Padé approximation via the Lanczos process', *IEEE Trans. Computer-Aided Design* **14**, 639–649.

P. Feldmann and R. W. Freund (1995*b*), Reduced-order modeling of large linear subcircuits via a block Lanczos algorithm, in *Proc. 32nd ACM/IEEE Design Automation Conference*, ACM, New York, pp. 474–479.

P. Feldmann and R. W. Freund (1997), Circuit noise evaluation by Padé approximation based model-reduction techniques, in *Technical Digest of the 1997 IEEE/ACM Int. Conf. on Computer-Aided Design*, IEEE Computer Society Press, Los Alamitos, CA, pp. 132–138.

L. Fortuna, G. Nunnari and A. Gallo (1992), *Model Order Reduction Techniques with Applications in Electrical Engineering*, Springer, London.

R. W. Freund (1993), Solution of shifted linear systems by quasi-minimal residual iterations, in *Numerical Linear Algebra* (L. Reichel, A. Ruttan and R. S. Varga, eds), W. de Gruyter, Berlin, Germany, pp. 101–121.

R. W. Freund (1995), Computation of matrix Padé approximations of transfer functions via a Lanczos-type process, in *Approximation Theory VIII, Vol. 1: Approximation and Interpolation* (C. Chui and L. Schumaker, eds), World Scientific, Singapore, pp. 215–222.

R. W. Freund (1997), Circuit simulation techniques based on Lanczos-type algorithms, in *Systems and Control in the Twenty-First Century* (C. I. Byrnes, B. N. Datta, D. S. Gilliam and C. F. Martin, eds), Birkhäuser, Boston, pp. 171–184.

R. W. Freund (1999*a*), Passive reduced-order models for interconnect simulation and their computation via Krylov-subspace algorithms, in *Proc. 36th ACM/IEEE Design Automation Conference*, ACM, New York, pp. 195–200.

R. W. Freund (1999*b*), Reduced-order modeling techniques based on Krylov subspaces and their use in circuit simulation, in *Applied and Computational Control, Signals, and Circuits* (B. N. Datta, ed.), Vol. 1, Birkhäuser, Boston, pp. 435–498.

R. W. Freund (2000*a*), Band Lanczos method (Section 7.10), in *Templates for the Solution of Algebraic Eigenvalue Problems: A Practical Guide* (Z. Bai, J. Demmel, J. Dongarra, A. Ruhe and H. van der Vorst, eds), SIAM Publications, Philadelphia, PA, pp. 205–216. Also available online from: `http://cm.bell-labs.com/cs/doc/99`.

R. W. Freund (2000*b*), 'Krylov-subspace methods for reduced-order modeling in circuit simulation', *J. Comput. Appl. Math.* **123**, 395–421.

R. W. Freund and P. Feldmann (1996*a*), Reduced-order modeling of large passive linear circuits by means of the SyPVL algorithm, in *Tech. Dig. 1996 IEEE/ACM International Conference on Computer-Aided Design*, IEEE Computer Society Press, Los Alamitos, CA, pp. 280–287.

R. W. Freund and P. Feldmann (1996*b*), 'Small-signal circuit analysis and sensitivity computations with the PVL algorithm', *IEEE Trans. Circuits and Systems, II: Analog and Digital Signal Processing* **43**, 577–585.

R. W. Freund and P. Feldmann (1997), The SyMPVL algorithm and its applications to interconnect simulation, in *Proc. 1997 International Conference on Simulation of Semiconductor Processes and Devices*, IEEE, Piscataway, NJ, pp. 113–116.

R. W. Freund and P. Feldmann (1998), Reduced-order modeling of large linear passive multi-terminal circuits using matrix-Padé approximation, in *Proc. Design, Automation and Test in Europe Conference 1998*, IEEE Computer Society Press, Los Alamitos, CA, pp. 530–537.

R. W. Freund and F. Jarre (2000), An extension of the positive real lemma to descriptor systems, Numerical Analysis Manuscript No. 00–3–09, Bell Laboratories, Murray Hill, NJ. Also available online from: `http://cm.bell-labs.com/cs/doc/00`.

R. W. Freund and F. Jarre (2003), Numerical computation of nearby positive real systems in the descriptor case, Numerical analysis manuscript, Bell Laboratories, Murray Hill, NJ. In preparation.

R. W. Freund and M. Malhotra (1997), 'A block QMR algorithm for non-Hermitian linear systems with multiple right-hand sides', *Linear Algebra Appl.* **254**, 119–157.

K. Gallivan, E. J. Grimme and P. Van Dooren (1994), 'Asymptotic waveform evaluation via a Lanczos method', *Appl. Math. Lett.* **7**, 75–80.

S. Glavaški, J. E. Marsden and R. M. Murray (1998), Model reduction, centering, and the Karhunen–Loève expansion, in *Proc. 37th IEEE Conference on Decision and Control*, IEEE, Piscataway, NJ, pp. 2071–2076.

K. Glover (1984), 'All optimal Hankel-norm approximations of linear multivariable systems and their l^∞-error bounds.', *Internat. J. Control* **39**, 1115–1193.

W. B. Gragg (1974), 'Matrix interpretations and applications of the continued fraction algorithm', *Rocky Mountain J. Math.* **4**, 213–225.

P. Holmes, J. L. Lumley and G. Berkooz (1996), *Turbulence, Coherent Structures, Dynamical Systems and Symmetry*, Cambridge University Press, Cambridge.

S.-Y. Kim, N. Gopal and L. T. Pillage (1994), 'Time-domain macromodels for VLSI interconnect analysis', *IEEE Trans. Computer-Aided Design* **13**, 1257–1270.

C. Lanczos (1950), 'An iteration method for the solution of the eigenvalue problem of linear differential and integral operators', *J. Res. Nat. Bur. Standards* **45**, 255–282.

R. Lozano, B. Brogliato, O. Egeland and B. Maschke (2000), *Dissipative Systems Analysis and Control*, Springer, London.

I. Masubuchi, Y. Kamitane, A. Ohara and N. Suda (1997), 'H_∞ control for descriptor systems: Matrix inequalities approach', *Automatica J. IFAC* **33**, 669–673.

B. C. Moore (1981), 'Principal component analysis in linear systems: Controllability, observability, and model reduction', *IEEE Trans. Automat. Control* **26**, 17–31.

A. Odabasioglu (1996), 'Provably passive RLC circuit reduction', MS thesis, Department of Electrical and Computer Engineering, Carnegie Mellon University.

A. Odabasioglu, M. Celik and L. T. Pileggi (1997), PRIMA: passive reduced-order interconnect macromodeling algorithm, in *Tech. Dig. 1997 IEEE/ACM International Conference on Computer-Aided Design*, IEEE Computer Society Press, Los Alamitos, CA, pp. 58–65.

D. P. O'Leary (1980), 'The block conjugate gradient algorithm and related methods', *Linear Algebra Appl.* **29**, 293–322.

L. T. Pileggi (1995), Coping with RC(L) interconnect design headaches, in *Tech. Dig. 1995 IEEE/ACM International Conference on Computer-Aided Design*, IEEE Computer Society Press, Los Alamitos, CA, pp. 246–253.

L. T. Pillage and R. A. Rohrer (1990), 'Asymptotic waveform evaluation for timing analysis', *IEEE Trans. Computer-Aided Design* **9**, 352–366.

V. Raghavan, R. A. Rohrer, L. T. Pillage, J. Y. Lee, J. E. Bracken and M. M. Alaybeyi (1993), AWE-inspired, in *Proc. IEEE Custom Integrated Circuits Conference*, pp. 18.1.1–18.1.8.

M. Rathinam and L. R. Petzold (2002), A new look at proper orthogonal decomposition. Submitted manuscript, University of California, Santa Barbara.

R. A. Rohrer and H. Nosrati (1981), 'Passivity considerations in stability studies of numerical integration algorithms', *IEEE Trans. Circuits and Systems* **28**, 857–866.

R. A. Rohrer, L. Nagel, R. Meyer and L. Weber (1971), 'Computationally efficient electronic-circuit noise calculations', *IEEE J. Solid-State Circuits* **6**, 204–213.

J. Roychowdhury (1999), 'Reduced-order modeling of time-varying systems', *IEEE Trans. Circuits and Systems, II: Analog and Digital Signal Processing* **46** 1273–1288.

A. E. Ruehli (1974), 'Equivalent circuit models for three-dimensional multiconductor systems', *IEEE Trans. Microwave Theory Tech.* **22**, 216–221.

T.-J. Su and R. R. Craig, Jr. (1991), 'Model reduction and control of flexible structures using Krylov vectors', *J. Guidance Control Dynamics* **14**, 260–267.

F. Tisseur and K. Meerbergen (2001), 'The quadratic eigenvalue problem', *SIAM Rev.* **43**, 235–286.

G. C. Verghese, B. C. Lévy and T. Kailath (1981), 'A generalized state-space for singular systems', *IEEE Trans. Automat. Control* **26**, 811–831.

C. de Villemagne and R. E. Skelton (1987), 'Model reductions using a projection formulation', *Internat. J. Control* **46**, 2141–2169.

J. Vlach and K. Singhal (1994), *Computer Methods for Circuit Analysis and Design*, 2nd edn, Van Nostrand Reinhold, New York.

J. L. Willems (1970), *Stability Theory of Dynamical Systems*, Wiley, New York.

K. Zhou, J. C. Doyle and K. Glover (1996), *Robust and Optimal Control*, Prentice-Hall, Upper Saddle River, NJ.

A. van der Ziel (1986), *Noise in Solid State Devices and Circuits*, Wiley, New York.

Acta Numerica (2003), pp. 321–398
DOI: 10.1017/S0962492902000132

A mathematical view of automatic differentiation

Andreas Griewank
Institute of Scientific Computing,
Department of Mathematics,
Technische Universität Dresden,
01052 Dresden, Germany
E-mail: griewank@math.tu-dresden.de

Automatic, or algorithmic, differentiation addresses the need for the accurate and efficient calculation of derivative values in scientific computing. To this end procedural programs for the evaluation of problem-specific functions are transformed into programs that also compute the required derivative values at the same numerical arguments in floating point arithmetic. Disregarding many important implementation issues, we examine in this article complexity bounds and other more mathematical aspects of the program transformation task sketched above.

CONTENTS

1. Introduction

Practically all calculus-based numerical methods for nonlinear computations are based on truncated Taylor expansions of problem-specific functions. Naturally we have to exempt from this blanket assertion those methods that are targeted for models defined by functions that are very rough, or even nondeterministic, as is the case for functions that can only be measured experimentally, or evaluated by Monte Carlo simulations. However, we should bear in mind that, in many other cases, the roughness and nondeterminacy may only be an artifact of the particular way in which the function is evaluated. Then a reasonably accurate evaluation of derivatives may well be possible using variants of the methodology described here. Consequently, we may view the subject of this article as an effort to extend the applicability of classical numerical methodology, dear to our heart, to the complex, large-scale models arising today in many fields of science and engineering. Sometimes, numerical analysts are content to verify the efficacy of their sophisticated methods on suites of academic test problems. These may have a very large number of variables and exhibit other complications, but they may still be comparatively easy to manipulate, especially as far as the calculation of derivatives is concerned. Potential users with more complex models may then give up on calculus-based models and resort to rather crude methods, possibly after drastically reducing the number of free variables in order to avoid the prohibitive runtimes resulting from excessive numbers of model reruns.

Not surprisingly, on the kind of real-life model indicated above, nothing can be achieved in an entirely automatic fashion. Therefore, the author much prefers the term 'algorithmic differentiation' and will refer to the subject from here on using the common acronym AD. Beyond this minor labelling issue lies the much more serious task of deciding which research and development activities ought to be described in a fair but focused survey on AD. To some, AD is an exercise in software development, about which they want to read (or write) nothing but a crisp online documentation. For others, AD has become the mainstay of their research activity with plenty of intriguing questions to resolve. As in all interdisciplinary endeavours, there are many close connections to other fields, especially numerical linear algebra, computer algebra, and compiler writing. It would be preposterous as well as imprudent to stake out an exclusive claim for any particular subject area or theoretical result. There has been a series of workshops and conferences focused on AD and its applications (Griewank and Corliss 1991, Berz, Bischof, Corliss and Griewank 1996, Corliss, Faure, Griewank, Hascoët and Naumann 2001), but nobody has seen the need for a dedicated journal. Results appear mostly in journals on optimization, or more generally, numerical analysis and scientific computing.

1.1. How numeric and how symbolic is AD?

There is a certain dichotomy in that the final results of AD are numerical derivative values, which are usually thought to belong to the domain of nonlinear analysis and optimization. On the other hand, the development of AD techniques and tools can for the most part stay blissfully oblivious to issues of floating point arithmetic, much like sparse matrix factorization in the positive definite case. Instead, we need to analyse and manipulate discrete objects like computer codes and related graphs, sometimes employing tools of combinatorial optimization. There are no division operations at all, and questions of numerical stability or convergence orders and rates arise mostly for dynamic codes that represent iterative solvers or adaptive discretizations.

The last assertion flies in the face of the notion that differentiation is an ill-conditioned process, which is firmly ingrained in the minds of numerical mathematicians. Whether or not this conviction is appropriate depends on the way in which the functions to be differentiated are provided by the *user*. If we merely have an oracle generating function values with prescribed accuracy, derivatives can indeed only be estimated quite inaccurately as divided differences; possibly averaged over a number of trial evaluations in situations where errors can be assumed to have a zero mean statistically. If, on the other hand, the *oracle* takes the form of a computer code for evaluating the function, then this code can often be analysed and transformed to yield an extended code that also evaluates desired derivatives. In both scenarios we have backward stability *à la* Wilkinson, in that the floating point values obtained can be interpreted as the exact derivatives of a 'slightly' perturbed problem with the same structure. The crucial difference is that in the first scenario there is not much structure to be preserved, as the functions may be subjected to discontinuous perturbations, generating slope changes of arbitrary size. In the second scenario the function is effectively prescribed as the composite of elementary functions, whose values and derivative are only subject to variations on the order of the machine accuracy.

More abstractly, we may simply state that, as a consequence of the chain rule, the composition operation

$$F \equiv F_2 \circ F_1 \quad \text{for} \quad F_i : \Omega_{i-1} \mapsto \Omega_i$$

is a jointly continuous mapping between Banach spaces of differentiable functions, *i.e.*, it belongs to

$$\mathcal{C}^1(\Omega_0, \Omega_1) \times \mathcal{C}^1(\Omega_1, \Omega_2) \mapsto \mathcal{C}^1(\Omega_0, \Omega_2).$$

In practical terms this means that we differentiate the composite function F by appropriately combining the derivatives of the two constituents F_1 and F_2, assuming that procedures for their evaluation are already in place. By doing this recursively, we effectively extend an original procedure for evaluating function values by themselves into one that also evaluates derivatives. This is essentially an algebraic manipulation, which again reflects the two-sided nature of AD, partly symbolic and partly numeric.

The term *numeric differentiation* is widely used to describe the common practice of approximating derivatives by divided differences (Anderssen and Bloomfield 1974). This approximation of differential quotients by difference quotients means geometrically the replacement of tangent slopes by secant slopes over a certain increment in each independent variable. It is well known that, even for the increment size that optimally balances truncation and rounding error, half of the significant digits are lost. Of course, the optimal increment typically differs for each variable/function pair and the situation is still more serious when it comes to approximating higher derivatives. No such difficulty occurs in AD, as no parameter needs to be selected and there is no significant loss of accuracy at all. Of course, the application of the chain rule does entail multiplications and additions of floating point numbers, which can only be performed with platform-dependent finite precision.

Since AD has really very little in common with divided differences, it is rather inappropriate to describe it as a 'halfway house' between numeric and symbolic differentiation, as is occasionally done in the literature. What then is the key feature that distinguishes AD from *fully symbolic differentiation*, as performed by most computer algebra (CA) systems? A short answer would be to say that AD applies the chain rule to floating point numbers rather than algebraic expressions. Indeed, no AD package yields tidy mathematical formulas for the derivatives of functions, no matter how algebraically simple they may be. All we get is more code, typically source or binary. It is usually not nice to look at and, as such, it never provides any analytical insight into the nature of the function and its derivatives. Of course, for multi-layered models, such insight by inspection is generally impossible anyway, as the direct expression of function values in terms of the independent variables leads to formulas that are already impenetrably complex. Instead, the aim in AD is the accurate evaluation of derivative values at a sequence of arguments, with an *a priori* bounded complexity in terms of the operation count, memory accesses, and memory size.

1.2. Various phases and costs of AD

Relative to the complexity of the function itself, the complexity growth is at worst linear in the number of independent or dependent variables and quadratic in the degree of the derivatives required. This moderate and predictable extra cost, combined with the achievement of good derivative accuracy and the absence of any free method parameter, has led to the wide-spread use of AD as built-in functionality in AMPL, GAMS, NEOS, and other modelling or optimization systems. In contrast, the complexity of symbolic manipulations typically grows exponentially with the depth of the expression tree or the computational graph that represent a function evaluation procedure. By making such blanket statements we are in danger of comparing apples to oranges.

In computing derivatives using either CA or AD we have to distinguish at least two distinct phases and their respective costs: first, a *symbolic phase* in which the given function specification is analysed and a procedure for evaluating the desired derivatives is prepared in a suitable fashion; second, a *numeric phase* where this evaluation is actually carried out at one or several numerical arguments. The more effort we invest in the symbolic phase the more runtime efficiency we can expect in the numeric phase, which will hopefully be applied sufficiently often to amortize the initial symbolic investment. For example, extensive symbolic preprocessing of a routine specifying a certain kind of finite element or the right-hand side of a stiff ODE is likely to pay off if the same routine and its derivative procedure are later called in the numeric phase at thousands of grid points or time-steps, respectively. On the other hand, investing much effort in the symbolic phase may not pay off, or simply be impossible, if the control flow of the original routine is very dynamic. For example, one may think of a recursive quadrature routine that adaptively subdivides the domains depending on certain error estimates that can vary dramatically from call to call. Then very little control flow information can be gleaned at compile-time and code optimization is nearly impossible.

Of course, we can easily conceive of a multi-phase scenario where the function specification is specialized at various levels and the corresponding derivative procedures are successively refined accordingly. For example, we may have an intermediate *qualitative phase* where, given the specification of certain parameters like mesh sizes and vector dimensions, the AD tool determines the sparsity pattern of a Jacobian matrix and correspondingly allocates suitable data structures in preparation for the subsequent numeric phase. Naturally, it may also output the sparsity pattern or provide other qualitative dependence information, such as the maximal rank of the Jacobian or even the algebraic multiplicity of certain eigenvalues.

Traditionally in AD the symbolic effort has been kept to the equivalent of a few compiler-type passes through the function specification, possibly including a final compilation by a standard compiler. Hence the symbolic effort is at most of the order of a single function evaluation, albeit probably with a rather large constant. However, this characteristic of all current AD tools may change when more extensive dependence analyses or combinatorial optimizations of the way in which the chain rule is applied are incorporated. For the most part these considerable extra efforts in the symbolic phase will reduce the resulting costs in the numerical phase only by constants but not by orders of magnitude. Some of them are rather mundane rearrangements or software modifications with no allure for the mathematician. Of course, such improvements can make all the difference for the practical feasibility of certain calculations.

This applies in particular to the calculation of gradients in the reverse mode of AD. Being a discrete analogue of adjoints in ODEs, this backward application of the chain rule yields all partials of a scalar function with respect to an arbitrary number of variables at the cost of a small multiple of the operation count for evaluating the function itself. If only multiplications are counted, this growth factor is at most 3, and for a typical evaluation code probably more like 2 on average. That kind of value seems reasonable to people who skilfully write adjoint code by hand (Zou, Vandenberghe, Pondeca and Kuo 1997), and it opens up the possibility of turning simulation codes into optimization codes with a growth of the total runtime by a factor in the tens rather than the hundreds. Largely because of memory effects, it is not uncommon that the runtime ratio for a single gradient achieved by current AD tools is of the order 10 rather than 2. Fortunately, the memory-induced cost penalty is the same if a handful of gradients, forming the Jacobian of a vector function, is evaluated simultaneously in what is called the *vector-reverse mode.*

In general, it is very important that the user or algorithm designer correctly identifies which derivative information is actually needed at any particular point in time, and then gets the AD tool to generate all of it jointly. In this way optimal use can be made of common subcalculations or memory accesses. The latter may well determine the wall clock time on a modern computer. Moreover, the term *derivative information* needs to be interpreted in a wider sense, not just denoting vectors, matrices or even tensors of partial derivatives, as is usually understood in hand-written formulas or the input to computer algebra systems. Such rectangular derivative arrays can often be contracted by certain weighting and direction vectors to yield derivative objects with fewer components. By building this contraction into the AD process all aspects of the computational costs can usually be significantly reduced. A typical scenario is the iterative calculation of approximate Newton steps, where only a sequence of Jacobian–vector, and possibly vector–Jacobian, products are needed, but never the Jacobian as such. In fact, as we will discuss in Section 6, it is not really clear what the 'Jacobian as such' really is. Instead we may have to consider various representations depending on the ultimate purpose of our numerical calculation. Even if this ultimate purpose is to compute exact Newton steps by a finite procedure, first calculating all Jacobian entries may not be a good idea, irrespective of whether it is sparse or not. In some sense Newton steps themselves become *derivative objects.* While that may seem a conceptual stretch, there are some other mathematical objects, such as Lipschitz constants, error estimates, and interval enclosures, which are naturally related to derivatives and can be evaluated by techniques familiar from AD.

1.3. Some historical remarks

Historically, there has been a particularly close connection between the reverse mode of AD and efforts to estimate the effects of rounding errors in evaluating functions or performing certain numerical algorithms. The adjoint quantities generated in the reverse mode for each intermediate variable of an evaluation process are simply the sensitivities of the weighted output with respect to perturbations of this particular variable. Hence we obtain the first-order Taylor expansion of the output error with respect to all intermediate errors, a linearized estimate apparently first published by Linnainmaa in 1972 (see Linnainmaa (1976)). Even earlier, in 1966, Volin and Ostrovski (see G. M. Ostrovskii and Borisov (1971)) suggested the reverse mode for the optimization of certain process models in chemical engineering. Interestingly, the first major publication dedicated to AD, namely the seminal book by Louis Rall (1983) did not recognize the reverse, or adjoint, mode as a variant of AD at all, but instead covered exclusively the forward or direct mode. This straightforward application of the chain rule goes back much further, at least to the early 1950s, when it was realized that computers could perform symbolic as well as numerical manipulations. In the past decade there has been growing interest in combinations of the forward and reverse mode, a wide range of possibilities, which had in principle already been recognized by Ostrovski *et al.* (see Volin and Ostrovskii (1985)).

Based on theoretical work by Cacuci, Weber, Oblow and Marable (1980), the first general-purpose tool implementing the forward and reverse mode was developed in the 1980s by Ed Oblow, Brian Worley and Jim Horwedel at Oak Ridge National Laboratory. Their system GRESS/ADGEN was successfully applied to many large scientific and industrial codes, especially from nuclear engineering (Griewank and Corliss 1991). Hence there was an early proof of concept, which did, however, have next to no impact in the numerical analysis community, where the notion that derivatives are always hard if not impossible to come by persisted for a long time, possibly to this day. Around the time of the first workshop on AD at Breckenridge in 1991, several system projects were started (*e.g.*, ADIFOR, Odyssée, TAMC, ADOL-C), though regrettably none with sufficient resources for the development of a truly professional tool. Currently several projects are under way, some to cover MATLAB, and others that aim at an even closer integration into a compilation platform for several source languages (Hague and Naumann 2001).

As suggested by the title, in this article we will concentrate on the mathematical questions of AD and leave the implementation issues largely aside. Following current practice in numerical linear algebra, we will measure computational complexity by counting fused multiply-adds and denote them simply by OPS. They can be performed in one or two cycles on modern super-scalar processors, and are essentially the only extra arithmetic operations introduced by AD. Furthermore, in Section 4 on reversal schedules, we will also emphasize the maximal memory requirement and the total number of memory accesses. They may dominate the runtime even though most of them occur in a strictly sequential fashion, thus causing only a minimal number of cache misses.

1.4. Structure of the article

The paper is organized as follows. In Section 2 we set up a framework of function evaluation procedures that is invariant with respect to adjoining, *i.e.*, application of reverse mode AD. To achieve this we consider from the beginning vector-valued and incremental elemental functions, whereas in Griewank (2000) and most other presentations of AD only scalar-valued assignments are discussed. These generalizations make the notation in some respects a little more complicated, and will therefore be suspended in the second part of Section 6. That final section discusses the rather intricate relationship between Jacobian matrices and computational graphs. It is in part speculative and meant to suggest directions of future research in AD. Section 3 reviews the basic and well-established techniques of AD, with methods for the evaluation of sparse Jacobians being merely sketched verbally. For further details on this and other aspects one may consult the author's book (Griewank 2000).

The following three sections treat rather different topics and can be read separately from each other. In many people's minds the main objection to the reverse mode in its basic form is its potentially very large demand for temporary memory. This may either exceed available storage or make the execution painfully slow owing to extensive data transfers to remote regions of the memory hierarchy. Therefore, in Section 4 we have elaborated checkpointing techniques for program reversals on serial and parallel machines in considerable detail. The emphasis is on the discrete optimization problem of checkpoint placement rather than their implementation from a computer science perspective. In Section 5 we will consider adjoints of iterative processes for the solution of linear or nonlinear state equations. In that case loop reversals can be avoided altogether using techniques that were also considered in Hascoët, Fidanova and Held (2001), Giles and Süli (2002) and Becker and Rannacher (2001).

Apart from second-order adjoints, which are covered in Sections 3 and 5, we will not discuss the evaluation of higher derivatives even though there are some interesting recent applications to DAEs and ODEs (see Pantelides (1988), Pryce (1998) and Röbenack and Reinschke (2000)). Explicit formulas for the higher derivatives of composite functions were published by Faa di Bruno (1856). Recursive formulas for the forward propagation of Taylor series with respect to a single variable were published by Moore (1979) and have been reproduced many times since: see, for instance, Kedem (1980), Rall (1983), Lohner (1992) and Griewank (2000). The key observation is that, because all elemental functions of interest, *e.g.*, $\exp(x)$, $\sin(x)$, *etc.*, are solutions of linear ODEs, univariate Taylor series can be propagated with a complexity that grows only quadratically with the degree d of the highest coefficient. Using FFT or other fast convolution algorithms we can reduce this complexity to order $d \ln(1 + d)$, a possibility that has not been exploited in practice. Following a suggestion by Rall, it was shown in Bischof, Corliss and Griewank (1993), Griewank, Utke and Walther (2000) and Neidinger (200x) that multivariate Taylor expansions can be computed rather efficiently using families of univariate polynomials. Similarly, the reverse propagation of Taylor polynomials does not introduce any really new aspects, and can be interpreted as performing the usual reverse mode in Taylor series arithmetic (Christianson 1992, Griewank 2000).

Like the kind of material presented, the style and depth of treatment in this article is rather heterogeneous. Many observations are just asserted verbally with references to the literature but others are formalized as lemmas or propositions. Proofs have been omitted, except for a new, shorter demonstration that binomial reversal schedules are optimal in Proposition 4.2, the proof of the folklore result, Proposition 6.3, concerning the generic rank of a Jacobian, and some simple observations regarding the new concept of Jacobian scarcity introduced in Section 6. For a more detailed exposition of the basic material the reader should consult the author's book (Griewank 2000), and for more recent results and application studies the proceedings volume edited by Corliss *et al.* (2001). A rather comprehensive bibliography on AD is maintained by Corliss (1991).

2. Evaluation procedures in incremental form

Rather than considering functions 'merely' as mathematical mappings we have to specify them by evaluation procedures and to a large extent identify the two. These conceptual codes are abstractions of actual computer codes, as they may be written in Fortran or C and their various extensions. Generally, the quality and complexity of the derived procedures for evaluating derivatives will reflect the properties of the underlying function evaluation procedure quite closely. The following formalism does not make any really substantial assumptions other than that the procedure represents a finite algorithm without branching of the control flow. We just have to be a little careful concerning the handling of intermediate quantities.

The basic assumption of AD is that the function to be differentiated is, at least conceptually, evaluated by a sequence of *elemental* statements, such as

$$v_i = \varphi_i(u_i) \quad \text{or} \quad v_i \mathrel{+}= \varphi_i(u_i) \quad \text{for all} \quad i \in \mathcal{I}. \tag{2.1}$$

In other words we have the evaluation of an elemental function φ_i followed by a standard assignment or an additive incrementation. Here the C-style notation $a \mathrel{+}= b$ abbreviates $a = a+b$ for any two compatible vectors a and b. We include incremental assignments and, later, allow overlap between the left-hand sides v_i, because both aspects occur frequently in adjoint programs. To obtain a unified notation for both kinds of assignment we introduce for each index $i \in \mathcal{I}$ a *flag* $\sigma_i \in \{0, 1\}$, and write

$$v_i = \sigma_i v_i + \varphi_i(u_i) \quad \text{for all} \quad i \in \mathcal{I}. \tag{2.2}$$

To make the adjoint procedure even more symmetric we may prefer the equivalent statement pair

$$\big[v_i \mathrel{*}= \sigma_i;\ v_i \mathrel{+}= \varphi_i(u_i)\big] \quad \text{for all} \quad i \in \mathcal{I}, \tag{2.3}$$

where $a \mathrel{*}= \alpha$ abbreviates $a = \alpha * a$ for any vector a and scalar α. In this way we have a fully incremental form, which turns out to be convenient for adjoining later on. Throughout we will neglect the cost of the additive and multiplicative incrementations. Normally the φ_i will be taken from a library of arithmetic operations and intrinsic functions. However, we may also include basic linear algebra subroutines (BLAS) and other library or user-defined functions, provided corresponding derivative procedures $\dot{\varphi}_i$ and $\bar{\varphi}_i$, as defined in Section 3, can be supplied.

The index set $\mathcal{I}$ is assumed to be partially ordered by an acyclic, nonreflexive precedence relation $j \prec i$, which indicates that φ_j must be applied before φ_i. Deviating from the more mathematical AD literature, including Griewank (2000), we will identify the indices $i \in \mathcal{I}$ with the whole statement (2.2) rather than just the output variable v_i. This approach is more in line with compiler literature (Hascoët 2001). The difference disappears whenever we exclude incremental statements. We may interpret $\mathcal{I}$ as the vertex set of the directed acyclic graph $\mathcal{G} = (\mathcal{I}, \mathcal{E})$ with edge set $\mathcal{E} \equiv \{(j, i) \in \mathcal{I} \times \mathcal{I} \big| \ j \prec i\}$. When the vertices of $\mathcal{G}$ are annotated with the elemental functions φ_i, it is often called the *computational graph*, a concept apparently due to Kantorovich (1957). As a small example we consider the scalar function

$$y = \exp(x_1) * \sin(x_1 + x_2). \tag{2.4}$$

It can be evaluated using 3 intermediates by the program listed in Figure 2.1; on the right is displayed the corresponding computational graph.

Program:

$v_3 = \exp(v_1)$

$v_4 = \sin(v_1 + v_2)$

$v_5 = v_3 * v_4$

Graph:

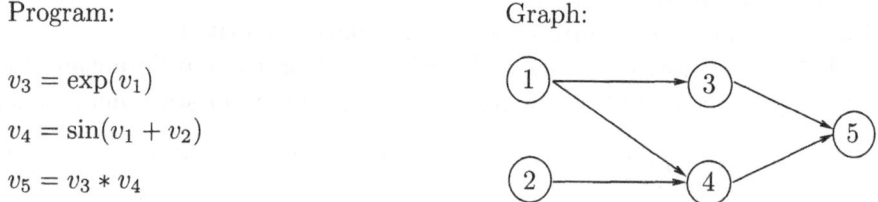

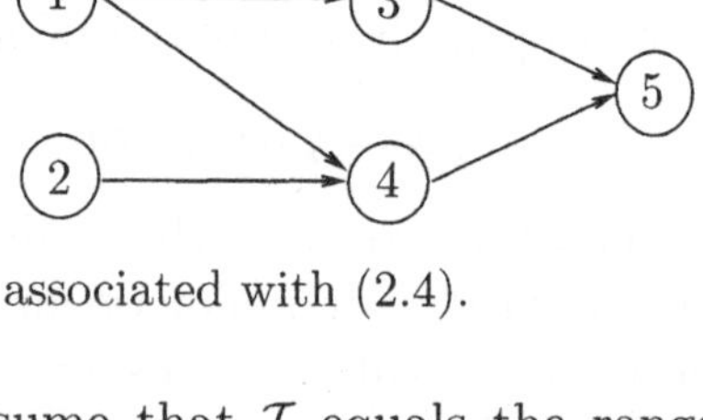

Figure 2.1. Program and graph associated with (2.4).

Without loss of generality, we may assume that $\mathcal{I}$ equals the range of integers $\mathcal{I} = [1 \ldots l]$, and consider (2.2) as a loop representing the succession of statements executed by a computer program for a given set of data.

In particular we consider the control flow to be fixed. The partial ordering $\prec$ of $\mathcal{I}$ allows us to discuss the runtime ratios between the derived and the original evaluation procedures on parallel as well as serial machines. We will relax the usual convention that the variables v_i are real scalars, and instead allow them to be elements of an arbitrary real Hilbert space $\mathcal{H}_i$ of dimension $m_i \equiv \dim(\mathcal{H}_i)$. We will refer to the *scalar case* whenever $m_i = 1$ for all $i = 1 \dots l$.

2.1. The state space and its transformations

In real computer programs several variables are often stored in the same location. Hence we do not require the $\mathcal{H}_i$ to be completely separate but to have a natural embedding into a common *state space* $\mathcal{H}$, such as

$$P_i \mathcal{H}_i \subset \mathcal{H} \quad \text{and} \quad v_i = P_i^T \mathbf{v} \quad \text{for} \quad \mathbf{v} \in \mathcal{H},$$

where P_i is orthogonal, in that $P_i^T P_i = I$ is the identity on $\mathcal{H}_i$. Here and throughout we will identify linear mappings between finite-dimensional spaces with their matrix description in terms of suitable orthonormal bases. Correspondingly we will denote the adjoint mapping of P as the transpose P^T. We will choose $\mathcal{H}$ minimal such that

$$\mathcal{H} \equiv P_1 \mathcal{H}_1 + P_2 \mathcal{H}_2 + \cdots + P_{l-1} \mathcal{H}_{l-1} + P_l \mathcal{H}_l.$$

In the simplest scenario $\mathcal{H}$ is the Cartesian product of all $P_i \mathcal{H}_i$, in which case we have a so-called *single assignment procedure*. Otherwise the variables v_i are said to *overlap*, and we have to specify more carefully what (2.2) and thus (2.3) actually mean. Namely, we assume that there is an underlying state vector $\mathbf{v} \in \mathcal{H}$, which is transformed according to

$$\Phi_i(\mathbf{v}) \equiv \left[I - (1 - \sigma_i)\, P_i P_i^T\right]\mathbf{v} + P_i\, \varphi_i(Q_i\, \mathbf{v}). \tag{2.5}$$

Here the *argument selections* Q_i denote orthogonal projections from $\mathcal{H}$ into the domains $\mathcal{D}_i$ of the elemental functions φ_i so that

$$u_i = Q_i\, \mathbf{v} \in \mathcal{D}_i = \operatorname{dom}(\varphi_i) \quad \text{with} \quad n_i = \dim(\mathcal{D}_i). \tag{2.6}$$

Rather than restricting ourselves to Cartesian projections that pick out coordinate components of $\mathbf{v}$, we allow general linear Q_i, and thus include the important concept of group partial separability (Conn, Gould and Toint 1992) in a natural way. Of course, the φ_i need not be well defined on the whole Euclidean space $\mathcal{D}_i$ in practice, but we will assume here that this difficulty does not occur at arguments of interest. For basic observations regarding exceptional arguments consider Sections 11.2 and 11.3 in Griewank (2000).

To distinguish them verbally from the underlying *elemental functions* φ_i we will refer to the Φ_i as *elemental transitions*. Given any initial state $\mathbf{u} \in \mathcal{H}$ we can now apply the Φ_i in any specified order and obtain a corresponding final state $\mathbf{v} \equiv \Phi(\mathbf{u}) \in \mathcal{H}$ where $\Phi \equiv \Phi_l \circ \Phi_{l-1} \circ \cdots \circ \Phi_1$ denotes the composition of the elemental transitions.

For example (2.4) we may set

$$\mathcal{H} = \mathbb{R}^5, \quad \mathcal{H}_i = \mathbb{R} \quad \text{and} \quad P_i = \mathbf{e}_i \quad \text{for} \quad i = 1 \ldots 5.$$

Here $\mathbf{e}_i \in \mathbb{R}^5$ denotes the ith Cartesian basis vector. Furthermore we select

$$Q_3 = (1\,0\,0\,0\,0), \quad Q_4 = (1\,1\,0\,0\,0), \quad Q_5 = \begin{pmatrix} 0 & 0 & 1 & 0 & 0 \\ 0 & 0 & 0 & 1 & 0 \end{pmatrix},$$

and may thus evaluate (2.4) by the transformations

$$\begin{aligned}
\mathbf{v}^{(0)} &= 0, \\
\mathbf{v}^{(1)} &= \left(I - \mathbf{e}_1\,\mathbf{e}_1^T\right)\mathbf{v}^{(0)} + \mathbf{e}_1\,x_1, \\
\mathbf{v}^{(2)} &= \left(I - \mathbf{e}_2\,\mathbf{e}_2^T\right)\mathbf{v}^{(1)} + \mathbf{e}_2\,x_2, \\
\mathbf{v}^{(3)} &= \left(I - \mathbf{e}_3\,\mathbf{e}_3^T\right)\mathbf{v}^{(2)} + \mathbf{e}_3\,\exp\left(Q_3\,\mathbf{v}^{(2)}\right), \\
\mathbf{v}^{(4)} &= \left(I - \mathbf{e}_4\,\mathbf{e}_4^T\right)\mathbf{v}^{(3)} + \mathbf{e}_4\,\sin\left(Q_4\,\mathbf{v}^{(3)}\right), \\
\mathbf{v}^{(5)} &= \left(I - \mathbf{e}_5\,\mathbf{e}_5^T\right)\mathbf{v}^{(4)} + \mathbf{e}_5\,\mathrm{prod}\left(Q_5\,\mathbf{v}^{(4)}\right),
\end{aligned}$$

where $I = I_5$ and $\mathrm{prod}(a, b) \equiv a * b$ for $(a, b) \in \mathbb{R}^2$.

2.2. No-overwrite conditions

Example (2.4) already shows some properties that we require in general. Namely, we normally impose the natural condition that the function φ_i may only be applied when its argument u_i has reached its final value. In other words we must have evaluated all φ_j that precede φ_i so that

$$Q_i\,P_j \neq 0 \quad \Rightarrow \quad j \prec i. \tag{2.7}$$

In particular we always have $Q_i\,P_i = 0$, as $\prec$ is assumed nonreflexive. In addition to these write–read dependences between the two statements φ_j and φ_i, we also have to worry about write–write dependences where several elemental functions 'overlap', in that equivalently

$$P_i^T P_j \neq 0 \quad \Leftrightarrow \quad P_j^T P_i \neq 0.$$

Overlapping really only makes sense if all but possibly one of the φ_i are incremental, *i.e.*, $\sigma_i = 1$, since otherwise values are overwritten before they are ever used. Moreover, since by their very nature the incremental φ_i with overlapping P_i do commute, their relative order does not matter, but they all must succeed a possible, nonincremental one. Formally we impose the condition

$$P_j^T P_i \neq 0 \quad \Rightarrow \quad \left(j \prec i \quad \text{and} \quad \sigma_i = 1\right) \quad \text{or} \quad \left(i \prec j \quad \text{and} \quad \sigma_j = 1\right). \tag{2.8}$$

From now on we will assume that we have an acyclic ordering satisfying conditions (2.7), (2.8) and furthermore make the monotonicity assumption $j \prec i \Rightarrow j < i$, where $<$ denotes the usual ordering of integers by size.

2.3. Independent and dependent variables

Even though they are patently obvious in many practical codes, we may characterize the independent and dependent variables in terms of the precedence relation $\prec$ as follows. All indices j that are minimal with respect to $\prec$ must have a trivial argument mapping $Q_j = 0$. Otherwise, by the assumed minimality of the state space we would have $Q_j P_k \neq 0$ for some k. All such minimal φ_j initialize v_j to some constant vector x_j unless j is incremental, which we will preclude by assumption. Furthermore, we assume without loss of generality that the minimal indices are given by $j = 1 \ldots n$ and may thus write

$$v_j = x_j \quad \text{for} \quad j = 1 \ldots n.$$

We consider

$$\mathbf{x} = (x_j)_{j=1\ldots n} \in X \equiv P_1 \mathcal{H}_1 \times P_2 \mathcal{H}_2 \times \cdots \times P_{n-1} \mathcal{H}_{n-1} \times P_n \mathcal{H}_n.$$

as the vector of *independent* variables. Similarly, we assume that the maximal indices $i \in \mathcal{I}$ with respect to $\prec$ are given by $i = l - m + 1 \ldots l$. The values v_i with $i > l - m$ do not impact any other elemental function and are therefore considered as *dependent* variables

$$y_i = v_{\hat{\imath}} \quad \text{with} \quad \hat{\imath} \equiv l - m + i \quad \text{for} \quad i = 1 \ldots m.$$

We consider

$$\mathbf{y} = (y_i)_{i=1\ldots m} \in Y \equiv P_{l-m+1} \mathcal{H}_{l-m+1} \times \cdots \times P_{l-1} \mathcal{H}_{l-1} \times P_l \mathcal{H}_l$$

as the vector of dependent variables. Combining our assumption on the independents and dependents in the following additional condition, we obtain

$$j \prec i \quad \Rightarrow \quad i > n \quad \text{and} \quad j \leq l - m. \tag{2.9}$$

Also, we exclude independent or dependent elementals from being incremental, so that

$$i \leq n \quad \text{or} \quad i > l - m \quad \Rightarrow \quad \sigma_i = 0.$$

Consequently, by (2.8) the $P_i \mathcal{H}_i$ for $i = 1 \ldots n$ and $i = l - m + 1 \ldots l$ must be mutually orthogonal, so that we have in fact $X \subset \mathcal{H}$ and $Y \subset \mathcal{H}$.

2.4. Four-part form and complexity

To ensure that the result of our evaluation procedure is uniquely defined even when there are incremental assignments, we will assume that the state

Table 2.1. Original function evaluation procedure.

$\mathbf{v}$	$=$	0	
v_i	$=$	x_i	for $i = 1 \ldots n$
v_i v_i	$*=$ $+=$	σ_i $\varphi_i(u_i)$	for $i = n+1 \ldots l$
y_{m-i}	$=$	v_{l-i}	for $i = m-1 \ldots 0$

vector $\mathbf{v}$ is initialized to zero. Hence we obtain the program structure displayed in Table 2.1.

With P_X and P_Y the orthogonal projections from $\mathcal{H}$ onto its subspaces X and Y, we obtain the mapping from $\mathbf{x} = \left(x_j\right)_{j=1}^{n}$ to $\mathbf{y} = \left(y_i\right)_{i=1}^{m}$ as the composite function

$$F \quad \equiv P_Y \circ \Phi_l \circ \Phi_{l-1} \circ \cdots \circ \Phi_2 \circ \Phi_1 \circ P_X^T, \tag{2.10}$$

where the transitions Φ_i are as defined in (2.5). The application of (2.10) to a particular vector $\mathbf{x} \in X$ can be viewed as the loop

$$\mathbf{v}^{(0)} = P_X^T \mathbf{x}, \quad \mathbf{v}^{(i)} = \Phi_i\left(\mathbf{v}^{(i-1)}\right) \quad \text{for} \quad i = 1 \ldots l, \quad \mathbf{y} = P_Y \, \mathbf{v}^{(l)}. \tag{2.11}$$

We may summarize the mathematical development in this section as follows.

General Assumption: Elemental Decomposition.

(i) The elemental functions $\varphi_i : \mathcal{D}_i \mapsto \mathcal{H}_i = P_i^T \mathcal{H}$ are $d \geq 1$ times continuously differentiable on their open domains $\mathcal{D}_i = Q_i \, \mathcal{H}$.

(ii) The pairs of linear mappings $\{P_i, Q_i\}$ and the partial ordering $\prec$ are consistent in that (2.7), (2.8), and (2.9) hold.

As an immediate consequence we state the following result without proof.

Proposition 2.1. Given the General Assumption, Table 2.1 yields, for any monotonic total ordering of $\mathcal{I}$, the same unique vector function defined in (2.10), that is,

$$X \ni \mathbf{x} \mapsto \mathbf{y} = F(\mathbf{x}) \in Y,$$

which is d times continuously differentiable, by the chain rule.

Only in Sections 3.6 and 5.4 will we need $d > 1$. The partial ordering in $\mathcal{I}$ is only important for parallel evaluation and reversal without a value stack, as described in Section 3.2.

As to the computational cost, we will assume that the operation count is additive in that

$$\texttt{OPS}\,(F(\mathbf{x})) = \sum_{i=1}^{l} \texttt{OPS}\,(\varphi_i(u_i)), \tag{2.12}$$

where we have neglected the cost of performing the incrementations and projections.

On a parallel machine elemental functions φ_i and φ_j that do not depend on each others' results can be executed simultaneously. It is customary to assume that a read conflict, *i.e.*, $Q_i\,Q_j^T \neq 0$, is no inhibition for concurrent execution, and we will assume here that the same is true for the incrementation of results, *i.e.*, $P_i^T P_j \neq 0$. This is a crucial assumption for showing that adjoints have the same degree of parallelism as the original evaluation procedure. Whether or not it is realistic depends on the computing platform and the nature of the elemental functions φ_i. If they are rather chunky, involving many internal calculations, the delays through incremental write conflict are indeed likely to be negligible. In any case, it makes no sense to schedule individual operations separately on a single processor machine, so that we may consider the φ_i to be more substantial subtasks in a parallel context. Then we can estimate the wall clock time by the longest, and thus critical path, *i.e.*,

$$\texttt{CLOCK}\,(F(\mathbf{x})) = \max_{\mathcal{P}\subset\mathcal{G}} \sum_{i\in\mathcal{P}} \texttt{OPS}\,(\varphi_i). \tag{2.13}$$

Here $\mathcal{P}$ denotes a directed path in $\mathcal{G}$ with $i \in \mathcal{P}$ ranging over its vertices. Throughout we will use CLOCK to denote the idealized wall clock time on a parallel machine with an unlimited supply of processors and zero communication costs. For practical implementations of AD with parallelism, see Carle and Fagan (1996), Christianson, Dixon and Brown (1997) and Mancini (2001).

3. Basic forward and reverse mode of AD

In this section we introduce the basic forward and reverse mode of AD and develop estimates for its computational complexity. They may be geometrically interpreted as the forward propagation of tangents or the backward propagation of normals or cotangents. Therefore they are often referred to as the *tangent* and the *cotangent mode*, respectively. Rather than propagating a single tangent or normal, we may also carry forward or back bundles of them in the so-called *vector mode*. It amortizes certain overhead costs and allows the efficient evaluation of sparse Jacobians by matrix compression. These aspects as well as techniques for propagating bit pattern will be sketched in Section 3.5. The section concludes with some remarks on higher-order adjoints and a brief section summary.

3.1. Forward differentiation

Differentiating (2.10), we obtain by the chain rule the Jacobian

$$F'(\mathbf{x}) \quad \equiv P_Y \ \Phi'_l \ \Phi'_{l-1} \ \cdots \ \Phi'_2 \ \Phi'_1 \ P_X^T, \tag{3.1}$$

where

$$\Phi'_i = I - (1-\sigma_i) P_i \, P_i^T + P_i \, \varphi'_i(u_i) \, Q_i. \tag{3.2}$$

Rather than computing the whole Jacobian explicitly, we usually prefer in AD to propagate directional derivatives along a smooth curve $\mathbf{x}(t) \subset X$ with $t \in (-\varepsilon, \varepsilon)$ for some $\varepsilon > 0$. Then $\mathbf{y}(t) \equiv F(\mathbf{x}(t)) \subset Y$ is also a smooth curve with the tangent

$$\dot{\mathbf{y}}(t) = \frac{\mathrm{d}}{\mathrm{d}t} \, F(\mathbf{x}(t)) = F'(\mathbf{x}(t)) \, \frac{\mathrm{d}}{\mathrm{d}t}\mathbf{x}(t) \ \subset Y. \tag{3.3}$$

Similarly all states $\mathbf{v}^{(i)} = \mathbf{v}^{(i)}(t)$ and intermediates $v_i = v_i(t) = P_i^T \mathbf{v}^{(i)}$ are differentiable functions of $t \in (-\varepsilon, \varepsilon)$ with the tangent values

$$\dot{v}_i \equiv \left. \frac{\mathrm{d}}{\mathrm{d}t} v_i(t) \right|_{t=0} .$$

Multiplying (3.1) by the input vector $\dot{\mathbf{x}} = \mathrm{d}\mathbf{x}(t)/\mathrm{d}t|_0$, we obtain the vector equation

$$\dot{\mathbf{y}} = P_Y \, \Phi'_l \left(\Phi'_{l-1} \left(\cdots \left(\Phi'_2 \left(\Phi'_1 \, P_X^T \, \dot{\mathbf{x}} \right) \right) \cdots \right) \right).$$

The bracketing is equivalent to the loop of matrix-vector products

$$\dot{\mathbf{v}}^{(0)} = P_X^T \, \dot{\mathbf{x}}, \ \dot{\mathbf{v}}^{(i)} = \Phi'_i(\mathbf{v}^{(i-1)}) \dot{\mathbf{v}}^{(i-1)} \ \text{ for } \ i = 1 \ldots l, \ \dot{\mathbf{y}} = P_Y \dot{\mathbf{v}}^{(l)}. \tag{3.4}$$

With $u_i = Q_i \mathbf{v}^{(i-1)}$, $\dot{u}_i = Q_i \, \dot{\mathbf{v}}^{(i-1)}$ and $\dot{\varphi}_i(u_i, \dot{u}_i) \equiv \varphi'_i(u_i)\dot{u}_i$ this may be rewritten as the so-called *tangent procedure* listed in Table 2.1, and it can now be stated formally.

Proposition 3.1. Given the General Assumption, the procedure listed in Table 3.1 yields the value $\dot{\mathbf{y}} = F'(\mathbf{x})\dot{\mathbf{x}}$ with F, as defined in Proposition 2.1.

Obviously Table 3.1 has the same form as Table 2.1 with $\mathcal{H}$ replaced by $\mathcal{H} \times \mathcal{H}$, Q_i by $Q_i \times Q_i$, and P_i by $P_i \times P_i$. In other words, all argument and value spaces have been doubled up by a derivative companion, but the flags σ_i, the precedence relation $\prec$ and thus the structure of the computational graph remain unchanged.

As we can see, the evaluation of the combined function

$$[F(\mathbf{x}), \dot{F}(\mathbf{x}, \dot{\mathbf{x}})] \equiv [F(\mathbf{x}), F'(\mathbf{x})\dot{\mathbf{x}}] \tag{3.5}$$

requires exactly one evaluation of the corresponding elemental combinations

$$[\varphi_i(u_i), \dot{\varphi}_i(u_i, \dot{u}_i)] \equiv [\varphi_i(u_i), \varphi'_i(u_i)\dot{u}_i].$$

Table 3.1. Tangent procedure derived from original procedure in Table 2.1.

$[\mathbf{v}, \dot{\mathbf{v}}]$	$=$	0	
$[v_i, \dot{v}_i]$	$=$	$[x_i, \dot{x}_i]$	for $i = 1 \ldots n$
$[v_i, \dot{v}_i]$	$*=$	σ_i	for $i = n+1 \ldots l$
$[v_i, \dot{v}_i]$	$+=$	$[\varphi_i(u_i), \dot{\varphi}_i(u_i, \dot{u}_i)]$	
$[y_{m-i}, \dot{y}_{m-i}]$	$=$	$[v_{l-i}, \dot{v}_{l-i}]$	for $i = m-1 \ldots 0$

We have bracketed $[\varphi_i, \dot{\varphi}_i]$ side-by-side to indicate that good use can often be made of common subexpressions in evaluating the function and its derivative. In any case we can show that

$$\frac{\mathtt{OPS}\big([F(\mathbf{x}), \dot{F}(\mathbf{x}, \dot{\mathbf{x}})]\big)}{\mathtt{OPS}\,\big(F(\mathbf{x})\big)} \leq \max_{1 \leq i \leq l} \frac{\mathtt{OPS}\,\big([\varphi_i(u_i), \dot{\varphi}_i(u_i, \dot{u}_i)]\big)}{\mathtt{OPS}\,\big(\varphi_i(u_i)\big)} \leq 3. \tag{3.6}$$

The upper bound 3 on the relative cost of performing a single tangent calculation is arrived at by taking the maximum over all elemental functions. It is actually attained for a single multiplication, which spawns two extra multiplications by the chain rule

$$v = \varphi(u, w) \equiv u * w \quad \rightarrow \quad \dot{v} = \dot{\varphi}(u, w, \dot{u}, \dot{w}) = u * \dot{w} + w * \dot{u}.$$

Since the data dependence relation $\prec$ applies to the $\dot{\varphi}_i$ in exactly the same way as to the φ_i, we obtain on a parallel machine

$$\mathtt{CLOCK}\big([F(\mathbf{x}), \dot{F}(\mathbf{x}, \dot{\mathbf{x}})]\big) \leq 3\,\mathtt{CLOCK}\big(F(\mathbf{x})\big),$$

where CLOCK is the idealized wall clock time introduced in (2.13).

The bound 3 is pessimistic as the cost ratio between φ and $\dot{\varphi}$ is more advantageous for most other elemental functions. On the other hand actual runtime ratios between codes representing Table 3.1 and Table 2.1 may well be worse on account of various effects including loss of vectorization. This gap between operation count and actual runtime is likely to be even more marked for the following adjoint calculation.

3.2. Reverse differentiation and adjoint vectors

Rather than propagating tangents forward we may propagate normals backward, that is, instead of computing $\dot{\mathbf{y}} = \dot{F}(\mathbf{x}, \dot{\mathbf{x}}) = F'(\mathbf{x})\dot{\mathbf{x}}$ we may evaluate $\bar{\mathbf{x}} = \bar{F}(\mathbf{x}, \bar{\mathbf{y}}) \equiv \bar{\mathbf{y}} F'(\mathbf{x})$. Here the dual vectors $\bar{\mathbf{y}} \in Y^* = Y^T$ and $\bar{\mathbf{x}} \in X^* = X^T$ are thought of as row vectors. Using the transpose of the

Jacobian product representation (3.1) we find that

$$\bar{\mathbf{x}}^T = P_X\,[\Phi_1']^T([\Phi_2']^T(\cdots([\Phi_{l-1}']^T([\Phi_l']^T\,P_Y^T\bar{\mathbf{y}}^T))\cdots)).$$

The bracketing is equivalent to the loop of vector-matrix products

$$\bar{\mathbf{v}}^{(l)} = \bar{\mathbf{y}}\,P_Y,\quad \bar{\mathbf{v}}^{(i-1)} = \bar{\mathbf{v}}^{(i)}\,\Phi_i'(\mathbf{v}^{(i-1)}) \quad \text{for} \quad i = l\ldots 1, \quad \bar{\mathbf{x}} = \bar{\mathbf{v}}^{(0)}\,P_X^T.$$

Formally we will write, in agreement with (3.2),

$$\bar{\mathbf{v}}^{(i-1)} = \bar{\Phi}_i\big(\mathbf{v}^{(i-1)}, \bar{\mathbf{v}}^{(i)}\big) \equiv \bar{\mathbf{v}}^{(i)}\Phi_i'\big(\mathbf{v}^{(i-1)}\big),$$

where, as ever, adjoints are interpreted as row vectors. With $\bar{v}_i \equiv \bar{\mathbf{v}}^{(i)}P_i$ and $\bar{u}_i = \bar{\mathbf{v}}^{(i)}Q_i^T$, we obtain from (3.2), using $Q_iP_i = 0$, the adjoint elemental function

$$\big[\bar{u}_i \mathrel{+}= \bar{v}_i\,\varphi_i'(u_i); \quad \bar{v}_i \mathrel{*}= \sigma_i\big]. \tag{3.7}$$

In other words the multiplicative statement $v_i \mathrel{*}= \sigma_i$ is self-adjoint in that, correspondingly, $\bar{v}_i \mathrel{*}= \sigma_i$. The incremental part $v_i \mathrel{+}= \varphi(u_i)$ generates the equally incremental statement

$$\bar{u}_i \mathrel{+}= \bar{\varphi}(u_i, \bar{v}_i) \equiv \bar{v}_i\,\varphi_i'(u_i). \tag{3.8}$$

Following the forward sweep of Table 2.1, we may then execute the so-called reverse sweep listed in Table 3.2. Now, using the assumptions on the partial ordering in $\mathcal{I}$, we obtain the following result.

Proposition 3.2. Given the General Assumption, the procedure listed in Table 3.1 yields the value $\bar{\mathbf{x}} = \bar{\mathbf{y}}\,F'(\mathbf{x})$, with F as defined in Proposition 2.1.

Proof. The only fact to ascertain is that the argument u_i of the $\bar{\varphi}_i$ defined in (3.8) still has the correct value when it is called up in the third line of Table 3.2. However, this follows from the definition of u_i in (2.6) and our assumption (2.7), so that none of the statements φ_j with $j > i$, nor of course the corresponding $\bar{\varphi}_j$, can alter u_i before it is used again by $\bar{\varphi}_i$. □

So far, we have treated the $\bar{v}_i$ merely as auxiliary quantities during reverse matrix multiplications. In fact, their final values can be interpreted as adjoint vectors in the sense that

$$\bar{v}_i \equiv \frac{\partial}{\partial v_i}\bar{\mathbf{y}}\,\mathbf{y}. \tag{3.9}$$

Here $\bar{\mathbf{y}}$ is considered constant and the notation $\frac{\partial}{\partial v_i}$ requires further explanation. Partial differentiation is normally defined with respect to one of several independent variables whose values uniquely determine the function being differentiated. However, here $\mathbf{y} = F(\mathbf{x})$ is fully determined, via the evaluation procedure of Table 2.1, as a function of $x_1 \ldots x_n$, with each v_i, for $i > n$, occurring merely as an intermediate variable. Its value, $v_i = v_i(\mathbf{x})$,

Table 3.2. Reverse sweep following forward sweep of Table 2.1.

$\bar{\mathbf{v}}$	$=$	0	
$\bar{v}_{l-m+i}$	$=$	$\bar{y}_i$	for $i = m \ldots 1$
$\bar{u}_i$	$\mathrel{+}=$	$\bar{\varphi}_i(u_i, \bar{v}_i)$	for $i = l \ldots n+1$
$\bar{v}_i$	$\mathrel{*}=$	σ_i	
$\bar{x}_j$	$=$	$\bar{v}_j$	for $j = n \ldots 1$

being also uniquely determined by $\mathbf{x}$, the intermediate v_i is certainly not another independent variable. To define $\bar{v}_i$ unambiguously, we should perturb the assignment (2.2) to

$$v_i = \sigma_i v_i + \varphi_i(u_i) + \delta_i \quad \text{for} \quad \delta_i \in \mathcal{H}_i,$$

and then set

$$\bar{v}_i = \nabla_{\delta_i} \bar{\mathbf{y}}\,\mathbf{y}\big|_{\delta_i=0} \in \mathcal{H}_i^* = \mathcal{H}_i. \tag{3.10}$$

Thus, $\bar{v}_i$ quantifies the sensitivity of $\bar{\mathbf{y}}\,\mathbf{y} = \bar{\mathbf{y}}\,F(\mathbf{x})$ to a small perturbation δ_i in the ith assignment. Clearly, δ_i can be viewed as a variable independent of $\mathbf{x}$, and we may allow one such perturbation for each v_i.

The resulting perturbation δ of $\bar{\mathbf{y}}\,\mathbf{y}$ can then be estimated by the first-order Taylor expansion

$$\delta \approx \sum_{i=1}^{l} \bar{v}_i \cdot \delta_i. \tag{3.11}$$

When v_i is scalar the δ_i may be interpreted as rounding errors incurred in evaluating $v_i = \varphi_i(u_i)$. Then we may assume $\delta_i = v_i\,\eta_i$ with all $|\,\eta_i| \le \eta$, the relative machine precision. Consequently, we obtain the approximate bound

$$|\,\delta\,| \lesssim \left|\sum_{i=1}^{l} \bar{v}_i\,v_i\,\eta_i\right| \le \eta \sum_{i=1}^{l} |\,\bar{v}_i\,v_i\,|. \tag{3.12}$$

This relation has been used by many authors for the estimation of round-off errors (Iri, Tsuchiya and Hoshi 1988). In particular, it led Seppo Linnainmaa to the first publication of the reverse mode in English (Linnainmaa 1983). The factor $\sum_{i=1}^{l} |\bar{v}_i v_i|$ has been extensively used by Stummel (1981) and others as a condition estimate for the function evaluation procedure. Braconnier and Langlois (2001) used the adjoints $\bar{v}_i$ to compensate evaluation errors.

In the context of AD we mostly regard adjoints as vectors rather than adjoint equations. They represent gradients, or more generally Fréchet derivatives, of the weighted final result $\bar{\mathbf{y}}\,\mathbf{y}$ with respect to all kinds of intermediate vectors that are generated in a well-defined hierarchical fashion. This procedural view differs strongly from the more global, equation-based concept prevalent in the literature (see, *e.g.*, Marcuk (1996)).

Viewing all u_i as given constants in Table 3.2, this procedure may by itself be identified with the form of Table 2.1. However, the roles of P_i and Q_i have been interchanged and the precedence relation $\prec$ must be reversed. To interpret the combination of Tables 2.1 and 3.2 as a single evaluation procedure, we introduce an isomorphic copy $\bar{\mathcal{I}}$ of $\mathcal{I}$, denoting the bijection between elements of $\mathcal{I}$ and $\bar{\mathcal{I}}$ by an overline:

$$i \in \mathcal{I} \;\Leftrightarrow\; \bar{\imath} \in \bar{\mathcal{I}} \;\Rightarrow\; \varphi_{\bar{\imath}}(u_i, \bar{v}_i) \equiv \bar{\varphi}_i(u_i, \bar{v}_i). \tag{3.13}$$

Identifying the Hilbert space $\mathcal{H}$ and its subspaces with their duals, we obtain the state space $\mathcal{H} \times \mathcal{H}$ for the combined procedure. Correspondingly the argument selections and value projections can be defined such that the computational graph for $\bar{F}(\mathbf{x}, \bar{\mathbf{y}}) = \bar{\mathbf{y}} F'(\mathbf{x})$ has the following ordering.

Lemma 3.3. With $\varphi_{\bar{\imath}}$ as defined in (3.13), the extension of the precedence relation $\prec$ to $\mathcal{I} \cup \bar{\mathcal{I}}$ is given by

$$\bar{\imath} \prec \bar{\jmath} \;\Leftrightarrow\; j \prec i \;\Leftrightarrow\; j \prec \bar{\imath}.$$

Proof. The extension can be written as

$$(Q_i, 0) \quad \text{for} \quad i \in \mathcal{I} \quad \text{and} \quad Q_{\bar{\imath}} \equiv \begin{pmatrix} Q_i & 0 \\ 0 & P_i^T \end{pmatrix} \quad \text{for} \quad \bar{\imath} \in \bar{\mathcal{I}}$$

and

$$\begin{pmatrix} P_i \\ 0 \end{pmatrix} \quad \text{for} \quad i \in \mathcal{I} \quad \text{and} \quad P_{\bar{\imath}} = \begin{pmatrix} 0 \\ Q_i^T \end{pmatrix} \quad \text{for} \quad \bar{\imath} \in \bar{\mathcal{I}}.$$

Hence we find, for the precedence between adjoint indices,

$$Q_{\bar{\imath}}\, P_{\bar{\jmath}} = P_i^T\, Q_j^T = (Q_j\, P_i)^T \;\Leftrightarrow\; \bar{\jmath} \prec \bar{\imath} \;\Leftrightarrow\; j \succ i.$$

For the precedence between direct and adjoint indices, we obtain

$$Q_{\bar{\imath}}\, P_j = Q_i\, P_j \quad \text{so that} \quad j \prec \bar{\imath} \;\Leftrightarrow\; j \prec i. \qquad \square$$

Hence we see that the graph with the vertex set $\mathcal{I} \cup \bar{\mathcal{I}}$ and the extended precedence relation consists of two halves whose internal dependencies are mirror images of each other. However, the connections between the two halves are not symmetric, in that $j \prec \bar{\imath}$ normally precludes rather than implies the relation $i \succ \bar{\jmath}$, as shown in Figure 3.1.

It is very important to notice that the adjoint operations are executed in reverse order. As we can see from Tables 2.1 and 3.2, the evaluation of the combined function $[F(\mathbf{x}), \bar{F}(\mathbf{x}, \bar{\mathbf{y}})]$ requires exactly one evaluation of the corresponding elemental combinations $\varphi_i(u_i)$ and $\bar{\varphi}_i(u_i, \bar{v}_i)$, though this time the two must be evaluated separately, *i.e.*, at different times. In any case, we find that

$$\frac{\mathtt{OPS}([F(\mathbf{x}),\bar F(\mathbf{x},\bar{\mathbf{y}})])}{\mathtt{OPS}(F(\mathbf{x}))} \le \max_{1\le i\le l} \frac{\mathtt{OPS}(\varphi_i(u_i)) + \mathtt{OPS}(\bar\varphi_i(u_i,\bar v_i))}{\mathtt{OPS}(\varphi_i(u_i))} \le 3. \qquad (3.14)$$

The upper bound 2 on the relative cost of performing a single reverse operation can again be found by taking the maximum over all elemental functions, which is actually obtained for a single multiplication,

$$v \mathrel{+}= \varphi(u,w) \equiv u * w \quad \rightarrow \quad [\bar u, \bar w] \mathrel{+}= [\bar v * w, \bar v * u],$$

only counting multiplications as elsewhere. For the estimate (2.13) of parallel execution time, our assumptions now allow us to conclude that

$$\mathtt{CLOCK}\left([F(\mathbf{x}),\bar F(\mathbf{x},\bar{\mathbf{y}})]\right) \le 3\,\mathtt{CLOCK}(F(\mathbf{x})). \qquad (3.15)$$

This estimate holds because every critical path in $\mathcal{I}\cup\bar{\mathcal{I}}$ must be the union of two paths $\mathcal{P}\subset\mathcal{I}$ and $\bar{\mathcal{P}}\subset\bar{\mathcal{I}}$, whose complexity is bounded by one and two times $\mathtt{CLOCK}(F(\mathbf{x}))$, respectively.

3.3. Symmetry of Hessian graphs

To make the extended graph fully symmetric we must be able to rewrite $\varphi_{\bar\imath}(u_i,\bar v_i)$ as $\varphi_{\bar\imath}(v_i,\bar v_i)$ so that

$$Q_{\bar\imath} \equiv \begin{pmatrix} P_i^T & 0 \\ 0 & P_i^T \end{pmatrix} \quad \text{and hence} \quad j \prec \bar\imath \;\Leftrightarrow\; P_i^T P_j \ne 0 \;\Leftrightarrow\; i \prec \bar\jmath.$$

In other words, we must pack all information needed for the adjoint operation into the value of the original one. For certain elementals such as $v = \varphi(u) \equiv \exp(u)$ this is very natural, for in that particular case $\varphi'(u) = v$. For others like $\varphi(u,w) = u * w$ we would have to append the partials $\partial\varphi/\partial u = w$ and $\partial\varphi/\partial w = u$ to the result, and thus return the vector value $v = (u * w, w, u)$. This is more or less what is done when a tangent linear code is first developed and then the partials are used in the cotangent linear, or adjoint code.

As an example we may again consider the function defined in (2.4). In Table 3.3 we have listed on the left the combination of the forward and reverse sweep without the inclusion of partials in the elemental values. On the right we have listed the extended symmetrized version.

The corresponding computational graphs are depicted in Figure 3.1. With the curved, dotted arcs included, we have the original version associated with the left-hand side of Table 3.3. The dotted arcs can be replaced by the direct, dashed arcs if the computational procedure is modified according to the right-hand side of Table 3.3. The superscripts denote additional components of the intermediate values, which contain the partial derivatives needed on the way back. Apart from its aesthetic appeal, the symmetry of the Hessian graph promises at least a halving of the operation count

Table 3.3. Adjoint procedure = forward sweep + reverse sweep.

Nonsymmetric adjoint		Adjoint with symmetric graph	
(v_1, v_2)	$= (x_1, x_2)$	(v_1, v_2)	$= (x_1, x_2)$
v_3	$= \exp(v_1)$	v_3	$= \exp(v_1)$
v_4	$= \sin(v_1+v_2)$	$(v_4^{(0)}, v_4^{(1)})$	$= (\sin(v_1+v_2), \cos(v_1+v_2))$
v_5	$= v_3 * v_4$	$(v_5^{(0)}, v_5^{(1)}, v_5^{(2)})$	$= (v_3 * v_4^{(0)}, v_4^{(0)}, v_3)$
y	$= v_5$	y	$= v_5^{(0)}$
$\bar{v}_5$	$= \bar{y}$	$\bar{v}_5^{(0)}$	$= \bar{y}$
$(\bar{v}_3, \bar{v}_4)$	$\mathrel{+}= \bar{v}_5 * (v_4, v_3)$	$(\bar{v}_3, \bar{v}_4)$	$\mathrel{+}= \bar{v}_5 * (v_5^{(1)}, v_5^{(2)})$
$(\bar{v}_1, \bar{v}_2)$	$\mathrel{+}= \bar{v}_4 * \cos(v_1+v_2) * (1, 1)$	$(\bar{v}_1, \bar{v}_2)$	$\mathrel{+}= \bar{v}_4 * v_4^{(1)} * (1, 1)$
$\bar{v}_1$	$\mathrel{+}= \bar{v}_3 * \exp(v_1)$	$\bar{v}_1$	$\mathrel{+}= \bar{v}_3 * v_3$
$(\bar{x}_1, \bar{x}_2)$	$= (\bar{v}_1, \bar{v}_2)$	$(\bar{x}_1, \bar{x}_2)$	$= (\bar{v}_1, \bar{v}_2)$

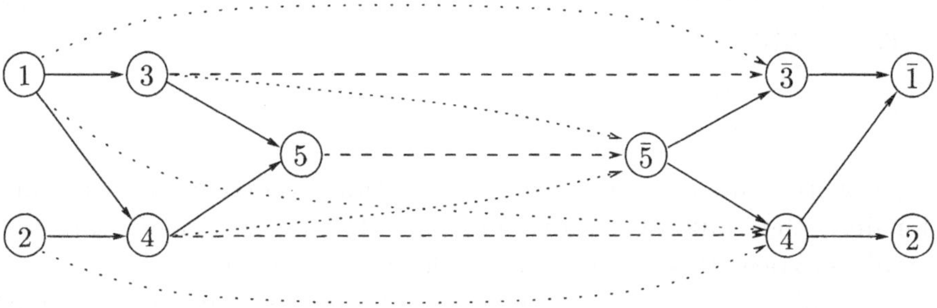

Figure 3.1. Graph corresponding to left $(\cdots)$ and right (- - -) part of Table 3.3.

and storage requirement for the accumulation task discussed in Section 6. A different but related symmetrization of Hessian graphs was developed by Dixon (1991).

3.4. The memory issue

The fact that gradients and other adjoint vectors $\bar{\mathbf{x}} = \bar{F}(\mathbf{x}, \bar{\mathbf{y}}) = \bar{\mathbf{y}}F'(\mathbf{x})$ can, according to (3.14) and (3.15), be computed with essentially the same operation count as the underlying $\mathbf{y} = F(\mathbf{x})$ is certainly impressive, and possibly still a little bit surprising. However, its practical benefit is sometimes in doubt on account of its potentially large memory requirement. Our no-overwrite conditions (2.7) and (2.8) mean that, except for incremental assignments, each elemental function requires new storage for its value. As we

Table 3.4. Direct/adjoint statement pairs.

Case	On forward sweep			On reverse sweep		
(i)	v_i	$\longrightarrow$	STACK	v_i	$\longleftarrow$	STACK
	v_i	$=$	$\varphi_i(u_i)$	$\bar{u}_i$	$+=$	$\bar{\varphi}_i(u_i, \bar{v}_i)$
(ii)	v_i	$+=$	$\varphi_i(u_i)$	$\bar{u}_i$	$+=$	$\bar{\varphi}_i(u_i, \bar{v}_i)$
				v_i	$-=$	$\varphi_i(u_i)$
(iii)	v_i	$=$	$\varphi_i(u_i)$	$\bar{u}_i$	$+=$	$\bar{\varphi}_i(v_i, \bar{v}_i)$
	v_i	$\longrightarrow$	STACK	v_i	$\longleftarrow$	STACK

noted before, the composition (2.10) always represents a well-defined vector function $\mathbf{y} = F(\mathbf{x})$, no matter how the projections Q_i and P_i are defined, provided the state $\mathbf{v}$ is initialized to zero. As a consequence the tangent procedure listed in Table 3.1 can also be applied without any difficulties, yielding consistent results.

Things are very different for the reverse mode. Here the $\bar{\varphi}_i(u_i, \bar{v}_i)$ or $\bar{\varphi}_i(v_i, \bar{v}_i)$ require the old values of u_i or v_i from the time $\varphi_i(u_i)$ itself was originally evaluated. One simple way to ensure this is to save the $v_i = P_i^T \mathbf{v}$ on a value stack during the forward sweep, and then to recover them on the way back just before or just after the form $\bar{\varphi}_i(u_i, \bar{v}_i)$ or $\bar{\varphi}_i(v_i, \bar{v}_i)$ is used, respectively. Then we may modify the statements in the main loops of Table 2.1 and Table 3.2 by one of the three pairs listed in Table 3.4.

The first case (i) describes a copy-on-write strategy for nonincremental statements where all pre-values are saved on a stack just before they are overwritten. On the reverse sweep the value is read back from the stack to restore the state $\mathbf{v}^{(i-1)}$ that was valid just before the original call to φ_i. No extra storage is necessary in case (ii) of additively incremental statement as the previous state can be restored by the corresponding decremental operation. This mechanism allows, for example, the adjoining of an LU factorization procedure with a stack whose size grows only quadratically rather than cubically in the matrix dimension. This growth rate can be made linear if we also choose to invert multiplicatively incremental operations like $v *= u$ and $v /= u$, provided we can ensure $u \neq 0$, which is of course a key point of pivoting in matrix factorization. In general, we face a choice between restoration from a stack and various modes of recomputation. For a more thorough discussion of these issues see Faure and Naumann (2001) and Giering and Kaminski (2001*b*). The main difference between cases (i) and (iii) is that, in the latter, the new value of v_i is needed for the adjoint $\bar{\varphi}(v_i, \bar{v}_i)$ so that restoration of the old value takes place afterwards.

If all elementals are treated according to (i) in Table 3.4, we obtain the bound

$$\texttt{MEM}\left(\bar{F}(\mathbf{x},\bar{\mathbf{y}})\right) \approx \texttt{MEM}\left(F\right) + \sum_{i=1}^{l} m_i \sim \texttt{OPS}\left(F(\mathbf{x})\right). \tag{3.16}$$

Here the proportionality relation assumes that $\texttt{OPS}(\varphi_i) \sim m_i \sim \texttt{MEM}(\varphi_i)$, which is reasonable under most circumstances. The situation is depicted schematically in Figure 3.2. The STACK of intermediates v_i is sometimes called a *trajectory* or *execution* log of the evaluation procedure.

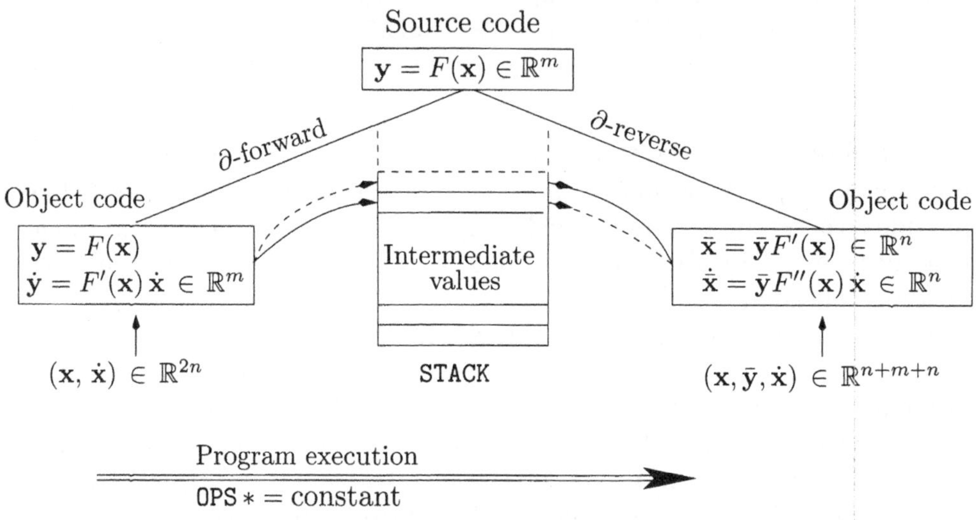

Figure 3.2. Practical execution of forward and reverse differentiation.

The situation is completely analogous to backward integration of the costate in the optimal control of ordinary differential equations (Campbell, Moore and Zhong 1994). More specifically, the problem

$$\min_{\mathbf{x}\in\mathcal{L}^\infty([0,T],\mathbb{R}^n)} \psi(\mathbf{v}(T)) \quad \text{such that} \quad \mathbf{v}(0) = \mathbf{v}_0 \in \mathbb{R}^n \tag{3.17}$$

and

$$\frac{\mathrm{d}}{\mathrm{d}t}\mathbf{v}(t) = \Phi(\mathbf{v}(t),\mathbf{x}(t)) \quad \text{for} \quad 0 \le t \le T \tag{3.18}$$

has the adjoint system

$$\frac{\mathrm{d}}{\mathrm{d}t}\bar{\mathbf{v}}(t) = -\bar{\mathbf{v}}(t)\,\Phi_{\mathbf{v}}(\mathbf{v}(t),\mathbf{x}(t)), \tag{3.19}$$

$$\bar{\mathbf{x}}(t) = \bar{\mathbf{v}}(t)\,\Phi_{\mathbf{x}}(\mathbf{v}(t),\mathbf{x}(t)), \tag{3.20}$$

with the terminal condition

$$\bar{\mathbf{v}}(T) = \nabla\psi(\mathbf{v}(T)). \tag{3.21}$$

Here the trajectory $\mathbf{v}(t)$ is a continuous solution path, whose points enter into the coefficients of the corresponding adjoint differential equation (3.19) for $\bar{\mathbf{v}}(t)$. Rather than storing the whole trajectory, we may instead store only certain checkpoints and repeat the forward simulation in pieces on the way back. This technique is the subject of the subsequent Section 4.

3.5. Propagating vectors or sparsity patterns

Rather than just evaluating one tangent $\dot{\mathbf{y}} = \dot{F}(\mathbf{x}, \dot{\mathbf{x}})$ or one adjoint $\bar{\mathbf{x}} = \bar{\mathbf{y}} F'(\mathbf{x})$, we frequently wish to evaluate several of them, or even the whole Jacobian $F'(\mathbf{x})$ simultaneously. While this bundling increases the memory requirement, it means that the general overhead, and especially the effort for evaluating the φ_i and their partial derivatives, may be better amortized. With $\dot{\mathbf{X}} \in X^p$ any set of p directions in X and $t \in \mathbb{R}^p$ a p-dimensional parameter, we may redefine for each intermediate v_i the derivative vector as

$$\dot{v}_i \equiv \nabla_t \, v_i\big(\mathbf{x} + \dot{\mathbf{X}}\, t\big)\Big|_{t=0} \in \mathcal{H}_i^p.$$

Then the tangent procedure Table 3.1 can be applied without any formal change, though the individual products $\dot{v}_i = \varphi(u_i, \dot{u}_i) = \varphi'(u_i)\dot{u}_i$ must now be computed for $\dot{u}_i \in \mathcal{D}_i^p$. The overall result of this *vector* forward mode is the matrix

$$\dot{\mathbf{Y}} = \dot{F}(\mathbf{x}, \dot{\mathbf{X}}) \equiv F'(\mathbf{x})\dot{\mathbf{X}} \in Y^p.$$

In terms of complexity we can derive the upper bounds

$$\texttt{OPS}\big(\dot{F}(\mathbf{x}, \dot{\mathbf{X}})\big) \le (1 + 2\,p)\,\texttt{OPS}(F(\mathbf{x})) \le p\,\texttt{OPS}\big(\dot{F}(\mathbf{x}, \dot{\mathbf{x}})\big), \qquad (3.22)$$

$$\texttt{MEM}\big(\dot{F}(\mathbf{x}, \dot{\mathbf{X}})\big) \le (1 + p)\,\texttt{MEM}\big(F(\mathbf{x})\big),$$

and

$$\texttt{CLOCK}\big(\dot{F}(\mathbf{x}, \dot{\mathbf{X}})\big) \le (1 + 2\,p)\,\texttt{CLOCK}(F(\mathbf{x})).$$

With a bit of optimism we could replace the factor 2 in (3.22) by 1, which would make the forward mode as expensive as one-sided differences. In fact, depending on the problem and the computing platform, the forward mode tool ADIFOR 2.0 typically achieves a factor somewhere between 0.5 and 2. Unfortunately, this cannot yet be said for reverse mode tools.

There we obtain, for given $\bar{\mathbf{Y}} \in (Y^*)^q$, the adjoint bundle

$$\bar{\mathbf{X}} = \bar{F}\big(\mathbf{x}, \bar{\mathbf{Y}}\big) \equiv \bar{\mathbf{Y}}\, F'(\mathbf{x}) \in \big(X^*\big)^q$$

at the temporal complexity

$$\texttt{OPS}\big(\bar{F}(\mathbf{x}, \bar{\mathbf{Y}})\big) \le (1 + 2\,q)\,\texttt{OPS}(F(\mathbf{x})) \le q\,\texttt{OPS}\big(\bar{F}(\mathbf{x}, \bar{\mathbf{y}})\big),$$

and

$$\texttt{CLOCK}\big(\bar{F}(\mathbf{x}, \bar{\mathbf{Y}})\big) \leq (1 + 2\,q)\,\texttt{CLOCK}(F(\mathbf{x})).$$

Since the trajectory size is independent of the adjoint dimension q, we obtain from (3.16) the spatial complexity

$$\texttt{MEM}(\bar{F}(\mathbf{x}, \bar{\mathbf{Y}})) \sim q\,\texttt{MEM}(F(\mathbf{x})) + \texttt{OPS}(F(\mathbf{x})) \sim \texttt{OPS}(F(\mathbf{x})).$$

The most important application of the vector mode is probably the efficient evaluation of sparse matrices by matrix compression. Here $\dot{\mathbf{X}} \in X^p$ is chosen as a *seed matrix* for a given sparsity pattern such that the resulting *compressed* Jacobian $F'(\mathbf{x})\dot{\mathbf{X}}$ allows the reconstruction of all nonzero entries in $F'(\mathbf{x})$. This technique apparently originated with the grouping proposal of Curtis, Powell and Reid (1974), where each row of $\dot{\mathbf{X}}$ contains exactly one nonzero element and the p columns of $\dot{\mathbf{Y}}$ are approximated by the divided differences

$$\frac{1}{\varepsilon}\left[F\big(\mathbf{x} + \varepsilon\,\dot{\mathbf{X}}\,\mathbf{e}_j\big) - F(\mathbf{x})\right] = \dot{\mathbf{Y}}\mathbf{e}_j + 0(\varepsilon) \quad \text{for} \quad j = 1 \ldots p.$$

Here $\mathbf{e}_j$ denotes the jth Cartesian basis vector in $\mathbb{R}^p$. In AD the matrix $\dot{\mathbf{Y}}$ is obtained with working accuracy, so that the conditioning of $\dot{\mathbf{X}}$ is not quite so critical. The reconstruction of $F'(\mathbf{x})$ from $F'(\mathbf{x})\dot{\mathbf{X}}$ relies on certain submatrices of $\dot{\mathbf{X}}$ being nonsingular. In the CPR approach they are permutations of identity matrices and thus ideally conditioned. However, there is a price to pay, namely the number of columns p must be greater or equal to the chromatic number of the column-incidence graph introduced by Coleman and Moré (1984). Any such colouring number is bounded below by $\hat{n} \leq n$, the maximal number of nonzeros in any row of the Jacobian. By a degree of freedom argument, we see immediately that $F'(\mathbf{x})$ cannot be reconstructed from $F'(\mathbf{x})\dot{\mathbf{X}}$ if $p < \hat{n}$, but $p = \hat{n}$ suffices for almost all choices of the seed matrix $\dot{\mathbf{X}}$. The gap between the chromatic number and $\hat{n}$ can be as large as $n - \hat{n}$, as demonstrated by an example of Hossain and Steihaug (1998). Whenever the gap is significant we should instead apply dense seeds $\dot{\mathbf{X}} \in X^{\hat{n}}$, which were proposed by Newsam and Ramsdell (1983). Rather than using the seemingly simple choice of Vandermonde matrices, we may prefer the much better conditioned Pascal or Lagrange seeds proposed by Hossain and Steihaug (2002) and Griewank and Verma (2003), respectively. In many applications sparsity can be enhanced by exploiting partial separability, which sometimes even allows the efficient calculation of dense gradients using the forward mode.

The compression techniques discussed above require the *a priori* knowledge of the sparsity pattern, which may be rather complicated and thus tedious for the user to supply. Then we may prefer to execute the forward mode with $p = n$ and the v_i stored and manipulated as dynamically sparse vectors (Bischof, Carle, Khademi and Mauer 1996). Excluding exact cancellations, we may conclude that the operation count for computing the whole Jacobians in this sparse vector mode is also bounded above by $(1 + 2\,\hat{n})\mathtt{OPS}(F(x))$. Unfortunately this bound may not be a very good indication of actual runtimes since the dynamic manipulation of sparse data structures typically incurs a rather large overhead cost. Alternatively we may propagate the sparsity pattern of the v_i as bit patterns encoded in $n/32$ integers (Giering and Kaminski 2001*a*). In this way the sparsity pattern of $F'(\mathbf{x})$ can be computed with about $n/32$ times the operation count of F itself and very little overhead. By so-called *probing algorithms* (Griewank and Mitev 2002) the cost factor $n/32$ can often be reduced to $O(\hat{n}\,\log n)$ for a seemingly large class of sparse matrices.

Throughout this subsection we have tacitly assumed that the sparsity pattern of $F'(\mathbf{x})$ is static, *i.e.*, does not vary as a function of the evaluation point $\mathbf{x}$. If it does, we have to either recompute the pattern at each new argument $\mathbf{x}$ or determine certain envelope patterns that are valid, at least in a certain subregion of the domain. All the techniques we have discussed here in their application to the Jacobian $F'(\mathbf{x})$ can be applied analogously to its transpose $F'(\mathbf{x})^T$ using the reverse mode (Coleman and Verma 1996). For certain matrices such as arrowheads, a combination of both modes is called for.

3.6. Higher-order adjoints

In Figure 3.2 we have already indicated that a combination of forward and reverse mode yields so-called *second-order adjoints* of the form

$$\dot{\bar{\mathbf{x}}} \equiv \dot{\bar{F}}(\mathbf{x}, \bar{\mathbf{y}}, \dot{\mathbf{x}}, \dot{\bar{\mathbf{y}}}) \equiv \bar{\mathbf{y}}F''(\mathbf{x})\dot{\mathbf{x}} + \dot{\bar{\mathbf{y}}}F'(\mathbf{x}) \in X = X^*$$

for any given $(\mathbf{x}, \dot{\mathbf{x}}) \in X^2$ and $(\bar{\mathbf{y}}, \dot{\bar{\mathbf{y}}}) \in (Y^*)^2$. In Figure 3.2 the vector $\dot{\bar{\mathbf{y}}} \in Y^*$ was assumed to vanish, and the abbreviation

$$\dot{\bar{F}}(\mathbf{x}, \bar{\mathbf{y}}, \dot{\mathbf{x}}, 0) \equiv \bar{\mathbf{y}}F''(\mathbf{x})\dot{\mathbf{x}} \equiv \nabla_{\mathbf{x}}\left[\bar{\mathbf{y}}F'(\mathbf{x})\dot{\mathbf{x}}\right] \in X^*$$

is certainly stretching conventional matrix notation. To obtain an evaluation procedure for $\dot{\bar{F}}$ we simply have to differentiate the combination of Table 2.1 and Table 3.2 once more in the forward mode. Consequently, composing (3.6) and (3.14) we obtain the bound

$$\mathtt{OPS}(\dot{\bar{F}}(\mathbf{x}, \bar{\mathbf{y}}, \dot{\mathbf{x}}, \dot{\bar{\mathbf{y}}})) \le 3\,\mathtt{OPS}(\bar{F}(\mathbf{x}, \bar{\mathbf{y}})) \le 9\,\mathtt{OPS}(F(\mathbf{x})).$$

Though based on a simple-minded count of multiplicative operations, this bound is, in our experience, not a bad estimate for actual runtimes.

In the context of optimization we may think of $F \equiv (f, c)$ as the combination of a scalar objective function $f : X \to \mathbb{R}$ and a vector constraint

$c : X \to \mathbb{R}^{m-1}$. With $\bar{\mathbf{y}}$ a vector of Lagrange multipliers $\bar{\mathbf{y}}_i$, the symmetric matrix $\bar{\mathbf{y}}F''(\mathbf{x}) = \sum_{i=1}^{m} \mathbf{y}_i \nabla^2 F_i$ is the Hessian of the Lagrangian function. One second-order adjoint calculation then yields exactly the information needed to perform one inner iteration step within a truncated Newton method applied to the KKT system

$$\bar{\mathbf{y}}F'(\mathbf{x}) = 0, \qquad c(\mathbf{x}) = 0.$$

Consequently the cost of executing one inner iteration step is roughly ten times that of evaluating $F = (f, c)$. Note in particular that a procedure for the evaluation of the whole constraint Jacobian ∇c need not be developed either automatically or by hand. Walther (2002) found that iteratively solving the linearized KKT system using exact second-order adjoints was much more reliable and also more efficient than the same method based on divided differences on the gradient of the Lagrangian function.

The brief description of second-order adjoints above begs the question of their relation to a nested application of the reverse mode. The answer is that $\dot{\bar{\mathbf{y}}} = \bar{\mathbf{y}}F''(\mathbf{x})\dot{\mathbf{x}}$ could also be computed in this way, but that the complication of such a nested reversal yields no advantage whatever. To see this we note first that, for a symmetric Jacobian $F'(\mathbf{x}) = F'(\mathbf{x})^T$, we obviously have $\bar{\mathbf{y}}F'(\mathbf{x}) = (F'(\mathbf{x})\,\dot{\mathbf{x}})^T$ if $\dot{\mathbf{x}} = \bar{\mathbf{y}}^T$. Hence the adjoint vector $\bar{\mathbf{x}} = \bar{\mathbf{y}}F'(\mathbf{x})$ can be computed by forward differentiation if $F'(\mathbf{x})$ is symmetric and thus actually the Hessian $\nabla^2\psi(\mathbf{x})$ of a scalar function $\psi(\mathbf{x})$ with the gradient $\nabla\psi(\mathbf{x}) = F(\mathbf{x})$. In other words, it never makes sense to apply the reverse mode to vector functions that are gradients. On the other hand, applying reverse differentiation to $\mathbf{y} = F(\mathbf{x})$ yields

$$\left[F(\mathbf{x})^T, \bar{\mathbf{y}}\,F'(\mathbf{x})\right]^T = \nabla_{\bar{\mathbf{y}},x}\,\bar{\mathbf{y}}\,F(\mathbf{x}).$$

This partitioned vector is the gradient of the Lagrangian $\bar{\mathbf{y}}F(\mathbf{x})$, so that a second differentiation may be carried out without loss of generality in the forward mode, which is exactly what we have done above. The process can be repeated to yield $\ddot{\bar{\mathbf{x}}} = \bar{\mathbf{y}}F'''(\mathbf{x})\dot{\mathbf{x}}\dot{\mathbf{x}}$ and so on by the higher forward differentiation techniques described in Chapter 10 of Griewank (2000).

3.7. *Section summary*

For functions $\mathbf{y} = F(\mathbf{x})$ evaluated by procedures of the kind specified in Section 2, we have derived procedures for computing corresponding tangents $F'(\mathbf{x})\dot{\mathbf{x}}$ and gradients $\bar{\mathbf{y}}F'(\mathbf{x})$, whose composition yields second-order adjoints $\bar{\mathbf{y}}F''(\mathbf{x})\dot{\mathbf{x}}$. All these derivative vectors are obtained with a small multiple of the operation count for $\mathbf{y}$ itself. Derivative matrices can be obtained by bundling these vectors, where sparsity can be exploited through matrix compression. By combining the well-known derivatives of elemental functions using the chain rule, the derived procedures can be generated line by line and, more generally, subroutine by subroutine. These simple observations form the core of AD, and everything else is questionable as to its relevance to AD and scientific computing in general.

4. Serial and parallel reversal schedules

The key difficulty in adjoining large evaluation programs is their reversal within a reasonable amount of memory. Performing the actual adjoint operations is a simple task compared to (re)producing all intermediate results in reverse order. The problem of reversing a program execution has received some perfunctory attention in the computer science literature (see, *e.g.*, Bennett (1973) and van der Snepscheut (1993)). The first authors to consider the problem from an AD point of view were apparently Volin and Ostrovskii (1985) and later Horwedel (1991). The problem of nested reversal that occurs in Pantoja's algorithm (Pantoja 1988) for computing Newton steps in discrete optimal control was first considered by Christianson (1999). No optimal reversal schedule has yet been devised for this discrete variant of the matrix Riccati equation. It might merely be a first example of computational schemes with an intricate, multi-directional information flow that can be treated by the checkpointing techniques developed in AD. In this survey we will examine only the case of discrete evolutions of the form (2.11). They may be interpreted as a chain of l subroutine calls. More generally we may consider the reversal of calling trees with arbitrary structure, whose depth in particular is a critical parameter for memory–runtime trade-offs (see Section 12.2 in Griewank (2000)).

Throughout this section we suppose that the transition functions Φ_i change all or at least most components of the state vector $\mathbf{v}^{(i-1)}$ so that l now represents the number of time-steps rather than elemental functions. Then the basic form of the reverse mode requires the storage of l times the full state vector $\mathbf{v}^{(i-1)}$ of dimension $N = \dim(\mathcal{H})$. Since both l and N may be rather large, their product may indicate an enormous memory requirement, which can however be avoided in one of two ways. First, without any conditions on the transition functions Φ_i we may apply checkpointing, as discussed in this section. The simple idea is to store only a selection of the intermediate state vectors on the first sweep through and then to recalculate the others using repeated partial forward sweeps. These alternative trade-offs between spatial and temporal complexities generate rather intricate serial or parallel reversal schedules, which are discussed in this section. For certain choices the resulting growth in memory and runtime or processor number is only logarithmically dependent on l.

Second, the dependence on l can be avoided completely if (2.12) is not a true iterative loop, but is instead either parallel or represents a contractive fixed-point iteration. In both cases the individual transitions Φ_i commute at least asymptotically and, as we will see in Section 5, they can be adjoined individually without the need for a global execution log.

4.1. Binary schedules for serial reversals

First we consider a simple reversal strategy, which already exhibits certain key characteristics, though it is never quite optimal. Suppose we have exactly $l = 16 = 2^4$ time-steps and enough memory to accommodate $s = 4$ states on a stack of checkpoints. On the first forward sweep we may then store the 0th, 8th, 12th, and 14th intermediate states in the checkpoints c_0, c_1, c_2, and c_3, respectively. This process is displayed in Figure 4.1, where the vertical axis represents *physical* time-steps and the horizontal axis *computational* time-steps or *cycles.*

After advancing to state 15 and then reversing the last step, *i.e.*, applying $\bar{\Phi}_{16}$, we may then restart the calculation at the last checkpoint c_3 containing the 14th state, and immediately perform $\bar{\Phi}_{15}$. The execution of these adjoint steps is represented by the little hooks at the end of each downward slanting run. They are assumed to take twice as long as the forward steps. Horizontal lines indicate checkpoint positions and slanted lines to the right represent forward transition steps, *i.e.*, applications of the Φ_i. After the successive reversals $\bar{\Phi}_{15}$, $\bar{\Phi}_{14}$, and $\bar{\Phi}_{13}$, the last two checkpoints can be abandoned and one of them subsequently overwritten by the 10th state, which can be reached by two steps from the 8th state, which resides in the second checkpoint.

Throughout we will let $\texttt{REPS}(l)$ denote the *repetition count*, *i.e.*, the total number of repeated forward steps needed by some reversal schedule. Rather than formalizing the concept of a reversal schedule as a sequence of certain

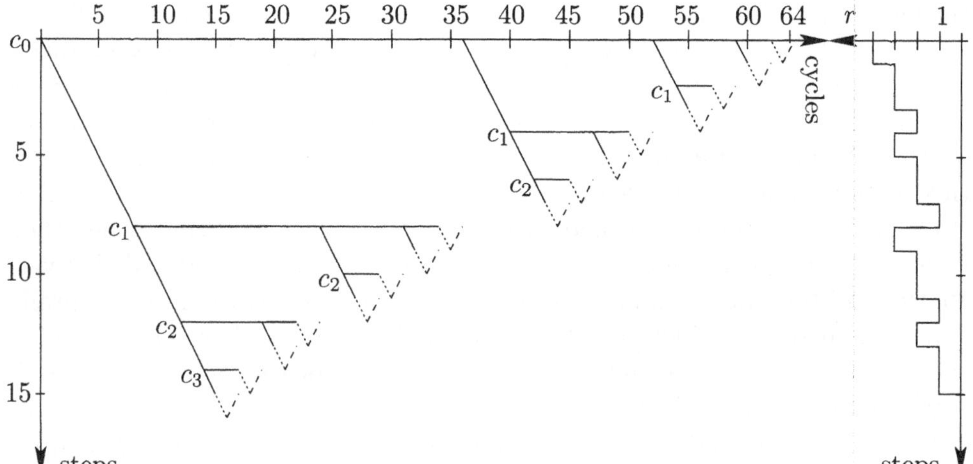

Figure 4.1. Binary serial reversal schedule for $l = 16 = 2^4$ and $s = 4$.

actions, we recommend thinking of them in terms of their graphical representation as depicted here in Figures 4.1 and 4.2. For a binary reversal of an even number of l time-steps we obtain the recurrence

$$\texttt{REPS}(2\,l) = 2\,\texttt{REPS}(l) + l \quad \text{with} \quad \texttt{REPS}(1) = 0.$$

Here l represents the cost of advancing to the middle and $2\,\texttt{REPS}(l)$ the cost of reversing the two halves. Consequently, we derive quite easily that, for all binary powers $l = 2^p$,

$$\texttt{REPS}(l) = l\,p/2 = l\,\log_2 l/2.$$

For $l = 16$ we obtain a total of $16 \cdot 4/2 = 32$ simple forward steps plus 16 adjoint steps, which adds up to a total runtime of $32 + 2 \cdot 16 = 64$ cycles, as depicted in Figure 4.1. The profile on the right margin indicates how often each one of the physical steps is repeated, a number r that varies here between 0 and 4. The total operation count for evaluating the adjoint $\bar{F}(\mathbf{x}, \bar{\mathbf{y}})$ is by

$$\texttt{REPS}(l)\;\texttt{OPS}(\Phi_i) + l\,\texttt{OPS}(\bar{\Phi}_i) \leq \big(\texttt{REPS}(l)/l + 3\big)\;\texttt{OPS}(F),$$

where the last inequality follows from the application of (3.14) to a single transition Φ_i. We also need $p = \log_2 l$ checkpoints, which is the maximal number of horizontal lines intersecting any imaginary vertical in Figure 4.1. Hence, both temporal and spatial complexity grow essentially by the same factor, namely,

$$2\,\frac{\texttt{OPS}(\bar{F}(\mathbf{x}, \bar{\mathbf{y}}))}{\texttt{OPS}(F(\mathbf{x}))} \approx \log_2 l \approx \frac{\texttt{MEM}(\bar{F}(\mathbf{x}, \bar{\mathbf{y}}))}{\texttt{MEM}(F(\mathbf{x}))}.$$

This common logarithmic factor must be compared with the ratios of 3 and l obtained for the operation count and memory requirement in basic reverse mode. Thus we have a drastic reduction in the memory requirement at the cost of a rather moderate increase in the operation count. On a sizable problem and a machine with a hierarchical memory the increase may not necessarily translate into an increased runtime, as data transfers to remote regions of memory can be avoided. Alternatively we may keep the wall clock time to its minimal value by performing the repeated forward sweeps using auxiliary processes on a parallel machine.

For example, if we have $\log_2 l$ processors as well as checkpoints we may execute the binary schedule depicted in Figure 4.2 for $l = 32 = 2^5$ in 64 cycles. We call such schedules *time-minimal* because both the initial forward and the subsequent reverse sweep proceed uninterrupted, requiring a total of $2l$ computational cycles. Here the maximal number of slanted lines intersecting an imaginary vertical line gives the number of processors required. There is a natural association between the checkpoint c_i and the

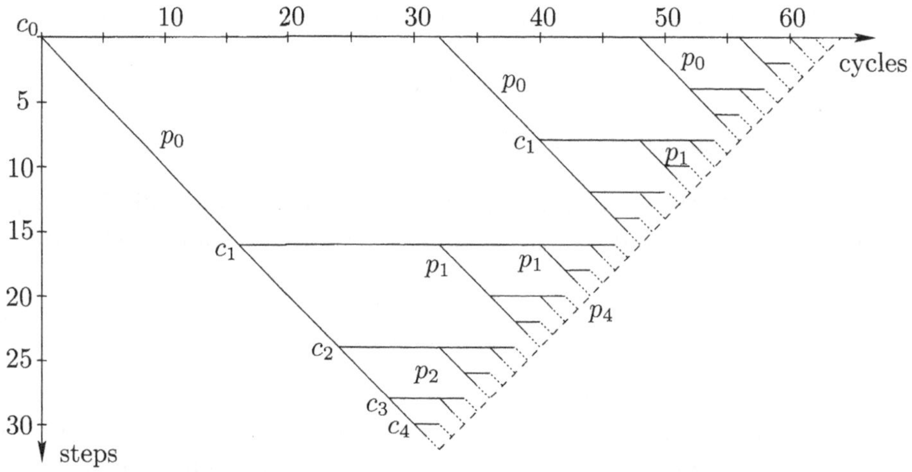

Figure 4.2. Binary parallel reversal schedule for $l = 32 = 2^5$.

processor p_i, which keeps restarting from the same state saved in c_i. For the optimal Fibonacci schedules considered below the task assignment for the individual processors is considerably more complicated. For the binary schedule executed in parallel we now have the complexity estimate

$$\frac{\mathtt{CLOCK}(\bar{F}(\mathbf{x}, \bar{\mathbf{y}}))}{\mathtt{CLOCK}(F(\mathbf{x}))} \approx 2 \quad \text{and} \quad \frac{\mathtt{PROCS}(\bar{F}(\mathbf{x}, \bar{\mathbf{y}}))}{\mathtt{PROCS}(F(\mathbf{x}))} \approx \log_2 l.$$

Here the original function evaluation may be carried out in parallel so that $\mathtt{PROCS}(F(x)) > 1$, but then $\log_2 l$ times as many processors must be available for the reversal with minimal wall clock time. If not quite as many are available, we may of course compress the sequential schedule somewhat along the computing axis without reaching the absolute minimal wall clock time. Such hybrid schemes will not be elaborated here.

4.2. Binomial schedules for serial reversal

While elegant in its simplicity, the binary schedule (like other schemes that recursively partition the remaining simulation lengths by a fixed proportion) is not optimal in either the serial or parallel context. More specifically, given memory for $\log_2 l$ checkpoints, we can construct serial reversal schedules that involve significantly fewer than $l \log_2 l/2$ repeated forward steps or time-minimal parallel reversal schedules that get by with fewer than $\log_2 l$ processors. The binary schedules have another major drawback, namely they cannot be implemented online for cases where the total number of steps to be taken is not known *a priori*. This situation may arise, for example, when a differential equation is integrated with adaptive step-size, even if the period of integration is fixed beforehand. Optimal online schedules have so far only been developed for multi-processors.

The construction of serially optimal schedules is possible by variations of dynamic programming. The key observation for this decomposition into smaller subschedules is that checkpoints are *persistent*, in that they exist until all later steps to the right of them have been reversed. In Figures 4.1 and 4.2 this means graphically that all horizontal lines continue until they (almost) reach the upward slanting line representing the step reversal.

Lemma 4.1. (Checkpoint persistence) Any reversal schedule can be modified without a reduction of its length l or an increase in its repetition count such that, once established at a state j, a checkpoint stays fixed and is not overwritten until the $(j+1)$st step has been reversed, *i.e.*, $\bar{\Phi}_{j+1}$ applied. Moreover, during the 'life-span' of the checkpoint at state j, all actions occur to the right, *i.e.*, concern only states $k \geq j$.

This checkpoint persistence principle is easy to prove (Griewank 2000) under the crucial assumption that all state vectors $\mathbf{v}^{(i)}$ are indeed of the same dimension and thus require the same amount of storage. While this assumption could be violated due to the use of adaptive grids, it would be hard to imagine that the variations in size could be so large that taking the upper bound of the checkpoint size as a uniform measure would result in very significant inefficiencies. Another uniformity assumption that we will use, but which can be quite easily relaxed, is that the computational effort for all individual transitions Φ_i is the same. Thus the optimal reversal schedules depend only on the total number l of steps to be taken.

In Figure 4.1 we see that the setting of the second checkpoint c_1 effectively splits the reversal into two subproblems. After storing the initial state into c_0 and then advancing to state eight, the remaining eight physical steps are reversed using just the three checkpoints c_1, c_2 and c_3. Subsequently the first eight steps are reversed also using only three checkpoints, namely c_0, c_1, c_2, although c_3 is again available. In fact, it would be better to set c_1 at a state $\check{l}$ that solves, for $s = 4$ and $l = 16$, the dynamic programming problem

$$\mathtt{REPS}(s,l) \equiv \min_{1<\check{l}<l} \left\{ \check{l} + \mathtt{REPS}(s-1, l-\check{l}) + \mathtt{REPS}(s,\check{l}) \right\}. \tag{4.1}$$

Here $\mathtt{REPS}(s,l)$ denotes the minimal repetition count for a reversal of l steps using just s checkpoints. The three terms on the right represent the effort of, respectively, advancing $\check{l}$ steps, reversing the right subchain $[\check{l} \dots l]$ using $s-1$ checkpoints, and finally reversing the left subchain $[0 \dots \check{l}]$, again using all s checkpoints. To derive a nearly explicit formula for the function $\mathtt{REPS}(s,l)$, we will consider values

$$l_r(s,l) \leq l \quad \text{for} \quad r \geq 0 < s.$$

They are defined as the number of steps $i = 1 \dots l$ that are repeated at most r

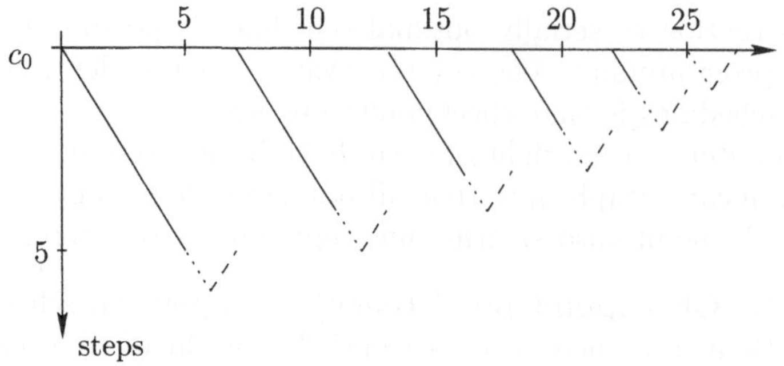

Figure 4.3. Reversal with a single checkpoint ($s = 1, l = 6$).

times, maximized over all reversals on the range $[0 \ldots l]$. By definition, the $l_r(s,l)$ are nondecreasing as functions of r, such that $l_{r+1}(s,l) \geq l_r(s,l)$. Moreover, as it turns out, these numbers have maxima

$$l_r(s) \equiv \max_{l \geq 0} l_r(s,l),$$

which are attained for all sufficiently large l.

For $s = 1$ this is easy to see. As there is just one checkpoint it must be set to the initial state, and the reversal schedule takes the trivial form depicted in Figure 4.3 for the case $l = 6$. Hence we see there is exactly one step that is never repeated, one that is repeated once, twice, and so on, so that

$$l_r(1) = r + 1 = l_r(1,l) \quad \text{for all} \quad 0 \leq r < l. \tag{4.2}$$

On the other hand, in any reversal schedule all but the final step must be repeated at least once, so that we have

$$l_0(s) = 1 = l_0(s,l) \quad \text{for all} \quad 0 \leq s < l. \tag{4.3}$$

The initial values (4.2) and (4.3) for $s = 1$ or $r = 0$ facilitate the proof of the binomial formula below by induction. The other assertions were originally established by Grimm, Pottier and Rostaing-Schmidt (1996).

Proposition 4.2.

(i) Lemma 4.1 implies that, for any reversal schedule,

$$l_r(s) \leq \beta(s,r) \equiv (s+r)!/(s!\,r!), \tag{4.4}$$

where l is the chain length and s the number of checkpoints.

(ii) The resulting repetition count is bounded below by

$$\texttt{REPS}(s,l) \geq l\, r_{\max} - \beta(s+1, r_{\max} - 1) \tag{4.5}$$

where $r_{\max} \equiv r_{\max}(s,l)$ is uniquely defined by

$$\beta(s, r_{\max} - 1) < l \leq \beta(s, r_{\max}).$$

(iii) REPS(s, l) does attain its lower bound (4.5), with (4.4) holding as equality for all $0 \leq r < r_{\max}$, and consequently

$$l_{r_{\max}}(s,l) = l - \beta(s, r_{\max} - 1,) = l - \sum_{i=0}^{r_{\max}-1} l_r(s,l).$$

Proof. It follows from checkpoint persistence that

$$l_r(s,l) \leq \max_{1<\check{l}<l} \{ l_{r-1}(s,\check{l}) + l_r(s-1, l-\check{l}) \},$$

because all steps in the left half $[0 \ldots \check{l}]$ get repeated one more time during the initial advance to $\check{l}$. Now, by taking the maximum over l and $\check{l} < l$, we obtain the recursive bound

$$l_r(s) \leq l_{r-1}(s) + l_r(s-1).$$

Together with the initial conditions (4.2) and (4.3), we immediately arrive at the binomial bound

$$l_r(s) \leq \beta(s,r) \equiv (s+r)!/(s!\, r!),$$

which establishes (i). To prove the remaining assertions let us suppose we have a given chain length l and some serial reversal schedule using s checkpoints that repeats $\Delta\tilde{l}_r \geq 0$ steps exactly r times for $r = 0, 1 \ldots l$. Looking for the most efficient schedule, we then obtain the following constrained minimization problem:

$$\text{Min} \sum_{r=0}^{\infty} r \Delta\tilde{l}_r \quad \text{such that} \quad l = \sum_{r=0}^{\infty} \Delta\tilde{l}_r \quad \text{and}$$
$$\tilde{l}_r \equiv \sum_{i=0}^{r} \Delta\tilde{l}_i \leq l_r(s) \quad \text{for all} \quad r \geq 0. \tag{4.6}$$

By replacing the right sides of the inequality constraints by their upper bounds $\beta(s,r)$, we relax the problem so that the minimal value can only go down. Also relaxing the integrality constraint on the variables $\Delta\tilde{l}_r$ for $r \geq 0$ we obtain a standard LP whose minimum is obtained at

$$\Delta\tilde{l}_r = \beta(s,r) - \beta(s,r-1) = \beta(s-1,r) \quad \text{for} \quad 0 \leq r < r_{\max} \quad \text{and}$$
$$\Delta\tilde{l}_{r_{\max}} = l - \beta(s, r_{\max}-1) \geq 0 \quad \text{where} \quad \beta(s, r_{\max}-1) < l \leq \beta(s, r_{\max}).$$

This solution is integral and the resulting lower bound on the cost is given by

$$\begin{aligned} \text{REPS}(s,l) &\geq \big[l - \beta(s, r_{\max}-1) \big]\, r_{\max} + \sum_{j=0}^{r_{\max}-1} j\, \beta(s-1,j) \\ &= r_{\max}\, l - \beta(s+1, r_{\max}-1) \leq r_{\max}\, l, \end{aligned}$$

where the equation follows from standard binomial identities (Knuth 1973). That the lower bound $\mathtt{REPS}(s,l)$ is actually attained is established by the following construction of suitable schedules by recursion.

When $s = 1$ we can only use the schedule displayed in Figure 4.3, where $\Delta\tilde{l}_r = 1$ for all $0 \le r < l$ so that $r_{\max} = l - 1$ and $\mathtt{REPS} = l\,(l-1) - \beta(2, r_{\max} - 1) = l\,(l-1)/2$. When $l = 1$ no step needs to be repeated at all, so that $\Delta\tilde{l}_0 = 1$, $r_{\max} = 0$ and thus $\mathtt{REPS} = 0 = \beta(s+1, r_{\max} - 1)$. As an induction hypothesis we suppose that a binomial schedule satisfying (4.4) and (4.5) as equalities exists for all pairs $\tilde{s} \ge 1 \le \tilde{l}$ that precede (s,l) in the lexicographic ordering. This induction hypothesis is trivially true for $s = 1$ and $l = 1$. There is a unique value $r_{\max}$ such that

$$\beta(s, r_{\max} - 2) + \beta(s-1, r_{\max} - 1) = \beta(s, r_{\max} - 1) < l \tag{4.7}$$

$$\text{and} \quad l \le \beta(s, r_{\max}) = \beta(s, r_{\max} - 1) + \beta(s-1, r_{\max}).$$

Then we can clearly partition l into $\check{l}$ and $l - \check{l}$ such that

$$\beta(s, r_{\max} - 2) < \check{l} \le \beta(s, r_{\max} - 1) \tag{4.8}$$

$$\text{and} \quad \beta(s-1, r_{\max} - 1) < l - \check{l} \le \beta(s-1, r_{\max}), \tag{4.9}$$

which means that the induction hypothesis is satisfied for the two subranges $[0 \ldots \check{l}]$ and $[\check{l} \ldots l]$. This concludes the proof by induction. □

Figure 4.4 displays the incremental counts

$$\Delta l_r(s) = l_r(s) - l_{r-1}(s) \approx \beta(s-1, r),$$

where the approximation holds as equality when $l_r(s)$ and $l_{r-1}(s)$ achieve their maximal values $\beta(s,r)$ and $\beta(s-1,r)$. The solid line represents an optimal schedule and the dashed line some other schedule for the given parameters $s = 3$ and $l = 68$. The area under the graphs is the same, as it must equal the number of steps, 68.

An optimal schedule for $l = 16$ and $s = 4$ is displayed in Figure 4.5. It achieves with $r_{\max} = 3$ the minimal repetition count $\mathtt{REPS}(3, 16) = 3 \cdot 16 - \beta(5,2) = 48 - 21 = 27$, which yields together with 16 adjoint steps a total number of 59 cycles. This compares with a total of 65 cycles for the binary schedule depicted in Figure 4.1, where the repetition count is 33 and the distribution amongst the physical steps less even. Of course, the discrepancy is larger on bigger problems.

The formula $\mathtt{REPS}(s,l) = l\, r_{\max} - \beta(s+1, r_{\max} - 1)$ shows that the number $r_{\max}$, which is defined as the maximal number of times that any step is repeated, gives a very good estimate of the temporal complexity growth between $\bar{F}(\mathbf{x}, \bar{\mathbf{y}})$ and $F(\mathbf{x})$, given s checkpoints. The resulting complexity is piecewise linear, as displayed in Figure 4.6. Not surprisingly, the more

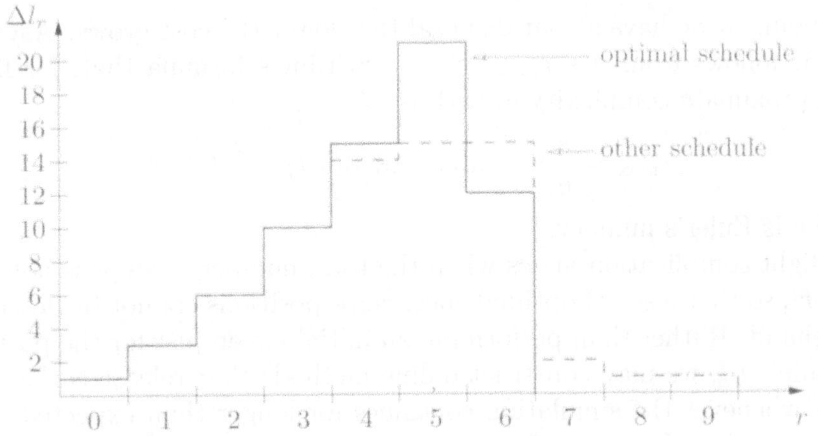

Figure 4.4. Repetition levels for $s = 3$ and $l = 68$.

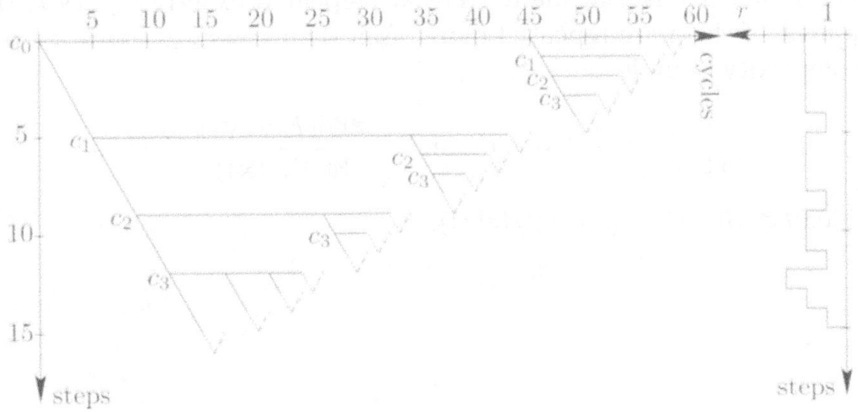

Figure 4.5. Binomial serial reversal schedule for $l = 16$ and $s = 4$.

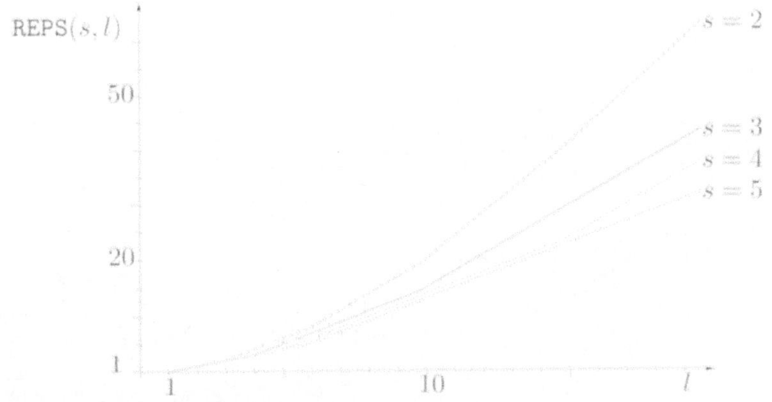

Figure 4.6. Repetition count for chain length l and checkpoint number s.

checkpoints s we have at our disposal the slower the cost grows. Asymptotically it follows from $\beta(s, r_{\max}) \approx l$ by Stirling's formula that, for fixed s, the approximate complexity growth is

$$r_{\max} \sim \frac{s}{\mathrm{e}} \sqrt[s]{l} \quad \text{and} \quad \mathtt{REPS}(s, l) \sim \frac{s}{\mathrm{e}} l^{1+1/s}, \tag{4.10}$$

where e is Euler's number.

A slight complication arises when the total number of steps is not known *a priori*, so that a set of optimal checkpoint positions cannot be determined beforehand. Rather than performing an initial sweep just for the purpose of determining l, we may construct online methods that release earlier checkpoints whenever the simulation continues for longer than expected. Sternberg (2002) found that the increase of the repeated step numbers compared with the (in hindsight) optimal schemes is rarely more than 5%.

A key advantage of the binomial reversal schedule compared with the simpler binary one is that many steps are repeated exactly $r_{\max}(s, l)$ times. Since none of them are repeated more often than $r = r_{\max}$ times, we obtain the complexity bounds

$$\frac{\mathtt{OPS}(\bar{F}(\mathbf{x}, \bar{\mathbf{y}}))}{\mathtt{OPS}(F(\mathbf{x}))} \leq 3 + r \quad \text{and} \quad \frac{\mathtt{MEM}(\bar{F}(\mathbf{x}, \bar{\mathbf{y}}))}{\mathtt{MEM}(F(\mathbf{x}))} \leq 3 + s \tag{4.11}$$

provided $l \leq \beta(s, r) \equiv (s + r)!/(s!\, r!)$.

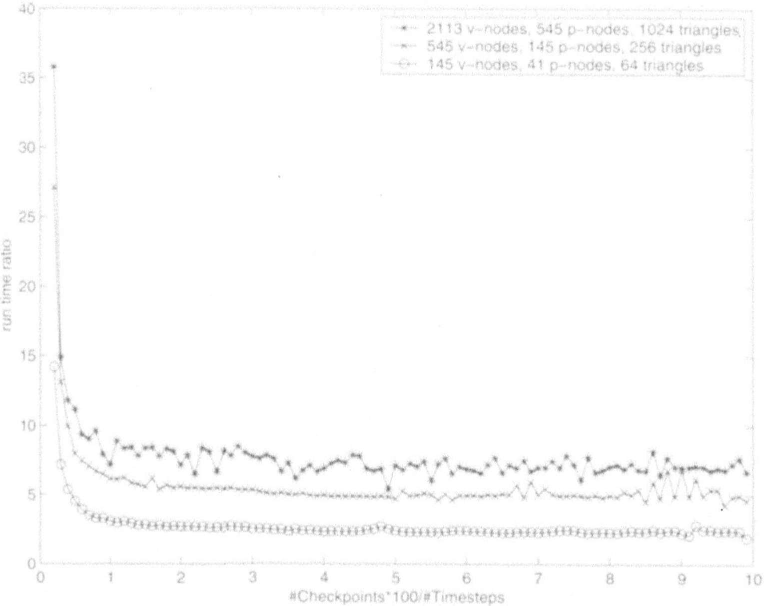

Figure 4.7. Runtime behaviour, 1000 time-steps.

The binomial reversal schedules were applied in an oceanographic study by Restrepo, Leaf and Griewank (1998), have been implemented in the software routine `revolve` (Griewank and Walther 2000), and are used in the Hilbert class library (Gockenbach, Petro and Symes 1999). Figure 4.7 reports the experimental runtime ratios on a linearized Navier–Stokes equation for the driven cavity problem. The problem was discretized using Taylor–Hood elements on 2113 velocity nodes and 545 pressure nodes. We need 38 kbytes to store one state: for details see Hinze and Walther (2002). Over 1000 time-steps the memory requirement for the basic approach of storing all intermediate states would be 38 Mbytes.

While this amount of storage may appear manageable it turns out that it can be reduced by a factor of about a hundred without any noticeable increase in runtime despite a growth factor of $r_{\max} = 8 \approx \frac{1}{\mathrm{e}}10(1000)^{1/10}$ in the repetition count. The total operation count ratio is somewhat smaller because of the constant effort for adjoint steps. By inspection of Figure 4.7 we also note that the runtime ratios are lower for coarser discretizations and thus smaller state space dimensions. This confirms the notion that the amount of data transfer from and to various parts of the memory hierarchy has become critical for the execution times.

The bound (4.11) is valid even when the assumption that $\mathtt{OPS}(\Phi_i)$ is the same for all i is not true. Only when these individual step costs vary by orders of magnitude is it worthwhile to minimize the temporal complexity of $\bar{F}(\mathbf{x}, \bar{\mathbf{y}})$ more carefully. The task of finding optimal reversal schedules for sequences of steps with nonuniform step costs is quite similar to the construction of optimal binary search trees considered by Knuth (1973). As in that setting, exact solutions can be computed with an effort of order l^2 and nearly optimal ones with an effort of order $l \log_2 l$.

The binomial reversal schedules can be generalized to the multi-step situation where

$$\mathbf{v}^{(i)} = \Phi_i(\mathbf{v}^{(i-1)}, \mathbf{v}^{(i-2)}, \ldots, \mathbf{v}^{(i-q)}) \quad \text{for some} \quad q > 1. \tag{4.12}$$

Then checkpoints consist of q consecutive states $[i-q, i)$ that are needed to advance repeatedly towards i and beyond. Under the critical uniformity assumption $\mathtt{MEM}(\Phi_i) = \mathtt{MEM}(\Phi_l)$ for $i = 1 \ldots l$, we can show that the checkpoint persistence principle still applies, so that dynamic programming is also applicable. More specifically, according to Theorem 3.1 in Walther (1999), we have for the minimal repetition count $\mathtt{REPS}(s, l, q)$

$$\lim_{l \to \infty} \frac{\mathtt{REPS}(s, l, q)}{l^{1+q/s}} = \left[\frac{(s/q)!}{q}\right]^{q/s} \approx \frac{s}{\mathrm{e}\, q^{(1+q/s)}}. \tag{4.13}$$

Exactly the same asymptotic complexity can be achieved by the one-step checkpointing schemes discussed above if they are adapted in the following way. Suppose the l original time-steps are interpreted as l/q *mega-time-*

steps between *mega-states* comprising q consecutive states $[qi-q, qi)$ for $i = 1 \ldots l/q$. Here we may have to increase l formally to the next integer multiple of q. While the number of time-steps is thus divided by q, the complexity of a mega-step is of course q times that of a normal step. Moreover, since the dimension of the state space has also grown q-fold, we have to assume that the number of checkpoints s is divisible by q so that we can now keep up to s/q mega-states in memory. Then, replacing REPS by REPS$/q$, l by l/q and s by s/q in (4.10) yields asymptotically the right-hand side of (4.13). Hence we may conclude that directly exploiting multi-step structure is only worthwhile when l/q is comparatively small. Otherwise, interpreting multi-step evolutions as one-step evolutions on mega-states yields nearly optimal reversal results.

4.3. Fibonacci schedules for parallel reversals

Just as in the serial case the binary schedules discussed in Section 4.1 are also quite good but not really optimal in the parallel scenario. Moreover, they cannot be conveniently updated when l changes in the bidirectional simulation scenario discussed below. First let us briefly consider the scenario which is depicted for $l = 8$ in Figure 4.8.

By inspection, the maximal numbers of horizontal and slanted lines intersecting any vertical line are 2 and 3, respectively. This can be executed using 2 checkpoints and 3 processors. In contrast, the binary schedule for that problem defined by the lowest quarter of Figure 4.2 requires 3 checkpoints and processors each. On larger problems the gap is considerably bigger, with the numbers of checkpoints and processors each being reduced by the factor $1/\log_2[(1+\sqrt{5})/2] \approx 1.44$.

The construction of optimal schedules is again recursive, though no decomposition into completely separate subtasks is possible. For our example the optimal position for setting checkpoint c_1 is state 5, so that we have two subschedules of length 5 and 3, represented by the two shaded triangles in the second layer of Figure 4.8 from the top. However, they must be performed at last in parts simultaneously, with computing resources being passed from the bottom left triangle to the top right triangle in a rather intricate manner.

As it turns out, it is best to count both checkpoints and processors together as resources, whose number is given by ϱ. Obviously, a processor can always transmute into a checkpoint by just standing still. The maximal length of any chain that can be reverted without any interruption using ϱ resources will be denoted by l_ϱ. The resources in use during any cycle are depicted in the histograms at the right-hand side of Figure 4.8. These resource profiles are partitioned into the subprofiles corresponding to the two subschedules, and the shaded area caused by additional activities linking

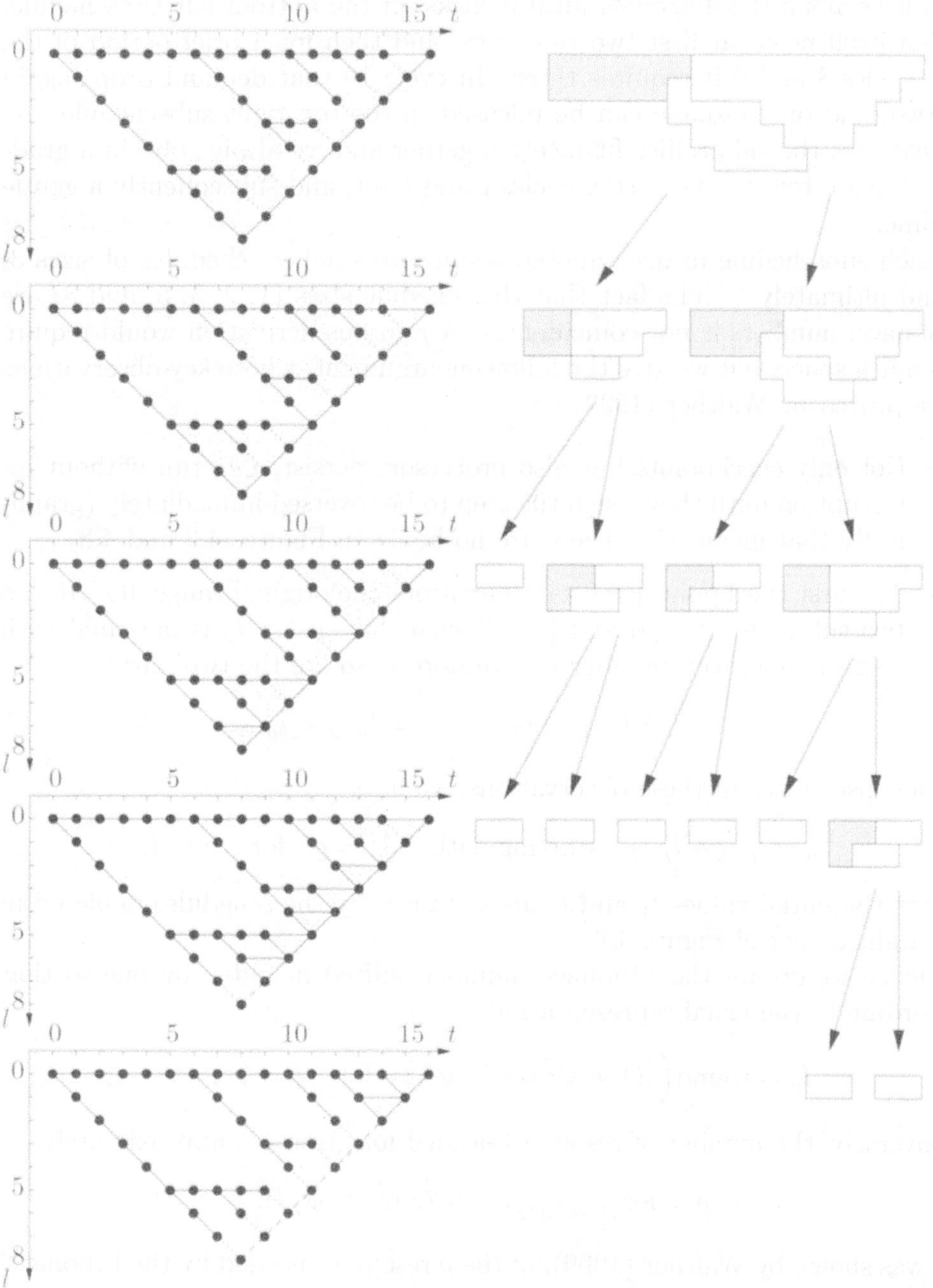

Figure 4.8. Optimal parallel reversal schedule with recursive decomposition into subschedules and corresponding resource profiles.

the two. At first, one checkpoint keeps the initial state, and one processor advances along the diagonals until it kicks off the bottom left subschedule, which itself needs at first two resources, and then for a brief period of the two cycles 8 and 9 it requires three. In cycle 10 that demand drops again to two, and one resource can be released to the top right subschedule. As we can see, the subprofiles fit nicely together and we always obtain a gradual increase toward the vertex cycles l and $l+1$, and subsequently a gentle decline.

Each subschedule in decomposed again into smaller schedules of sizes 3, 2 and ultimately 1. The fact that the schedule sizes (1, 2, 3, 5, and 8) are Fibonacci numbers is not coincidental. A rigorous derivation would require too much space, but we give the following argument, whose key observations were proved by Walther (1999).

- Not only checkpoints but also processors persist, *i.e.*, run without interruption until they reach the step to be reversed immediately (graphically that means that there are no hooks in Figures 4.2 and 4.8).
- The first checkpoint position $\check{l}$ partitions the original range $[0 \ldots l]$ into two subranges $[0 \ldots \check{l}]$ and $[\check{l} \ldots l]$ such that, if $l = l_\varrho$ is maximal with respect to a given number ϱ of resources, so are the two parts

$$\check{l} = l_{\varrho-1} \quad \text{and} \quad l - \check{l} = l_{\varrho-2} < l_{\varrho-1}.$$

As a consequence of these observations, we have

$$l_\varrho = l_{\varrho-1} + l_{\varrho-2} \quad \text{starting with} \quad l_\varrho = \varrho \quad \text{for} \quad \varrho \leq 3,$$

where the initial values l_1 and l_2 are obtained by the schedules depicted in the right corner of Figure 4.8.

Hence we obtain the Fibonacci number shifted in index by one so that according to the usual representation,

$$l_\varrho = \text{round}\Big(\big[(1+\sqrt{5})/2\big]^\rho/\sqrt{5}\Big) \quad \text{for} \quad \varrho = 1, 2 \ldots.$$

Conversely, the number of resources needed for given l is approximately

$$\varrho \approx \log_{(1+\sqrt{5})/2} l \approx 0.7202\ (2 \log_2 l).$$

As was shown by Walther (1999), of the ϱ resources needed by the Fibonacci schedule at most one more than half must be processors. Consequently, compared with the binary schedule the optimal parallel schedule achieves a reduction in both resources by about 28%. While this gain is not really enormous, the difference between the binary and Fibonacci reversals becomes even more pronounced if we consider the scenario where l is not known *a priori*.

Imagine we have a simulation that runs forward in time with a fair bit of diffusion or energy dissipation, so that we cannot integrate the model equations backwards. In order to be able to run the 'movie' backwards nevertheless, we take checkpoints at selected intermediate states and then repeat partial forward runs using auxiliary processors. The reversal is supposed to start instantaneously and to run at exactly the same speed as the forward simulation. Moreover, the user can switch directions at any time. As it turns out, the Fibonacci schedules can be updated adaptively so that ϱ resources, with roughly half of them processors, suffice as long as l does not exceed l_ϱ. When that happens, one extra resource must be added so that $\varrho += 1$.

An example using maximally 5 processors and 3 checkpoints is depicted in Figure 4.9. First the simulation is advanced from state 0 to state $48 < 55 = l_9$. In the 48th transition step 6 checkpoints have been set and 2 processors are running. Then the 'movie' is run backwards at exactly the same speed until state $32 < 34 = l_8$, at which point 5 processors are running and 3 checkpoints are in memory. Then we advance again until state 40, reverse once more and so on. The key property of the Fibonacci numbers used in these bidirectional simulation schedules is that, by the removal of a checkpoint between gaps whose size equals neighbouring Fibonacci numbers, the size of the new gap equals the next-larger Fibonacci numbers.

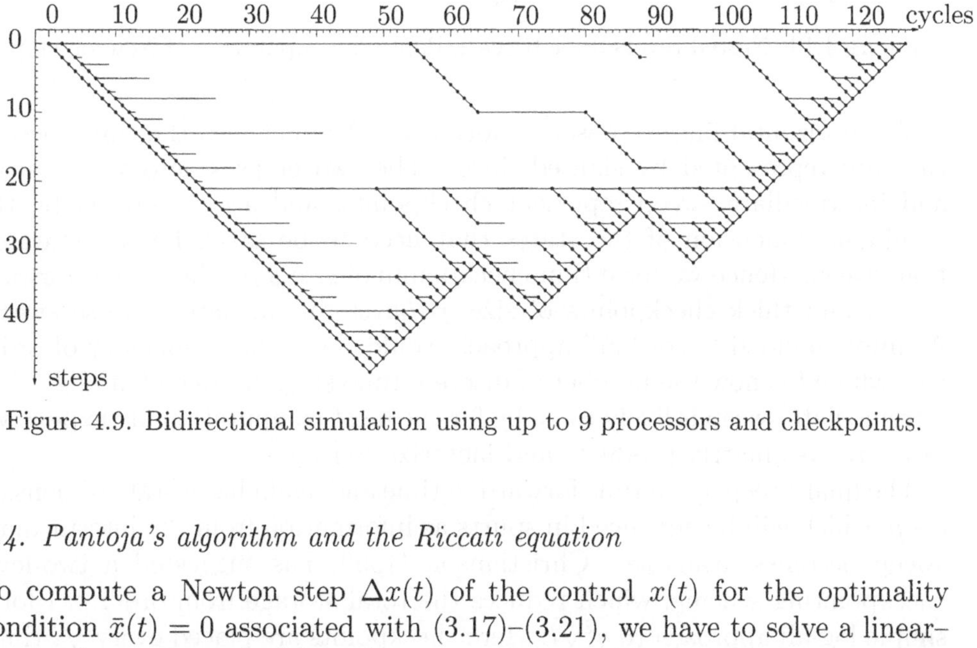

Figure 4.9. Bidirectional simulation using up to 9 processors and checkpoints.

4.4. Pantoja's algorithm and the Riccati equation

To compute a Newton step $\Delta x(t)$ of the control $x(t)$ for the optimality condition $\bar{x}(t) = 0$ associated with (3.17)–(3.21), we have to solve a linear–quadratic regulator problem that approximates the given nonlinear control problem along the current trajectory. Its solution requires the integration of the matrix Riccati equation (Maurer and Augustin 2001, Caillau and

Noailles 2001), backward from a terminal condition involving the objective Hessian $\nabla^2\psi$. After this intermediate, reverse sweep, a third, forward sweep yields the control correction $\Delta x(t)$. Pantoja's algorithm (Pantoja 1988) may be viewed as a discrete variant of these coupled ODEs (Dunn and Bertsekas 1989). It was originally derived as a clever way of computing the Newton step on the equality-constrained finite-dimensional problem obtained by discretizing (3.17)–(3.21). Here, we are interested primarily in the dimensions of the state spaces for the various evaluation sweeps and the information that flows between them. Assuming that the control dimension $p = \dim(\mathbf{x})$ is much smaller than the original state space dimension $q = \dim(\mathbf{v})$, we arrive at the scenario sketched in Figure 4.10. Here $B(t)$ is a $p \times q$ matrix path that must be communicated from the intermediate to the third sweep.

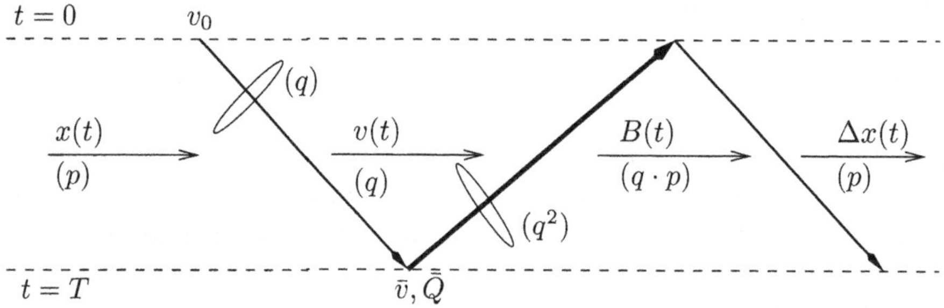

Figure 4.10. Nested reversal for Riccati/Pantoja computation of Newton step.

The horizontal lines represent information flow between the three sweeps that are represented by slanted lines. The two ellipses across the initial and intermediate sweep represent checkpoints, and are annotated by the dominant dimension of the states that need to be saved for a restart on that sweep. Hence we have thin checkpoints of size q on the initial, forward sweep, and thick checkpoints of size q^2 on the immediate, reverse sweep. A simple-minded 'record all' approach would thus require memory of order $l\,q^2$, where l is now the number of discrete time-steps between 0 and T. The intermediate sweep is likely to be by far the most computationally expensive, as it involves matrix products and factorizations.

The final sweep again runs forward in time and again has a state dimension $q+p$, which will be advanced in spurts as information from the intermediate sweep becomes available. Christianson (1999) has suggested a two-level checkpointing scheme, which reduces the total storage from order $q^2\,l$ for a simple log-all approach to $q^2\sqrt{l}$. Here checkpoints are placed every $\sqrt{l}$ time-steps along the initial and the intermediate sweep. The subproblems of length $\sqrt{l}$ are then treated with complete logging of all intermediate states. More sophisticated serial and parallel reversal schedules are currently under development and will certainly achieve a merely logarithmic dependence of the temporal and spatial complexity on l.

4.5. Section summary

The basic reverse mode makes both the operation count $\mathtt{OPS}(\bar{F}(\mathbf{x},\bar{\mathbf{y}}))$ and the memory requirement $\mathtt{MEM}(\bar{F}(\mathbf{x},\bar{\mathbf{y}}))$ for adjoints proportional to the operation count $\mathtt{OPS}(F(\mathbf{x})) \geq \mathtt{MEM}(F(\mathbf{x}))$. By serial checkpointing on explicitly time-dependent problems, both ratios

$$\mathtt{OPS}(\bar{F}(\mathbf{x},\bar{\mathbf{y}}))/\mathtt{OPS}(F(\mathbf{x})) \quad \text{and} \quad \mathtt{MEM}(\bar{F}(\mathbf{x},\bar{\mathbf{y}}))/\mathtt{MEM}(F(\mathbf{x}))$$

can be made proportional to the logarithm of the number of time-steps, which equals roughly $\mathtt{OPS}(F(\mathbf{x}))/\mathtt{MEM}(F(\mathbf{x}))$. A similar logarithmic number of checkpoints and processors suffices to maintain the minimal wall clock ratio

$$\mathtt{CLOCK}(F(\mathbf{x},\bar{\mathbf{y}}))/\mathtt{CLOCK}(F(\mathbf{x})) \in [1,2]$$

that is obtained theoretically when all intermediate results can be stored. In practice the massive data movements may well lead to significant slowdown. Parallel reversal schedules can be implemented online so that, at any point in time the direction of simulation can be reversed without any delay and repeatedly.

5. Differentiating iterative solvers

Throughout the previous sections we implicitly made the crucial assumption that the trajectory traced out by the forward simulation actually matters. Otherwise the taking of checkpoints, and multiple restarts, would not be worth the trouble at all.

Many large-scale computations involve pseudo-time-stepping for solving systems of equations. The resulting solution vector can then be used to evaluate certain performance measures whose adjoints might be required for design optimization. In that case, a mechanical application of the adjoining process developed in the previous sections may still yield reasonably accurate reduced gradients. However, the accuracy achieved is nearly impossible to control and the cost is unnecessarily high, especially in terms of memory space. With regard to derivative accuracy, we should first recall that the derivatives obtained in the forward mode are, up to rounding errors, identical to the ones obtained in the (mechanical) reverse mode, while the corresponding computational cost can differ widely, of course. Therefore we begin this section with an analysis of the application of the forward mode to a contractive fixed-point iteration.

5.1. Black box differentiation in forward mode

Let us consider a loop of the form (2.11) with the vector $\mathbf{v}^{(i)} = (\mathbf{x}, \mathbf{z}_i, \mathbf{y}) \in X \times Z \times Y = H$ partitioned into the vector of independent variables $\mathbf{x}$, the proper state vector $\mathbf{z}_i$ and the vector of dependent variables $\mathbf{y}$. While $\mathbf{x}$ is assumed constant throughout the iteration, each vector $\mathbf{y}$ is a simple output, or function, and may therefore not affect any of the other values. Then we may write

$$\mathbf{z}_0 = \mathbf{z}_0(\mathbf{x}), \quad \mathbf{z}_i = G(\mathbf{x}, \mathbf{z}_{i-1}) \text{ for } i = 1 \ldots l, \quad \mathbf{y} = F(\mathbf{x}, \mathbf{z}_l) \tag{5.1}$$

for suitable functions

$$G : X \times Z \mapsto Z \quad \text{and} \quad F : X \times Z \mapsto Y.$$

The statement $\mathbf{z}_0 = \mathbf{z}_0(\mathbf{x})$ is meant to indicate that $\mathbf{z}_0$ is suitably initialized given the value of $\mathbf{x}$. In many applications there might be an explicit dependence of G on the iterations counter i so that, in effect, $\mathbf{z}_i = G^{(i)}(\mathbf{x}, \mathbf{z}_{i-1})$, a situation that was considered in Campbell and Hollenbeck (1996) and Griewank and Faure (2002). For simplicity we will assume here that the iteration is stationary in that $G^{(i)} = G$, an assumption that was also made in the fundamental papers by Gilbert (1992) and Christianson (1994). To assure convergence to a fixed point we make the following assumption.

Contractivity Assumption. The equation

$$\mathbf{z} = G(\mathbf{x}, \mathbf{z})$$

has a solution $\mathbf{z}_* = \mathbf{z}_*(\mathbf{x})$, and there are constants $\delta > 0$ and $\rho < 1$ such that

$$\|\mathbf{z} - \mathbf{z}_*(\mathbf{x})\| \le \delta \;\Rightarrow\; \|G_\mathbf{z}(\mathbf{x}, \mathbf{z})\| \le \rho < 1, \tag{5.2}$$

where $G_\mathbf{z} \equiv \partial G/\partial \mathbf{z}$, represents the Jacobian of G with respect to $\mathbf{z}$ and $\|G_\mathbf{z}\|$ denotes a matrix norm that must be consistent with a corresponding vector norm $\|\mathbf{z}\|$ on Z.

According to Ostrowski's theorem (see Propositions 10.1.3 and 10.1.4 in Ortega and Rheinboldt (1970)), it follows immediately that the initial condition $\|\mathbf{z}_0 - \mathbf{z}_*\| < \delta$ implies Q-linear convergence in that

$$Q\{\mathbf{z}_i - \mathbf{z}_*\} \equiv \limsup_{i \to \infty} \|\mathbf{z}_i - \mathbf{z}_*\| / \|\mathbf{z}_{i-1} - \mathbf{z}_*\| \le \rho. \tag{5.3}$$

Another consequence of our contractivity assumption is that 1 is not an eigenvalue of $G_\mathbf{z}(\mathbf{x}, \mathbf{z}_*)$, so that, by the implicit function theorem, the locally unique solution $\mathbf{z}_* = \mathbf{z}_*(\mathbf{x})$ has the derivative

$$\partial \mathbf{z}_* / \partial \mathbf{x} = [I - G_\mathbf{z}(\mathbf{x}, \mathbf{z}_*)]^{-1} G_\mathbf{x}(\mathbf{x}, \mathbf{z}_*). \tag{5.4}$$

For a given tangent $\dot{\mathbf{x}}$, we obtain the directional derivative $\dot{\mathbf{z}}_* \equiv [\partial \mathbf{z}_* / \partial \mathbf{x}] \dot{\mathbf{x}}$.

We continue to use such restrictions to one-dimensional subspaces of domain and (dual) range in order to stay within matrix-vector notation as much as possible. The resulting vector

$$\dot{\mathbf{y}}_* = \dot{F}(\mathbf{x}, \mathbf{z}_*, \dot{\mathbf{x}}, \dot{\mathbf{z}}_*) \equiv F_{\mathbf{x}}(\mathbf{x}, \mathbf{z}_*)\dot{\mathbf{x}} + F_{\mathbf{z}}(\mathbf{x}, \mathbf{z}_*)\dot{\mathbf{z}}_*$$

represents a directional derivative $f'(\mathbf{x})\dot{\mathbf{x}}$ of the *reduced function*

$$f(\mathbf{x}) \equiv F(\mathbf{x}, \mathbf{z}_*(\mathbf{x})). \tag{5.5}$$

Now the crucial question is whether and how this implicit derivative can be evaluated at least approximately by AD. The simplest approach is to differentiate the loop (5.1) in black box fashion.

Provided not only G but also F is differentiable in some neighbourhood of the fixed point $(\mathbf{x}, \mathbf{z}_*(\mathbf{x}))$, we obtain by differentiation in the forward mode

$$\dot{\mathbf{z}}_0 = 0, \quad \dot{\mathbf{z}}_i = \dot{G}(\mathbf{x}, \mathbf{z}_{i-1}, \dot{\mathbf{x}}, \dot{\mathbf{z}}_{i-1}) \text{ for } i = 1 \ldots l, \quad \dot{\mathbf{y}} = \dot{F}(\mathbf{x}, \mathbf{z}_l, \dot{\mathbf{x}}, \dot{\mathbf{z}}_l), \tag{5.6}$$

where

$$\dot{G}(\mathbf{x}, \mathbf{z}, \dot{\mathbf{x}}, \dot{\mathbf{z}}) \equiv G_{\mathbf{x}}(\mathbf{x}, \mathbf{z})\dot{\mathbf{x}} + G_{\mathbf{z}}(\mathbf{x}, \mathbf{z})\dot{\mathbf{z}} \tag{5.7}$$

and $\dot{F}$ is defined analogously. Since the derivative recurrence (5.6) is merely a linearization of the original iteration (5.1), the contractivity is inherited and we obtain for any initial $\dot{\mathbf{z}}_0$ the R-linear convergence result

$$R\{\dot{\mathbf{z}}_i - \dot{\mathbf{z}}_*\} \equiv \limsup_{i\to\infty} \sqrt[i]{\|\dot{\mathbf{z}}_i - \dot{\mathbf{z}}_*\|} \leq \rho. \tag{5.8}$$

This result was originally established by Gilbert (1992) and later generalized by Griewank, Bischof, Corliss, Carle and Williamson (1993) to a much more general class of Newton-like methods. We may abbreviate (5.8) to $\dot{\mathbf{z}}_i = \dot{\mathbf{z}}_* + \tilde{O}(\rho^i)$ for ease of algebraic manipulation, and it is an immediate consequence for the reduced function $f(\mathbf{x})$ defined in (5.5) that

$$F(\mathbf{x}, \mathbf{z}_i) = f(\mathbf{x}) + \tilde{O}(\rho^i) \quad \text{and} \quad \dot{F}(\mathbf{x}, \mathbf{z}_i, \dot{\mathbf{x}}, \dot{\mathbf{z}}_i) = f'(\mathbf{x})\dot{\mathbf{x}} + \tilde{O}(\rho^i).$$

As discussed by Ortega and Rheinboldt (1970), R-linear convergence is a little weaker than Q-linear convergence, in that successive discrepancies $\|\dot{\mathbf{z}}_i - \dot{\mathbf{z}}_*\|$ need not go down monotonically but must merely decline on average by the factor ρ. This lack of monotonicity comes about because the leading term on the right-hand side of (5.6), as defined in (5.7), may also approach its limit $G_{\mathbf{x}}(\mathbf{x}, \mathbf{z}_*)\dot{\mathbf{x}}$ in an irregular fashion. A very similar effect occurs for Lagrange multipliers in nonlinear programming. Since (5.3) implies by the triangle inequality

$$\limsup_{i\to\infty} \|\mathbf{z}_i - \mathbf{z}_*\| \,/\, \|\mathbf{z}_i - \mathbf{z}_{i-1}\| \leq \frac{\rho}{1-\rho},$$

we may use the last step-size $\|\mathbf{z}_i - \mathbf{z}_{i-1}\|$ as an estimate for the remaining discrepancy $\|\mathbf{z}_i - \mathbf{z}_*\|$. In contrast, this is not possible for R-linearly converging sequences, so the last step-size $\|\dot{\mathbf{z}}_i - \dot{\mathbf{z}}_{i-1}\|$ does not provide a reliable indication of the remaining discrepancy $\|\dot{\mathbf{z}}_i - \dot{\mathbf{z}}_*\|$.

In order to gauge whether the current value $\dot{\mathbf{z}}_i$ is a reasonable approximation to $\dot{\mathbf{z}}_*$, we recall that the latter is a solution of the *direct sensitivity* equation

$$[I - G_{\mathbf{z}}(\mathbf{x}, \mathbf{z}_*)]\dot{\mathbf{z}} = G_{\mathbf{x}}(\mathbf{x}, \mathbf{z}_*)\dot{\mathbf{x}}, \tag{5.9}$$

which is a mere rewrite of (5.4). Hence it follows under our assumptions that, for any candidate pair $(\mathbf{z}, \dot{\mathbf{z}})$,

$$\big(\|\mathbf{z} - \mathbf{z}_*\| + \|\dot{\mathbf{z}} - \dot{\mathbf{z}}_*\|\big) = O\big(\|\mathbf{z} - G(\mathbf{x}, \mathbf{z})\| + \|\dot{\mathbf{z}} - \dot{G}(\mathbf{x}, \mathbf{z}, \dot{\mathbf{x}}, \dot{\mathbf{z}})\|\big), \tag{5.10}$$

with $\dot{G}$ as defined in (5.7). The directional derivative $\dot{G}(\mathbf{x}, \mathbf{z}, \dot{\mathbf{x}}, \dot{\mathbf{z}})$ can be computed quite easily in the forward mode of automatic differentiation so that the residual on the right-hand side of (5.10) is constructively available.

Hence we may execute (5.1) and (5.6) simultaneously, and stop the iteration when not only $\|\mathbf{z}_i - G(\mathbf{x}, \mathbf{z}_i)\|$ but also $\|\dot{\mathbf{z}}_i - \dot{G}(\mathbf{x}, \mathbf{z}_i, \dot{\mathbf{x}}, \dot{\mathbf{z}}_i)\|$ is sufficiently small. The latter condition requires a modification of the stopping criterion, but otherwise the directional derivative $\dot{\mathbf{z}}_i \approx \dot{\mathbf{z}}_* = \dot{\mathbf{z}}_*(\mathbf{x}, \dot{\mathbf{x}})$ and the resulting $\dot{\mathbf{y}}_i = \dot{F}(\mathbf{x}, \mathbf{z}_i, \dot{\mathbf{x}}, \dot{\mathbf{z}}_i) \approx \dot{\mathbf{y}}_*$ can be obtained by black box differentiation of the original iterative solver.

Sometimes G takes the form

$$G(\mathbf{x}, \mathbf{z}) = \mathbf{z} - P(\mathbf{x}, \mathbf{z}) \cdot H(\mathbf{x}, \mathbf{z}),$$

with $H(\mathbf{x}, \mathbf{z}) = 0$ the state equation to be solved and $P(\mathbf{x}, \mathbf{z}) \approx H_{\mathbf{z}}^{-1}(\mathbf{x}, \mathbf{z})$ a suitable preconditioner. The optimal choice $P(\mathbf{x}, \mathbf{z}) = H_{\mathbf{z}}^{-1}(\mathbf{x}, \mathbf{z})$ represents Newton's method but can rarely be realized exactly on large-scale problems. Anyway we find

$$\dot{\mathbf{z}} - \dot{G}(\mathbf{x}, \mathbf{z}, \dot{\mathbf{x}}, \dot{\mathbf{z}}) = \dot{\mathbf{z}} - P(\mathbf{x}, \mathbf{z}) \left[H_{\mathbf{x}}(\mathbf{x}, \mathbf{z})\, \dot{\mathbf{x}} + H_{\mathbf{z}}(\mathbf{x}, \mathbf{z})\, \dot{\mathbf{z}}\right] - \dot{P}(\mathbf{x}, \mathbf{z})\, H(\mathbf{x}, \mathbf{z}),$$

where

$$\dot{P}(\mathbf{x}, \mathbf{z}) = P_{\mathbf{x}}(\mathbf{x}, \mathbf{z})\, \dot{\mathbf{x}} + P_{\mathbf{z}}(\mathbf{x}, \mathbf{z})\, \dot{\mathbf{z}},$$

provided $P(\mathbf{x}, \mathbf{z})$ is differentiable at all. As $H(\mathbf{x}, \mathbf{z}_i)$ converges to zero, the second term $\dot{P}(\mathbf{x}, \mathbf{z})\, H(\mathbf{x}, \mathbf{z}_i)$ could, and probably should, be omitted when evaluating $\dot{G}(\mathbf{x}, \mathbf{z}_i, \dot{\mathbf{x}}, \dot{\mathbf{z}}_i)$. This applies in particular when $P(\mathbf{x}, \mathbf{z})$ is not even continuously differentiable, for example due to pivoting in a preconditioner based on an incomplete LU factorization. Setting $\dot{P} = 0$ does not affect the R-linear convergence result (5.8), and may reduce the cost of the derivative iteration. However, it does require separating the preconditioner P from the residual H, which may not be a simple task if the whole iteration function G is incorporated in a legacy code. In Figure 5.1 the curves labelled 'state equation residual' and 'direct derivative residual' represent $\|\mathbf{z}_i - G(\mathbf{x}, \mathbf{z}_i)\|$ and $\|\dot{\mathbf{z}}_i - \dot{G}(\mathbf{x}, \mathbf{z}_i, \dot{\mathbf{x}}, \dot{\mathbf{z}}_i)\|$, respectively. The other two residuals are explained in the following subsection.

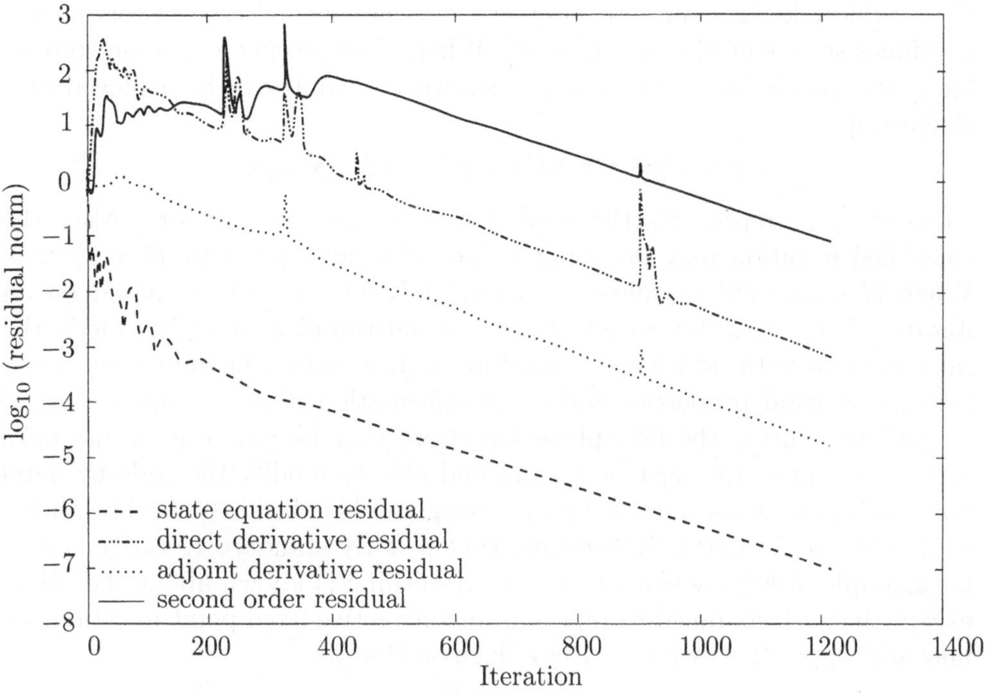

Figure 5.1. History of residuals on $2D$ Euler solver, from Griewank and Faure (2002).

While the contractivity assumption (5.2) seems quite natural, it is by no means always satisfied. For example, conjugate gradient and other general Krylov space methods cannot be written in the form (5.1) with an iteration function $G(\mathbf{x}, \mathbf{z})$ that has a bounded derivative with respect to $\mathbf{z}$. The problem is that the current residual $H(\mathbf{z}, \mathbf{x})$ often occurs as a norm or in inner products in denominators, so that $G_{\mathbf{z}}(\mathbf{z}, \mathbf{x})$ turns out to be unbounded in the vicinity of a fixed point $\mathbf{z}_* = \mathbf{z}_*(\mathbf{x})$. The same is also true for quasi-Newton methods based on secant updating, but a rather careful analysis showed that (5.8) is still true if (5.2) is satisfied initially (Griewank *et al.* 1993). In practice R-linear convergence of the derivatives has been observed for other Krylov subspace methods, but we know of no proof that this must occur under suitable assumptions.

5.2. Two-phase and adjoint differentiation

Many authors (Ball 1969, Cappelaere, Elizondo and Faure 2001) have advocated the following approach, for which directives are included in TAMC (Giering and Kaminski 2000) and possibly other AD tools. Rather than differentiating the fixed point iteration (5.1) itself, we may let it run undifferentiated until the desired solution accuracy has been obtained, and then solve the sensitivity equation (5.9) in a second phase. Owing to its linearity, this problem looks somewhat simpler than the original task of solving the nonlinear state equation $\mathbf{z} = G(\mathbf{x}, \mathbf{z})$. When G represents a Newton step we have asymptotically $\rho = 0$, and the solution of (5.9) can be achieved in a single step

$$\dot{\mathbf{z}} = \dot{G}(\mathbf{x}, \mathbf{z}, \dot{\mathbf{x}}, 0) = -P(\mathbf{x}, \mathbf{z})\, H_{\mathbf{x}}(\mathbf{x}, \mathbf{z})\, \dot{\mathbf{x}}$$

where $\mathbf{z} \approx \mathbf{z}_*$ represents the final iterate of the state vector. Also, the simplified iteration may be applied since $\dot{P}$ is multiplied by $H(\mathbf{x}, \mathbf{z}) \approx 0$. When G represents an inexact version of Newton's method based on an iterative linear equation solver, it seems a natural idea to apply exactly the same method to the sensitivity equation (5.9). This may be quite economical because spectral properties and other information, that is known *a priori* or gathered during the first phase iteration, may be put to good use once more. Of course, we must be willing and able to modify the code by hand unless AD provides a suitable tool (Giering and Kaminski 2000). The ability to do this would normally presume that a fairly standard iterative solver, for example of Krylov type, is in use. If nothing is known about the solver except that it is assumed to represent a contractive fixed point iteration, we may still apply (5.6) with $\mathbf{z}_{i-1} = \mathbf{z}$ fixed so that

$$\dot{\mathbf{z}}_i = G_{\mathbf{x}}(\mathbf{x}, \mathbf{z})\dot{\mathbf{x}} + G_{\mathbf{z}}(\mathbf{x}, \mathbf{z})\dot{\mathbf{z}}_i \quad \text{for} \quad i = 1 \dots l. \tag{5.11}$$

Theoretically, the AD tool could exploit the constancy of $\mathbf{z}$ to avoid the repeated evaluations of certain intermediates. In effect we apply the linearization of the last state space iteration to propagate derivatives forward. The idea of just exploiting the linearization of the last step has actually been advocated more frequently for evaluating adjoints of iterative solvers (Giering and Kaminski 1998, Christianson 1994).

To elaborate on this we first need to derive the adjoint of the fixed point equation $\mathbf{z} = G(\mathbf{z}, \mathbf{x})$. It follows from (5.4) by the chain rule that the total derivative of the reduced response function defined in (5.5) is given by

$$f'(\mathbf{x}) = F_{\mathbf{x}} + F_{\mathbf{z}} \left[I - G_{\mathbf{z}}(\mathbf{x}, \mathbf{z}_*)\right]^{-1} G_{\mathbf{x}}(\mathbf{x}, \mathbf{z}_*).$$

Applying a weighting functional $\bar{\mathbf{y}}$ to $\mathbf{y}$, we obtain the adjoint vector

$$\bar{\mathbf{x}}_* = \bar{\mathbf{y}} f'(\mathbf{x}) = \bar{\mathbf{y}}\, F_{\mathbf{x}}(\mathbf{x}, \mathbf{z}_*) + \bar{\mathbf{g}}_* G_{\mathbf{x}}(\mathbf{x}, \mathbf{z}_*), \tag{5.12}$$

where

$$\bar{\mathbf{g}}_* \equiv \bar{\mathbf{z}}_* \left[I - G_{\mathbf{z}}(\mathbf{x}, \mathbf{z}_*)\right]^{-1} \quad \text{with} \quad \bar{\mathbf{z}}_* \equiv \bar{\mathbf{y}}\, F_{\mathbf{z}}(\mathbf{x}, \mathbf{z}_*). \tag{5.13}$$

While the definition of $\bar{\mathbf{z}}_*$ follows our usual concept of an adjoint vector, the role of $\bar{\mathbf{g}}_* \in Z$ warrants some further explanation. Suppose we introduce an additive perturbation $\mathbf{g}$ of G so that we have the system

$$\mathbf{z} = G(\mathbf{z}, \mathbf{x}) + \mathbf{g}, \quad \mathbf{y} = F(\mathbf{z}, \mathbf{x}).$$

Then it follows from the implicit function theorem that $\bar{\mathbf{g}}_*$, as given by (5.13), is exactly the gradient of $\bar{\mathbf{y}}\,\mathbf{y}$ with respect to $\mathbf{g}$ evaluated at $\mathbf{g} = 0$. In other words $\bar{\mathbf{g}}_*$ is the vector of Lagrange multipliers of the constraint $\mathbf{z} = G(\mathbf{x}, \mathbf{z})$ given the objective function $\bar{\mathbf{y}}\,\mathbf{y}$. From (5.13) it follows directly that $\bar{\mathbf{g}}_*$ is the unique solution of the adjoint sensitivity equation

$$\bar{\mathbf{g}}\big[I - G_{\mathbf{z}}(\mathbf{x}, \mathbf{z}_*)\big] = \bar{\mathbf{z}}_*. \tag{5.14}$$

Under our contractivity assumption (5.2) on $G_{\mathbf{z}}$, the corresponding fixed point iteration

$$\bar{\mathbf{g}}_{i+1} = \bar{\mathbf{z}}_* + \bar{\mathbf{g}}_i\, G_{\mathbf{z}}(\mathbf{x}, \mathbf{z}_*) \tag{5.15}$$

is also convergent, whence

$$Q\{\bar{\mathbf{g}}_i - \bar{\mathbf{g}}_*\} = \limsup_{i\to\infty} \|\bar{\mathbf{g}}_i - \bar{\mathbf{g}}_*\| / \|\bar{\mathbf{g}}_{i-1} - \bar{\mathbf{g}}_*\| \le \rho. \tag{5.16}$$

In the notation of Section 3, the right-hand side of (5.15) can be evaluated as

$$\bar{\mathbf{g}}_i G_{\mathbf{z}}(\mathbf{x}, \mathbf{z}_*) \equiv \bar{G}(\mathbf{x}, \mathbf{z}_*, \bar{\mathbf{g}}_i)$$

by a reverse sweep on the procedure for evaluating $G(\mathbf{x}, \mathbf{z}_*)$. In many applications, this adjoining of the iteration function $G(\mathbf{x}, \mathbf{z})$ is no serious problem, and requires only a moderate amount of memory space. When $G = I - P\,H$ with $P = H_{\mathbf{z}}^{-1}$ representing Newton's method, we have $G_{\mathbf{z}}(\mathbf{x}, \mathbf{z}_*) = 0$, and the equation (5.15) reduces to $\mathbf{g}_1 = \bar{\mathbf{z}}_* = \mathbf{g}_*$. The relation between (5.15) and mechanical adjoining of original fixed point iteration was analysed carefully in Giles (2001) for linear time-variant systems and their discretizations.

Both (5.11) and (5.15) assume that we really have a contractive, single-step iteration. If in fact $\mathbf{z}_i = G^{(i)}(\mathbf{x}, \mathbf{z}_i)$ and each individual $G^{(i)}$ only contracts the solution error on some subspace, as occurs for example in multi-grid methods, then the linearization of what happens to be the last iteration will not provide a convergent solver for the direct or adjoint sensitivity equation.

5.3. Adjoint-based error correction

Whenever we have computed approximate solutions $\mathbf{z}$ and $\bar{\mathbf{z}}$ to the state equation (5.2) and the adjoint sensitivity equation (5.14), we have in analogy to (5.10) the error bound

$$\big(\|\mathbf{z} - \mathbf{z}_*\| + \|\bar{\mathbf{z}} - \bar{\mathbf{z}}_*\|\big) = O\big(\|\mathbf{z} - G\,(\mathbf{x}, \mathbf{z})\| + \|\bar{\mathbf{z}} - \bar{G}\,(\mathbf{x}, \mathbf{z}, \bar{\mathbf{x}}, \bar{\mathbf{z}})\|\big). \tag{5.17}$$

Using the approximate adjoint solution $\bar{\mathbf{g}}$ we may improve the estimate for the weighted response $\bar{\mathbf{y}}\,F(\mathbf{x}, \mathbf{z}_*) = \bar{\mathbf{y}}\,f(\mathbf{x})$ by the correction $\bar{\mathbf{g}}(G(\mathbf{z}, \mathbf{x}) - \mathbf{z})$. More specifically, we have for Lipschitz continuously differentiable F and G the Taylor expansion

$$F(\mathbf{x}, \mathbf{z}_*) = F(\mathbf{x}, \mathbf{z}) + F_{\mathbf{z}}(\mathbf{x}, \mathbf{z})(\mathbf{z}_* - \mathbf{z}) + O\big(\|\mathbf{z} - \mathbf{z}_*\|^2\big)$$

and

$$0 = \mathbf{z}_* - G(\mathbf{x}, \mathbf{z}_*)) = \mathbf{z} - G(\mathbf{x}, \mathbf{z}) + \big[I - G_\mathbf{z}(\mathbf{x}, \mathbf{z})\big](\mathbf{z}_* - \mathbf{z}) + O\big(\|\mathbf{z} - \mathbf{z}_*\|^2\big).$$

By subtracting $\bar{\mathbf{g}}$ times the second equation from the first, we obtain the estimate

$$\begin{aligned}\bar{\mathbf{y}}\, F(\mathbf{x}, \mathbf{z}) &- \bar{\mathbf{g}}\big[\mathbf{z} - G(\mathbf{x}, \mathbf{z})\big] - \bar{\mathbf{y}}\, f(\mathbf{x}) \\ &= O\big(\|\bar{\mathbf{g}}\big[I - G_\mathbf{z}(\mathbf{x}, \mathbf{z})\big] - \bar{\mathbf{z}}\|\; \|\mathbf{z} - \mathbf{z}_*\| + \|\mathbf{z} - \mathbf{z}_*\|^2\big).\end{aligned}$$

Now, if $\mathbf{z} = \mathbf{z}_i$ and $\bar{\mathbf{g}} = \bar{\mathbf{g}}_i$ are generated according to (5.1) and (5.15), we derive from (5.3) and (5.16) that

$$\bar{\mathbf{y}}\, F(\mathbf{x}, \mathbf{z}_i) - \bar{\mathbf{g}}_i\big[\mathbf{z}_i - G(\mathbf{x}, \mathbf{z}_i)\big] = \bar{\mathbf{y}}\, f(\mathbf{x}) + \tilde{O}(\rho^{2i}).$$

Hence the corrected estimate converges twice as fast as the uncorrected one. It has been shown by Griewank and Faure (2002) that, for linear G and F and zero initializations $\mathbf{z}_0 = 0$ and $\bar{\mathbf{g}}_0 = 0$, the ith corrected estimate is exactly equal to the $2\,i$th uncorrected value $\bar{\mathbf{y}}\, F(\mathbf{x}, \mathbf{z}_i)$. The same effect occurs when the initial $\mathbf{z}_0$ is quite good compared to $\bar{\mathbf{z}}_0$, since the adjoint iteration (5.15) is then almost linear. This effect is displayed in Figure 5.2, where the normal (uncorrected) value lags behind the double (adjoint corrected) one by almost exactly the factor 2. Here the weighted response $\bar{\mathbf{y}}F$ was the drag coefficient of the NACA0012 airfoil.

In our setting the discrepancies $\mathbf{z} - \mathbf{z}_*$ and $\bar{\mathbf{g}} - \mathbf{g}_*$ come about through iterative equation solving. The same duality arguments apply if $\mathbf{z}_*$ and $\bar{\mathbf{g}}_*$ are solutions of operator equations that are approximated by solutions $\mathbf{z}$ and $\bar{\mathbf{g}}$ of corresponding discretizations. Under suitable conditions elaborated

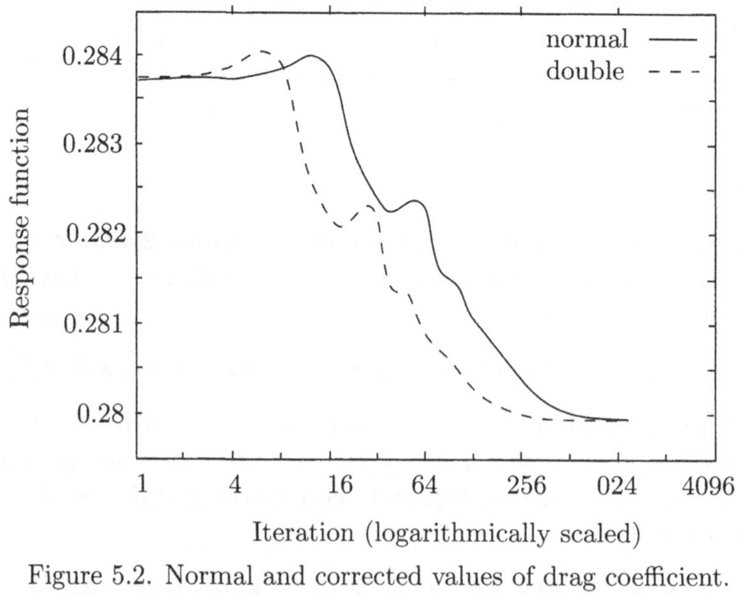

Figure 5.2. Normal and corrected values of drag coefficient.

in Giles and Pierce (2001), Giles and Süli (2002) and Becker and Rannacher (2001), the adjoint correction technique then doubles the order of convergence with respect to the mesh-width. In both scenarios, solving the adjoint equation provides accurate sensitivities of the weighted response with respect to solution inaccuracies. For discretized PDEs this information may then be used to selectively refine the grid where solution inaccuracies have the largest effect on the weighted response (Becker and Rannacher 2001).

5.4. Grey box differentiation and one-shot optimization

The two-phase approach considered above seems inappropriate in all situations where the design vector $\mathbf{x}$ is not constant, but updated repeatedly to arrive at some desirable value of the response $f(\mathbf{x}) = F(\mathbf{x}, \mathbf{z}_*(\mathbf{x}))$, for example a minimum. Then it makes more sense to apply simultaneously with the state space iteration (5.1) the adjoint iteration

$$\bar{\mathbf{g}}_{i+1} = \bar{\mathbf{z}}_i + \bar{\mathbf{g}}_i G_{\mathbf{z}}(\mathbf{x}, \mathbf{z}_i) = \bar{\mathbf{y}}\, F_{\mathbf{z}}(\mathbf{x}, \mathbf{z}_i) + \bar{\mathbf{g}}_i G_{\mathbf{z}}(\mathbf{x}, \mathbf{z}_i). \tag{5.18}$$

As we observed for the forward derivative recurrence (5.6), the continuing variation in $\mathbf{z}_i$ destroys the proof of Q-linear convergence, but we have

$$R\{\bar{\mathbf{g}}_i - \bar{\mathbf{g}}_*\} \equiv \limsup_{i\to\infty} \sqrt[i]{\|\bar{\mathbf{g}}_i - \bar{\mathbf{g}}_*\|} \le \rho, \tag{5.19}$$

analogous to (5.8). The $\bar{\mathbf{g}}_i$ allow the computation of the approximate reduced gradients

$$\bar{\mathbf{x}}_i \equiv \bar{\mathbf{y}}\, F_{\mathbf{x}}(\mathbf{x}, \mathbf{z}_i) + \bar{\mathbf{g}}_i\, G_{\mathbf{x}}(\mathbf{x}, \mathbf{z}_i) = \bar{\mathbf{x}}_* + \tilde{O}(\rho^i). \tag{5.20}$$

Differentiating (5.18) once more, we obtain the second-order adjoint iteration

$$\dot{\bar{\mathbf{g}}}_{i+1} = \bar{\mathbf{y}}\dot{F}_{\mathbf{z}}(\mathbf{x}, \mathbf{z}_i, \dot{\mathbf{x}}, \dot{\mathbf{z}}_i) + \bar{\mathbf{g}}_i\dot{G}_{\mathbf{z}}(\mathbf{x}, \mathbf{z}_i, \dot{\mathbf{x}}, \dot{\mathbf{z}}_i) + \dot{\bar{\mathbf{g}}}_i G_{\mathbf{z}}(\mathbf{x}, \mathbf{z}_i), \tag{5.21}$$

where

$$\dot{F}_{\mathbf{z}}(\mathbf{x}, \mathbf{z}_i, \dot{\mathbf{x}}, \dot{\mathbf{z}}_i) \equiv F_{\mathbf{z}\mathbf{x}}(\mathbf{x}, \mathbf{z}_i)\dot{\mathbf{x}} + F_{\mathbf{z}\mathbf{z}}(\mathbf{x}, \mathbf{z}_i)\dot{\mathbf{z}}_i$$

and $\dot{G}_{\mathbf{z}}$ is defined analogously. The vector $\dot{\bar{\mathbf{g}}}_i \in X^*$ obtained from (5.21) may then be used to calculate

$$\begin{aligned}\dot{\bar{\mathbf{x}}}_i &= \bar{\mathbf{y}}\dot{F}_{\mathbf{x}}(\mathbf{x}, \mathbf{z}_i, \dot{\mathbf{x}}, \dot{\mathbf{z}}_i) + \dot{\bar{\mathbf{g}}}_i G_{\mathbf{x}}(\mathbf{x}, \mathbf{z}_i) + \bar{\mathbf{g}}_i\dot{G}_{\mathbf{x}}(\mathbf{x}, \mathbf{z}_i, \dot{\mathbf{x}}, \dot{\mathbf{z}}_i)\\ &= \dot{\bar{\mathbf{x}}}_* + \tilde{O}(\rho^i) = \bar{\mathbf{y}} f''(\mathbf{x})\dot{\mathbf{x}} + \tilde{O}(\rho^i),\end{aligned}$$

where $\dot{F}_{\mathbf{x}}$ and $\dot{G}_{\mathbf{x}}$ are also defined by analogy with $\dot{F}_{\mathbf{z}}$. The vector $\dot{\bar{\mathbf{x}}}$ represent a first-order approximation to the product of the reduced Hessian $\bar{\mathbf{y}} f''(\mathbf{x}) = \nabla^2_{\mathbf{x}}[\bar{\mathbf{y}} f(\mathbf{x})]$ with the direction $\dot{\mathbf{x}}$.

While the right-hand side of (5.21) looks rather complicated, it can be evaluated as a second-order adjoint of the vector function

$$E(\mathbf{x}, \mathbf{z}) \equiv [F(\mathbf{x}, \mathbf{z}), G(\mathbf{x}, \mathbf{z})],$$

whose first-order adjoint $\bar{E}$ appears on the right-hand side of (5.18).

Overall we have the following derivative enhanced iterations for $i = 0, 1 \ldots$:

$$\begin{aligned}
(\mathbf{y}, \mathbf{z}) &= \mathbf{E}\,(\mathbf{x}, \mathbf{z}) && \text{(original)},\\
(\bar{\mathbf{x}}, \bar{\mathbf{g}}) &= \bar{\mathbf{E}}\,(\mathbf{x}, \mathbf{z}, \bar{\mathbf{y}}, \bar{\mathbf{g}}) && \text{(adjoint)},\\
(\dot{\mathbf{y}}, \dot{\mathbf{z}}) &= \dot{\mathbf{E}}\,(\mathbf{x}, \mathbf{z}, \dot{\mathbf{x}}, \dot{\mathbf{z}}) && \text{(direct)},\\
(\dot{\bar{\mathbf{x}}}, \dot{\bar{\mathbf{g}}}) &= \dot{\bar{\mathbf{E}}}\,(\mathbf{x}, \mathbf{z}, \bar{\mathbf{y}}, \bar{\mathbf{g}}, \dot{\mathbf{x}}, \dot{\mathbf{z}}, \dot{\bar{\mathbf{g}}}) && \text{(second)}.
\end{aligned}$$

Here we have omitted the indices, because the new versions of $\mathbf{y}, \mathbf{z}, \bar{\mathbf{x}}, \bar{\mathbf{g}}, \dot{\mathbf{y}}, \dot{\mathbf{z}}$, $\dot{\bar{\mathbf{x}}}$ and $\dot{\bar{\mathbf{g}}}$ can immediately overwrite the old ones. Under our contractivity assumption all converge with the same R-factor ρ. Norms of the discrepancies between the left- and right-hand sides can be used to measure the progress of the iteration. Their plot in Figure 5.1 confirms the asymptotic result, but also suggests that higher derivatives lag behind lower derivatives, as was to be expected.

We have named this section 'grey box differentiation' because the generation of the nonincremental adjoint statement $(\bar{\mathbf{x}}, \bar{\mathbf{g}}) = \bar{E}(\mathbf{x}, \mathbf{z}, \bar{\mathbf{y}}, \bar{\mathbf{g}})$ from the original $(\mathbf{y}, \mathbf{z}) = E(\mathbf{x}, \mathbf{z})$ does not follow the usual recipe of adjoint generation, first and foremost because the outer loop need not be reversed, and consequently the old versions of the various variables need not be saved on a stack. Also, the adjoints are not updated incremently by += but properly assigned new values. There are many possible variations of the derived fixed point iterations. In 'one shot' optimization we try to solve simultaneously the stationarity condition $\bar{\mathbf{x}} = \nabla f(\mathbf{x}) = 0$, possibly by another fixed point iteration

$$\mathbf{x} = \mathbf{x} - Q^{-1}\bar{\mathbf{x}} \quad \text{with} \quad Q \approx \nabla^2 f(\mathbf{x}).$$

The reduced Hessian approximation Q might be based on $\mathtt{dim}(X)$ second-order adjoints of the form $\dot{\bar{\mathbf{x}}}$, with $\dot{\mathbf{x}}$ ranging over a basis in X. Alternatively, we may use secant updates or apply Uzawa-like algorithms. Some of these variants are employed in Newman, Hou, Jones, Taylor and Korivi (1992), Keyes, Hovland, McInnes and Samyono (2001), Forth and Evans (2001), Mohammadi and Pironneau (2001), Venditti and Darmofal (2000), Hinze and Slawig (2002) and Courty, Dervieux, Koobus and Hascoët (2002).

6. Jacobian matrices and graphs

The forward and reverse mode are nearly optimal choices when it comes to evaluating a single tangent $\dot{\mathbf{y}} = F'(\mathbf{x})\dot{\mathbf{x}}$ or a single gradient $\bar{\mathbf{x}} = \bar{\mathbf{y}}F'(\mathbf{x})$, respectively. In contrast, when we wish to evaluate a Jacobian with $m > 1$ rows and $n > 1$ columns, we observe in this section that the forward and reverse mode are just two extreme options from a wide range of possible ways to apply the chain rule. This generalization opens up the chance for a very significant reduction in operation count, although the search for an elimination ordering with absolutely minimal operation count is a hard combinatorial problem.

6.1. Nonincremental structure and Bauer's formula

In this subsection we will exclude all incremental statements so that there is a 1–1 correspondence between the φ_i and their results v_i. They must then, by (2.8), all be independent of each other in that $P_i^T P_j = 0$ if $i \neq j$. Theoretically any evaluation procedure with incremental statements can be rewritten by combining all contributions to a certain value in one statement. For example, the reverse sweep of the adjoint procedure on the right-hand side of Table 3.3 can be combined to the procedure listed in Figure 6.1.

$$\begin{aligned}
\bar{v}_5 &= \bar{y} \\
\bar{v}_4 &= \bar{v}_5 * v_3 \\
\bar{v}_3 &= \bar{v}_5 * v_4 \\
\bar{v}_2 &= \bar{v}_4 * \cos(v_1 + v_2) \\
\bar{v}_1 &= \bar{v}_4 * \cos(v_1 + v_2) + \bar{v}_3 * v_3
\end{aligned}$$

Figure 6.1. Nonincremental adjoint.

On larger codes this elimination of incremental statements may be quite laborious and change the structure significantly. Without incremental assignments we have the orthogonal decomposition

$$\mathcal{H} = \mathcal{H}_1 \oplus \mathcal{H}_2 \oplus \cdots \oplus \mathcal{H}_l \quad \text{and} \quad v_i \in \mathcal{H}_i \quad \text{for} \quad i = 1 \ldots l.$$

Consequently we have

$$Q_i = Q_i \sum_{j \prec i} P_j P_j^T = \sum_{j \prec i} Q_i P_j P_j^T$$

and may write

$$v_i = \varphi_i \left(\sum_{j \prec i} Q_i P_j v_j \right).$$

Thus we obtain the partial derivatives

$$C_{ij} = \frac{\partial \varphi_i}{\partial v_j} = \varphi_i'(u_i) Q_i P_j \in \mathbb{R}^{m_i \times m_j}.$$

The interpretation of the C_{ij} as $m_i \times m_j$ matrices again assumes a fixed orthonormal basis for all $\mathcal{H}_i \triangleq \mathbb{R}^{m_i}$ and thus for their Cartesian product $\mathcal{H}$. Since C_{ij} is nontrivial exactly when $j \prec i$, we may attach these matrices as

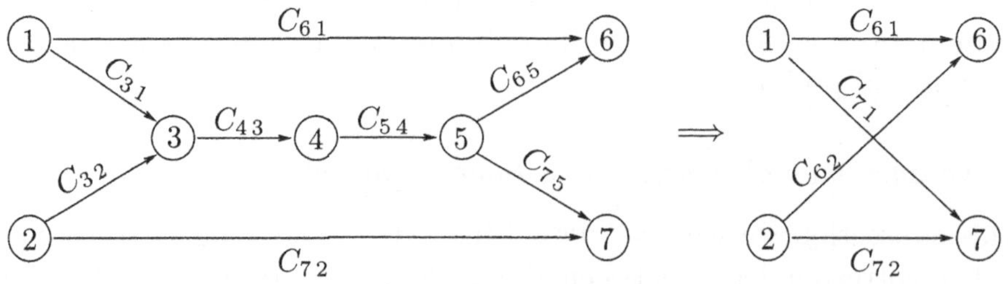

Figure 6.2. Jacobian vertex accumulation on a simple graph.

labels to the edges of the computation graph, as sketched in Figure 6.2 for a particular case with two independents and two dependents.

Obviously the local derivatives C_{ij} uniquely determine the overall Jacobian $F'(\mathbf{x})$, whose calculation from the C_{ij} we will call *accumulation.* Formally we may use the following explicit expression for the individual Jacobian blocks, which can be derived from the chain rule by induction on l.

Lemma 6.1. (Bauer's formula) The derivative of any dependent variable $y_i = v_{l-m+i}$ with respect to any independent variable $x_j = v_j$ is given by

$$\frac{\partial y_i}{\partial x_j} = \sum_{\mathcal{P}\in[j\to\hat{\imath}]} \prod_{(\tilde{\jmath},\tilde{\imath})\subset\mathcal{P}} C_{\tilde{\imath}\tilde{\jmath}}, \tag{6.1}$$

where $[j \to \hat{\imath}]$ denotes the set of all paths connecting j to $\hat{\imath} \equiv l - m + i$ and the pair $(\tilde{\jmath}, \tilde{\imath})$ ranges over all edges in $\mathcal{P}$ in descending order from left to right.

The order of the factors C_{ij} along any one of the paths matters only if some of the vertex dimensions $m_i = \dim(v_i)$ are greater than 1. For the example depicted in Figure 6.2 we find, according to Lemma 6.1,

$$\frac{\partial y_1}{\partial x_1} = \frac{\partial v_6}{\partial v_1} \equiv C_{61} = C_{61} + C_{65}\, C_{54}\, C_{43}\, C_{31},$$
$$\frac{\partial y_1}{\partial x_2} = \frac{\partial v_6}{\partial v_2} \equiv C_{62} = C_{65}\, C_{54}\, C_{43}\, C_{32},$$
$$\frac{\partial y_2}{\partial x_1} = \frac{\partial v_7}{\partial v_1} \equiv C_{71} = C_{75}\, C_{54}\, C_{43}\, C_{31},$$
$$\frac{\partial y_2}{\partial x_2} = \frac{\partial v_7}{\partial v_2} \equiv C_{72} = C_{72} + C_{75}\, C_{54}\, C_{43}\, C_{32}.$$

In the scalar case the number of multiplications or additions needed to accumulate all entries of the Jacobian by Bauer's formula is given by

$$\mathtt{SIZE}\,(F(\mathbf{x})) = \sum_{j=1}^{n} \sum_{i=1}^{m} \sum_{\mathcal{P}\in[j\to\hat{\imath}]} |\mathcal{P}|,$$

where $|\mathcal{P}|$ counts the number of vertices in $\mathcal{P}$. It is not hard to see that this often enormous number is proportional to the length of the formulas that express each y_i directly in terms of all x_j without any named intermediates.

The expression (6.1) has the flavour of a determinant, which is no coincidence since in the scalar case it is in fact a determinant, as we shall see in Section 6.3. Naturally, an explicit evaluation is usually very wasteful for there are often common subexpressions, such as $C_{54}\,C_{43}$, that should probably be calculated first. Actually this is not necessarily so, unless we assume all intermediates v_i to be scalars. For instance, if the two independents v_1, v_2 and dependents v_6, v_7 were scalars but the three intermediates v_3, v_4 and v_5 were vectors of length d, then the standard matrix product $C_{54}\,C_{43}$ of two $d \times d$ matrices would cost d^3 multiplications; whereas a total of 8 matrix-vector multiplications would suffice to compute the four Jacobian entries by bracketing their expressions from left to right or *vice versa.*

6.2. Jacobian accumulation by vertex elimination

In graph terminology we can interpret the calculation of the product $C_{53} \equiv C_{54}\,C_{43}$ as the elimination of the vertex ④ in Figure 6.2. Then we have afterwards the simplified graph depicted on the left-hand side of Figure 6.3.

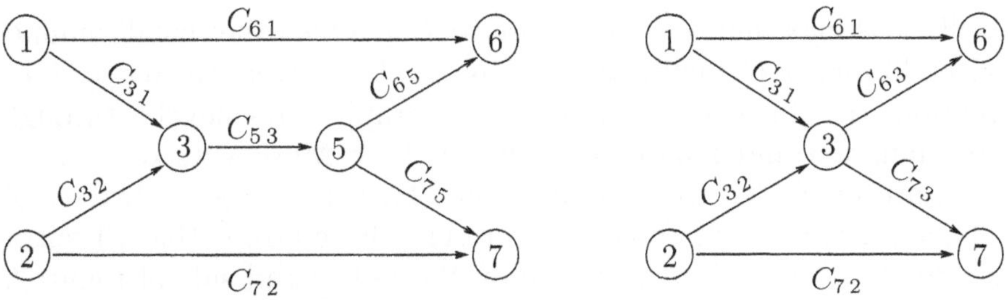

Figure 6.3. Successive vertex eliminations on problem displayed in Figure 6.2.

Subsequently we may eliminate vertex ⑤ by setting $C_{63} \equiv C_{65}\,C_{53}$ and $C_{73} = C_{75}\,C_{53}$. Finally, eliminating vertex ③ we arrive at the labels C_{61}, C_{62}, C_{71}, and C_{72}, which are exactly the (block) elements of the Jacobian $F'(\mathbf{x})$ represented as a bipartite graph in Figure 6.2.

In general, elimination of an intermediate vertex v_j from the computational graph requires the incrementation

$$C_{ik} \mathrel{+}= C_{ij}\,C_{jk} \in \mathbb{R}^{m_k \times m_i}$$

for all predecessor–successor pairs (k, i) with $k \prec j \prec i$. If a direct edge (i, k) did not exist beforehand it must now be introduced with the initial value $C_{ik} = C_{ij}\,C_{jk}$. In other words, the precedence relation $\prec$ and the graph structure must be updated such that all pairs (k, i) with $k \prec j \prec i$

are directly connected by an edge, afterwards the vertex j and its edges are deleted. Again assuming elementary matrix–matrix arithmetic, we have the total elimination cost

$$\mathtt{MARK}(v_j) \equiv \mathtt{OPS}\big(\mathtt{Elim}(v_j)\big) = m_j \left(\sum_{k \prec j \prec i} m_k \, m_i \right).$$

When all vertices are scalars $\mathtt{MARK}(v_j)$ reduces to the Markowitz degree familiar from sparse Gaussian elimination (Rose and Tarjan 1978). There as here, the overall objective is to minimize the vertex accumulation cost

$$\mathtt{VACC}\big(F'(\mathbf{x})\big) \equiv \sum_{n<j\leq l-m} \mathtt{MARK}(v_j),$$

where the degree $\mathtt{MARK}(v_j)$ needs to be computed just before the elimination of the jth vertex. This accumulation effort is likely to dominate the overall operation count compared to the effort for evaluating the elemental partials

$$\mathtt{OPS}\left(F'(\mathbf{x})\right) - \mathtt{VACC}\left(F'(\mathbf{x})\right) \equiv \sum_{i=1}^{l} \mathtt{OPS}\left(\{C_{ij}\}_{j \prec i}\right) \leq q \, \mathtt{OPS}\left(F(\mathbf{x})\right),$$

where

$$q \equiv \max_i \big\{ \mathtt{OPS}\left(\{C_{ij}\}_{j \prec i}\right) / \mathtt{OPS}\left(\varphi_i\right) \big\}.$$

Typically this maximal ratio q is close to 2 or some other small number for any given library of elementary functions φ_i. In contrast to sparse Gaussian elimination, the linearized computational graph stays acyclic throughout, and the minimal and maximal vertices, whose Markowitz degrees vanish trivially, are precluded from elimination. They form the vertex set of the final bipartite graph whose edges represent the nonzero (block) entries of the accumulated Jacobian, as shown on the right-hand side of Figure 6.2.

For the small example Figure 6.2 with scalar vertices, the elimination order ④, ⑤ and ③ discussed above requires $1 + 2 + 4 = 7$ multiplications. In contrast, elimination in the natural order ③, ④ and ⑤, or its exact opposite ⑤, ④ and ③, requires $2+2+4 = 8$ multiplications. Without too much difficulty it can be seen that, in general, eliminating all intermediate vertices in the orders $n+1, n+2, \ldots, (l-1), l$ or $l, l-1, \ldots, n+1, n$ is mathematically equivalent to applying the sparse vector forward or the sparse reverse mode discussed in Section 3.5, respectively. Thus our tiny example demonstrates that, already, neither of the two standard modes of AD needs to be optimal in terms of the operation count for Jacobian accumulation.

As in the case of sparse Gaussian elimination (Rose and Tarjan 1978), it has been shown that minimizing the fill-in during the accumulation of Jacobian is an NP-hard problem and the same is probably true for minimizing the operation count. In any case no convincing heuristics for selecting one of the $(l-m-n)!$ vertex elimination orderings have yet been developed. A good reason for this lack of progress may be the observation that, on one hand, Jacobian accumulation can be performed in more general ways, and on the other hand, it may not be a good idea in the first place. These two aspects are discussed in the remainder of this section.

Recently John Reid has observed that, even though Jacobian accumulation involves no divisions, certain vertex elimination orderings may lead to numerical instabilities. He gave an example similar to the one depicted in Figure 6.4, where the arcs are annotated by constant partials u and h, whose values are assumed to be rather close to 2 and -1, respectively.

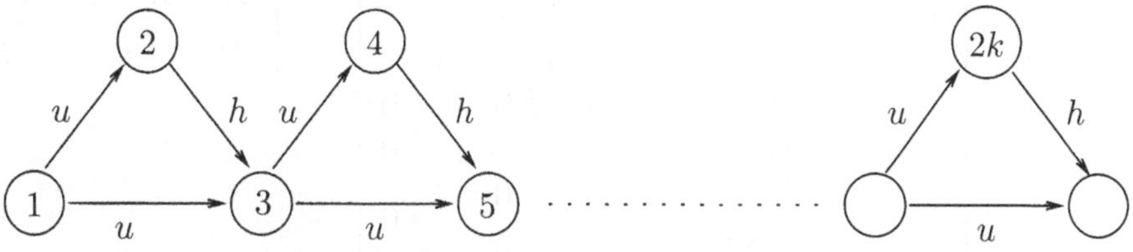

Figure 6.4. Reid's example of potential accumulation instability.

When the intermediate vertices $2, 3, 4, \ldots, 2k-1, 2k$ are eliminated forward or backward, all newly calculated partial derivatives are close to zero, as should be the final result $c_{2k+1,1}$. However, if we first eliminate only the odd vertices $3, 5, \ldots, 2k-1$, the arc between vertex 1 and vertex $2k+1$ temporarily reaches the value $u^k \approx 2^k$. The subsequent elimination of the even intermediate vertices $2, 4, 6, \ldots, 2k$ theoretically balances out this enormous value to nearly zero. However, in floating point arithmetic errors are certain to be amplified enormously. Hence, it is clear that numerical stability must be a concern in the study of suitable elimination orderings. In this particular case the numerically unstable, prior elimination of the odd vertices is also the one that makes the least sense in terms of minimizing fill-in and operation count. More specifically, the number of newly allocated and computed arcs grows quadratically with the depth k of the computational graph. Practically any other elimination order will result in a temporal and spatial complexity that is linear in k.

6.3. Jacobians as Schur complements

Because we have excluded incremental assignments and imposed the single assignment condition (2.7), the structure of the evaluation procedure Table 2.1 is exactly reflected in the nonlinear system

$$0 = E(\mathbf{x}; \mathbf{v}) = (\varphi_i(u_i) - v_i)_{i=1\ldots l},$$

where, as before, $\varphi_i(u_i) = \varphi_i(\,) = x_i$ for $i = 1 \ldots n$. By our assumptions on the data dependence $\prec$, the system is triangular and its $l \times l$ (block)

Jacobian has the structure

$$E'(\mathbf{x};\mathbf{v}) = \Big(C_{ij} - \delta_{ij} I\Big)_{i=1\ldots l}^{j=1\ldots l} \equiv C - I \tag{6.2}$$

$$= \left[\begin{array}{cccc|cccc|cccc}
-I & 0 & \cdots & 0 & 0 & & \cdots & 0 & 0 & & \cdots & 0 \\
0 & \ddots & & \vdots & \vdots & & & \vdots & \vdots & & & \vdots \\
\vdots & & \ddots & 0 & \vdots & & & \vdots & \vdots & & & \vdots \\
0 & \cdots & 0 & -I & 0 & & \cdots & 0 & 0 & & \cdots & 0 \\
\hline
\mathrm{x} & \cdots & & \mathrm{x} & -I & 0 & \cdots & 0 & 0 & & \cdots & 0 \\
\vdots & & & \vdots & \mathrm{x} & \ddots & & \vdots & \vdots & & & \vdots \\
\vdots & & & \vdots & \vdots & & \ddots & 0 & \vdots & & & \vdots \\
\mathrm{x} & \cdots & & \mathrm{x} & \mathrm{x} & \cdots & \mathrm{x} & -I & 0 & & \cdots & 0 \\
\hline
\mathrm{x} & \cdots & & \mathrm{x} & \mathrm{x} & \cdots & & \mathrm{x} & -I & 0 & \cdots & 0 \\
\vdots & & & \vdots & \vdots & & & \vdots & 0 & \ddots & & \vdots \\
\vdots & & & \vdots & \vdots & & & \vdots & \vdots & & \ddots & 0 \\
\mathrm{x} & \cdots & & \mathrm{x} & \mathrm{x} & \cdots & & \mathrm{x} & 0 & \cdots & 0 & -I
\end{array}\right] = \begin{bmatrix} -I & 0 & 0 \\ B & L-I & 0 \\ R & T & -I \end{bmatrix}.$$

Here δ_{ij} is the Kronecker delta and $C_{ij} \neq 0 \Longleftrightarrow j \prec i \Longrightarrow j < i$. The last relation implies in particular that the matrix L is strictly lower-triangular. Applying the implicit function theorem to $E(\mathbf{x};\mathbf{v}) = 0$, we obtain the derivative

$$\begin{aligned} F'(\mathbf{x}) &\equiv R + T(I-L)^{-1}B \\ &= R + T[(I-L)^{-1}B] = R + [T(I-L)^{-1}]B. \end{aligned} \tag{6.3}$$

The two bracketings on the last line represent two particular ways of accumulating the Jacobian $F'(\mathbf{x})$. One involves the solution of n linear systems in the unary lower-triangular matrix $(L - I)$ and the other requires the solution of m linear systems in its transpose $(L-I)^T$. As observed in Chapter 8 of Griewank (2000), the two alternatives correspond once more to the forward and reverse mode of AD, and their relative cost need not be determined by the ratio m/n nor even by the sparsity structure of the Jacobian $F'(\mathbf{x})$ alone. Instead, what matters is the sparsity of the extended Jacobian $E'(\mathbf{v}, \mathbf{x})$, which is usually too huge to deal with in many cases. In the scalar case with $R = 0$ we can rewrite (6.3) for any two Cartesian basis vectors $e_i \in \mathbb{R}^m$ and $e_j \in \mathbb{R}^n$ by Cramer's rule as

$$e_i^T F'(\mathbf{x})\, e_j = \det\begin{bmatrix} I-L & B\mathbf{e}_j\, \mathbf{e}_i^T \\ -T & I \end{bmatrix}. \tag{6.4}$$

Thus we see that Bauer's formula given in Lemma 6.1 indeed represents the determinant of a sparse matrix. Moreover, we may derive the following alternative interpretation of Jacobian accumulation procedures.

6.4. Accumulation by edge elimination

Eliminating a certain vertex j in the computational graph, as discussed in the previous section, corresponds to using the $-I$ in the jth diagonal block of (6.2) to eliminate all other (block) elements in its row or column. This can be done for all $n > j \leq l-m$ in arbitrary order until the lower-triangular part L has been zeroed out completely, and the bottom left block R now contains the Jacobian $F'(\mathbf{x})$. Rather than eliminating all subdiagonal blocks in a row or column at once, we may also just zero out one of them, say C_{ij}. This can be interpreted as back- or front-elimination of the edge (j,i) from the computational graph, and requires the updates

$$C_{ik} \mathrel{+}= C_{ij}\, C_{jk} \quad \text{for all} \quad k \prec j \tag{6.5}$$

or

$$C_{hj} \mathrel{+}= C_{hi}\, C_{ij} \quad \text{for all} \quad h \succ i, \tag{6.6}$$

respectively. It is easy to check that, after the precedence relation $\prec$ has been updated accordingly, Bauer's formulas is unaffected. Front-eliminating all edges incoming to a node j or back-eliminating all outgoing edges is equivalent to the vertex elimination of j. The back-elimination (6.5) or the front-elimination (6.6) of the edge (j,i) may actually increase the number of nonzero (block-)elements in (6.2), which equals the number of edges in the computational graph. However, we can show that the sum over the length of all directed path in the graph and also the number of nonzeros in the subdiagonals of (6.2) are monotonically decreasing. Therefore, elimination of edges in any order must again lead eventually to the same bipartite graph representing the Jacobian.

Let us briefly consider edge elimination for the 'lion' example of Naumann displayed in Figure 6.5. It can be checked quite easily that eliminating the two intermediate vertices ① and ② in either order requires a total of 12 multiplications. However, if the edge c_{84} is first back-eliminated, the total effort is reduced by one multiplication. Hence we see that the generalization from vertex to edge elimination makes a further reduction in the operation count possible. However, it is not yet really known how to make good use of this extra freedom, and there is the danger that a poor choice of the order in which the edges are eliminated can lead to an exponential effort overall. This cannot happen in vertex elimination for which $\mathtt{VACC}(F'(\mathbf{x}))$ always has the cubic bound $\left[l \max_{1\leq i\leq l}(m_i)\right]^3$.

The same upper bound applies to edge elimination if we ensure that no edge is eliminated and reintroduced repeatedly through fill-in. To exclude this possibility, edge (j,i) may only be eliminated when it is final, *i.e.*, no other path connects the origin j to the destination i. Certain edge elimination orderings correspond to breaking down the product representation (3.1) further into a matrix chain, where originally each factor corresponds

Program:

$v_3 = c_{21} * v_1 + c_{32} * v_2$
$v_4 = c_{43} * v_3$
$v_5 = c_{54} * v_4$
$v_6 = c_{64} * v_4$
$v_7 = c_{74} * v_4$
$v_8 = c_{84} * v_4 + c_{83} * v_3$

Graph:

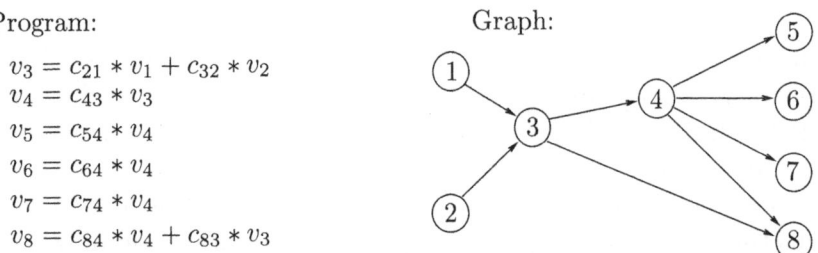

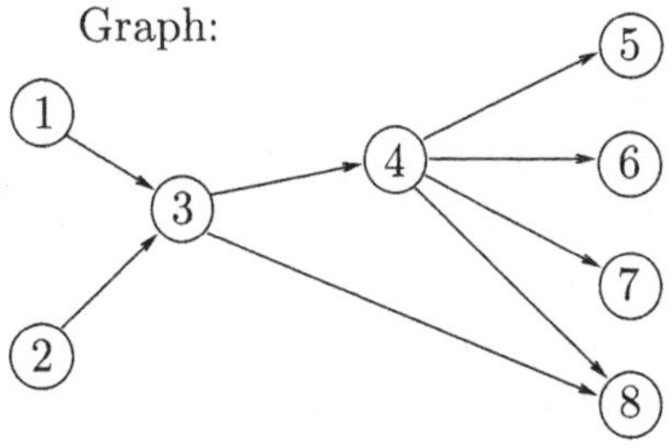

Figure 6.5. The 'lion' example from Naumann (1999) with $l = 8$.

to a single elemental partial C_{ij}. Many of these extremely sparse matrices commute and can thus be reordered. Moreover, the resulting chain can be bracketed in many different ways. The bracketing can be optimized by dynamic programming (Griewank and Naumann 2003*b*), which is classical when all matrix factors are dense (Aho, Hopcroft and Ullman 1974). Various elimination strategies were tested on the stencil generated by the Roe flux for hyperbolic PDEs (Tadjouddine, Forth and Pryce 2001).

6.5. Face elimination and its optimality

Rather than combining one edge with all its successors or with all its predecessors, we may prefer to combine only two edges at a time. In other words, we just wish to update $C_{ik} \mathrel{+}= C_{ij}\, C_{jk}$ for a single triple $k \prec j \prec i$, which is called a 'face' in Griewank and Naumann (2003*a*). Unfortunately, this kind of modification cannot be represented in a simple way on the level of the computational graph. Instead we have to perform *face eliminations* on the line graph whose vertices $\textcircled{j,i}$ correspond initially to the edges (j,i) of the original graph appended by two extra sets of minimal vertices o_j for $j = 1 \ldots n$ and maximal vertices d_i for $i = 1 \ldots m$. They represent edges $o_j = (-\infty, j)$ connecting a common source $(-\infty)$ to the independent vertices of the original graphs and edges $d_i = (\hat{\imath}, \infty)$ connecting the dependent vertices $\hat{\imath} = l - m + i$ for $i = 1 \ldots m$ to a common sink $(+\infty)$. A line edge connects two line vertices $\textcircled{k,i}$ and $\textcircled{j,h}$ exactly when $j = i$. Now the vertices rather than the edges are labelled by the values C_{ij}. Without a formal definition we display in Figure 6.6 the line graph corresponding to the 'lion' example defined in Figure 6.5. It is easy to check that line graphs of DAGs (directed acyclic graphs) are also acyclic. They have certain special properties (Griewank and Naumann 2003*a*), which may, however, be lost for a while by face elimination, as discussed below.

Bauer's formula (6.1) may be rewritten in terms of the line graph as

$$\frac{\partial y_i}{\partial x_j} = \sum_{\mathcal{P} \in [o_j \to d_i]} \; \prod_{\textcircled{\tilde{\jmath},\tilde{\imath}} \subset \mathcal{P}} C_{\tilde{\imath}\tilde{\jmath}}, \tag{6.7}$$

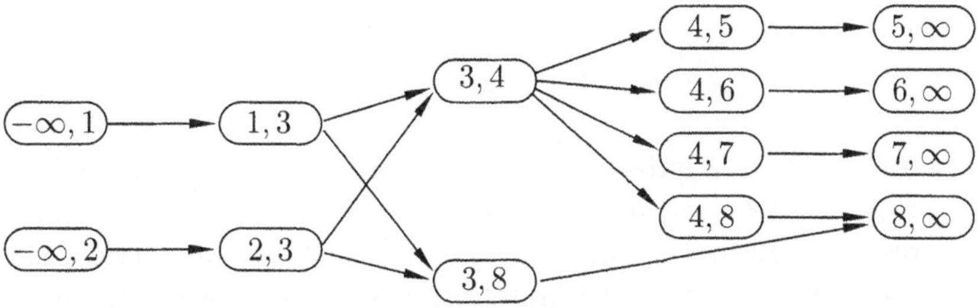

Figure 6.6. Line graph for Naumann's 'lion' example.

where $[o_j \to d_i]$ represents the set of all paths $\mathcal{P}$ connecting o_j to d_i, and the $(\tilde{j},\tilde{i})$ range over all vertices belonging to path $\mathcal{P}$ in descending order. The elimination of a face (k,j), (j,i) with $k \neq -\infty$ and $i \neq +\infty$ can now proceed by introducing a new line vertex with the value $C_{ij}\,C_{jk}$, and connecting it to all predecessors of (k,j) and all successors of (j,i). It is very easy to see that this modification leaves (6.7) valid and again reduces the total sum of the length of all maximal paths in the line graph. After the successive elimination of all interior edges of the line graph whose vertices correspond initially to edges in the computational graph, we arrive at a tripartite graph. The latter can be interpreted as the line graph of a bipartite graph, whose edge values are the vertex labels of the central layer of the tripartite graph. Naumann (2001) has constructed an example where accumulation by face elimination reduces the number of multiplications below the minimal count achievable by edge elimination. It is believed that face elimination on the scalar line graph is the most general procedure for accumulating Jacobians.

Conjecture 6.2. Given a scalar computational graph, consider a fixed sequence of multiplications and additions to compute the Jacobian entries $\partial y_i/\partial x_j$ given by (6.1) and (6.7) for arbitrary real values of the C_{ij} for $j \prec i$. Then the number of multiplications required is no smaller than that required by some face elimination sequence on the computational graph, that is,

$$\texttt{ACC}\big(F'(\mathbf{x})\big) = \texttt{FAAC}\big(F'(\mathbf{x})\big) \leq \texttt{EACC}\big(F'(\mathbf{x})\big) \leq \texttt{VACC}\big(F'(\mathbf{x})\big)$$

where the inequalities hold strictly for certain example problems, and FAAC, EACC denote the minimal operation count achievable by face and edge elimination, respectively.

Proof. We prove the assertion under the additional assumptions that all $c_{ij} = C_{ij}$ are scalar, and that the computational graph is absorption-free in that any two vertices j and i are connected by at most one directed path. The property is inherited by the line graph. Then the accumulation procedure involves only multiplications, and must generate a sequence of partial products of the form

$$c_{\mathcal{P}} \equiv \prod_{(j,i) \subset \mathcal{P}} c_{ij}, \tag{6.8}$$

where $\mathcal{P} \subset \mathcal{P}^c \subset \mathcal{G}$ is a subset of a connected path $\mathcal{P}^c$. Multiplying coefficients c_{ij} that are not contained in a common path makes no sense, since the resulting products cannot occur in any one of the Jacobian entries. Now suppose we modify the original accumulation procedure by keeping all products in factored form, that is,

$$c_{\mathcal{P}_k} = \prod_{(j,i) \in \mathcal{P}_k} c_{ij} \quad \text{for} \quad k = 0, 1, \ldots, \bar{k},$$

where $\mathcal{P} = \bigcup_{k=0}^{\bar{k}} \mathcal{P}_k$ is the decomposition of $\mathcal{P}$ into its $\bar{k} + 1$ maximal connected subcomponents. We show, by induction on the number of elements in $\mathcal{P}$, that this partitioned representation can be obtained by exactly $\bar{k}$ fewer multiplications than $c_{\mathcal{P}}$ itself. The assertion is trivially true for all individual arcs, which form a single element path with $\bar{k} = 0$. Now suppose we multiply two partial products $c_{\mathcal{P}}$ and $c_{\mathcal{P}'}$ with $\mathcal{P} \cap \mathcal{P}' = \emptyset$. Then the new partial path

$$\mathcal{P} \cup \mathcal{P}' = \bigcup_{k=0}^{\bar{k}} \mathcal{P}_k \cup \bigcup_{k'=0}^{\bar{k}'} \mathcal{P}'_{k'}$$

must be partitioned again into maximal connected components. Before these mergers we have $\bar{k} + \bar{k}' + 2$ subpaths, and, by induction hypothesis, $\bar{k} + \bar{k}' + 1$ saved multiplications. Whenever two subpaths $\mathcal{P}_k$ and $\mathcal{P}'_{k'}$ are adjacent we merge them, expending one multiplication and eliminating one gap. Hence the number of connected components minus one stays exactly the same as the number of saved multiplications. In summary, we find that we may rewrite the accumulation procedure such that all $\mathcal{P}$ successively occurring in (6.8) are, in fact, connected paths. Furthermore, without loss of generality we may assume that they are ordered according to their length. Now we will show that each of them can be interpreted as a face elimination on a line graph, whose initial structure was defined above. The first accumulation step must be of the form $c_{ijk} = c_{ij}\, c_{jk}$ with $k \prec j$ and $j \prec i$. This simplification may be interpreted as a face elimination on the line graph, which stays absorption-free. Hence we face the same problem as before, but for a graph of somewhat reduced total length. Consequently, the assertion follows by induction on the total length, *i.e.*, the sum of the lengths of all maximal paths. □

According to the conjecture, accumulating Jacobians with minimal multiplication count requires the selection of an optimal face elimination sequence of which there exists a truly enormous number, even for comparatively small initial line graphs. Before attacking such a difficult combinatorial problem we may pause to question whether the accumulation of Jacobians is really such an inevitable task as it may seem at first. The answer is in general 'no'. To demonstrate this we make two rather simple but fundamental observations. Firstly, Newton steps may sometimes be calculated more cheaply by a finite algorithm (of infinite precision) that does not involve accumulating the corresponding square Jacobian. Secondly, the representation of the Jacobian even as a sparse matrix, *i.e.*, a rectangular array of numbers, may be inappropriate, because it hides structure and wastes storage.

6.6. Jacobians with singular minors

Neither in Section 3 nor in Sections 6.4 and 6.5 have we obtained a definite answer concerning the cost ratio

$$\texttt{OPS}\big([F(\mathbf{x}), F'(\mathbf{x})]\big)/\texttt{OPS}(F(\mathbf{x})) \in \big[1, 3\min(\hat{n}, \hat{m})\big].$$

Here $\hat{n} \leq n$ and $\hat{m} \leq m$ denote, as in Section 3.5, the maximal number of nonzeros per row and column, respectively. A very simple dense example for which the upper bound is attained up to the constant factor 6 is

$$F(\mathbf{x}) = \frac{1}{2}\,(\mathbf{a}^T\mathbf{x})^2\mathbf{b} \quad \text{with} \quad \mathbf{a} \in \mathbb{R}^n, \quad \mathbf{b} \in \mathbb{R}^m. \tag{6.9}$$

Here we have $\texttt{OPS}(F(\mathbf{x})) = n + m$, but $F'(\mathbf{x}) = \mathbf{b}(\mathbf{a}^T\mathbf{x})\mathbf{a}^T \in \mathbb{R}^{m\times n}$ has in general mn distinct entries so that

$$\texttt{OPS}(F'(\mathbf{x}))/\texttt{OPS}(F(\mathbf{x})) \geq mn/(m+n) \geq \min(m, n)/2,$$

for absolutely any method of calculating the Jacobians as a rectangular array of reals.

The troubling aspect of this example is that we can hardly imagine a numerical purpose where it would make sense to accumulate the Jacobian – quite the opposite. If we treat the inner product $z = \mathbf{a}^T\mathbf{x} = \sum_{i=1}^n a_i x_i$ as a single elemental operation, the computational graph of (6.9) takes for $n = 3$ and $m = 2$ the form sketched in Figure 6.7. Here z is also the derivative of $\frac{1}{2}z^2$.

The elimination of the two intermediate nodes in Figure 6.7 yields an $m \times n$ matrix with mn nonzero elements. Owing to rounding errors this matrix will typically have full rank, whereas the unaccumulated representation $F'(\mathbf{x}) = \mathbf{b}(\mathbf{a}^T\mathbf{x})\mathbf{a}^T$ reveals that $\operatorname{rank}(F'(\mathbf{x})) \leq 1$ for all $\mathbf{x} \in \mathbb{R}^n$. Thus we conclude that important structural information can be lost during the accumulation process. Even if we consider the $n + m$ components of $\mathbf{a}$ and $\mathbf{b}$ as free parameters, the set

$$\texttt{Reach}\{F'(\mathbf{x})\} \equiv \{F'(\mathbf{x}) : \mathbf{x}, \mathbf{a} \in \mathbb{R}^n, \; \mathbf{b} \in \mathbb{R}^m\} \subset \mathbb{R}^{m\times n}$$

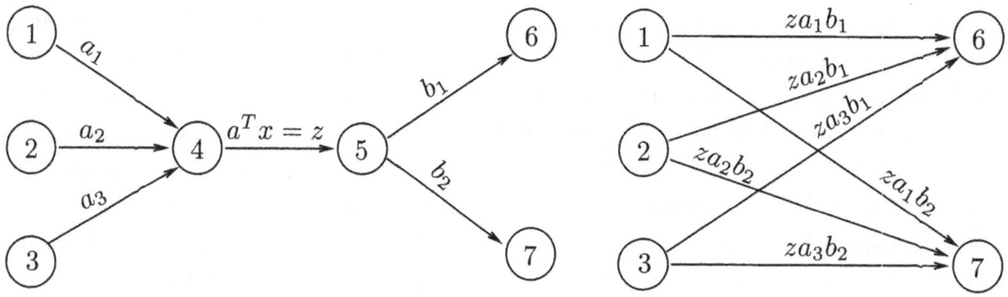

Figure 6.7. Vertex accumulation of a Jacobian on a simple graph.

is only an $(n+m-1)$-dimensional submanifold of the set of all real $m \times n$ matrices. We can check without too much difficulty that, in this case, $\texttt{Reach}\{F'(\mathbf{x})\}$ is characterized uniquely as the set of $m \times n$ matrices $A = (a_{ij})$ for which all 2×2 submatrices are singular, that is,

$$0 = \det \begin{pmatrix} a_{ik} & a_{ij} \\ a_{hk} & a_{hj} \end{pmatrix} \quad \text{for all} \quad 1 \leq i < h \leq m, \ \ 1 \leq k < j \leq n.$$

6.7. Generic rank

In the computational graph displayed in Figure 6.7, all 2×2 matrices correspond to a subgraph with the structure obtained overall for $n = 2 = m$. It is obvious that such matrices are singular since $z = \mathbf{a}^T \mathbf{x}$ is the single degree of freedom into which all the independent variables $x_1 \ldots x_n$ are mapped. More generally, we obtain the following fundamental result.

Proposition 6.3. Let the edge values $\{c_{ij}\}$ of a certain linearized computational graph be considered as free parameters. Then the rank of the matrix A with elements

$$a_{ij} = \sum_{\mathcal{P} \in [j \to i]} \ \prod_{(\tilde{\jmath},\tilde{\imath}) \in \mathcal{P}} c_{\tilde{\imath}\tilde{\jmath}}, \quad 1 \leq i \leq m, \ \ 1 \leq j \leq n,$$

is, for almost all combinations $\{c_{ij}\}_{j \prec i}$, equal to the size r of a maximal match, that is, the largest number of disjoint paths connecting the roots to the leaves of the graph.

Proof. Suppose we split all interior vertices into an input port and an output port connected by an internal edge of capacity 1. Furthermore we connect all minimal vertices by edges of unit capacity to a common source, and analogously all maximal ones to a common sink. All other edges are given infinite capacity. Then the solution of the max-flow (Tarjan 1983) problem is integral and represents exactly a maximal match as defined above. The corresponding min-cut solution consists of a set of internal edges of finite and thus unit capacity, whose removal would split the modified graph into two halves. The values v_j at these split vertices form a vector $\mathbf{z}$ of reals that depends differentiable on $\mathbf{x}$ and determines differentiably $\mathbf{y}$. Hence we have a decomposition

$$\mathbf{y} = F(\mathbf{x}) = H(\mathbf{z}) \quad \text{with} \quad \mathbf{z} = G(\mathbf{x}),$$

so that

$$\begin{aligned} \operatorname{rank}\big(F'(\mathbf{x})\big) &= \operatorname{rank}\big(H'(\mathbf{z})\, G'(\mathbf{x})\big) \\ &\leq \max\big\{\operatorname{rank}\big(H'(\mathbf{z})\big),\ \operatorname{rank}\big(G'(\mathbf{x})\big)\big\} \leq r \equiv \dim(\mathbf{z}). \end{aligned}$$

To attain the upper bound, we simply set $c_{ij} = 1$ for every edge (j, i) in the maximal match, all others being set to zero. Then the resulting matrix A is a permutation of a diagonal matrix with the number of nonzero elements being equal to its rank r. Because the determinant corresponding to the permuted diagonal has the value 1 for the special choice of the edge values given above, and it is a polynomial in the c_{ij}, it must in fact be nonzero for almost all combinations of the c_{ij}. This completes the proof. □

As we have seen in the proof, generic rank can be determined by a max flow computation for which very efficient algorithms are now available (Tarjan 1983). There are other structural properties which, like rank, can be deduced from the structure of the computational graph, such as the multiplicity of the zero eigenvalue (Röbenack and Reinschke 2001). All this kind of structural information is obliterated when we accumulate the Jacobian into a rectangular array of numbers. Therefore, in the remainder of this section we hope to motivate and initiate an investigation into the properties of Jacobian graphs and their optimal representation.

6.8. Scarce Jacobians and their representation

A simple criterion for the loss of information in accumulating Jacobians is a count on the number of degrees of freedom. In the graph Figure 6.3 of the rank-one example (6.8) we have $n + m + 1$ free edge values whose accumulation into dense $m \times n$ matrices cannot possibly reach all of $\mathbb{R}^{m \times n}$ when $n > 1 < m$. In some sense that is nothing special, since, for sufficiently smooth F, the set of reachable Jacobians

$$\big\{F'(\mathbf{x}) \in \mathbb{R}^{m \times n} : \mathbf{x} \in \mathbb{R}^n\big\}$$

forms by its very definition an n-dimensional manifold embedded in the $m\,n$ dimensional linear space $\mathbb{R}^{m \times n}$. Some of that structure may be directly visible as *sparsity*, where certain Jacobian entries are zero or otherwise constant. Then the set of reachable Jacobians is in fact contained in an affine subspace of $\mathbb{R}^{m \times n}$, which often has only a dimension of order $\mathcal{O}(n + m)$. It is well understood that such sparsity structure can be exploited for storing, factoring and otherwise manipulating these matrices economically, especially when the dimensions m and n are rather large. We contend here that a similar effect can occur when the computational graph rather than the Jacobian is sparse in a certain sense.

More specifically, we will call the computational graph and the resulting Jacobian *scarce* if the matrices $(\partial y_i/\partial x_j)_{j=1\ldots n,i=1\ldots m}$ defined by Bauer's formula (6.7) for arbitrary values of the elemental partials $C_{\bar{\imath}\bar{\jmath}}$ do not range over all of $\mathbb{R}^{m\times n}$. For the example considered in Section 6.7, we obtain exactly the set of matrices whose rank equals one or zero, which is a smooth manifold of dimension $m+n-1$. Naturally, this manifold need not be regular and, owing to its homogeneity, there are always bifurcations at the origin. The difference between $n\,m$ and the dimension of the manifold may be defined as the *degree of scarcity*. The dimension of the manifold is bounded above but generally not equal to the number of arcs in the computational graph. The discrepancy is exactly two for the example above.

A particularly simple kind of scarcity is sparsity, where the number of zeros yields the degree of sparsity, provided the other entries are free and independent of each other. It may then still make sense to accumulate the Jacobian, that is, represent it as set of partial derivatives. We contend that this is not true for Jacobians that are scarce but not sparse, such as the rank-one example.

In the light of Proposition 6.3 and our observation on the rank-one example, we might think that the scarcity structure can always be characterized in terms of a collection of vanishing subdeterminants. This conjecture is refuted by the following example, which is also representative for a more significant class of problems for which full accumulation seems a rather bad idea.

Suppose we have a 3×3 grid of values v_j for $j=1\ldots 9$, as displayed in Figure 6.8, that are subject to two successive transformations. Each time the new value is a function of the old values at the same place and its neighbours to the west, south and southwest. At the southern and western boundary we continue the dependence periodically so that the new value of v_4, for example, depends on the old values of v_4, v_3, v_1, and v_6. This dependency is displayed in Figure 6.8. Hence the 9×9 Jacobians of each of the two transitions contain exactly four nonzero elements in each row. Consequently we have $4\cdot 9\cdot 2 = 72$ free edge values but $9\cdot 9 = 81$ entries in the product of the two matrices, which is the Jacobian of the final state with respect to the initial state.

Since by following the arrows in Figure 6.8 we can get from any grid point to any other point in two moves, the Jacobian is dense. Therefore the degree of scarcity defined above is at least $81-72=9$. However, the scarcity of this Jacobian cannot be explained in terms of singular submatrices. Any such square submatrix would be characterized by two sets $J_i \subset \{1,2,\ldots,9\}$ for $i=0,2$ with J_0 selecting the indices of columns (= independent variables) and J_2 the rows (= dependent variables) of the submatrix. It is then not too difficult to find an intermediate set J_1 such that each element in J_1 is

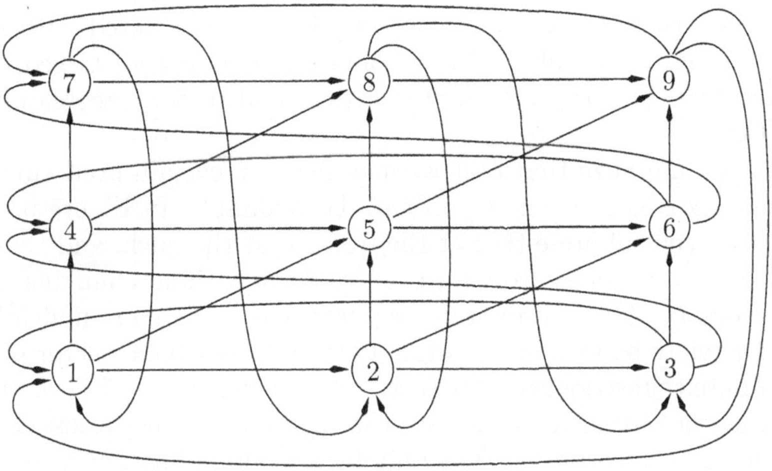

Figure 6.8. Update dependences on the upwinding example.

the middle of a directed path from an element in J_0 to an element in J_2 with all these paths being disjoint. Then it follows from Proposition 6.3 that the submatrix has generically full rank, and hence its determinant cannot vanish.

The example sketched in Figure 6.8 may be viewed as prototypical of upwind discretizations of a time-dependent PDE on a two-dimensional domain. There the Jacobian of the final state with respect to the initial state is the product of transformation matrices that are block-tridiagonal for reasonably regular discretizations. If there are just enough time-steps such that information can traverse the whole domain, the Jacobian will be dense but still scarce, as shown in Section 8.3 of Griewank (2000). It was also observed in that book that the greedy Markowitz heuristic has significantly worse performance than forward and reverse. We may view this as yet another indication that accumulating Jacobians may not be such a good idea in the first place.

That leaves the question of determining what is the best representation of scarce Jacobians whose full accumulation is redundant by definition of scarcity. Suppose we have evaluated all local partials c_{ij} and are told that a rather large number of Jacobian–vector or vector–Jacobian products must be computed subsequently. This requires execution of the forward loop

$$\dot{v}_i = \sum_{j \prec i} c_{ij} * \dot{v}_j \quad \text{for} \quad i = n+1 \dots l \tag{6.10}$$

or the (nonincremental) reverse loop

$$\bar{v}_j = \sum_{i \succ j} \bar{v}_i * c_{ij} \quad \text{for} \quad j = l-m \dots 1, \tag{6.11}$$

which follow from Tables 3.1 and 3.2, provided all v_i are scalar and disjoint. In either loop there is exactly one multiplication associated with each arc value c_{ij}, so that the number of arcs is an exact measure for the cost of computing Jacobian–vector products $\dot{\mathbf{y}} = F'(\mathbf{x})\dot{\mathbf{x}}$ and vector–Jacobian products $\bar{\mathbf{x}} = \bar{\mathbf{y}}F'(\mathbf{x})$.

In order to minimize that cost we may perform certain preaccumulations by eliminating edges or whole vertices. For example, in Figure 6.7 we can eliminate the central arc either at the front or at the back, and thus reduce the number of edge values from $n+m+1$ to $n+m$ without changing the reach of the Jacobian. This elimination requires here $\min(n, m)$ multiplications and will therefore be worthwhile even if only one Jacobian–vector product is to be computed subsequently. Further vertex, edge, or even face eliminations would make the size of the edge set grow again, and cause a loss of scarcity in the sense discussed above. The only further simplification respecting the Jacobian structure would be to normalize any one of the edge values to 1 and thus save additional multiplications during subsequent vector product calculations. Unfortunately, at least this last simplification is clearly not unique and it would appear that there is in general no unique minimal representation of a computational graph. As might have been expected, there is a rather close connection between the avoidance of fill-in and the maintenance of scarcity.

Proposition 6.4.

(i) If the front- or back-elimination of an edge in a computational graph does not increase the total number of edges, then the degree of scarcity remains constant.

(ii) If the elimination of a vertex would lead to a reduction in the total number of edges, then at least one of its edges can be eliminated via (i) without loss of scarcity.

Proof. Without loss of generality we may consider back-elimination of an edge (j, i), as depicted in Figure 6.9. The no fill-in condition requires that all predecessors of ⓙ but possibly one, say (k_0), are already directly connected to ⓘ. Now we have to show that the values of the new arcs after the elimination can be chosen such that they reproduce any possible sensitivities in the subgraph displayed in Figure 6.9. Since the arc from ⓙ to ⓗ and possibly other successors of ⓙ remain unchanged, we must also keep all arcs c_{jk_t} for $t = 0 \ldots s$ unaffected. If also $k_0 \prec i$ there is no difficulty at all, as we may scale c_{ij} to one without loss of generality so that the new arc values are given by $\tilde{c}_{k_t} = c_{ik_t} + c_{jk_t}$, which is obviously reversible. Otherwise we have

$$\tilde{c}_{ik_0} = c_{ij} * c_{jk_0} \quad \text{and} \quad \tilde{c}_{ik_t} = c_{ik_t} + c_{ij} * c_{jk_t} \quad \text{for} \quad k = 1 \ldots s.$$

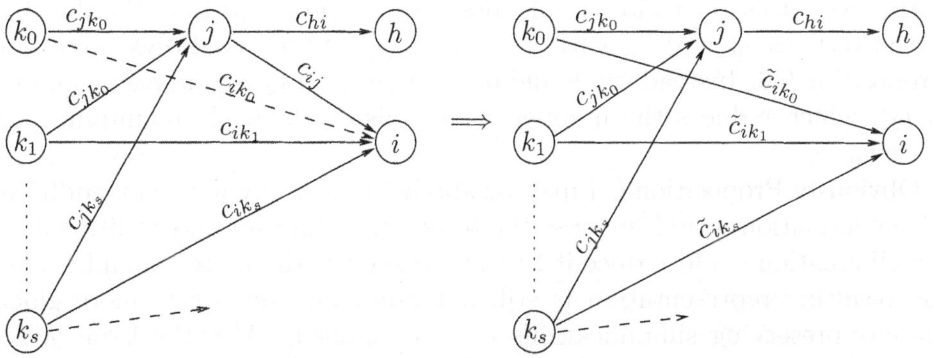

Figure 6.9. Scarcity-preserving back-elimination of (j, i).

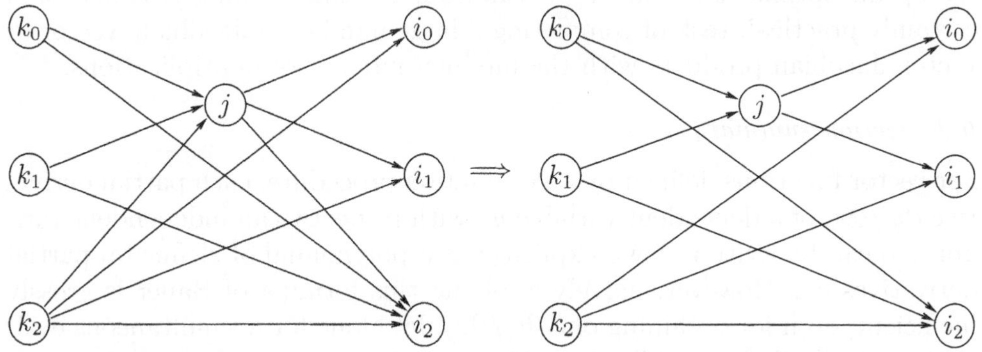

Figure 6.10. Scarcity-preserving elimination of (j, i_2) and (k_2, j).

Since we may assume without loss of generality that $c_{jk_0} \neq 0$, this relation may be reversed so that the degree of scarcity is indeed maintained as asserted.

To prove the second assertion, let us assume that the vertex ⓙ to be eliminated has $(p+1)$ predecessors and $(s+1)$ successors, which span together with ⓙ a subgraph of $p+s+3$ vertices. After the elimination of ⓙ they form a dense bipartite graph with $(s+1)(p+1)$ edges. Because of the negative fill-in, the number of direct arcs between the predecessors and successors of ⓙ is at least

$$(s+1)(p+1) - (s+1+p+1) + 1 = s\,p.$$

From this it can easily be shown by contradiction that at least one predecessor of ⓙ is connected to all but possibly one of its successors, so that its link to ⓙ can be front-eliminated. □

The application of the proposition to the 3×3 subgraph depicted in Figure 6.10 shows that there are $6+5$ edges in the original graph on the left which would be reduced to 9 by the elimination of the central node.

However, this operation would destroy the property that the Jacobian of the vertices $\textcircled{i_0}$ and $\textcircled{i_1}$ with respect to $\textcircled{k_0}$ and $\textcircled{k_1}$ is always singular by Proposition 6.3. Instead we should front-eliminate (k_2, j) and back-eliminate (j, i_2), which reduces the number of arcs also to 9, while maintaining the scarcity.

Obviously Proposition 6.4 may be applied repeatedly until any additional edge elimination would increase the total edge count and no additional vertex elimination could reduce it further. However, there are examples where the resulting representation is still not minimal, and other, more global, scarcity-preserving simplifications may be applied. Also the issue of normalization, that is, the question of which arc values can be set to 1 to save extra multiplications, appears to be completely open. Hence we cannot yet set up an optimal Jacobian representation for the seemingly simple, and certainly practical, task of completing a large number of Jacobian–vector or vector–Jacobian products with the minimal number of multiplications.

6.9. Section summary

For vector functions defined by an evaluation procedure, each partial derivative $\partial y_i/\partial x_j$ of a dependent variable y_i, with respect to an independent variable x_j, can be written down explicitly as a polynomial of elemental partial derivatives $c_{\bar{i}\bar{j}}$. However, naively applying this formula of Bauer is grossly inefficient, even for obtaining one $\partial y_i/\partial x_j$, let alone for a simultaneous evaluation of all of them. The identification of common subexpressions can be interpreted as vertex and edge elimination on the computational graph or face elimination on the associated line graph. The latter approach is believed to be the most general, and thus potentially most efficient, procedure. In contrast to vertex and edge elimination, face elimination can apparently *not* be interpreted in terms of linear algebra operations on the extended Jacobian (6.2). In any case, heuristics to determine good, if suboptimal, elimination sequences remain to be developed.

The accumulation of Jacobians as rectangular arrays of numbers appears to be a bad idea when they are scarce without being sparse, as defined in Section 6.8. Then there is a certain intrinsic relation between the matrix entries which is lost by accumulation. Instead we should strive to simplify the computational graph by suitable eliminations until a minimal representation is reached. This would, for example, be useful for the subsequent evaluation of Jacobian–vector or vector–Jacobian products with minimal costs. More importantly, structural properties like the generic rank should be respected. Thus we face the challenge of developing a theory of canonical representations of linear computational graphs, including in particular their composition.

Acknowledgements
The author is greatly indebted to many collaborators and co-authors, as well as the AD community as a whole. To name some would mean to disregard others, and quite a few are likely to disagree with my view of the field anyway. The initial typesetting of the manuscript was done by Sigrid Eckstein, who took particular care in preparing the many tables and figures.

REFERENCES

A. Aho, J. Hopcroft and J. Ullman (1974), *The Design and Analysis of Computer Algorithms*, Addison-Wesley, Reading, MA.

R. Anderssen and P. Bloomfield (1974), 'Numerical differentiation proceedings for non-exact data', *Numer. Math.* **22**, 157–182.

W. Ball (1969), *Material and Energy Balance Computations*, Wiley, pp. 560–566.

R. Becker and R. Rannacher (2001), An optimal control approach to error control and mesh adaptation in finite element methods, in *Acta Numerica*, Vol. 10, Cambridge University Press, pp. 1–102.

C. Bennett (1973), 'Logical reversibility of computation', *IBM J. Research Development* **17**, 525–532.

M. Berz, C. Bischof, G. Corliss and A. Griewank, eds (1996), *Computational Differentiation: Techniques, Applications, and Tools*, SIAM, Philadelphia, PA.

C. Bischof, G. Corliss and A. Griewank (1993), 'Structured second- and higher-order derivatives through univariate Taylor series', *Optim. Methods Software* **2**, 211–232.

C. Bischof, A. Carle, P. Khademi and A. Mauer (1996), 'The ADIFOR 2.0 system for the automatic differentiation of Fortran 77 programs', *IEEE Comput. Sci. Engr.* **3**, 18–32.

T. Braconnier and P. Langlois (2001), From rounding error estimation to automatic correction with AD, in Corliss *et al.* (2001), Chapter 42, pp. 333–339.

D. Cacuci, C. Weber, E. Oblow and J. Marable (1980), 'Sensitivity theory for general systems of nonlinear equations', *Nucl. Sci. Engr.* **88**, 88–110.

J.-B. Caillau and J. Noailles (2001), Optimal control sensitivity analysis with AD, in Corliss *et al.* (2001), Chapter 11, pp. 105–111.

S. Campbell and R. Hollenbeck (1996), Automatic differentiation and implicit differential equations, in Berz *et al.* (1996), pp. 215–227.

S. Campbell, E. Moore and Y. Zhong (1994), 'Utilization of automatic differentiation in control algorithms', *IEEE Trans. Automatic Control* **39**, 1047–1052.

B. Cappelaere, D. Elizondo and C. Faure (2001), Odyssée versus hand differentiation of a terrain modelling application, in Corliss *et al.* (2001), Chapter 7, pp. 71–78.

A. Carle and M. Fagan (1996), Improving derivative performance for CFD by using simplified recurrences, in Berz *et al.* (1996), pp. 343–351.

B. Christianson (1992), 'Automatic Hessians by reverse accumulation', *IMA J. Numer. Anal.* **12**, 135–150.

B. Christianson (1994), 'Reverse accumulation and attractive fixed points', *Optim. Methods Software* **3**, 311–326.

B. Christianson (1998), 'Reverse accumulation and implicit functions', *Optim. Methods Software* **9**, 307–322.

B. Christianson (1999), 'Cheap Newton steps for optimal control problems: Automatic differentiation and Pantoja's algorithm', *Optim. Methods Software* **10**, 729–743.

B. Christianson, L. Dixon and S. Brown (1997), Automatic differentiation of computer programs in a parallel computing environment, in *Applications of High Performance Computing in Engineering V* (H. Power and J. C. Long, eds), Computational Mechanics Publications, Southampton, pp. 169–178.

T. Coleman and J. Moré (1984), 'Estimation of sparse Jacobian matrices and graph coloring problems', *SIAM J. Numer. Anal.* **20**, 187–209.

T. Coleman and A. Verma (1996), Structure and efficient Jacobian calculation, in Berz *et al.* (1996), pp. 149–159.

A. Conn, N. Gould and P. Toint (1992), *LANCELOT, a Fortran Package for Large-Scale Nonlinear Optimization (Release A)*, Vol. 17 of *Computational Mathematics*, Springer, Berlin.

G. Corliss (1991), Automatic differentiation bibliography, in Griewank and Corliss (1991), pp. 331–353.

G. Corliss, C. Faure, A. Griewank, L. Hascoët and U. Naumann, eds (2001), *Automatic Differentiation: From Simulation to Optimization*, series in *Computer and Information Science*, Springer, New York.

F. Courty, A. Dervieux, B. Koobus and L. Hascoët (2002), Reverse automatic differentiation for optimum design: From adjoint state assembly to gradient computation, Technical report, INRIA Sophia-Antipolis.

A. Curtis, M. Powell and J. Reid (1974), 'On the estimation of sparse Jacobian matrices', *J. Inst. Math. Appl.* **13**, 117–119.

L. C. W. Dixon (1991), Use of automatic differentiation for calculating Hessians and Newton steps, in Griewank and Corliss (1991), pp. 114–125.

J. Dunn and D. Bertsekas (1989), 'Efficient dynamic programming implementations of Newton's method for unconstrained optimal control problems', *J. Optim. Theory Appl.* **63**, 23–38.

Faa di Bruno (1856), 'Note sur une nouvelle formule de calcule differentiel', *Quar. J. Math.* **1**, 359–360.

C. Faure and U. Naumann (2001), Minimizing the tape size, in Corliss *et al.* (2001), Chapter 34, pp. 279–284.

S. A. Forth and T. Evans (2001), Aerofoil optimisation via AD of a multigrid cell-vertex Euler flow solver, in Corliss *et al.* (2001), Chapter 17, pp. 149–156.

Y. M. V. G. M. Ostrovskii and W. W. Borisov (1971), 'Über die Berechnung von Ableitungen', *Wissenschaftliche Zeitschrift der Technischen Hochschule für Chemie* **13**, 382–384.

R. Giering and T. Kaminski (1998), 'Recipes for adjoint code construction', *ACM Trans. Math. Software* **24**, 437–474.

R. Giering and T. Kaminski (2000), On the performance of derivative code generated by TAMC. Manuscript, FastOpt, Hamburg, Germany. Submitted to *Optim. Methods Software* See `www.FastOpt.de/papers/perftamc.ps.gz`.

R. Giering and T. Kaminski (2001*a*), 'Automatic sparsity detection'. Draft.

R. Giering and T. Kaminski (2001*b*), Generating recomputations in reverse mode AD, in Corliss *et al.* (2001), Chapter 33, pp. 271–278.

J. C. Gilbert (1992), 'Automatic differentiation and iterative processes', *Optim. Methods Software* **1**, 13–21.

M. Giles (2001), On the iterative solution of adjoint equations, in Corliss *et al.* (2001), pp. 145–151.

M. B. Giles and N. A. Pierce (2001), 'An introduction to the adjoint approach to design', *Flow, Turbulence and Combustion* **65**, 393–415.

M. B Giles and E. Süli (2002), Adjoint methods for PDEs: *a posteriori* error analysis and postprocessing by duality, in *Acta Numerica*, Vol. 11, Cambridge University Press, pp. 145–236.

M. Gockenbach, M. J. Petro and W. Symes (1999), 'C++ classes for linking optimization with complex simulations', *ACM Trans. Math. Software* **25**, 191–212.

A. Griewank (2000), *Evaluating Derivatives, Principles and Techniques of Algorithmic Differentiation*, Vol. 19 of *Frontiers in Applied Mathematics*, SIAM, Philadelphia.

A. Griewank and G. Corliss, eds (1991), *Automatic Differentiation of Algorithms: Theory, Implementation, and Application*, SIAM, Philadelphia, PA.

A. Griewank and C. Faure (2002), 'Reduced functions, gradients and Hessians from fixed point iteration for state equations', *Numer. Alg.* **30**, 113–139.

A. Griewank and C. Mitev (2002), 'Detecting Jacobian sparsity patterns by Bayesian probing', *Math. Prog.* **93**, 1–25.

A. Griewank and U. Naumann (2003*a*), Accumulating Jacobians by vertex, edge or face elimination, in *Proceedings of the 6th African Conference on Research in Computer Science* (L. Andriamampianina, B. Philippe, E. Kamgnia and M. Tchuente, eds), Imprimerie Saint Paul, Camerun and INRIA, France, pp. 375–383.

A. Griewank and U. Naumann (2003*b*), 'Accumulating Jacobians as chained sparse matrix products', *Math. Prog.* **95**, 555–571.

A. Griewank and A. Verma (2003), 'On Newsam–Ramsdell-seeds for calculating sparse Jacobians'. In preparation.

A. Griewank and A. Walther (2000), 'Revolve: An implementation of checkpointing for the reverse or adjoint mode of computational differentiation', *Trans. Math. Software* **26**, 19–45.

A. Griewank, C. Bischof, G. Corliss, A. Carle and K. Williamson (1993), 'Derivative convergence for iterative equation solvers', *Optim. Methods Software* **2**, 321–355.

A. Griewank, J. Utke and A. Walther (2000), 'Evaluating higher derivative tensors by forward propagation of univariate Taylor series', *Math. Comp.* **69**, 1117–1130.

J. Grimm, L. Pottier and N. Rostaing-Schmidt (1996), Optimal time and minimum space-time product for reversing a certain class of programs, in Berz *et al.* (1996), pp. 95–106.

S. Hague and U. Naumann (2001), Present and future scientific computation environments, in Corliss *et al.* (2001), Chapter 5, pp. 55–62.

L. Hascoët (2001), The data-dependence graph of adjoint programs, Research Report 4167, INRIA, Sophia-Antipolis, France. See `www.inria.fr/rrrt/rr-4167.html`.

L. Hascoët, S. Fidanova and C. Held (2001), Adjoining independent computations, in Corliss *et al.* (2001), Chapter 35, pp. 285–290.

M. Hinze and T. Slawig (2002), Adjoint gradients compared to gradients from algorithmic differentiation in instantaneous control of the Navier–Stokes equations, Preprint, TU Dresden.

M. Hinze and A. Walther (2002), An optimal memory-reduced procedure for calculating adjoints of the instationary Navier–Stokes equations, Preprint MATH-NM-06-2002, TU Dresden.

J. Horwedel (1991), GRESS: A preprocessor for sensitivity studies on Fortran programs, in Griewank and Corliss (1991), pp. 243–250.

A. S. Hossain and T. Steihaug (1998), 'Computing a sparse Jacobian matrix by rows and columns', *Optim. Methods Software* **10**, 33–48.

S. Hossain and T. Steihaug (2002), Sparsity issues in the computation of Jacobian matrices, Technical Report 223, Department of Computer Science, University of Bergen, Norway.

M. Iri, T. Tsuchiya and M. Hoshi (1988), 'Automatic computation of partial derivatives and rounding error estimates with applications to large-scale systems of nonlinear equations', *Comput. Appl. Math.* **24**, 365–392. Original Japanese version appeared in *J. Information Processing* **26** (1985), 1411–1420.

L. Kantorovich (1957), 'Ob odnoj matematicheskoj simvolike, udobnoj pri provedenii vychislenij na mashinakh', *Doklady Akademii Nauk. SSSR* **113**, 738–741.

G. Kedem (1980), 'Automatic differentiation of computer programs', *ACM Trans. Math. Software* **6**, 150–165.

D. Keyes, P. Hovland, L. McInnes and W. Samyono (2001), Using automatic differentiation for second-order matrix-free methods in PDE-constrained optimization, in Corliss *et al.* (2001), Chapter 3, pp. 33–48.

D. Knuth (1973), *The Art of Computer Programming*, Vol. 1 of *Computer Science and Information Processing*, Addison-Wesley, MI.

S. Linnainmaa (1976), 'Taylor expansion of the accumulated rounding error', *BIT* (*Nordisk Tidskrift for Informationsbehandling*) **16**, 146–160.

S. Linnainmaa (1983), 'Error linearization as an effective tool for experimental analysis of the numerical stability of algorithms', *BIT* **23**, 346–359.

R. Lohner (1992), Verified computing and programs in PASCAL-XSC, with examples, Habilitationsschrift, Institute for Applied Mathematics, University of Karlsruhe.

M. Mancini (2001), A parallel hierarchical approach for automatic differentiation, in Corliss *et al.* (2001), Chapter 27, pp. 225–230.

G. Marcuk (1996), *Adjoint Equations and Perturbation Algorithms in Nonlinear Problems*, CRC Press, Boca Raton, FL.

H. Maurer and D. Augustin (2001), Sensitivity analysis and real-time control of parametric optimal control problems using boundary value methods, in *Online Optimization of Large Scale Systems* (M. Grötschel, S. Krumke and J. Rambau, eds), Springer, Berlin/Heidelberg/New York, Chapter I, pp. 17–55.

B. Mohammadi and O. Pironneau (2001), *Applied Shape Optimization for Fluids*, series in *Numerical Mathematics and Scientific Computation*, Clarendon Press, Oxford.

R. Moore (1979), *Methods and Applications of Interval Analysis*, SIAM, Philadelphia, PA.

U. Naumann (1999), Efficient calculation of Jacobian matrices by optimized application of the chain rule to computational graphs, PhD thesis, Institute of Scientific Computing, Germany.

U. Naumann (2001), Elimination techniques for cheap Jacobians, in Corliss *et al.* (2001), Chapter 29, pp. 241–246.

R. Neidinger (200x), Directions for computing multivariate Taylor series. Technical report, Davidson College, Davidson. To appear in *Math. Comp.*

P. Newman, G.-W. Hou, H. Jones, A. Taylor and V. Korivi (1992), Observations on computational methodologies for use in large-scale, gradient-based, multidisciplinary design incorporating advanced CFD codes, Technical Memorandum 104206, NASA Langley Research Center. AVSCOM Technical Report 92-B-007.

G. Newsam and J. Ramsdell (1983), 'Estimation of sparse Jacobian matrices', *SIAM J. Algebraic Discrete Methods* **4**, 404–417.

J. Ortega and W. Rheinboldt (1970), *Iterative Solution of Nonlinear Equations in Several Variables*, Academic Press, New York.

C. Pantelides (1988), 'The consistent initialization of differential-algebraic systems', *SIAM J. Sci. Statist. Comput.* **9**, 213–231.

J. D. O. Pantoja (1988), 'Differential dynamic programming and Newton's method', *Internat. J. Control* **47**, 1539–1553.

J. Pryce (1998), 'Solving high-index DAEs by Taylor series', *Numer. Alg.* **19**, 195–211.

L. Rall (1983), Differentiation and generation of Taylor coefficients in Pascal-SC, in *A New Approach to Scientific Computation* (U. Kulisch and W. L. Miranker, eds), Academic Press, New York, pp. 291–309.

J. Restrepo, G. Leaf and A. Griewank (1998), 'Circumventing storage limitations in variational data assimilation studies', *SIAM J. Sci. Comput.* **19**, 1586–1605.

K. Röbenack and K. Reinschke (2000), 'Graph-theoretic characterization of structural controllability for singular DAE', *Z. Angew. Math. Mech.* **8 Suppl. 1**, 849–850.

K. Röbenack and K. Reinschke (2001), Nonlinear observer design using automatic differentiation, in Corliss *et al.* (2001), Chapter 15, pp. 133–138.

D. Rose and R. Tarjan (1978), 'Algorithmic aspects of vertex elimination on directed graphs', *SIAM J. Appl. Math.* **34**, 177–197.

J. Sternberg (2002), Adaptive Umkehrschemata für Schrittfolgen mit nichtuniformen Kosten, Diploma Thesis, Institute of Scientific Computing, Germany.

F. Stummel (1981), 'Optimal error estimates for Gaussian elimination in floating-point arithmetic', *GAMM, Numerical Analysis* **62**, 355–357.

M. Tadjouddine, S. A. Forth and J. Pryce (2001), AD tools and prospects for optimal AD in CFD flux Jacobian calculations, in Corliss *et al.* (2001), Chapter 30, pp. 247–252.

R. Tarjan (1983), 'Data structures and network algorithms', *CBMS-NSF Reg. Conf. Ser. Appl. Math.* **44**, 131.

J. van der Snepscheut (1993), *What Computing Is All About*, Suppl. 2 of *Texts and Monographs in Computer Science*, Springer, Berlin.

D. Venditti and D. Darmofal (2000), 'Adjoint error estimation and grid adaptation for functional outputs: application to quasi-one-dimensional flow', *J. Comput. Phys.* **164**, 204–227.

Y. Volin and G. Ostrovskii (1985), 'Automatic computation of derivatives with the use of the multilevel differentiating technique, I: Algorithmic basis', *Comput. Math. Appl.* **11**, 1099–1114.

A. Walther (1999), Program reversal schedules for single- and multi-processor machines, PhD thesis, Institute of Scientific Computing, Germany.

A. Walther (2002), 'Adjoint based truncated Newton methods for equality constrained optimization'.

X. Zou, F. Vandenberghe, M. Pondeca and Y.-H. Kuo (1997), Introduction to adjoint techniques and the MM5 adjoint modeling system, NCAR Technical Note NCAR/TN-435-STR, Mesocale and Microscale Meteorology Division, National Center for Atmospheric Research, Boulder, Colorado.

Acta Numerica (2003), pp. 399–450
DOI: 10.1017/S0962492902000144

Geometric numerical integration illustrated by the Störmer–Verlet method

Ernst Hairer
Section de Mathématiques,
Université de Genève, Switzerland
E-mail: Ernst.Hairer@math.unige.ch

Christian Lubich
Mathematisches Institut,
Universität Tübingen, Germany
E-mail: Lubich@na.uni-tuebingen.de

Gerhard Wanner
Section de Mathématiques,
Université de Genève, Switzerland
E-mail: Gerhard.Wanner@math.unige.ch

The subject of geometric numerical integration deals with numerical integrators that preserve geometric properties of the flow of a differential equation, and it explains how structure preservation leads to improved long-time behaviour. This article illustrates concepts and results of geometric numerical integration on the important example of the Störmer–Verlet method. It thus presents a cross-section of the recent monograph by the authors, enriched by some additional material.

After an introduction to the Newton–Störmer–Verlet–leapfrog method and its various interpretations, there follows a discussion of geometric properties: reversibility, symplecticity, volume preservation, and conservation of first integrals. The extension to Hamiltonian systems on manifolds is also described. The theoretical foundation relies on a backward error analysis, which translates the geometric properties of the method into the structure of a modified differential equation, whose flow is nearly identical to the numerical method. Combined with results from perturbation theory, this explains the excellent long-time behaviour of the method: long-time energy conservation, linear error growth and preservation of invariant tori in near-integrable systems, a discrete virial theorem, and preservation of adiabatic invariants.

CONTENTS

1. The Newton–Störmer–Verlet–leapfrog method

We start by considering systems of second-order differential equations

$$\ddot{q} = f(q), \tag{1.1}$$

where the right-hand side $f(q)$ does not depend on $\dot{q}$. Many problems in astronomy, molecular dynamics, and other areas of physics are of this form.

1.1. Two-step formulation

If we choose a step size h and grid points $t_n = t_0 + nh$, the most natural discretization of (1.1) is

$$q_{n+1} - 2q_n + q_{n-1} = h^2 f(q_n), \tag{1.2}$$

which determines q_{n+1} whenever q_{n-1} and q_n are known. Geometrically, this amounts to determining an interpolating parabola which, in the mid-point, assumes the second derivative prescribed by equation (1.1); see Figure 1.1, left.

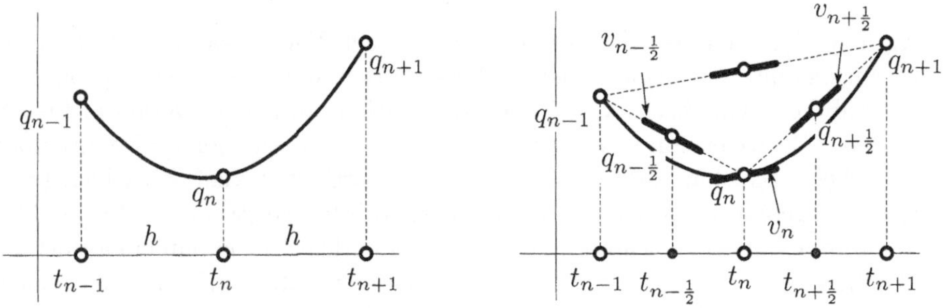

Figure 1.1. Method (1.2): two-step formulation (left); one-step formulations (right).

1.2. *One-step formulations*

Introducing the velocity $\dot{q} = v$ turns equation (1.1) into a first-order system of doubled dimension

$$\dot{q} = v, \qquad \dot{v} = f(q), \tag{1.3}$$

an equation in the so-called *phase space*. In analogy to this, we introduce discrete approximations of v and q as follows:

$$v_n = \frac{q_{n+1} - q_{n-1}}{2h}, \qquad v_{n-\frac{1}{2}} = \frac{q_n - q_{n-1}}{h}, \qquad q_{n-\frac{1}{2}} = \frac{q_n + q_{n-1}}{2}, \tag{1.4}$$

where some derivatives, in order to preserve second-order and symmetry, are evaluated on the *staggered grid* $t_{n-\frac{1}{2}}, t_{n+\frac{1}{2}}, \ldots$; see Figure 1.1, right. Inserting these expressions into the method (or simply looking at the picture) we see that method (1.2) can now be interpreted as a *one-step method* $\Phi_h : (q_n, v_n) \mapsto (q_{n+1}, v_{n+1})$, given by

$$\text{(A)} \qquad \begin{aligned} v_{n+\frac{1}{2}} &= v_n + \tfrac{h}{2}\, f(q_n), \\ q_{n+1} &= q_n + h\, v_{n+\frac{1}{2}}, \\ v_{n+1} &= v_{n+\frac{1}{2}} + \tfrac{h}{2}\, f(q_{n+1}). \end{aligned} \tag{1.5}$$

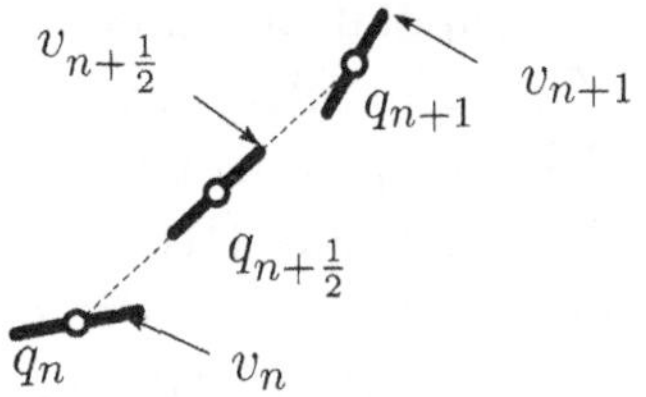

There is a *dual* variant of the method on the staggered grid $(v_{n-\frac{1}{2}}, q_{n-\frac{1}{2}}) \mapsto (v_{n+\frac{1}{2}}, q_{n+\frac{1}{2}})$ as follows:

$$\text{(B)} \qquad \begin{aligned} q_n &= q_{n-\frac{1}{2}} + \tfrac{h}{2}\, v_{n-\frac{1}{2}}, \\ v_{n+\frac{1}{2}} &= v_{n-\frac{1}{2}} + h\, f(q_n), \\ q_{n+\frac{1}{2}} &= q_n + \tfrac{h}{2}\, v_{n+\frac{1}{2}}. \end{aligned} \tag{1.6}$$

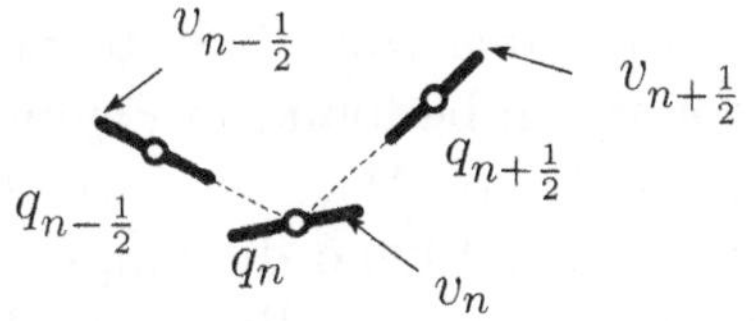

For both arrays (A) and (B), we can concatenate, in the actual step-by-step procedure, the last line of the previous step with the first line of the subsequent step. Both schemes then turn into the same method, where the q-values are evaluated on the original grid, and the v-values are evaluated on the staggered grid:

$$\begin{aligned} v_{n+\frac{1}{2}} &= v_{n-\frac{1}{2}} + h\, f(q_n), \\ q_{n+1} &= q_n + h\, v_{n+\frac{1}{2}}. \end{aligned} \tag{1.7}$$

This is the computationally most economic implementation, and numerically more stable than (1.2); see Hairer, Nørsett and Wanner (1993, p. 472).

1.3. Historical remarks

> Isn't that ingenious? I borrowed it straight from Newton. It comes right out of the *Principia*, diagram and all. R. Feynman (1965, p. 43)

The above schemes are known in the literature under various names. In particular, in molecular dynamics they are often called the *Verlet method* (Verlet 1967) and have become by far the most widely used integration scheme in this field.

Another name for this method is the *Störmer method*, since C. Störmer, in 1907, used higher-order variants of it for his computations of the motion of ionized particles in the earth's magnetic field (aurora borealis); see, *e.g.*, Hairer *et al.* (1993, Section III.10). Sometimes it is also called the *Encke method*, because J. F. Encke, around 1860, did extensive calculations for the perturbation terms of planetary orbits, which obey systems of second-order differential equations of precisely the form (1.1). Mainly in the context of partial differential equations of wave propagation, this method is called the *leapfrog method.* In yet another context, this formula is the basic method for the GBS extrapolation scheme, as it was proposed, for the case of equation (1.1), by Gragg in 1965; see Hairer *et al.* (1993, p. 294*f.*). Furthermore, the scheme (1.7) is equivalent to Nyström's method of order 2; see Hairer *et al.* (1993, p. 362, formula (III.1.13′)).

A curious fact is that Professor Loup Verlet, who later became interested in the history of science, discovered precisely 'his' method in several places in the classical literature, for example, in the calculations of logarithms and astronomical tables by J. B. Delambre in 1792: this paper was translated and discussed in McLachlan and Quispel (2002, Appendix C). Even more spectacular is the finding that the 'Verlet method' was used in Newton's *Principia* from 1687 to prove Kepler's second law. An especially clear account can be found in Feynman's *Messenger Lecture* from 1964; see Feynman (1965, p. 41), from which we reproduce with pleasure[1] two of Feynman's original hand drawings.

The argument is as follows: if there are no forces, the body advances with uniform speed, and the radius vector covers equal areas in equal times, simply because the two triangles Sun-1-2 and Sun-2-3 have the same base and common altitudes (see Figure 1.2). If the gravitational force acts at the midpoint, the planet is deviated in such a way that the top of the second triangle moves parallel to the sun ray (see Figure 1.3). Hence, the triangle Sun-2-4 also has the same area. The whole procedure (uniform motion on half the interval, then a 'kick' to the velocity in the direction of the Sun, and another uniform motion on the second half) is precisely variant (B) of the Störmer–Verlet scheme.

[1] ... and with permission of the publisher

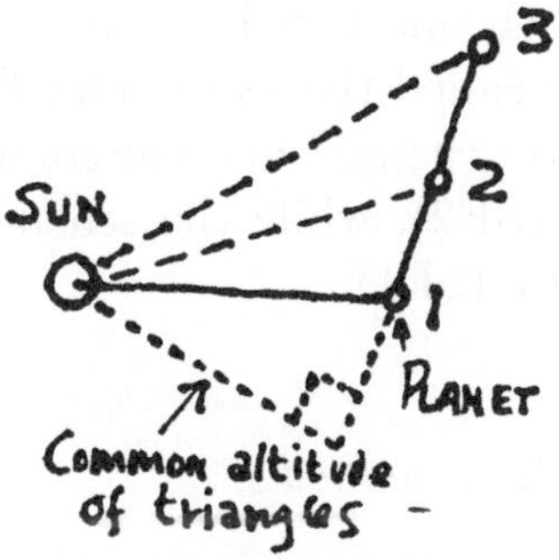

Figure 1.2. Uniform motion of a planet
(drawing by R. Feynman).

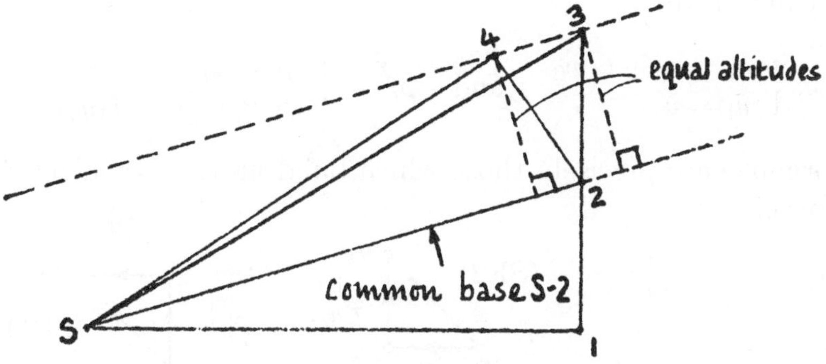

Figure 1.3. Gravitation acting at mid-point
(drawing by R. Feynman).

1.4. Interpretation as composition method (symplectic Euler)

We can go a step further and split the formulae in the middle of the schemes (1.5) and (1.6). We then arrive at the schemes $(v_n, q_n) \mapsto (v_{n+\frac{1}{2}}, q_{n+\frac{1}{2}})$ given by

$$\text{(SE1)} \qquad \begin{aligned} v_{n+\frac{1}{2}} &= v_n + \tfrac{h}{2}\, f(q_n), \\ q_{n+\frac{1}{2}} &= q_n + \tfrac{h}{2}\, v_{n+\frac{1}{2}}, \end{aligned} \tag{1.8}$$

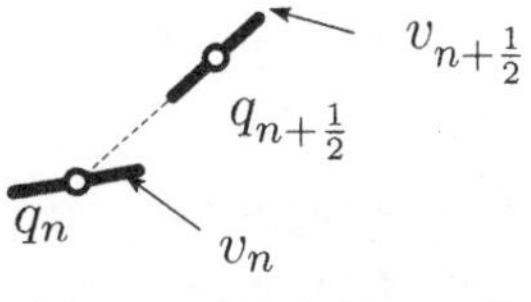

as well as the *adjoint* scheme $(v_{n+\frac{1}{2}}, q_{n+\frac{1}{2}}) \mapsto (v_{n+1}, q_{n+1})$ obtained by formally replacing the subscript n by $n+1$ and h by $-h$,

$$\text{(SE2)} \qquad \begin{aligned} q_{n+1} &= q_{n+\frac{1}{2}} + \tfrac{h}{2}\, v_{n+\frac{1}{2}}, \\ v_{n+1} &= v_{n+\frac{1}{2}} + \tfrac{h}{2}\, f(q_{n+1}). \end{aligned} \tag{1.9}$$

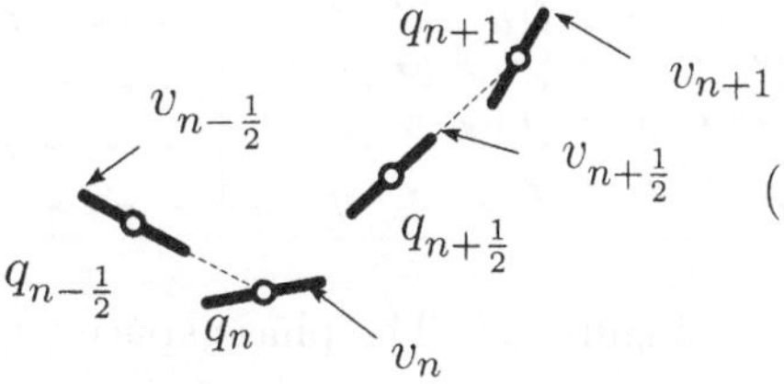

Both these schemes, in which one variable is used at the old value and the other at the new value, are called the *symplectic Euler method.*

We thus see that the above scheme (A) is the composition of the symplectic Euler schemes (SE1) with (SE2), while the scheme (B) is the composition of method (SE2) followed by (SE1).

1.5. Interpretation as splitting method

We consider the vector field $(v, f(q))$ of (1.3) 'split' as the sum of two vector fields $(v, 0)$ and $(0, f(q))$, as indicated in Figure 1.4. The *exact* flows $\varphi_t^{[1]}$ and $\varphi_t^{[2]}$ of these two vector fields, which both have a constant time derivative, are easily obtained:

$$\varphi_t^{[1]} : \begin{cases} q_1 = q_0 + t \cdot v_0 \\ v_1 = v_0 \end{cases} \quad \text{and} \quad \varphi_t^{[2]} : \begin{cases} q_1 = q_0 \\ v_1 = v_0 + t \cdot f(q_0). \end{cases} \tag{1.10}$$

These formulae are precisely those which build up the formulae (SE1) and (SE2) above:

$$(\text{SE2}) = \varphi_{h/2}^{[2]} \circ \varphi_{h/2}^{[1]}, \qquad (\text{SE1}) = \varphi_{h/2}^{[1]} \circ \varphi_{h/2}^{[2]}. \tag{1.11}$$

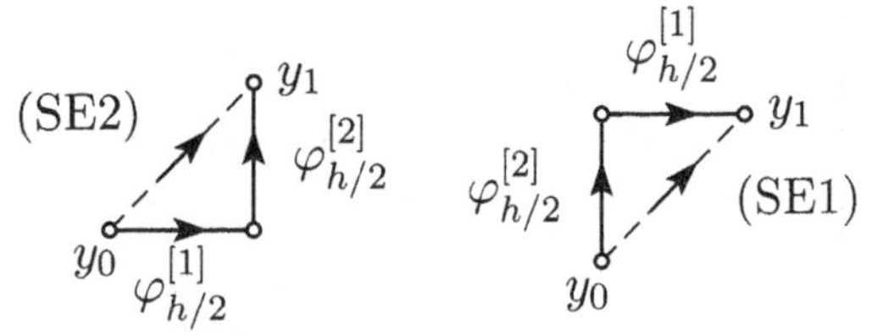

For the two versions of the Störmer–Verlet method we thus obtain the diagrams

$$(\text{A}) = (\text{SE2}) \circ (\text{SE1}), \qquad (\text{B}) = (\text{SE1}) \circ (\text{SE2}), \tag{1.12}$$

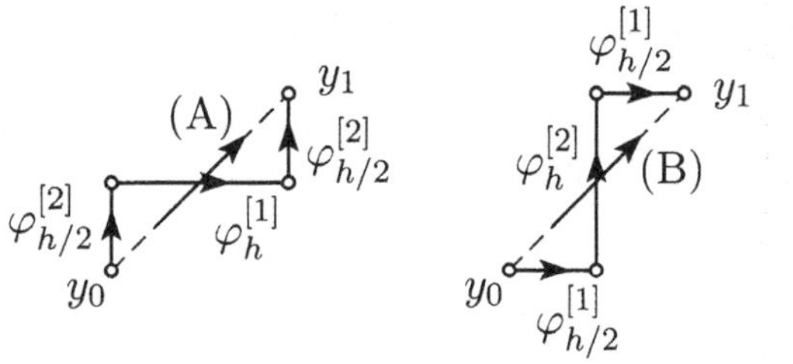

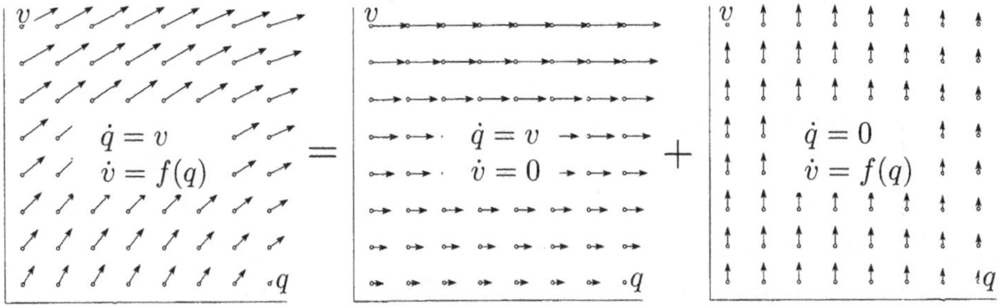

Figure 1.4. The phase space vector field split into two fields.

or, in explicit formulae,

$$\Phi_h^{(A)} = \varphi_{h/2}^{[2]} \circ \varphi_h^{[1]} \circ \varphi_{h/2}^{[2]},$$
$$\Phi_h^{(B)} = \varphi_{h/2}^{[1]} \circ \varphi_h^{[2]} \circ \varphi_{h/2}^{[1]}. \tag{1.13}$$

This way of composing the flows of split vector fields is often referred to as *Strang splitting*, after Strang (1968). For a careful survey of splitting methods we refer to McLachlan and Quispel (2002).

1.6. Interpretation as variational integrator

A further approach to the Störmer–Verlet method is obtained by discretizing Hamilton's principle. This variational principle states that the motion of a mechanical system between any two positions $q(t_0) = q_0$ and $q(t_N) = q_N$ is such that the action integral

$$\int_{t_0}^{t_N} L\big(q(t), \dot{q}(t)\big)\,\mathrm{d}t \quad \text{is minimized,} \tag{1.14}$$

where $L(q, v)$ is the *Lagrangian* of the system. Typically, it is the difference between the kinetic and the potential energy, that is,

$$L(q, v) = \frac{1}{2} v^T M v - U(q), \tag{1.15}$$

with a symmetric positive definite mass matrix M. When M does not depend on q, the Euler–Lagrange equations of this variational problem, $\frac{\mathrm{d}}{\mathrm{d}t}\frac{\partial L}{\partial v} = \frac{\partial L}{\partial q}$, reduce to the second-order differential equation $M\ddot{q} = -\nabla U(q)$.

We now approximate $q(t)$ by a piecewise linear function, interpolating grid values (t_n, q_n) for $n = 0, 1, \ldots, N$, and the action integral by the trapezoidal rule. We then require that $q_1, \ldots, q_{N-1}$ be such that, instead of (1.14),

$$\sum_{n=0}^{N-1} S_h(q_n, q_{n+1}) \quad \text{is minimized,} \tag{1.16}$$

where

$$S_h(q_n, q_{n+1}) = \frac{h}{2} L\Big(q_n, \frac{q_{n+1} - q_n}{h}\Big) + \frac{h}{2} L\Big(q_{n+1}, \frac{q_{n+1} - q_n}{h}\Big). \tag{1.17}$$

The requirement that the gradient with respect to q_n be zero, yields the discrete Euler–Lagrange equations

$$\nabla_Q S_h(q_{n-1}, q_n) + \nabla_q S_h(q_n, q_{n+1}) = 0$$

for $n = 1, \ldots, N-1$, where the partial gradients ∇_q, ∇_Q refer to $S_h = S_h(q, Q)$. In the case of the Lagrangian (1.15) these equations reduce to

$$M(q_{n+1} - 2q_n + q_{n-1}) + h^2 \nabla U(q_n) = 0, \tag{1.18}$$

which is just the two-step formulation (1.2) of the Störmer–Verlet method, with $f(q) = -M^{-1}\nabla U(q)$.

This variational interpretation of the Störmer–Verlet method was given by MacKay (1992). A comprehensive survey of variational integrators can be found in Marsden and West (2001).

1.7. *Numerical example*

We choose the Kepler problem

$$\ddot{q}_1 = -\frac{q_1}{(q_1^2+q_2^2)^{3/2}}, \qquad \ddot{q}_2 = -\frac{q_2}{(q_1^2+q_2^2)^{3/2}}. \tag{1.19}$$

As initial values we take

$$q_1(0) = 1-e, \quad q_2(0) = 0, \quad \dot{q}_1(0) = 0, \quad \dot{q}_2(0) = \sqrt{\frac{1+e}{1-e}}, \tag{1.20}$$

with $e = 0.6$. The period of the exact solution is 2π. Figure 1.5 presents the numerical values of the Störmer–Verlet method for two different step sizes. These solutions are compared to those of the explicit midpoint rule in Runge's one-step formulation; see Hairer, Lubich and Wanner (2002, p. 24, Figure 1.2. and equation (1.3)). This second method is of the same order and for the first steps it behaves very similarly to the Störmer–Verlet scheme (the first step is even identical!), but it deteriorates significantly as the integration interval increases. The explanation of this strange difference is the subject of the theories below.

1.8. *Extension to general partitioned problems*

For the extension of the above formulae to the more general system

$$\dot{q} = g(q,v), \qquad \dot{v} = f(q,v), \tag{1.21}$$

we follow the ideas of De Vogelaere (1956). This is a marvellous paper, short, clear, elegant, written in one week, submitted for publication – and never published. We first extend the formulae (1.8) and (1.9), by taking over the missing arguments from one equation to the other. This gives

$$\text{(SE1)} \qquad \begin{aligned} v_{n+\frac12} &= v_n + \tfrac{h}{2} f\big(q_n, v_{n+\frac12}\big), \\ q_{n+\frac12} &= q_n + \tfrac{h}{2} g\big(q_n, v_{n+\frac12}\big), \end{aligned} \tag{1.22}$$

and

$$\text{(SE2)} \qquad \begin{aligned} q_{n+1} &= q_{n+\frac12} + \tfrac{h}{2} g\big(q_{n+1}, v_{n+\frac12}\big), \\ v_{n+1} &= v_{n+\frac12} + \tfrac{h}{2} f\big(q_{n+1}, v_{n+\frac12}\big). \end{aligned} \tag{1.23}$$

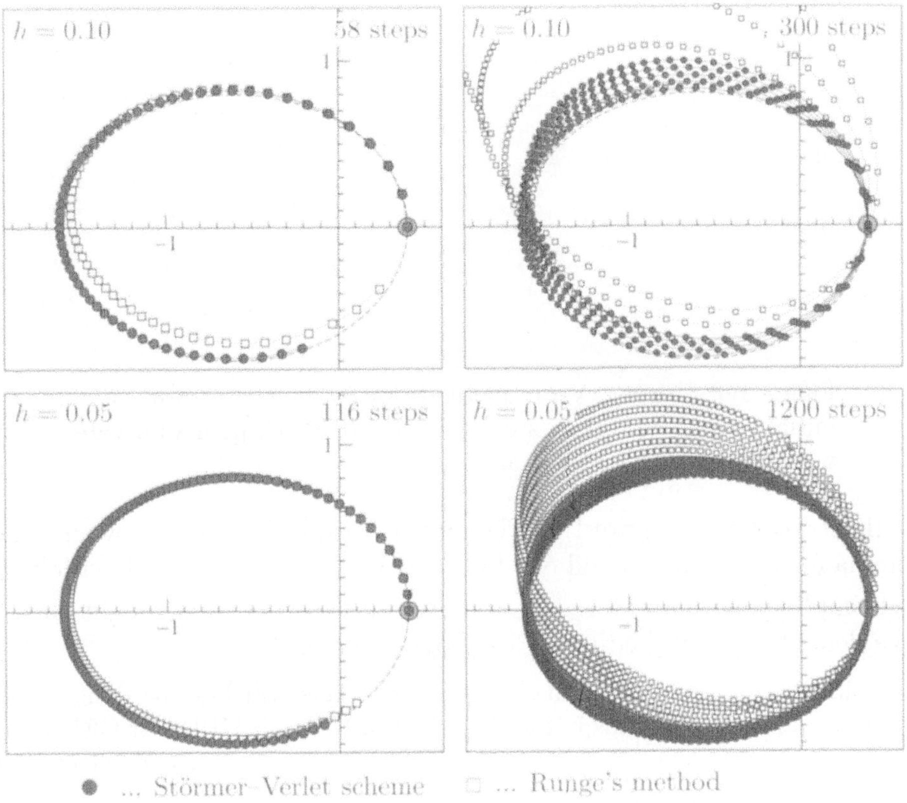

Figure 1.5. Kepler problem: the dashed line is the exact solution.

In each of these algorithms the derivative evaluations of both formulae are taken at the same point. The extensions of the Störmer–Verlet schemes are now obtained by composition, in the same way as in Section 1.4:

$$(\mathrm{A}) = (\mathrm{SE2})\circ(\mathrm{SE1}) \quad \begin{aligned} v_{n+\frac{1}{2}} &= v_n + \tfrac{h}{2}\, f\big(q_n, v_{n+\frac{1}{2}}\big), \\ q_{n+1} &= q_n + \tfrac{h}{2}\Big(g\big(q_n, v_{n+\frac{1}{2}}\big) + g\big(q_{n+1}, v_{n+\frac{1}{2}}\big)\Big), \\ v_{n+1} &= v_{n+\frac{1}{2}} + \tfrac{h}{2}\, f\big(q_{n+1}, v_{n+\frac{1}{2}}\big), \end{aligned} \tag{1.24}$$

and, for the dual version,

$$(\mathrm{B}) = (\mathrm{SE1})\circ(\mathrm{SE2}) \quad \begin{aligned} q_n &= q_{n-\frac{1}{2}} + \tfrac{h}{2}\, g\big(q_n, v_{n-\frac{1}{2}}\big), \\ v_{n+\frac{1}{2}} &= v_{n-\frac{1}{2}} + \tfrac{h}{2}\Big(f\big(q_n, v_{n-\frac{1}{2}}\big) + f\big(q_n, v_{n+\frac{1}{2}}\big)\Big), \\ q_{n+\frac{1}{2}} &= q_n + \tfrac{h}{2}\, g\big(q_n, v_{n+\frac{1}{2}}\big). \end{aligned} \tag{1.25}$$

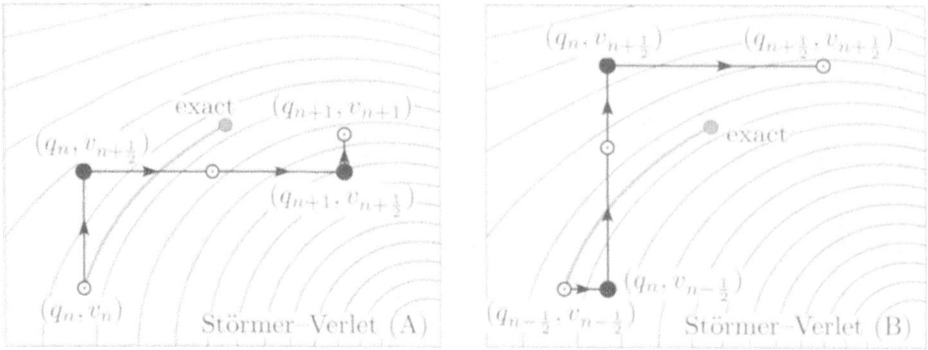

Figure 1.6. Störmer–Verlet methods for $\dot{q} = v$, $\dot{v} = -\sin q - v^2/5$, initial values $(-1.8, 0.3)$, step size $h = 1.5$. Black points indicate where the vector field is evaluated.

For illustrations see Figure 1.6. The first equation of (1.24) is now an *implicit* formula for $v_{n+\frac{1}{2}}$, the second one for q_{n+1}, while only the last one is explicit. Such implicit methods were not common in the 1950s and might then not have delighted journal editors – or programmers:

> No detailed example or discussion is given. This will best be done by those working on these problems in the Brookhaven, Harwell, MURA or CERN group.
>
> De Vogelaere (1956)

2. Geometric properties

We study geometric properties of the flow of differential equations which are preserved by the Störmer–Verlet method. The properties discussed are reversibility, symplecticity, and volume preservation.

2.1. Symmetry and reversibility

The Störmer–Verlet method is *symmetric* with respect to changing the direction of time: in its one-step formulation (1.5), replacing h by $-h$ and exchanging the subscripts $n \leftrightarrow n+1$ (*i.e.*, reflecting time at the centre $t_{n+1/2}$) gives the same method again. Similarly, the replacements $h \leftrightarrow -h$ and $n - \frac{1}{2} \leftrightarrow n + \frac{1}{2}$ leave the formulation (1.6) unchanged. In terms of the numerical one-step map $\Phi_h : (q_n, v_n) \mapsto (q_{n+1}, v_{n+1})$, this symmetry can be stated more formally as

$$\Phi_h = \Phi_{-h}^{-1}. \tag{2.1}$$

Such a relation does not hold for the symplectic Euler methods (1.8) and (1.9), where the above time-reflection transforms (SE1) to (SE2) and *vice versa*.

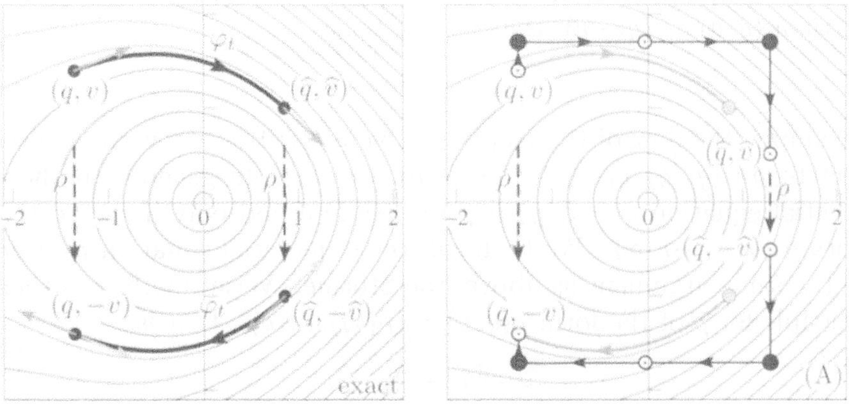

Figure 2.1. A reversible system (left) and the symmetric Störmer–Verlet method (right); the same equation as in Figure 1.6.

The time-symmetry of the Störmer–Verlet method implies an important geometric property of the numerical map in the phase space, namely *reversibility*, to which we turn next. The importance of this property in numerical analysis was first emphasized by Stoffer (1988).

The system (1.3) has the property that inverting the direction of the initial velocity does not change the solution trajectory, it just inverts the direction of motion. The flow φ_t thus satisfies that

$$\varphi_t(q,v) = (\widehat{q},\widehat{v}) \quad \text{implies} \quad \varphi_t(\widehat{q},-\widehat{v}) = (q,-v), \tag{2.2}$$

and we call it *reversible* with respect to the reflection $\rho : (q,v) \mapsto (q,-v)$. This property is illustrated in Figure 2.1, left. The numerical one-step map Φ_h of the Störmer–Verlet method satisfies similarly

$$\Phi_h(q,v) = (\widehat{q},\widehat{v}) \quad \text{implies} \quad \Phi_h(\widehat{q},-\widehat{v}) = (q,-v), \tag{2.3}$$

for all q, v and all h; see Figure 2.1, right. This holds because practically all numerical methods for (1.3), and in particular the Störmer–Verlet method and the symplectic Euler methods, are such that

$$\Phi_h(q,v) = (\widehat{q},\widehat{v}) \quad \text{implies} \quad \Phi_{-h}(q,-v) = (\widehat{q},-\widehat{v}), \tag{2.4}$$

as is readily seen from the defining formulae such as (1.5). The symmetry (2.1) of the Störmer–Verlet method is therefore equivalent to the reversibility (2.3). Let us summarize these considerations.

Theorem 2.1. The Störmer–Verlet method applied to the second-order differential equation (1.1) is both symmetric and reversible, *i.e.*, its one-step map satisfies (2.1) and (2.3).

In some situations, the flow is *ρ-reversible* with respect to involutions ρ other than $(q,v) \mapsto (q,-v)$, that is, it satisfies

$$\rho \circ \varphi_t = \varphi_t^{-1} \circ \rho. \tag{2.5}$$

For example, the flow of the Kepler problem (1.19) is ρ-reversible also with respect to $\rho : (q_1,q_2,v_1,v_2) \mapsto (q_1,-q_2,-v_1,v_2)$. In general, the flow of a differential equation $\dot{y} = F(y)$ is ρ-reversible if and only if the vector field satisfies $\rho \circ F = -F \circ \rho$. We then call the differential equation *ρ-reversible.*

By the same argument as above, the Störmer–Verlet method is then also ρ-reversible for ρ of the form $\rho(q,v) = (\rho_1(q), \rho_2(v))$, that is,

$$\rho \circ \Phi_h = \Phi_h^{-1} \circ \rho. \tag{2.6}$$

2.2. Hamiltonian systems and symplecticity

We now turn to the important class of *Hamiltonian systems*

$$\dot{p} = -\nabla_q H(p,q), \quad \dot{q} = \nabla_p H(p,q), \tag{2.7}$$

where $H(p,q)$ is an arbitrary scalar function of the variables (p,q). When the Hamiltonian is of the form

$$H(p,q) = \frac{1}{2}\, p^T M^{-1} p + U(q), \tag{2.8}$$

with a positive definite mass matrix M and a potential $U(q)$, then the system (2.7) turns into the second-order differential equation (1.3) upon expressing the momenta $p = Mv$ in terms of the velocities and setting $f(q) = -M^{-1}\nabla U(q)$. Equation (2.8) expresses the total energy H as the sum of kinetic and potential energy.

A characteristic geometric property of Hamiltonian systems is that the flow φ_t is *symplectic*, that is, the derivative $\varphi_t' = \partial\varphi_t/\partial(p,q)$ of the flow satisfies, for all (p,q) and t where $\varphi_t(p,q)$ is defined,

$$\varphi_t'(p,q)^T J\, \varphi_t'(p,q) = J \qquad \text{with } J = \begin{pmatrix} 0 & I \\ -I & 0 \end{pmatrix}, \tag{2.9}$$

where I is the identity matrix of the dimension of p or q; see, *e.g.*, Arnold (1989, p. 204) or Hairer *et al.* (2002, p. 172).

The relation (2.9) is formally similar to orthogonality (which it would be if J were replaced by the identity matrix), but, unlike orthogonality, it is not related to the conservation of lengths but of *areas* in phase space. In fact, for systems with one degree of freedom (*i.e.*, $p,q \in \mathbb{R}$), equation (2.9) expresses that the flow preserves the area of sets of initial values in the (p,q)-plane; see Figure 2.2, left. For higher-dimensional systems, symplecticity (2.9) means that the flow preserves the sum of the oriented areas of the projections of $\varphi_t(A)$ onto the (p_i,q_i)-coordinate planes, for any two-dimensional bounded manifold of initial values A; see, *e.g.*, Hairer *et al.* (2002, p. 171*f.*) for a justification of this interpretation.

The Störmer–Verlet method (1.24) applied to the Hamiltonian system (2.7) reads

$$\text{(A)}\qquad \begin{aligned} p_{n+\frac{1}{2}} &= p_n - \tfrac{h}{2}\,\nabla_q H\big(p_{n+\frac{1}{2}}, q_n\big), \\ q_{n+1} &= q_n + \tfrac{h}{2}\Big(\nabla_p H\big(p_{n+\frac{1}{2}}, q_n\big) + \nabla_p H\big(p_{n+\frac{1}{2}}, q_{n+1}\big)\Big), \\ p_{n+1} &= p_{n+\frac{1}{2}} - \tfrac{h}{2}\,\nabla_q H\big(p_{n+\frac{1}{2}}, q_{n+1}\big), \end{aligned} \qquad (2.10)$$

and a similar formula for variant (B). In the particular case of the Hamiltonian (2.8), the method reduces to the Störmer–Verlet method (1.5) with $f(q) = -M^{-1}\nabla U(q)$, upon setting $p_n = Mv_n$.

A numerical method is called *symplectic* if, for Hamiltonian systems (2.7), the Jacobian of the numerical flow $\Phi_h : (p_n, q_n) \mapsto (p_{n+1}, q_{n+1})$ satisfies condition (2.9), that is, if

$$\Phi_h'(p,q)^T J\, \Phi_h'(p,q) = J \qquad (2.11)$$

for all (p, q) and all step sizes h.

Symplecticity of numerical methods was first considered by De Vogelaere (1956), but was not followed up until Ruth (1983) and Feng (1985). In the late 1980s, the results of Lasagni (1988), Sanz-Serna (1988), and Suris (1988) started off an avalanche of papers on symplectic numerical methods. Sanz-Serna and Calvo (1994) was the first book dealing with this subject.

Theorem 2.2. The Störmer–Verlet method applied to a Hamiltonian system is symplectic.

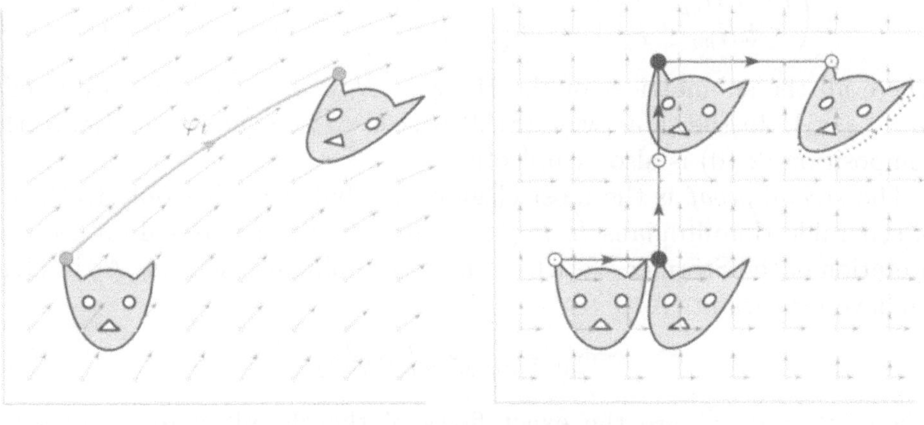

Figure 2.2. Symplecticity of the Störmer–Verlet method for a separable Hamiltonian.

We give four different proofs of this result, which all correspond to different interpretations of the method: as a composition method, as a splitting method, as a variational integrator, and using generating functions. Each of these interpretations lends itself to generalizations to other symplectic integrators, of higher order and/or for constrained Hamiltonian systems. Yet another proof is based on the preservation of quadratic invariants and will be mentioned in Section 3 below. The second proof applies only to Hamiltonians of the special form (2.8), the third proof is formulated for such Hamiltonians for convenience.

The historically *first proof*, due to De Vogelaere (1956), uses the interpretation of the Störmer–Verlet method as the composition of the symplectic Euler method

$$\text{(SE1)} \qquad \begin{aligned} p_{n+\frac{1}{2}} &= p_n - \tfrac{h}{2}\,\nabla_q H\big(p_{n+\frac{1}{2}}, q_n\big), \\ q_{n+\frac{1}{2}} &= q_n + \tfrac{h}{2}\,\nabla_p H\big(p_{n+\frac{1}{2}}, q_n\big), \end{aligned} \tag{2.12}$$

and its adjoint

$$\text{(SE2)} \qquad \begin{aligned} q_{n+1} &= q_{n+\frac{1}{2}} + \tfrac{h}{2}\,\nabla_p H\big(p_{n+\frac{1}{2}}, q_{n+1}\big), \\ p_{n+1} &= p_{n+\frac{1}{2}} - \tfrac{h}{2}\,\nabla_q H\big(p_{n+\frac{1}{2}}, q_{n+1}\big). \end{aligned} \tag{2.13}$$

The method (SE1) is indeed symplectic, as is seen by direct verification of the symplecticity condition

$$\left(\frac{\partial(p_{n+1/2}, q_{n+1/2})}{\partial(p_n, q_n)}\right)^T J \left(\frac{\partial(p_{n+1/2}, q_{n+1/2})}{\partial(p_n, q_n)}\right) = J.$$

The matrix of partial derivatives is obtained from differentiating equation (2.12):

$$\begin{pmatrix} I + hH_{qp}^T & 0 \\ -hH_{pp} & I \end{pmatrix} \left(\frac{\partial(p_{n+1/2}, q_{n+1/2})}{\partial(p_n, q_n)}\right) = \begin{pmatrix} I & -hH_{qq} \\ 0 & I + hH_{qp} \end{pmatrix},$$

where all the submatrices of the Hessian, H_{qp}, H_{pp}, *etc.*, are evaluated at $(p_{n+1/2}, q_n)$. In the same way, (SE2) is seen to be symplectic. Hence their composition (2.10) is also symplectic.

The *second proof* is the most elegant one, but it applies only to the case of separable Hamiltonians $H(p,q) = T(p) + U(q)$. It is based on the interpretation of the Störmer–Verlet method as a splitting method. As in (1.13), we have for variant (A)

$$\Phi_h = \varphi_{h/2}^U \circ \varphi_h^T \circ \varphi_{h/2}^U, \tag{2.14}$$

where φ_t^T and φ_t^U are the exact flows of the Hamiltonian systems with Hamiltonian $T(p) = \frac{1}{2}p^T M^{-1} p$ and $U(q)$, *i.e.*, $\dot p = 0$, $\dot q = M^{-1}p$ and $\dot p = -\nabla U(q)$, $\dot q = 0$, respectively, corresponding to the splitting $H(p,q) =$

$T(p)+U(q)$ of the Hamiltonian (2.8) into kinetic and potential energy. Since the flows of Hamiltonian systems are symplectic, so is their composition (2.14). This is illustrated in Figure 2.2, right. Variant (B) has the flows of T and U interchanged in (2.14), and is thus likewise symplectic.

The *third proof* uses the interpretation of the Störmer–Verlet method as a variational integrator (see Section 1.6). The symplecticity of variational integrators derives from non-numerical work by Maeda (1980) and Veselov (1991). Using (1.4) and the first line of (1.5), we have for $S_h(q,Q)$ of (1.17), in the case of the Lagrangian (1.15) which corresponds to the Hamiltonian (2.8),

$$-\nabla_q S_h(q_n, q_{n+1}) = M\,\frac{q_{n+1}-q_n}{h} + \frac{h}{2}\nabla U(q_n) = Mv_n = p_n \tag{2.15}$$

and similarly

$$\nabla_Q S_h(q_n, q_{n+1}) = M\,\frac{q_{n+1}-q_n}{h} - \frac{h}{2}\nabla U(q_{n+1}) = Mv_{n+1} = p_{n+1}. \tag{2.16}$$

Given (p_n, q_n), the first of the above two equations determines q_{n+1}, and the second one p_{n+1}. The one-step map $\Phi_h : (p_n, q_n) \mapsto (p_{n+1}, q_{n+1})$ of the Störmer–Verlet method is thus generated by the scalar-valued function S_h via (2.15) and (2.16). The desired result then follows from the fact that a map $(p,q) \mapsto (P,Q)$ generated by

$$-\nabla_q S(q,Q) = p, \quad \nabla_Q S(q,Q) = P,$$

is symplectic for *any* function S. This is verified by directly checking the symplecticity condition. Differentiation of the above equations gives the following relations for the matrices of partial derivatives P_p, P_q, Q_p, Q_q:

$$\begin{aligned} S_{qq} + S_{qQ}Q_q &= 0, & S_{qQ}Q_p &= I, \\ S_{Qq} + S_{QQ}Q_q &= P_q, & S_{QQ}Q_p &= P_p. \end{aligned}$$

These equations yield

$$\begin{pmatrix} P_p & P_q \\ Q_p & Q_q \end{pmatrix}^T \begin{pmatrix} 0 & I \\ -I & 0 \end{pmatrix} \begin{pmatrix} P_p & P_q \\ Q_p & Q_q \end{pmatrix} = \begin{pmatrix} 0 & I \\ -I & 0 \end{pmatrix}$$

after multiplying out, as is required for symplecticity. This completes the third proof of symplecticity of the Störmer–Verlet method.

A *fourth proof* of the symplecticity is based on ideas of Lasagni (1988). A step of the Störmer–Verlet method can be generated by a function $\widehat{S}_h(p_1, q_0)$ in the same way as the symplectic Euler method:

$$\begin{aligned} p_1 &= p_0 - \nabla_q \widehat{S}_h(p_1, q_0), \\ q_1 &= q_0 + \nabla_p \widehat{S}_h(p_1, q_0). \end{aligned} \tag{2.17}$$

As we have seen in the first proof, such maps are symplectic. The generating function is simply $\widehat{S}_h = hH$ for the symplectic Euler method. For the Störmer–Verlet method $\widehat{S}_h$ is obtained as

$$\begin{aligned}\widehat{S}_h(p_1, q_0) = \frac{h}{2}\Big(H(p_{1/2}, q_0) + H(p_{1/2}, q_1)\Big) \\ - \frac{h^2}{4}\nabla_q H(p_{1/2}, q_1)^T\Big(\nabla_p H(p_{1/2}, q_0) + \nabla_p H(p_{1/2}, q_1)\Big),\end{aligned} \tag{2.18}$$

where q_1 and $p_{1/2}$ are defined by the Störmer–Verlet formulae and are now considered as functions of (p_1, q_0). We do not give the computational details, which can be found in Hairer *et al.* (2002, Section VI.5) for a more general class of symplectic integrators.

2.3. Volume preservation

The flow φ_t of a system of differential equations $\dot{y} = F(y)$ with divergence-free vector field ($\operatorname{div} F(y) = 0$ for all y) satisfies $\det \varphi_t'(y) = 1$ for all y. It therefore preserves volume in phase space: for every bounded open set Ω, and for every t for which $\varphi_t(y)$ exists for all $y \in \Omega$,

$$\operatorname{vol}(\varphi_t(\Omega)) = \operatorname{vol}(\Omega).$$

The vector field $(v, f(q))$ of a second-order differential equation (1.3), written as a first-order system, is divergence-free. The same is true for Hamiltonian vector fields $(-\nabla_q H(p,q), \nabla_p H(p,q))$.

The Störmer–Verlet method preserves volume,

$$\operatorname{vol}(\Phi_h(\Omega)) = \operatorname{vol}(\Omega),$$

in the following two situations.

For the method (2.10), applied to a Hamiltonian system (2.7), this follows from its symplecticity (2.11), which implies $\det \Phi_h'(p,q) = 1$ for all (p,q).

For partitioned differential equations of the form

$$\dot{q} = g(v), \qquad \dot{v} = f(q), \tag{2.19}$$

the method (1.24) can be interpreted as the splitting (1.13), where $\varphi_t^{[1]}$ and $\varphi_t^{[2]}$ are the exact flows of $\dot{q} = g(v)$, $\dot{v} = 0$ and $\dot{q} = 0$, $\dot{v} = f(q)$, respectively. Since the vector fields of these flows are divergence-free, they are volume-preserving and so is their composition.

The same idea allows us to extend the Störmer–Verlet method to a volume-preserving algorithm for systems partitioned into the *three* equations

$$\dot{x} = a(y,z), \qquad \dot{y} = b(x,z), \qquad \dot{z} = c(x,y), \tag{2.20}$$

for which the diagonal blocks of the Jacobian are zero. We split them symmetrically, giving

$$\varphi_{h/2}^{[1]} \circ \varphi_{h/2}^{[2]} \circ \varphi_h^{[3]} \circ \varphi_{h/2}^{[2]} \circ \varphi_{h/2}^{[1]}, \tag{2.21}$$

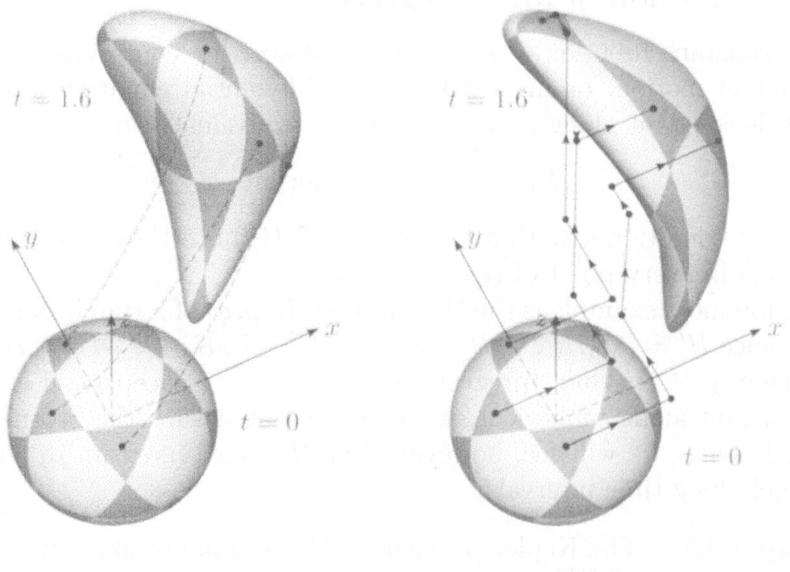

Figure 2.3. Volume-preserving deformation of the ball of radius 0.9, centred at the origin, by the ABC flow (left) and by method (2.22) (right).

where $\varphi_t^{[1]}$ is the (volume-preserving) flow of $\dot{x} = a(y,z)$, $\dot{y} = 0$, $\dot{z} = 0$ and similarly for $\varphi_t^{[2]}$ and $\varphi_t^{[3]}$. Written out, this becomes

$$\begin{aligned} x_{n+\frac{1}{2}} &= x_n + \frac{h}{2}\, a(y_n, z_n), \\ y_{n+\frac{1}{2}} &= y_n + \frac{h}{2}\, b\big(x_{n+\frac{1}{2}}, z_n\big), \\ z_{n+1} &= z_n + h\, c\big(x_{n+\frac{1}{2}}, y_{n+\frac{1}{2}}\big), \\ y_{n+1} &= y_{n+\frac{1}{2}} + \frac{h}{2}\, b\big(x_{n+\frac{1}{2}}, z_{n+1}\big), \\ x_{n+1} &= x_{n+\frac{1}{2}} + \frac{h}{2}\, a(y_{n+1}, z_{n+1}). \end{aligned} \tag{2.22}$$

An illustration of this algorithm, applied to the ABC-flow

$$\begin{aligned} \dot{x} &= A \sin z + C \cos y, \\ \dot{y} &= B \sin x + A \cos z, \\ \dot{z} &= C \sin y + B \cos x, \end{aligned}$$

is presented in Figure 2.3 for $A = 1/2$, $B = C = 1$.

More ingenuity is necessary if the system is divergence-free with nonzero elements on the diagonal of the Jacobian. Feng and Shang (1995) give a volume-preserving extension of the above scheme to the general case.

3. Conservation of first integrals

A non-constant function $I(y)$ is a *first integral* (or conserved quantity, or constant of motion, or invariant) of the differential equation $\dot{y} = F(y)$ if $I(y(t))$ is constant along every solution, or equivalently, if

$$I'(y)F(y) = 0 \qquad \text{for all} \quad y. \tag{3.1}$$

The latter condition says that the gradient $\nabla I(y)$ is orthogonal to the vector field $F(y)$ in every point of the phase space.

The foremost example is the Hamiltonian $H(p,q)$ of a Hamiltonian system (2.7): since $H' = (\nabla_p H^T, \nabla_q H^T)$ and $\nabla_p H^T(-\nabla_q H) + \nabla_q H^T\, \nabla_p H = 0$, the total energy H is a first integral. Apart from very exceptional cases, H is *not* constant along numerical solutions computed with the Störmer–Verlet method. Later we will see, however, that H is conserved up to $\mathcal{O}(h^2)$ over extremely long time intervals.

Example 3.1. The Kepler problem (1.19) is Hamiltonian with $H(p,q) = \frac{1}{2}(p_1^2 + p_2^2) - 1/\sqrt{q_1^2 + q_2^2}$. In addition to the Hamiltonian, this system has the following conserved quantities, as can be easily checked: the angular momentum $L = q_1p_2 - q_2p_1$, and the nonzero components of the Runge–Lenz–Pauli vector

$$\begin{pmatrix} A_1 \\ A_2 \\ 0 \end{pmatrix} = \begin{pmatrix} p_1 \\ p_2 \\ 0 \end{pmatrix} \times \begin{pmatrix} 0 \\ 0 \\ q_1p_2 - q_2p_1 \end{pmatrix} - \frac{1}{\sqrt{q_1^2 + q_2^2}} \begin{pmatrix} q_1 \\ q_2 \\ 0 \end{pmatrix}.$$

Figure 3.1 shows the behaviour of these quantities along a numerical solution of the Störmer–Verlet method. The method preserves the angular momentum exactly (see Section 1.3), and there are only small errors in the Hamiltonian along the numerical solution, but no drift. There is, however, a linear drift in the Runge–Lenz–Pauli vector. In contrast, for explicit Runge–Kutta methods, none of the first integrals is preserved, and there is a drift away from the constant value for all of them.

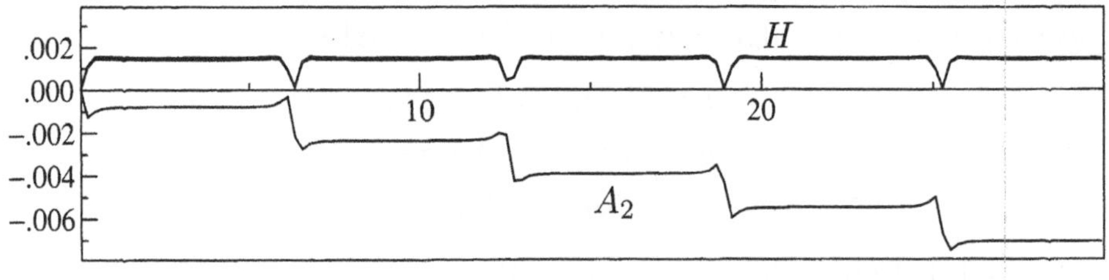

Figure 3.1. The Hamiltonian and the second component of the Runge–Lenz–Pauli vector along the numerical solution of the Störmer–Verlet method with step size $h = 0.02$.

Example 3.2. (Conservation of total linear and angular momentum of N-body systems) A system of N particles interacting pairwise, with potential forces depending on the distances of the particles, is formulated as a Hamiltonian system with total energy

$$H(p,q) = \frac{1}{2}\sum_{i=1}^{N} \frac{1}{m_i} p_i^T p_i + \sum_{i=2}^{N}\sum_{j=1}^{i-1} V_{ij}\big(\|q_i - q_j\|\big). \tag{3.2}$$

Here $q_i, p_i \in \mathbb{R}^3$ represent the position and momentum of the ith particle of mass m_i, and $V_{ij}(r)$ $(i > j)$ is the interaction potential between the ith and jth particle. The equations of motion read

$$\dot{q}_i = \frac{1}{m_i} p_i, \quad \dot{p}_i = \sum_{j=1}^{N} \nu_{ij}\,(q_i - q_j)$$

where, for $i > j$, we have $\nu_{ij} = \nu_{ji} = -V_{ij}'(r_{ij})/r_{ij}$ with $r_{ij} = \|q_i - q_j\|$, and ν_{ii} is arbitrary, say $\nu_{ii} = 0$. The conservation of the total *linear momentum* $P = \sum_{i=1}^{N} p_i$ and the total *angular momentum* $L = \sum_{i=1}^{N} q_i \times p_i$ is a consequence of the symmetry relation $\nu_{ij} = \nu_{ji}$:

$$\frac{\mathrm{d}}{\mathrm{d}t}\sum_{i=1}^{N} p_i = \sum_{i=1}^{N}\sum_{j=1}^{N} \nu_{ij}(q_i - q_j) = 0,$$

$$\frac{\mathrm{d}}{\mathrm{d}t}\sum_{i=1}^{N} q_i \times p_i = \sum_{i=1}^{N} \frac{1}{m_i} p_i \times p_i + \sum_{i=1}^{N}\sum_{j=1}^{N} q_i \times \nu_{ij}(q_i - q_j) = 0.$$

The exact preservation of *linear* first integrals, such as the total linear momentum, is common to most numerical integrators.

Theorem 3.3. The Störmer–Verlet method preserves linear first integrals.

Proof. Let the linear first integral be $I(q,v) = b^T q + c^T v$, so that $b^T v + c^T f(q) = 0$ for all q, v. Necessarily then, $c^T f(q) = 0$ for all q, and $b = 0$. Multiplying the formulae for v in (1.5) by c^T thus yields $c^T v_1 = c^T v_0$. □

Quadratic first integrals are not generally preserved by the Störmer–Verlet method, as the following example shows.

Example 3.4. Consider the harmonic oscillator, which has the Hamiltonian $H(p,q) = \frac{1}{2}p^2 + \frac{1}{2}\omega^2 q^2$ $(p, q \in \mathbb{R})$. Applying the Störmer–Verlet method gives

$$\begin{pmatrix} p_{n+1} \\ \omega q_{n+1} \end{pmatrix} = A(h\omega) \begin{pmatrix} p_n \\ \omega q_n \end{pmatrix} \tag{3.3}$$

with the propagation matrix

$$A(h\omega) = \begin{pmatrix} 1 - \frac{h^2\omega^2}{2} & -\frac{h\omega}{2}\left(1 - \frac{h^2\omega^2}{4}\right) \\ \frac{h\omega}{2} & 1 - \frac{h^2\omega^2}{2} \end{pmatrix}. \tag{3.4}$$

Since $A(h\omega)$ is not an orthogonal matrix, $H(p,q)$ is not preserved along numerical solutions. Notice, however, that the characteristic polynomial is $\lambda^2 - (2 - h^2\omega^2)\lambda + 1$, so that the eigenvalues are of modulus one if (and only if) $|h\omega| \le 2$. The matrix V of eigenvectors is close to the identity for small $h\omega$, and the norm of $V^{-1}(p_n, \omega q_n)^T$ is conserved.

The Störmer–Verlet method does, however, preserve an important subclass of quadratic first integrals, and in particular the total angular momentum of N-body systems. As we have seen in Section 1.3, Newton was aware that the method preserves angular momentum in the Kepler problem and used this fact to prove Kepler's second law. In the following result C is a constant square matrix and c a constant vector.

Theorem 3.5. The Störmer–Verlet method preserves quadratic first integrals of the form $I(q,v) = v^T(Cq + c)$ (or $I(p,q) = p^T(Bq + b)$ in the Hamiltonian case).

Proof. By (3.1), $f(q)^T(Cq+c) + v^T Cv = 0$ for all q, v. Writing the Störmer–Verlet method as the composition of the two symplectic Euler methods (1.8) and (1.9), we obtain for the first half-step

$$\begin{aligned} v_{n+1/2}^T(Cq_{n+1/2} + c) &= v_n^T(Cq_n + c) \\ &\quad + \frac{h}{2}\big(f(q_n)^T(Cq_n + c) + v_{n+1/2}^T C v_{n+1/2}\big), \end{aligned}$$

where we notice that the term in the second line vanishes. For the second half-step we obtain in the same way $v_{n+1}^T(Cq_{n+1} + c) = v_{n+1/2}^T(Cq_{n+1/2} + c)$, and the result follows. □

The most important source of first integrals of Hamiltonian systems is *Noether's theorem*, which states that continuous symmetries yield first integrals: if the associated Lagrangian is invariant under the flow α_s of the vector field $a(q)$, that is, $L(\alpha_s(q), \alpha_s'(q)v) = L(q,v)$ for all real s near 0 and all (q,v), then $I(p,q) = p^T a(q)$ is a first integral; see, *e.g.*, Arnold (1989, p. 88). For Hamiltonian systems of the form (2.8), where the associated Lagrangian is (1.15), it can be shown that $a(q)$ must be linear: $a(q) = Bq + b$, with MB skew-symmetric. Hence, for Hamiltonian systems (2.7) with a Hamiltonian of the form (2.8), all first integrals originating from Noether's theorem are preserved by the Störmer–Verlet method.

Theorem 3.5 yields yet another proof (and further insight) of the symplecticity of the Störmer–Verlet method, following an argument by Bochev and

Scovel (1994): consider the Hamiltonian system $\dot{p} = -\nabla U(q)$, $\dot{q} = M^{-1}p$ together with its variational equation

$$\dot{Y} = \begin{pmatrix} 0 & -\nabla^2 U(q) \\ M^{-1} & 0 \end{pmatrix} Y \qquad \text{with } Y = \begin{pmatrix} P_p & P_q \\ Q_p & Q_q \end{pmatrix}.$$

The derivative of the flow is then $\varphi_t'(p,q) = Y(t)$ corresponding to the initial conditions p, q and $Y(0) = I$. The derivative $\Phi_h'(p,q)$ of the numerical solution with respect to the initial values equals the result Y_1 obtained by applying the method to the combined system of the Hamiltonian system together with its variational equation, partitioned into (p, P_p, P_q) and (q, Q_p, Q_q). Symplecticity means that the components of $Y^T J Y$ are first integrals. Since they are of the mixed quadratic type considered above, Theorem 3.5 shows that they are preserved by the Störmer–Verlet method: $Y_1^T J Y_1 = Y_0^T J Y_0$, which is just the symplecticity ${\Phi_h'}^T J \Phi_h' = J$.

4. Backward error analysis

The theoretical foundation of geometric integrators is mainly based on a backward interpretation which considers the numerical approximation as the exact solution of a modified problem. Such an interpretation has been intuitively used in the physics literature, *e.g.*, Ruth (1983). A rigorous formulation evolved around 1990, beginning with the papers by Feng (1991), McLachlan and Atela (1992), Sanz-Serna (1992) and Yoshida (1993). Exponentially small error bounds and applications of backward error analysis to explain the long-time behaviour of numerical integrators were subsequently given by Benettin and Giorgilli (1994), Hairer and Lubich (1997), and Reich (1999*a*). We explain the essential ideas and we illustrate them for the Störmer–Verlet method.

4.1. Construction of the modified equation

The idea of backward error analysis applies to general ordinary differential equations and to general numerical integrators, and a restriction to special methods for second-order problems would hide the essentials. We therefore consider the differential equation

$$\dot{y} = F(y), \tag{4.1}$$

and a numerical one-step method $y_{n+1} = \Phi_h(y_n)$. The idea consists in searching and studying a *modified differential equation*

$$\dot{y} = F(y) + hF_2(y) + h^2 F_3(y) + \cdots, \tag{4.2}$$

such that the exact time-h flow $\widetilde{\varphi}_h(y)$ of (4.2) is equal to the numerical flow $\Phi_h(y)$. Unfortunately, the series in (4.2) cannot be expected to converge in general, and the precise statement has to be formulated as follows.

Theorem 4.1. Consider (4.1) with an infinitely differentiable vector field $F(y)$, and assume that the numerical method admits a Taylor series expansion of the form

$$\Phi_h(y) = y + hF(y) + h^2 D_2(y) + h^3 D_3(y) + \cdots \tag{4.3}$$

with smooth $D_j(y)$. Then there exist unique vector fields $F_j(y)$ such that, for any $N \geq 1$,

$$\Phi_h(y) = \widetilde{\varphi}_{h,N}(y) + \mathcal{O}(h^{N+1}),$$

where $\widetilde{\varphi}_{t,N}$ is the exact flow of the truncated modified equation

$$\dot{y} = F(y) + hF_2(y) + \cdots + h^{N-1} F_N(y). \tag{4.4}$$

Proof. Disregarding convergence issues, we expand the exact flow of (4.2) into a Taylor series (using the notation $\widetilde{y}(t) = \widetilde{\varphi}_t(y)$)

$$\begin{aligned}\widetilde{\varphi}_h(y) &= y + h\dot{\widetilde{y}}(0) + \frac{h^2}{2!}\ddot{\widetilde{y}}(0) + \frac{h^3}{3!}\widetilde{y}^{(3)}(0) + \cdots \\ &= y + h\big(F(y) + hF_2(y) + h^2 F_3(y) + \cdots\big) \\ &\quad + \frac{h^2}{2!}\big(F'(y) + hF_2'(y) + \cdots\big)\big(F(y) + hF_2(y) + \cdots\big) + \cdots\end{aligned} \tag{4.5}$$

and we compare like powers of h in the expressions (4.5) and (4.3). This yields recurrence relations for the functions $F_j(y)$, namely,

$$\begin{aligned}F_2(y) &= D_2(y) - \tfrac{1}{2!}F'F(y), \\ F_3(y) &= D_3(y) - \tfrac{1}{3!}\big(F''(F,F)(y) + F'F'F(y)\big) - \tfrac{1}{2!}\big(F'F_2(y) + F_2'F(y)\big),\end{aligned} \tag{4.6}$$

and uniquely defines the functions $F_j(y)$ in a constructive manner. □

4.2. Modified equation of the Störmer–Verlet method

Putting $y = (q,v)^T$ and $F(y) = (v, f(q))^T$, the differential equation (1.3) is of the form (4.1). For the Störmer–Verlet scheme (1.5) we have

$$\Phi_h(q,v) = \begin{pmatrix} q + hv + \frac{h^2}{2}f(q) \\ v + \frac{h}{2}f(q) + \frac{h}{2}f\big(q + hv + \frac{h^2}{2}f(q)\big)\end{pmatrix}. \tag{4.7}$$

Expanding this function into a Taylor series we get (4.3) with

$$D_2(q,v) = \frac{1}{2}\begin{pmatrix} f(q) \\ f'(q)v\end{pmatrix}, \quad D_3(q,v) = \frac{1}{4}\begin{pmatrix} 0 \\ f'(q)f(q) + f''(q)(v,v)\end{pmatrix}, \quad \cdots$$

and the functions $F_j(q,v)$ can be computed as in the proof of Theorem 4.1. Since the Störmer–Verlet method is of second order, the function $D_2(q,v)$

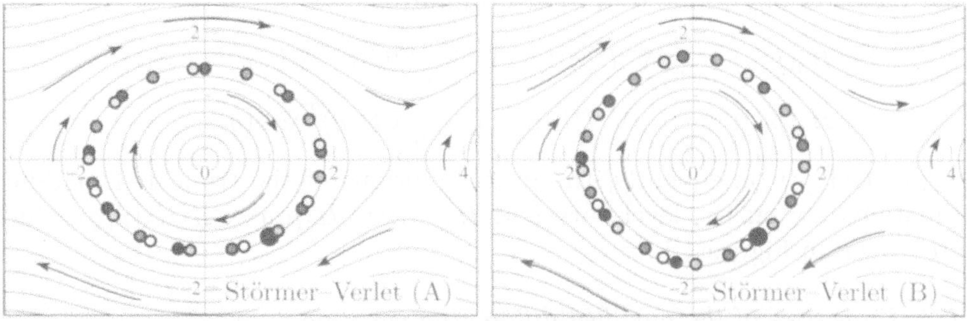

Figure 4.1. Numerical solution with step size $h = 0.9$ for the two versions of the Störmer–Verlet method compared to the exact flow of their modified differential equations truncated after the $\mathcal{O}(h^2)$ term.

has to coincide with the h^2-coefficient of the exact solution and we have $F_2(q, v) = 0$. We then get

$$F_3(q,v) = \frac{1}{12}\begin{pmatrix} -2\,f'(q)v \\ f'(q)f(q) + f''(q)(v,v) \end{pmatrix}, \tag{4.8}$$

and for the next function we obtain $F_4(q, v) = 0$. The vanishing of this function follows from the symmetry of the method (*cf.* Section 4.3). For larger (odd) j the functions $F_j(q, v)$ become more and more complicated, and higher derivatives of $f(q)$ are involved. The explicit formula for $F_3(q, v)$ also shows that the modified differential equation (4.2) is no longer a second-order equation like (1.3).

A similar computation for the version (B) of the Störmer–Verlet method (see (1.6)) gives

$$F_3(q,v) = \frac{1}{24}\begin{pmatrix} 2\,f'(q)v \\ -4\,f'(q)f(q) - f''(q)(v,v) \end{pmatrix}, \tag{4.9}$$

and, obviously, also $F_2(q, v) = F_4(q, v) = 0$.

As a concrete example consider the pendulum equation for which $f(q) = -\sin q$. The two pictures of Figure 4.1 show the exact flow of the modified differential equations (truncated after the $\mathcal{O}(h^2)$ term) corresponding to the two versions (1.5) and (1.6) of the Störmer–Verlet scheme together with the numerical solution for the initial value $(p_0, q_0) = (1.0, -1.2)$. The shade of the numerical approximations (dark grey to light grey) indicates the increasing time. We observe a surprisingly good agreement.

In both cases the solutions of the modified equation are periodic, and the numerical approximation lies near a closed curve, so that correct qualitative behaviour is obtained. This is explained by the fact that for $f(q) = -\nabla U(q)$

the vector fields (4.8) and (4.9) are Hamiltonian with

$$H_3(p,q) = \frac{1}{12}\,\nabla^2 U(q)(p,p) + \frac{1}{24}\,\nabla U(q)^T \nabla U(q) \qquad \text{and}$$
$$H_3(p,q) = -\frac{1}{24}\,\nabla^2 U(q)(p,p) - \frac{1}{6}\,\nabla U(q)^T \nabla U(q),$$

respectively. Consequently, the exact solutions of the truncated modified equation stay on the level curves of $\widetilde{H}(p,q) = H(p,q) + h^2 H_3(p,q)$ which are drawn in Figure 4.1.

4.3. Properties of the modified differential equation

In Section 4.4 we shall see that the numerical solution is extremely close to the exact solution of a truncated modified equation. To study properties of the numerical solution it is therefore justified to investigate instead the corresponding properties of the modified differential equation.

It follows from the definition of the modified equation that for methods of *order r*, that is, $\Phi_h(y) = \varphi_h(y) + \mathcal{O}(h^{r+1})$, we have

$$F_j(y) = 0 \qquad \text{for} \;\; j = 2, \ldots, r.$$

Furthermore, if the leading term of the *local truncation error* is $E_{r+1}(y)$, that is, $\Phi_h(y) = \varphi_h(y) + h^{r+1}E_{r+1}(y) + \mathcal{O}(h^{r+2})$, then

$$F_{r+1}(y) = E_{r+1}(y).$$

By Theorem 2.1 the Störmer–Verlet method is *symmetric*. For such methods the modified equation has an expansion in even powers of h, that is,

$$F_{2j}(y) = 0 \quad \text{for} \;\; j = 1, 2, \ldots. \tag{4.10}$$

This can be proved as follows: to indicate the h-dependence of the vector field (4.2), we let $\widetilde{\varphi}_{t,h}(y)$ denote the (formal) flow of (4.2). Backward error analysis tells us that $\Phi_h(y) = \widetilde{\varphi}_{h,h}(y)$. We thus have $\Phi_{-h}(y) = \widetilde{\varphi}_{-h,-h}(y)$ and, by the group property of the exact flow, $\Phi_{-h}^{-1}(y) = \widetilde{\varphi}_{h,-h}(y)$. The symmetry condition (2.1) thus implies that $\widetilde{\varphi}_{t,h}(y) = \widetilde{\varphi}_{t,-h}(y)$ for $t = h$, and the computation of (4.5) shows that this is only possible if (4.10) holds.

Geometric properties of a numerical method have their counterparts in the modified equation. Let us explain this for the properties discussed in Sections 2 and 3.

Theorem 4.2. (Reversible systems) If the Störmer–Verlet method (1.5) is applied to a differential equation (1.3), then every truncation of the modified differential equation is reversible with respect to the reflection $\rho(q,v) = (q,-v)$.

Theorem 4.3. (Hamiltonian systems) If the Störmer–Verlet method (2.10) is applied to a Hamiltonian system, then every truncation of the modified differential equation is Hamiltonian.

Theorem 4.4. (Divergence-free systems) If the Störmer–Verlet method (1.24) is applied to a divergence-free system of the form (2.19), then every truncation of the modified differential equation is divergence-free.

Theorem 4.5. (First integrals) If the Störmer–Verlet method (1.24) is applied to a differential equation with a first integral of the form $I(q, v) = v^T(Cq + c)$, then every truncation of the modified differential equation has $I(q, v)$ as a first integral.

The proofs are based on an induction argument. Since they are all very similar (see Hairer *et al.* (2002, Chapter IX)), we only present the proof of Theorem 4.3, for the case where the Hamiltonian $H(p, q)$ is defined on a simply connected domain. This proof was first given by Benettin and Giorgilli (1994) and Tang (1994), and its ideas can be traced back to Moser (1968).

Proof. With $y = (p, q)$, the Hamiltonian system (2.7) is written more compactly as $\dot{y} = J^{-1}\nabla H(y)$ with J of (2.9). We will show that all the coefficient functions of the modified equation can be written as

$$F_j(y) = J^{-1}\nabla H_j(y). \tag{4.11}$$

Assume, by induction, that (4.11) holds for $j = 1, 2, \ldots, N$ (this is satisfied for $N = 1$, because $F_1(y) = F(y) = J^{-1}\nabla H(y)$). We have to prove the existence of a Hamiltonian $H_{N+1}(y)$. The idea is to consider the truncated modified equation (4.4), which is then a Hamiltonian system with Hamiltonian $H(y) + hH_2(y) + \cdots + h^{N-1}H_N(y)$. Its flow $\varphi_{N,t}(y)$, compared to that of (4.2) and thus to the one-step map Φ_h of the Störmer–Verlet method, satisfies

$$\Phi_h(y) = \varphi_{N,h}(y) + h^{N+1}F_{N+1}(y) + \mathcal{O}(h^{N+2}),$$

and also

$$\Phi_h'(y) = \varphi_{N,h}'(y) + h^{N+1}F_{N+1}'(y) + \mathcal{O}(h^{N+2}).$$

By Theorem 2.2 and by the induction hypothesis, Φ_h and $\varphi_{N,h}$ are symplectic transformations. This, together with $\varphi_{N,h}'(y) = I + \mathcal{O}(h)$, therefore implies

$$J = \Phi_h'(y)^T J \Phi_h'(y) = J + h^{N+1}\big(F_{N+1}'(y)^T J + JF_{N+1}'(y)\big) + \mathcal{O}(h^{N+2}).$$

Consequently, the matrix $JF_{N+1}'(y)$ is symmetric. The function $JF_{N+1}(y)$ is therefore the gradient of some scalar function $H_{N+1}(y)$, which proves (4.11) for $j = N + 1$. □

The last argument of this proof requires that the domain be simply connected. For general domains, we have to use the representation (2.17) with the help of the generating function (2.18). We refer to Section IX.3.2 of Hairer *et al.* (2002) for details of the proof.

4.4. Exponentially small error estimates

Theorem 4.1 proves a statement that is valid for all $N \geq 1$, and it is natural to ask which choice of N gives the best estimate.

Example 4.6. Consider the simple differential equation $\ddot{q} = f(t)$ (which becomes autonomous after adding $\ddot{t} = 0$). If we try to compute the modified equation for the Störmer–Verlet method, we are readily convinced that its q-component is of the form

$$\ddot{q}(t) = f(t) + h^2 b_2 \ddot{f}(t) + h^4 b_4 f^{(4)}(t) + h^6 b_6 f^{(6)}(t) + \cdots . \tag{4.12}$$

Putting $f(t) = \mathrm{e}^t$, the solution of this modified equation is

$$\widetilde{q}(t) = C_1 + tC_2 + (1 + b_2 h^2 + b_4 h^4 + b_6 h^6 + \cdots)\, \mathrm{e}^t,$$

and inserted into (1.2) we obtain

$$(1 + b_2 h^2 + b_4 h^4 + b_6 h^6 + \cdots)(\mathrm{e}^{-h} - 2 + \mathrm{e}^h) = h^2. \tag{4.13}$$

This shows that $1 + b_2 h^2 + b_4 h^4 + \cdots$ is analytic in a disc of radius 2π centred at the origin. Consequently, the coefficients behave like $b_{2k} \approx \text{Const}\,(2\pi)^{-2k}$ for $k \to \infty$.

Now consider functions $f(t)$ whose derivatives grow like $f^{(k)}(t) \approx k!\, M\, R^{-k}$. This is the case for analytic $f(t)$ with finite poles. The individual terms of the modified equation (4.12) then behave like

$$h^{2k} b_{2k} f^{(2k)}(t) \approx \text{Const}\, \frac{h^{2k}(2k)!}{(R \cdot 2\pi)^{2k}} \approx \text{Const}\, \sqrt{4\pi k} \left(\frac{h \cdot 2k}{R \cdot 2\pi\, \mathrm{e}} \right)^{2k} \tag{4.14}$$

(using Stirling's formula). Even for very small step sizes h this expression is unbounded for $k \to \infty$, so that the series (4.12) cannot converge. However, formula (4.14) tells us that the terms of the series decrease until $2k$ approaches the value $2\pi R/h$, and then they tend rapidly to ∞. It is therefore natural to truncate the modified equation after N terms, where $N \approx 2\pi R/h$.

To find a reasonably good truncation index N for general differential equations, we have to know estimates for all derivatives of $F(y)$ and of the coefficient functions $D_j(y)$ of the Taylor expansion of the numerical flow. One convenient way of doing this is to assume analyticity of these functions.

Exponentially small error bounds were first derived by Benettin and Giorgilli (1994). The following estimates are from Hairer *et al.* (2002, p. 306).

Theorem 4.7. Let $F(y)$ be analytic in $B_{2R}(y_0)$, let the coefficients $D_j(y)$ of the method (4.3) be analytic in $B_R(y_0)$, and assume that

$$\|F(y)\| \leq M \quad \text{and} \quad \|D_j(y)\| \leq \mu M \left(\frac{2\kappa M}{R} \right)^{j-1} \tag{4.15}$$

hold for $y \in B_{2R}(y_0)$ and $y \in B_R(y_0)$, respectively. If $h \leq h_0/4$ with

$h_0 = R/(\mathrm{e}\eta M)$ and $\eta = 2\max\big(\kappa, \mu/(2\ln 2 - 1)\big)$, then there exists $N = N(h)$ (namely N equal to the largest integer satisfying $hN \le h_0$) such that the difference between the numerical solution $y_1 = \Phi_h(y_0)$ and the exact solution $\widetilde{\varphi}_{N,t}(y_0)$ of the truncated modified equation (4.4) satisfies

$$\|\Phi_h(y_0) - \widetilde{\varphi}_{N,h}(y_0)\| \le h\gamma M \mathrm{e}^{-h_0/h},$$

where $\gamma = \mathrm{e}(2 + 1.65\eta + \mu)$ depends only on the method.

The proof of this theorem is technical and long; see Hairer *et al.* (2002, Section IX.7) for details. We just explain how the assumptions can be checked for the Störmer–Verlet method (4.7).

We let $y = (q, v)^T$, $F(y) = (v, f(q))^T$, and we consider the scaled norm $\|y\| = \|q\| + h\|v\|$. The quantities R and M are then given by the problem. The computation of the beginning of Section 4.2 shows that the functions $D_j(q, v)$ are composed of derivatives of $f(q)$ so that they are analytic on the same domain as $f(q)$. To find the constants μ and κ in (4.15), we use $\|F(y)\| = \|v\| + h\|f(q)\| \le M$ for $\|q - q_0\| + h\|v - v_0\| \le 2R$, and we estimate

$$\left\| \Phi_h(q,v) - \begin{pmatrix} q + hv \\ v + \frac{h}{2} f(q) \end{pmatrix} \right\| = \left\| \begin{pmatrix} \frac{h^2}{2} f(q) \\ \frac{h}{2} f(q + hv + \frac{h^2}{2} f(q)) \end{pmatrix} \right\| \le h M$$

for $\|q - q_0\| + h\|v - v_0\| \le R$ and for $hM \le R$. This follows from the fact that the argument of f satisfies $\|q + hv + \frac{h^2}{2} f(q) - q_0\| \le R + hM \le 2R$. Considered as a function of h, $\Phi_h(q, v)$ is analytic in the complex disc $|h| \le R/M$. Cauchy's estimate therefore yields

$$\|D_j(q,v)\| = \frac{1}{j!} \left\| \frac{\mathrm{d}^j}{\mathrm{d}h^j} \left(\Phi_h(q,v) - \begin{pmatrix} q + hv \\ v + \frac{h}{2} f(q) \end{pmatrix} \right) \Big|_{h=0} \right\| \le M \left(\frac{M}{R} \right)^{j-1}$$

for $j \ge 2$. This proves the estimates (4.15) with $\mu = 1$ and $\kappa = 1/2$.

5. Long-time behaviour of numerical solutions

In this section we show how the geometric properties of Section 2 turn into favourable long-term behaviour. Most of the results are obtained with the help of backward error analysis.

5.1. Energy conservation

We have seen in Example 3.4 that the total energy $H(p, q)$ of a Hamiltonian system is not preserved exactly by the Störmer–Verlet method. In that example it is, however, approximately preserved. Also for the Kepler problem, Figure 3.1 indicates no drift in the energy. As the following theorem shows, the Hamiltonian is in fact approximately preserved over very long times for general Hamiltonian systems.

Theorem 5.1. The total energy along a numerical solution (p_n, q_n) of the Störmer–Verlet method satisfies

$$|H(p_n, q_n) - H(p_0, q_0)| \le Ch^2 + C_N h^N t \quad \text{for} \quad 0 \le t = nh \le h^{-N}$$

for arbitrary positive integer N. The constants C and C_N are independent of t and h. C_N depends on bounds of derivatives of H up to $(N+1)$th order in a region that contains the numerical solution values (p_n, q_n).

We give two different proofs of this result, the first one based on the symplecticity, the second one on the symmetry of the method. When the Hamiltonian is analytic, both proofs can be refined to yield an estimate $Ch^2 + C_0 e^{-c/h} t$ over exponentially long times $t \le e^{c/h}$, with c proportional to $1/\Omega$, where Ω is an upper bound of $\|M^{-1/2}\nabla^2 U(q) M^{-1/2}\|^{1/2}$, *i.e.*, of the highest frequency in the linearized system.

The *first proof* uses the symplecticity of the Störmer–Verlet method via backward error analysis, in an argument due to Benettin and Giorgilli (1994). It applies to general symplectic methods for general (smooth) Hamiltonian systems (2.7). We know from Theorem 4.3 that the modified differential equation, truncated after N terms, is again Hamiltonian, with a modified Hamiltonian $\widetilde{H}$ that is $\mathcal{O}(h^2)$ close to the original Hamiltonian H in a neighbourhood of the numerical solution values. Consider now $\widetilde{H}$ along the numerical solution. We write the deviation of $\widetilde{H}$ as a telescoping sum

$$\widetilde{H}(p_n, q_n) - \widetilde{H}(p_0, q_0) = \sum_{j=0}^{n-1} \bigl(\widetilde{H}(p_{j+1}, q_{j+1}) - \widetilde{H}(p_j, q_j)\bigr).$$

By construction of the modified equation, we have for its flow $\widetilde{\varphi}_h(p_j, q_j) = (p_{j+1}, q_{j+1}) + \mathcal{O}(h^{N+1})$. On the other hand, the flow $\widetilde{\varphi}_t$ preserves the modified Hamiltonian, and hence

$$\widetilde{H}(p_{j+1}, q_{j+1}) - \widetilde{H}(p_j, q_j) = \widetilde{H}(p_{j+1}, q_{j+1}) - \widetilde{H}(\widetilde{\varphi}_h(p_j, q_j)) = \mathcal{O}(h^{N+1}).$$

Inserting this estimate in the above sum yields the result.

The *second proof* uses only the symmetry of the Störmer–Verlet method. It was given in Hairer and Lubich (2000*b*) because its arguments extend to numerical energy conservation in oscillatory systems when the product of the step size with the highest frequencies is bounded away from 0 (see Section 5.4). Backward error analysis, or the asymptotic h^2-expansion of the numerical solution, shows that there exists, for every n, a function $q^n(t)$ with $q^n(0) = q_n$ and $q^n(-h) = q_{n-1} + \mathcal{O}(h^{N+1})$ satisfying

$$q^n(t+h) - 2q^n(t) + q^n(t-h) = h^2 f(q^n(t)) + \mathcal{O}(h^{N+2}) \tag{5.1}$$

for t in some fixed interval around 0. The functions $q^n(t+h)$ and $q^{n+1}(t)$ agree up to $\mathcal{O}(h^{N+1})$, as do their kth derivatives multiplied with h^k, for $k \le N$. By Taylor expansion in (5.1),

$$\sum_{l=1}^{N/2} \frac{2}{(2l)!} \frac{\mathrm{d}^{2l} q^n}{\mathrm{d}t^{2l}}(t)\, h^{2l-2} = f(q^n(t)) + \mathcal{O}(h^N). \tag{5.2}$$

Because of the symmetry of the method, only even-order derivatives of $q^n(t)$ (and even powers of the step size) are present in (5.2).

We multiply (5.2) with $\dot{q}^n(t)^T M$ and integrate over t. The key observation is now that the product of $\dot{q}^n(t)$ with an *even*-order derivative of $q^n(t)$ is a total differential (we omit the superscript n in the following formula):

$$\dot{q}^T M q^{(2l)} = \frac{\mathrm{d}}{\mathrm{d}t} \mathcal{A}_l[q]$$

with

$$\mathcal{A}_l[q] = \Big(\dot{q}^T M q^{(2l-1)} - \ddot{q}^T M q^{(2l-2)} + \cdots \mp (q^{(l-1)})^T M q^{(l+1)} \pm \frac{1}{2} (q^{(l)})^T M q^{(l)}\Big).$$

In particular, $\mathcal{A}_1[q] = \frac{1}{2}\dot{q}^T M \dot{q}$. Moreover, for $f(q) = -M^{-1}\nabla U(q)$ we clearly have $\dot{q}^T M f(q) = -(\mathrm{d}/\mathrm{d}t) U(q)$. For the energy functional

$$\mathcal{H}[q](t) = \sum_{l=1}^{N/2} \frac{2}{(2l)!} \mathcal{A}_l[q](t)\, h^{2l-2} + U(q(t))$$

we thus obtain $(\mathrm{d}/\mathrm{d}t)\mathcal{H}[q^n](t) = \mathcal{O}(h^N)$, and hence

$$\mathcal{H}[q^n](h) - \mathcal{H}[q^n](0) = \mathcal{O}(h^{N+1}). \tag{5.3}$$

Since the functions $q^n(t+h)$ and $q^{n+1}(t)$, together with their kth derivatives scaled by h^k $(k \le N)$, are equal up to $\mathcal{O}(h^{N+1})$, we further have

$$\mathcal{H}[q^{n+1}](0) - \mathcal{H}[q^n](h) = \mathcal{O}(h^{N+1}). \tag{5.4}$$

Moreover, with $p^n(t) = M\dot{q}^n(t)$ we have

$$\mathcal{H}[q^n](0) = H(p^n(0), q^n(0)) + \mathcal{O}(h^2) = H(p_n, q_n) + \mathcal{O}(h^2), \tag{5.5}$$

where the last equation follows by noting

$$p_n = M\, \frac{q_{n+1} - q_{n-1}}{2h} = p^n(0) + \mathcal{O}(h^2).$$

Hence, from (5.3)–(5.5),

$$\begin{aligned} H(p_n, q_n) - H(p_0, q_0) &= \mathcal{H}[q^n](0) - \mathcal{H}[q^0](0) + \mathcal{O}(h^2) \\ &= \mathcal{O}(nh^{N+1}) + \mathcal{O}(h^2), \end{aligned}$$

which completes the proof.

5.2. Linear error growth for integrable systems

General Hamiltonian systems may have extremely complicated dynamics, and little can be said about the long-time behaviour of their discretizations apart from the long-time near-conservation of the total energy considered above. At the other end, the simplest conceivable dynamics – uniform motion on a Cartesian product of circles – appears in integrable Hamiltonian systems. Their practical interest lies in the fact that many physical systems are perturbations of integrable systems, with planetary motion as the classical example and historical driving force.

A Hamiltonian system (2.7) is *integrable* if there exists a symplectic transformation

$$(p, q) = \psi(a, \theta) \tag{5.6}$$

to *action-angle variables* (a, θ), defined for actions $a = (a_1, \dots, a_d)$ in some open set of $\mathbb{R}^d$ and for angles θ on the whole d-dimensional torus

$$\mathbb{T}^d = \mathbb{R}^d/(2\pi\mathbb{Z}^d) = \{(\theta_1, \dots, \theta_d);\ \theta_i \in \mathbb{R} \bmod 2\pi\},$$

such that the Hamiltonian in these variables depends only on the actions

$$H(p, q) = H(\psi(a, \theta)) = K(a). \tag{5.7}$$

In the action-angle variables, the equations of motion are simply

$$\dot{a} = 0, \qquad \dot{\theta} = \omega(a), \tag{5.8}$$

with the *frequencies* $\omega = (\omega_1, \dots, \omega_d)^T = \nabla_a K$ (note that $\nabla_\theta K = 0$). This has a quasi-periodic (or possibly periodic) flow:

$$\varphi_t(a, \theta) = (a,\ \theta + \omega(a)t). \tag{5.9}$$

For every a, the torus $\{(a, \theta) : \theta \in \mathbb{T}^d\}$ is thus invariant under the flow. We express the actions and angles in terms of the original variables (p, q) via the inverse transform of (5.6) as

$$(a, \theta) = (I(p, q), \Theta(p, q)),$$

and note that the components of $I = (I_1, \dots, I_d)$ are first integrals of the integrable system.

Integrability of a Hamiltonian system is an exceptional property: the system has d independent first integrals $I_1, \dots, I_d$ whose Poisson brackets vanish pairwise, that is,

$$\{I_i, I_j\} = \nabla_q I_i^T \nabla_p I_j - \nabla_p I_i^T \nabla_q I_j = 0 \quad \text{for all } i, j.$$

The solution trajectories of the Hamiltonian systems with Hamiltonian I_i exist for all time (in the action-angle variables, their flow is simply $\varphi_t^{[i]}(a, \theta) =$

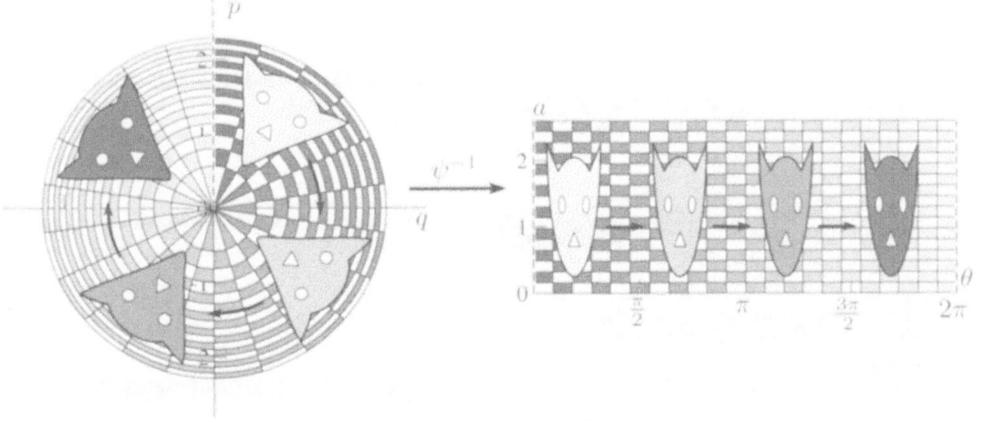

Figure 5.1. Transformation to action-angle variables.

$(a, \theta + te_i)$ with e_i denoting the ith unit vector of $\mathbb{R}^d$), and the level sets of I are compact (the invariant tori $\{a = \text{Const}\, \theta \in \mathbb{T}^d\}$). Conversely, the *Arnold–Liouville theorem* (Arnold 1963) states that every Hamiltonian system having d first integrals with the above properties can be transformed to action-angle variables with a Hamiltonian depending only on the actions.

Example 5.2. The harmonic oscillator $H(p,q) = \frac{1}{2}p^2 + \frac{1}{2}q^2$ is integrable, with the transformation to action-angle coordinates given by

$$\begin{pmatrix} p \\ q \end{pmatrix} = \begin{pmatrix} \sqrt{2a}\,\cos\theta \\ \sqrt{2a}\,\sin\theta \end{pmatrix},$$

with $a = H(p,q)$: see Figure 5.1. Here, the action-angle coordinates are symplectic polar coordinates.

Example 5.3. The Kepler problem, with $H(p,q) = \frac{1}{2}(p_1^2+p_2^2)-(q_1^2+q_2^2)^{-\frac{1}{2}}$ in the range $H < 0$, is integrable with actions $a_1 = 1/\sqrt{-2H}$ and $a_2 = L$ (the angular momentum, $L = q_1p_2 - q_2p_1$). The frequencies are $\omega_1 = \omega_2 = 2\pi/T$, where $T = 2\pi/(-2H)^{3/2}$ is the period of a trajectory with total energy H.

Example 5.4. A further celebrated example of an integrable system is the Toda lattice (Toda 1970, Flaschka 1974), which describes a system of particles on a line interacting with exponential forces. The Hamiltonian is

$$H(p,q) = \sum_{k=1}^{d} \left(\frac{1}{2}p_k^2 + \exp(q_k - q_{k+1}) \right)$$

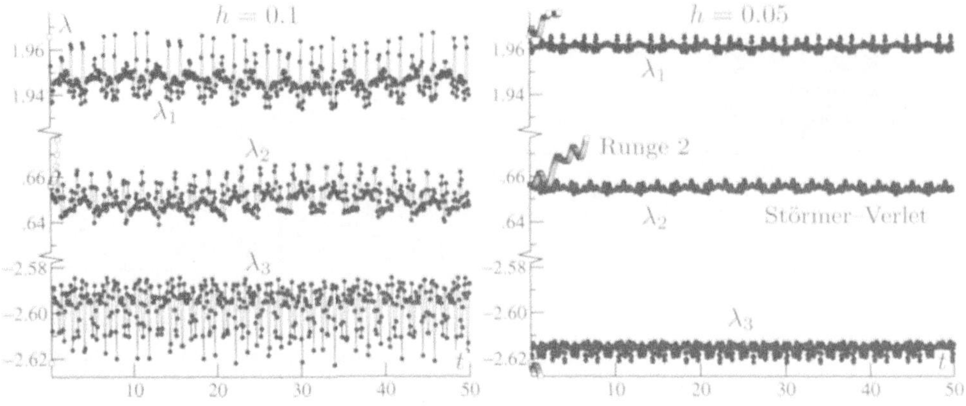

Figure 5.2. Toda eigenvalues along the numerical solution.

with periodic extension $q_{d+1} = q_1$. The eigenvalues of the matrix

$$L = \begin{pmatrix} a_1 & b_1 & & & b_d \\ b_1 & a_2 & b_2 & 0 & \\ & b_2 & \ddots & \ddots & \\ & 0 & \ddots & a_{d-1} & b_{d-1} \\ b_d & & & b_{d-1} & a_d \end{pmatrix}, \qquad \begin{aligned} a_k &= -\tfrac{1}{2}p_k, \\ b_k &= \tfrac{1}{2}\exp\bigl(\tfrac{1}{2}(q_k - q_{k+1})\bigr) \end{aligned}$$

are first integrals whose Poisson brackets vanish pairwise.

We consider the case $d = 3$ and choose initial values $q_0 = (1, 2, -1)^T$ and $p_0 = (-1.5, 1, 0.5)^T$. Figure 5.2 shows the eigenvalues of L along the numerical solution of the Störmer–Verlet and the second-order Runge method obtained with step sizes $h = 0.1$ (left) and $h = 0.05$ (right) on the interval $0 \le t \le 50$. Not only the Hamiltonian (Theorem 5.1), but all d first integrals of the integrable system are well approximated over long times with an error of size $\mathcal{O}(h^2)$. This is explained by Theorem 5.5 below.

The global error in (p, q) is plotted in Figure 5.3. We observe a linear error growth for the Störmer–Verlet method, in contrast to a quadratic error growth for the second-order Runge method.

The study of the error behaviour of the numerical method combines *backward error analysis*, by means of which the numerical map is interpreted as being essentially the time-h flow of a modified Hamiltonian system, and the *perturbation theory* of integrable systems, a rich mathematical theory originally developed for problems of celestial mechanics (Poincaré 1892/1893/1899, Siegel and Moser 1971). The effect of a small perturbation

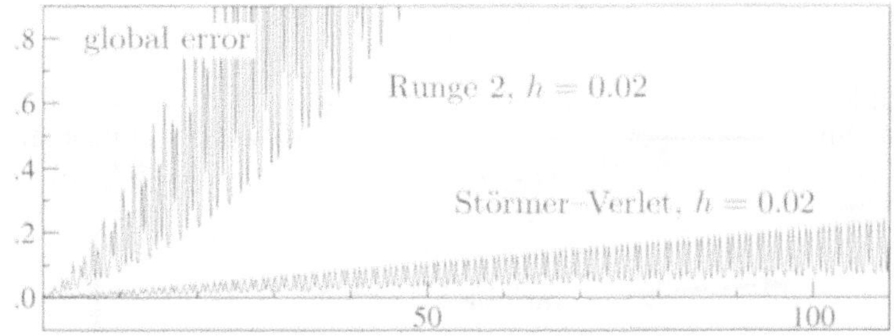

Figure 5.3. Global error of the Störmer–Verlet and the second-order Runge method on the Toda lattice.

of an integrable system is well under control in subsets of the phase space where the frequencies ω satisfy Siegel's *diophantine condition*:

$$|k \cdot \omega| \geq \gamma |k|^{-\nu} \quad \text{for all} \quad k \in \mathbb{Z}^d \tag{5.10}$$

for some positive constants γ and ν, with $|k| = \sum_i k_i$. For $\nu > d-1$, almost all frequencies (in the sense of the Lebesgue measure) satisfy (5.10) for some $\gamma > 0$. For any choice of γ and ν the complementary set is, however, open and dense in $\mathbb{R}^d$.

For general numerical integrators applied to integrable systems (or perturbations thereof) the error grows quadratically with time, and there is a linear drift away from the first integrals I_i. For symplectic methods such as the Störmer–Verlet method there is linear error growth and long-time near-preservation of the first integrals I_i, as is shown by the following result from Hairer *et al.* (2002, Section X.3).

Theorem 5.5. Consider applying the Störmer–Verlet method to an integrable system (2.7) with real-analytic Hamiltonian. Suppose that $\omega^* \in \mathbb{R}^d$ satisfies the diophantine condition (5.10). Then there exist positive constants C, c and h_0 such that the following holds for all step sizes $h \leq h_0$: every numerical solution (p_n, q_n) starting with frequencies $\omega_0 = \omega(I(p_0, q_0))$ such that $\|\omega_0 - \omega^*\| \leq c|\log h|^{-\nu-1}$, satisfies

$$\begin{aligned} \|(p_n, q_n) - (p(t), q(t))\| &\leq C\, t\, h^2, \\ \|I(p_n, q_n) - I(p_0, q_0)\| &\leq C\, h^2, \end{aligned} \qquad \text{for } t = nh \leq h^{-2}.$$

The constants h_0, c, C depend on d, γ, ν and on bounds of the Hamiltonian.

The basic steps of the proof are summarized in Figure 5.4. By backward error analysis, the numerical method coincides, up to arbitrary order in h, with the flow of the modified differential equation, which is a Hamiltonian

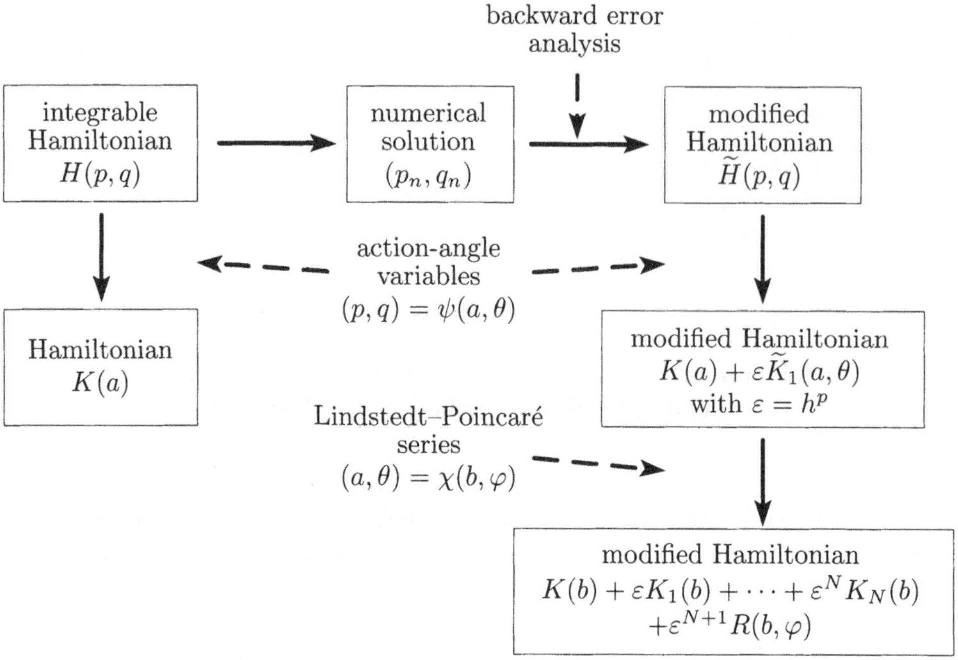

Figure 5.4. Transformations in the proof of Theorem 5.5.

perturbation of size $\varepsilon = h^2$ of the original, integrable system. We are thus in the realm of classical perturbation theory. In addition to the transformation to the action-angle variables (a,θ), which gives the modified Hamiltonian in the form $K(a)+\varepsilon\widetilde{K}_1(a,\theta)$, we use a further symplectic coordinate transformation $(a,\theta)=\chi(b,\varphi)$ which eliminates, up to high-order terms in ε, the dependence on the angles in the modified Hamiltonian. This transformation is $\mathcal{O}(\varepsilon)$ close to the identity. It is constructed as

$$b = a - \nabla_\theta S(b,\theta), \qquad \varphi = \theta + \nabla_b S(b,\theta),$$

where the generating function $S(b,\theta)$ is given by a Lindstedt–Poincaré perturbation series,

$$S(b,\theta) = \varepsilon S_1(b,\theta) + \varepsilon^2 S_2(b,\theta) + \cdots + \varepsilon^N S_N(b,\theta).$$

The error propagation is then studied in the (b,φ)-variables, with the result

$$\begin{aligned} \|b(t)-b_0\| &\le C\,t\,\varepsilon^{N+1}, \\ \|\varphi(t)-\varphi_0-\omega_\varepsilon(b_0)\,t\| &\le C\,(t+t^2)\,\varepsilon^N, \end{aligned} \qquad \text{for } t^2 \le 1/\varepsilon^N,$$

with $\omega_\varepsilon(b)=\omega(b)+\mathcal{O}(\varepsilon)$. Transforming back to the original variables (p,q) finally yields the stated result.

Theorem 5.5 admits extensions in several directions. It is just one of a series of results on the long-time behaviour of geometric integrators.

- The theorem does not apply directly to the Kepler problem, which has two identical frequencies $\omega_1 = \omega_2 = (-2H)^{3/2}$. However, since the angular momentum $a_2 = L$ is preserved exactly by the Störmer–Verlet method, it turns out that the modified Hamiltonian written in the action-angle variables of the Kepler problem is independent of the angle θ_2. Only the angle θ_1 must therefore be eliminated via the perturbation series, and this involves only the single frequency ω_1 for which the diophantine condition is trivially satisfied. The proof and result of Theorem 5.5 thus extend to the Kepler problem.
- The linear error growth remains intact when the method is applied to *perturbed integrable systems* $H(p,q) + \varepsilon G(p,q)$ with a perturbation parameter of size $\varepsilon = \mathcal{O}(h^\alpha)$ for some positive exponent α.
- Under stronger conditions on the initial values or on the system, the near-preservation of the action variables along the numerical solution holds for times that are exponentially long in a negative power of the step size (Hairer and Lubich 1997, Moan 2002). For a Cantor set of initial values and a Cantor set of step sizes this holds even perpetually, in view of the existence of invariant tori of the numerical integrator close to the invariant tori of the integrable system (Shang 1999, 2000).
- Perturbed integrable systems have KAM tori, *i.e.*, deformations of the invariant tori of the integrable system corresponding to diophantine frequencies ω, which are invariant under the flow of the perturbed system. If the method is applied to such a perturbed integrable system, then the numerical method has tori which are near-invariant over exponentially long times (Hairer and Lubich 1997). For a Cantor set of non-resonant step sizes there are even truly invariant tori on which the numerical one-step map reduces to rotation by $h\omega$ in suitable coordinates (Hairer *et al.* 2002, p. 371).
- There is a completely analogous theory for *integrable reversible systems* (Hairer *et al.* 2002, Chapter XI). These are differential equations with reversible flow (2.2), which are transformed to the form (5.8) by a transformation $(q,v) = (\mu(a,\theta), \nu(a,\theta))$ that preserves reversibility, *i.e.*, μ is odd in θ and ν is even in θ. In that theory, only the reversibility of the numerical method comes into play, not the symplecticity. There is again linear error growth, long-time near-preservation of the action variables, and an abundance of invariant tori.
- For dissipatively perturbed integrable systems, where only one torus survives the perturbation and becomes weakly attractive, the existence of a nearby invariant torus of the numerical method is shown under weak assumptions on the step size in Stoffer (1998) and Hairer and Lubich (1999).

5.3. Statistical behaviour

> The equation of motion of a system of 864 particles interacting through a Lennard-Jones potential has been integrated for various values of the temperature and density, relative, generally, to a fluid state. The equilibrium properties have been calculated and are shown to agree very well with the corresponding properties of argon. L. Verlet (1967)

In molecular dynamics, it is the computation of statistical or thermodynamic quantities, such as temperature, which is of interest, rather than single trajectories. The success of the Störmer–Verlet method in this field lies in the observation that the method is apparently able to reproduce the correct statistical behaviour over long times. Since Verlet (1967), this has been confirmed in countless computational experiments. Backward error analysis gives indications as to why this might be so, but to our knowledge there are as yet no rigorous mathematical results in the literature explaining the favourable statistical behaviour.

In the following we derive a result which is a discrete analogue of the virial theorem of statistical mechanics; *cf.* Abraham and Marsden (1978, p. 243) and Gallavotti (1999, p. 129). It comes as a consequence of the long-time near-conservation of energy. Consider the Poisson bracket $\{F,H\} = \nabla_q F^T \nabla_p H - \nabla_p F^T \nabla_q H$ of an arbitrary differentiable function $F(p,q)$ with the Hamiltonian. Along every solution $(p(t),q(t))$ of the Hamiltonian system we have

$$\{F,H\}(p(t),q(t)) = \frac{\mathrm{d}}{\mathrm{d}t} F(p(t),q(t)),$$

and hence the time average of the Poisson bracket along a solution is

$$\frac{1}{T}\int_0^T \{F,H\}(p(t),q(t))\,\mathrm{d}t = \frac{1}{T}\big(F(p(T),q(T)) - F(p(0),q(0))\big).$$

If F is bounded along the solution, this shows that the average is of size $\mathcal{O}(1/T)$ as $T \to \infty$. In particular, this condition is satisfied if the energy level set $\{(p,q) : H(p,q) = H(p(0),q(0))\}$ is compact.

Example 5.6. For a separable Hamiltonian (2.8) the choice $F(p,q) = p^T q$ yields the virial theorem of Clausius (Gallavotti 1999, p. 129),

$$\lim_{T\to\infty} \frac{1}{T}\int_0^T p(t)^T M^{-1} p(t)\,\mathrm{d}t = \lim_{T\to\infty} \frac{1}{T}\int_0^T q(t)^T \nabla U(q(t))\,\mathrm{d}t,$$

i.e., the time average of twice the kinetic energy equals that of the *virial function* $q^T \nabla U(q)$.

For the numerical discretization there is the following result.

Theorem 5.7. Let $H(p,q)$ be a real-analytic Hamiltonian for which

$$K_\delta = \{(p,q) : |H(p,q) - H_0| \le \delta\} \quad \text{is compact}$$

for some $\delta > 0$. Let $F(p,q)$ be any smooth real-valued function, bounded by μ on K_δ. Then the numerical solution (p_n, q_n) obtained by the Störmer–Verlet method satisfies

$$\left| \frac{1}{N} \sum_{n=0}^{N}{}' \{F,H\}(p_n,q_n) \right| \le \frac{2\mu}{Nh} + Ch^2 \qquad \text{for } Nh \le \mathrm{e}^{c/h} \text{ and } h \le h_0, \tag{5.11}$$

where the prime on the sum indicates that the first and last term are taken with weight $\frac{1}{2}$. The constants $C, c, h_0 > 0$ depend on bounds of H on a complex neighbourhood of K_δ and on bounds of the first three derivatives of F on K_δ, but they are independent of h and $(p_0, q_0) \in K_{\delta/2}$.

In particular, the left-hand side of (5.11) is $\mathcal{O}(h^2)$ for $h^{-2} \le Nh \le \mathrm{e}^{c/h}$.

Proof. By Theorem 5.1 and the remark thereafter, we know that

$$y_n := (p_n, q_n) \in K_\delta \qquad \text{for } nh \le \mathrm{e}^{c/h}.$$

Since $y_{n+1} = \varphi_h(y_n) + \mathcal{O}(h^3)$ and $\{F,H\}(\varphi_t(y)) = \frac{\mathrm{d}}{\mathrm{d}t} F(\varphi_t(y))$, we have

$$\begin{aligned} \frac{h}{2}\{F,H\}(y_n) + \frac{h}{2}\{F,H\}(y_{n+1}) &= \int_0^h \{F,H\}\big(\varphi_t(y_n)\big)\,\mathrm{d}t + \mathcal{O}(h^3) \\ &= F(y_{n+1}) - F(y_n) + \mathcal{O}(h^3). \end{aligned}$$

Hence

$$\frac{1}{N} \sum_{n=0}^{N}{}' \{F,H\}(y_n) = \frac{1}{Nh}\big(F(y_N) - F(y_0)\big) + \mathcal{O}(h^2),$$

which yields the stated estimate. □

Example 5.8. We give a numerical experiment with a small-scale version of Verlet's argon model. It considers N_A atoms interacting by the Lennard-Jones potential

$$V(r) = 4\varepsilon\left(\left(\frac{\sigma}{r}\right)^{12} - \left(\frac{\sigma}{r}\right)^{6}\right).$$

The Hamiltonian of the system is (3.2) with $V_{ij} = V$. We choose $N_A = 7$

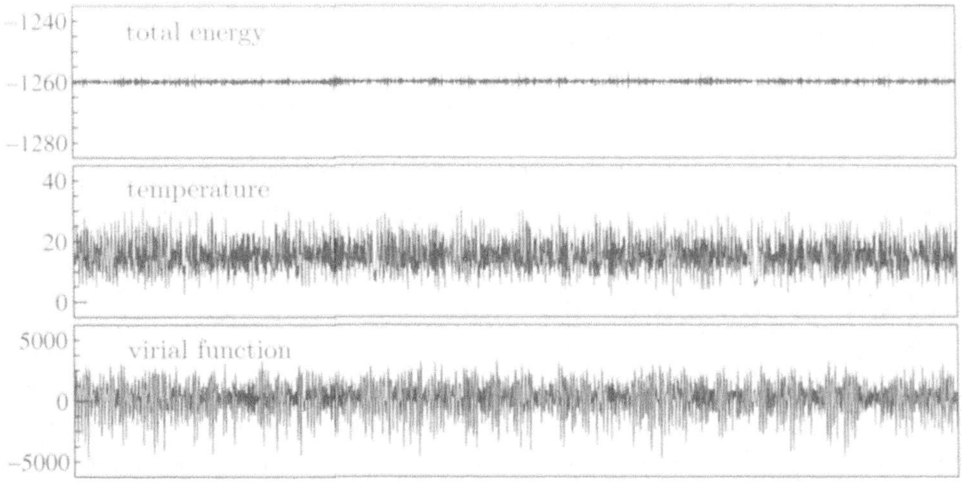

Figure 5.5. Computed total energy, temperature and virial function of the argon crystal, 10 000 steps of size $h = 40$ [fsec].

and the data of Biesiadecki and Skeel (1993); see also Hairer *et al.* (2002, p. 15). Figure 5.5 shows the Hamiltonian, the temperature

$$T(p) = \frac{1}{N_A k_B} \frac{1}{2m} \sum_{i=1}^{N_A} \|p_i\|^2$$

(k_B is Boltzmann's constant), and the virial function

$$C(q) = \sum_{i=2}^{N_A} \sum_{j=1}^{i-1} V'(r_{ij}) r_{ij}$$

(with $r_{ij} = \|q_i - q_j\|$) over an interval of length $4 \cdot 10^5$ [fsec], obtained by the Störmer–Verlet method with step size $h = 40$ [fsec]. The units in the figure are such that $k_B = 1$. The size of the oscillations in the Hamiltonian is proportional to h^2, whereas that in the temperature and in the virial function is independent of h. At the end of the integration (after 10 000 steps) the averages of twice the kinetic energy and of the virial function are 217.1 and 217.8, respectively.

5.4. Oscillatory differential equations

Nonlinear mass-spring models have traditionally been very useful in explaining various phenomena of more complicated 'real' physical systems. We have already mentioned the Toda lattice. An equally famous problem is the Fermi–Pasta–Ulam model (Fermi, Pasta and Ulam 1955, Ford 1992), where a nonlinear perturbation to a primarily linear problem is studied over long times. Here we use a variant of this problem for gaining insight into the long-time energy behaviour of the Störmer–Verlet method applied to oscillatory systems with multiple time scales. We are interested in using step sizes h for which the product with the highest frequency ω in the system is bounded away from zero. (Values of $h\omega \approx 1/2$ are routinely used in molecular dynamics.) In this situation, backward error analysis is no longer applicable, since the 'exponentially small' error terms are then of size $\mathcal{O}(\mathrm{e}^{-c/h\omega}) = \mathcal{O}(1)$.

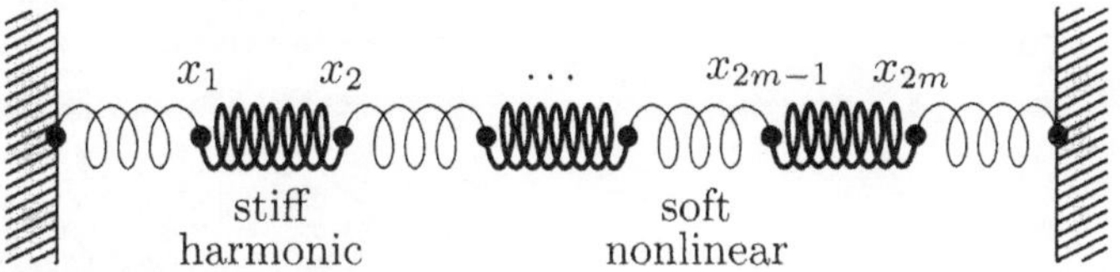

Figure 5.6. Chain with alternating soft nonlinear and stiff linear springs.

Example 5.9. Consider a chain of $2m$ mass points, connected with alternating soft nonlinear and stiff linear springs, and fixed at the end points; see Galgani, Giorgilli, Martinoli and Vanzini (1992) and Figure 5.6. The variables $x_1, \ldots, x_{2m}$ stand for the displacements of the mass points. In terms of the new variables

$$q_i = (x_{2i} + x_{2i-1})/\sqrt{2}, \qquad q_{m+i} = (x_{2i} - x_{2i-1})/\sqrt{2}$$

(which represent a scaled displacement and a scaled expansion/compression of the ith stiff spring) and the momenta $p_i = \dot{q}_i$, the motion is described by a Hamiltonian system with

$$\begin{aligned} H(p,q) = \frac{1}{2}\sum_{i=1}^{2m} p_i^2 + \frac{\omega^2}{2}\sum_{i=1}^{m} q_{m+i}^2 + \frac{1}{4}\Bigg((q_1 - q_{m+1})^4 \\ + \sum_{i=1}^{m-1} (q_{i+1} - q_{m+i+1} - q_i - q_{m+i})^4 + (q_m + q_{2m})^4 \Bigg), \end{aligned}$$

where $\omega \gg 1$ is a large parameter. Here we assume cubic nonlinear springs, but the special form of the nonlinearity is not important.

For an illustration we consider $m = 3$ and choose $\omega = 30$. In Figure 5.7 we have plotted the following quantities as functions of time: the Hamiltonian H (actually we plot $H - 0.8$ for graphical reasons), the oscillatory energy I defined as

$$I = I_1 + I_2 + I_3 \quad \text{with } I_j = \frac{1}{2}p_{m+j}^2 + \frac{1}{2}\omega^2 q_{m+j}^2,$$

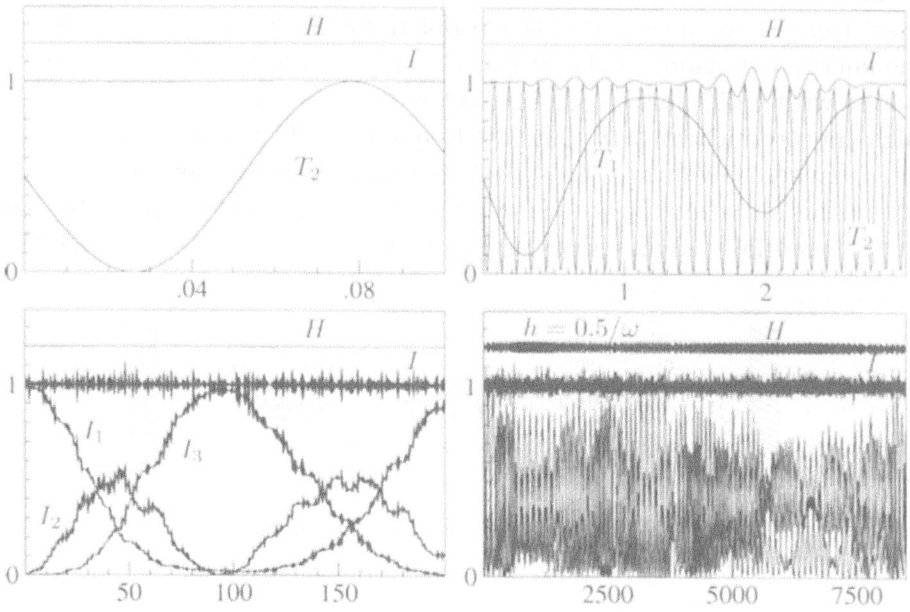

Figure 5.7. Different time scales in a Fermi–Pasta–Ulam problem, and energy conservation of the Störmer–Verlet method (last picture).

and the kinetic energies of the mass centre motion and of the relative motion of masses joined by a stiff spring,

$$T_1 = \frac{1}{2}(p_1^2 + p_2^2 + p_3^2), \qquad T_2 = \frac{1}{2}(p_4^2 + p_5^2 + p_6^2).$$

The system has different dynamics on several time scales: on the fast scale ω^{-1} the motion is nearly harmonic in the stiff springs, on scale ω^0 there is the motion of the soft springs driven by the nonlinearity, and on the slow scale ω there is an energy exchange between the stiff linear springs. For the first three pictures the solutions were computed with high accuracy.

In the last picture we show the results obtained by the Störmer–Verlet method with step size $h = 0.5/\omega$. We note that both H and I are approximately conserved over long times. For fixed ω the size of the oscillations in H is proportional to h^2. However, the oscillations remain of the same size if h decreases and ω increases such that $h\omega$ remains constant. The oscillations in I are of size $\mathcal{O}(\omega^{-1})$ uniformly for $h \to 0$.

The equations of motion for the above example are of the form

$$\ddot{q} = -\Omega^2 q - \nabla U(q) \qquad \text{with } \Omega = \begin{pmatrix} 0 & 0 \\ 0 & \omega I \end{pmatrix}, \tag{5.12}$$

with a single high frequency $\omega \gg 1$ and with a smooth potential $U(q)$ whose derivatives are bounded independently of ω. In addition to the total energy as a conserved quantity,

$$H(p,q) = \frac{1}{2} p^T p + \frac{1}{2} q^T \Omega^2 q + U(q),$$

the system has an adiabatic invariant: over times exponentially long in ω, the *oscillatory energy*

$$I(p,q) = \frac{1}{2} p^T \begin{pmatrix} 0 & 0 \\ 0 & I \end{pmatrix} p + \frac{\omega^2}{2} q^T \begin{pmatrix} 0 & 0 \\ 0 & I \end{pmatrix} q \tag{5.13}$$

is preserved up to $\mathcal{O}(\omega^{-1})$. This holds uniformly for all initial values for which the total energy is bounded by a constant independent of ω, *i.e.*, for bounded (p,q) with $(0\ \ I)q = \mathcal{O}(\omega^{-1})$.

Now consider applying the Störmer–Verlet method to such a system. The step size is then restricted to $h\omega < 2$ for linear stability, as Example 3.4 shows. The Hamiltonian $H(p_n,q_n)$ and the oscillatory energy $I(p_n,q_n)$ of (5.13) oscillate rapidly, but stay within an $\mathcal{O}((h\omega)^2)$ band over long times. The oscillations do not become smaller when h is decreased but ω is increased such that their product $h\omega$ is kept fixed. Nevertheless, the following result shows that the *time averages* of the total and oscillatory energies

$$\overline{H}_n = \frac{h}{T} \sum_{|jh| \le T/2} H(p_{n+j}, q_{n+j}),$$

$$\overline{I}_n = \frac{h}{T} \sum_{|jh| \le T/2} I(p_{n+j}, q_{n+j}),$$

for an arbitrary fixed $T > 0$, remain constant up to $\mathcal{O}(h)$ over long times even when $h\omega$ is bounded away from zero, but within the range of linear stability.

Theorem 5.10. Let the Störmer–Verlet method be applied to the problem (5.12) with a step size h for which $0 < c_0 \le h\omega \le c_1 < 2$. Let $\widetilde{\omega}$ be defined by the relation $\sin(\frac{1}{2}h\widetilde{\omega}) = \frac{1}{2}h\omega$ and suppose $|\sin(\frac{1}{2}kh\widetilde{\omega})| \ge c\sqrt{h}$ for $k = 1, \ldots, N$ for some $N \ge 2$ and $c > 0$. Suppose further that the total energy at the initial value (p_0, q_0) is bounded independently of ω, and that the numerical solution values q_n stay in a region where all derivatives of the potential U are bounded. Then, the time averages of the total and the oscillatory energy along the numerical solution satisfy

$$\begin{aligned} \overline{H}_n &= \overline{H}_0 + \mathcal{O}(h), \\ \overline{I}_n &= \overline{I}_0 + \mathcal{O}(h), \end{aligned} \qquad \text{for } 0 \le nh \le h^{-N+1}.$$

The constants symbolized by $\mathcal{O}$ are independent of n, h, ω with the above conditions.

It should, however, be noted that the time averages $\overline{H}_n$ and $\overline{I}_n$ do not, in general, remain $\mathcal{O}(h)$ close to the initial values $H(p_0, q_0)$ and $I(p_0, q_0)$.

The estimates of Theorem 5.10 can be improved to $\mathcal{O}(h^2)$ if a weighted time average is taken, replacing the characteristic function of the interval $[-T/2, T/2]$ by a smooth windowing function with bounded support, and if the oscillatory energy I is replaced by $J(p,q) = I(p,q) + q^T\left(\begin{smallmatrix}0 & 0\\ 0 & I\end{smallmatrix}\right)\nabla U(q)$, which is preserved up to $\mathcal{O}(\omega^{-2})$ over exponentially long time intervals.

For $h\omega \to 0$, the long-time near-preservation of the adiabatic invariant I can be shown using backward error analysis (Reich 1999*b*), but this argument breaks down for $h\omega$ bounded away from zero as in Theorem 5.10.

We comment only briefly on the proof of Theorem 5.10; see Hairer *et al.* (2002, Chapter XIII) for the full proof. It is based on representing the numerical solution locally (on bounded intervals) by a *modulated Fourier expansion*

$$q_n = \sum_{|k|<N} z_k(t)\mathrm{e}^{\mathrm{i}k\widetilde{\omega}t} + \mathcal{O}(h^N) \qquad \text{for } t = nh,$$

where the coefficients $z_k(t)$ together with all their derivatives (up to some arbitrarily fixed order) are bounded by $\mathcal{O}(\widetilde{\omega}^{-|k|})$. A similar representation holds for p_n. The expansion coefficients $y_k(t) = z_k(t)\mathrm{e}^{\mathrm{i}k\widetilde{\omega}t}$ satisfy a system of equations similar in structure to (5.2) (but of higher dimension). This permits us to use similar arguments to the second proof of Theorem 5.1 to infer the existence of certain modified energies H^* and I^*, which the numerical method preserves up to $\mathcal{O}(h)$ over times h^{-N+1}. Finally, the time averages $\overline{H}_n$ and $\overline{I}_n$ can be expressed, up to $\mathcal{O}(h)$, in terms of these modified energies.

6. Constrained Hamiltonian systems

A minimal set of coordinates of a mechanical system is often difficult to find. The minimal coordinates may be defined only implicitly, or frequent changes of charts are necessary along a solution of the system. In this situation it is favourable to formulate the problem as a Hamiltonian system with constraints.

6.1. Formulation as differential-algebraic equations

We consider a mechanical system with coordinates $q \in \mathbb{R}^d$ that are subject to constraints $g(q) = 0$. The equations of motion are then of the form

$$\begin{aligned} \dot{p} &= -\nabla_q H(p,q) - \nabla_q g(q)\lambda, \\ \dot{q} &= \nabla_p H(p,q), \qquad 0 = g(q), \end{aligned} \tag{6.1}$$

where the Hamiltonian $H(p,q)$ is usually given by (2.8). Here, p and q are vectors in $\mathbb{R}^d$, $g(q) = \big(g_1(q),\dots,g_m(q)\big)^T$ is the vector of constraints, and $\nabla_q g = \big(\nabla_q g_1,\dots,\nabla_q g_m\big)$ is the transposed Jacobian matrix of $g(q)$.

To compute the Lagrange multiplier λ, we differentiate the constraint $0 = g\big(q(t)\big)$ with respect to time. This yields the so-called hidden constraint

$$0 = \nabla_q g(q)^T \nabla_p H(p,q), \tag{6.2}$$

which is an invariant of the flow of (6.1). A further differentiation gives

$$\begin{aligned} 0 = {} & \frac{\partial}{\partial q}\big(\nabla_q g(q)^T \nabla_p H(p,q)\big)\nabla_p H(p,q) \\ & - \nabla_q g(q)^T \nabla_p^2 H(p,q)\big(\nabla_q H(p,q) + \nabla_q g(q)\lambda\big), \end{aligned} \tag{6.3}$$

which allows us to express λ in terms of (p,q), if the matrix

$$\nabla_q g(q)^T \nabla_p^2 H(p,q) \nabla_q g(q) \qquad \text{is invertible} \tag{6.4}$$

($\nabla_p^2 H$ denotes the Hessian matrix of H). Inserting the so-obtained function $\lambda(p,q)$ into (6.1) gives the ordinary differential equation

$$\begin{aligned} \dot p &= -\nabla_q H(p,q) - \nabla_q g(q)\lambda(p,q), \\ \dot q &= \nabla_p H(p,q) \end{aligned} \tag{6.5}$$

for (p,q), which is well defined on the domain where $H(p,q)$ and $g(q)$ are defined, and not only for $g(q) = 0$. The standard theory for ordinary differential equations can be used to deduce existence and uniqueness of the solution. Important properties of the system (6.1) are the following.

- Whenever the initial values satisfy $(p_0,q_0) \in \mathcal{M}$ with

$$\mathcal{M} = \big\{(p,q) \ : \ g(q) = 0,\ \nabla_q g(q)^T \nabla_p H(p,q) = 0\big\}, \tag{6.6}$$

  the solution stays on the manifold $\mathcal{M}$ for all t; hence, the flow of (6.1) is a mapping $\varphi_t : \mathcal{M} \to \mathcal{M}$.

- The flow φ_t is a symplectic transformation on $\mathcal{M}$, which means that

$$(\varphi_t'(p,q)\xi)^T J\, \varphi_t'(p,q)\eta = \xi^T J\,\eta \quad \text{for} \quad \xi,\eta \in T_{(p,q)}\mathcal{M}. \tag{6.7}$$

  Here, $T_{(p,q)}\mathcal{M}$ denotes the tangent space of $\mathcal{M}$ at $(p,q) \in \mathcal{M}$, and the product $\varphi_t'(p,q)\xi$ has to be interpreted as the directional derivative on the manifold.

- For Hamiltonians satisfying

$$H(-p,q) = H(p,q),$$

  the flow φ_t is ρ-reversible for $\rho(p,q) = (-p,q)$ in the sense that (2.5) holds for $(p,q) \in \mathcal{M}$.

The first of these properties follows from the definition of $\lambda(p,q)$. For $(p_0,q_0)\in\mathcal{M}$, a first integration of (6.3) gives (6.2) and a second integration yields $g(q)=0$ along the solution of (6.5).

To prove the symplecticity, we consider the (unconstrained) Hamiltonian system with $K(p,q)=H(p,q)+g(q)^T\lambda(p,q)$. Its flow is symplectic and coincides with that of (6.5) on the manifold $\mathcal{M}$.

The reversibility is a consequence of the fact that $H(-p,q)=H(p,q)$ implies $\lambda(-p,q)=\lambda(p,q)$. The flow of (6.5) and hence also its restriction onto $\mathcal{M}$ is thus ρ-reversible.

Example 6.1. (Kepler and two-body problems on the sphere) Following Kozlov and Harin (1992), we consider a particle moving on the unit sphere attracted by a fixed point a on the sphere. The potential is given as a fundamental solution of the Laplace–Beltrami equation on the sphere:

$$U(q,a)=-\frac{\cos\vartheta}{\sin\vartheta},\qquad \cos\vartheta=\langle q,a\rangle. \tag{6.8}$$

The Kepler problem on the sphere is then of the form (6.1) with

$$H(p,q)=\frac{1}{2}p^Tp+U(q,a),\qquad g(q)=q^Tq-1.$$

The left picture of Figure 6.1 shows the solution corresponding to the point $a=(0.3\sqrt{2},0.3\sqrt{2},0.8)^T$ and to initial values given in spherical coordinates by $\varphi_0=1$, $\theta_0=1.1$ and $\dot\varphi_0=1.2$, $\dot\theta_0=-1.1$. The point a and the initial value are indicated by a larger symbol in Figure 6.1.

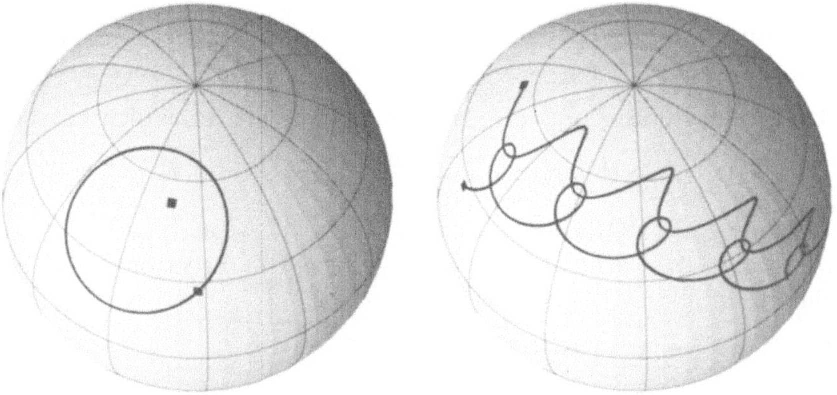

Figure 6.1. Solutions of the Kepler problem (left) and the two-body problem on the sphere (right).

Whereas in Euclidean space the two-body problem reduces to the Kepler problem, this is not the case on the sphere. For the two-body problem the Hamiltonian is

$$H(p_1, p_2, q_1, q_2) = \frac{1}{2}\, p_1^T p_1 + \frac{1}{2}\, p_2^T p_2 + U(q_1, q_2)$$

with $U(q_1, q_2)$ given by (6.8). The constraints are $g_i(q_1, q_2) = q_i^T q_i - 1$ for $i = 1, 2$. The solution with initial values $\varphi_{10} = -0.3$, $\theta_{10} = 1.1$, $\varphi_{20} = -0.8$, $\theta_{20} = 0.6$ and $\dot\varphi_{10} = 0.9$, $\dot\theta_{10} = -0.5$, $\dot\varphi_{20} = 0.3$, $\dot\theta_{20} = -0.1$ is plotted in Figure 6.1 (right).

Example 6.2. (Rigid body) The motion of a rigid body with a fixed point chosen at the origin can be described by an orthogonal matrix $Q(t)$. Letting I_1, I_2, I_3 denote the moments of inertia of the body, its kinetic energy is

$$T = \frac{1}{2}\big(I_1\Omega_1^2 + I_2\Omega_2^2 + I_3\Omega_3^2\big),$$

where the angular velocity $\Omega = (\Omega_1, \Omega_2, \Omega_3)^T$ of the body is defined by

$$\widehat{\Omega} = \begin{pmatrix} 0 & -\Omega_3 & \Omega_2 \\ \Omega_3 & 0 & -\Omega_1 \\ -\Omega_2 & \Omega_1 & 0 \end{pmatrix} = Q^T \dot Q,$$

(Arnold 1989, Chapter 6). In terms of Q, the kinetic energy on the manifold $O(3) = \{Q \,|\, Q^T Q = I\}$ becomes

$$T = \frac{1}{2}\,\mathrm{trace}\,(\widehat{\Omega} D \widehat{\Omega}^T) = \frac{1}{2}\,\mathrm{trace}\,(Q^T \dot Q D \dot Q^T Q) = \frac{1}{2}\,\mathrm{trace}\,(\dot Q D \dot Q^T),$$

where $D = \mathrm{diag}\,(d_1, d_2, d_3)$ is given by the relations $I_1 = d_2 + d_3$, $I_2 = d_3 + d_1$, and $I_3 = d_1 + d_2$. With $P = \partial T/\partial \dot Q = \dot Q D$, we are thus concerned with

$$H(P, Q) = \frac{1}{2}\,\mathrm{trace}\,(P D^{-1} P^T) + U(Q),$$

and the constrained Hamiltonian system becomes

$$\begin{aligned} \dot P &= -\nabla_Q U(Q) - Q\Lambda, \\ \dot Q &= P D^{-1}, \qquad 0 = Q^T Q - I, \end{aligned} \tag{6.9}$$

where Λ is a symmetric matrix consisting of Lagrange multipliers. This is of the form (6.1) and satisfies the regularity condition (6.4).

6.2. Development of the Rattle algorithm

The most important numerical algorithm for the solution of constrained Hamiltonian systems is an adaptation of the Störmer–Verlet method. Its historical development is in three main steps.

First step. For Hamiltonians $H(p,q) = \frac{1}{2} p^T M^{-1} p + U(q)$ with constant mass matrix M (*cf.* Section 2.2), the problem is a second-order differential equation $M\ddot{q} = -\nabla_q U(q) - \nabla_q g(q)\lambda$ with constraint $g(q) = 0$. The most natural extension of (1.2) is

$$\begin{aligned} q_{n+1} - 2q_n + q_{n-1} &= -h^2 M^{-1}\big(\nabla_q U(q_n) + \nabla_q g(q_n)\lambda_n\big), \\ 0 &= g(q_{n+1}). \end{aligned} \tag{6.10}$$

This algorithm (called *Shake*) was originally proposed by Ryckaert, Ciccotti and Berendsen (1977) for computations in molecular dynamics. The p-components, not used in the recursion, are approximated by $p_n = M(q_{n+1} - q_{n-1})/2h$.

Second step. A one-step formulation of this method, obtained by a formal analogy to formula (1.5), reads

$$\begin{aligned} p_{n+1/2} &= p_n - \frac{h}{2}\big(\nabla_q U(q_n) + \nabla_q g(q_n)\lambda_n\big), \\ q_{n+1} &= q_n + hM^{-1}p_{n+1/2}, \qquad 0 = g(q_{n+1}), \\ p_{n+1} &= p_{n+1/2} - \frac{h}{2}\big(\nabla_q U(q_{n+1}) + \nabla_q g(q_{n+1})\lambda_{n+1}\big). \end{aligned} \tag{6.11}$$

This formula cannot be implemented, because λ_{n+1} is not yet available at this step (it is computed together with q_{n+2}). As a remedy, Andersen (1983) suggests replacing the last line in (6.11) with the projection step

$$\begin{aligned} p_{n+1} &= p_{n+1/2} - \frac{h}{2}\big(\nabla_q U(q_{n+1}) + \nabla_q g(q_{n+1})\mu_n\big), \\ 0 &= \nabla_q g(q_{n+1})^T M^{-1} p_{n+1}. \end{aligned} \tag{6.12}$$

This modification, called *Rattle*, is motivated by the fact that the numerical approximation (p_{n+1}, q_{n+1}) lies on the solution manifold $\mathcal{M}$.

Third step. Jay (1994) and Reich (1993) observed independently that the Rattle method can be interpreted as a partitioned Runge–Kutta method and thus allows the extension to general Hamiltonians

$$\begin{aligned} p_{n+1/2} &= p_n - \frac{h}{2}\big(\nabla_q H(p_{n+1/2}, q_n) + \nabla_q g(q_n)\lambda_n\big), \\ q_{n+1} &= q_n + \frac{h}{2}\big(\nabla_p H(p_{n+1/2}, q_n) + \nabla_p H(p_{n+1/2}, q_{n+1})\big), \\ 0 &= g(q_{n+1}), \\ p_{n+1} &= p_{n+1/2} - \frac{h}{2}\big(\nabla_q H(p_{n+1/2}, q_{n+1}) + \nabla_q g(q_{n+1})\mu_n\big), \\ 0 &= \nabla_q g(q_{n+1})^T \nabla_p H(p_{n+1}, q_{n+1}) \end{aligned} \tag{6.13}$$

whenever $(p_n, q_n) \in \mathcal{M}$. The first three equations of (6.13) determine $(p_{n+1/2}, q_{n+1}, \lambda_n)$, whereas the remaining two are equations for (p_{n+1}, μ_n).

For a sufficiently small step size, these equations have a locally unique solution (Hairer *et al.* 2002, p. 214).

Example 6.3. (Kepler problem on the sphere) We apply the Rattle method with a large step size $h = 0.07$ to the problem of Example 6.1. The numerical solution, plotted in Figure 6.2, shows a precession as it appears in computations with symplectic integrators for the Kepler problem in Euclidean space; see Figure 1.5. We remark that the value of the Hamiltonian along the numerical solution oscillates around the correct value and the energy error remains bounded by 0.114 on very long time intervals.

Since the constraint $g(q)$ is quadratic and the Hamiltonian is separable, the formulae (6.13) are explicit with exception of the computation of λ_n, for which a scalar quadratic equation needs to be solved.

Example 6.4. (Rigid body) The Rattle method (6.13) applied to (6.9) yields

$$\begin{aligned} P_{1/2} &= P_0 - \frac{h}{2}\nabla_Q V(Q_0) - \frac{h}{2}Q_0\Lambda_1, \\ Q_1 &= Q_0 + hP_{1/2}D^{-1}, \qquad Q_1^T Q_1 = I, \\ P_1 &= P_{1/2} - \frac{h}{2}\nabla_Q V(Q_1) - \frac{h}{2}Q_1\Lambda_2, \quad D^{-1}P_1^T Q_1 + Q_1^T P_1 D^{-1} = 0, \end{aligned} \tag{6.14}$$

where both Λ_1 and Λ_2 are symmetric matrices. For consistent initial values, Q_0 is orthogonal and $Q_0^T P_0 D^{-1} = \widehat{\Omega}_0$ is skew-symmetric. Working with

$$\widehat{\Omega}_0 = Q_0^T \dot{Q}_0 = Q_0^T P_0 D^{-1}, \quad \widehat{\Omega}_{1/2} = Q_0^T P_{1/2} D^{-1}, \quad \widehat{\Omega}_1 = Q_1^T P_1 D^{-1},$$

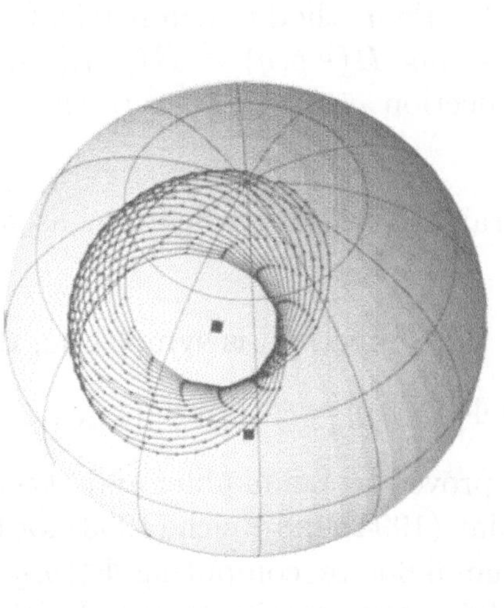

Figure 6.2. Numerical solution of the Kepler problem on the sphere, obtained with the Rattle method using step size $h = 0.07$.

instead of P_0, $P_{1/2}$, P_1, the equations (6.14) become the following integrator $(Q_0, \widehat{\Omega}_0) \mapsto (Q_1, \widehat{\Omega}_1)$:

- find an orthogonal matrix $I + h\widehat{\Omega}_{1/2}$ such that

$$\widehat{\Omega}_{1/2} = \widehat{\Omega}_0 - \frac{h}{2} Q_0^T \nabla_Q V(Q_0) D^{-1} - \frac{h}{2} \Lambda_1 D^{-1}$$

  holds with a symmetric matrix Λ_1;
- compute $Q_1 = Q_0(I + h\widehat{\Omega}_{1/2})$;
- compute a skew-symmetric matrix $\widehat{\Omega}_1$ such that

$$\widehat{\Omega}_1 = \widehat{\Omega}_{1/2} - \frac{h}{2} Q_1^T \nabla_Q V(Q_1) D^{-1} - (\widehat{\Omega}_{1/2} + \widehat{\Omega}_{1/2}^T) - \frac{h}{2} \Lambda_2 D^{-1}$$

  holds with a symmetric matrix Λ_2.

This algorithm for the simulation of the heavy top is proposed in McLachlan and Scovel (1995). An efficient implementation uses the representation of the appearing orthogonal matrices by quaternions (Hairer 2003).

6.3. Geometric properties of Rattle

For consistent initial values $(p_n, q_n) \in \mathcal{M}$, the Rattle method (6.13) yields an approximation (p_{n+1}, q_{n+1}) which is again on $\mathcal{M}$. We thus have a numerical flow $\Phi_h : \mathcal{M} \to \mathcal{M}$. The geometric properties of Section 2 for the Störmer–Verlet method extend to this algorithm.

Theorem 6.5. The Rattle method is symmetric, that is, $\Phi_h = \Phi_{-h}^{-1}$ on $\mathcal{M}$. For Hamiltonians satisfying $H(-p,q) = H(p,q)$, the method is reversible with respect to the reflection $\rho(p,q) = (-p,q)$, that is, it satisfies $\rho \circ \Phi_h = \Phi_h^{-1} \circ \rho$ on $\mathcal{M}$.

The proof is by straightforward verification, as for the Störmer–Verlet method.

Theorem 6.6. The Rattle method is symplectic, that is,

$$(\Phi_h'(p,q)\xi)^T J \, \Phi_h'(p,q)\eta = \xi^T J \, \eta \quad \text{for} \quad \xi, \eta \in T_{(p,q)}\mathcal{M}. \tag{6.15}$$

This result was first proved by Leimkuhler and Skeel (1994) for the method (6.11)–(6.12), and by Jay (1994) and Reich (1993) for the general case (6.13).

One proof of Theorem 6.6 is by computing $\Phi_h'(p,q)\xi$ using implicit differentiation, and by verifying the identity (6.15). Further proofs are based on the interpretation as a variational integrator (Marsden and West 2001), and on explicit formulae of a generating function as in (2.18); see Hairer (2003).

7. Geometric integration beyond Störmer–Verlet

In this article we deliberately considered only the Störmer–Verlet method and a few selected geometric properties. Even within the class of ordinary differential equations, we have not mentioned important topics of geometric integration, such as

- higher-order methods, for instance, symmetric composition, partitioned Runge–Kutta, and linear multistep methods,
- the structure-preserving use of variable step sizes,
- differential equations with further geometric properties such as differential equations on Lie groups, problems with multiple time scales, *etc.*

The reader will find more on these topics in the monographs by Sanz-Serna and Calvo (1994) and Hairer *et al.* (2002), in the special journal issue Budd and Iserles (1999), and in the survey articles by Iserles, Munthe-Kaas, Nørsett and Zanna (2000), Marsden and West (2001) and McLachlan and Quispel (2002).

REFERENCES

R. Abraham and J. E. Marsden (1978), *Foundations of Mechanics*, 2nd edn, Benjamin/Cummings, Reading, MA.

H. C. Andersen (1983), Rattle: A 'velocity' version of the Shake algorithm for molecular dynamics calculations, *J. Comput. Phys.* **52**, 24–34.

V. I. Arnold (1963), Small denominators and problems of stability of motion in classical and celestial mechanics, *Russian Math. Surveys* **18**, 85–191.

V. I. Arnold (1989), *Mathematical Methods of Classical Mechanics*, 2nd edn, Springer, New York.

G. Benettin and A. Giorgilli (1994), On the Hamiltonian interpolation of near to the identity symplectic mappings with application to symplectic integration algorithms, *J. Statist. Phys.* **74**, 1117–1143.

J. J. Biesiadecki and R. D. Skeel (1993), Dangers of multiple time step methods, *J. Comput. Phys.* **109**, 318–328.

P. B. Bochev and C. Scovel (1994), On quadratic invariants and symplectic structure, *BIT* **34**, 337–345.

C. J. Budd and A. Iserles, eds (1999), *Geometric Integration: Numerical Solution of Differential Equations on Manifolds*, special issue of *R. Soc. Lond. Philos. Trans. Ser. A* **357**(1754).

R. De Vogelaere (1956), Methods of integration which preserve the contact transformation property of the Hamiltonian equations, Report No. 4, Department of Mathematics, University of Notre Dame, IN.

K. Feng (1985), On difference schemes and symplectic geometry, in *Proc. Fifth Intern. Symposium on Differential Geometry and Differential Equations, August 1984, Beijing*, pp. 42–58.

K. Feng (1991), Formal power series and numerical algorithms for dynamical systems, in *Proc. International Conference on Scientific Computation, Hangzhou, China* (T, Chan and Z.-C. Shi, eds), Vol. 1 of *Series on Appl. Math.*, pp. 28–35.

K. Feng and Z. Shang (1995), Volume-preserving algorithms for source-free dynamical systems, *Numer. Math.* **71**, 451–463.

E. Fermi, J. Pasta and S. Ulam (1955), Studies of nonlinear problems, Los Alamos Report No. LA-1940. Later published in *E. Fermi: Collected Papers*, Vol. II, Chicago University Press (1965), pp. 978–988.

R. Feynman (1965), *The Character of Physical Law*, first published by the BBC (1965). MIT Press (1967).

H. Flaschka (1974), The Toda lattice, II: Existence of Integrals, *Phys. Rev. B* **9**, 1924–1925.

J. Ford (1992), The Fermi–Pasta–Ulam problem: Paradox turns discovery, *Physics Reports* **213**, 271–310.

L. Galgani, A. Giorgilli, A. Martinoli and S. Vanzini (1992), On the problem of energy equipartition for large systems of the Fermi–Pasta–Ulam type: Analytical and numerical estimates, *Physica D* **59**, 334–348.

G. Gallavotti (1999), *Statistical Mechanics: A Short Treatise*, Springer, Berlin.

E. Hairer (2003), Global modified Hamiltonian for constrained symplectic integrators, *Numer. Math.* To appear.

E. Hairer and Ch. Lubich (1997), The life-span of backward error analysis for numerical integrators, *Numer. Math.* **76**, 441–462.

E. Hairer and Ch. Lubich (1999), Invariant tori of dissipatively perturbed Hamiltonian systems under symplectic discretization, *Appl. Numer. Math.* **29**, 57–71.

E. Hairer and Ch. Lubich (2000*a*), Long-time energy conservation of numerical methods for oscillatory differential equations, *SIAM J. Numer. Anal.* **38**, 414–441.

E. Hairer and Ch. Lubich (2000*b*), Energy conservation by Störmer-type numerical integrators, in *Numerical Analysis 1999* (G. F. Griffiths and G. A. Watson, eds), CRC Press LLC, pp. 169–190.

E. Hairer, S. P. Nørsett and G. Wanner (1993), *Solving Ordinary Differential Equations I: Nonstiff Problems*, Springer, Heidelberg.

E. Hairer, Ch. Lubich and G. Wanner (2002), *Geometric Numerical Integration: Structure-Preserving Algorithms for Ordinary Differential Equations*, Springer, Berlin.

A. Iserles, H. Z. Munthe-Kaas, S. P. Nørsett and A. Zanna (2000), Lie-group methods, in *Acta Numerica*, Vol. 9, Cambridge University Press, pp. 215–365.

L. Jay (1994), Runge–Kutta type methods for index three differential-algebraic equations with applications to Hamiltonian systems, Thesis No. 2658, University of Genève.

V. V. Kozlov and A. O. Harin (1992), Kepler's problem in constant curvature spaces, *Celestial Mech. Dynam. Astronom.* **54**, 393–399.

F. M. Lasagni (1988), Canonical Runge–Kutta methods, *ZAMP* **39**, 952–953.

B. J. Leimkuhler and R. D. Skeel (1994), Symplectic numerical integrators in constrained Hamiltonian systems, *J. Comput. Phys.* **112**, 117–125.

R. MacKay (1992), Some aspects of the dynamics of Hamiltonian systems, in *The Dynamics of Numerics and the Numerics of Dynamics* (D. S. Broomhead and A. Iserles, eds), Clarendon Press, Oxford, pp. 137–193.

S. Maeda (1980), Canonical structure and symmetries for discrete systems, *Math. Japonica* **25**, 405–420.

J. E. Marsden and M. West (2001), Discrete mechanics and variational integrators, in *Acta Numerica*, Vol. 10, Cambridge University Press, pp. 357–514.

R. I. McLachlan and P. Atela (1992), The accuracy of symplectic integrators, *Nonlinearity* **5**, 541–562.

R. I. McLachlan and G. R. W. Quispel (2002), Splitting methods, in *Acta Numerica*, Vol. 11, Cambridge University Press, pp. 341–434.

R. I. McLachlan and C. Scovel (1995), Equivariant constrained symplectic integration, *J. Nonlinear Sci.* **5**, 233–256.

P. C. Moan (2002), On backward error analysis and Nekhoroshev stability in the numerical analysis of conservative systems of ODEs, PhD thesis, University of Cambridge.

J. Moser (1968), Lectures on Hamiltonian systems, *Mem. Amer. Math. Soc.* **81**, 1–60.

H. Poincaré (1892/1893/1899), *Les Méthodes Nouvelles de la Mécanique Céleste, Tome I–III*, Gauthier-Villars, Paris.

S. Reich (1993), Symplectic integration of constrained Hamiltonian systems by Runge–Kutta methods, Technical Report 93-13, Department of Computer Science, University of British Columbia.

S. Reich (1999*a*), Backward error analysis for numerical integrators, *SIAM J. Numer. Anal.* **36**, 1549–1570.

S. Reich (1999*b*), Preservation of adiabatic invariants under symplectic discretization, *Appl. Numer. Math.* **29**, 45–56.

R. D. Ruth (1983), A canonical integration technique, *IEEE Trans. Nuclear Science* **NS-30**, 2669–2671.

J.-P. Ryckaert, G. Ciccotti and H. J. C. Berendsen (1977), Numerical integration of the cartesian equations of motion of a system with constraints: Molecular dynamics of n-alkanes, *J. Comput. Phys.* **23**, 327–341.

J. M. Sanz-Serna (1988), Runge–Kutta schemes for Hamiltonian systems, *BIT* **28**, 877–883.

J. M. Sanz-Serna (1992), Symplectic integrators for Hamiltonian problems: An overview, in *Acta Numerica*, Vol. 1, Cambridge University Press, pp. 243–286.

J. M. Sanz-Serna and M. P. Calvo (1994), *Numerical Hamiltonian Problems*, Chapman and Hall, London.

Z. Shang (1999), KAM theorem of symplectic algorithms for Hamiltonian systems, *Numer. Math.* **83**, 477–496.

Z. Shang (2000), Resonant and diophantine step sizes in computing invariant tori of Hamiltonian systems, *Nonlinearity* **13**, 299–308.

C. L. Siegel and J. K. Moser (1971), Lectures on Celestial Mechanics, Vol. 187 of *Grundlehren der Mathematischen Wissenschaft*, Springer, Heidelberg.

D. Stoffer (1988), On reversible and canonical integration methods, SAM-Report No. 88-05, ETH Zürich.

D. Stoffer (1998), On the qualitative behaviour of symplectic integrators, III: Perturbed integrable systems, *J. Math. Anal. Appl.* **217**, 521–545.

G. Strang (1968), On the construction and comparison of difference schemes, *SIAM J. Numer. Anal.* **5**, 506–517.

Y. B. Suris (1988), On the conservation of the symplectic structure in the numerical solution of Hamiltonian systems, in *Numerical Solution of Ordinary Differential Equations* (S. S. Filippov, ed.), Keldysh Institute of Applied Mathematics, USSR Academy of Sciences, Moscow, pp. 148–160. (In Russian.)

Y.-F. Tang (1994), Formal energy of a symplectic scheme for Hamiltonian systems and its applications, I, *Comput. Math. Appl.* **27**, 31–39.

M. Toda (1970), Waves in nonlinear lattice, *Progr. Theor. Phys. Suppl.* **45**, 174–200.

L. Verlet (1967), Computer 'experiments' on classical fluids, I: Thermodynamical properties of Lennard-Jones molecules, *Phys. Rev.* **159**, 98–103.

A. P. Veselov (1991), Integrable maps, *Russ. Math. Surv.* **46**, 1–51.

H. Yoshida (1993), Recent progress in the theory and application of symplectic integrators, *Celestial Mech. Dynam. Astronom.* **56**, 27–43.

Acta Numerica (2003), pp. 451–512
DOI: 10.1017/S0962492902000156

Entropy stability theory for difference approximations of nonlinear conservation laws and related time-dependent problems

Eitan Tadmor*
Department of Mathematics,
Center for Scientific Computation
and Mathematical Modeling (CSCAMM)
and
Institute for Physical Science & Technology (IPST),
University of Maryland,
College Park, MD 20742, USA
E-mail: tadmor@cscamm.umd.edu

We study the entropy stability of difference approximations to nonlinear hyperbolic conservation laws, and related time-dependent problems governed by additional dissipative and dispersive forcing terms. We employ a comparison principle as the main tool for entropy stability analysis, comparing the entropy production of a given scheme against properly chosen *entropy-conservative* schemes.

To this end, we introduce general families of entropy-conservative schemes, interesting in their own right. The present treatment of such schemes extends our earlier recipe for construction of entropy-conservative schemes, introduced in Tadmor (1987*b*). The new families of entropy-conservative schemes offer two main advantages, namely, (i) their numerical fluxes admit an explicit, closed-form expression, and (ii) by a proper choice of their path of integration in phase space, we can distinguish between different families of waves within the same computational cell; in particular, entropy stability can be enforced on rarefactions while keeping the sharp resolution of shock discontinuities.

A comparison with the numerical viscosities associated with entropy-conservative schemes provides a useful framework for the construction and analysis of entropy-stable schemes. We employ this framework for a detailed study of entropy stability for a host of first- and second-order accurate schemes. The comparison approach yields a precise characterization of the entropy stability of semi-discrete schemes for both scalar problems and systems of equations.

* Research was supported by NSF grants DMS01-07917 and DMS01-07428 and by ONR grant N00014-91-J-1076.

We extend these results to fully discrete schemes. Here, spatial entropy dissipation is balanced by the entropy production due to time discretization with a sufficiently small time-step, satisfying a suitable CFL condition. Finally, we revisit the question of entropy stability for fully discrete schemes using a different approach based on *homotopy* arguments. We prove entropy stability under optimal CFL conditions.

CONTENTS

1. Introduction

We discuss the stability of difference approximations to conservation laws and related time-dependent problems. The related problems we have in mind are governed by additional dissipative and dispersive forcing terms. Our main focus, however, is devoted to nonlinear convection governed by hyperbolic systems of conservation laws. In the linear hyperbolic framework, L^2-stability is sought as a discrete analogue for the *a priori* energy estimates available in the differential set-up, *e.g.*, Richtmyer and Morton (1967) and Gustafsson, Kreiss and Oliger (1995); consult the recent *Acta Numerica* review by Kreiss and Lorenz (1998). In the present context of nonlinear problems dominated by nonlinear convection, we seek *entropy stability* as a discrete analogue for the corresponding statement in the differential set-up. The prototype one-dimensional problem consists of systems of conservation laws, $\mathbf{u}_t+\mathbf{f}(\mathbf{u})_x = 0$. A distinctive feature of this problem is the spontaneous formation of shock discontinuities. The entropy condition plays a decisive role in the theory and numerics of such problems (Lax 1972, Smoller 1983, Dafermos 2000). It requires $\mathbf{u}$ to satisfy the additional inequality, $U(\mathbf{u})_t + F(\mathbf{u})_x \leq 0$, for all admissible entropy pairs $(U(\mathbf{u}), F(\mathbf{u}))$. It follows that the total amount of entropy, $\int U(\mathbf{u}(\cdot,t)\,\mathrm{d}x$, does not increase in time. This is a generalization of the (weighted) L^2-energy bound encountered in the linear case. The possibility of strict inequality reflects entropy decay due to concentration along shock discontinuities.

We consider difference approximations of the general conservative form,

$$\frac{\mathrm{d}}{\mathrm{d}t}\mathbf{u}_\nu(t) + \frac{\mathbf{f}_{\nu+\frac{1}{2}} - \mathbf{f}_{\nu-\frac{1}{2}}}{\Delta x_\nu} = 0.$$

Here $\mathbf{u}_\nu(t)$ is the numerical solution computed at discrete grid lines (x_ν, t), and $\mathbf{f}_{\nu+\frac{1}{2}} \sim \mathbf{f}$ is a numerical flux based on a stencil of neighbouring grid values, $\mathbf{u}_{\nu-p+1}, \ldots, \mathbf{u}_{\nu+p}$. We enquire when such schemes are entropy-stable in the sense of satisfying the corresponding discrete entropy inequality,

$$\frac{\mathrm{d}}{\mathrm{d}t}U(\mathbf{u}_\nu(t)) + \frac{F_{\nu+\frac{1}{2}} - F_{\nu-\frac{1}{2}}}{\Delta x_\nu} \leq 0.$$

So far we have specified semi-discrete schemes based on spatial differencing. We will address the question of entropy stability for the semi-discrete as well as the fully discrete case, taking into account additional temporal discretization. The extension to the multidimensional set-up and a host of related problems with additional dissipative and dispersive terms can be handled in a straightforward manner.

We distinguish between three main tools of the trade in the analysis of entropy stability: comparison arguments, a homotopy approach and kinetic formulations. We will discuss the first two and refer the reader to Bouchut (2002), Makridakis and Perthame (2003) and the references therein for recent contributions regarding the third. Most of our discussion will be devoted to the main approach, based on a *comparison principle*: we compare the amount of entropy dissipation produced by a given scheme against a properly chosen entropy-stable reference. The entropy stability of solutions to monotone schemes, for example, Harten, Hyman and Lax (1976), is carried out by a comparison with the (entropy-stable) constant solution (Crandall and Majda 1980). The class of entropy-stable E-schemes (Osher 1984) is characterized by having more numerical viscosity than the entropy-stable Godunov scheme (Tadmor 1984*b*). And we mention in passing the kinetic approach presented in Makridakis and Perthame (2003), which is based on comparison of the corresponding pseudo-Maxwellians.

In Tadmor (1987*b*), the question of entropy stability was addressed by the construction of certain *entropy-conservative* schemes, interesting for their own sake. We begin, in Section 3, with the construction of these entropy-conservative schemes. There are two main ingredients: (i) the use of entropy variables, outlined in Section 2, and (ii) the choice of certain paths of integration in phase space of these entropy variables. In the scalar case, the numerical fluxes are path-independent, and entropy-conservative schemes are unique (for a given entropy pair). In Section 4 we study a host of instructive scalar examples whose entropy stability is verified by comparison with entropy-conservative ones. These include the first-order Engquist–Osher, the optimality of the Godunov scheme and the second-order Lax–Wendroff, as well as other centred schemes. In Section 5 we turn our attention to systems. Here we revisit the construction of entropy-conservative schemes in terms of numerical fluxes which are integrated along straight line paths in phase space. A comparison of numerical viscosities provides a detailed study of entropy stability for Rusanov, Lax–Friedrichs and the family of Roe-type schemes, as well as second-order extensions. In Section 6 we present the general framework, introducing new families of entropy-conservative fluxes subject to the choice of path of integration in phase space. These new entropy-conservative schemes offer two main advantages: (i) their numerical fluxes admit an explicit, closed-form expression, and, more importantly, (ii) by a proper choice of the path of integration (aligned with the eigen-directions of $\mathbf{f_u}$), one can distinguish between different families of waves within the same cell, $[x_\nu, x_{\nu+1}]$. In particular, entropy stability can be enforced on rarefactions while keeping the sharp resolution of shock discontinuities. In Section 7 we extend our discussion to fully discrete schemes,

$$\mathbf{u}_\nu^{n+1} \equiv \mathbf{u}_\nu(t^n + \Delta t) = \mathbf{u}_\nu^n - \frac{\Delta t}{\Delta x_\nu}\Big[\mathbf{f}_{\nu+\frac{1}{2}}(\overline{\mathbf{u}}^{n+\frac{1}{2}}) - \mathbf{f}_{\nu-\frac{1}{2}}(\overline{\mathbf{u}}^{n+\frac{1}{2}})\Big].$$

There are three prototype examples. In the fully implicit case where we set $\overline{\mathbf{u}}^{n+\frac{1}{2}} := \mathbf{u}^{n+1}$, additional entropy dissipation is introduced by the time discretization and hence this implicit backward Euler scheme is entropy-stable whenever the semi-discrete scheme is. In the case of Crank–Nicolson time discretization, a proper (possibly nonlinear) choice of intermediate values $\overline{\mathbf{u}}^{n+\frac{1}{2}}$ inherits the same unconditional entropy stability properties of the semi-discrete problem associated with the numerical flux $\mathbf{f}_{\nu+\frac{1}{2}}$; finally, the fully explicit case, $\overline{\mathbf{u}}^{n+\frac{1}{2}} := \mathbf{u}^n$, yields entropy production which needs to be balanced by entropy dissipation on the spatial part. This balance is achieved for a mesh ratio satisfying a suitable Courant–Friedrichs–Lewy (CFL) condition, $\frac{\Delta t}{\Delta x}\|\mathbf{f_u}\| \leq \text{Const}$.

In Section 8 we revisit the question of entropy stability using a completely different approach, based on *homotopy* arguments. The results apply to semi- and fully discrete approximations of scalar and systems of conservation laws. We prove the entropy stability for a large class of first-order schemes, this time under an optimal CFL condition. For second-order scalar extensions we refer to Nessyahu and Tadmor (1990, Appendix). The homotopy argument was introduced by Lax (1971) in the context of the Lax–Friedrichs scheme.

The entropy stability study is based on comparison with entropy-conservative schemes. The entropy-stable schemes discussed so far were limited by the use of second-order accurate entropy-conservative schemes as a reference for a comparison. We conclude, in Section 9, with higher-order extensions. We recast the original entropy-conservative schemes in their piecewise linear finite element formulation (Tadmor 1986*b*). Higher orders with larger stencils follow from piecewise polynomials of higher degrees. A general framework for such high-order entropy-conservative schemes was recently introduced in LeFloch, Mercier and Rohde (2002), and should serve as the starting point for the corresponding higher-order entropy stability analysis.

Our discussion on entropy stability theory is tied to several topics which we were unable to explore in the present framework, and we conclude this Introduction by mentioning a few of the items that were left out.

- *Entropy-conservative schemes* play an essential role in our discussion below as the main reference for calibrating entropy stability. Entropy-conservative schemes are interesting in their own right in the context of zero dispersion limits, and completely integrable systems (consult, for example, Lax, Levermore and Venakidis (1993) and Deift and McLaughlin (1998)), with much recent renewed interest (Abramov, Kovačič and Majda 2003, Abramov and Majda 2002). Entropy-conservative schemes are also sought in the context of energy conservation for long-term shock-free integration: for example, Arakawa (1966). Let us mention the related class of completely conservative schemes developed by the school of A. A. Samarskii and co-workers: consult, for example, Moskalkov (1980) and the references therein.
- *Entropy stability* serves as an essential guideline in the design of new computationally reliable difference schemes. Much of our discussion below is devoted to the development of a general framework for proving the entropy stability of such schemes. As an alternative approach, we mention the design of entropy *corrections* for existing schemes. For early numerical simulations with entropy corrections along these lines, we refer to Khalfallah and Lerat (1988) and Kaddouri (1993), for example. The new class of entropy-conservative/entropy-stable schemes explored in Section 6 offers a challenging new set-up for revisiting numerical simulations with entropy corrected schemes.
- *Entropy variables* are essential for symmetrization, and hence for the sense of ordering required for the comparison approach in verifying entropy stability. Entropy variables are essential for the weak finite element formulation as briefly outlined in Section 9. We refer to the streamline diffusion of Hughes, Johnson and collaborators (Hughes, Franca and Mallet 1986, Johnson and Szepessy 1986) as an example of a successful class of entropy variables-based finite element methods (FEM) for treating convection-dominated problems.

- *Compensated compactness.* Quantifying the amount of entropy dissipation (consult Corollary 5.1 below) enables us to convert the entropy stability statement into a convergence proof by compensated compactness arguments (Tartar 1975, DiPerna 1983, Chen 2000). This methodology was applied to different classes of discrete methods: for instance, FEM streamline diffusion (Johnson, Szepessy and Hansbo 1990), the spectral viscosity method (Tadmor 1989), and multidimensional finite volume methods (LeVeque 2002). The present framework should pave the way for a systematic development of a convergence theory for a large class of entropy-stable finite difference approximations (for scalar and 2×2 systems).
- *Nonclassical shocks* and a host of nonlinear phenomena are governed by a borderline balance between dissipative and dispersive forces: we refer, for example, to the recent phase transitions studies of LeFloch and co-workers (LeFloch 2002). The numerical simulation in those regimes becomes possible by carefully tuning the amount of entropy dissipation/dispersion added to the entropy-conservative schemes.
- *Boundary conditions.* Once the entropy-conservative schemes are introduced, the question of entropy stability is answered by summation by parts, carried out in phase space of entropy variables. This reveals the skew-symmetry of the spatial operators (Tadmor 1984*a*), while retaining the conservative form. Consequently, summation by parts along these lines should in principle enable us to treat the question of entropy stability in the presence of boundaries: consult Olsson (1995), for example.

2. The entropy variables

We consider systems of conservation laws of the form

$$\frac{\partial}{\partial t}\mathbf{u} + \frac{\partial}{\partial x}\mathbf{f}(\mathbf{u}) = 0, \quad (x,t) \in \mathbb{R} \times [0,\infty), \tag{2.1}$$

where $\mathbf{f}(\mathbf{u}) = (f_1(\mathbf{u}), \ldots, f_N(\mathbf{u}))^\top$ are smooth flux functions of the N-vector of conservative variables[1] $\mathbf{u}(x,t) = (u_1(x,t), \ldots, u_N(x,t))^\top$. We assume that system (2.1) is equipped with a convex *entropy function*, $U(\mathbf{u})$, such that

$$U_{\mathbf{uu}}A = [U_{\mathbf{uu}}A]^\top, \quad A(\mathbf{u}) := \mathbf{f}_{\mathbf{u}}(\mathbf{u}). \tag{2.2}$$

Thus, the Hessian of an entropy function symmetrizes the system (2.1) upon multiplication 'on the left' (Friedrichs and Lax 1971). An alternative

[1] Here and below, scalars are distinguished from vectors, which are denoted by **bold** letters.

procedure, which respects both strong and weak solutions of (2.1), is to symmetrize 'on the right', where (2.2) is replaced by the equivalent statement

$$A(U_{\mathbf{uu}})^{-1} = \left[A(U_{\mathbf{uu}})^{-1}\right]^\top. \tag{2.3}$$

To this end, Mock (1980) (see also Godunov (1961)) suggested the following procedure. Define the *entropy variables*

$$\mathbf{v} \equiv \mathbf{v}(\mathbf{u}) := \nabla_{\mathbf{u}} U(\mathbf{u}). \tag{2.4}$$

Thanks to the convexity of $U(\mathbf{u})$, the mapping $\mathbf{u} \to \mathbf{v}$ is one-to-one and hence we can make the change of variables $\mathbf{u} = \mathbf{u}(\mathbf{v})$, which puts the system (2.1) into its equivalent symmetric form

$$\frac{\partial}{\partial t}\mathbf{u}(\mathbf{v}) + \frac{\partial}{\partial x}\mathbf{g}(\mathbf{v}) = 0, \quad \mathbf{g}(\mathbf{v}) := \mathbf{f}(\mathbf{u}(\mathbf{v})). \tag{2.5}$$

Here, $\mathbf{u}(\cdot)$ and $\mathbf{g}(\cdot)$ become the temporal and spatial fluxes in the independent entropy variables, $\mathbf{v}$, and the system (2.5) is symmetric in the sense that the Jacobians of these fluxes are, namely

$$H(\mathbf{v}) := \mathbf{u}_{\mathbf{v}}(\mathbf{v}) = H^\top(\mathbf{v}) > 0 \quad \text{and} \quad B(\mathbf{v}) := \mathbf{g}_{\mathbf{v}}(\mathbf{v}) = B^\top(\mathbf{v}). \tag{2.6}$$

Indeed, (2.2) holds if and only if there exists an entropy flux function, $F = F(\mathbf{u})$, such that the following compatibility relation holds:

$$U_{\mathbf{u}}^\top \mathbf{f}_{\mathbf{u}} = F_{\mathbf{u}}^\top. \tag{2.7}$$

Consequently, we have

$$\mathbf{u}(\mathbf{v}) = \nabla_{\mathbf{v}}\phi(\mathbf{v}), \quad \phi(\mathbf{v}) := \langle \mathbf{v}, \mathbf{u}(\mathbf{v})\rangle - U(\mathbf{u}(\mathbf{v})) \tag{2.8}$$

$$\mathbf{g}(\mathbf{v}) = \nabla_{\mathbf{v}}\psi(\mathbf{v}), \quad \psi(\mathbf{v}) := \langle \mathbf{v}, \mathbf{g}(\mathbf{v})\rangle - F(\mathbf{u}(\mathbf{v})), \tag{2.9}$$

where $\langle\cdot,\cdot\rangle$ denotes the usual Euclidean inner product. Hence the Jacobians $H(\mathbf{v})$ and $B(\mathbf{v})$ in (2.6) are symmetric, being the Hessians of $\phi(\mathbf{v})$ and $\psi(\mathbf{v})$. The latter, so-called potential functions, $\phi(\mathbf{v})$ and $\psi(\mathbf{v})$, are significant tools in our discussion below. Observe that the symmetry of $B = AH$ amounts to the symmetrization 'on the right' indicated in (2.3).

Entropy functions play an important role in the stability theory of PDEs dominated by the nonlinear convection of the type (2.1). We provide below a brief overview and refer the reader to a detailed account in Volpert (1967), Kružkov (1970), Friedrichs and Lax (1971), Lax (1972), Tartar (1975), DiPerna (1983), Smoller (1983), Majda (1984), Serre (1999), Dafermos (2000) and LeFloch (2002). We first recall that 'physically relevant' solutions of (2.1), are those arising as vanishing viscosity limits, $\mathbf{u} = \lim_{\epsilon\downarrow 0}\mathbf{u}^\epsilon$, where

$$\mathbf{u}_t^\epsilon + \mathbf{f}(\mathbf{u}^\epsilon)_x = \epsilon(P\mathbf{u}_x^\epsilon)_x. \tag{2.10}$$

Here $P = P(\mathbf{u}, \mathbf{u}_x)$ is any admissible viscosity matrix which is H-symmetric (compare (2.3)), that is,

$$PH = \big[PH\big]^\top \geq 0, \quad H = (U_{\mathbf{uu}})^{-1}, \tag{2.11}$$

so that integration of (2.10) against $U_{\mathbf{u}}^\top$ yields

$$\begin{aligned}\frac{\partial}{\partial t}U(\mathbf{u}^\epsilon) + \frac{\partial}{\partial x}F(\mathbf{u}^\epsilon) &= \left\langle U_{\mathbf{u}}^\top, \frac{\partial}{\partial t}\mathbf{u} + \frac{\partial}{\partial x}\mathbf{f}(\mathbf{u})\right\rangle \\ &= -\epsilon\langle U_{\mathbf{uu}}\mathbf{u}_x^\epsilon, P\mathbf{u}_x^\epsilon\rangle \\ &= -\epsilon\big\langle(H^{-1}\mathbf{u}_x^\epsilon), PH(H^{-1}\mathbf{u}_x^\epsilon)\big\rangle \leq 0. \end{aligned} \tag{2.12}$$

Passing to the limit we obtain the *entropy inequality*

$$\frac{\partial}{\partial t}U(\mathbf{u}) + \frac{\partial}{\partial x}F(\mathbf{u}) \leq 0. \tag{2.13}$$

The passage to the limit on the right of (2.12) is understood weakly, in the sense of measures; the passage inside the nonlinear terms on the left, however, requires strong limits: consult the recent breakthrough of Bianchini and Bressan (Bianchini and Bressan 2003, Bressan 2003). The possibility of a strictly negative measure on the left of (2.13) is due to concentration of entropy dissipation along shock discontinuities on the right of (2.12).

The entropy inequality (2.13) is necessary in order to single out a unique, 'physically relevant' solution among the possibly many weak solutions of (2.1). In this context it is important whether (2.1) is endowed with a sufficiently 'rich' family of entropy pairs, (U, F): consult, for example, Serre (1991). How 'rich' is the family of such entropy functions? In the scalar case, $N = 1$, scalar Jacobians are symmetric and hence *every* convex U serves as an entropy function. This is the starting point for the L^1-stability theory of Kružkov (1970) for general scalar equations; we postpone this discussion to the end of this section. If the $N \times N$ system happens to be symmetric to begin with, then we can use the identity as a symmetrizer in (2.2), $U_{\mathbf{uu}} = I_N$, and hence the usual 'energy', $U(\mathbf{u}) = |\mathbf{u}|^2/2$, is an entropy function (Godunov 1961). In this case, integration of (2.13) yields the entropy bound

$$\int_x U(\mathbf{u}(x,t))\,\mathrm{d}x \leq \int_x U(\mathbf{u}(x,0))\,\mathrm{d}x, \tag{2.14}$$

which is the usual L^2-stability statement familiar from the linear theory of symmetric hyperbolic systems. Thus, entropy stability could be viewed as a nonlinear extension of the L^2 linear stability set-up to general, non-symmetric $N \times N$ systems. For 2×2 systems, the symmetrizing requirement from an entropy function, (2.2), amounts to a second-order linear hyperbolic equation and Lax (1971) has shown how to construct a family of entropy functions in this case. For general $N \geq 3$ equations, (2.2) is over-determined.

Nevertheless, most physically relevant systems are equipped with (at least one) entropy pair. The canonical example is of course the following one.

Example 2.1. (Euler equations) We consider entropy solutions, $\mathbf{u} = (\rho, m, E)^\top$ of the Euler equations

$$\frac{\partial}{\partial t}\begin{bmatrix}\rho\\ m\\ E\end{bmatrix} + \frac{\partial}{\partial x}\begin{bmatrix} m\\ qm+p\\ q(E+p)\end{bmatrix} = 0. \tag{2.15}$$

These equations govern inviscid polytropic gas dynamics, asserting the conservation of the density ρ, the momentum m, and the total energy E. Here q and p are, respectively, the velocity $q := \frac{m}{\rho}$ and the pressure $p = (\gamma-1)\cdot\left[E - \frac{m^2}{2\rho}\right]$ (where γ is the adiabatic exponent). Harten has shown that this system of equations is equipped with a family of entropy pairs, (U, F). These pairs take the form

$$U(\mathbf{u}) = -\rho h(S), \quad F(\mathbf{u}) = -mh(S) \tag{2.16}$$

(Harten 1983*b*; consult also Tadmor (1986*a*)). Here S stands for the nondimensional specific entropy

$$S = \ell n(p\rho^{-\gamma}), \tag{2.17}$$

and $h = h(S)$ is any scalar function satisfying

$$h' - \gamma h'' > 0, \quad h' > 0, \tag{2.18}$$

so that the requirement for $U(\mathbf{u})$ to be convex is met (Harten 1983*b*). The corresponding entropy variables are given by

$$\mathbf{v} \equiv \begin{bmatrix} v_1\\ v_2\\ v_3\end{bmatrix} = (1-\gamma)\cdot\frac{h'(S)}{p}\cdot\begin{bmatrix} E + \frac{p}{\gamma-1}\left(\frac{h(S)}{h'(S)} - \gamma - 1\right)\\ -m\\ \rho\end{bmatrix}, \tag{2.19}$$

with the corresponding potential pairs, $(\phi, \psi) = (\gamma-1)h'(S)(\rho, m)$. A particularly convenient form to work with is determined by $h(S) = \frac{\gamma+1}{\gamma-1}\cdot e^{\frac{S}{\gamma+1}}$. With this choice we find the entropy pair

$$U(\mathbf{u}) = \frac{\gamma+1}{1-\gamma}\cdot(\rho p)^{\frac{1}{\gamma+1}}, \quad F(\mathbf{u}) = \frac{\gamma+1}{1-\gamma}q\cdot(\rho p)^{\frac{1}{\gamma+1}}, \tag{2.20}$$

with the corresponding entropy variables, $\mathbf{v} = \mathbf{v}(\mathbf{u})$, given by

$$\mathbf{v} \equiv \begin{bmatrix} v_1\\ v_2\\ v_3\end{bmatrix} = -(\rho p)^{-\frac{\gamma}{\gamma+1}}\cdot\begin{bmatrix} E\\ -m\\ \rho\end{bmatrix}. \tag{2.21}$$

The inverse mapping, $\mathbf{v} \to \mathbf{u}$, is easily obtained as

$$\mathbf{u} = \begin{bmatrix} \rho \\ m \\ E \end{bmatrix} = -(\rho p)^{\frac{\gamma}{\gamma+1}} \begin{bmatrix} v_3 \\ -v_2 \\ v_1 \end{bmatrix}, \tag{2.22}$$

where

$$\rho p = \left[(\gamma - 1) \left(v_1 v_3 - \frac{v_2^2}{2} \right) \right]^{\frac{1+\gamma}{1-\gamma}}. \tag{2.23}$$

Godunov has studied the special choice $h(S) = S$, which leads to the canonical 'physical' entropy pair

$$U(\mathbf{u}) = -\rho S, \quad F(\mathbf{u}) = -mS. \tag{2.24}$$

Expressed in terms of the absolute temperature, T, the entropy variables in this case read

$$\mathbf{v} \equiv \begin{bmatrix} v_1 \\ v_2 \\ v_3 \end{bmatrix} = -\frac{c_v}{T} \begin{bmatrix} T(S-\gamma) + \frac{q^2}{2} \\ -q \\ 1 \end{bmatrix}, \qquad T := (\gamma - 1) \cdot c_v \cdot \frac{p}{\rho}, \tag{2.25}$$

and the inverse mapping $\mathbf{v} \to \mathbf{u}$ can be found in Harten (1983*b*). We conclude this example with several remarks.

(1) The Euler equations (2.15) provide us with an example which shows how the 'richness' of the entropy pairs can be used for a stability statement: using the one-parameter family[2] $(-\rho(S-c)^-, -m(S-c)^-)$, which is admissible by (2.18), we obtain a minimum entropy principle, $S(x,t) \geq \min_y S(y,0)$ (consult Tadmor (1986*a*)).

(2) We note that the family of admissible entropy pairs, (2.16), (2.17), (2.18), becomes smaller once we seek further symmetrization of the viscous Navier–Stokes terms along the lines of (2.11) (consult Hughes *et al.* (1986)).

(3) Finally, we call attention to the fact that, with the particular choice of entropy pair $(U,F) = (-\rho S, -mS)$ (consult (2.24), (2.25)), the corresponding potential pair (ϕ, ψ) turns out to be the density and momentum components of the flow, $(\phi(\mathbf{v}), \psi(\mathbf{v})) = (\gamma - 1)(\rho, m)$. Hence, in view of (2.8), (2.9), Euler equations can be rewritten in the intriguing form

$$\frac{\partial}{\partial t}[\nabla_{\mathbf{v}} \rho] + \frac{\partial}{\partial x}[\nabla_{\mathbf{v}} m] = 0. \tag{2.26}$$

[2] The superscript $^+$ (respectively $^-$) denotes the positive (respectively negative) part of the indicated scalar.

We close the section with the promised discussion on the entropy stability of the scalar case. We start with the following result, extending the penetrating scalar arguments of Kružkov (1970), which demonstrates how the 'richness' of the family of entropy pairs is converted into a stability statement.

Theorem 2.2. (Tadmor 1997, Theorem 2.1) Assume the system (2.1) is endowed with an N-parameter family of entropy pairs, $(U(\mathbf{u};\mathbf{c}),\ F(\mathbf{u};\mathbf{c}))$, $\mathbf{c} \in \mathbb{R}^N$, satisfying the symmetry property

$$U(\mathbf{u};\mathbf{c}) = U(\mathbf{c};\mathbf{u}), \quad F(\mathbf{u};\mathbf{c}) = F(\mathbf{c};\mathbf{u}). \tag{2.27}$$

Let $\mathbf{u}^1, \mathbf{u}^2$ be two entropy solutions of (2.1). Then the following *a priori* estimate holds:

$$\int_x U(\mathbf{u}^1(x,t);\mathbf{u}^2(x,t))\,\mathrm{d}x \le \int_x U(\mathbf{u}^1(x,0);\mathbf{u}^2(x,0))\,\mathrm{d}x. \tag{2.28}$$

Sketch of proof. Let $\mathbf{u}^1(x,t)$ be an entropy solution of (2.1) satisfying the entropy inequality (2.13). We employ the latter with the entropy pair $(U(\mathbf{u};\mathbf{c}), F(\mathbf{u};\mathbf{c}))$ parametrized with $\mathbf{c} = \mathbf{u}^2(y,\tau)$. This tells us that $\mathbf{u}^1(x,t)$ satisfies

$$\partial_t U(\mathbf{u}^1(x,t);\mathbf{u}^2(y,\tau)) + \partial_x F(\mathbf{u}^1(x,t);\mathbf{u}^2(y,\tau)) \le 0. \tag{2.29}$$

Let φ_δ denote a symmetric C_0^∞ unit mass mollifier that converges to Dirac mass in $\mathbb{R}$ as $\delta \downarrow 0$; set $\phi_\delta(x-y,t-\tau) := \varphi_\delta(\frac{x-y}{2})\varphi_\delta(\frac{t-\tau}{2})$ as an approximate Dirac mass in $\mathbb{R}\times\mathbb{R}^+$. 'Multiplication' of the (distributional) entropy inequality (2.13) by $\phi_\delta(x-y,t-\tau)$ yields

$$\begin{aligned}\partial_t(\phi_\delta U(\mathbf{u}^1;\mathbf{u}^2)) &+ \partial_x(\phi_\delta F(\mathbf{u}^1;\mathbf{u}^2)) \\ &\le (\partial_t\phi_\delta)U(\mathbf{u}^1;\mathbf{u}^2) + (\partial_x\phi_\delta)F(\mathbf{u}^1;\mathbf{u}^2).\end{aligned} \tag{2.30}$$

A dual manipulation, this time with (y,τ) as the primary integration variables of $\mathbf{u}^2(y,\tau)$ and (x,t) parametrizing $\mathbf{c} = \mathbf{u}^1(x,t)$, yields

$$\begin{aligned}\partial_\tau(\phi_\delta U(\mathbf{u}^2;\mathbf{u}^1)) &+ \partial_y(\phi_\delta F(\mathbf{u}^2;\mathbf{u}^1)) \\ &\le (\partial_\tau\phi_\delta)U(\mathbf{u}^2;\mathbf{u}^1) + (\partial_y\phi_\delta)F(\mathbf{u}^2;\mathbf{u}^1).\end{aligned} \tag{2.31}$$

We now add the last two inequalities: by the symmetry property (2.27), the sum of the right-hand sides of (2.30) and (2.31) vanishes; whereas by sending δ to zero, the sum of the left-hand sides of (2.30) and (2.31) amounts to

$$\partial_t U(\mathbf{u}^1(x,t);\mathbf{u}^2(x,t)) + \partial_x F(\mathbf{u}^1(x,t);\mathbf{u}^2(x,t)) \le 0.$$

The result follows by spatial integration. □

Let us point out that the elegance of the last result is confronted with the difficulty of satisfying the symmetry property (2.27). Thus, for example, $N \times N$ symmetric hyperbolic systems are endowed with the N-parameter family of entropies $U(\mathbf{u}, \mathbf{c}) = |\mathbf{u}-\mathbf{c}|^2/2$, but (2.27) fails for the corresponding entropy fluxes, $F(\mathbf{u}, \mathbf{c}) = \langle \mathbf{u} - \mathbf{c}, f(\mathbf{u})\rangle - \int^{\mathbf{u}} \langle \mathbf{f}(\mathbf{w}), d\mathbf{w}\rangle$. The favourable situation occurs in the scalar case where each convex U serves as an entropy function. In particular, Kružkov (1970) set the one-parameter family

$$U(u) = |u - c|, \qquad F(u) = \operatorname{sgn}(u - c)(f(u) - f(c)).$$

The symmetry requirement (2.27) holds and (2.28) leads to the following L^1-stability estimate.

Corollary 2.3. (Kružkov 1970) If u^1, u^2 are two entropy solutions of the scalar conservation law (2.1) subject to L^1 initial data, then

$$\|u^2(\cdot, t) - u^1(\cdot, t)\|_{L^1(x)} \leq \|u^2(\cdot, 0) - u^1(\cdot, 0)\|_{L^1(x)}. \tag{2.32}$$

Thus there exists a unique (entropy) solution operator associated with the scalar conservation law (2.1), $\mathcal{S}(t) : u(\cdot, 0) \mapsto u(\cdot, t)$, which is conservative and, according to Corollary 2.3, is also L^1-contractive, and hence by the Crandall–Tartar lemma (Crandall and Tartar 1980), $\mathcal{S}$ is order-preserving, $u^2(\cdot, 0) \geq u^1(\cdot, 0) \Longrightarrow \mathcal{S}(t)u^2(\cdot, 0) \geq \mathcal{S}(t)u^1(\cdot, 0)$. There is a parallel discrete theory for so-called *monotone schemes* which respect a similar discrete property of order preserving. The entropy stability of such schemes goes back to the pioneering work of Harten *et al.* (1976). We will not be able to expand on the details in the limited framework of this review, but let us mention the elegant approach of Crandall and Majda (1980), which clarified the entropy stability of monotone schemes in terms of a *comparison* with the constant solution. Sanders (1983) generalized the result to variable grids and we refer to Godlewski and Raviart (1996), Kröner (1997), Tadmor (1998) and LeVeque (2002) and the references therein for a series of later works, with particular emphasis on multidimensional extensions. Monotone schemes are at most first-order accurate (Harten *et al.* 1976); indeed, being entropy-stable with respect to all convex entropies, monotone schemes are necessarily limited to first-order accuracy (Osher and Tadmor 1988). This limitation led to systematic development of high-resolution schemes which circumvent this first-order limitation. For a brief overview of the convergence analysis of such schemes, we refer to Tadmor (1998). Our discussion below focuses on the question of entropy stability of such first-order as well as higher-order resolution schemes, in the context of both scalar and systems of conservation laws.

3. Entropy-conservative and entropy-stable schemes

We consider semi-discrete conservative schemes of the form

$$\frac{\mathrm{d}}{\mathrm{d}t}\mathbf{u}_\nu(t) = -\frac{1}{\Delta x_\nu}\Big[\mathbf{f}_{\nu+\frac{1}{2}} - \mathbf{f}_{\nu-\frac{1}{2}}\Big], \tag{3.1}$$

serving as consistent approximations to (2.1). Here, $\mathbf{u}_\nu(t)$ denotes the discrete solution along the grid line (x_ν, t) with $\Delta x_\nu := \frac{1}{2}(x_{\nu+1} - x_{\nu-1})$ being the variable meshsize, and $\mathbf{f}_{\nu+\frac{1}{2}}$ being the Lipschitz-continuous numerical flux consistent with the differential flux, that is,

$$\mathbf{f}_{\nu+\frac{1}{2}} = \mathbf{f}(\mathbf{u}_{\nu-p+1}, \ldots, \mathbf{u}_{\nu+p}), \quad \mathbf{f}(\mathbf{u}, \mathbf{u}, \ldots, \mathbf{u}) \equiv \mathbf{f}(\mathbf{u}). \tag{3.2}$$

The numerical flux, $\mathbf{f}(\cdot, \cdot, \ldots, \cdot)$, involves a stencil of $2p$ neighbouring grid values, and as such could be clearly distinguished from the (same notation of) the differential flux, $\mathbf{f}(\cdot)$. The difference schemes (3.1) and (3.2) are conservative in the sense of Lax and Wendroff (1960), namely, the change of total mass,

$$\sum_{\nu=-L}^{R} \mathbf{u}_\nu(t)\Delta x_\nu,$$

is solely due to the flux through the local neighbourhoods of the *arbitrary* boundaries at x_{-L} and x_R.

We are concerned here with the question of *entropy stability* of such schemes. To this end, let (U, F) be an entropy pair associated with the system (2.1). We ask whether the scheme (3.1) is *entropy-stable* with respect to such a pair, in the sense of satisfying a discrete entropy inequality analogous to (2.13), that is,

$$\frac{\mathrm{d}}{\mathrm{d}t}U(\mathbf{u}_\nu(t)) + \frac{1}{\Delta x_\nu}\Big[F_{\nu+\frac{1}{2}} - F_{\nu-\frac{1}{2}}\Big] \leq 0. \tag{3.3}$$

Here, $F_{\nu+\frac{1}{2}}$ is a consistent numerical entropy flux

$$F_{\nu+\frac{1}{2}} = F(\mathbf{u}_{\nu-p+1}, \ldots, \mathbf{u}_{\nu+p}), \quad F(\mathbf{u}, \mathbf{u}, \ldots, \mathbf{u}) = F(\mathbf{u}). \tag{3.4}$$

If, in particular, equality holds in (3.3), we say that the scheme (3.1) is *entropy-conservative.*

The answer to this question of entropy stability provided in Tadmor (1987*b*) consists of two main ingredients: (i) the use of the entropy variables and (ii) the comparison with appropriate *entropy-conservative* schemes. We conclude this section with a brief overview.

By making the changes of variables $\mathbf{u}_\nu = \mathbf{u}(\mathbf{v}_\nu)$, the scheme (3.1) recasts into the equivalent form

$$\frac{\mathrm{d}}{\mathrm{d}t}\mathbf{u}_\nu(t) = -\frac{1}{\Delta x_\nu}\Big[\mathbf{g}_{\nu+\frac{1}{2}} - \mathbf{g}_{\nu-\frac{1}{2}}\Big], \quad \mathbf{u}_\nu(t) = \mathbf{u}(\mathbf{v}_\nu(t)), \tag{3.5}$$

with a numerical flux

$$\mathbf{g}_{\nu+\frac{1}{2}} = \mathbf{g}(\mathbf{v}_{\nu-p+1}, \dots, \mathbf{v}_{\nu+p}) := \mathbf{f}(\mathbf{u}(\mathbf{v}_{\nu-p+1}), \dots, \mathbf{u}(\mathbf{v}_{\nu+p})), \tag{3.6}$$

consistent with the differential flux, that is,

$$\mathbf{g}(\mathbf{v}, \mathbf{v}, \dots, \mathbf{v}) = \mathbf{g}(\mathbf{v}) \equiv \mathbf{f}(\mathbf{u}(\mathbf{v})). \tag{3.7}$$

Define

$$F_{\nu+\frac{1}{2}} := \frac{1}{2}\Big\langle [\mathbf{v}_\nu + \mathbf{v}_{\nu+1}], \mathbf{g}_{\nu+\frac{1}{2}} \Big\rangle - \frac{1}{2}\big[\psi(\mathbf{v}_\nu) + \psi(\mathbf{v}_{\nu+1})\big]. \tag{3.8}$$

Then the following identity holds:

$$\begin{aligned} &\frac{\mathrm{d}}{\mathrm{d}t} U(\mathbf{u}_\nu(t)) + \frac{1}{\Delta x_\nu}\Big[F_{\nu+\frac{1}{2}} - F_{\nu-\frac{1}{2}}\Big] \qquad (3.9) \\ &= \frac{1}{2}\Big[\Big\langle \Delta\mathbf{v}_{\nu+\frac{1}{2}}, \mathbf{g}_{\nu+\frac{1}{2}} \Big\rangle - \Delta\psi_{\nu+\frac{1}{2}}\Big] + \frac{1}{2}\Big[\Big\langle \Delta\mathbf{v}_{\nu-\frac{1}{2}}, \mathbf{g}_{\nu-\frac{1}{2}} \Big\rangle - \Delta\psi_{\nu-\frac{1}{2}}\Big] \end{aligned}$$

(Tadmor 1987*b*, Section 4; see also Osher (1984)). Here $\Delta\psi_{\nu+\frac{1}{2}} := \psi(\mathbf{v}_{\nu+1}) - \psi(\mathbf{v}_\nu)$ denotes the difference of entropy flux potential, (2.9), of two neighbouring grid values $\mathbf{v}_\nu$ and $\mathbf{v}_{\nu+1}$. Thanks to (2.9), $F_{\nu+\frac{1}{2}}$ is a consistent entropy flux and this brings us to the next result.

Theorem 3.1. (Tadmor 1987*b*, Theorem 5.2) The conservative scheme (3.5) is entropy-stable (respectively, entropy-conservative) if, and for three-point schemes ($p = 1$) only if,

$$\Big\langle \Delta\mathbf{v}_{\nu+\frac{1}{2}}, \mathbf{g}_{\nu+\frac{1}{2}} \Big\rangle \leq \Delta\psi_{\nu+\frac{1}{2}}, \tag{3.10}$$

and, respectively,

$$\Big\langle \Delta\mathbf{v}_{\nu+\frac{1}{2}}, \mathbf{g}_{\nu+\frac{1}{2}} \Big\rangle = \Delta\psi_{\nu+\frac{1}{2}}. \tag{3.11}$$

4. The scalar problem

We discuss the entropy stability of *scalar* schemes of the form (see (3.5))

$$\frac{\mathrm{d}}{\mathrm{d}t} u_\nu(t) = -\frac{1}{\Delta x_\nu}\Big[g_{\nu+\frac{1}{2}} - g_{\nu-\frac{1}{2}}\Big], \quad u_\nu(t) \equiv u(v_\nu(t)). \tag{4.1}$$

For a more convenient formulation, let us define for $\Delta v_{\nu+\frac{1}{2}} \neq 0$

$$Q_{\nu+\frac{1}{2}} = \frac{f(u_\nu) + f(u_{\nu+1}) - 2g_{\nu+\frac{1}{2}}}{\Delta v_{\nu+\frac{1}{2}}}, \quad \Delta v_{\nu+\frac{1}{2}} := v_{\nu+1} - v_\nu. \tag{4.2}$$

Our scheme recasts into the equivalent *viscosity form*

$$\frac{\mathrm{d}}{\mathrm{d}t}u_\nu(t) = -\frac{1}{2\Delta x_\nu}\Big[f(u_{\nu+1}) - f(u_{\nu-1})\Big] + \frac{1}{2\Delta x_\nu}\Big[Q_{\nu+\frac{1}{2}}\Delta v_{\nu+\frac{1}{2}} - Q_{\nu-\frac{1}{2}}\Delta v_{\nu-\frac{1}{2}}\Big], \tag{4.3}$$

which reveals the role of $Q_{\nu+\frac{1}{2}}$ as the numerical viscosity coefficient (*e.g.*, Tadmor (1984*b*)).

According to (3.11), scalar entropy-conservative schemes are uniquely determined by the numerical flux $g_{\nu+\frac{1}{2}} = g^*_{\nu+\frac{1}{2}}$, that is,

$$g^*_{\nu+\frac{1}{2}} := \frac{\Delta\psi_{\nu+\frac{1}{2}}}{\Delta v_{\nu+\frac{1}{2}}} \equiv \int_{\xi=-\frac{1}{2}}^{\frac{1}{2}} g\Big(v_{\nu+\frac{1}{2}}(\xi)\Big)\,\mathrm{d}\xi,$$
$$v_{\nu+\frac{1}{2}}(\xi) := \frac{1}{2}(v_\nu + v_{\nu+1}) + \xi\Delta v_{\nu+\frac{1}{2}}. \tag{4.4}$$

Noting that

$$g^*_{\nu+\frac{1}{2}} = \int_{\xi=-\frac{1}{2}}^{\frac{1}{2}} \frac{\mathrm{d}}{\mathrm{d}\xi}(\xi)\cdot g\Big(v_{\nu+\frac{1}{2}}(\xi)\Big)\,\mathrm{d}\xi, \tag{4.5}$$

we find upon integration by parts that entropy-conservative schemes admit the viscosity form (4.3), with a viscosity coefficient $Q_{\nu+\frac{1}{2}} = Q^*_{\nu+\frac{1}{2}}$ given by[3]

$$Q^*_{\nu+\frac{1}{2}} = \int_{\xi=-\frac{1}{2}}^{\frac{1}{2}} 2\xi g'\Big(v_{\nu+\frac{1}{2}}(\xi)\Big)\,\mathrm{d}\xi. \tag{4.6}$$

The entropy-conservative scheme then takes the form

$$\begin{aligned}\frac{\mathrm{d}}{\mathrm{d}t}u_\nu(t) &= -\frac{1}{\Delta x_\nu}\Big[g^*_{\nu+\frac{1}{2}} - g^*_{\nu-\frac{1}{2}}\Big]\\ &= -\frac{1}{2\Delta x_\nu}\big[f(u_{\nu+1}) - f(u_{\nu-1})\big]\\ &\quad + \frac{1}{2\Delta x_\nu}\Big[Q^*_{\nu+\frac{1}{2}}\Delta v_{\nu+\frac{1}{2}} - Q^*_{\nu-\frac{1}{2}}\Delta v_{\nu-\frac{1}{2}}\Big].\end{aligned} \tag{4.7}$$

The entropy stability portion of Theorem 3.1 can now be restated in the following form.

Corollary 4.1. (Tadmor 1987*b*, Theorem 5.1) The conservative scheme (4.7) and (4.3) is entropy-stable, if – and for three-point schemes ($p = 1$) only if – it contains more viscosity than the entropy-conservative one (4.6),

[3] We use primes to indicate differentiation with respect to primary dependent variables, *e.g.*, $\mathbf{g}' = \mathbf{g_v}(\mathbf{v}), \mathbf{f}'' = \mathbf{f_{uu}}(\mathbf{u})$, *etc.*

that is,

$$Q^*_{\nu+\frac{1}{2}} \leq Q_{\nu+\frac{1}{2}}. \tag{4.8}$$

The rest of this section is devoted to examples demonstrating applications of the last corollary.

Example 4.2. (Entropy-conservative schemes) We begin with two examples of entropy-conservative schemes, interesting in their own right, which played a significant role in studying zero dispersion phenomena: see, *e.g.*, Lax (1986) and collaborators.

We consider the inviscid Burgers' equation, $u_t + (\frac{1}{2}u^2)_x = 0$, and we seek a semi-discrete scheme that conserves the logarithmic entropy $U(u) = -\ln u$. The entropy flux in this case is $F(u) = -u$. Using the entropy variable $v(u) = -1/u$, we compute the entropy flux potential

$$\psi(v) = vf(u(v)) - F(u(v)) = -\frac{1}{2v},$$

which in turn yields the entropy-conservative flux

$$g^*_{\nu+\frac{1}{2}} = \frac{\psi(v_{\nu+1} - \psi(v_\nu)}{v_{\nu+1} - v_\nu} = \frac{1}{2}\frac{1}{v_\nu v_{\nu+1}} = \frac{1}{2}u_\nu u_{\nu+1}.$$

This yields the entropy-conservative centred schemes

$$\frac{\mathrm{d}}{\mathrm{d}t}u_\nu(t) = u_\nu(t)\frac{u_{\nu+1}(t) - u_{\nu-1}(t)}{2\Delta x_\nu},$$

studied in Goodman and Lax (1988), Hou and Lax (1991) and Levermore and Liu (1996), among others.

Next, we consider $u_t + (\mathrm{e}^u)_x = 0$ and we seek the semi-discrete scheme that conserves the exponential entropy, $U(u) = \mathrm{e}^u$. The entropy flux is $F(u) = \frac{1}{2}\mathrm{e}^{2u}$. Using the corresponding entropy variable $v(u) = \mathrm{e}^u$, we compute the entropy flux potential

$$\psi(v) = vf(u(v)) - F(u(v)) = \frac{1}{2}v^2,$$

which in turn yields the entropy-conservative flux

$$g^*_{\nu+\frac{1}{2}} = \frac{\psi(v_{\nu+1}) - \psi(v_\nu)}{v_{\nu+1} - v_\nu} = \frac{1}{2}(v_\nu + v_{\nu+1}) = \frac{1}{2}\left[\mathrm{e}^{u_\nu} + \mathrm{e}^{u_{\nu+1}}\right].$$

This yields the entropy-conservative centred schemes

$$\frac{\mathrm{d}}{\mathrm{d}t}u_\nu(t) = \frac{\mathrm{e}^{u_{\nu+1}(t)} - \mathrm{e}^{u_{\nu-1}(t)}}{2\Delta x_\nu}$$

associated with Toda flow: consult Lax *et al.* (1993), Levermore and Liu (1996), Deift and McLaughlin (1998), and the references therein.

We continue with a series of entropy-stable examples.

Example 4.3. (Engquist and Osher 1980) Using the estimate

$$Q^*_{\nu+\frac{1}{2}} \leq \int_{\xi=-\frac{1}{2}}^{\frac{1}{2}} |g'\big(v_{\nu+\frac{1}{2}}(\xi)\big)|\,\mathrm{d}\xi = \int |f'\Big(u\big(v_{\nu+\frac{1}{2}}(\xi)\big)\Big)| \left|\frac{\mathrm{d}u(v_{\nu+\frac{1}{2}}(\xi))}{\Delta v_{\nu+\frac{1}{2}}}\right|$$
$$= \frac{1}{\Delta v_{\nu+\frac{1}{2}}}\left[\int_{u_\nu}^{u_{\nu+1}} |f'(u)|\,\mathrm{d}u\right] =: Q^{EO}_{\nu+\frac{1}{2}}, \qquad (4.9)$$

we obtain an upper bound $Q^{EO}_{\nu+\frac{1}{2}}$, which is the viscosity coefficient associated with the entropy-stable Engquist–Osher (EO) scheme (Engquist and Osher 1980).

The quantity inside the brackets on the right-hand side of (4.9) is independent of different choices for entropy variables. Consequently, the entropy stability of the EO scheme is uniform with respect to all admissible entropy pairs, (U, F). This raises the question of the minimal amount of viscosity required to maintain such uniformity.

Example 4.4. (Godunov 1959) We rewrite the second term on the right-hand side of the schemes (4.3) as

$$\frac{1}{2\Delta x_\nu}\left[\left(Q_{\nu+\frac{1}{2}}\frac{\Delta v_{\nu+\frac{1}{2}}}{\Delta u_{\nu+\frac{1}{2}}}\right)\Delta u_{\nu+\frac{1}{2}} - \left(Q_{\nu-\frac{1}{2}}\frac{\Delta v_{\nu-\frac{1}{2}}}{\Delta u_{\nu-\frac{1}{2}}}\right)\Delta u_{\nu-\frac{1}{2}}\right],$$

thus normalizing their viscous part by using the conservative variables as our fixed scale. Since in the scalar case *all* convex functions, $U(u)$, are admissible entropy functions, it follows that, for an entropy stability which is uniform with respect to every such U, we need to *maximize* the corresponding entropy viscous factors $Q^*_{\nu+\frac{1}{2}}(\Delta v_{\nu+\frac{1}{2}}/\Delta u_{\nu+\frac{1}{2}})$,

$$\sup_v\left[\frac{f(u_\nu)+f(u_{\nu+1})-2g^*_{\nu+\frac{1}{2}}}{\Delta u_{\nu+\frac{1}{2}}}\right], \quad g^*_{\nu+\frac{1}{2}} = \int_{\xi=-\frac{1}{2}}^{\frac{1}{2}} f\big(u(v_{\nu+\frac{1}{2}}(\xi))\big)\,\mathrm{d}\xi,$$

where the supremum is taken over *all increasing* $v = v(u)$. This yields Godunov's viscosity coefficient (Osher 1985, Tadmor 1984*b*)

$$Q^G_{\nu+\frac{1}{2}} = \max_{\min(u_\nu,u_{\nu+1})\leq u\leq\max(u_\nu,u_{\nu+1})}\left[\frac{f(u_\nu)+f(u_{\nu+1})-2f(u)}{\Delta u_{\nu+\frac{1}{2}}}\right]. \qquad (4.10)$$

Thus, the scalar schemes which are uniformly entropy-stable with respect to all convex entropies are precisely those that contain at least as much numerical viscosity as the Godunov scheme does. These so-called E-schemes were first identified in Osher (1984); see also Tadmor (1984*b*).

The E-schemes are only first-order accurate: consult, *e.g.*, Lemma 4.5 below. Corollary 4.1 enables us to verify the entropy stability of second-order accurate schemes as well. To this end we recall from (4.6) that

$$Q^*_{\nu+\frac{1}{2}} = \int_{\xi=-\frac{1}{2}}^{\frac{1}{2}} 2\xi g'\Big(v_{\nu+\frac{1}{2}}(\xi)\Big)\,\mathrm{d}\xi, \quad v_{\nu+\frac{1}{2}}(\xi) = \frac{1}{2}(v_\nu + v_{\nu+1}) + \xi\Delta v_{\nu+\frac{1}{2}}.$$

Integration by parts yields

$$\begin{aligned} Q^*_{\nu+\frac{1}{2}} &= \int_{\xi=-\frac{1}{2}}^{\frac{1}{2}} \frac{\mathrm{d}}{\mathrm{d}\xi}\left(\xi^2 - \frac{1}{4}\right) g'\Big(v_{\nu+\frac{1}{2}}(\xi)\Big)\,\mathrm{d}\xi \\ &= \int_{\xi=-\frac{1}{2}}^{\frac{1}{2}} \left(\frac{1}{4} - \xi^2\right) \frac{\mathrm{d}}{\mathrm{d}\xi} g'\Big(v_{\nu+\frac{1}{2}}(\xi)\Big)\,\mathrm{d}\xi, \end{aligned} \tag{4.11}$$

and hence the entropy-conservative viscosity coefficient $Q^*_{\nu+\frac{1}{2}}$ takes the form

$$Q^*_{\nu+\frac{1}{2}} = \int_{\xi=-\frac{1}{2}}^{\frac{1}{2}} \left(\frac{1}{4} - \xi^2\right) g''\Big(v_{\nu+\frac{1}{2}}(\xi)\Big)\,\mathrm{d}\xi \cdot \Delta v_{\nu+\frac{1}{2}}. \tag{4.12}$$

Thus, the viscosity coefficients of the entropy-conservative schemes are in fact of order $\mathcal{O}(|\Delta v_{\nu+\frac{1}{2}}|)$, and this implies their second-order accuracy in view of the following lemma.

Lemma 4.5. Consider the conservative schemes (4.3) with viscosity coefficient, $Q_{\nu+\frac{1}{2}}$, such that $\big(Q_{\nu+\frac{1}{2}}/\Delta v_{\nu+\frac{1}{2}}\big)$ is Lipschitz-continuous. Then these schemes are second-order accurate, in the sense that their local truncation error is of the order

$$\mathcal{O}\Big[|x_{\nu+1} - x_\nu|^2 + |x_\nu - x_{\nu-1}|^2 + |x_{\nu+1} - 2x_\nu + x_{\nu-1}|\Big].$$

Verification of this lemma is straightforward and therefore omitted.

Example 4.6. (Second-order accurate schemes) Using the simple upper bound

$$Q^*_{\nu+\frac{1}{2}} \le \frac{1}{6} \max_{\min(v_\nu, v_{\nu+1}) \le v \le \max(v_\nu, v_{\nu+1})} |g''(v)| \cdot \left|\Delta v_{\nu+\frac{1}{2}}\right|, \tag{4.13}$$

we obtain a viscosity coefficient on the right of (4.13) which, according to Corollary 4.1 and Lemma 4.5, maintains both entropy stability and second-order accuracy. Viscosity terms similar to this were previously derived in a number of special cases, dealing with the entropy stability question of second-order schemes, such as (generalized) van Leer's MUSCL scheme (van Leer 1977, Osher 1985, Lions and Souganidis 1985, Yang 1996*a*) as well as other high-resolution schemes (Majda and Osher 1978, Majda and Osher 1979, Harten 1983*a*, Harten and Hyman 1983, Nessyahu and Tadmor 1990). We remark that the careful calculations required in those derivations are due to the delicate balance of the cubic order of entropy loss, which should match the third-order dissipation in this case.

An instructive example of using the above arguments of entropy stability is provided in the genuinely nonlinear case, where $f(u)$ is, say, convex. A quadratic entropy stability is sufficient in this case, to single out the unique physically relevant solution (Szepessy 1989, Chen 2000). In particular, the choice of the quadratic entropy function $U(u) = \frac{1}{2}u^2$ leads to entropy variables that coincide with the conservative ones, $g(v) = f(u)$. The last three examples of this section deal with this important special case.

Example 4.7. (Lax and Wendroff 1960) By convexity, the entropy-conservative viscosity coefficient in (4.12),

$$Q^*_{\nu+\frac{1}{2}} = \int_{\xi=-\frac{1}{2}}^{\frac{1}{2}} \left(\frac{1}{4} - \xi^2\right) f''\left(u_{\nu+\frac{1}{2}}(\xi)\right) \mathrm{d}\xi \cdot \Delta u_{\nu+\frac{1}{2}}$$

is negative whenever $\Delta u_{\nu+\frac{1}{2}}$ is negative, and hence numerical viscosity is required only in the case of rarefactions where $\Delta u_{\nu+\frac{1}{2}} > 0$. To see how much viscosity is required in this case, we use the fact that the integrand on the right of Q^* is positive, leading to the upper bound

$$Q^*_{\nu+\frac{1}{2}} \leq \frac{1}{4} \int_{\xi=-\frac{1}{2}}^{\frac{1}{2}} f''\left(u_{\nu+\frac{1}{2}}(\xi)\right) \mathrm{d}\xi \cdot \Delta u_{\nu+\frac{1}{2}} = \frac{1}{4}\left[a(u_{\nu+1}) - a(u_\nu)\right]^+. \quad (4.14)$$

The resulting viscosity coefficient on the right is the second-order accurate viscosity originally proposed by Lax and Wendroff (1960),

$$Q^{LW}_{\nu+\frac{1}{2}} = \frac{1}{4}\left[a(u_{\nu+1}) - a(u_\nu)\right]^+, \quad a(u) = f'(u). \quad (4.15)$$

Example 4.8. (Centred schemes) According to (2.9) with $g(v) = f(u)$, the entropy flux potential is given by the primitive of $f(\cdot)$, and by (3.10), entropy stability is guaranteed if

$$\Delta u_{\nu+\frac{1}{2}} \cdot f_{\nu+\frac{1}{2}} \leq \int_{u_\nu}^{u_{\nu+1}} f(u)\,\mathrm{d}u.$$

In the rarefaction case, $\Delta u_{\nu+\frac{1}{2}} > 0$, the integral on the right approximated from below by the midpoint rule; in the case of a shock, $\Delta u_{\nu+\frac{1}{2}} < 0$, signs are reversed and we can instead use the trapezoidal rule. Thus we derive a second-order accurate entropy stable scheme (4.1), whose simple numerical flux is given by the centred numerical flux

$$g_{\nu+\frac{1}{2}} = f_{\nu+\frac{1}{2}} = \begin{cases} f\Big(\dfrac{u_\nu + u_{\nu+1}}{2}\Big), & \Delta u_{\nu+\frac{1}{2}} > 0 \\[2ex] \dfrac{f(u_\nu) + f(u_{\nu+1})}{2}, & \Delta u_{\nu-\frac{1}{2}} \leq 0. \end{cases} \tag{4.16}$$

We conclude this section with the following example.

Example 4.9. A well-known 'trick' for deriving a quadratic entropy-conservative approximation in the particular case of the inviscid Burgers' equation

$$\frac{\partial}{\partial t}u + \frac{\partial}{\partial x}\left[\frac{1}{2}u^2\right] = 0, \tag{4.17}$$

is based on centred differencing of its equivalent skew-adjoint form (Tadmor 1984*a*)

$$\frac{\partial}{\partial t}u + \frac{1}{3}\frac{\partial}{\partial x}[u^2] + \frac{1}{3}u\frac{\partial}{\partial x}[u] = 0,$$

which yields

$$\frac{\mathrm{d}}{\mathrm{d}t}u_\nu(t) = -\frac{1}{3}\frac{1}{2\Delta x_\nu}\left[u^2_{\nu+1} - u^2_{\nu-1}\right] - \frac{1}{3}u_\nu\frac{1}{2\Delta x_\nu}\left[u_{\nu+1} - u_{\nu-1}\right].$$

In fact, there is more than just a 'trick' here: the resulting scheme is simply a special case of our entropy-conservative recipe (4.6)

$$\begin{aligned}\frac{\mathrm{d}}{\mathrm{d}t}u_\nu(t) = &-\frac{1}{2\Delta x_\nu}\left[\frac{1}{2}u^2_{\nu+1} - \frac{1}{2}u^2_{\nu-1}\right] \\ &+ \frac{1}{2\Delta x_\nu}\left[\frac{1}{6}\Big(\Delta u_{\nu+\frac{1}{2}}\Big)^2 - \frac{1}{6}\Big(\Delta u_{\nu-\frac{1}{2}}\Big)^2\right].\end{aligned} \tag{4.18}$$

If we exclude negative viscosity, however, then according to (4.12) the least viscous entropy-stable approximation of Burgers' equation (4.17) is given by

$$\begin{aligned}\frac{\mathrm{d}}{\mathrm{d}t}u_\nu(t) = &-\frac{1}{2\Delta x_\nu}\left[\frac{1}{2}u^2_{\nu+1} - \frac{1}{2}u^2_{\nu-1}\right] \\ &+ \frac{1}{2\Delta x_\nu}\left[\frac{1}{6}\Big(\Delta u_{\nu+\frac{1}{2}}\Big)^+\Delta u_{\nu+\frac{1}{2}} - \frac{1}{6}\Big(\Delta u_{\nu-\frac{1}{2}}\Big)^+\Delta u_{\nu-\frac{1}{2}}\right].\end{aligned} \tag{4.19}$$

5. Systems of conservation laws

We study the entropy stability of the semi-discrete schemes that are consistent with the *system* of conservation laws (2.5). The schemes assume the following viscosity form:

$$\frac{\mathrm{d}}{\mathrm{d}t}\mathbf{u}_\nu(t) = -\frac{1}{2\Delta x_\nu}\left[\mathbf{f}(\mathbf{u}_{\nu+1}) - \mathbf{f}(\mathbf{u}_{\nu-1})\right] + \frac{1}{2\Delta x_\nu}\left[Q_{\nu+\frac{1}{2}}\Delta\mathbf{v}_{\nu+\frac{1}{2}} - Q_{\nu-\frac{1}{2}}\Delta\mathbf{v}_{\nu-\frac{1}{2}}\right]. \tag{5.1}$$

Difference schemes that admit the viscosity form (5.1) are precisely the so-called essentially three-point schemes (Harten (1983*a*), Tadmor (1987*b*, Lemma 5.1)), namely, difference schemes whose numerical flux, $\mathbf{f}(\cdot,\cdot,\cdots\cdot)$ satisfies the restricted consistency relation

$$\mathbf{f}(\mathbf{u}_{\nu-p+1},\ldots,\mathbf{u}_\nu = \mathbf{u}_{\nu+1} = \mathbf{u},\ldots,\mathbf{u}_{\nu+p}) = \mathbf{f}(\mathbf{u}).$$

This is the case of (5.1), with

$$\mathbf{f}(\ldots,\mathbf{u}_\nu,\mathbf{u}_{\nu+1},\ldots) = \frac{1}{2}\left[\mathbf{f}(\mathbf{u}_\nu) + \mathbf{f}(\mathbf{u}_{\nu+1})\right] - \frac{1}{2}Q_{\nu+\frac{1}{2}}(\mathbf{v}_{\nu+1} - \mathbf{v}_\nu),$$
$$\mathbf{v}_\nu = \mathbf{v}(\mathbf{u}_\nu).$$

A couple of remarks are in order.

(1) The class of essentially three-point schemes includes classical schemes based on three-point stencils ($p = 1$), as well as most modern high-resolution schemes (van Leer 1977, Harten 1983*a*); consult Godlewski and Raviart (1996), Kröner (1997), LeVeque (1992, 2002) and the references therein.

(2) The use of essentially three-point stencils in this section is linked to the specific second-order entropy-conservative schemes discussed below. Extensions to higher orders and larger stencils were carried out by LeFloch and Rohde (2000, Section 4); consult Section 9 below.

To extend our scalar entropy stability analysis to systems of conservation laws we proceed as before, by comparison with certain entropy-conservative schemes. Unlike the scalar problem, however, we now have more than one way to meet the entropy conservation requirement (3.11). The various ways differ in their choice of the path of integration in phase space. In this section, we restrict our attention to the simplest choice along the *straight path* $\mathbf{v}_{\nu+\frac{1}{2}}(\xi) = \frac{1}{2}(\mathbf{v}_\nu+\mathbf{v}_{\nu+1})+\xi\Delta\mathbf{v}_{\nu+\frac{1}{2}}$. The corresponding entropy-conservative flux is given by

$$\mathbf{g}^*_{\nu+\frac{1}{2}} = \int_{\xi=-\frac{1}{2}}^{\frac{1}{2}} \mathbf{g}\Big(\mathbf{v}_{\nu+\frac{1}{2}}(\xi)\Big)\,\mathrm{d}\xi,$$
$$\mathbf{v}_{\nu+\frac{1}{2}}(\xi) := \frac{1}{2}(\mathbf{v}_\nu + \mathbf{v}_{\nu+1}) + \xi\Delta\mathbf{v}_{\nu+\frac{1}{2}}. \tag{5.2}$$

Indeed, the entropy conservation requirement (3.1) is fulfilled in this case, since, in view of (2.9),

$$\left\langle \Delta \mathbf{v}_{\nu+\frac{1}{2}}, \mathbf{g}^*_{\nu+\frac{1}{2}} \right\rangle = \int_{\xi=-\frac{1}{2}}^{\frac{1}{2}} \left\langle \Delta \mathbf{v}_{\nu+\frac{1}{2}}, \mathbf{g}\left(\mathbf{v}_{\nu+\frac{1}{2}}(\xi)\right) \right\rangle \mathrm{d}\xi$$
$$= \int_{\mathbf{v}_\nu}^{\mathbf{v}_{\nu+1}} \langle \mathrm{d}\mathbf{v}, \mathbf{g}(\mathbf{v}) \rangle = \Delta \psi_{\nu+\frac{1}{2}}.$$

The entropy-conservative flux (5.2) was introduced in Tadmor (1986*b*, 1987*b*). As before (see (4.5), (4.6)), we integrate by parts to find

$$\begin{aligned} \mathbf{g}^*_{\nu+\frac{1}{2}} &= \int_{\xi=-\frac{1}{2}}^{\frac{1}{2}} \frac{\mathrm{d}}{\mathrm{d}\xi}(\xi) \mathbf{g}\left(\mathbf{v}_{\nu+\frac{1}{2}}(\xi)\right) \mathrm{d}\xi \\ &= \xi \mathbf{g}\left(\mathbf{v}_{\nu+\frac{1}{2}}(\xi)\right)\Big|_{\xi=-\frac{1}{2}}^{\frac{1}{2}} - \int_{\xi=-\frac{1}{2}}^{\frac{1}{2}} \xi \mathbf{g}_{\mathbf{v}}\left(\mathbf{v}_{\nu+\frac{1}{2}}(\xi)\right) \frac{\mathrm{d}\mathbf{v}_{\nu+\frac{1}{2}}(\xi)}{\mathrm{d}\xi} \mathrm{d}\xi \\ &= \frac{1}{2}[\mathbf{f}(\mathbf{u}_\nu) + \mathbf{f}(\mathbf{u}_{\nu+1})] - \int_{\xi=-\frac{1}{2}}^{\frac{1}{2}} \xi B\left(\mathbf{v}_{\nu+\frac{1}{2}}(\xi)\right) \mathrm{d}\xi \Delta \mathbf{v}_{\nu+\frac{1}{2}}. \end{aligned} \tag{5.3}$$

Thus, the entropy-conservative scheme (5.2) admits the equivalent viscosity form

$$\begin{aligned} \frac{\mathrm{d}}{\mathrm{d}t}\mathbf{u}_\nu(t) = &-\frac{1}{2\Delta x_\nu}\left[\mathbf{f}(\mathbf{u}_{\nu+1}) - \mathbf{f}(\mathbf{u}_{\nu-1})\right] \\ &+ \frac{1}{2\Delta x_\nu}\left[Q^*_{\nu+\frac{1}{2}} \Delta \mathbf{v}_{\nu+\frac{1}{2}} - Q^*_{\nu-\frac{1}{2}} \Delta \mathbf{v}_{\nu-\frac{1}{2}}\right], \end{aligned} \tag{5.4}$$

with a numerical viscosity matrix coefficient, $Q^*_{\nu+\frac{1}{2}}$, given by

$$Q^*_{\nu+\frac{1}{2}} := \int_{\xi=-\frac{1}{2}}^{\frac{1}{2}} 2\xi B\left(\mathbf{v}_{\nu+\frac{1}{2}}(\xi)\right) \mathrm{d}\xi, \quad B(\mathbf{v}) = \mathbf{g}_{\mathbf{v}}(\mathbf{v}). \tag{5.5}$$

The entropy stability portion of Theorem 3.1 can now be conveniently interpreted as follows.

Corollary 5.1. The conservative scheme (5.1) is entropy-stable if – and for three-point schemes ($p = 1$) only if – it contains more viscosity than the entropy-conservative one (5.4), (5.5), that is,

$$\left\langle \Delta \mathbf{v}_{\nu+\frac{1}{2}}, Q^*_{\nu+\frac{1}{2}} \Delta \mathbf{v}_{\nu+\frac{1}{2}} \right\rangle \leq \left\langle \Delta \mathbf{v}_{\nu+\frac{1}{2}}, Q_{\nu+\frac{1}{2}} \Delta \mathbf{v}_{\nu+\frac{1}{2}} \right\rangle. \tag{5.6}$$

Indeed, we can provide a precise measure for the amount of entropy dissipation in terms of the *dissipation matrix* $D_{\nu+\frac{1}{2}} \equiv D_{\nu+\frac{1}{2}}(\mathbf{v}(t)) := Q_{\nu+\frac{1}{2}} - Q^*_{\nu+\frac{1}{2}}$

(Tadmor 1987*b*, Theorem 5.2)

$$\frac{\mathrm{d}}{\mathrm{d}t}U(\mathbf{u}_\nu(t)) + \frac{1}{\Delta x_\nu}\left[F_{\nu+\frac{1}{2}} - F_{\nu-\frac{1}{2}}\right] \tag{5.7}$$
$$= -\frac{1}{4\Delta x_\nu}\left[\left\langle \Delta\mathbf{v}_{\nu-\frac{1}{2}}, D_{\nu-\frac{1}{2}}\Delta\mathbf{v}_{\nu-\frac{1}{2}}\right\rangle + \frac{1}{4}\left\langle \Delta\mathbf{v}_{\nu+\frac{1}{2}}, D_{\nu+\frac{1}{2}}\Delta\mathbf{v}_{\nu+\frac{1}{2}}\right\rangle\right]$$

Here, $F_{\nu+\frac{1}{2}}$ stands for the entropy flux (see (3.8))

$$F_{\nu+\frac{1}{2}} = \frac{1}{2}\left\langle \mathbf{v}_\nu + \mathbf{v}_{\nu+1}, \mathbf{g}^*_{\nu+\frac{1}{2}}\right\rangle - \frac{1}{2}\left[\psi(\mathbf{v}_\nu) + \psi(\mathbf{v}_{\nu+1})\right]$$
$$- \frac{1}{4\Delta x_\nu}\left\langle \mathbf{v}_\nu + \mathbf{v}_{\nu+1}, D_{\nu+\frac{1}{2}}\Delta\mathbf{v}_{\nu+\frac{1}{2}}\right\rangle. \tag{5.8}$$

The entropy-conservative flux (5.2), and likewise its corresponding viscosity coefficient in (5.5), cannot be evaluated in a closed form. However, Corollary 5.1 enables us to verify entropy stability by *comparison*, $Q^*_{\nu+\frac{1}{2}} \leq \operatorname{Re} Q_{\nu+\frac{1}{2}}$, with the usual ordering between symmetric matrices. We note in passing that $Q^*_{\nu+\frac{1}{2}}$ is symmetric (since $B(\cdot)$ is) and that, in the generic case, the viscosity coefficient $Q_{\nu+\frac{1}{2}}$ is also symmetric. The following examples demonstrate this point.

Example 5.2. (Rusanov 1961, Lax 1954) We seek a scalar viscosity coefficient, $p_{\nu+\frac{1}{2}}I_N$, which guarantees the entropy stability of the scheme

$$\frac{\mathrm{d}}{\mathrm{d}t}\mathbf{u}_\nu(t) = -\frac{1}{2\Delta x_\nu}\left[\mathbf{f}(\mathbf{u}_{\nu+1}) - f(\mathbf{u}_{\nu-1})\right]$$
$$+ \frac{1}{2\Delta x_\nu}\left[p_{\nu+\frac{1}{2}}\Delta\mathbf{u}_{\nu+\frac{1}{2}} - p_{\nu-\frac{1}{2}}\Delta\mathbf{u}_{\nu-\frac{1}{2}}\right]. \tag{5.9}$$

Using (2.6) we have

$$\Delta\mathbf{u}_{\nu+\frac{1}{2}} = \int_{\xi=-\frac{1}{2}}^{\frac{1}{2}} \frac{\mathrm{d}}{\mathrm{d}\xi}\mathbf{u}\left(\mathbf{v}_{\nu+\frac{1}{2}}(\xi)\right)\mathrm{d}\xi = \int_{\xi=-\frac{1}{2}}^{\frac{1}{2}} H\left(\mathbf{v}_{\nu+\frac{1}{2}}(\xi)\right)\mathrm{d}\xi\cdot\Delta\mathbf{v}_{\nu+\frac{1}{2}}, \tag{5.10}$$

and hence the viscous part of the scheme (5.9) can be interpreted in terms of the entropy variables (rather than the conservative ones), as

$$p_{\nu+\frac{1}{2}}\Delta\mathbf{u}_{\nu+\frac{1}{2}} = Q_{\nu+\frac{1}{2}}\Delta\mathbf{v}_{\nu+\frac{1}{2}}, \tag{5.11}$$

where

$$Q_{\nu+\frac{1}{2}} = p_{\nu+\frac{1}{2}}\int_{\xi=-\frac{1}{2}}^{\frac{1}{2}} H(\xi)\,\mathrm{d}\xi, \quad H(\xi) \equiv H\left(\mathbf{v}_{\nu+\frac{1}{2}}(\xi)\right). \tag{5.12}$$

Recalling (5.5), we conclude that entropy stability is guaranteed by Corollary 5.1, provided $p_{\nu+\frac{1}{2}}$ is chosen so that the inequality

$$2\xi B(\xi) \leq p_{\nu+\frac{1}{2}} H(\xi), \quad B(\xi) \equiv B\Big(\mathbf{v}_{\nu+\frac{1}{2}}(\xi)\Big), \quad -\frac{1}{2} \leq \xi \leq \frac{1}{2},$$

holds. To this end, multiply both sides by $H^{-\frac{1}{2}}(\xi)$; by congruence, we end up with the equivalent inequality[4]

$$2\xi \sup_{\lambda} \lambda\big[H^{-\frac{1}{2}}(\xi)B(\xi)H^{-\frac{1}{2}}(\xi)\big] \leq p_{\nu+\frac{1}{2}} I_N. \tag{5.13}$$

We recall that $B = \mathbf{g}_\mathbf{v} = \mathbf{f}_\mathbf{u}\mathbf{u}_\mathbf{v} = AH$, that is,

$$B(\xi) = A(\xi)H(\xi), \quad A(\xi) \equiv A\Big(\mathbf{u}\Big(\mathbf{v}_{\nu+\frac{1}{2}}(\xi)\Big)\Big). \tag{5.14}$$

Hence (5.13) holds and entropy stability follows, for any scalar $p_{\nu+\frac{1}{2}}$ satisfying

$$\begin{aligned} p_{\nu+\frac{1}{2}} &\geq \max_{\lambda, |\xi| \leq \frac{1}{2}} \Big|2\xi\lambda\Big[H^{-\frac{1}{2}}(\xi)A(\xi)H^{\frac{1}{2}}(\xi)\Big]\Big| \\ &= \max_{\lambda, |\xi| \leq \frac{1}{2}} \Big|\lambda\Big[A\Big(\mathbf{u}\Big(\mathbf{v}_{\nu+\frac{1}{2}}(\xi)\Big)\Big)\Big]\Big|. \end{aligned} \tag{5.15}$$

The cell-dependent viscosity factor on the right corresponds to the Rusanov scheme (Rusanov 1961; see also Richtmyer and Morton (1967, Section 2)), while a uniform viscosity factor, satisfying

$$p_{\nu+\frac{1}{2}} \equiv p \geq \max_{\lambda, \mathbf{u}} |\lambda[A(\mathbf{u})]|,$$

corresponds to a Lax–Friedrichs viscosity (Friedrichs 1954, Lax 1954). Both schemes are entropy-stable with respect to any entropy pair associated with equation (2.1).

The last example was restricted to first-order accurate schemes. Yet Corollary 5.1 can be used to maintain both entropy stability and second-order accuracy, as was done in the scalar case. To this end, we proceed as follows. Using (5.14) we can rewrite the quantity on the left of (5.6) as

$$\begin{aligned} &\Big\langle \Delta\mathbf{v}_{\nu+\frac{1}{2}}, Q^*_{\nu+\frac{1}{2}}\Delta\mathbf{v}_{\nu+\frac{1}{2}}\Big\rangle \\ &= \int_{\xi=-\frac{1}{2}}^{\frac{1}{2}} 2\xi\Big\langle H^{\frac{1}{2}}(\xi)\Delta\mathbf{v}_{\nu+\frac{1}{2}}, H^{-\frac{1}{2}}(\xi)A(\xi)H^{\frac{1}{2}}(\xi)\cdot H^{\frac{1}{2}}(\xi)\Delta\mathbf{v}_{\nu+\frac{1}{2}}\Big\rangle \,\mathrm{d}\xi. \end{aligned} \tag{5.16}$$

Let $\{a_k(\xi), \mathbf{r}_k(\xi)\}_{k=1}^N$ be the eigenpairs of $A(\xi)$, that is,

$$a_k(\xi) \equiv a^{(k)}_{\nu+\frac{1}{2}}\Big(\mathbf{u}\Big(\mathbf{v}_{\nu+\frac{1}{2}}(\xi)\Big)\Big), \qquad \mathbf{r}_k(\xi) \equiv \mathbf{r}^{(k)}_{\nu+\frac{1}{2}}\Big(\mathbf{u}\Big(\mathbf{v}_{\nu+\frac{1}{2}}(\xi)\Big)\Big).$$

[4] Here and below, $\lambda_k[\cdot]$ denotes the kth eigenvalue of a matrix.

Since $H^{-\frac{1}{2}}(\xi)\mathbf{r}_k(\xi)$ are the eigenvectors of the matrix $H^{-\frac{1}{2}}(\xi)A(\xi)H^{\frac{1}{2}}(\xi)$, and since, by (5.14),

$$H^{-\frac{1}{2}}(\xi)A(\xi)H^{\frac{1}{2}}(\xi) \equiv H^{-\frac{1}{2}}(\xi)B(\xi)H^{-\frac{1}{2}}(\xi), \quad B(\xi) \equiv B\Big(\mathbf{v}_{\nu+\frac{1}{2}}(\xi)\Big), \tag{5.17}$$

is a symmetric matrix, it follows after normalization that $\{H^{-\frac{1}{2}}(\xi)\mathbf{r}_k(\xi)\}$ form an orthonormal system, that is,

$$\Big\langle H^{-\frac{1}{2}}(\xi)\mathbf{r}_k(\xi), H^{-\frac{1}{2}}(\xi)\mathbf{r}_j(\xi)\Big\rangle = \delta_{jk}. \tag{5.18}$$

We expand $H^{\frac{1}{2}}(\xi)\Delta\mathbf{v}_{\nu+\frac{1}{2}}$ and substitute the expansion into the right-hand side of (5.16) to find

$$\Big\langle \Delta\mathbf{v}_{\nu+\frac{1}{2}}, Q^*_{\nu+\frac{1}{2}}\Delta\mathbf{v}_{\nu+\frac{1}{2}}\Big\rangle = \sum_{k=1}^{N}\int_{\xi=-\frac{1}{2}}^{\frac{1}{2}} 2\xi a_k(\xi)\cdot\Big|\Big\langle \mathbf{r}_k(\xi), \Delta\mathbf{v}_{\nu+\frac{1}{2}}\Big\rangle\Big|^2\,\mathrm{d}\xi. \tag{5.19}$$

Finally, we integrate by parts along the lines of (4.11), arriving at

$$\begin{aligned}
&\Big\langle \Delta\mathbf{v}_{\nu+\frac{1}{2}}, Q^*_{\nu+\frac{1}{2}}\Delta\mathbf{v}_{\nu+\frac{1}{2}}\Big\rangle \\
&\quad= \sum_{k=1}^{N}\int_{\xi=-\frac{1}{2}}^{\frac{1}{2}} \Big(\frac{1}{4}-\xi^2\Big)\frac{\mathrm{d}}{\mathrm{d}\xi}a_k(\xi)\cdot\Big|\Big\langle \mathbf{r}_k(\xi), \Delta\mathbf{v}_{\nu+\frac{1}{2}}\Big\rangle\Big|^2\,\mathrm{d}\xi \\
&\qquad+ \sum_{k=1}^{N}\int_{\xi=-\frac{1}{2}}^{\frac{1}{2}} \Big(\frac{1}{4}-\xi^2\Big)a_k(\xi)\cdot\frac{\mathrm{d}}{\mathrm{d}\xi}\Big|\Big\langle \mathbf{r}_k(\xi), \Delta\mathbf{v}_{\nu+\frac{1}{2}}\Big\rangle\Big|^2\,\mathrm{d}\xi.
\end{aligned} \tag{5.20}$$

We compute

$$\frac{\mathrm{d}}{\mathrm{d}\xi}a_k(\xi)\cdot\Big|\Big\langle \mathbf{r}_k(\xi), \Delta\mathbf{v}_{\nu+\frac{1}{2}}\Big\rangle\Big|^2 = \Big\langle \nabla_{\mathbf{v}}a_k(\xi), \Delta\mathbf{v}_{\nu+\frac{1}{2}}\Big\rangle\cdot\Big|\Big\langle \mathbf{r}_k(\xi), \Delta\mathbf{v}_{\nu+\frac{1}{2}}\Big\rangle\Big|^2$$

and

$$\frac{\mathrm{d}}{\mathrm{d}\xi}\Big|\Big\langle \mathbf{r}_k(\xi), \Delta\mathbf{v}_{\nu+\frac{1}{2}}\Big\rangle\Big|^2 = 2\cdot\Big\langle \mathbf{r}_k(\xi), \Delta\mathbf{v}_{\nu+\frac{1}{2}}\Big\rangle\cdot\Big\langle \Delta\mathbf{v}_{\nu+\frac{1}{2}}, \nabla_{\mathbf{v}}\mathbf{r}_k(\xi)\Delta\mathbf{v}_{\nu+\frac{1}{2}}\Big\rangle,$$

and since both terms are of order $\mathcal{O}\big(|\Delta\mathbf{v}_{\nu+\frac{1}{2}}|^3\big)$, it follows that the quantity on the right of (5.20) does not exceed

$$\Big|\Big\langle \Delta\mathbf{v}_{\nu+\frac{1}{2}}, Q^*_{\nu+\frac{1}{2}}\Delta\mathbf{v}_{\nu+\frac{1}{2}}\Big\rangle\Big| \le C_{\nu+\frac{1}{2}}\cdot\big|\Delta\mathbf{v}_{\nu+\frac{1}{2}}\big|^3. \tag{5.21}$$

Thus, the entropy-conservative schemes (5.2) dissipate entropy at a cubic rate and are therefore second-order accurate: consult Lemma 4.5. Comparison of this in light of Corollary 5.1 yields the following entropy stability criterion which respects second-order accuracy.

Theorem 5.3. The conservative scheme

$$\begin{aligned}\frac{\mathrm{d}}{\mathrm{d}t}\mathbf{u}_\nu(t) = &-\frac{1}{2\Delta x_\nu}\left[\mathbf{f}(\mathbf{u}_{\nu+1}) - \mathbf{f}(\mathbf{u}_{\nu-1})\right] \\ &+ \frac{1}{2\Delta x_\nu}\left[Q_{\nu+\frac{1}{2}}\Delta\mathbf{v}_{\nu+\frac{1}{2}} - Q_{\nu-\frac{1}{2}}\Delta\mathbf{v}_{\nu-\frac{1}{2}}\right]\end{aligned} \tag{5.22}$$

is entropy-stable if the eigenvalues of (the symmetric part of) its viscosity coefficient matrix, $\mathrm{Re}\, Q_{\nu+\frac{1}{2}}$, satisfy

$$\min_\lambda \lambda\left[\mathrm{Re}\, Q_{\nu+\frac{1}{2}}\right] \geq C_{\nu+\frac{1}{2}} \cdot \left|\Delta\mathbf{v}_{\nu+\frac{1}{2}}\right|. \tag{5.23}$$

Next, we would like to convert this entropy stability criterion to difference schemes which are written in the standard form

$$\begin{aligned}\frac{\mathrm{d}}{\mathrm{d}t}\mathbf{u}_\nu(t) = &-\frac{1}{2\Delta x_\nu}\left[\mathbf{f}(\mathbf{u}_{\nu+1}) - \mathbf{f}(\mathbf{u}_{\nu-1})\right] \\ &+ \frac{1}{2\Delta x_\nu}\left[P_{\nu+\frac{1}{2}}\Delta\mathbf{u}_{\nu+\frac{1}{2}} - P_{\nu-\frac{1}{2}}\Delta\mathbf{u}_{\nu-\frac{1}{2}}\right].\end{aligned} \tag{5.24}$$

Thus, the viscous part is expressed entirely in terms of the conservative variables, $\mathbf{u}$, instead of the entropy variables, $\mathbf{v}$, used in (5.22). From the corresponding differential set-up, (2.10), we already know that the admissible viscosity matrices for this formulation are those Ps for which $PU_{\mathbf{uu}}^{-1}$ are symmetric positive definite, (2.11), $PH - HP^\top = 0$; consequently, a discrete analogue should hold, at least to leading order, that is,

$$\left\|P_{\nu+\frac{1}{2}}H_{\nu+\frac{1}{2}} - H_{\nu+\frac{1}{2}}P^\top_{\nu+\frac{1}{2}}\right\| \leq \delta_{\nu+\frac{1}{2}}\left|\Delta\mathbf{u}_{\nu+\frac{1}{2}}\right|. \tag{5.25}$$

Here, $H_{\nu+\frac{1}{2}}$ can be any first-order symmetric approximation to the inverse Hessian, $H = U_{\mathbf{uu}}^{-1}$,

$$\begin{aligned}H_{\nu+\frac{1}{2}} &= \int_{\xi=-\frac{1}{2}}^{\frac{1}{2}} H\left(\mathbf{v}_{\nu+\frac{1}{2}}(\xi)\right)\mathrm{d}\xi + \mathcal{O}\left(\left|\Delta\mathbf{u}_{\nu+\frac{1}{2}}\right|\right), \\ 0 &< \frac{1}{K}\cdot I_N \leq H_{\nu+\frac{1}{2}} \leq K\cdot I_N.\end{aligned} \tag{5.26}$$

Theorem 5.4. Consider the conservative difference scheme (5.24) with numerical viscosity coefficient $P_{\nu+\frac{1}{2}}$, which is essentially H-symmetric (5.25), (5.26). The scheme is entropy-stable if the eigenvalues of its viscosity coefficient matrix $\lambda(P_{\nu+\frac{1}{2}})$ satisfy

$$\min_\lambda \lambda\left[P_{\nu+\frac{1}{2}}\right] \geq \gamma_{\nu+\frac{1}{2}}\left|\Delta\mathbf{u}_{\nu+\frac{1}{2}}\right|. \tag{5.27}$$

Here $\gamma_{\nu+\frac{1}{2}}$ is a suitably large constant depending on (5.21), (5.25), (5.26) for which

$$\gamma_{\nu+\frac{1}{2}} \geq K\Big(C_{\nu+\frac{1}{2}} + \delta_{\nu+\frac{1}{2}}\Big). \tag{5.28}$$

Proof. Using (5.10), the scheme (5.24) meets the desired form in (5.22) with

$$Q_{\nu+\frac{1}{2}} = P_{\nu+\frac{1}{2}}H, \quad H \equiv H_{\nu+\frac{1}{2}} = \int_{\xi=-\frac{1}{2}}^{\frac{1}{2}} H\Big(\mathbf{v}_{\nu+\frac{1}{2}}(\xi)\Big)\,\mathrm{d}\xi.$$

Next, we invoke the identity

$$H^{-\frac{1}{2}}P_{\nu+\frac{1}{2}}H^{\frac{1}{2}} = H^{-\frac{1}{2}}\Big(\operatorname{Re}Q_{\nu+\frac{1}{2}}\Big)H^{-\frac{1}{2}} + H^{-\frac{1}{2}}\left(\frac{P_{\nu+\frac{1}{2}}H - HP^{\top}_{\nu+\frac{1}{2}}}{2}\right)H^{-\frac{1}{2}}.$$

According to (5.27), the eigenvalues of the matrix on the left are bounded from below by $\gamma_{\nu+\frac{1}{2}} \cdot |\Delta\mathbf{u}_{\nu+\frac{1}{2}}|$. Hence, by (5.25), (5.26), the same is true for the eigenvalues of the first matrix on the right; more precisely, we have

$$\lambda\Big[H^{-\frac{1}{2}}\Big(\operatorname{Re}Q_{\nu+\frac{1}{2}}\Big)H^{-\frac{1}{2}}\Big] \geq \Big(\gamma_{\nu+\frac{1}{2}} - K\delta_{\nu+\frac{1}{2}}\Big)\cdot\big|\Delta\mathbf{u}_{\nu+\frac{1}{2}}\big|.$$

Multiplying on both sides by $H^{\frac{1}{2}}$ we find, on account of (5.28),

$$\lambda\Big[\operatorname{Re}Q_{\nu+\frac{1}{2}}\Big] \geq \left(\frac{1}{K}\gamma_{\nu+\frac{1}{2}} - \delta_{\nu+\frac{1}{2}}\right)\cdot\big|\Delta\mathbf{u}_{\nu+\frac{1}{2}}\big| \geq C_{\nu+\frac{1}{2}}\cdot\big|\Delta\mathbf{v}_{\nu+\frac{1}{2}}\big|,$$

and entropy stability follows from Theorem 5.3. □

Equipped with Theorem 5.4 we turn to the following example.

Example 5.5. (Second-order accurate scalar numerical viscosity) We re-examine Example 5.2, considering the case of scalar viscosity in (5.9), where $p_{\nu+\frac{1}{2}} = p_{\nu+\frac{1}{2}}I_N$. By Theorem 5.4, any scalar satisfying

$$p_{\nu+\frac{1}{2}} \geq KC_{\nu+\frac{1}{2}}\cdot\big|\Delta\mathbf{u}_{\nu+\frac{1}{2}}\big| \tag{5.29}$$

will guarantee entropy stability as well as maintain second-order accuracy.

Example 5.6. (Roe-type schemes) We consider the class of schemes based on Roe's decomposition (Roe 1981). To this end, we introduce a Lipschitz-continuous averaged Jacobian, the so-called Roe matrix, $\overline{A}_{\nu+\frac{1}{2}}$, satisfying

$$\Delta\mathbf{f}_{\nu+\frac{1}{2}} \equiv \overline{A}_{\nu+\frac{1}{2}}\Delta\mathbf{u}_{\nu+\frac{1}{2}}, \qquad \Delta\mathbf{f}_{\nu+\frac{1}{2}} := \mathbf{f}(\mathbf{u}_{\nu+1}) - \mathbf{f}(\mathbf{u}_{\nu}), \tag{5.30}$$

and having a complete real eigensystem. Roe (1981) constructed such a matrix for the Euler equations (2.15); Harten and Lax (1981) have shown its existence in the general case, namely

$$\overline{A}_{\nu+\frac{1}{2}} = B_{\nu+\frac{1}{2}} \cdot H^{-1}_{\nu+\frac{1}{2}}, \tag{5.31}$$

where $B_{\nu+\frac{1}{2}}$ and $H_{\nu+\frac{1}{2}}$ are defined by the cell averages

$$B_{\nu+\frac{1}{2}} = \int_{\xi=-\frac{1}{2}}^{\frac{1}{2}} B\left(\mathbf{v}_{\nu+\frac{1}{2}}(\xi)\right) \mathrm{d}\xi, \quad H_{\nu+\frac{1}{2}} = \int_{\xi=-\frac{1}{2}}^{\frac{1}{2}} H\left(\mathbf{v}_{\nu+\frac{1}{2}}(\xi)\right) \mathrm{d}\xi. \tag{5.32}$$

Given a Roe matrix, the viscosity coefficient in (5.24), $P_{\nu+\frac{1}{2}}$, is then set to be

$$P_{\nu+\frac{1}{2}} = p\left(\overline{A}_{\nu+\frac{1}{2}}\right). \tag{5.33}$$

Here $p(\cdot)$ is an appropriate viscosity function which is computed according to the spectral decomposition of $\overline{A}_{\nu+\frac{1}{2}}$, namely,

$$p\left(\overline{A}_{\nu+\frac{1}{2}}\right) = R_{\nu+\frac{1}{2}} \cdot \begin{bmatrix} p(\overline{a}_1) & & \\ & \ddots & \\ & & p(\overline{a}_N) \end{bmatrix} \cdot R^{-1}_{\nu+\frac{1}{2}}. \tag{5.34}$$

where $\{(\overline{a})_1^N, R_{\nu+\frac{1}{2}}\}$ is the eigensystem of $\overline{A}$, that is,

$$\overline{A}_{\nu+\frac{1}{2}} = R_{\nu+\frac{1}{2}} \cdot \begin{bmatrix} \overline{a}_1 & & \\ & \ddots & \\ & & \overline{a}_N \end{bmatrix} \cdot R^{-1}_{\nu+\frac{1}{2}}, \quad \overline{a}_k := \lambda_k\left[\overline{A}_{\nu+\frac{1}{2}}\right].$$

For a given system, the possibly various choices of a Roe matrices are within $\mathcal{O}(|\Delta\mathbf{u}_{\nu+\frac{1}{2}}|)$ of each other; since the set-up of Theorem 5.4 is invariant under such perturbations we can discuss without restriction the one choice given in (5.31)–(5.32). With this choice of a Roe matrix we have

$$P_{\nu+\frac{1}{2}} = p\left(\overline{A}_{\nu+\frac{1}{2}}\right) = H^{\frac{1}{2}}_{\nu+\frac{1}{2}} \cdot p\left[H^{-\frac{1}{2}}_{\nu+\frac{1}{2}} \cdot \int_{\xi=-\frac{1}{2}}^{\frac{1}{2}} B\left(\mathbf{v}_{\nu+\frac{1}{2}}(\xi)\right) \mathrm{d}\xi \cdot H^{-\frac{1}{2}}_{\nu+\frac{1}{2}}\right] \cdot H^{-\frac{1}{2}}_{\nu+\frac{1}{2}},$$

and hence, by the symmetry of B, it follows that P is H-symmetric, so that (5.25) holds with $\delta_{\nu+\frac{1}{2}} = 0$. Theorem 5.4 applies and we are led to the following.

Theorem 5.7. The conservative Roe-type scheme (5.24), (5.33), (5.34) is entropy-stable, provided that its viscosity function $p(\cdot)$ satisfies

$$p(\overline{a}_k) \geq KC_{\nu+\frac{1}{2}} \cdot \left|\Delta\mathbf{u}_{\nu+\frac{1}{2}}\right|, \quad \overline{a}_k := \lambda_k\left[\overline{A}_{\nu+\frac{1}{2}}\right]. \tag{5.35}$$

There are various ways of choosing a viscosity function $p(\cdot)$ satisfying (5.35), which give rise to either first- or second-order accurate entropy-stable schemes. We start by discussing the pros and cons of some first-order choices.

The original choice of Roe (1981) employs the viscosity function

$$p(\overline{a}_k) = |\overline{a}_k|. \tag{5.36}$$

It has the desirable property that discrete steady shocks are perfectly resolved on the grid. With this choice, the entropy stability requirement (5.35) reads

$$|\overline{a}_k| \geq KC_{\nu+\frac{1}{2}} \cdot \left|\Delta \mathbf{u}_{\nu+\frac{1}{2}}\right|,$$

and it is fulfilled as long as we are away from sonic points. Yet this requirement may be violated in sonic neighbourhoods where $\overline{a}_k \approx 0$; indeed, Roe's scheme is the canonical example of an entropy-unstable scheme, for it admits steady expansion shocks. Theorem 5.7 suggests a simple modification – first proposed by Osher (1985, Theorem 3.3) – in order to avoid such instability.

Example 5.8. (First-order entropy fix of the Roe scheme) The Roe scheme (5.24), (5.33), (5.34), is entropy-stable with a viscosity function

$$p(\overline{a}_k) = \max\left\{|\overline{a}_k|, KC_{\nu+\frac{1}{2}} \cdot \left|\Delta \mathbf{u}_{\nu+\frac{1}{2}}\right|\right\}. \tag{5.37}$$

The slightly more viscous modification of Harten (1983*a*) takes the form

$$p(\overline{a}_k) = \max\left\{|\overline{a}_k|, \varepsilon\right\}, \quad \left|\Delta \mathbf{u}_{\nu+\frac{1}{2}}\right| \ll 1. \tag{5.38}$$

In these cases entropy stability is achieved by adding viscosity near sonic points, regardless of whether they occur in rarefaction or shock waves. This is done at the expense of destroying the sharp steady shock resolution of Roe's original scheme (5.36).

However, we can do better with regard to Roe-type schemes, by sharpening the general sufficient entropy stability condition (5.27) which led us to Theorem 5.7. To this end, we first note that the eigensystem of a Roe matrix in (5.30), $\overline{A}_{\nu+\frac{1}{2}}$, is within $\mathcal{O}(|\Delta \mathbf{u}_{\nu+\frac{1}{2}}|^2)$ from the eigensystem of the exact mid-value Jacobian, say $A(\xi = 0)$, for example,

$$|\overline{\mathbf{r}}_k - \mathbf{r}_k(\xi = 0)| + |\overline{a}_k - a_k(\xi = 0)| \leq \text{Const}\left|\Delta \mathbf{u}_{\nu+\frac{1}{2}}\right|^2. \tag{5.39}$$

By virtue of (5.39) we can obtain rather detailed information about the entropy dissipation rate of the Roe-type schemes (5.24), (5.33), (5.34).

Let $\Delta a_k(\mathbf{u}_\nu)$ denote the jump in the kth eigenvalue

$$\Delta a_k(\mathbf{u}_\nu) = \lambda_k(A(\mathbf{u}_{\nu+1})) - \lambda_k(A(\mathbf{u}_\nu)).$$

In the Appendix we prove the following theorem.

Theorem 5.9. The conservative Roe-type scheme (5.24), (5.33), (5.34) is entropy-stable if its viscosity function, $p(\cdot)$, satisfies

$$p(\overline{a}_k) \geq \frac{1}{6}\left[\Delta a_k(\mathbf{u}_\nu) + \varepsilon_k|\overline{a}_k| + \left(1 + \frac{|\overline{a}_k|}{\varepsilon_k}\right)\text{Const}\left|\Delta \mathbf{u}_{\nu+\frac{1}{2}}\right|^2\right]. \tag{5.40}$$

Here $\varepsilon_k > 0$ are arbitrary parameters at our disposal.

Remark. We note that the essential ingredient of Theorem 5.9 is not the Roe averaging property (5.30), but the requirement that the eigensystem of $\overline{A}_{\nu+\frac{1}{2}}$ be within $\mathcal{O}(|\Delta \mathbf{u}_{\nu+\frac{1}{2}}|^2)$ of the eigensystem of $A(\xi = 0)$, (5.39). Hence, Theorem 5.9 and its consequences apply to other Roe averages, for example, $\overline{A}_{\nu+\frac{1}{2}} = A[\frac{1}{2}(\mathbf{u}_\nu + \mathbf{u}_{\nu+1})]$ or $\overline{A}_{\nu+\frac{1}{2}} = \frac{1}{2}\left[A(\mathbf{u}_\nu) + A(\mathbf{u}_{\nu+1})\right]$.

We shall apply Theorem 5.9 to hyperbolic systems which contain either *genuinely nonlinear* (GNL) or linearly degenerate waves. Lax (1957) has shown that any two nearby states in such systems, $\mathbf{u}_\nu$ and $\mathbf{u}_{\nu+1}$, can be connected by a certain continuous path in phase space; the jump from $\mathbf{u}_\nu$ to $\mathbf{u}_{\nu+1}$ is resolved into a succession of k-waves, $k = 1, 2, \ldots, N$, each of which is either a k-shock, a k-contact or a k-rarefaction, depending on whether a_k increases, remains constant or decreases, respectively, along the corresponding kth subpath. In this section we answer the entropy stability question by comparison with the entropy-conservative schemes (5.2) which are based upon integration along a simple straight path in phase space. Therefore, these schemes do not resolve the full structure in phase space of the solution path to the Riemann problem described above. Instead, we shall confine ourselves to identify each cell with one dominant k-wave. This fact of one dominant wave per cell is certainly the case with GNL scalar problems and, as observed by Harten (1983*a*), is also valid in actual computations with the gas dynamics system (2.15). We will refine our stability analysis in Section 6 below, in terms of new entropy conservative schemes which do take into account different subpaths in phase space.

Choosing $\varepsilon_k = 6$ in (5.40), then the following entropy-stable modification of the first-order Roe-type scheme is obtained.

Example 5.10. (Modified Roe scheme revisited) The conservative Roe-type scheme (5.24), (5.33), (5.34) is entropy-stable with viscosity function

$$p(\overline{a}_k) = |\overline{a}_k| + \left[\frac{1}{6}\Delta a_k(\mathbf{u}_\nu) + \text{Const}\left|\Delta \mathbf{u}_{\nu+\frac{1}{2}}\right|^2\right]^+. \tag{5.41}$$

This choice of viscosity function was suggested in Harten and Hyman (1983, Appendix A) and numerical simulations were carried out in Kaddouri (1993), for example. To gain a better insight into this choice, we shall distinguish between three different cases.

Case I. $\Delta a_k(\mathbf{u}_\nu) \le -\text{Const}\big|\Delta\mathbf{u}_{\nu+\frac{1}{2}}\big| + \mathcal{O}\big(\big|\Delta\mathbf{u}_{\nu+\frac{1}{2}}\big|^2\big)$.
In this case the jump Δa_k is dominated by a k-shock, and with sufficiently small variation, no additional viscosity is required in (5.36), *i.e.*, (5.41) is reduced to $p(\overline{a}_k) = |\overline{a}_k|$. Thus, (5.41) retains the perfect resolution of (sufficiently weak) discrete steady shocks.

Case II. $\Delta a_k(\mathbf{u}_\nu) = \mathcal{O}\big(\big|\Delta\mathbf{u}_{\nu+\frac{1}{2}}\big|^2\big)$.
In this case the jump Δa_k is essentially due to the k-contact field and/or the balance between the other fields. Here, a minimal amount of viscosity is required near sonic points $p(\overline{a}_k) = |\overline{a}_k| + \text{Const}\big|\Delta\mathbf{u}_{\nu+\frac{1}{2}}\big|^2$.

Case III. Finally, in all other cases we shall identify the jump $\Delta a_k(\mathbf{u}_\nu)$ as dominated by a k-rarefaction, and as expected, $\mathcal{O}\big(\big|\Delta\mathbf{u}_{\nu+\frac{1}{2}}\big|\big)$ amount of dissipation is required near sonic points, $p(\overline{a}_k) = |\overline{a}_k| + \text{Const}\big|\Delta\mathbf{u}_{\nu+\frac{1}{2}}\big|$.

This concludes our discussion of the first-order accurate Roe-type schemes and we turn to the second-order case. Choosing $\varepsilon_k \sim \big|\Delta\mathbf{u}_{\nu+\frac{1}{2}}\big|$ in (5.40), we find $p(\overline{a}_k) = \Delta a_k(\mathbf{u}_\nu) + \text{Const}\big(|\overline{a}_k|\Delta\mathbf{u}_\nu + |\Delta\mathbf{u}_\nu|^2\big)$. Thus, if we set

$$p(\overline{a}_k) = \text{Const}\big|\Delta\mathbf{u}_{\nu+\frac{1}{2}}\big|, \tag{5.42}$$

then the resulting Roe-type scheme (5.24), (5.33), (5.34) is second-order accurate by Lemma 4.5, and it is entropy-stable for sufficiently large Const $\ge KC_{\nu+\frac{1}{2}}$: consult Theorem 5.7.

The examples studied so far are based on *a priori* (positive) bounds for the entropy-conservative viscosity. We close this section with the following example, which shows how to enforce entropy stability *a posteriori* by carefully removing any viscosity production. We start with an essentially three-point scheme in its conservative variables formulation (5.24),

$$\begin{aligned}\frac{\mathrm{d}}{\mathrm{d}t}\mathbf{u}_\nu(t) &= -\frac{1}{2\Delta x_\nu}\big[\mathbf{f}(\mathbf{u}_{\nu+1}) - \mathbf{f}(\mathbf{u}_{\nu-1})\big]\\ &\quad + \frac{1}{2\Delta x_\nu}\Big[P_{\nu+\frac{1}{2}}\Delta\mathbf{u}_{\nu+\frac{1}{2}} - P_{\nu-\frac{1}{2}}\Delta\mathbf{u}_{\nu-\frac{1}{2}}\Big],\end{aligned} \tag{5.43}$$

and we compare it with the corresponding formulation of the conservative scheme (5.4),

$$\begin{aligned}\frac{\mathrm{d}}{\mathrm{d}t}\mathbf{u}_\nu(t) &= -\frac{1}{2\Delta x_\nu}\big[\mathbf{f}(\mathbf{u}_{\nu+1}) - \mathbf{f}(\mathbf{u}_{\nu-1})\big]\\ &\quad + \frac{1}{2\Delta x_\nu}\Big[P^*_{\nu+\frac{1}{2}}\Delta\mathbf{u}_{\nu+\frac{1}{2}} - P^*_{\nu-\frac{1}{2}}\Delta\mathbf{u}_{\nu-\frac{1}{2}}\Big],\end{aligned}$$

$$\text{where } P^*_{\nu+\frac{1}{2}} := Q^*_{\nu+\frac{1}{2}}(H^{-1})^*, \quad (H^{-1})^* := \int_{\xi=-\frac{1}{2}}^{\frac{1}{2}} H^{-1}\Big(\mathbf{u}_{\nu+\frac{1}{2}}(\xi)\Big)\,\mathrm{d}\xi.$$

According to (5.7) which we express in terms of the conservative variables, the entropy production of (5.43) is quantified by

$$e_{\nu+\frac{1}{2}} := \left\langle \Delta \mathbf{v}_{\nu+\frac{1}{2}}, E_{\nu+\frac{1}{2}} \Delta \mathbf{u}_{\nu+\frac{1}{2}} \right\rangle, \quad E_{\nu+\frac{1}{2}} := P^*_{\nu+\frac{1}{2}} - P_{\nu+\frac{1}{2}},$$

and we arrive at the following example.

Example 5.11. (Khalfallah and Lerat 1988) The scheme (5.43) becomes entropy-stable if we add a minimal amount of (scalar) viscosity correction, $p^c_{\nu+\frac{1}{2}}$, replacing $P_{\nu+\frac{1}{2}}$ with

$$P_{\nu+\frac{1}{2}} \longrightarrow P_{\nu+\frac{1}{2}} + p^c_{\nu+\frac{1}{2}} I_{N\times N}, \quad p^c := \frac{(e_{\nu+\frac{1}{2}})^+}{\left\langle \Delta \mathbf{v}_{\nu+\frac{1}{2}}, \mathbf{u}_{\nu+\frac{1}{2}} \right\rangle}. \tag{5.44}$$

We note that the quantity on the right is well defined since the denominator does not vanish $\langle (H^{-1})^* \mathbf{u}, \mathbf{u} \rangle > 0$. The correction preserves second-order accuracy, and Lerat and his co-workers (Khalfallah and Lerat 1988) report on successful applications of such entropy correction in numerical simulations of fluid dynamics problems. Let us point out two limitations to the present approach: (i) we need to compute the entropy-conservative term $P^* \mathbf{u} = Q^* \mathbf{v}$, which might not be readily available, and (ii), as before, the entropy correction does not distinguish between different waves within the same cell. Both points are addressed in the context of the new entropy-conservative schemes introduced in the next section: consult (6.12) below, for example.

6. Entropy-conservative schemes revisited

Our study of entropy stability is based on comparison with entropy-conservative schemes. In the scalar case, entropy-conservative schemes are unique (for a given entropy pair). For systems, there are various choices for numerical fluxes which meet the entropy conservation requirement (3.11). In Section 5 we restricted our attention to just one such choice. In this section we present the general framework.

The entropy-conservative schemes treated in Section 5 are based on integration along a straight path in phase space. Consequently, the corresponding entropy stability analysis, for instance, Example 5.10, took into account only one dominant wave per cell. In contrast, in this section we introduce a new general family of entropy-conservative schemes which are based on different paths in phase space. This enables us to enforce entropy stability by fine-tuning the amount of numerical viscosity along each subpath carrying different intermediate waves. Moreover, the straight path integration of the entropy-conservative flux (5.2) does not admit a closed form, whereas the new family of entropy-conservative schemes enjoys an *explicit, closed-form formulation.* To this end, at each cell consisting of two neighbouring values

$\mathbf{v}_\nu$ and $\mathbf{v}_{\nu+1}$, we let $\{\mathbf{r}^j_{\nu+\frac12}\}_{j=1}^N$ be an *arbitrary* set of N linearly independent N-vectors, and let $\{\boldsymbol{\ell}^j_{\nu+\frac12}\}_{j=1}^N$ denote the corresponding orthogonal set, $\langle \boldsymbol{\ell}^j_{\nu+\frac12}, \mathbf{r}^k_{\nu+\frac12}\rangle = \delta_{jk}$. Next, we introduce the intermediate states, $\{\mathbf{v}^j_{\nu+\frac12}\}_{j=1}^N$, starting with $\mathbf{v}^1_{\nu+\frac12} = \mathbf{v}_\nu$, and followed by

$$\mathbf{v}^{j+1}_{\nu+\frac12} = \mathbf{v}^j_{\nu+\frac12} + \left\langle \boldsymbol{\ell}^j_{\nu+\frac12}, \Delta\mathbf{v}_{\nu+\frac12}\right\rangle \mathbf{r}^j_{\nu+\frac12}, \quad j = 1, 2, \ldots, N, \tag{6.1}$$

thus defining a path in phase space, connecting $\mathbf{v}_\nu$ to $\mathbf{v}_{\nu+1}$,

$$\mathbf{v}^{N+1}_{\nu+\frac12} = \mathbf{v}^1_{\nu+\frac12} + \sum_{j=1}^N \left\langle \boldsymbol{\ell}^j_{\nu+\frac12}, \Delta\mathbf{v}_{\nu+\frac12}\right\rangle \mathbf{r}^j_{\nu+\frac12} = \mathbf{v}_\nu + \Delta\mathbf{v}_{\nu+\frac12} \equiv \mathbf{v}_{\nu+1}. \tag{6.2}$$

Since the mapping $\mathbf{u} \mapsto \mathbf{v}$ is one-to-one, the path is mirrored in the usual phase space of conservative variables, $\{\mathbf{u}^j_{\nu+\frac12} := \mathbf{u}(\mathbf{v}^j_{\nu+\frac12})\}_{j=1}^{N+1}$, starting with $\mathbf{u}^1_{\nu+\frac12} = \mathbf{u}_\nu$ and ending with $\mathbf{u}^{N+1}_{\nu+\frac12} = \mathbf{u}_{\nu+1}$. Equipped with this notation we turn to our next result.

Theorem 6.1. The conservative scheme

$$\frac{\mathrm{d}}{\mathrm{d}t}\mathbf{u}_\nu(t) = -\frac{1}{\Delta x_\nu}\left[\mathbf{g}^*_{\nu+\frac12} - \mathbf{g}^*_{\nu-\frac12}\right],$$

with a numerical flux $\mathbf{g}^*_{\nu+\frac12}$ given by

$$\mathbf{g}^*_{\nu+\frac12} = \sum_{j=1}^N \frac{\psi(\mathbf{v}^{j+1}_{\nu+\frac12}) - \psi(\mathbf{v}^j_{\nu+\frac12})}{\langle \boldsymbol{\ell}^j_{\nu+\frac12}, \Delta\mathbf{v}_{\nu+\frac12}\rangle} \boldsymbol{\ell}^j_{\nu+\frac12}, \tag{6.3}$$

is an entropy-conservative approximation consistent with (2.5). Here, ψ is the entropy flux potential associated with the conserved entropy pair (U, F).

Remark. We note that the quantities on the right of (6.3) are well defined: consult (6.6) below.

Proof. The entropy conservation requirement (3.11) follows directly from (6.2) for

$$\begin{aligned}\left\langle \Delta\mathbf{v}_{\nu+\frac12}, \mathbf{g}^*_{\nu+\frac12}\right\rangle &= \sum_{j=1}^N \frac{\psi(\mathbf{v}^{j+1}_{\nu+\frac12}) - \psi(\mathbf{v}^j_{\nu+\frac12})}{\langle \boldsymbol{\ell}^j_{\nu+\frac12}, \Delta\mathbf{v}_{\nu+\frac12}\rangle}\left\langle \boldsymbol{\ell}^j_{\nu+\frac12}, \Delta\mathbf{v}_{\nu+\frac12}\right\rangle \\ &= \sum_{j=1}^N \psi\left(\mathbf{v}^{j+1}_{\nu+\frac12}\right) - \psi\left(\mathbf{v}^j_{\nu+\frac12}\right) \\ &= \psi\left(\mathbf{v}^{N+1}_{\nu+\frac12}\right) - \psi\left(\mathbf{v}^1_{\nu+\frac12}\right) = \Delta\psi_{\nu+\frac12}.\end{aligned}$$

It remains to verify the consistency relation (3.7). Let

$$\mathbf{v}^{j+\frac{1}{2}}_{\nu+\frac{1}{2}}(\xi) := \frac{1}{2}\left(\mathbf{v}^{j}_{\nu+\frac{1}{2}} + \mathbf{v}^{j+1}_{\nu+\frac{1}{2}}\right) + \xi\left\langle \boldsymbol{\ell}^{j}_{\nu+\frac{1}{2}}, \Delta\mathbf{v}_{\nu+\frac{1}{2}}\right\rangle \mathbf{r}^{j}_{\nu+\frac{1}{2}}, \quad -\frac{1}{2} \le \xi \le \frac{1}{2}, \tag{6.4}$$

denote the straight subpath connecting $\mathbf{v}^{j}_{\nu+\frac{1}{2}}$ and $\mathbf{v}^{j+1}_{\nu+\frac{1}{2}}$; then we can use (2.9) to express the ψ-potential jump between two consecutive intermediate states as

$$\begin{aligned}\psi\left(\mathbf{v}^{j+1}_{\nu+\frac{1}{2}}\right) - \psi\left(\mathbf{v}^{j}_{\nu+\frac{1}{2}}\right) &= \int_{\xi=-\frac{1}{2}}^{\frac{1}{2}} \frac{\mathrm{d}}{\mathrm{d}\xi}\psi\left(\mathbf{v}^{j+\frac{1}{2}}_{\nu+\frac{1}{2}}(\xi)\right)\mathrm{d}\xi \\ &= \left\langle \int_{\xi=-\frac{1}{2}}^{\frac{1}{2}} \mathbf{g}\left(\mathbf{v}^{j+\frac{1}{2}}_{\nu+\frac{1}{2}}(\xi)\right)\mathrm{d}\xi, \mathbf{r}^{j}_{\nu+\frac{1}{2}}\right\rangle \left\langle \boldsymbol{\ell}^{j}_{\nu+\frac{1}{2}}, \Delta\mathbf{v}_{\nu+\frac{1}{2}}\right\rangle.\end{aligned} \tag{6.5}$$

Inserting this into (6.3), we find that the entropy-conservative flux can be equivalently written as

$$\mathbf{g}^{*}_{\nu+\frac{1}{2}} = \sum_{j=1}^{N}\left\langle \int_{\xi=-\frac{1}{2}}^{\frac{1}{2}} \mathbf{g}\left(\mathbf{v}^{j+\frac{1}{2}}_{\nu+\frac{1}{2}}(\xi)\right)\mathrm{d}\xi, \mathbf{r}^{j}_{\nu+\frac{1}{2}}\right\rangle \boldsymbol{\ell}^{j}_{\nu+\frac{1}{2}}, \tag{6.6}$$

and consistency is now obvious:

$$\mathbf{g}^{*}(\mathbf{v},\mathbf{v}) = \sum_{j=1}^{N}\left\langle \mathbf{g}(\mathbf{v}), \mathbf{r}^{j}_{\nu+\frac{1}{2}}\right\rangle \boldsymbol{\ell}^{j}_{\nu+\frac{1}{2}} = \mathbf{g}(\mathbf{v}). \tag{6.7}$$

□

Remark. We note that if we let $\left\{\mathbf{r}^{j+\frac{1}{2}}_{\nu+\frac{1}{2}}\right\}_{j=1}^{N+1}$ collapse into the same direction of $\Delta\mathbf{v}_{\nu+\frac{1}{2}}$, then the new entropy-conservative flux (6.5) collapses into the entropy-conservative flux of the 'first kind' studied earlier in Section 5.

As before, the new entropy-conservative schemes admit a viscosity form, subject to the phase space path. Considering a typical subpath factor on the right of (6.6), we integrate by parts along the lines of (5.3), to obtain

$$\begin{aligned}\int_{\xi=-\frac{1}{2}}^{\frac{1}{2}} \frac{\mathrm{d}}{\mathrm{d}\xi}(\xi)\left\langle \mathbf{g}\left(\mathbf{v}^{j+\frac{1}{2}}_{\nu+\frac{1}{2}}(\xi)\right)\mathrm{d}\xi, \mathbf{r}^{j}_{\nu+\frac{1}{2}}\right\rangle \\ = \frac{1}{2}\left\langle \mathbf{f}\left(\mathbf{u}^{j}_{\nu+\frac{1}{2}}\right) + \mathbf{f}\left(\mathbf{u}^{j+1}_{\nu+\frac{1}{2}}\right), \mathbf{r}^{j}_{\nu+\frac{1}{2}}\right\rangle \\ + \int_{\xi=-\frac{1}{2}}^{\frac{1}{2}} \xi\left\langle \mathbf{r}^{j}_{\nu+\frac{1}{2}}, B\left(\mathbf{v}^{j+\frac{1}{2}}_{\nu+\frac{1}{2}}(\xi)\right)\mathbf{r}^{j}_{\nu+\frac{1}{2}}\right\rangle \left\langle \boldsymbol{\ell}^{j}_{\nu+\frac{1}{2}}, \Delta\mathbf{v}_{\nu+\frac{1}{2}}\right\rangle.\end{aligned}$$

This yields the following family of entropy-conservative schemes.

Corollary 6.2. Given a complete path in phase space,

$$\left\{\mathbf{u}^j_{\nu+\frac{1}{2}} := \mathbf{u}\left(\mathbf{v}^j_{\nu+\frac{1}{2}}\right)\right\}_{j=1}^{N+1},$$

associated with left and right orthogonal sets $\left\langle \boldsymbol{\ell}^j_{\nu+\frac{1}{2}}, \mathbf{r}^k_{\nu+\frac{1}{2}}\right\rangle = \delta_{jk}$, where $\mathbf{r}^j_{\nu+\frac{1}{2}}$ is in the direction of $\mathbf{v}^{j+1}_{\nu+\frac{1}{2}} - \mathbf{v}^j_{\nu+\frac{1}{2}}$. Then we have the following entropy-conservative scheme:

$$\begin{aligned}
\frac{\mathrm{d}}{\mathrm{d}t}\mathbf{u}_\nu(t) = -\frac{1}{2\Delta x_\nu}&\left[\sum_{j=1}^{N}\left\langle \mathbf{f}\left(\mathbf{u}^j_{\nu+\frac{1}{2}}\right) + \mathbf{f}\left(\mathbf{u}^{j+1}_{\nu+\frac{1}{2}}\right), \mathbf{r}^j_{\nu+\frac{1}{2}}\right\rangle \boldsymbol{\ell}^j_{\nu+\frac{1}{2}}\right.\\
&\left.-\sum_{j=1}^{N}\left\langle \mathbf{f}\left(\mathbf{u}^j_{\nu-\frac{1}{2}}\right) + \mathbf{f}\left(\mathbf{u}^{j+1}_{\nu-\frac{1}{2}}\right), \mathbf{r}^j_{\nu-\frac{1}{2}}\right\rangle \boldsymbol{\ell}^j_{\nu-\frac{1}{2}}\right]\\
+\frac{1}{2\Delta x_\nu}&\left[\sum_{j=1}^{N}\left\langle \mathbf{r}^j_{\nu+\frac{1}{2}}, Q^{j+\frac{1}{2},*}_{\nu+\frac{1}{2}}\mathbf{r}^j_{\nu+\frac{1}{2}}\right\rangle\left\langle \boldsymbol{\ell}^j_{\nu+\frac{1}{2}}, \Delta\mathbf{v}_{\nu+\frac{1}{2}}\right\rangle \boldsymbol{\ell}^j_{\nu+\frac{1}{2}}\right.\\
&\left.-\sum_{j=1}^{N}\left\langle \mathbf{r}^j_{\nu-\frac{1}{2}}, Q^{j+\frac{1}{2},*}_{\nu-\frac{1}{2}}\mathbf{r}^j_{\nu-\frac{1}{2}}\right\rangle\left\langle \boldsymbol{\ell}^j_{\nu-\frac{1}{2}}, \Delta\mathbf{v}_{\nu-\frac{1}{2}}\right\rangle \boldsymbol{\ell}^j_{\nu-\frac{1}{2}}\right],
\end{aligned}$$

$$Q^{j+\frac{1}{2},*}_{\nu+\frac{1}{2}} := \int_{\xi=-\frac{1}{2}}^{\frac{1}{2}} 2\xi B\left(\mathbf{v}^{j+\frac{1}{2}}_{\nu+\frac{1}{2}}(\xi)\right)\mathrm{d}\xi. \tag{6.8}$$

The viscosity form of the entropy-conservative scheme outlined in Corollary 6.2 is a refinement of the entropy-conservative schemes (5.4). In particular, we can revisit the examples of entropy-stable recipes outlined in Section 5, using the two ingredients of (i) comparison with entropy-conservative schemes, and (ii) a proper choice of path in phase space. We continue with a discussion of these two ingredients.

(i) *Comparison.* We seek appropriate viscosity amplitudes, $q^{j+\frac{1}{2}}_{\nu+\frac{1}{2}}$, which upper-bound the amount of entropy-conservative viscosities on each subpath in phase space, $\mathbf{v}^{j+\frac{1}{2}}_{\nu+\frac{1}{2}}(\xi)$, so that (compare Corollary 5.1)

$$\left\langle \mathbf{r}^j_{\nu+\frac{1}{2}}, Q^{j+\frac{1}{2},*}_{\nu+\frac{1}{2}}\mathbf{r}^j_{\nu+\frac{1}{2}}\right\rangle \le q^{j+\frac{1}{2}}_{\nu+\frac{1}{2}}. \tag{6.9}$$

A straightforward argument along the lines of our previous results yields the following result.

Theorem 6.3. The semi-discrete scheme

$$\frac{d}{dt}\mathbf{u}_\nu(t) = -\frac{1}{2\Delta x_\nu}\left[\sum_{j=1}^{N}\left\langle \mathbf{f}\left(\mathbf{u}^j_{\nu+\frac{1}{2}}\right) + \mathbf{f}\left(\mathbf{u}^{j+1}_{\nu+\frac{1}{2}}\right), \mathbf{r}^j_{\nu+\frac{1}{2}}\right\rangle \boldsymbol{\ell}^j_{\nu+\frac{1}{2}}\right.$$
$$\left. -\left\langle \mathbf{f}\left(\mathbf{u}^j_{\nu-\frac{1}{2}}\right) + \mathbf{f}\left(\mathbf{u}^{j+1}_{\nu-\frac{1}{2}}\right), \mathbf{r}^j_{\nu-\frac{1}{2}}\right\rangle \boldsymbol{\ell}^j_{\nu-\frac{1}{2}}\right]$$
$$+\frac{1}{2\Delta x_\nu}\left[\sum_{j=1}^{N} q^{j+\frac{1}{2}}_{\nu+\frac{1}{2}}\left\langle \boldsymbol{\ell}^j_{\nu+\frac{1}{2}}, \Delta \mathbf{v}_{\nu+\frac{1}{2}}\right\rangle \boldsymbol{\ell}^j_{\nu+\frac{1}{2}}\right.$$
$$\left. -\sum_{j=1}^{N} q^{j+\frac{1}{2}}_{\nu-\frac{1}{2}}\left\langle \boldsymbol{\ell}^j_{\nu-\frac{1}{2}}, \Delta \mathbf{v}_{\nu-\frac{1}{2}}\right\rangle \boldsymbol{\ell}^j_{\nu-\frac{1}{2}}\right], \tag{6.10}$$

is entropy-stable if it contains more numerical viscosity than the entropy-conservative one in the sense that (6.9) holds.

(ii) *Choice of path.* The new ingredient here is the choice of a proper subpath in phase space. We demonstrate the advantage of using such a subpath in the context of second-order accurate reformulation of the conservative schemes outlined in Corollary 6.2. Let

$$\left\{\mathbf{w}^k(\mathbf{v}(\xi)) = \mathbf{w}^k\left(\mathbf{v}^{j+\frac{1}{2}}_{\nu+\frac{1}{2}}(\xi)\right)\right\}$$

be the orthonormal eigensystem of the symmetric $B = B\left(\mathbf{v}^{j+\frac{1}{2}}_{\nu+\frac{1}{2}}(\xi)\right)$,

$$B\left(\mathbf{v}^{j+\frac{1}{2}}_{\nu+\frac{1}{2}}(\xi)\right)\mathbf{w}^k(\mathbf{v}(\xi)) = b_k(\mathbf{v}(\xi))\mathbf{w}^k(\mathbf{v}(\xi)), \quad b_k(\mathbf{v}(\xi)) := \lambda_k\left(B\left(\mathbf{v}^{j+\frac{1}{2}}_{\nu+\frac{1}{2}}\right)\right).$$

Expanding $\mathbf{r}^j_{\nu+\frac{1}{2}} = \sum_k \langle \mathbf{w}^k(\mathbf{v}(\xi)), \mathbf{r}^j_{\nu+\frac{1}{2}}\rangle \mathbf{w}^k(\mathbf{v}(\xi))$, we rewrite the amount of entropy-conservative viscosity corresponding to a typical subpath on the left of (6.9)

$$\left\langle \mathbf{r}^j_{\nu+\frac{1}{2}}, Q^{j+\frac{1}{2},*}_{\nu+\frac{1}{2}} \mathbf{r}^j_{\nu+\frac{1}{2}}\right\rangle = \int_{\xi=-\frac{1}{2}}^{\frac{1}{2}} 2\xi\left\langle \mathbf{r}^j_{\nu+\frac{1}{2}}, B\left(\mathbf{v}^{j+\frac{1}{2}}_{\nu+\frac{1}{2}}(\xi)\right)\mathbf{r}^j_{\nu+\frac{1}{2}}\right\rangle d\xi$$
$$= \sum_{k=1}^{N}\int_{\xi=-\frac{1}{2}}^{\frac{1}{2}} 2\xi b_k(\mathbf{v}(\xi))\left\langle \mathbf{w}^k(\mathbf{v}(\xi)), \mathbf{r}^j_{\nu+\frac{1}{2}}\right\rangle^2 d\xi.$$

Simple upper bounds, for instance, $2\xi b_k(\mathbf{v}(\xi)) \leq \sup_\xi |b_k(\mathbf{v}(\xi))|$, characterize the first-order Roe-type schemes. For second-order accuracy, we perform one more integration by parts along the lines of (5.20):

$$\left\langle \mathbf{r}^j_{\nu+\frac{1}{2}}, Q^{j+\frac{1}{2},*}_{\nu+\frac{1}{2}} \mathbf{r}^j_{\nu+\frac{1}{2}} \right\rangle$$
$$= \sum_{k=1}^{N} \int_{\xi=-\frac{1}{2}}^{\frac{1}{2}} \left(\frac{1}{4} - \xi^2\right) \left[\left\langle \nabla_{\mathbf{v}} b_k(\mathbf{v}(\xi)), \mathbf{r}^j_{\nu+\frac{1}{2}} \right\rangle \left\langle \mathbf{w}^k(\mathbf{v}(\xi)), \mathbf{r}^j_{\nu+\frac{1}{2}} \right\rangle^2 d\xi \right.$$
$$\left. + 2 b_k(\mathbf{v}(\xi)) \left\langle \mathbf{r}^j_{\nu+\frac{1}{2}}, \nabla_{\mathbf{v}} \mathbf{w}^k(\mathbf{v}(\xi)) \mathbf{r}^j_{\nu+\frac{1}{2}} \right\rangle \right] d\xi. \tag{6.11}$$

Here, second-order accuracy is reflected by viscosity amplitudes of order $\mathcal{O}(|\Delta \mathbf{v}_{\nu+\frac{1}{2}}|)$ along each subpath (being entropy-conservative, the amount of entropy dissipation is zero). How should we choose an appropriate subpath? To simplify matters we consider the symmetric case where the entropy and conservative variables coincide, $B(\mathbf{v}) = A(\mathbf{u})$. We let $\{\mathbf{u}^j_{\nu+\frac{1}{2}}\}_{j=1}^N$ be the breakpoints along the path of (approximate) solutions to the Riemann problem. It is well known (Lax 1957) that each subpath is directed along the eigensystem of $A(\mathbf{u}^j_{\nu+\frac{1}{2}})$, that is, $\mathbf{u}^{j+1}_{\nu+\frac{1}{2}} - \mathbf{u}^j_{\nu+\frac{1}{2}} \sim \mathbf{r}^j_{\nu+\frac{1}{2}}$, so $\mathbf{w}^k \sim \mathbf{r}^k_{\nu+\frac{1}{2}}$ is the normalized eigensystem of A. With this choice, all but one of the terms on the right of (6.11) vanish to higher order (in $|\Delta \mathbf{u}_{\nu+\frac{1}{2}}|$) and the leading term governing entropy dissipation is given by

$$\left\langle \mathbf{r}^j_{\nu+\frac{1}{2}}, Q^{j+\frac{1}{2},*}_{\nu+\frac{1}{2}} \mathbf{r}^j_{\nu+\frac{1}{2}} \right\rangle \approx \int_{\xi=-\frac{1}{2}}^{\frac{1}{2}} \left(\frac{1}{4} - \xi^2\right) \left\langle \nabla_{\mathbf{u}} a_j\left(\mathbf{u}^{j+\frac{1}{2}}_{\nu+\frac{1}{2}}(\xi)\right), \mathbf{r}^j_{\nu+\frac{1}{2}} \right\rangle d\xi.$$

The last expression captures the essence of the entropy-conservative schemes that balance between entropy dissipation along j-shocks, where

$$\left\langle \nabla_{\mathbf{u}} a_j(\mathbf{u}(\xi)), \mathbf{r}^j_{\nu+\frac{1}{2}} \right\rangle > 0,$$

and the entropy production along j-rarefactions, where

$$\left\langle \nabla_{\mathbf{v}} a_j(\mathbf{u}(\xi)), \mathbf{r}^j_{\nu+\frac{1}{2}} \right\rangle < 0.$$

To enforce entropy stability, we need to increase the amount of numerical viscosity. The use of different subpaths allows us to stabilize rarefactions while avoiding spurious entropy dissipation with shocks. A detailed study for the general nonsymmetric case requires lengthy calculations, and can be carried out along the lines of the Appendix. Here we note a simple entropy-stable correction by turning off the entropy production along the rarefactions, leading to viscosity amplitude, $q^{j+\frac{1}{2}}_{\nu+\frac{1}{2}}$, acting along the j-wave,

$$q^{j+\frac{1}{2}}_{\nu+\frac{1}{2}} = \int_{\xi=-\frac{1}{2}}^{\frac{1}{2}} \left(\frac{1}{4} - \xi^2\right) \left\langle \nabla_{\mathbf{u}} a_j\left(\mathbf{u}^{j+\frac{1}{2}}_{\nu+\frac{1}{2}}(\xi)\right), \mathbf{r}^j_{\nu+\frac{1}{2}} \right\rangle^+ d\xi. \tag{6.12}$$

We conclude this section with the following two corollaries.

Corollary 6.4. The difference scheme (6.10), (6.12) is a second-order accurate entropy-stable approximation of (2.1). No artificial dissipation is added in shocks and, in particular, it has the desirable property of keeping the sharpness of shock profiles.

Next, we note that if the path connecting $\mathbf{u}^j_{\nu+\frac{1}{2}}$ and $\mathbf{u}^{j+1}_{\nu+\frac{1}{2}}$ is chosen along the (approximate) Riemann solution, then the integrand on the right of (6.12) does not change sign. A simple upper bound of the entropy-conservative amplitude on the right of (6.12) along the lines of Example 4.7 yields an entropy-stable Lax–Wendroff-type viscosity

$$\int_{\xi=-\frac{1}{2}}^{\frac{1}{2}} \left(\frac{1}{4} - \xi^2\right)\left\langle \nabla_{\mathbf{u}} a_j(\mathbf{u}(\xi)), \mathbf{r}^j_{\nu+\frac{1}{2}}\right\rangle^+ \mathrm{d}\xi \leq \frac{1}{4}\frac{\left[a_j(\mathbf{u}^{j+1}_{\nu+\frac{1}{2}}) - a_j(\mathbf{u}^j_{\nu+\frac{1}{2}})\right]^+}{\left\langle \boldsymbol{\ell}^j_{\nu+\frac{1}{2}}, \Delta v_{\nu+\frac{1}{2}}\right\rangle}. \tag{6.13}$$

This yields our next result.

Corollary 6.5. The following Lax–Wendroff-type difference scheme is a second-order accurate entropy-stable approximation of (2.1):

$$\begin{aligned}
\frac{\mathrm{d}}{\mathrm{d}t}\mathbf{u}_\nu(t) = -\frac{1}{2\Delta x_\nu}&\left[\sum_{j=1}^N \left\langle \mathbf{f}\left(\mathbf{u}^j_{\nu+\frac{1}{2}}\right) + \mathbf{f}\left(\mathbf{u}^{j+1}_{\nu+\frac{1}{2}}\right), \mathbf{r}^j_{\nu+\frac{1}{2}}\right\rangle \boldsymbol{\ell}^j_{\nu+\frac{1}{2}}\right.\\
&\left. -\left\langle \mathbf{f}\left(\mathbf{u}^j_{\nu-\frac{1}{2}}\right) + \mathbf{f}\left(\mathbf{u}^{j+1}_{\nu-\frac{1}{2}}\right), \mathbf{r}^j_{\nu-\frac{1}{2}}\right\rangle \boldsymbol{\ell}^j_{\nu-\frac{1}{2}}\right]\\
+\frac{1}{8\Delta x_\nu}&\left[\sum_{j=1}^N \left[a_j\left(\mathbf{u}^{j+1}_{\nu+\frac{1}{2}}\right) - a_j\left(\mathbf{u}^j_{\nu+\frac{1}{2}}\right)\right]^+ \boldsymbol{\ell}^j_{\nu+\frac{1}{2}}\right.\\
&\left. -\sum_{j=1}^N \left[a_j\left(\mathbf{u}^{j+1}_{\nu-\frac{1}{2}}\right) - a_j\left(\mathbf{u}^j_{\nu-\frac{1}{2}}\right)\right]^+ \boldsymbol{\ell}^j_{\nu-\frac{1}{2}}\right].
\end{aligned} \tag{6.14}$$

No artificial dissipation is added in shocks and in particular, it has the desirable property of keeping the sharpness of shock profiles.

7. Entropy stability of fully discrete schemes

In this section we study the time discretizations of the semi-discrete entropy-stable schemes

$$\frac{\mathrm{d}}{\mathrm{d}t}\mathbf{u}_\nu(t) = -\frac{1}{\Delta x_\nu}\left[\mathbf{g}_{\nu+\frac{1}{2}} - \mathbf{g}_{\nu-\frac{1}{2}}\right] \tag{7.1}$$

with essentially three-point numerical flux

$$\mathbf{g}_{\nu+\frac{1}{2}}(\mathbf{v}(t)) = \frac{1}{2}[\mathbf{f}(\mathbf{u}_\nu(t)) + \mathbf{f}(\mathbf{u}_{\nu+1}(t))] - \frac{1}{2}Q_{\nu+\frac{1}{2}}\Delta\mathbf{v}_{\nu+\frac{1}{2}}. \tag{7.2}$$

According to Corollary 5.1, the semi-discrete scheme is entropy-stable if it contains more numerical viscosity than entropy-conservative schemes, namely $Q^*_{\nu+\frac{1}{2}} \leq \operatorname{Re} Q_{\nu+\frac{1}{2}}$. We recall (5.7), which allows us to measure the amount of entropy dissipation in (7.1) in terms of the dissipation matrices

$$D_{\nu+\frac{1}{2}} \equiv D_{\nu+\frac{1}{2}}(\mathbf{v}(t)) := Q_{\nu+\frac{1}{2}} - Q^*_{\nu+\frac{1}{2}}. \tag{7.3}$$

The spatial part of (7.1) satisfies

$$\left\langle \mathbf{v}_\nu(t), \left[\mathbf{g}_{\nu+\frac{1}{2}} - \mathbf{g}_{\nu-\frac{1}{2}}\right]\right\rangle = \left[F_{\nu+\frac{1}{2}} - F_{\nu-\frac{1}{2}}\right] + \frac{1}{\Delta x_\nu}\mathcal{E}^{(x)}_\nu(\mathbf{v}(t)), \tag{7.4}$$

with $F_{\nu+\frac{1}{2}}$ being the entropy flux specified by (5.8) and $\mathcal{E}^{(x)}_\nu$ denoting the amount of entropy dissipation due to spatial discretization in (7.1), given by

$$\mathcal{E}^{(x)}_\nu := \frac{1}{4}\left\langle \Delta\mathbf{v}_{\nu-\frac{1}{2}}, D_{\nu-\frac{1}{2}}\Delta\mathbf{v}_{\nu-\frac{1}{2}}\right\rangle + \frac{1}{4}\left\langle \Delta\mathbf{v}_{\nu+\frac{1}{2}}, D_{\nu+\frac{1}{2}}\Delta\mathbf{v}_{\nu+\frac{1}{2}}\right\rangle \geq 0. \tag{7.5}$$

We note in passing that use of the dissipation matrix $D_{\nu+\frac{1}{2}}$ is restricted here to entropy-conservative schemes of the 'first kind' discussed in Section 5, and we can use a similar, refined argument with the entropy-conservative schemes of the 'second kind' in Section 6, leading to the corresponding generalization of the fully discrete entropy stability analysis presented below.

To discretize in time, we introduce a local time step, $t^{n+1} = t^n + \Delta t^{n+\frac{1}{2}}$. We shall use superscripts to denote dependence on the time level, for instance, $\mathbf{u}^n_\nu = \mathbf{u}(\mathbf{v}(x_\nu, t^n)), \mathbf{g}^n_{\nu+\frac{1}{2}} = \mathbf{g}_{\nu+\frac{1}{2}}(\mathbf{v}(t^n))$, *etc.* To simplify notation, we suppress the variability of the time step and grid cell width, abbreviating $\Delta t^{n+\frac{1}{2}}/\Delta x_{\nu+\frac{1}{2}} = \frac{\Delta t}{\Delta x}$. We shall study the entropy stability of the fully discrete schemes in terms of three prototype examples, which demonstrate the balance between the entropy dissipation from spatial stencil vs. the entropy dissipation/production due to the time discretization. We begin with the following.

Example 7.1. (Implicit backward Euler (BE) time discretization) We discretize (7.1) by the backward Euler scheme

$$\mathbf{u}^{n+1}_\nu = \mathbf{u}^n_\nu - \frac{\Delta t}{\Delta x}\left[\mathbf{g}_{\nu+\frac{1}{2}}(\mathbf{v}^{n+1}) - \mathbf{g}_{\nu-\frac{1}{2}}(\mathbf{v}^{n+1})\right], \quad \mathbf{v}^{n+1} = \mathbf{v}(\mathbf{u}(t^{n+1})). \tag{7.6}$$

We claim that the fully implicit time discretization in (7.6) is unconditionally entropy-stable. Indeed, implicit time discretization is responsible for additional entropy dissipation. For a quantitative measure of this statement, we invoke the identity

$$
\begin{aligned}
U(\mathbf{u}(\mathbf{v}_\nu^{n+1})) - U(\mathbf{u}(\mathbf{v}_\nu^{n})) &\equiv \int_{\xi=-\frac{1}{2}}^{\frac{1}{2}} \frac{\mathrm{d}}{\mathrm{d}\xi} U\Big(\mathbf{u}\Big(\mathbf{v}_\nu^{n+\frac{1}{2}}(\xi)\Big)\Big)\,\mathrm{d}\xi \\
&= \int_{\xi=-\frac{1}{2}}^{\frac{1}{2}} \Big\langle \mathbf{v}_\nu^{n+\frac{1}{2}}(\xi), H\Big(\mathbf{v}_\nu^{n+\frac{1}{2}}(\xi)\Big)\Delta\mathbf{v}_\nu^{n+\frac{1}{2}} \Big\rangle\,\mathrm{d}\xi,
\end{aligned} \tag{7.7}
$$

where the following abbreviation is used:

$$
\mathbf{v}_\nu^{n+\frac{1}{2}}(\xi) = \frac{1}{2}\big(\mathbf{v}_\nu^{n+1} + \mathbf{v}_\nu^{n}\big) + \xi\Delta\mathbf{v}_\nu^{n+\frac{1}{2}}, \quad \Delta\mathbf{v}_\nu^{n+\frac{1}{2}} := \mathbf{v}_\nu^{n+1} - \mathbf{v}_\nu^{n}. \tag{7.8}
$$

Rearranging the last term on the right of (7.7), we find that time discretization yields

$$
\big\langle \mathbf{v}_\nu^{n+1}, \mathbf{u}_\nu^{n+1} - \mathbf{u}_\nu^{n} \big\rangle = U\big(\mathbf{u}_\nu^{n+1}\big) - U\big(\mathbf{u}_\nu^{n}\big) + \mathcal{E}_\nu^{BE}\big(\mathbf{v}^{n+\frac{1}{2}}\big), \tag{7.9}
$$

where $\mathcal{E}_\nu^{BE}$ measures the entropy dissipation due to time discretization by backward Euler differencing:

$$
\mathcal{E}_\nu^{BE}\big(\mathbf{v}^{n+\frac{1}{2}}\big) := \int_{\xi=-\frac{1}{2}}^{\frac{1}{2}} \Big(\frac{1}{2} - \xi\Big)\Big\langle \Delta\mathbf{v}_\nu^{n+\frac{1}{2}}, H\big(\mathbf{v}_\nu^{n+\frac{1}{2}}(\xi)\big)\Delta\mathbf{v}_\nu^{n+\frac{1}{2}} \Big\rangle\,\mathrm{d}\xi \geq 0. \tag{7.10}
$$

Returning to (7.6), we multiply by $\mathbf{v}_\nu^{n+1}$ and obtain entropy dissipation from both the spatial discretization (7.4), (7.5) and time discretization (7.9), (7.10)

$$
\begin{aligned}
&U\big(\mathbf{u}_\nu^{n+1}\big) - U\big(\mathbf{u}_\nu^{n}\big) + \frac{\Delta t}{\Delta x}\Big[F_{\nu+\frac{1}{2}}^{n+1} - F_{\nu-\frac{1}{2}}^{n+1}\Big] \\
&\quad = \big\langle \mathbf{v}_\nu^{n+1}, \mathbf{u}_\nu^{n+1} - \mathbf{u}_\nu^{n} \big\rangle + \frac{\Delta t}{\Delta x}\Big\langle \mathbf{v}_\nu^{n+1}, \Big[\mathbf{g}_{\nu+\frac{1}{2}}\big(\mathbf{v}^{n+1}\big) - \mathbf{g}\big(\mathbf{v}^{n}\big)\Big]\Big\rangle \\
&\qquad - \frac{\Delta t}{\Delta x}\mathcal{E}_\nu^{(x)}\big(\mathbf{v}^{n+1}\big) - \mathcal{E}_\nu^{BE}\big(\mathbf{v}^{n+\frac{1}{2}}\big) \\
&\quad = -\frac{\Delta t}{\Delta x}\mathcal{E}_\nu^{(x)}\big(\mathbf{v}^{n+1}\big) - \mathcal{E}_\nu^{BE}\big(\mathbf{v}^{n+\frac{1}{2}}\big) \leq 0.
\end{aligned} \tag{7.11}
$$

Entropy stability is enhanced by fully implicit time discretization. In contrast, explicit time discretization, discussed in the next example, leads to entropy production. Thus, the entropy stability of explicit schemes hinges on a delicate balance between temporal entropy production and spatial entropy dissipation.

Example 7.2. (Explicit forward Euler (FE) time discretization) We discretize (7.1) by the forward Euler scheme

$$
\mathbf{u}_\nu^{n+1} = \mathbf{u}_\nu^{n} - \frac{\Delta t}{\Delta x}\Big[\mathbf{g}_{\nu+\frac{1}{2}}\big(\mathbf{v}^{n}\big) - \mathbf{g}_{\nu-\frac{1}{2}}\big(\mathbf{v}^{n}\big)\Big]. \tag{7.12}
$$

Now, the identity (7.7), (7.8) can be put into the equivalent form

$$\langle \mathbf{v}_\nu^n, \mathbf{u}_\nu^{n+1} - \mathbf{u}_\nu^n \rangle = U(\mathbf{u}_\nu^{n+1}) - U(\mathbf{u}_\nu^n) - \mathcal{E}_\nu^{FE}(\mathbf{v}^{n+\frac{1}{2}}), \tag{7.13}$$

with entropy production $\mathcal{E}_\nu^{FE}(\mathbf{v}^{n+\frac{1}{2}})$ given by

$$\mathcal{E}_\nu^{FE}(\mathbf{v}^{n+\frac{1}{2}}) := \int_{\xi=-\frac{1}{2}}^{\frac{1}{2}} \left(\frac{1}{2}+\xi\right)\left\langle \Delta\mathbf{v}_\nu^{n+\frac{1}{2}}, H\big(\mathbf{v}_\nu^{n+\frac{1}{2}}(\xi)\big)\Delta\mathbf{v}_\nu^{n+\frac{1}{2}}\right\rangle \mathrm{d}\xi \geq 0. \tag{7.14}$$

We multiply (7.12) by $\mathbf{v}_\nu^n$, and together with the spatial dissipation of entropy quantified in (7.5), we arrive at

$$U(\mathbf{u}_\nu^{n+1}) - U(\mathbf{u}_\nu^n) + \frac{\Delta t}{\Delta x}\left[F_{\nu+\frac{1}{2}}^n - F_{\nu-\frac{1}{2}}^n\right] = -\frac{\Delta t}{\Delta x}\mathcal{E}_\nu^{(x)}(\mathbf{v}^n) + \mathcal{E}_\nu^{FE}(\mathbf{v}^{n+\frac{1}{2}}). \tag{7.15}$$

To study the entropy stability of (7.12), we therefore need to upper-bound the entropy production $\mathcal{E}_\nu^{FE}$, in terms of the spatial dissipation matrices $D_{\nu\pm\frac{1}{2}}$, which are responsible for the entropy dissipation in (7.5). We proceed as follows. From (7.14) we have

$$\begin{aligned}\mathcal{E}_\nu^{FE}(\mathbf{v}^{n+\frac{1}{2}}) &\leq \int_{\xi=-\frac{1}{2}}^{\frac{1}{2}} \left(\frac{1}{2}+\xi\right)\left\langle \Delta\mathbf{v}_\nu^{n+\frac{1}{2}}, H\big(\mathbf{v}_\nu^{n+\frac{1}{2}}(\xi)\big)\Delta\mathbf{v}_\nu^{n+\frac{1}{2}}\right\rangle \mathrm{d}\xi \\ &\leq \frac{K}{2}|\Delta\mathbf{v}_\nu^{n+\frac{1}{2}}|^2 \leq \frac{K^3}{2}|\Delta\mathbf{u}_\nu^{n+\frac{1}{2}}|^2,\end{aligned} \tag{7.16}$$

where K^2 is the condition number of H: see (5.26). To upper-bound the time differences, $\Delta\mathbf{u}_\nu^{n+\frac{1}{2}}$, we recall $\mathbf{g}(\mathbf{v}_{\nu+1}^n) - \mathbf{g}(\mathbf{v}_\nu^n) = B_{\nu+\frac{1}{2}}\Delta\mathbf{v}_{\nu+\frac{1}{2}}^n$ with $B_{\nu+\frac{1}{2}}$ given in (5.32) as $B_{\nu+\frac{1}{2}} = \int B\big(\mathbf{v}_{\nu+\frac{1}{2}}^n(\xi)\big)\,\mathrm{d}\xi$. This enables us to rewrite the discrete forward Euler scheme (7.12) in the equivalent incremental form

$$\begin{aligned}\mathbf{u}_\nu^{n+1} - \mathbf{u}_\nu^n &= \frac{\Delta t}{2\Delta x}\Big[\Big(\mathbf{g}(\mathbf{v}_{\nu+1}^n) - \mathbf{g}(\mathbf{v}_\nu^n)\Big) + Q_{\nu+\frac{1}{2}}\Delta\mathbf{v}_{\nu+\frac{1}{2}}^n \\ &\qquad + \Big(\mathbf{g}(\mathbf{v}_\nu^n) - \mathbf{g}(\mathbf{v}_{\nu-1}^n)\Big) + Q_{\nu-\frac{1}{2}}\Delta\mathbf{v}_{\nu-\frac{1}{2}}^n\Big] \\ &= \frac{\Delta t}{2\Delta x}\Big[\Big(B_{\nu+\frac{1}{2}} + Q_{\nu+\frac{1}{2}}\Big)\Delta\mathbf{v}_{\nu+\frac{1}{2}}^n + \Big(B_{\nu-\frac{1}{2}} + Q_{\nu-\frac{1}{2}}\Big)\Delta\mathbf{v}_{\nu-\frac{1}{2}}^n\Big].\end{aligned}$$

Finally, we recall the viscosity matrix $Q_{\nu+\frac{1}{2}} = Q_{\nu+\frac{1}{2}}^* + D_{\nu+\frac{1}{2}}$. This enables us to rewrite the last expression as

$$\mathbf{u}_\nu^{n+1} - \mathbf{u}_\nu^n = \frac{\Delta t}{2\Delta x}\Big[\Big(\widetilde{B}_{\nu+\frac{1}{2}} + D_{\nu+\frac{1}{2}}\Big)\Delta\mathbf{v}_{\nu+\frac{1}{2}}^n + \Big(\widetilde{B}_{\nu-\frac{1}{2}} + D_{\nu-\frac{1}{2}}\Big)\Delta\mathbf{v}_{\nu-\frac{1}{2}}^n\Big], \tag{7.17}$$

where

$$\widetilde{B} := B + Q^* = \int_{\xi=-\frac{1}{2}}^{\frac{1}{2}} (1+2\xi)B\left(\mathbf{v}^n_{\nu+\frac{1}{2}}\right) \mathrm{d}\xi = B_{\nu+\frac{1}{2}} + \mathcal{O}\left(|\Delta \mathbf{v}_{\nu+\frac{1}{2}}|\right).$$

Squaring (7.17) we find

$$|\Delta \mathbf{u}_\nu^{n+\frac{1}{2}}|^2 \le \frac{1}{2}\left(\frac{\Delta t}{\Delta x}\right)^2 \left[\left\langle \Delta \mathbf{v}_{\nu+\frac{1}{2}}, \left(\widetilde{B}_{\nu+\frac{1}{2}} + D_{\nu+\frac{1}{2}}\right)^2 \Delta \mathbf{v}_{\nu+\frac{1}{2}}\right\rangle \right.$$
$$\left. + \left\langle \Delta \mathbf{v}_{\nu-\frac{1}{2}}, \left(\widetilde{B}_{\nu-\frac{1}{2}} + D_{\nu-\frac{1}{2}}\right)^2 \Delta \mathbf{v}_{\nu-\frac{1}{2}}\right\rangle\right]. \tag{7.18}$$

Compared with the spatial entropy dissipation in (7.5), we find that the forward Euler scheme is entropy-stable, $-\frac{\Delta t}{\Delta x}\mathcal{E}_\nu^{(x)}(\mathbf{v}^n) + \mathcal{E}_\nu^{FE}(\mathbf{v}^{n+\frac{1}{2}}) \le 0$, provided D is sufficiently large that

$$K^3\left(\frac{\Delta t}{\Delta x}\right)^2 \left(\widetilde{B}_{\nu+\frac{1}{2}} + D_{\nu+\frac{1}{2}}\right)^2 \le \frac{\Delta t}{\Delta x} D_{\nu+\frac{1}{2}}. \tag{7.19}$$

We consider the two prototype examples of centred and upwind schemes. If we set $D_{\nu+\frac{1}{2}} = \frac{\Delta x}{2\Delta t} I_{N\times N}$ we obtain the centred *modified Lax–Friedrichs* (MLxF) scheme (*e.g.*, Tadmor (1984*b*))

$$\mathbf{u}_\nu^{n+1} = \frac{1}{4}\left(\mathbf{u}^n_{\nu+1} + 2\mathbf{u}^n_\nu + \mathbf{u}^n_{\nu-1}\right) + \frac{\Delta t}{2\Delta x}\left[\mathbf{f}\left(\mathbf{u}^n_{\nu+1}\right) - \mathbf{f}\left(\mathbf{u}^n_{\nu-1}\right)\right]. \tag{7.20}$$

To simplify matters, we consider the symmetric case, where the Bs are turned into As, and (7.19) with condition number $K = 1$ yields the entropy stability of the MLxF for sufficiently small CFL number

$$\frac{\Delta t}{\Delta x} \max_\lambda |\lambda(\widetilde{A})| \le \frac{\sqrt{2}-1}{2}.$$

Similarly, the viscosity coefficient matrix, $D_{\nu+\frac{1}{2}} = |\widetilde{A}_{\nu+\frac{1}{2}}|$ leads to the upwind scheme

$$\mathbf{u}_\nu^{n+1} = \mathbf{u}^n_\nu - \frac{\Delta t}{2\Delta x}\left[\mathbf{f}\left(\mathbf{u}^n_{\nu+1}\right) - \mathbf{f}\left(\mathbf{u}^n_{\nu-1}\right)\right]$$
$$+ \frac{1}{2\Delta x_\nu}\left[\left(|A_{\nu+\frac{1}{2}} + Q^*_{\nu+\frac{1}{2}}| + Q^*_{\nu+\frac{1}{2}}\right)\Delta \mathbf{v}^n_{\nu+\frac{1}{2}}\right.$$
$$\left. - \left(|A_{\nu-\frac{1}{2}} + Q^*_{\nu+\frac{1}{2}}| + Q^*_{\nu-\frac{1}{2}}\right)\Delta \mathbf{v}^n_{\nu-\frac{1}{2}}\right]. \tag{7.21}$$

According to (7.19), entropy stability follows under the CFL condition

$$\frac{\Delta t}{\Delta x} \max_{\lambda,\nu} \left|\lambda\left(\widetilde{A}_{\nu+\frac{1}{2}}\right)\right| \le 1/4, \quad \widetilde{A} := A + Q^*.$$

We conclude this fully explicit example with several remarks.

(1) *CFL optimality.* In both examples of the centred and upwind schemes, entropy stability is obtained under less than optimal CFL conditions, which is due to less than optimal bounds on the entropy production rate, $\mathcal{E}_\nu^{FE}$. In particular, the resulting entropy stability condition (7.19) excludes the entropy stability of second-order fully discrete schemes, which are identified with Lax–Wendroff (LxW) dissipation matrices of order $D_{\nu+\frac{1}{2}} \sim \frac{\Delta t}{\Delta x}\tilde{A}^2_{\nu+\frac{1}{2}}$.

(2) *Entropy stability of Lax–Wendroff scheme.* For first-order accurate schemes, sharp CFL entropy stability conditions would follow from an alternative approach discussed in Section 8 below. The question of entropy stability for second-order fully discrete schemes, however, is more delicate. It would be desirable to refine the above arguments to obtain an improved CFL condition, which in particular entertains the second-order case. For a systematic approach to enforcing entropy stability of the second-order scalar LxW scheme we refer to Majda and Osher (1978, 1979). We note the limitation that entropy stability places on fully discrete forward Euler time discretization, namely, higher-order accuracy requires spatial stencils with more than three points (Schonbek 1982). Part of the difficulty is due to lack of fully discrete entropy-conservative schemes (LeFloch and Rohde 2000, Theorem 6.1). This requires entropy production bounds of the kind discussed in the current example. Sharp entropy production bounds in the scalar case can be found in Chalons and LeFloch (2001*b*).

(3) *Entropy stability with distinguished waves.* Finally, we remark that an extension based on entropy-conservative schemes of the 'second kind' discussed in Section 6 would lead to an entropy stability statement under a refinement of the CFL statement (7.19), similar to the semi-discrete discussion in Section 6.

Example 7.3. (Crank–Nicolson time discretization) The fully explicit Euler time discretization does not conserve entropy except in the case of linear fluxes (LeFloch and Rohde 2000). Consequently, both the fully explicit and fully implicit Euler differencing do not respect (nonlinear) entropy conservation, independent of the spatial discretization. Fully discrete entropy conservation is offered by Crank–Nicolson time differencing. In its standard version, for example, Richtmyer and Morton (1967) and Gustafsson *et al.* (1995), time is replaced by divided differences centred at $t^{n+\frac{1}{2}} := \frac{1}{2}(t^n + t^{n+1})$ and spatial terms are evaluated at the mid-value, $\frac{1}{2}(\mathbf{v}^n + \mathbf{v}^{n+1})$. In the present nonlinear context, the mid-value should be weighted by the specific entropy function we are dealing with. We set

$$\overline{\mathbf{v}}^{n+\frac{1}{2}} := \int_{\xi=-\frac{1}{2}}^{\frac{1}{2}} \mathbf{v}\left(\frac{1}{2}(\mathbf{u}^n + \mathbf{u}^{n+1}) + \xi\Delta\mathbf{u}^{n+\frac{1}{2}}\right) \mathrm{d}\xi, \quad \Delta\mathbf{u}^{n+\frac{1}{2}} = \mathbf{u}^{n+1} - \mathbf{u}^n, \tag{7.22}$$

and we discretize (7.1) by the (generalized) Crank–Nicolson scheme

$$\mathbf{u}_\nu^{n+1} = \mathbf{u}_\nu^n - \frac{\Delta t}{\Delta x}\left[\mathbf{g}_{\nu+\frac{1}{2}}\left(\overline{\mathbf{v}}^{n+\frac{1}{2}}\right) - \mathbf{g}_{\nu-\frac{1}{2}}\left(\overline{\mathbf{v}}^{n+\frac{1}{2}}\right)\right]. \tag{7.23}$$

Noting that $\langle \overline{\mathbf{v}}^{n+\frac{1}{2}}, \mathbf{u}^{n+1} - \mathbf{u}^n \rangle = U(\mathbf{u}^{n+1}) - U(\mathbf{u}^n)$, we conclude the following.

Corollary 7.4. The Crank–Nicolson scheme (7.22), (7.23) is entropy-stable (and, respectively, entropy-conservative), if and only if the semi-discrete scheme associated with the numerical flux $\mathbf{g}(\cdot)$ is entropy-stable (respectively, entropy-conservative).

Observe that in the symmetric case, $\overline{\mathbf{v}}^{n+\frac{1}{2}} = \frac{1}{2}(\mathbf{u}^n + \mathbf{u}^{n+1})$, and (7.23) recovers the standard differencing centred around $t^{n+\frac{1}{2}}$.

We conclude this section by referring the reader to the recent work of LeFloch and his co-workers (LeFloch and Rohde 2000, LeFloch *et al.* 2002) for a general framework along these lines for entropy stability of fully discrete schemes.

8. Entropy stability by the homotopy approach

We study the cell entropy inequality for general difference schemes written in their viscosity form corresponding to (5.24):

$$\begin{aligned}\mathbf{u}_\nu^{n+1} = \mathbf{u}_\nu^n &- \frac{\Delta t}{2\Delta x}\left[\mathbf{f}(\mathbf{u}_{\nu+1}^n) - \mathbf{f}(\mathbf{u}_{\nu-1}^n)\right] \\ &+ \frac{\Delta t}{2\Delta x}\left[P_{\nu+\frac{1}{2}}(\mathbf{u}_{\nu+1}^n - \mathbf{u}_\nu^n) - P_{\nu-\frac{1}{2}}(\mathbf{u}_\nu^n - \mathbf{u}_{\nu-1}^n)\right].\end{aligned} \tag{8.1}$$

We decompose $\mathbf{u}_\nu^{n+1} = \left(\mathbf{u}_{\nu+\frac{1}{2}}^{n+1} + \mathbf{u}_{\nu-\frac{1}{2}}^{n+1}\right)/2$ where

$$\begin{aligned}\mathbf{u}_{\nu+\frac{1}{2}}^{n+1} &:= \mathbf{u}_\nu^n - \frac{\Delta t}{\Delta x}\left[\mathbf{f}(\mathbf{u}_{\nu+1}^n) - \mathbf{f}(\mathbf{u}_\nu^n)\right] + \frac{\Delta t}{\Delta x}P_{\nu+\frac{1}{2}}(\mathbf{u}_{\nu+1}^n - \mathbf{u}_\nu^n), \\ \mathbf{u}_{\nu-\frac{1}{2}}^{n+1} &:= \mathbf{u}_\nu^n - \frac{\Delta t}{\Delta x}\left[\mathbf{f}(\mathbf{u}_\nu^n) - \mathbf{f}(\mathbf{u}_{\nu-1}^n)\right] - \frac{\Delta t}{\Delta x}P_{\nu+\frac{1}{2}}(\mathbf{u}_\nu^n - \mathbf{u}_{\nu-1}^n),\end{aligned}$$

and we study the entropy inequality for each term. This decomposition into left- and right-handed stencils in the context of cell entropy inequality was first introduced in Tadmor (1984*b*). We begin by considering $\mathbf{u}^{n+1}_{\nu+\frac{1}{2}}$. To this end, we set $\mathbf{u}^n_{\nu+\frac{1}{2}}(s) := \mathbf{u}^n_\nu + s(\mathbf{u}^n_{\nu+1} - \mathbf{u}^n_\nu)$ and the following inequality is sought (here and below, $\Delta\mathbf{u} := \mathbf{u}^n_{\nu+1} - \mathbf{u}^n_\nu$):

$$\mathcal{I}_+ := U\Big(\mathbf{u}^{n+1}_{\nu+\frac{1}{2}}\Big) - U(\mathbf{u}^n_\nu) + \frac{\Delta t}{\Delta x}\big[F(\mathbf{u}^n_{\nu+1}) - F(\mathbf{u}^n_\nu)\big]$$
$$- \frac{\Delta t}{\Delta x}\int_{s=0}^{1}\Big\langle U'\Big(\mathbf{u}^n_{\nu+\frac{1}{2}}(s)\Big), P_{\nu+\frac{1}{2}}\Delta\mathbf{u}\Big\rangle\,\mathrm{d}s \le 0. \tag{8.2}$$

We refer to the last statement as a *quasi-cell entropy inequality* since the last expression on the right is not conservative. To verify (8.2) we proceed as follows. We set

$$\mathbf{u}^{n+1}_{\nu+\frac{1}{2}}(s) := \mathbf{u}^n_\nu - \frac{\Delta t}{\Delta x}\Big[\mathbf{f}\Big(\mathbf{u}^n_{\nu+\frac{1}{2}}(s)\Big) - \mathbf{f}(\mathbf{u}^n_\nu)\Big] + \frac{\Delta t}{\Delta x}P_{\nu+\frac{1}{2}}\Big(\mathbf{u}^n_{\nu+\frac{1}{2}}(s) - \mathbf{u}^n_\nu\Big).$$

Noting that $\mathbf{u}^{n+1}_{\nu+\frac{1}{2}}(0) = \mathbf{u}^n_\nu$ and $\mathbf{u}^{n+1}_{\nu+\frac{1}{2}}(1) = \mathbf{u}^{n+1}_{\nu+\frac{1}{2}}$, we compute

$$\begin{aligned}
&U\Big(\mathbf{u}^{n+1}_{\nu+\frac{1}{2}}\Big) - U(\mathbf{u}^n_\nu)\\
&\quad= \int_{s=0}^{1}\frac{\mathrm{d}}{\mathrm{d}s}U\Big(\mathbf{u}^{n+1}_{\nu+\frac{1}{2}}(s)\Big)\,\mathrm{d}s\\
&\quad= \int_{s=0}^{1}\Big\langle U'\Big(\mathbf{u}^{n+1}_{\nu+\frac{1}{2}}(s)\Big), \Big(-\frac{\Delta t}{\Delta x}A\Big(\mathbf{u}^{n+1}_{\nu+\frac{1}{2}}(s)\Big) + \frac{\Delta t}{\Delta x}P_{\nu+\frac{1}{2}}\Big)\Delta\mathbf{u}\Big\rangle\,\mathrm{d}s
\end{aligned}$$

and

$$\begin{aligned}
\frac{\Delta t}{\Delta x}\big[F(\mathbf{u}^n_{\nu+1}) - F(\mathbf{u}^n_\nu)\big] &= \frac{\Delta t}{\Delta x}\int_{s=0}^{1}\Big\langle F'\Big(\mathbf{u}^n_{\nu+\frac{1}{2}}(s)\Big), (\mathbf{u}^n_{\nu+1} - \mathbf{u}^n_\nu)\Big\rangle\,\mathrm{d}s\\
&= \frac{\Delta t}{\Delta x}\int_{s=0}^{1}\Big\langle U'\Big(\mathbf{u}^n_{\nu+\frac{1}{2}}(s)\Big)A\Big(\mathbf{u}^n_{\nu+\frac{1}{2}}(s)\Big), \Delta\mathbf{u}\Big\rangle\,\mathrm{d}s.
\end{aligned}$$

Adding the last two equalities yields

$$\mathcal{I}_+ = \int_{s=0}^{1}\Big\langle U'\Big(\mathbf{u}^{n+1}_{\nu+\frac{1}{2}}(s)\Big) - U'\Big(\mathbf{u}^n_{\nu+\frac{1}{2}}(s)\Big),$$
$$-\frac{\Delta t}{\Delta x}\Big(P_{\nu+\frac{1}{2}} - A\Big(\mathbf{u}^n_{\nu+\frac{1}{2}}(s)\Big)\Big)\Delta\mathbf{u}\Big\rangle\,\mathrm{d}s. \tag{8.3}$$

Next, we introduce

$$\mathbf{u}^n_{\nu+\frac{1}{2}}(r,s) := \mathbf{u}^n_{\nu+\frac{1}{2}}(s) + r\Big(\mathbf{u}^n_\nu - \mathbf{u}^n_{\nu+\frac{1}{2}}(s)\Big) \equiv \mathbf{u}^n_\nu + s(1-r)\big(\mathbf{u}^n_{\nu+1} - \mathbf{u}^n_\nu\big), \tag{8.4}$$

and we set

$$\mathbf{u}^{n+1}_{\nu+\frac{1}{2}}(r,s) = \mathbf{u}^{n}_{\nu+\frac{1}{2}}(r,s) - \frac{\Delta t}{\Delta x}\Big(\mathbf{f}\Big(\mathbf{u}^{n}_{\nu+\frac{1}{2}}(s)\Big) - \mathbf{f}\Big(\mathbf{u}^{n}_{\nu+\frac{1}{2}}(r,s)\Big)\Big) + \frac{\Delta t}{\Delta x}P_{\nu+\frac{1}{2}}\Big(\mathbf{u}^{n}_{\nu+\frac{1}{2}}(s) - \mathbf{u}^{n}_{\nu+\frac{1}{2}}(r,s)\Big)$$

so that $\mathbf{u}^{n+1}_{\nu+\frac{1}{2}}(0,s) = \mathbf{u}^{n}_{\nu+\frac{1}{2}}(s)$ and $\mathbf{u}^{n+1}_{\nu+\frac{1}{2}}(1,s) = \mathbf{u}^{n+1}_{\nu+\frac{1}{2}}(s)$. This then yields

$$\begin{aligned} U'\Big(\mathbf{u}^{n+1}_{\nu+\frac{1}{2}}(s)\Big) - U'\Big(\mathbf{u}^{n}_{\nu+1}(s)\Big) &= \int_{r=0}^{1} \frac{\mathrm{d}}{\mathrm{d}r}U'\Big(\mathbf{u}^{n+1}_{\nu+\frac{1}{2}}(r,s)\Big)\,\mathrm{d}r \\ &= -s\int_{r=0}^{1} U''\Big(\mathbf{u}^{n+1}_{\nu+\frac{1}{2}}(r,s)\Big)\,\mathrm{d}r\Big(I + \frac{\Delta t}{\Delta x}A\Big(\mathbf{u}^{n}_{\nu+\frac{1}{2}}(r,s)\Big) - \frac{\Delta t}{\Delta x}P_{\nu+\frac{1}{2}}\Big)\Delta\mathbf{u}. \end{aligned}$$

Inserting the last expression into the right-hand side of (8.3) we end up with

$$\begin{aligned} \mathcal{I}_{+} = -\int_{r,s=0}^{1} s\Big\langle\Big(I + \frac{\Delta t}{\Delta x}A\Big(\mathbf{u}^{n}_{\nu+\frac{1}{2}}(r,s)\Big) - \frac{\Delta t}{\Delta x}P_{\nu+\frac{1}{2}}\Big)\Delta\mathbf{u}, \\ U''\Big(\mathbf{u}^{n+1}_{\nu+\frac{1}{2}}(r,s)\Big)\Big(-\frac{\Delta t}{\Delta x}A\Big(\mathbf{u}^{n}_{\nu+\frac{1}{2}}(s)\Big) + \frac{\Delta t}{\Delta x}P_{\nu+\frac{1}{2}}\Big)\Delta\mathbf{u}\Big\rangle\,\mathrm{d}r\,\mathrm{d}s. \end{aligned} \tag{8.5}$$

To continue, we focus our attention on two prototype cases.

(i) *The scalar case.* The positivity of the last expression on the right of (8.5) follows from a CFL condition

$$\frac{\Delta t}{\Delta x}a\Big(u^{n}_{\nu+\frac{1}{2}}(s)\Big) \leq \frac{\Delta t}{\Delta x}p_{\nu+\frac{1}{2}} \leq I + \frac{\Delta t}{\Delta x}a\Big(u^{n}_{\nu+\frac{1}{2}}(r,s)\Big). \tag{8.6}$$

In a similar manner, the CFL condition

$$-\frac{\Delta t}{\Delta x}a\Big(u^{n}_{\nu-\frac{1}{2}}(s)\Big) \leq \frac{\Delta t}{\Delta x}p_{\nu-\frac{1}{2}} \leq I - \frac{\Delta t}{\Delta x}a\Big(u^{n}_{\nu-\frac{1}{2}}(r,s)\Big)$$

yields the quasi-cell entropy inequality

$$\begin{aligned} \mathcal{I}_{-} := U\Big(u^{n+1}_{\nu-\frac{1}{2}}\Big) - U(u^{n}_{\nu}) + \frac{\Delta t}{\Delta x}\Big(F(u^{n}_{\nu}) - F(\mathbf{u}^{n}_{\nu-1})\Big) \\ + \frac{\Delta t}{\Delta x}\int_{s=0}^{1}\Big\langle U'\Big(u^{n}_{\nu-\frac{1}{2}}(s)\Big), p_{\nu-\frac{1}{2}}(u^{n}_{\nu} - u^{n}_{\nu-1})\Big\rangle\,\mathrm{d}s \leq 0. \end{aligned} \tag{8.7}$$

Again, the last expression is nonconservative, but together with (8.2) we end up with the cell entropy inequality.

Corollary 8.1. Consider the fully discrete scalar scheme (8.1) and assume the CFL condition

$$\frac{\Delta t}{\Delta x}\Big|a\Big(u^{n}_{\nu+\frac{1}{2}}(s)\Big)\Big| \leq \frac{\Delta t}{\Delta x}p_{\nu+\frac{1}{2}} \leq I - \frac{\Delta t}{\Delta x}\Big|a\Big(u^{n}_{\nu+\frac{1}{2}}(r,s)\Big)\Big| \tag{8.8}$$

is fulfilled. Then the following cell entropy inequality holds:

$$\begin{aligned}
U\left(u_\nu^{n+1}\right) &\le \frac{1}{2}\left(U\left(u_{\nu+\frac{1}{2}}^{n+1}\right) + U\left(u_{\nu-\frac{1}{2}}^{n+1}\right)\right)\\
&\le U\left(u_\nu^n\right) - \frac{\Delta t}{\Delta x}\left[F\left(u_{\nu+1}^n\right) - F\left(u_{\nu-1}^n\right)\right]\\
&\quad + \frac{\Delta t}{2\Delta x}\left[\int_{s=0}^1 \left\langle U'\left(u_{\nu+\frac{1}{2}}^n(s)\right), p_{\nu+\frac{1}{2}}\left(u_{\nu+1}^n - u_\nu^n\right)\right\rangle ds\right.\\
&\quad \left. - \int_{s=0}^1 \left\langle U'\left(u_{\nu-\frac{1}{2}}^n(s)\right), p_{\nu-\frac{1}{2}}\left(u_\nu^n - u_{\nu-1}^n\right)\right\rangle ds\right].
\end{aligned}$$

Next, we extend our discussion to systems of conservation laws.

(ii) *Symmetric systems* of conservation laws with the quadratic entropy, $U(\mathbf{u}) = |\mathbf{u}|^2/2$. We start by setting $C(s) := P_{\nu+\frac{1}{2}} - A(\mathbf{u}_{\nu+\frac{1}{2}}^n(s))$, and noting the (r,s)-variables in (8.4), we find that

$$A\left(\mathbf{u}_{\nu+\frac{1}{2}}^n(r,s)\right) - P_{\nu+\frac{1}{2}} = -C((1-r)s)).$$

Change of variables, $t := (1-r)s$, in (8.5) then yields

$$\mathcal{I}_+ = \int_{s=0}^1 \int_{t=0}^s \left\langle \left(I - \frac{\Delta t}{\Delta x}C(t)\right)\Delta\mathbf{u}, \frac{\Delta t}{\Delta x}C(s)\Delta\mathbf{u}\right\rangle dt\, ds. \tag{8.9}$$

We now make the first requirement of positivity, assuming $C(\cdot) \ge 0$; then the positivity of $\mathcal{I}_+$ follows if and only if the corresponding eigenvalues satisfy $\lambda\left[C(s)\left(I - \frac{\Delta t}{\Delta x}C(t)\right)\right] \ge 0$. But $C(s)\left(I - \frac{\Delta t}{\Delta x}C(t)\right)$ is similar to $C^{\frac{1}{2}}(s)\left(I - \frac{\Delta t}{\Delta x}C(t)\right)C^{\frac{1}{2}}$, which is congruent to, and hence by Sylvester's theorem has the same number of nonnegative eigenvalues as, $I - \frac{\Delta t}{\Delta x}C(t)$. This leads to the second requirement, $\frac{\Delta t}{\Delta x}\lambda(C(\cdot)) \le 1$. Recall that $C(s) = P_{\nu+\frac{1}{2}} - A\left(\mathbf{u}_{\nu+\frac{1}{2}}^n(s)\right)$ is symmetric, and hence the last two requirements amount to the same CFL condition we met earlier in connection with the scalar case (8.6):

$$\frac{\Delta t}{\Delta x}A\left(\mathbf{u}_{\nu+\frac{1}{2}}^n(s)\right) \le \frac{\Delta t}{\Delta x}P_{\nu+\frac{1}{2}} \le I + \frac{\Delta t}{\Delta x}A\left(\mathbf{u}_{\nu+\frac{1}{2}}(r,s)\right).$$

In a similar manner, the CFL condition

$$-\frac{\Delta t}{\Delta x}A\left(\mathbf{u}_{\nu+\frac{1}{2}}^n(s)\right) \le \frac{\Delta t}{\Delta x}P_{\nu+\frac{1}{2}} \le I - \frac{\Delta t}{\Delta x}A\left(\mathbf{u}_{\nu+\frac{1}{2}}(r,s)\right)$$

yields the quasi-cell entropy inequality for $\mathbf{u}_{\nu-\frac{1}{2}}^{n+1}$, and the following conclusion.

Corollary 8.2. Consider the fully discrete scheme (8.1) consistent with the symmetric system (2.1) and assume the CFL condition[5]

$$\frac{\Delta t}{\Delta x}\left|A\left(\mathbf{u}^n_{\nu+\frac{1}{2}}(s)\right)\right| \leq \frac{\Delta t}{\Delta x}P_{\nu+\frac{1}{2}} \leq I - \frac{\Delta t}{\Delta x}\left|A\left(\mathbf{u}^n_{\nu+\frac{1}{2}}(r,s)\right)\right| \tag{8.10}$$

is fulfilled. Then the following cell entropy inequality holds for the quadratic entropy pair $U(\mathbf{u}) = |\mathbf{u}|^2/2, F(\mathbf{u}) = \int^{\mathbf{u}} \mathbf{f}(\mathbf{w})\,\mathrm{d}\mathbf{w} - \langle \mathbf{u}, \mathbf{f}(\mathbf{u})\rangle$:

$$\begin{aligned}
U(\mathbf{u}^{n+1}_\nu) &\leq \frac{1}{2}\left(U\left(\mathbf{u}^{n+1}_{\nu+\frac{1}{2}}\right) + U\left(\mathbf{u}^{n+1}_{\nu-\frac{1}{2}}\right)\right)\\
&\leq U(\mathbf{u}^n_\nu) - \frac{\Delta t}{2\Delta x}\left(F(\mathbf{u}^n_{\nu+1}) - F(\mathbf{u}^n_{\nu-1})\right)\\
&\quad + \frac{\Delta t}{2\Delta x}\left(\int_{s=0}^1 \left\langle U'\left(\mathbf{v}^n_{\nu+\frac{1}{2}}(s)\right), P_{\nu+\frac{1}{2}}(\mathbf{u}^n_{\nu+1} - \mathbf{u}^n_\nu)\right\rangle \mathrm{d}s\right.\\
&\quad \left. - \int_{s=0}^1 \left\langle U'\left(\mathbf{u}^n_{\nu-\frac{1}{2}}(s)\right), P_{\nu-\frac{1}{2}}(\mathbf{u}^n_\nu - \mathbf{u}^n_{\nu-1})\right\rangle \mathrm{d}s\right).
\end{aligned}$$

We demonstrate the application of Corollaries 8.1 and 8.2 with two prototype examples of centred and upwind schemes.

Example 8.3. (Modified Lax–Friedrichs scheme) Here we set $P_{\nu+\frac{1}{2}} = \frac{\Delta x}{2\Delta t}I_{N\times N}$, leading to the *modified* Lax–Friedrichs scheme (7.20)

$$\mathbf{u}^{n+1}_\nu = \frac{1}{4}(\mathbf{u}^n_{\nu-1} - 2\mathbf{u}^n_\nu + \mathbf{u}^n_{\nu+1}) + \frac{\Delta t}{2\Delta x}\left[\mathbf{f}(\mathbf{u}^n_{\nu+1}) - \mathbf{f}(\mathbf{u}^n_{\nu-1})\right].$$

The modified Lax–Friedrichs scheme is entropy-stable with respect to the quadratic entropy function (for symmetric systems) and for all convex entropies (for scalar equations), provided the CFL condition (8.8), (8.10) holds, which amounts to

$$\frac{\Delta t}{2\Delta x}\sup_{s,\lambda}\left|\lambda\left(A\left(\mathbf{u}_{\nu+\frac{1}{2}}(s)\right)\right)\right| \leq \frac{1}{2}.$$

A linearized von Neumann stability analysis reveals that this CFL condition is sharp.

Remark. The original homotopy argument in this context of entropy stability is due to Lax (1971), where he proves the entropy stability of the Lax–Friedrichs (LxF) scheme, corresponding to $P_{\nu+\frac{1}{2}} = \frac{\Delta x}{\Delta t}I_{N\times N}$. In this

[5] We recall (consult (5.34)) that $|A|$ stands for the absolute value of A, defined by its spectral decomposition

$$|A| = R\begin{bmatrix} |a_1| & & \\ & \ddots & \\ & & |a_N| \end{bmatrix}R^{-1}.$$

special case of a two-point LxF stencil, we can apply the homotopy argument on the full stencil of the scheme. In the general case of essentially three-point schemes, (8.1), we follow the decomposition into left- and right-handed stencils and, as in Tadmor (1984*b*), this restricts the maximal viscosity coefficient to that of the modified LxF scheme. For a recent extension to entropy stability under an optimal CFL condition in the scalar case, consult Makridakis and Perthame (2003).

We now turn to discussion of entropy stability with the minimal amount of viscosity.

Example 8.4. (Upwind scheme) We set $P_{\nu+\frac{1}{2}} = p\big(A\big(\mathbf{u}^n_{\nu+\frac{1}{2}}(s)\big)\big)$ with $p(\cdot)$ being any viscosity function satisfying $p(\cdot) \geq |\cdot|$: consult Example 5.6. The typical example is the upwind scheme

$$\begin{aligned}\mathbf{u}^{n+1}_\nu &= \mathbf{u}^n_\nu - \frac{\Delta t}{2\Delta x}\big[\mathbf{f}(\mathbf{u}^n_{\nu+1}) - \mathbf{f}(\mathbf{u}^n_{\nu-1})\big] \qquad (8.11)\\ &+ \frac{\Delta t}{2\Delta x}\Big[\Big(\sup_s \Big|A\Big(\mathbf{u}^n_{\nu+\frac{1}{2}}(s)\Big)\Big|\Big)\Delta\mathbf{u}^n_{\nu+\frac{1}{2}} - \Big(\sup_s \Big|A\Big(\mathbf{u}^n_{\nu-\frac{1}{2}}(s)\Big)\Big)\Big|\Delta\mathbf{u}^n_{\nu-\frac{1}{2}}\Big].\end{aligned}$$

We find that the upwind scheme is entropy-stable for the quadratic entropy function (for symmetric systems) and for all convex entropies (for scalar equations), provided the CFL condition (8.8), (8.10) holds, which amounts to

$$\frac{\Delta t}{2\Delta x}\sup_{s,\lambda}\Big|\lambda\Big(A\Big(\mathbf{u}_{\nu+\frac{1}{2}}(s)\Big)\Big)\Big| \leq 1. \qquad (8.12)$$

Again, a linearized von Neumann stability analysis reveals that the CFL condition is sharp.

We conclude this section with several remarks.

(1) *Comparison with Roe scheme.* Consider the numerical viscosity of the Roe-type scheme (5.24), (5.33), $P_{\nu+\frac{1}{2}} = |\overline{A}_{\nu+\frac{1}{2}}|$. A comparison with the upwind scheme (8.11), $P_{\nu+\frac{1}{2}} = \sup_s \big|A\big(\mathbf{u}^n_{\nu-\frac{1}{2}}(s)\big)\big|$, reveals that the entropy stability of the latter is explained by taking into account *all intermediate* values between two neighbouring values, $\mathbf{u}_\nu$ and $\mathbf{u}_{\nu+1}$. An alternative entropy correction discussed in Example 5.10 adds the additional term of order $\mathcal{O}(|\Delta\mathbf{u}|)$ to compensate for the missing intermediate values.

(2) *Extensions.* The entropy stability results in this section are based on a homotopy approach. Our initial point was the essentially three-point scheme (8.1). The same homotopy approach refinement, starting with

the entropy-conservative schemes of the 'second kind' in Theorem 6.1, would yield a refinement of the above entropy stability statements. In particular, (i) the intermediate values sought in the upwind viscosities, (8.12), would be confined to each subpath, $\{\mathbf{u}^j_{\nu+\frac{1}{2}}\}_{j=1}^N$. This leads to an entropy stability criterion which distinguishes between shock, rarefaction and contact waves; and (ii) starting the present homotopy approach with the entropy variables, rather than the entropy-conservative formulation in (8.1), enables us to extend the above entropy stability statements to the general nonsymmetric case.

(3) *Second-order accuracy.* The CFL conditions (8.8) and (8.10) are restricted to first-order accurate schemes. A similar, more careful computation along these lines enables us to treat the entropy stability of *second-order* accurate schemes. For a scalar entropy stability analysis along the lines of the second-order accurate Nessyahu–Tadmor central scheme, we refer to Nessyahu and Tadmor (1990, Appendix).

(4) *The scalar case.* More could be said on the scalar case, and we should mention a considerable amount of work in this direction. In particular, the entropy stability of fully discrete second-order schemes was systematically analysed in Osher and Tadmor (1988) (see the follow-up in Aiso (1993)). A key to enforcing entropy stability in this case is the use of *all* intermediate values within critical cells: consult Coquel and LeFloch (1995), Bouchut, Bourdarias and Perthame (1996), Johnson and Szepessy (1986) and Yang (1996c). Otherwise, entropy stability is enforced for a single entropy, as in Osher and Tadmor (1988), for example, or high-order accuracy should be given up (Yang 1996*b*).

9. Higher-order extensions

We generalize the construction of second-order entropy-conservative schemes to higher orders. To this end, we revisit the original derivation of the second-order entropy-conservative schemes (Tadmor 1986*b*), using finite element discretization. We begin with the weak formulation of the systems of conservation laws (2.6),

$$\int_\Omega \left\langle \mathbf{w}(x,t), \frac{\partial}{\partial t}\mathbf{u}(\mathbf{v}) \right\rangle = \int_\Omega \left\langle \frac{\partial}{\partial x}\mathbf{w}(x,t), \mathbf{g}(\mathbf{v}) \right\rangle \mathrm{d}x\,\mathrm{d}t, \quad \Omega \subset \mathbb{R}\times(0,T), \tag{9.1}$$

where $\mathbf{w}(\cdot)$ is an arbitrary $C_0^\infty(\Omega)$ test function. The key point is the use of the entropy variables, which enables us to use the standard finite element framework where both the primary computed solution $\mathbf{v}$ and the test function $\mathbf{w}$ belong to the same finite-dimensional scale of spaces. In particular, let the trial solution $\hat{\mathbf{v}} = \sum_\mu \mathbf{v}_\mu(t)\hat{H}_\mu(x)$ be chosen from the typical finite element space spanned by the C^0 'hat functions'

$$\hat{H}_\mu(x) = \begin{bmatrix} \frac{x-x_{\mu-1}}{x_\mu - x_{\mu-1}}, & x_{\mu-1} \le x \le x_\mu, \\ \frac{x_{\mu+1}-x}{x_{\mu+1}-x_\mu}, & x_\mu \le x \le x_{\mu+1}. \end{bmatrix}$$

Testing (9.1) against $\mathbf{w}(x) = w(x) = \hat{H}_\nu(x)$, the right-hand side of (9.1) yields

$$\int_{x_{\nu-1}}^{x_{\nu+1}} \frac{\partial}{\partial x}\hat{H}_\nu(x)\mathbf{g}\left(\sum_\mu \mathbf{v}_\mu(t)\hat{H}_\mu(x)\right) \mathrm{d}x\, \mathrm{d}t = \\ -\left[\int_{\xi=-\frac{1}{2}}^{\frac{1}{2}} \mathbf{g}\left(\mathbf{v}_{\nu+\frac{1}{2}}(\xi)\right) \mathrm{d}\xi - \int_{\xi=-\frac{1}{2}}^{\frac{1}{2}} \mathbf{g}\left(\mathbf{v}_{\nu-\frac{1}{2}}(\xi)\right) \mathrm{d}\xi\right], \qquad (9.2)$$

where we employed a change of variables, expressed in terms of the usual $\mathbf{v}_{\nu+\frac{1}{2}} = \frac{1}{2}(\mathbf{v}_\nu + \mathbf{v}_{\nu+1}) + \xi\Delta\mathbf{v}_{\nu+\frac{1}{2}}$. A second-order mass lumping on the left of (9.1) leads to

$$\int_{x_{\nu-1}}^{x_{\nu+1}} \hat{H}_\nu \frac{\partial}{\partial t}\mathbf{u}\left(\sum_\mu \mathbf{v}_\mu(t)\hat{H}_\mu(x)\right) \mathrm{d}x\, \mathrm{d}t \\ = \Delta x_\nu \frac{\mathrm{d}}{\mathrm{d}t}\mathbf{u}(\mathbf{v}_\nu(t)) + \mathcal{O}\left(\left|\mathbf{v}_{\nu+\frac{1}{2}}\right|\right)^2. \qquad (9.3)$$

Equating (9.2) and (9.3) while neglecting the quadratic error term, we end up with the entropy-conservative scheme (5.2):

$$\frac{\mathrm{d}}{\mathrm{d}t}\mathbf{u}_\nu(t) = -\frac{1}{\Delta x_\nu}\left[\mathbf{g}^*_{\nu+\frac{1}{2}} - \mathbf{g}^*_{\nu-\frac{1}{2}}\right], \quad \mathbf{g}^*_{\nu+\frac{1}{2}} = \int_{\xi=-\frac{1}{2}}^{\frac{1}{2}} \mathbf{g}\left(\mathbf{v}_{\nu+\frac{1}{2}}(\xi)\right) \mathrm{d}\xi.$$

Mass lumping preserves the entropy conservation induced by (and, in a sense, built into) the weak formulation (9.1), upon choosing $\hat{\mathbf{w}}(x,t) = \hat{\mathbf{v}}(x,t)$,

$$0 = \int_\Omega \left[\left\langle \hat{\mathbf{v}}(x,t), \frac{\mathrm{d}}{\mathrm{d}t}\mathbf{u}(\hat{\mathbf{v}})\right\rangle - \left\langle \frac{\partial}{\partial x}\hat{\mathbf{v}}(x,t)\mathbf{g}(\hat{\mathbf{v}}(x,t))\right\rangle\right] \mathrm{d}x\, \mathrm{d}t \\ = \int_\Omega \left[\frac{\partial}{\partial t}U(\mathbf{u}(\hat{\mathbf{v}}(x,t))) + \frac{\partial}{\partial x}F(\mathbf{u}(\hat{\mathbf{v}}(x,t)))\right] \mathrm{d}x\, \mathrm{d}t. \qquad (9.4)$$

Using higher-order piecewise polynomial finite element building blocks will lead to entropy-conservative schemes of any desired order. We note in passing that, with the increased order, the size of the stencil increases. For example, piecewise quadratic splines would lead to five-point entropy-conservative stencils of the following form.

Theorem 9.1. (LeFloch and Rohde 2000, Section 3) Consider the semi-discrete scheme

$$\frac{\mathrm{d}}{\mathrm{d}t}\mathbf{u}_\nu(t) = -\frac{1}{\Delta x_\nu}\left[\mathbf{g}^*_{\nu+\frac{1}{2}} - \mathbf{g}^*_{\nu-\frac{1}{2}}\right], \tag{9.5}$$

with a numerical flux, $\mathbf{g}_{\nu+\frac{1}{2}} = \mathbf{g}(\mathbf{v}_{\nu-1}, \mathbf{v}_\nu, \mathbf{v}_{\nu+1}, \mathbf{v}_{\nu+2})$, given by

$$\begin{aligned}\mathbf{g}_{\nu+\frac{1}{2}} = &\int_{\xi=-\frac{1}{2}}^{\frac{1}{2}} \mathbf{g}\Big(\mathbf{v}_{\nu+\frac{1}{2}}(\xi)\Big)\,\mathrm{d}\xi \\ &- \frac{1}{12}\left[Q^{**}_{\nu+\frac{3}{2}}(\mathbf{v}_{\nu+2} - \mathbf{v}_{\nu+1}) - Q^{**}_{\nu-\frac{1}{2}}(\mathbf{v}_\nu - \mathbf{v}_{\nu-1})\right].\end{aligned} \tag{9.6}$$

Here, $Q^{**}_{\nu+\frac{1}{2}}$ is a secondary viscosity coefficient depending on

$$Q^{**}_{\nu+\frac{1}{2}} = Q^{**}(\mathbf{v}_{\nu-1}, \mathbf{v}_\nu, \mathbf{v}_{\nu+1}) \tag{9.7}$$

The resulting five-point scheme (9.5), (9.6), (9.7) is entropy-conservative, and it is of (at least) third-order accuracy provided $Q^{**}(\mathbf{v}, \mathbf{v}, \mathbf{v}) = B(\mathbf{v})$.

We observe that the higher-order accuracy is intimately linked to the wider stencil, beyond the essentially three-point schemes discussed in Section 5. Once more, these wider stencils could serve as the starting point for an entropy stability theory for higher (than second)-order entropy-stable schemes.

REFERENCES

R. Abramov and A. Majda (2002), 'Discrete approximations with additional conserved quantities: Deterministic and statistical behavior', preprint.

R. Abramov, G. Kovačič and A. Majda (2003), 'Hamiltonian structure and statistically relevant conserved quantities for the truncated Burgers–Hopf equation', *Comm. Pure Appl. Math.* **56**, 1–46.

H. Aiso (1993), 'Admissibility of difference approximations for scalar conservation laws', *Hiroshima Math. J.* **23**, 15–61.

A. Arakawa (1966), 'Computational design for long-term numerical integration of the equations of fluid motion: Two-dimensional incompressible flow, Part I', *J. Comput. Phys.* **1**, 119–143.

S. Bianchini and A. Bressan (2003), 'Vanishing viscosity solutions of nonlinear hyperbolic systems', *Ann. Math.* To appear.

F. Bouchut (2002), 'Entropy satisfying flux vector splitting and kinetic BGK models', *Numer. Math.*, DOI http://dx.doi.org/10.1007/s00211-002-0426-9.

F. Bouchut, Ch. Bourdarias and B. Perthame (1996), 'A MUSCL method satisfying all the numerical entropy inequalities', *Math. Comp.* **65**, 1439–1461

A. Bressan (2003), Viscosity solutions of nonlinear hyperbolic systems, in *Hyperbolic Problems: Theory, Numerics and Applications, Proc. 9th Hyperbolic Conference* (T. Hou and E. Tadmor, eds), Springer. To appear.

C. Chalons and P. LeFloch (2001*a*), 'High-order entropy-conservative schemes and kinetic relations for van der Waals fluids', *J. Comput. Phys.* **168**, 184–206.

C. Chalons and P. LeFloch (2001*b*), 'A fully discrete scheme for diffusive-dispersive conservation law', *Numer. Math.* **89**, 493–509.

G.-Q. Chen (2000), 'Compactness methods and nonlinear hyperbolic conservation laws: Some current topics on nonlinear conservation laws', in Vol. 15 of *AMS/IP Stud. Adv. Math.*, Amer. Math. Soc., Providence, RI, pp. 33–75.

F. Coquel and P. LeFloch (1995), 'An entropy satisfying MUSCL scheme for systems of conservation laws', *CR Acad. Sci. Paris, Série 1* **320**, 1263–1268.

M. G. Crandall and A. Majda, (1980) 'Monotone difference approximations for scalar conservation laws', *Math. Comp.* **34**, 1–21.

M. G. Crandall and L. Tartar (1980), 'Some relations between non-expansive and order preserving mapping', *Proc. Amer. Math. Soc.* **78**, 385–390.

C. Dafermos (2000) *Hyperbolic Conservation Laws in Continuum Mechanics*, Springer.

P. Deift and K. T. R. McLaughlin, (1998) 'A continuum limit of the Toda lattice', *Mem. Amer. Math. Soc.* **131**, 1–216.

R. DiPerna (1979), 'Uniqueness of solutions to hyperbolic conservation laws', *Indiana University Math. J.* **28**, 137–188.

R. J. DiPerna (1983), 'Convergence of approximate solutions to conservation laws', *Arch. Rational Mech. Anal.* **82**, 27–70.

B. Engquist and S. Osher (1980), 'Stable and entropy condition satisfying approximations for transonic flow calculations', *Math. Comp.* **34**, 44–75.

L. P. Franca, I. Harari, T. J. R. Hughes, M. Mallet, F. Shakib, T. E. Spelce, F. Chalot and T. E. Tezduyar (1986), A Petrov–Galerkin finite element method for the compressible Euler and Navier–Stokes equations, in *Numerical Methods for Compressible Flows: Finite Difference, Element and Volume Techniques* (T. E. Tezduyar and T. J. R. Hughes, eds), *Amer. Soc. Mech. Eng.*, AMD, Vol. 78, pp. 19–44.

K. O. Friedrichs (1954), 'Symmetric hyperbolic linear differential equations', *Comm. Pure Appl. Math.* **7**, 345–392.

K. O. Friedrichs and P. D. Lax (1971), 'Systems of conservation laws with a convex extension', *Proc. Nat. Acad. Sci. USA* **68**, 1686–1688.

E. Godlewski and P.-A. Raviart (1996), *Numerical Approximation of Hyperbolic Systems of Conservation Laws*, Springer.

S. K. Godunov (1959), 'A difference scheme for numerical computation of discontinuous solutions of fluid dynamics', *Mat. Sb.* **47**, 271–306.

S. K. Godunov (1961), 'An interesting class of quasilinear systems', *Dokl. Acad. Nauk. SSSR* **139**, 521–523.

J. Goodman and P. D. Lax (1981), 'On dispersive difference schemes, I', *Comm. Pure Appl. Math.* **41**, 591–613.

B. Gustafsson, H.-O. Kreiss and J. Oliger (1995), *Time Dependent Problems and Difference Methods*, Wiley-Interscience.

A. Harten (1983*a*), 'High-resolution scheme for hyperbolic conservation laws', *J. Comput. Phys.* **49**, 357–393.

A. Harten (1983*b*), 'On the symmetric form of systems of conservation laws with entropy', *J. Comput. Phys.* **49**, 151–164.

A. Harten and J. M. Hyman (1983), 'Self-adjusting grid method for one-dimensional hyperbolic conservation laws', *J. Comput. Phys.* **50**, 235–269.

A. Harten and P. D. Lax (1981), 'A random choice finite difference scheme for hyperbolic conservation laws', *SIAM J. Numer. Anal.* **18**, 289–315.

A. Harten, J. M. Hyman, and P. D. Lax (1976), 'On finite difference approximations and entropy conditions for shocks', *Comm. Pure Appl. Math.*, **29**, 297–322.

A. Harten, B. Engquist, S. Osher, and S. Chakravarthy (1987), 'Uniformly high-order accurate essentially non-oscillatory schemes, III', *J. Comput. Phys.* **71**, 231–303.

T. Hou and P. Lax (1991), 'Dispersive approximations in fluid dynamics', *Comm. Pure Appl. Math.* **44**, 1–40.

T. J. R. Hughes, L. P. Franca, and M. Mallet (1986), 'A new finite-element formulation for computational fluid dynamics, I: Symmetric forms of the compressible Euler and Navier–Stokes equations and the second law of thermodynamics', *Comput. Methods Appl. Mech. Eng.* **54**, 223–234.

G.-S. Jiang and C. -W. Shu (1994), 'On a cell entropy inequality for discontinuous Galerkin method', *Math. Comp.* **62**, 531–538.

C. Johnson and A. Szepessy (1986), 'A shock-capturing streamline diffusion finite element method for a nonlinear hyperbolic conservation laws', Technical Report 1986-V9, Mathematics Department, Chalmers University of Technology, Goteborg.

C. Johnson, A. Szepessy, and P. Hansbo (1990), 'On the convergence of shock-capturing streamline diffusion finite element methods for hyperbolic conservation laws' *Math. Comp.* **54**, 107–130.

L. Kaddouri (1993), 'Une méthode d'éléments finis discontinus pour les équations d'Euler des fluides compressibles', Thesis, Université Paris 6.

K. Khalfallah and A. Lerat (1988), 'Correction d'entropie pour des schémas numeriques approchant un système hyperbolique', *Note CR Acad. Sci.*

H.-O. Kreiss and J. Lorenz (1998), 'Stability for time-dependent differential equations', in *Acta Numerica*, Vol. 7, Cambridge University Press, pp. 203–286.

D. Króner (1997), *Numerical Schemes for Conservation Laws*, Wiley-Teubner, Stuttgart.

S. N. Kružkov (1970), 'First order quasilinear equations in several independent variables', *Math. USSR Sbornik* **10**, 217–243.

P. D. Lax (1954), 'Weak solutions of non-linear hyperbolic equations and their numerical computations', *Comm. Pure Appl. Math.* **7**, 159–193.

P. D. Lax (1957), 'Hyperbolic systems of conservation laws, II', *Comm. Pure Appl. Math.* **10**, 537–566.

P. D. Lax (1971), Shock waves and entropy, in *Contributions to Nonlinear Functional Analysis*, (E. A. Zarantonello, ed.), Academic Press, New York, pp. 603–634.

P. D. Lax (1972), *Hyperbolic Systems of Conservation Laws and the Mathematical Theory of Shock Waves*, Vol. 11 of *SIAM Regional Conference Lectures in Applied Mathematics.*

P. D. Lax (1986), 'On dispersive difference schemes', *Physica* **18D**, 250–254.

P. D. Lax and B. Wendroff (1960), 'Systems of conservation laws', *Comm. Pure Appl. Math.* **13**, 217–237.

P. D. Lax, D. Levermore and S. Venakidis (1993), The generation and propagation of oscillations in dispersive IVPs and their limiting behavior, in *Important Developments in Soliton Theory 1980–1990* (T. Fokas and V. E. Zakharov, eds), Springer, Berlin.

B. van Leer (1977), 'Towards the ultimate conservative difference scheme, III: Upstream-centered finite-difference schemes for ideal compressible flow', *J. Comput. Phys.* **23**, 263–275.

P. LeFloch (2002), *Hyperbolic Systems of Conservation Laws: The Theory of Classical and Nonclassical Shock Waves*, Springer, *Lectures in Mathematics, ETH Zürich.*

P. LeFloch and C. Rohde (2000), 'High-order schemes, entropy inequalities and nonclassical shocks', *SIAM J. Numer. Anal.* **37**, 2023–2060.

P. LeFloch, J. M. Mercier and C. Rohde (2002), 'Fully discrete, entropy conservative schemes of arbitrary order', *SIAM J. Numer. Anal.* **40**, 1968–1992.

R. LeVeque (1992), *Numerical Methods for Conservation Laws*, Birkhäuser, Basel, *Lectures in Mathematics.*

R. LeVeque (2002), *Finite Volume Methods for Hyperbolic Problems*, Cambridge University Press, *Texts in Applied Mathematics.*

D. Levermore and J.-G. Liu (1996), 'Large oscillations arising from a dispersive numerical scheme', *Phys. D* **99**, 191–216.

P. L. Lions and Souganidis (1995), 'Convergence of MUSCL and filtered schemes for scalar conservation laws and Hamilton–Jacobi equations', *Numer. Math.* **69**, 441–470.

A. Majda (1984), *Compressible Fluid Flow and Systems of Conservation Laws in Several Space Variables*, Springer, New York.

A. Majda and S. Osher (1978), 'A systematic approach for correcting nonlinear instabilities: The Lax–Wendroff scheme for scalar conservation laws', *Numer. Math.* **30**, 429–452.

A. Majda and S. Osher (1979), 'Numerical viscosity and the entropy condition', *Comm. Pure Appl. Math.* **32**, 797–838.

C. Makridakis and B. Perthame (2003), 'Sharp CFL, discrete kinetic formulation and entropy schemes for scalar conservation laws', *SIAM J. Numer. Anal.* To appear.

M. Merriam (1989), 'Towards a rigorous approach to artificial dissipation', AIAA-89-0471, Reno, Nevada.

M. S. Mock (1978), 'Some higher order difference schemes enforcing an entropy inequality', *Michigan Math. J.* **25**, 325–344.

M. S. Mock (1980), 'Systems of conservation of mixed type', *J. Diff. Eqns* **37**, 70–88.

M. Moskalkov (1980), 'Completely conservative schemes for gas dynamics', *USSR Comput. Maths. Math. Phys.* **20**, 162–170.

H. Nessyahu and E. Tadmor (1990), 'Non-oscillatory central differencing for hyperbolic conservation laws'. *J. Comput. Phys.* **87**, 408–463.

J. von Neumann and R. D. Richtmyer (1950), 'A method for the numerical calculation of hydrodynamic shocks', *J. Appl. Phys.* **21**, 232–237.

P. Olsson (1995), 'Summation by parts, projections, and stability, III', RIACS Technical report, 95.06.

S. Osher (1984), 'Riemann solvers, the entropy condition, and difference approximations', *SIAM J. Numer. Anal.* **21**, 217–235.

S. Osher (1985), 'Convergence of generalized MUSCL schemes', *SIAM J. Numer. Anal.* **22**, 947–961.

S. Osher and S. Chakravarthy (1984), 'High resolution schemes and the entropy condition', *SIAM J. Numer. Anal.* **21**, 955–984.

S. Osher and F. Solomon (1982), 'Upwind difference schemes for hyperbolic conservation laws', *Math. Comp.* **38**, 339–374.

S. Osher and E. Tadmor (1988), 'On the convergence of difference approximations to scalar conservation laws', *Math. Comp.* **50**, 19–51.

R. D. Richtmyer and K.W. Morton (1967), *Difference Methods for Initial-Value Problems*, Interscience, 2nd edn.

P. L. Roe (1981), 'Approximate Riemann solvers, parameter vectors and difference schemes', *J. Comput. Phys.* **43**, 357–372.

V. V. Rusanov (1961), 'Calculation of interaction of non-steady shock-waves with obstacles', *J. Comput. Math. Phys., USSR* **1**, 267–279.

R. Sanders (1983), 'On convergence of monotone finite difference schemes with variable spatial differencing', *Math. Comp.* **40**, 91–106.

D. Serre (1991), 'Richness and the classification of quasilinear hyperbolic systems', *Multidimensional Hyperbolic Problems and Computations*, Minneapolis, MN, *Inst. Math. Appl.*, Vol 29, Springer, pp. 315–333.

D. Serre (1999), *Systems of Conservation Laws, 1: Hyperbolicity, Entropies, Shock Waves* (English translation), Cambridge University Press.

M. Schonbek (1982), 'Convergence of solutions to nonlinear dispersive equations', *Comm. PDEs* **7**, 959–1000.

J. Smoller (1983), *Shock Waves and Reaction Diffusion Equations*, Springer, Berlin.

T. Sonar (1992), 'Entropy production in second-order three-point schemes', *Numer. Math.* **62**, 371–390.

A. Szepessy (1989), 'An existence result for scalar conservation laws using measure valued solutions', *Comm. PDEs* **14**, 1329–1350.

E. Tadmor (1984*a*), 'Skew-adjoint form for systems of conservation form', *J. Math. Anal. Appl.* **103**, 428–442.

E. Tadmor (1984*b*), 'Numerical viscosity and the entropy condition for conservative difference schemes', *Math. Comp.* **43**, 369–381.

E. Tadmor (1986*a*), 'A minimum entropy principle in the gas dynamics equations', *Appl. Numer. Math.* **2**, 211–219.

E. Tadmor (1986*b*), Entropy conservative finite element schemes, in *Numerical Methods for Compressible Flows: Finite Difference Element and Volume Techniques, Proc. Winter Annual Meeting of the Amer. Soc. Mech. Eng. AMD*, Vol. 78 (T. E. Tezduyar and T. J. R. Hughes, eds), pp. 149–158.

E. Tadmor (1987*a*), 'Entropy functions for symmetric systems of conservation laws', *J. Math. Anal. Appl.* **121**, 355–359.

E. Tadmor (1987*b*), 'The numerical viscosity of entropy stable schemes for systems of conservation laws, I', *Math. Comp.* **49**, 91–103.

E. Tadmor (1989), 'Convergence of spectral methods for nonlinear conservation laws', *SIAM J. Numer. Anal.* **26**, 30–44.

E. Tadmor (1997), Approximate solution of nonlinear conservation laws and related equations, in *Recent Advances in Partial Differential Equations and Applications, Proc. 1996 Venice Conference in Honor of Peter D. Lax and Louis Nirenberg on their 70th Birthday* (R. Spigler and S. Venakides, ed.), *AMS Proceedings Symp. Appl. Math.* Vol. 54, Providence, RI, pp. 321–368.

E. Tadmor (1998), Approximate solutions of nonlinear conservation laws, in *Advanced Numerical Approximation of Nonlinear Hyperbolic Equations, Lecture Notes from CIME Course Cetraro, Italy, 1997* (A. Quarteroni, ed.), Vol. 1697 of *Lecture Notes in Mathematics*, Springer, pp. 1–150.

L. Tartar (1975), Compensated compactness and applications to partial differential equations, in *Nonlinear Analysis and Mechanics, Heriot-Watt Symposium, Vol. 4* (R. J. Knopps, ed.), Vol. 39 of *Research Notes in Mathematics*, Pitman Press, pp. 136–211.

A. I. Vol'pert (1967), 'The spaces BV and quasilinear equations', *Math. USSR-Sb.* **2**, 225–267.

H. Yang (1996*a*), 'On wavewise entropy inequalities for high-resolution schemes, I: The semidiscrete case' *Math. Comp.* **65**, 45–67. Supplement *Math. Comp.* **65** S1–S13.

H. Yang (1996*b*), 'Convergence of Godunov type schemes' *Appl. Math. Letters* **9**, 63–67.

H. Yang (1998), 'On wavewise entropy inequalities for high-resolution schemes, II: Fully discrete MUSCL schemes with exact evolution in small time', *SIAM J. Numer. Anal.* **36**, 1–31.

Appendix: Entropy stability of Roe-type schemes

We consider Roe-type schemes of the form

$$\begin{aligned}\frac{\mathrm{d}}{dt}\mathbf{u}_\nu(t) = &-\frac{1}{2\Delta x_\nu}\left[f(\mathbf{u}_{\nu+1}) - f(\mathbf{u}_{\nu-1})\right] \\ &+ \frac{1}{2\Delta x_\nu}\left[p_{\nu+\frac{1}{2}}\Delta\mathbf{u}_{\nu+\frac{1}{2}} - p_{\nu-\frac{1}{2}}\Delta\mathbf{u}_{\nu-\frac{1}{2}}\right], \quad p_{\nu+\frac{1}{2}} = p\left(\overline{A}_{\nu+\frac{1}{2}}\right).\end{aligned} \tag{A.1}$$

Once more (consult Example 5.2), we use (5.10) to rewrite the viscous part of (A.1) in terms of the entropy variables, obtaining

$$\begin{aligned}\frac{\mathrm{d}}{dt}\mathbf{u}_\nu(t) = &-\frac{1}{2\Delta x_\nu}\left[\mathbf{f}(\mathbf{u}_{\nu+1}) - \mathbf{f}(\mathbf{u}_{\nu-1})\right] \\ &+ \frac{1}{2\Delta x_\nu}\left[Q_{\nu+\frac{1}{2}}\Delta\mathbf{v}_{\nu+\frac{1}{2}} - Q_{\nu-\frac{1}{2}}\Delta\mathbf{v}_{\nu-\frac{1}{2}}\right],\end{aligned} \tag{A.2}$$

where

$$Q_{\nu+\frac{1}{2}} = \int_{\xi=-\frac{1}{2}}^{\frac{1}{2}} p\left(\overline{A}_{\nu+\frac{1}{2}}\right)H(\xi)\,\mathrm{d}\xi, \quad H(\xi) \equiv H\left(\mathbf{v}_{\nu+\frac{1}{2}}(\xi)\right). \tag{A.3}$$

Corollary 5.1 suggests that the entropy dissipation of these schemes should be measured by the quantity $\langle \Delta\mathbf{v}_{\nu+\frac{1}{2}}, Q_{\nu+\frac{1}{2}}\Delta\mathbf{v}_{\nu+\frac{1}{2}}\rangle$. A lower bound for the latter is provided in the following lemma.

Lemma A1. Let $\{\overline{\mathbf{r}}_k, \overline{a}_k\}$ be the eigensystem of $\overline{A}_{\nu+\frac{1}{2}}$, and assume that

$$|\overline{\mathbf{r}}_k - \mathbf{r}_k(\xi = 0)| + |\overline{a}_k - a_k(\xi = 0)| \leq \text{Const}\left|\Delta\mathbf{v}_{\nu+\frac{1}{2}}\right|^2. \tag{A.4}$$

Then we have

$$\left\langle \Delta\mathbf{v}_{\nu+\frac{1}{2}}, Q_{\nu+\frac{1}{2}}\Delta\mathbf{v}_{\nu+\frac{1}{2}}\right\rangle \geq \sum_{k=1}^{N}\left[p(\overline{a}_k) - \text{Const}\left|\Delta\mathbf{v}_{\nu+\frac{1}{2}}\right|^2\right]\int_{\xi=-\frac{1}{2}}^{\frac{1}{2}}\left|\left\langle \mathbf{r}_k(\xi), \Delta\mathbf{v}_{\nu+\frac{1}{2}}\right\rangle\right|^2 \mathrm{d}\xi. \tag{A.5}$$

Proof. Using the orthonormal system $\{H^{-\frac{1}{2}}(\xi)\mathbf{r}_k(\xi)\}$ in (5.18), we can expand the right-hand side of the equality (A.3), which we rewrite as

$$\left\langle \Delta\mathbf{v}_{\nu+\frac{1}{2}}, Q_{\nu+\frac{1}{2}}\Delta\mathbf{v}_{\nu+\frac{1}{2}}\right\rangle = \int_{\xi=-\frac{1}{2}}^{\frac{1}{2}}\left\langle \Delta\mathbf{v}_{\nu+\frac{1}{2}}, p\left(\overline{A}_{\nu+\frac{1}{2}}\right)H^{\frac{1}{2}}(\xi)\cdot H^{\frac{1}{2}}(\xi)\Delta\mathbf{v}_{\nu+\frac{1}{2}}\right\rangle \mathrm{d}\xi,$$

and find

$$\left\langle \Delta\mathbf{v}_{\nu+\frac{1}{2}}, Q_{\nu+\frac{1}{2}}\Delta\mathbf{v}_{\nu+\frac{1}{2}}\right\rangle = \sum_{k=1}^{N}\int_{\xi=-\frac{1}{2}}^{\frac{1}{2}}\left\langle \Delta\mathbf{v}_{\nu+\frac{1}{2}}, p\left(A_{\nu+\frac{1}{2}}\right)\mathbf{r}_k(\xi)\right\rangle \alpha_k(\xi)\,\mathrm{d}\xi, \tag{A.6}$$

where $\alpha_k(\xi)$ abbreviates $\alpha_k(\xi) := \langle \mathbf{r}_k(\xi), \Delta\mathbf{v}_{\nu+\frac{1}{2}}\rangle$.

Consider the quantities on the right of (A.6): their dependence on ξ is reflected through their dependence on $\mathbf{v}_{\nu+\frac{1}{2}}(\xi) = \frac{1}{2}(\mathbf{v}_\nu + \mathbf{v}_{\nu+1}) + \xi\Delta v_{\nu+\frac{1}{2}}$; for such quantities we have

$$\left|\frac{\mathrm{d}^s X(\xi)}{\mathrm{d}\xi^s} \equiv \frac{\mathrm{d}^s}{\mathrm{d}\xi^s}X\left(\mathbf{v}_{\nu+\frac{1}{2}}(\xi)\right)\right| = \mathcal{O}\left(\left|\Delta\mathbf{v}_{\nu+\frac{1}{2}}\right|^s\right). \tag{A.7}$$

By Taylor's theorem,

$$\mathbf{r}_k(\xi) = \mathbf{r}_k(0) + \xi\dot{\mathbf{r}}_k(0) + \frac{\xi^2}{2}\ddot{\mathbf{r}}_k(\theta\xi), \quad \text{for some } \theta \in [0,1].$$

Here and below, $(\dot{\ })$ denotes ξ-differentiation, $\frac{\mathrm{d}}{\mathrm{d}\xi}(\ )$. In view of (A.7), $|\ddot{\mathbf{r}}_k| \leq \text{Const}\left|\Delta\mathbf{v}_{\nu+\frac{1}{2}}\right|^2$, and together with assumption (A.4) we have

$$\mathbf{r}_k(\xi) = \overline{\mathbf{r}}_k + \xi\dot{\mathbf{r}}_k(0) + J_k, \quad |J_k| \leq \text{Const}\left|\Delta\mathbf{v}_{\nu+\frac{1}{2}}\right|^2. \tag{A.8}$$

Inserting this into (A.6) we conclude that

$$\begin{aligned}
&\left\langle \Delta\mathbf{v}_{\nu+\frac{1}{2}}, Q_{\nu+\frac{1}{2}}\Delta\mathbf{v}_{\nu+\frac{1}{2}}\right\rangle \\
&\quad = \sum_{k=1}^{N}\int_{\xi=-\frac{1}{2}}^{\frac{1}{2}} \left\langle \Delta\mathbf{v}_{\nu+\frac{1}{2}}, p\left(\overline{A}_{\nu+\frac{1}{2}}\right)\overline{\mathbf{r}}_k\right\rangle \alpha_k(\xi)\,\mathrm{d}\xi \\
&\qquad + \sum_{k=1}^{N}\int_{\xi=-\frac{1}{2}}^{\frac{1}{2}} \xi\left\langle \Delta\mathbf{v}_{\nu+\frac{1}{2}}, p\left(\overline{A}_{\nu+\frac{1}{2}}\right)\dot{\mathbf{r}}_k(0)\right\rangle \alpha_k(\xi)\,\mathrm{d}\xi \\
&\qquad + \sum_{k=1}^{N}\int_{\xi=-\frac{1}{2}}^{\frac{1}{2}} \left\langle \Delta\mathbf{v}_{\nu+\frac{1}{2}}, p\left(\overline{A}_{\nu+\frac{1}{2}}\right)J_k\right\rangle \alpha_k(\xi)\,\mathrm{d}\xi. \qquad \text{(A.9)}
\end{aligned}$$

Finally, taking into account (A.8), we have

$$p\left(\overline{A}_{\nu+\frac{1}{2}}\right)\overline{\mathbf{r}}_k = p(\overline{a}_k)\mathbf{r}_k = p(\overline{a}_k)\mathbf{r}_k(\xi) - \xi p(\overline{a}_k)\dot{\mathbf{r}}_k(0) - p(\overline{a}_k)J_k;$$

we substitute the last three terms into three summations on the right of (A.9), respectively, and end up with

$$\begin{aligned}
&\left\langle \Delta\mathbf{v}_{\nu+\frac{1}{2}}, Q_{\nu+\frac{1}{2}}\Delta\mathbf{v}_{\nu+\frac{1}{2}}\right\rangle \\
&\quad = \sum_{k=1}^{N} p(\overline{a}_k)\cdot\int_{\xi=-\frac{1}{2}}^{N} \alpha_k^2(\xi)\,\mathrm{d}\xi \\
&\qquad + \sum_{k=1}^{N}\int_{\xi=-\frac{1}{2}}^{\frac{1}{2}} \xi\left\langle \Delta\mathbf{v}_{\nu+\frac{1}{2}}, \left[p\left(\overline{A}_{\nu+\frac{1}{2}}\right) - p(\overline{a}_k)\cdot I_N\right]\dot{\mathbf{r}}_k(0)\right\rangle \alpha_k(\xi)\,\mathrm{d}\xi \\
&\qquad + \sum_{k=1}^{N}\int_{\xi=-\frac{1}{2}}^{\frac{1}{2}} \left\langle \Delta\mathbf{v}_{\nu+\frac{1}{2}}, \left[p\left(\overline{A}_{\nu+\frac{1}{2}}\right) - p(\overline{a}_k)\cdot I_N\right]J_k\right\rangle \alpha_k(\xi)\,\mathrm{d}\xi \\
&\quad = \sum_{k=1}^{N} p\left(\overline{a}_k\right)\int_{\xi=-\frac{1}{2}}^{\frac{1}{2}} \left|\left\langle \mathbf{r}_k(\xi), \Delta\mathbf{v}_{\nu+\frac{1}{2}}\right\rangle\right|^2 \mathrm{d}\xi + II + III. \qquad \text{(A.10)}
\end{aligned}$$

Since $J_k\alpha_k(\xi)$ is of order $\mathcal{O}\left(\left|\Delta\mathbf{v}_{\nu+\frac{1}{2}}\right|^3\right)$, we have

$$|III| \leq \mathrm{Const}_{III}\cdot\left|\Delta\mathbf{v}_{\nu+\frac{1}{2}}\right|^4; \qquad \text{(A.11)}$$

integration by parts of the second summation on the right of (A.10) yields

$$II = \sum_{k=1}^{N}\int_{\xi=-\frac{1}{2}}^{\frac{1}{2}} \left(\frac{1}{4}-\xi^2\right)\frac{\mathrm{d}}{\mathrm{d}\xi}\left\{\left\langle \Delta\mathbf{v}_{\nu+\frac{1}{2}}, \left[p\left(\overline{A}_{\nu+\frac{1}{2}}\right) - p(\overline{a}_k)I_N\right]\cdot\mathbf{r}_k(0)\right\rangle \alpha_k(\xi)\right\}\mathrm{d}\xi.$$

Since $\dot{\mathbf{r}}_k(0)\alpha_k(\xi)$ is of order $\mathcal{O}(|\Delta\mathbf{v}_{\nu+\frac{1}{2}}|^2)$, the expansion inside the curly brackets is $\mathcal{O}(|\Delta\mathbf{v}_{\nu+\frac{1}{2}}|^3)$, and its ξ-derivative gives us (consult (A.7))

$$|II| \leq \mathrm{Const}_{II} \cdot |\Delta\mathbf{v}_{\nu+\frac{1}{2}}|^4. \tag{A.12}$$

The result (A.5) now follows, noting that $H(\xi) \leq K \cdot I_N$ (see (5.26)), and hence

$$\begin{aligned} |\Delta\mathbf{v}_{\nu+\frac{1}{2}}|^2 &\leq \frac{1}{K}\int_{\xi=-\frac{1}{2}}^{\frac{1}{2}} |H^{\frac{1}{2}}(\xi)\Delta\mathbf{v}_{\nu+\frac{1}{2}}|^2 \, d\xi \\ &\leq \frac{1}{K}\sum_{k=1}^{N}\int_{\xi=-\frac{1}{2}}^{\frac{1}{2}} |\langle \mathbf{r}_k(\xi), \Delta\mathbf{v}_{\nu+\frac{1}{2}}\rangle|^2 \, d\xi; \end{aligned} \tag{A.13}$$

consequently, (A.11), (A.12), and (A.13) imply

$$II + III \geq -\sum_{k=1}^{N}\frac{1}{K}(\mathrm{Const}_{II} + \mathrm{Const}_{III}) \cdot |\Delta\mathbf{v}_{\nu+\frac{1}{2}}|^2 \cdot \int_{\xi=-\frac{1}{2}}^{\frac{1}{2}} |\langle \mathbf{r}_k(\xi), \Delta\mathbf{v}_{\nu+\frac{1}{2}}\rangle|^2 \, d\xi$$

and together with (A.10), this amounts to having (A.5). □

Next we turn to bounding the viscosity part of the entropy-conservative scheme (5.4), (5.5) from above, in the spirit of Lemma A1.

Lemma A2. Let $Q^*_{\nu+\frac{1}{2}}$ be the viscosity matrix associated with the entropy-conservative scheme (5.2), (5.3). Let

$$\Delta a_k(\mathbf{u}_\nu) \equiv a_k(\mathbf{u}_{\nu+1}) - a_k(\mathbf{u}_\nu)$$

denote the jump of the kth eigenvalue, $\lambda_k(A(\mathbf{u}))$, from the state on the left, $\mathbf{u}_\nu$, to its right neighbour $\mathbf{u}_{\nu+1}$. Then, for arbitrary $\varepsilon_k > 0$, we have

$$\begin{aligned} &\langle \Delta\mathbf{v}_{\nu+\frac{1}{2}}, Q^*_{\nu+\frac{1}{2}}\Delta\mathbf{v}_{\nu+\frac{1}{2}}\rangle \\ &\quad \leq \sum_{k=1}^{N}\frac{1}{6}\left[\Delta a_k(\mathbf{u}_\nu) + \varepsilon_k|\bar{a}_k| + \left(1 + \frac{|\bar{a}_k|}{\varepsilon_k}\right)\mathrm{Const}|\Delta\mathbf{v}_{\nu+\frac{1}{2}}|^2\right] \\ &\qquad \times \int_{\xi=-\frac{1}{2}}^{\frac{1}{2}} |\langle \mathbf{r}_k(\xi), \Delta\mathbf{v}_{\nu+\frac{1}{2}}\rangle|^2 \, d\xi. \end{aligned} \tag{A.14}$$

Proof. According to (5.19) we have

$$\langle \Delta\mathbf{v}_{\nu+\frac{1}{2}}, Q^*_{\nu+\frac{1}{2}}\Delta\mathbf{v}_{\nu+\frac{1}{2}}\rangle = \sum_{k=1}^{N}\int_{\xi=-\frac{1}{2}}^{\frac{1}{2}} 2\xi a_k(\xi)\alpha_k^2(\xi) \, d\xi.$$

Changing variables, $\xi \to -\xi$, and averaging, we can rewrite this as

$$\left\langle \Delta \mathbf{v}_{\nu+\frac{1}{2}}, Q^*_{\nu+\frac{1}{2}} \Delta \mathbf{v}_{\nu+\frac{1}{2}} \right\rangle = \sum_{k=1}^{N} \int_{\xi=-\frac{1}{2}}^{\frac{1}{2}} \xi \left[a_k(\xi)\alpha_k^2(\xi) - a_k(-\xi)\alpha_k^2(-\xi) \right] \mathrm{d}\xi. \tag{A.15}$$

By Taylor's expansion,

$$\alpha_k^2(\pm\xi) = \alpha_k^2(0) \pm \xi \alpha_k(0)\dot{\alpha}_k(0) + \frac{\xi^2}{2} \frac{\mathrm{d}^2}{\mathrm{d}\xi^2}(\alpha_k^2(\overline{\xi})). \tag{A.16}$$

In view of (A.7), $\frac{\mathrm{d}^2}{\mathrm{d}\xi^2}(\alpha_k^2(\overline{\xi}))$ is of order $\mathcal{O}(|\Delta \mathbf{v}_{\nu+\frac{1}{2}}|^4)$, and therefore

$$\begin{aligned} &\left[a_k(\xi)\alpha_k^2(\xi) - a_k(-\xi)\alpha_k^2(-\xi) \right] \\ &\quad = \left[a_k(\xi) - a_k(-\xi) \right] \alpha_k^2(0) \\ &\qquad + \left[a_k(\xi) + a_k(-\xi) \right] 2\xi \alpha_k(0)\dot{\alpha}_k(0) + \mathcal{O}\left(|\Delta \mathbf{v}_{\nu+\frac{1}{2}}|^4 \right). \end{aligned} \tag{A.17}$$

Moreover, we have

$$\alpha_k(\pm\xi) = \alpha_k(0) \pm \xi \dot{a}_k(0) + \frac{\xi^2}{2}\ddot{a}_k, \quad |\ddot{a}_k| \leq \text{Const}|\Delta \mathbf{v}_{\nu+\frac{1}{2}}|^2;$$

we substitute this into the right-hand side of (A.17): since $\dot{a}_k(0), \alpha_k^2(0)$ and $\alpha_k(0)\dot{\alpha}_k(0)$ are of order $\mathcal{O}(|\Delta \mathbf{v}_{\nu+\frac{1}{2}}|), \mathcal{O}(|\Delta \mathbf{v}_{\nu+\frac{1}{2}}|^2)$, and $\mathcal{O}(|\Delta \mathbf{v}_{\nu+\frac{1}{2}}|^3)$ respectively, we obtain

$$\begin{aligned} &\left[a_k(\xi)\alpha_k^2(\xi) - a_k(-\xi)\alpha_k^2(-\xi) \right] \\ &\quad = 2\xi \dot{a}_k(0)\alpha_k^2(0) + 4\xi a_k(0)\alpha_k(0)\dot{\alpha}_k(0) + \mathcal{O}\left(|\Delta \mathbf{v}_{\nu+\frac{1}{2}}|^4 \right). \end{aligned} \tag{A.18}$$

Inserting the latter expression into (A.15), we find after integration that

$$\begin{aligned} &\left\langle \Delta \mathbf{v}_{\nu+\frac{1}{2}}, Q^*_{\nu+\frac{1}{2}} \Delta \mathbf{v}_{\nu+\frac{1}{2}} \right\rangle \\ &\quad \leq \sum_{k=1}^{N} \frac{1}{6} \left[\dot{a}_k(0)\alpha_k^2(0) + 2a_k(0)\alpha_k(0)\dot{\alpha}_k(0) \right] + \text{Const}\left(|\Delta \mathbf{v}_{\nu+\frac{1}{2}}|^4 \right). \end{aligned}$$

In view of (A.4), we can replace $a_k(0)$ by $\overline{a}_k$ and end up with

$$\begin{aligned} &\left\langle \Delta \mathbf{v}_{\nu+\frac{1}{2}}, Q^*_{\nu+\frac{1}{2}} \Delta \mathbf{v}_{\nu+\frac{1}{2}} \right\rangle \\ &\quad \leq \sum_{k=1}^{N} \frac{1}{6} \left[\dot{a}_k(0)\alpha_k^2(0) + 2\overline{a}_k\alpha_k(0)\dot{\alpha}_k(0) \right] + \text{Const}\left(|\Delta \mathbf{v}_{\nu+\frac{1}{2}}|^4 \right). \end{aligned} \tag{A.19}$$

We upper-bound the second term by Cauchy–Schwartz:

$$2\overline{a}_k\alpha_k(0)\dot{\alpha}_k(0) \leq \varepsilon_k|\overline{a}_k|\alpha_k^2(0) + \frac{1}{\varepsilon_k}|\overline{a}_k|(\dot{\alpha}_k(0))^2,$$

and since $(\dot{\alpha}_k(0))^2 \le \mathrm{Const}\left|\Delta \mathbf{v}_{\nu+\frac{1}{2}}\right|^4$, (A.19) gives us

$$\begin{aligned}
&\left\langle \Delta \mathbf{v}_{\nu+\frac{1}{2}}, Q^*_{\nu+\frac{1}{2}} \Delta \mathbf{v}_{\nu+\frac{1}{2}} \right\rangle \qquad \text{(A.20)} \\
&\quad \le \sum_{k=1}^{N} \frac{1}{6}\left[\dot{a}_k(0) + \varepsilon_k |a_k|\right]\alpha_k^2(0) + \left(1 + \frac{|\overline{a}_k|}{\varepsilon_k}\right)\mathrm{Const}\left(\left|\Delta \mathbf{v}_{\nu+\frac{1}{2}}\right|^4\right).
\end{aligned}$$

Finally, by (A.16) we have

$$\alpha_k^2(0) = \int_{\xi=-\frac{1}{2}}^{\frac{1}{2}} \left|\left\langle \mathbf{r}_k(\xi), \Delta \mathbf{v}_{\nu+\frac{1}{2}} \right\rangle\right|^2 \mathrm{d}\xi + \mathcal{O}\left(\left|\Delta \mathbf{v}_{\nu+\frac{1}{2}}\right|^4\right),$$

and, according to (A.13),

$$\left|\Delta \mathbf{v}_{\nu+\frac{1}{2}}\right|^2 \le \frac{1}{K} \sum_{k=1}^{N} \int_{\xi=-\frac{1}{2}}^{\frac{1}{2}} \left|\left\langle \mathbf{r}_k(\xi), \Delta \mathbf{v}_{\nu+\frac{1}{2}} \right\rangle\right|^2 \mathrm{d}\xi.$$

Using this together with (A.20) implies

$$\begin{aligned}
&\left\langle \Delta \mathbf{v}_{\nu+\frac{1}{2}}, Q^*_{\nu+\frac{1}{2}} \Delta \mathbf{v}_{\nu+\frac{1}{2}} \right\rangle \\
&\quad \le \sum_{k=1}^{N} \frac{1}{6}\left[\dot{a}_k(0) + \varepsilon_k |\overline{a}_k| + \left(1 + \frac{|\overline{a}_k|}{\varepsilon_k}\right)\mathrm{Const}\left|\Delta \mathbf{v}_{\nu+\frac{1}{2}}\right|^2\right] \\
&\qquad \times \int_{\xi=-\frac{1}{2}}^{\frac{1}{2}} \left|\left\langle \mathbf{r}_k(\xi), \Delta \mathbf{v}_{\nu+\frac{1}{2}} \right\rangle\right|^2 \mathrm{d}\xi, \qquad \text{(A.21)}
\end{aligned}$$

and the result (A.14) follows, noting that $\dot{a}_k(0) = \Delta a_k(\mathbf{u}_\nu) + \mathcal{O}\left(\left|\Delta \mathbf{v}_{\nu+\frac{1}{2}}\right|^2\right)$. □

We close this section with the following proof.

Proof of Theorem 5.9. Comparing (A.5) and (A.14), we conclude from Corollary 5.1 that the Roe-type scheme (A.1), (A.4) is entropy-stable provided that

$$p(\overline{a}_k) \ge \frac{1}{6}\left[\Delta a_k(\mathbf{u}_\nu) + \varepsilon_k |\overline{a}_k| + \left(1 + \frac{|\overline{a}_k|}{\varepsilon_k}\right)\mathrm{Const}\left|\Delta \mathbf{v}_{\nu+\frac{1}{2}}\right|^2\right], \qquad \text{(A.22)}$$

and since, by (5.10), (5.26),

$$\frac{1}{K}\left|\Delta \mathbf{v}_{\nu+\frac{1}{2}}\right| \le \left|\Delta \mathbf{u}_{\nu+\frac{1}{2}} = H_{\nu+\frac{1}{2}} \Delta \mathbf{v}_{\nu+\frac{1}{2}}\right| \le K\left|\Delta \mathbf{v}_{\nu+\frac{1}{2}}\right|,$$

it follows that (A.22) is equivalent to (5.40). □